Graduate Texts in Mathematics 266

Graduate Texts in Mathematics

Graduate Texts in Mathematics bridge the gap between passive study and creative understanding, offering graduate-level introductions to advanced topics in mathematics. The volumes are carefully written as teaching aids and highlight characteristic features of the theory. Although these books are frequently used as textbooks in graduate courses, they are also suitable for individual study.

For further volumes:
http://www.springer.com/series/136

Jean-Paul Penot

Calculus Without Derivatives

 Springer

Jean-Paul Penot
Laboratoire Jacques-Louis Lions
Université Pierre et Marie Curie
Paris, France

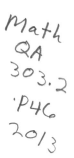

ISSN 0072-5285
ISBN 978-1-4614-4537-1   ISBN 978-1-4614-4538-8 (eBook)
DOI 10.1007/978-1-4614-4538-8
Springer New York Heidelberg Dordrecht London

Library of Congress Control Number: 2012945592

Mathematics Subject Classification (2010): 49J52, 49J53, 58C20, 54C60, 52A41, 90C30

Printed on acid-free paper

Springer is part of Springer Science+Business Media (www.springer.com)

To the memory of my parents,
with the hope that their thirst
for culture and knowledge
will be transmitted
to the reader through this book

Preface

A famous Indian saying can be approximatively phrased in the following way: "Our earth is not just a legacy from our parents; it is a loan from our children."

In mathematical analysis, a precious legacy has been given to us: differential calculus and integral calculus are tools that play an important role in the present state of knowledge and technology. They even gave rise to a philosophical opinion, often called determinism, that amounts to saying that any phenomenon can be predicted, provided one knows its rules and the initial conditions. Such a triumphant claim has been mitigated by modern theories such as quantum mechanics. The "fuzziness" one meets in this book presents some analogy with modern mechanics. In some sense, it is the best we can leave to our children in case they have to deal with rough data.

In the middle of the nineteenth century, Weierstrass made clear the fact that not all functions are differentiable. He even proved that there are continuous functions of one real variable that are nowhere differentiable. Although such "exotic" functions are not negligible, it appears that most nonsmooth functions that are met in concrete mathematical problems have a behavior that is not beyond the reach of analysis.

It is the purpose of the present book to show that an organized bundle of knowledge can be applied to situations in which differentiability is not present.

In favorable cases, such as pointwise maxima of finite families of differentiable functions or sums of convex functions with differentiable functions, a rather simple apparatus allows us to extend in a unified way the rules known in the realms of convex analysis and differentiable analysis. The pioneers in this restricted framework were Pshenichnii, Ioffe, and Tikhomirov (and later on, Demy'anov, Janin, among others). For general functions, more subtle constructions must be devised.

Already at this elementary stage, a combination of geometrical and analytical viewpoints gives greater and more incisive insight. Such a unified viewpoint is one of the revolutionary characteristics of nonsmooth analysis: functions, sets, mappings, and multimappings (or correspondences) can be considered to be equally important, and the links between them allow us to detect fruitful consequences. Historically, geometrical concepts (tangent and normal cones with Bouligand, Severi, Choquet, Dubovitskii-Milyutin, ...) appeared earlier than analytical notions (generalized directional derivatives, subdifferentials with Clarke, Ioffe, Kruger, Mordukhovich, . . .).

On the other hand, the variety of situations and needs has led to different approaches. In our opinion, it would not be sensible to leave the reader with the impression that a single type of answer or construction can meet all the needs one may encounter (it is not even the case with smooth calculus). It is our purpose to give the reader the ability to choose an appropriate scheme depending on the specificities of the problem at hand. Quite often, the problem itself leads to an adapted space. In turn, the space often commands the choice of the subdifferential as a manageable substitute for the derivative.

In this book we endeavor to present a balanced picture of the most elementary attempts to replace a derivative with a one-sided generalized derivative called a subdifferential. This means that instead of associating a linear form to a function at some reference point in order to summarize some information about the behavior of the function around that point, one associates a bunch of linear forms. Of course, the usefulness of such a process relies on the accurateness of the information provided by such a set of linear forms. It also relies on the calculus rules one can design. These two requirements appear to be somewhat antagonistic. Therefore, it may be worthwhile to dispose of various approaches satisfying at least one of these two requirements.

In spite of the variety of approaches, we hope that our presentation here will give an impression of unity. We do not consider the topic as a field full of disorder. On the contrary, it has its own methods, and its various achievements justify a comprehensive approach that has not yet been presented. Still, we do not look for completeness; we rather prefer to present significant tools and methods. The references, notes, exercises, and supplements we present will help the reader to get a more thorough insight into the subject.

In writing a book, one has to face a delicate challenge: either follow a tradition or prepare for a more rigorous use. Our experience with texts that were written about a lifetime ago showed us that the need for rigor and precision has increased and is likely to increase more. Thus, we have avoided some common abuses such as confusing a function with its value, a sequence with its general term, a space with its dual, the gradient of a function with its derivative, the adjoint of a continuous linear map with its transpose. That choice may lead to unusual expressions. But in general, we have made efforts to reach as much simplicity as possible in proofs, terminology and notation, even if some proofs remain long. Moreover, we have preferred suggestive names (such as allied, coherence, gap, soft) to complicated expressions or acronyms, and we have avoided a heavy use of multiple indices, of Greek letters (and also of Cyrillic, Gothic, Hebrew fonts). It appears to us that sophisticated notation blossoms when the concepts are fresh and still obscure; as soon as the concepts appear as natural and simple, the notation tends to get simpler too. Of course, besides mathematicians who are attached to traditions, there are some others who implicitly present themselves as magicians or learned people and like to keep sophisticated notation.

Let us present in greater detail the analysis that served as a guideline for this book.

The field of mathematics offers a number of topics presenting beautiful results. However, many of them are rather remote from practical applications. This fact makes them not too attractive to many students. Still, they are proposed in many courses because they are considered either as important from a theoretical viewpoint or precious for the formation of minds.

It is the purpose of this book to present fundamental aspects of analysis that have close connections with applications. There is no need to insist on the success of analysis. So many achievements of modern technology rely on methods or results from mathematical analysis that it is difficult to imagine what our lives would be like if the consequences of the so-called infinitesimal analysis of Fermat, Leibniz, Newton, Euler and many others would be withdrawn from us.

However, the classical differential calculus is unable to handle a number of problems in which order plays a key role; J.-J. Moreau called them "unilateral problems," i.e., one-sided problems. Usually, they are caused by constraints or obstacles.

A few decades ago, some tools were designed to study such problems. They are applied in a variety of fields, such as economics, mechanics, optimization, numerical analysis, partial differential equations. We believe that this rich spectrum of applications can be attractive for the reader and deserves a sequel to this book with complementary references, since here we do not consider applications as important as those in optimal control theory and mathematical analysis. Also, we do not consider special classes of functions or sets, and we do not even evoke higher-order notions, although considering second-order generalized derivatives of nondifferentiable functions can be considered a feat!

Besides some elements of topology and functional analysis oriented to our needs, we gather here three approaches: differential calculus, convex analysis, and nonsmooth analysis. The third of these is the most recent, but it is becoming a classical topic encompassing the first two.

The novelty of a joint presentation of these topics is justified by several arguments. First of all, since nonsmooth analysis encompasses both convex analysis and differential calculus, it is natural to present these two subjects as the two basic elements on which nonsmooth analysis is built. They both serve as an introduction to the newest topic. Moreover, they are both used as ingredients in the proofs of calculus rules in the nonsmooth framework. On the other hand, nonsmooth analysis represents an incentive to enrich convex analysis (and maybe differential calculus too, as shown here by the novelty of incorporating directional smoothness in the approach). As an example, we mention the relationship between the subdifferential of the distance function to a closed convex set C at some point z out of C and the normal cone to the set C at points of C that almost minimize the distance to z (Exercises 6 and 7 of Sect. 7.1 of the chapter on convex analysis). Another example is the fuzzy calculus that is common to convex analysis and nonsmooth analysis and was prompted by the last domain.

In this book, we convey some ideas that are simple enough but important. First we want to convince the reader that approximate calculus rules are almost as useful as exact calculus rules. They are more realistic, since from a numerical viewpoint, only approximate values of functions and derivatives can be computed (apart from some special cases).

Second, we stress the idea that basic notions, methods, or results such as variational principles, methods of error bounds, calmness, and metric regularity properties offer powerful tools in analysis. They are of interest in themselves, and we are convinced that they may serve as a motivated approach to the study of metric spaces, whereas such a topic is often considered very abstract by students.

The penalization method is another example illustrating our attempt. It is a simple idea that in order to ensure that a constraint (for instance a speed limit or an environmental constraint) is taken into account by an agent, a possible method consists in penalizing the violation of this constraint. The higher the penalty, the better the behavior. We believe that such methods related to the experience of the reader may enhance his or her interest in mathematics. They are present in the roots of nonsmooth calculus rules and in the study of partial differential equations.

Thus, the contents of the first part can be used for at least three courses besides nonsmooth analysis: metric and topological notions, convex analysis, and differential calculus. These topics are also deeply linked with optimization questions and geometric concepts.

In the following chapters dealing with nonsmooth analysis, we endeavor to present a view encompassing the main approaches, whereas most of the books on that topic focus on a particular theory. Indeed, we believe that it is appropriate to deal with nonsmooth problems with an open mind. It is often the nature of the problem that suggests the choice of the spaces. In turn, the choice of the nonsmooth concepts (normal cones, subdifferentials, etc.) depends on the properties of the chosen spaces and on the objectives of the study. Some concepts are accurate, but are lacking good calculus rules; some enjoy nice convexity or duality properties but are not so precise. We would like to convince the reader that such a variety is a source of richness rather than disorder.

The quotation below would be appropriate if in the present case it corresponded to what actually occurred. But the truth is that the book would never had been written if Alexander Ioffe had not suggested the idea to the author and contributed to many aspects of it. The author expresses his deepest gratitude to him. He also wants to thank the many colleagues and friends, in particular, D. Azé, A. Dontchev, E. Giner, A. Ioffe, M. Lassonde, K. Nachi, L. Thibault, who made useful criticisms or suggestions, and he apologizes to those who are not given credit or given not enough credit.

> N'écrire jamais rien qui de soi ne sortit,
> Et modeste d'ailleurs, se dire mon petit,
> Soit satisfait des fleurs, des fruits, même des feuilles,
> Si c'est dans ton jardin à toi que tu les cueilles!
> ...Ne pas monter bien haut, peut-être, mais tout seul!

Edmond Rostand, *Cyrano de Bergerac*, Acte II, Scène 8

Never to write anything that does not proceed from the heart,
and, moreover, to say modestly to myself, "My dear,
be content with flowers, with fruits, even with leaves,
if you gather them in your own garden!"
... Not to climb very high perhaps, but to climb all alone!

Pau and Paris, France   Jean-Paul Penot

Contents

Notation

i.e.	That is
f.i.	For instance
iff	If and only if
$:=$	Equality by definition
$\forall$	For all, for every
$\exists$	There exists
$\in$	Belongs to
$\subset$	Included in
$\cap$	Intersection
$\cup$	Union
∞	Infinity
$\varnothing$	The empty set
$X \setminus Y$	The set of elements of X not in Y
$x \to_f \bar{x}$	Stands for $x \to \bar{x}$ with $f(x) \to f(\bar{x})$
$\to_S$	Converges while remaining in S
$t(\in T) \to 0$	t converges to 0 while remaining in T
$\overset{*}{\to}$	Converges in the weak* topology
$(x_n^*) \to^* x^*$	(x_n^*) is bounded and has x^* as a weak* cluster point
$\mathbb{N}$	The set of natural numbers
$\mathbb{N}_k$	The set $\{1, \ldots, k\}$
$\mathbb{P}$	The set of positive real numbers
$\mathbb{Q}$	The set of rational numbers
$\mathbb{R}$	The set of real numbers
$\mathbb{R}_+$	The set of nonnegative real numbers
$\mathbb{R}_-$	The set of nonpositive real numbers
$\overline{\mathbb{R}}_+$	The set $\mathbb{R}_+ \cup \{+\infty\}$ of nonnegative extended real numbers
$\overline{\mathbb{R}}$	The set $\mathbb{R} \cup \{-\infty, +\infty\}$ of extended real numbers
$\mathbb{R}_\infty$	The set $\mathbb{R} \cup \{+\infty\}$
Δ_m	The canonical m-simplex $\{(t_1, \ldots, t_m) \in \mathbb{R}_+^m : t_1 + \cdots + t_m = 1\}$
$B(x, r)$	The open ball with center x and radius r in a metric space
$B[x, r]$	The closed ball with center x and radius r in a metric space

B_X	The closed unit ball in a normed space X
S_X	The unit sphere of the normed space X
$B(x, \varepsilon, f)$	The set $B(x, \varepsilon) \cap f^{-1}(B(f(x), \varepsilon))$
$\wedge_Z f$	The stabilized infimum of f on Z
$\wedge(f, g)$	The stabilized infimum of $f + g$
F^{-1}	The inverse of a multimap F
$C(X, Y)$	The set of continuous maps from X to another topological space Y
$L(X, Y)$	The set of continuous linear maps from X to another normed space Y
X^*	The topological dual space $L(X, \mathbb{R})$ of a normed space X
$\langle \cdot, \cdot \rangle$	The coupling function between a normed space and its dual
$(\cdot \mid \cdot)$	The scalar product
A^T	The transpose of the continuous linear map A
A^*	The adjoint of the continuous linear map A between Hilbert spaces
$\mathrm{int}\, S$	The interior of the subset S of a topological space
$\mathrm{cl}\, S$	The closure of the subset S of a topological space
$\mathrm{bdry}\, S$	The boundary of the subset S of a topological space
$\mathrm{co}\, S$	The convex hull of the subset S of a linear space
$\overline{\mathrm{co}}\, S$	The closed convex hull of the subset S of a topological linear space
$\overline{\mathrm{co}}^*\, S$	The weak* closed convex hull of the subset S of a dual space
$\mathrm{span}\, S$	The smallest linear subspace containing S
S^0	The polar set $\{x^* \in X^* : \forall x \in S \langle x^*, x \rangle \leq 1\}$ of S
$\mathrm{diam}\,(S)$	The diameter $\sup\{d(w, x) : w, x \in B\}$ of S
I_S	The identity map from S to S
ι_S	The indicator function of the subset S
σ_S	The support function of the subset S
$\mathscr{G}_\delta$	The family of countable intersections of open subsets of a topological space X
$\mathscr{F}(X)$	The set of lower semicontinuous proper functions on X
$\mathscr{L}(X)$	The space of locally Lipschitzian functions on a metric space X
$\mathscr{N}(x)$	The family of neighborhoods of x in a topological space
$\mathscr{P}(X)$	The set of subsets of the set X
$\mathscr{S}(X)$	The family of separable linear subspaces of a Banach space X
$\Gamma(X)$	The set of closed convex proper convex functions on a t.v.s. X
f' or Df	The derivative of a function f or a map f
∇f	The gradient of a function f
$\|\nabla\| f(x)$	The slope of f at x
$\partial f(x)$	A subdifferential of f at x or the Moreau–Rockafellar subdifferential
$\partial_C f(x)$	The circa-subdifferential or Clarke subdifferential of f at x
$\partial_D f(x)$	The directional or Dini–Hadamard subdifferential of f at x
$\partial_F f(x)$	The firm or Fréchet subdifferential of f at x
$\partial_G f(x)$	The graded subdifferential or Ioffe subdifferential of f at x
$\partial_H f(x)$	The Hadamard (viscosity) subdifferential of f at x
$\partial_h f(x)$	The limiting Hadamard subdifferential of f at x
$\partial_L f(x)$	The limiting (firm) subdifferential of f at x
$\partial_\ell f(x)$	The limiting directional subdifferential of f at x
$\partial_{MR} f(x)$	The Moreau–Rockafellar subdifferential of f at x

Chapter 1
Metric and Topological Tools

I do not know what I may appear to the world, but to myself I seem to have been only like a boy playing on the sea-shore, and diverting myself in now and then finding a smoother pebble or a prettier shell than ordinary, whilst the great ocean of truth lay all undiscovered before me.

—Isaac Newton

We devote this opening chapter to some preliminary material dealing with sets, set-valued maps, convergences, estimates, and well-posedness.

A mastery of set theory (or rather calculus with standard operations dealing with sets) and of set-valued maps is necessary for nonsmooth analysis. In fact, one of the most attractive features of nonsmooth analysis consists in easy and frequent passages from sets to functions and vice versa. Moreover, several concepts of nonsmooth analysis become clear when one has some knowledge of set convergence. As an example, recall that the tangent to a curve C at some $x_0 \in C$ is defined as the limit of a secant passing through x_0 and another point x of C as $x \to x_0$ in C.

In this first chapter we gather some basic material that will be used in the rest of the book. Part of it is in standard use. However, we present it for the convenience of the reader. It can serve as a refresher for various notions used in the sequel; it also serves to fix notation and terminology. Thus, parts of it can be skipped by the learned reader. Still, some elements of the chapter are not so classical, although widely used in the field of nonsmooth analysis.

The most important results for further use are the Ekeland variational principle expounded in Sect. 1.5 along with a convenient decrease principle, and the application to metric regularity made in Sect. 1.6. The general variational principle of Deville–Godefroy–Zizler obtained in Sect. 1.7 as a consequence of a study of well-posedness will be given a smooth version in Chap. 2. These variational principles are such important tools for nonsmooth analysis that we already display some applications and present in supplements and exercises several variants of interest.

J.-P. Penot, *Calculus Without Derivatives*, Graduate Texts in Mathematics 266, DOI 10.1007/978-1-4614-4538-8_1, © Springer Science+Business Media New York 2013

Among the direct consequences of smooth and nonsmooth variational principles are the study of conditioning of minimization problems, which is tied to the study of error bounds and sufficient conditions in order to get metric regularity. All these applications are cornerstones of optimization theory and nonsmooth analysis. For obtaining calculus rules, variational principles will be adjoined with penalization techniques in order to obtain decoupling processes. These techniques are displayed in Sects. 1.6.4 and 1.6.5 and are rather elementary. These preparations will open an easy route to calculus. But the reader is already provided with tools that give precious information without using derivatives.

1.1 Convergences and Topologies

1.1.1 Sets and Orders

A knowledge of basic set theory is desirable for the reading of the present book, as in various branches of analysis. We assume that the reader has such a familiarity with the standard uses of set theory. But we recall here some elements related to orders, because Zorn's lemma yields (among many other results) the Hahn–Banach theorem, which has itself numerous versions adapted to different situations.

Recall that a *preorder* or partial order or preference relation on a set X is a relation A between elements of X, often denoted by $\leq$, with $A(x) := \{y \in X : x \leq y\}$ that is reflexive ($x \leq x$ or $x \in A(x)$ for all $x \in X$) and transitive ($A \circ A \subset A$ i.e., $x \leq y$, $y \leq z \Rightarrow x \leq z$ for $x, y, z \in X$). One also writes $y \geq x$ instead of $x \leq y$ or $y \in A(x)$ and one reads, y is above x or y is preferred to x. A preorder is an *order* whenever it is *antisymmetric* in the sense that $x \leq y$, $y \leq x \Rightarrow x = y$ for every $x, y \in X$. Two elements x, y of a preordered set $(X, \leq)$ are said to be *comparable* if either $x \leq y$ or $y \leq x$. If such is the case for all pairs of elements of X, one says that $(X, \leq)$ is *totally ordered*. That is not always the case (think of the set $X := \mathscr{P}(S)$ of subsets of a set S with the inclusion or of a modern family with the order provided by authority). Given a subset S of $(X, \leq)$, an element m of X is called an *upper bound* (resp. a *lower bound*) of S if one has $s \leq m$ (resp. $m \leq s$) for all $s \in S$. A preordered set $(I, \leq)$ is *directed* if every finite subset F of I has an upper bound. A subset J of a preordered set $(I, \leq)$ is said to be *cofinal* if for all $i \in I$ there exists $j \in J$ such that $j \geq i$. A map $f : H \to I$ between two preordered spaces is said to be *filtering* if for all $i \in I$ there exists $h \in H$ such that $f(k) \geq i$ whenever $k \in H$ satisfies $k \geq h$. A subset C of $(X, \leq)$ that is totally ordered for the induced preorder is called a *chain*. A preorder on X is said to be *upper inductive* (resp. *lower inductive*) if every chain C has an upper bound (resp. a lower bound). Recall that an element $\bar{x}$ of a preordered space $(X, \leq)$ is said to be *maximal* if for every $x \in X$ such that $\bar{x} \leq x$ one has $x \leq \bar{x}$; it is called *minimal* if it is maximal for the reverse preorder. Zorn's lemma can be stated as follows; it is known to be equivalent to a number of other axioms, such as the axiom of choice, that seem to be very natural axioms. We shall not deal with such aspects of the foundations of mathematics.

Theorem 1.1 (Zorn's lemma or Zorn's axiom). *Every preordered set whose preorder is upper (resp. lower) inductive has at least one maximal (resp. minimal) element.*

Exercises

1. Show that a subset C of a preordered space $(X, \leq)$ is a chain if and only if $C \times C \subset A \cup A^{-1}$, where $A := \{(x,y) : x \leq y\}$, $A^{-1} := \{(x,y) : (y,x) \in A\}$.

2. Let $(I, \leq)$ be a directed set. Show that if $J \subset I$ is not cofinal, then $I \setminus J$ is cofinal.

3. Let $(X, \leq)$ be a preordered space. Check that the relation $<$ defined by $x < y$ if $x \leq y$ and not $y \leq x$ is transitive.

4. A map $f : H \to I$ between two preordered spaces is said to be *homotone* (resp. *antitone*) if $f(h) \leq f(h')$ (resp. $f(h') \leq f(h)$) when $h \leq h'$. It is *isotone* if it is a bijection such that f and f^{-1} are homotone. Show that if f is a homotone bijection, if $(H, \leq)$ is totally ordered, and if $(I, \leq)$ is ordered, then f is isotone.

5. Show that a homotone map $f : H \to I$ between two preordered spaces is filtering if and only if $f(H)$ is cofinal.

6. Let J be a subset of a preordered space $(I, \leq)$. An element s of I is a *supremum* of I if $s \in M := \{m \in I : \forall j \in J, j \leq m\}$ and for all $m \in M$ one has $s \leq m$. Give an example of a subset J of a preordered space $(I, \leq)$ having more than one supremum. Check that when $\leq$ is an order, a subset of I has at most one supremum.

7. Check that when a subset J of a preordered space $(I, \leq)$ has a greatest element k, then k is a supremum of J and for every supremum s of J one has $s \leq k$. Note that when a supremum s of J belongs to J, then s is a greatest element of J.

1.1.2 A Short Refresher About Topologies and Convergences

Most of the sequel takes place in normed spaces. However, it may be useful to use the concepts of metric spaces and to have some notions of general topology. In particular, we will use weak* topologies on dual Banach spaces. We will not attempt to give an axiomatic definition of convergence (however, see Exercise 4). But it is useful to master some notions of topology. Pointwise convergence of functions cannot enter the framework of normed spaces or even metric spaces.

A topology on a set X is obtained by selecting a family of subsets called the family of closed subsets having a stability property in terms of convergence.

Equivalently, one usually defines a *topology* on X as the data of a family $\mathcal{O}$ of so-called *open* subsets that satisfies the following two requirements:

(O1) The union of any subfamily of $\mathcal{O}$ belongs to $\mathcal{O}$.

(O2) The intersection of any finite subfamily of $\mathcal{O}$ belongs to $\mathcal{O}$.

By convention, we admit that these two conditions include the requirements that X and the empty set $\varnothing$ belong to $\mathcal{O}$. A topological space $(X, \mathcal{O})$ is also denoted by X if the choice of the topology $\mathcal{O}$ is unambiguous. A subset F of X is declared to be *closed* if $X \setminus F$ belongs to $\mathcal{O}$. The *closure* $\mathrm{cl}(S)$ of a subset S of a topological space $(X, \mathcal{O})$ is the intersection of the family of all closed subsets of X containing S. It is clearly the smallest closed subset of $(X, \mathcal{O})$ containing S. The *interior* $\mathrm{int}(T)$ of a subset T of $(X, \mathcal{O})$ is the set $X \setminus \mathrm{cl}(S)$, where $S := X \setminus T$. It is clearly the union of all the open subsets of $(X, \mathcal{O})$ contained in T. A subset D of $(X, \mathcal{O})$ is said to be *dense* in a subset E of X if $D \subset E$ and if E is contained in the closure of D. A topological space is said to be *separable* if it contains a countable dense subset.

A map $f : (X, \mathcal{O}) \to (X', \mathcal{O}')$ between two topological spaces is said to be *continuous* if for every $O' \in \mathcal{O}'$ its inverse image $f^{-1}(O') := \{x \in X : f(x) \in O'\}$ belongs to $\mathcal{O}$. The composition of two continuous maps is clearly continuous.

A topology $\mathcal{O}'$ on X is said to be *weaker* than a topology $\mathcal{O}$ if the *identity map* $I_X : (X, \mathcal{O}) \to (X, \mathcal{O}')$ is continuous, i.e., if any member of $\mathcal{O}'$ is in $\mathcal{O}$, i.e., if $\mathcal{O}' \subset \mathcal{O}$. Given a family $\mathcal{G}$ of subsets of a set X, there is a topology $\mathcal{O}$ on X that is the weakest among those containing $\mathcal{G}$. Then one says that $\mathcal{G}$ *generates* $\mathcal{O}$. If $\mathcal{B} \subset \mathcal{O}$ is such that every element of $\mathcal{O}$ is an union of elements of $\mathcal{B}$, one says that $\mathcal{B}$ is a *base* of $\mathcal{O}$. It is easy to check that when $\mathcal{G}$ generates $\mathcal{O}$, the family $\mathcal{B}$ of finite intersections of elements of $\mathcal{G}$ is a base of $\mathcal{O}$. A subset V of a topological space $(X, \mathcal{O})$ is a *neighborhood* of some $\bar{x} \in X$ if there exists some $U \in \mathcal{O}$ such that $\bar{x} \in U \subset V$. A family $\mathcal{U}$ of subsets of X is a *base of neighborhoods* of $\bar{x}$ if $\mathcal{U}$ is contained in the family $\mathcal{N}(\bar{x})$ of neighborhoods of $\bar{x}$ and if for every $V \in \mathcal{N}(\bar{x})$ there exists some $U \in \mathcal{U}$ such that $U \subset V$. Given $\mathcal{B} \subset \mathcal{O}$, we see that $\mathcal{B}$ is a base of $\mathcal{O}$ iff (if and only if) for all $x \in X$, $\mathcal{B}(x) := \{U \in \mathcal{B} : x \in U\}$ is a base of neighborhoods of x.

The notion of continuity can be localized by using neighborhoods or neighborhood bases. A map $f : (X, \mathcal{O}) \to (X', \mathcal{O}')$ is said to be *continuous at* $\bar{x} \in X$ if for every neighborhood V' of $f(\bar{x})$ in $(X', \mathcal{O}')$ there exists some $V \in \mathcal{N}(\bar{x})$ such that $f(V) \subset V'$. One can easily show that f is continuous if and only if it is continuous at every point of X.

To a topology $\mathcal{O}$ on X, one can associate a convergence for nets and sequences in X. Recall that a *net* (or generalized sequence) $(x_i)_{i \in I}$ in X is a mapping $i \mapsto x_i$ from a directed (preordered) set I to X. A *subnet* of a net $(x_i)_{i \in I}$ is a net $(y_j)_{j \in J}$ such that there exists a mapping $\theta : J \to I$ that is filtering and such that $y_j = x_{\theta(j)}$ for all $j \in J$. Note that in contrast to what occurs for subsequences, one takes for J a directed set that may differ from I. It is often of the form $J := I \times K$, where K is another directed set, or a subset of $I \times K$. One says that $(x_i)_{i \in I}$ *converges* to some $\bar{x} \in X$ if for every $V \in \mathcal{N}(\bar{x})$ one can find some $i_V \in I$ such that $x_i \in V$ for all $i \geq i_V$. Then one writes $(x_i)_{i \in I} \to \bar{x}$ or $\bar{x} = \lim_{i \in I} x_i$. One says that $(x_i)_{i \in I}$ has a *cluster point* $\bar{x} \in X$ if for every $V \in \mathcal{N}(\bar{x})$ and every $h \in I$ one can find some $i \in I$ such that $i \geq h$

and $x_i \in V$. One can show that $\bar{x} \in X$ is a cluster point of $(x_i)_{i \in I}$ if and only if there exists a subnet of $(x_i)_{i \in I}$ that converges to $\bar{x}$. The "if" condition is immediate. For the necessary condition one can take $J := \{(i, V) \in I \times \mathcal{N}(\bar{x}) : x_i \in V\}$, a cofinal subset of $I \times \mathcal{N}(\bar{x})$ for the product order, and define $\theta : J \to I$ by $\theta(i, V) := i$. The topology $\mathcal{O}$ on X is uniquely determined by its associated convergence for nets: a subset C of X is closed iff it contains the limits of its convergent nets. In general $\mathcal{O}$ is not determined by the convergence of sequences. The aim of the following proposition is to give the reader some familiarity with nets.

Proposition 1.2. *A map $f : X \to Y$ between two topological spaces is continuous at $\bar{x} \in X$ if and only if for every net $(x_i)_{i \in I}$ of X converging to $\bar{x}$, the net $(f(x_i))_{i \in I}$ converges to $f(\bar{x})$. Sequences can be used when $\mathcal{N}(\bar{x})$ has a countable base.*

Proof. Necessity is immediate. Let us show sufficiency. Suppose f is not continuous at $\bar{x}$. Then there exists $V \in \mathcal{N}(f(\bar{x}))$ such that for all $U \in \mathcal{N}(\bar{x})$ there exists some $x_U \in U$ with $f(x_U) \notin V$. Then for the net $(x_U)_{U \in \mathcal{N}(\bar{x})}$ we have $(x_U)_{U \in \mathcal{N}(\bar{x})} \to \bar{x}$, but $(f(x_U))_{U \in \mathcal{N}(\bar{x})}$ does not converge to $f(\bar{x})$.   □

Let X or (X, d) be a *metric space* i.e., a pair formed by a space X and a function $d : X \times X \to \mathbb{R}_+$ such that for all $x, x', x'' \in X$, one has $d(x, x') = d(x', x)$, $d(x, x') = 0$ iff $x = x'$, and the so-called triangle inequality $d(x, x'') \le d(x, x') + d(x', x'')$. Then the function d, called a *metric*, induces a topology $\mathcal{O}$ on X defined by $G \in \mathcal{O}$ iff for all $x \in G$ there exists some $r > 0$ such that the *open ball* $B(x, r) := \{x' \in X : d(x, x') < r\}$ is contained in G. Thus $\mathcal{O}$ is the topology generated by the family of open balls. In fact, this family is a base of $\mathcal{O}$ and for all $\bar{x} \in X$, the family of open balls centered at $\bar{x}$ is a base of neighborhoods of $\bar{x}$. In the sequel, the *closed ball* with center x and radius $r \in \mathbb{R}_+$ is the set

$$B[x, r] := \{x' \in X : d(x, x') \le r\}.$$

The family of closed balls centered at x with positive radius is also a base of neighborhoods of x. Thus continuity can be expressed with the help of ε's and δ's. A map $f : (X, d) \to (X', d')$ is said to be *Lipschitzian* if there exists some $c \in \mathbb{R}_+$ such that $d'(f(x_1), f(x_2)) \le cd(x_1, x_2)$ for all $x_1, x_2 \in X$. The constant c is called a Lipschitz constant (or rate, or rank). The least such constant is called the (exact) *Lipschitz rate* of f. If this rate is 1, f is said to be *nonexpansive*. For $x \in X$ the Lipschitz rate of f at x is the infimum of the Lipschitz rates of the restrictions of f to the neighborhoods of x (and $+\infty$ if there is no neighborhood of x on which f is Lipschitzian). If for all $x \in X$ there is a neighborhood V of x such that the restriction $f \mid V$ is Lipschitzian, f is said to be *locally Lipschitzian*. On the product $Z := X \times Y$ of two metric spaces (X, d_X), (Y, d_Y) a metric d is called a *product metric* if the canonical projections $p_X : Z \to X$, $p_Y : Z \to Y$ and the insertions $j_b : x \mapsto (x, b)$, $j_a : y \mapsto (a, y)$ are nonexpansive.

The structure of metric spaces is richer than the structure of topological spaces. In particular, one has the notion of a Cauchy sequence: a sequence (x_n) of (X, d) is called a *Cauchy sequence* if $(d(x_n, x_p)) \to 0$ as $n, p \to +\infty$. A metric space is said to be *complete* if its Cauchy sequences are convergent. Since a sequence (x_n) of (X,d) converges to some $x \in X$ iff $(d(x_n,x)) \to 0$ as $n \to +\infty$, convergence is reduced to convergence of real numbers.

Whereas the product X of a family of metric spaces (X_s,d_s) ($s \in S$, an arbitrary set) cannot be provided with a (sensible) metric in general, a *product* of topological spaces $(X_s,\mathcal{O}_s)$ ($s \in S$) can always be endowed with a topology $\mathcal{O}$ that makes the projections $p_s : X \to X_s$ continuous and that is as weak as possible. It is the topology generated by the sets $p_s^{-1}(O_s)$ for $s \in S$, $O_s \in \mathcal{O}_s$. Its associated convergence is componentwise convergence. When $(X_s,\mathcal{O}_s) := (Y,\mathcal{O}_Y)$ for all $s \in S$, identifying the product X with the set Y^S of maps from S to Y, the convergence associated with the product topology $\mathcal{O}$ on X coincides with *pointwise convergence*: $(f_i)_{i \in I} \to f$ in Y^S if for all $s \in S$, $(f_i(s))_{i \in I} \to f(s)$. When $\mathcal{O}_Y$ is the topology associated with a metric d_Y on Y, a stronger convergence can be defined on Y^S: it is the so-called *uniform convergence* for which $(f_i)_{i \in I} \to f$ iff $(d_\infty(f_i,f)) := (\sup_{s \in S} d_Y(f_i(s),f(s))) \to 0$.

In metric spaces one has notions of uniformity that are more demanding than their topological counterparts. In particular, a map $f : (X,d_X) \to (Y,d_Y)$ is *uniformly continuous* if for all $\varepsilon \in \mathbb{P} := (0,+\infty)$ one can find some $\delta \in \mathbb{P}$ such that $d_Y(f(x),f(x')) < \varepsilon$ whenever $x,x' \in X$ satisfy $d_X(x,x') < \delta$. One can show that f is uniformly continuous iff there exists a *modulus* μ (i.e., a function $\mu : \mathbb{R}_+ \to \overline{\mathbb{R}}_+ := \mathbb{R}_+ \cup \{+\infty\}$ continuous at 0 with $\mu(0) = 0$) such that $d_Y(f(x),f(x')) \leq \mu(d_X(x,x'))$ for all $x,x' \in X$. Such a modulus is called a *modulus of uniform continuity*. The example of $f : x \mapsto x^2$ shows that uniform continuity is more exacting than continuity.

Given two topological spaces $(W,\mathcal{O})$, $(X',\mathcal{O}')$, a subset X of W, and $w \in \mathrm{cl}(X)$, one says that $f : X \to X'$ has a limit $\bar{x}'$ as $x \to_X w$ (i.e., $x \to w$ with $x \in X$), or that f converges to $\bar{x}'$ as $x \to_X w$, and one writes $\bar{x}' = \lim_{x \to_X w} f(x)$, if for every $V' \in \mathcal{N}(\bar{x}')$ there exists $V \in \mathcal{N}(w)$ such that $f(V \cap X) \subset V'$. If $X = W$, one just writes $\bar{x}' = \lim_{x \to w} f(x)$. Thus, f is continuous at w if and only if f has the limit $f(w)$ as $x \to w$. Given another map $g : X \to Y$ with values in another topological space $(Y,\mathcal{G})$ and some $\bar{y} \in Y$, one says that f has a limit $\bar{x}'$ as $g(x) \to \bar{y}$, or that f converges to $\bar{x}'$ as $g(x) \to \bar{y}$, and one writes $\bar{x}' = \lim_{g(x) \to \bar{y}} f(x)$, if for every $V' \in \mathcal{N}(\bar{x}')$ there exists $W \in \mathcal{N}(\bar{y})$ such that $f(x) \in V'$ for all $x \in g^{-1}(W)$. Taking for g the canonical injection of X into $(Y,\mathcal{G}) = (W,\mathcal{O})$, one recovers the preceding notion of limit.

Given a directed set $(I,\leq)$, let $I_\infty := I \cup \{\infty\}$, where ∞ is an additional element satisfying $i \leq \infty$ for all $i \in I$. Then one can endow I_∞ with the topology $\mathcal{O}$ defined by $G \in \mathcal{O}$ if either G is contained in I or there exists some $h \in I$ such that $i \in G$ for all $i \in I_\infty$ such that $i \geq h$. Given a topological space $(X,\mathcal{O})$, $x \in X$, and a net $(x_i)_{i \in I}$ of X, one easily checks that $(x_i)_{i \in I} \to x$ if and only if the map $f : I_\infty \to X$ given by $f(i) := x_i$, $f(\infty) := x$ is continuous at ∞, if and only if f has limit x as $i \to_I \infty$.

A topological space $(X,\mathcal{O})$ is said to be *Hausdorff* (or T_2) if for every $x,x' \in X$ with $x \neq x'$ one can find $V \in \mathcal{N}(x)$, $V' \in \mathcal{N}(x')$ such that $V \cap V' = \varnothing$. Then the limit of a net of X is unique. A topological space $(X,\mathcal{O})$ is said to be *regular* (or T_3) if every $x \in X$ has a base of neighborhoods that are closed. Hausdorff *topological groups*, i.e., groups endowed with a topology for which the operation and the inversion are continuous, and metric spaces are regular. A topological space $(X, \mathcal{O})$ is said to be *compact* if it is Hausdorff and if every net of X has a convergent subnet. Equivalently, $(X, \mathcal{O})$ is compact if for every *covering* (i.e., a family of subsets whose union is X) of X by open sets has a finite subfamily that is still a covering of X. Another characterization (obtained by taking complements) is that every family $(C_i)_{i \in I}$ that has the finite intersection property has a nonempty intersection; here $(C_i)_{i \in I}$ has the *finite intersection property* if for every finite subset J of I the intersection of the family $(C_j)_{j \in J}$ is nonempty. If (X, d) is a metric space, then X is compact if and only if it is complete and precompact. *Precompact* means that for every $\varepsilon > 0$ one can find a finite subset F of X such that $X = B(F, \varepsilon) := \{x \in X : \exists a \in F, d(x, a) < \varepsilon\}$.

We admit the following famous theorem, whose proof requires Zorn's lemma when the product has an infinite number of factors.

Theorem 1.3 (Tykhonov). *The product of a family of compact topological spaces is compact.*

We define the *lower limit* (resp. *upper limit*) of a net $(r_i)_{i \in I}$ of real numbers by

$$\liminf_{i \in I} r_i := \sup_{h \in I} \inf_{i \in I, i \geq h} r_i \quad (\text{resp. } \limsup_{i \in I} r_i := \inf_{h \in I} \sup_{i \in I, i \geq h} r_i).$$

These substitutes for the limit always exist in $\overline{\mathbb{R}} := \mathbb{R} \cup \{-\infty, +\infty\}$. One can show that $\liminf_{i \in I} r_i$ is the least cluster point of the net $(r_i)_{i \in I}$ in the compact space $\overline{\mathbb{R}}$. A similar characterization holds for $\limsup_{i \in I} r_i$.

Exercises

1. Prove the assertions of this section given without proofs. In particular, prove Tykhonov's theorem first for a product of two spaces, then in the general case.

2. Let (X, d) be a metric space. Check that the functions $d' := \min(d, 1)$ and $d'' := d/(1 + d)$ are bounded metrics inducing on X the topology associated with d.

3. (Urysohn's theorem) Let S be a closed subset of a metric space (X, d) and let $f : S \to \mathbb{R}$ be a continuous function. Prove that there exists a continuous function $g : X \to \mathbb{R}$ such that $g(x) = f(x)$ for $x \in S$ and such that $g(X) \subset [\alpha, \beta]$ if $f(S) \subset [\alpha, \beta]$.

4. (Convergence space) Show that the convergence of nets in a topological space $(X, \mathcal{O})$ satisfies the following properties:

(a) For every $x \in X$ the constant net with value x converges to x.
(b) If $(x_i)_{i \in I} \to x$ and if $(x_j)_{j \in J}$ is a subnet of $(x_i)_{i \in I}$, then $(x_j)_{j \in J} \to x$.
(c) If $x \in X$ and $(x_i)_{i \in I}$ is a net of X such that for every subnet $(x_j)_{j \in J}$ of $(x_i)_{i \in I}$ there exists a subnet $(x_k)_{k \in K}$ of $(x_j)_{j \in J}$ that converges to x, then $(x_i)_{i \in I} \to x$.

These three properties can be taken for axioms of convergence spaces (also called spaces with limits). Check these axioms for the convergence defined in Sect. 1.3.1 on the power set $\mathscr{P}(X)$ of a topological space X.

5. Show that a compact space is regular.

6. Show that $\bar{x} \in X$ is in the closure of a subset S of a topological space X iff there exists a net $(x_i)_{i \in I}$ of S that converges to $\bar{x}$. Show that $\bar{x}$ is in the interior of S iff for every net $(x_i)_{i \in I} \to \bar{x}$ there exists $h \in I$ such that $x_i \in S$ for all $i \geq h$.

7. Prove **Lebesgue's lemma**: given a *sequentially compact* subset K of a metric space (X, d), i.e., a subset such that every sequence of K has a subsequence converging in K and given a family $(O_i)_{i \in I}$ of open subsets of X whose union contains K, there exists some $r > 0$ such that for all $x \in K$ the ball $B(x, r)$ is contained in some O_i.

8. Show that a metric space is separable iff it has a countable base of open sets.

9. (The window lemma) Let K (resp. L) be a compact subset of a topological space X (resp. Y) and let W be an open subset of $X \times Y$ containing $K \times L$. Prove that one can find an open subset U of X containing K and an open subset V of Y containing L such that $U \times V \subset W$. Give an interpretation in terms of building.

10. Let X, Y be topological spaces and let $C := C(X, Y)$ be the set of continuous maps from X to Y. The *compact-open topology* on C is the topology generated by the sets $W(K, G) := \{f \in C : f(K) \subset G\}$, where G is an open subset of Y and K is a compact subset of X.

(a) Show that a topology $\mathscr{T}$ on C such that the evaluation map $e : C \times X \to Y$ given by $e(f, x) := f(x)$ is continuous is finer than the compact-open topology. [Hint: Given $f \in W(K, G)$ use the preceding exercise to find some open neighborhood V of f for $\mathscr{T}$ such that $V \times K \subset e^{-1}(G)$.]

(b) Check that when X is locally compact the evaluation map e is continuous.

(c) Assuming that Y is a metric space, compare the compact-open topology with the topology of uniform convergence on compact subsets of X.

1.1.3 Weak Topologies

As mentioned above, we will work essentially in *normed spaces*, i.e., linear spaces equipped with a norm (see below for the definition if you have forgotten it) or even in *Banach spaces*, i.e., complete normed spaces. We assume the reader is familiar with such a framework, which is an important class of metric spaces, in which the metric d associated to a norm $\|\cdot\|$ on a linear space X is given by $d(x, y) = \|x - y\|$. A linear map $\ell : X \to Y$ between two normed spaces $(X, \|\cdot\|_X)$, $(Y, \|\cdot\|_Y)$ is continuous (and in fact Lipschitzian) if and only if there exists some $c \in \mathbb{R}_+$ such that $\|\ell(x)\|_Y \leq c \|x\|_X$ for all $x \in X$. The least such constant is called the *norm* of ℓ. It turns the space

$L(X,Y)$ of continuous linear maps from X to Y into a normed space that is complete whenever Y is complete. In particular, the (topological) *dual space* $X^* := L(X,\mathbb{R})$ of a normed space X is always a Banach space. A norm $\|\cdot\|$ on the product $X \times Y$ of two normed spaces is a *product norm* if its associated metric is a product metric. That amounts to the following inequalities for all $(x,y) \in X \times Y$:

$$\|(x,y)\|_\infty := \max(\|x\|_X, \|y\|_Y) \le \|(x,y)\| \le \|(x,y)\|_1 := \|x\|_X + \|y\|_Y.$$

In the infinite-dimensional case, it appears that compact subsets are scarce. A natural means to get a richer family of compact subsets on a normed space $(X, \|\cdot\|)$ is to weaken the topology: then there will be more convergent nets, and since open covers will be not as rich, finding finite subfamilies will be easier. The drawbacks are that continuity of maps issued from X will be lost in general and that no norm will be available to define the weakened topology if X is infinite-dimensional. A partial remedy for the first inconvenience will be proposed in the next subsection. Now, the lack of a norm will not be too dramatic if one realizes that the structure of a *topological linear space* is preserved. That means that the two operations $(x,y) \mapsto x + y$ and $(\lambda, x) \mapsto \lambda x$ will be continuous for the new topology. One will even have a family of seminorms defining the topology, a *seminorm* on a linear space X being a function $p : X \to \mathbb{R}_+$ that is *subadditive* (i.e., such that $p(x+y) \le p(x) + p(y)$ for all $(x,y) \in X \times X$) and *absolutely homogeneous* (i.e., such that $p(\lambda x) = |\lambda| p(x)$ for all $(\lambda, x) \in \mathbb{R} \times X$) or *positively homogeneous* (i.e., such that $p(\lambda x) = \lambda p(x)$ for all $(\lambda, x) \in \mathbb{R}_+ \times X$) and *even* (i.e., such that $p(-x) = p(x)$ for all $x \in X$). Note that a seminorm p is a *norm* iff $p^{-1}(0) = \{0\}$. The topology associated with a family $(p_i)_{i \in I}$ of seminorms on X is the topology generated by the family of semiballs $B_i(a,r) := \{x \in X : p_i(x-a) < r\}$ for all $a \in X, r \in \mathbb{P}, i \in I$. Such a topology is clearly compatible with the operations on X, so that X becomes a topological linear space. It is even a *locally convex topological linear space* in the sense that each point has a base of neighborhoods that are convex. One can show that this property is equivalent to the existence of a family of seminorms defining the topology.

On the (topological) *dual space* X^* of a topological linear space X, i.e., on the space of continuous linear forms on X, a natural family of seminorms is the family $(p_x)_{x \in X}$ given by $p_x(f) := |f(x)|$ or, with a notation we will use frequently, $p_x(x^*) := |\langle x^*, x \rangle|$. Then a net $(f_i)_{i \in I}$ of X^* converges to some $f \in X^*$ if and only if for all $x \in X$, $(f_i(x))_{i \in I} \to f(x)$; then we write $(f_i)_{i \in I} \overset{*}{\to} f$. Thus, the obtained topology on X^*, called the *weak* topology*, is just the topology w^* induced by pointwise convergence. It is the weakest topology on X^* for which the evaluations $f \mapsto f(x)$ are continuous for all $x \in X$. Although this topology is poor, it preserves some continuity properties. In particular, if X and Y are normed spaces and if $A \in L(X,Y)$, its *transpose* (often called the adjoint) $A^\mathsf{T} : Y^* \to X^*$ defined by $A^\mathsf{T}(y^*) := y^* \circ A$ for $y^* \in Y^*$ or

$$\langle A^\mathsf{T}(y^*), x \rangle = \langle y^*, A(x) \rangle, \qquad (x, y^*) \in X \times Y^*,$$

is not just continuous for the topologies induced by the dual norms (the so-called strong topologies); it is also continuous for the weak* topologies: when $(y_i^*)_{i \in I} \xrightarrow{*} y^*$ one has $(A^{\mathsf{T}}(y_i^*))_{i \in I} \xrightarrow{*} A^{\mathsf{T}}(y^*)$, since for all $x \in X$ one has $(A^{\mathsf{T}}(y_i^*)(x))_{i \in I} = (y_i^*(A(x)))_{i \in I} \to y^*(A(x))$. Note that when X and Y are *Hilbert spaces* (i.e., Banach spaces whose norms derive from scalar products), so that they can be identified with their dual spaces, A^{T} corresponds to the *adjoint* $A^* : Y \to X$ of A characterized by $(A^*(y) \mid x)_X = (y \mid Ax)_Y$ for all $x \in X, y \in Y$, $(\cdot \mid \cdot)_X$ (resp. $(\cdot \mid \cdot)_Y$) denoting the scalar product in X (resp. Y).

Let us show that there are sufficiently many linear forms on X^* that are continuous for the weak* topology.

Proposition 1.4. *The set of continuous linear forms on X^* endowed with the weak* topology can be identified with X.*

Proof. By definition, for all $x \in X$, the linear form $e_x : x^* \mapsto \langle x^*, x \rangle$ on X^* is continuous for the weak* topology. Let us show that every continuous linear form f on (X^*, w^*) coincides with some e_x. We can find $\delta > 0$ and a finite family $(a_1, \ldots, a_m)$ in X such that $|f(x^*)| < 1$ for all $x^* \in X^*$ satisfying $p_{a_i}(x^*) := |\langle x^*, a_i \rangle| < \delta$ for $i \in \mathbb{N}_m := \{1, \ldots, m\}$. Setting $x_i := a_i/\delta$, we get $|f(x^*)| \le \max_{1 \le i \le k} |\langle x^*, x_i \rangle|$, since otherwise, by homogeneity, we can find $x^* \in X^*$ such that $|f(x^*)| = 1$ and $\max_{1 \le i \le m} |\langle x^*, x_i \rangle| < 1$, a contradiction to the choice of a_i and x_i. Changing the indexing if necessary, we may suppose that for some $k \in \mathbb{N}_m$, $x_1, \ldots, x_k$ form a basis of the linear space spanned by $x_1, \ldots, x_m$. Let $A := (x_1, \ldots, x_k) : X^* \to \mathbb{R}^k$. Then denoting by N the kernel of A and by $p : X^* \to X^*/N$ the canonical projection, f can be factorized into $f = g \circ p$ for some linear form g on X^*/N. There is also an isomorphism $B : X^*/N \to \mathbb{R}^k$ such that A is factorized into $A = B \circ p$. Then $f = g \circ B^{-1} \circ A$, whence $f(x^*) = c_1 x^*(x_1) + \cdots + c_k x^*(x_k)$ for all $x^* \in X^*$, where $c_1, \ldots, c_k$ are the components of $g \circ B^{-1}$ in $(\mathbb{R}^k)^*$. Thus $f = e_x$ for $x := c_1 x_1 + \cdots + c_k x_k \in X$. □

The following result shows that in introducing the weak* topology we have attained our aim of obtaining sufficiently many compact subsets.

Theorem 1.5 (Alaoglu–Bourbaki). *Every weak* closed, bounded subset of the dual space X^* is weak* compact, i.e., is compact for the weak* topology.*

Proof. It suffices to show that the closed unit ball $B^* := B_{X^*}$ of X^* is weak* compact. To do so, let us denote by S the closed unit sphere S_X of X, by H the space of positively homogeneous functions on X, and by H_S the space of all the restrictions to S of the elements of H. The restriction operator $r : H \to H_S$ is then a bijection, with inverse given by $r^{-1}(h)(x) = th(t^{-1}x)$ for $x \in X \setminus \{0\}, t := \|x\|, r^{-1}(h)(0) = 0$. Then r and r^{-1} are continuous for the pointwise convergence topologies on H and H_S, and for this topology H_S is homeomorphic to the product space $\mathbb{R}^S$. The subset B^* of H is easily seen to be closed for the pointwise convergence topology on H. Moreover, $r(B^*)$ is contained in $[-1, 1]^S$, which is compact, by Tikhonov's theorem. Thus, $r(B^*)$ and B^* are compact in H_S and H respectively. It follows that B^* is compact in X^* endowed with the weak* topology. □

The *weak topology* on X is the topology induced by the embedding of X into $X^{**} := (X^*)^*$ given by $j(x)(x^*) := x^*(x)$ for $x \in X, x^* \in X^*$. Thus, it is the topology induced by the family $(p_f)_{f \in X^*}$ of seminorms on X given by $p_f(x) := |f(x)|$ for $x \in X$, $f \in X^*$. The definition of the dual norm on X^{**} shows that for all $x \in X$ one has $\|j(x)\|_{X^{**}} \leq \|x\|$; in fact, Corollary 1.172 of Subsection 1.4.2 shows that equality holds, so that j is a *monometry*, i.e., an isometry from X onto its image $j(X)$. When j is surjective (hence an isometry), X is said to be *reflexive*. In general, the weak topology does not provide compact subsets as easily as does the weak* topology. However, when X is reflexive, since then the weak topology coincides with the weak* topology obtained by considering X as the dual of X^*, we do get a rich family of compact subsets. We state this fact in the following corollary; its second assertion depends on another consequence of the Hahn–Banach theorem, asserting that closed convex subsets of a Banach space are weakly closed. It will be displayed later.

Corollary 1.6. *Every bounded weakly closed subset of a reflexive Banach space X is weakly compact. In particular, every bounded closed convex subset of X is weakly compact.*

When X is finite-dimensional, the weak* topology on X^* coincides with the strong topology (and similarly, the weak topology of X coincides with the topology associated with the norm). In fact, a net $(f_i)_{i \in I}$ of X^* converges to some $f \in X^*$ if and only if for every element b of a base of X the net $(f_i(b))_{i \in I}$ converges to $f(b)$, and this is enough to imply the convergence for the dual norm. If X is infinite-dimensional, the weak (resp. weak*) topology never coincides with the strong topology (the one induced by the norm or dual norm). This stems from the fact that no neighborhood V of 0 for the weak* topology is bounded, since it contains the intersection of the kernels of a finite family of linear forms.

Let us give without proofs some results of interest that are outside the scope of our purposes, although they have some bearing on our study in the reflexive case. We refer to [329, 337, 376, 507] for the proofs.

Theorem 1.7 (Kaplanski). *If $\bar{x}$ is in the weak closure w-cl(S) of a subset S of a Banach space X, then there exists a countable subset D of S such that $\bar{x} \in$ w-cl(D).*

Theorem 1.8 (Eberlein–Šmulian). *For a subset S of a Banach space X the following assertions are equivalent:*

(a) w-cl(S) *is weakly compact;*
(b) Every sequence of S has a weakly convergent subsequence;
(c) Every sequence of S has a weak cluster point.

Theorem 1.9. *A normed space X is reflexive if and only if its closed unit ball B_X is weakly sequentially compact in the sense that every sequence of B_X has a weakly convergent subsequence in B_X.*

For the next result (whose proof is given in [329, p. 426] for instance (f.i.)), we recall that a topological space is said to be *metrizable* if its topology can be defined by a metric.

Theorem 1.10. (a) *The topology induced by the weak topology on every bounded subset of a Banach space X is metrizable if and only if X^* is separable.*
(b) *The topology induced by the weak* topology on every bounded subset of the dual X^* of a Banach space X is metrizable if and only if X is separable.*

In fact, if X is separable and if $D := \{x_n : n \in \mathbb{N}\}$ is dense in B_X, then the topology induced by the weak* topology on a bounded subset of X^* coincides with the topology induced by the norm $\|\cdot\|_D$ on X^* defined by

$$\|x^*\|_D := \sum_{n \geq 0} 2^{-n} |x^*(x_n)| .$$

It may be useful to consider a topology $\mathscr{W}_b$ or bw* on the dual X^* of a normed space X, called the *bounded weak* topology*, that is weaker than the strong (or norm) topology and stronger than the weak* topology. It is the strongest topology agreeing with the weak* topology on weak* compact subsets of X^*. Thus, a subset of $(X^*, \mathscr{W}_b)$ is closed if and only if for all $r > 0$ its intersection with rB_{X^*} is weak* closed. For convex subsets, no change occurs, thanks to the following result, due to Banach and Dieudonné when C is a linear subspace.

Theorem 1.11 (Krein–Šmulian). *A convex subset C of the dual of a Banach space X is bw*-closed if and only if it is weak* closed.*

Theorem 1.12. *A linear form f on the dual X^* of a Banach space X is continuous for the bw*-topology if and only if it is continuous for the weak* topology, if and only if it is of the form $x^* \mapsto x^*(x)$ for some $x \in X$.*

Theorem 1.13. *The bounded weak* topology $\mathscr{W}_b$ on the dual X^* of a Banach space X coincides with the topology of uniform convergence on compact subsets of X.*

Thus, the bw*-topology is defined by the family of seminorms $x^* \mapsto p_K(x^*) := \sup\{|x^*(x)| : x \in K\}$, where K belongs to the family $\mathscr{K}(X)$ of compact subsets of X or to the family $\mathscr{K}_s(X)$ of symmetric $(-K = K)$ compact subsets of X. Note that when $K \in \mathscr{K}_s(X)$, p_K is the support function of K: $p_K(x^*) = \sup\{x^*(x) : x \in K\}$. Since the family $\mathscr{K}_s(X)$ is stable by dilations, we get that a basis of neighborhoods of 0 for the bw*-topology is formed by the family of *polar sets*

$$K^0 := \{x^* \in X^* : x^*(x) \leq 1 \,\forall x \in K\}, \qquad K \in \mathscr{K}_s(X).$$

Proof. First, let us prove that for every compact subset K of X the set K^0 is a neighborhood of 0 for the topology $\mathscr{W}_b$. It suffices to show that

$$V := \{x^* \in X^* : \forall x \in K \; x^*(x) < 1\}$$

is open for $\mathscr{W}_b$, since $0 \in V \subset K^0$. This means that for all $r > 0$, $V \cap rB_{X^*}$ is open in the topology on rB_{X^*} induced by the weak* topology. Let $\bar{x}^* \in V \cap rB_{X^*}$. There exists $s > 0$ such that $\bar{x}^*(x) \leq 1 - s$ for all $x \in K$. Let F be a finite subset of X such that $2K$ is contained in the union of the open balls $B(a, s/2r)$ for $a \in F$. Let $x^* \in (\bar{x}^* + sF^0) \cap rB_{X^*}$, $x^* \neq \bar{x}^*$ and let $w^* := x^* - \bar{x}^* \in sF^0 \cap 2rB_{X^*}$. Given $x \in K$, we pick $a \in F$ such that $\|2x - a\| < s/2r$, so that

$$x^*(x) = \bar{x}^*(x) + (1/2)w^*(2x) \leq 1 - s + (1/2)w^*(a) + (1/2)w^*(2x - a)$$

$$< 1 - s + (1/2)s + (1/2)\|w^*\|(s/2r) \leq 1.$$

Thus $(\bar{x}^* + (s^{-1}F)^0) \cap rB_{X^*} \subset V$. This shows that $V \cap rB_{X^*}$ is a neighborhood of $\bar{x}^*$ in the topology on rB_{X^*} induced by the weak* topology, hence that $V \cap rB_{X^*}$ is open in that topology. Therefore V is open for $\mathscr{W}_b$.

Now, given an open neighborhood V of 0 for the bw*-topology $\mathscr{W}_b$ on X^*, let us find $K \in \mathscr{K}(X)$ such that $K^0 \subset V$. By definition of $\mathscr{W}_b$, for all $n \in \mathbb{N} \setminus \{0\}$, there exists a finite subset $F_n \subset X$ such that $F_n^0 \cap nB_{X^*} \subset V$. Let us first show that there is a finite subset A_n of $(1/n)B_X$ such that $F_n^0 \cap A_n^0 \cap (n+1)B_{X^*} \subset V$. Suppose the contrary. Then the family of sets

$$F_n^0 \cap A^0 \cap (n+1)B_{X^*} \cap (X \setminus V),$$

for A in the family of finite subsets of $(1/n)B_X$, has the finite intersection property (we use the fact that $B^0 \cap C^0 = (B \cup C)^0$). Since $(n+1)B_{X^*} \cap (X \setminus V)$ is a weak* closed subset of the weak* compact set $(n+1)B_{X^*}$, the intersection of this family is nonempty. Take x^* in this intersection. Then $x^* \in F_n^0 \cap (n+1)B_{X^*}$, $x^* \in X \setminus V$, and $x^*(x) \leq 1$ for all $x \in (1/n)B_X$. Thus $\|x^*\| \leq n$ and $x^* \in F_n^0 \cap nB_{X^*} \subset V$, a contradiction to $x^* \in X \setminus V$.

Setting $F_{n+1} := F_n \cup A_n$, starting with $F_0 := \{0\}$, we construct inductively a sequence (F_n) of finite subsets of X such that $F_{n+1} \setminus F_n \subset (1/n)B_X$ and $F_n^0 \cap nB_{X^*} \subset V$ for all n. The union K of the family (F_n) is easily shown to be compact, and one has $K^0 \subset V$. □

Exercises

1. The purpose of this exercise is to show that weak continuity of continuous maps cannot be expected in general in an infinite-dimensional Banach space X.

(a) Prove that the unit sphere S_X of X is dense in the closed unit ball B_X for the weak topology σ.

(b) Check the continuity of the retraction $r : X \to B_X$ given by $r(x) := x/\max(\|x\|, 1)$.

(c) Given $x \in X$ such that $\|x\| = 1/2$, let $(x_i)_{i \in I}$ be a net of S_X weakly convergent to x. Observe that $(r(2x_i))_{i \in I} = (x_i)_{i \in I}$ weakly converges to x and not to $r(2x) = 2x$.

2. Show that a Banach space X is reflexive if and only if X^* is reflexive.

3. Show that a closed subspace of a reflexive Banach space is reflexive.

4. A *cone* of a linear space is a subset stable by the homotheties $h_t : x \mapsto tx$ for all $t > 0$. Show that a closed cone Q of a normed space X is weakly *locally compact* (i.e., for each point $\bar{x}$ of Q there exists a weak neighborhood V of $\bar{x}$ such that $Q \cap V$ is weakly compact) if and only if there exists a neighborhood U of 0 such that $Q \cap U$ is weakly compact. [Hint: Let $f_1, \ldots, f_n \in X^*$ be such that $U_1 \cap \cdots \cap U_n \subset U$ for $U_i := f_i^{-1}((-\infty, 1])$. Given $\bar{x} \in Q$, let $t > \max(1, f_1(\bar{x}), \ldots, f_n(\bar{x}))$. Let $V_i := f_i^{-1}((-\infty, t])$ for $i = 1, \ldots, n$. Then $V := V_1 \cap \cdots \cap V_n$ is a weak neighborhood of $\bar{x}$ and $Q \cap V = t(Q \cap t^{-1}V) \subset t(Q \cap U)$, which is weakly compact.]

5. Prove a similar result for a weak* closed cone of a dual space endowed with the weak* topology.

6. (a) Show that the polar set P^0 of a cone P of a normed space X is a cone and is given by $P^0 = \{x^* \in X^* : \langle x^*, x \rangle \le 0 \ \forall x \in P\}$.

    (b) A *base* of a convex cone Q is a convex subset C of Q such that $0 \notin C$ and $Q = \mathbb{R}_+ C$. Show that a closed convex cone P of a Banach space has a nonempty interior if and only if its polar cone $Q := P^0$ has a weak* compact base.

7. (a) Check that the polar cone Q of the cone $P := \{0\} \times \mathbb{R}_+ \subset \mathbb{R}^2$ is locally compact but does not have a compact base.

    (b) Prove that if Q is a weak* closed convex cone of the dual of a Banach space, Q has a weak* compact base if and only if it is locally compact (see [336]).

8. Show that a bilinear map $b : X \times Y \to Z$ between normed spaces (i.e., b is linear with respect to each of its two variables) is continuous if and only if there exists some $c \in \mathbb{R}_+$ such that $\|b(x,y)\| \le c \|x\| \cdot \|y\|$ for all $x \in X$, $y \in Y$. The least such constant is called the norm of b. In particular, the evaluation $e : X \times L(X,Y) \to Y$ given by $e(x, \ell) := \ell(x)$ is continuous.

9. Check that a continuous linear map $A : X \to X^*$ from a normed space X into its dual defines a continuous bilinear map $b : X \times X \to \mathbb{R}$ by setting $b(x, y) := \langle Ax, y \rangle$ for $x, y \in X$ and that every continuous bilinear map $b : X \times X \to \mathbb{R}$ is obtained in that way.

10. A function $q : X \to \mathbb{R}$ on a normed space is said to be *quadratic* if there exists a symmetric bilinear map $b : X \times X \to \mathbb{R}$ such that $q(x) = b(x,x)$ for all $x \in X$. Check that such a bilinear map is unique. Prove that b is continuous if and only if q is continuous. [Hint: Define b by $b(x, y) := \frac{1}{2}[q(x + y) - q(x) - q(y)]$ for $x, y \in X$.]

11. Prove the *Cauchy–Schwarz inequality*: for every bilinear form b whose associated quadratic form q is nonnegative one has $|b(x,y)|^2 \le q(x)q(y)$ for all $x, y \in X$.

12. Deduce from the Cauchy–Schwarz inequality that if q is the quadratic form associated with a symmetric, continuous, linear map $A : X \to X^*$ and if q is nonnegative (then A is called *positive semidefinite*), then $q^{-1}(0) = N(A)$, the kernel of A.

13. Let $A : X \to X^*$ be a *positive definite* symmetric continuous linear map in the sense that A is an isomorphism and is positive semidefinite. Show that there is some positive constant α such that $\langle Ax, x \rangle \geq \alpha \|x\|^2$ for all $x \in X$. [Hint: Let $c := \|A^{-1}\|^{-1}$. Given $x \in X$ and $r \in (0, 1)$, pick $y \in B_X$ such that $\langle Ax, y \rangle \geq rc \|x\|$ and use the Cauchy–Schwarz inequality to deduce that $r^2 c^2 \|x\|^2 \leq \|A\| \langle Ax, x \rangle$; conclude that one can take $\alpha = c^2 \|A\|^{-1}$.]

14. Prove the *Lax–Milgram theorem*: if a continuous bilinear form b on a Banach space X is such that for some $\alpha > 0$ one has $b(x, x) \geq \alpha \|x\|^2$ for all $x \in X$, then the associated linear map $A : X \to X^*$ is an isomorphism. Observe that in fact X is a Hilbert space for an equivalent norm.

15. Given a positive semidefinite symmetric map $A : X \to X^*$, show that there is a unique positive semidefinite symmetric map $B : X \to X^*$ such that $B \circ B = A$ and $A \circ B = B \circ A$. It is called the *square root* of A. [Hint: Use a power series expansion or spectral theory.]

16. Using an orthonormal base $(e_n)_{n \geq 0}$ of a separable Hilbert space, show that the quadratic form $x \mapsto q(x) := \|x\|^2$ is not continuous for the weak topology. [Hint: Check that $(e_n)_{n \geq 0} \to 0$ for the weak topology.]

17. *Opial's inequality.* Let (x_n) be a sequence of a Hilbert space that weakly converges to x. Show that $\liminf_n \|x_n - w\|^2 \geq \liminf_n \|x_n - x\|^2 + \|w - x\|^2$ for all $w \in X$.

18. Show that a topological linear space is a regular space when it is Hausdorff.

1.1.4 Semicontinuity of Functions and Existence Results

A large part of nonsmooth analysis differs from traditional analysis in that it is a unilateral analysis, or in other words, a one-sided analysis. The fact that it often arises from minimization problems explains this viewpoint: for such problems, points at which the maximum of the objective function is attained are not of interest. In order to deal with such problems, one has to use a one-sided counterpart to continuity: for several questions, a continuity assumption would not be realistic, while a lower semicontinuity hypothesis can be satisfied. Moreover, in several situations, one is interested in having no abrupt decrease of the values of the function, whereas a sudden increase is not an embarrassment (think about your income ...). Thus, a notion of one-sided continuity is in order. A precise definition is as follows.

Definition 1.14. A function $f : X \to \overline{\mathbb{R}} := \mathbb{R} \cup \{-\infty, +\infty\}$ on a topological space X is said to be *lower semicontinuous* at some $\bar{x} \in X$ if for every real number $r < f(\bar{x})$ there exists some member V of the family $\mathcal{N}(\bar{x})$ of neighborhoods of $\bar{x}$ such that $r < f(v)$ for all $v \in V$.

The function f is said to be lower semicontinuous on some subset S of X if f is lower semicontinuous at each point of S.

The function f is said to be *outward continuous* at $\bar{x}$ whenever $-f$ is lower semicontinuous at $\bar{x}$.

We denote by $\mathscr{F}(X)$ the set of lower semicontinuous functions on X that are proper, i.e., with values in $\mathbb{R} \cup \{+\infty\}$ and taking at least one finite value.

We observe that f is automatically lower semicontinuous at $\bar{x}$ when $f(\bar{x}) = -\infty$; when $f(\bar{x}) = +\infty$ the lower semicontinuity of f means that the values of f remain as large as required, provided one stays in some small neighborhood of $\bar{x}$. When $f(\bar{x})$ is finite, the definition amounts to assigning to each $\varepsilon > 0$ a neighborhood V_ε of $\bar{x}$ such that $f(v) > f(\bar{x}) - \varepsilon$ for all $v \in V_\varepsilon$. Thus, lower semicontinuity allows sudden upward changes of the value of f but excludes sudden downward changes. Obviously, f is continuous at $\bar{x}$ iff it is both lower semicontinuous and outward continuous at $\bar{x}$.

Example. The *indicator function* ι_A of a subset A of X, defined by $\iota_A(x) = 0$ for $x \in A$, $\iota_A(x) = +\infty$ for $x \in X \setminus A$, is lower semicontinuous iff A is closed, as is easily seen. Such a function is of great use in optimization theory and nonsmooth analysis.

Example. The *characteristic function* χ_A or 1_A of $A \subset X$, defined by $\chi_A(x) = 1$ for $x \in A$, $\chi_A(x) = 0$ for $x \in X \setminus A$, is lower semicontinuous iff A is open. Such a function, of crucial importance in integration theory, is seldom used in optimization.

Example: The length function. Given a metric space (M, d), let $X := C(T, M)$ be the space of continuous maps from $T := [0, 1]$ to M. Given a finite subdivision $\sigma := \{t_0 = 0 < t_1 < \cdots < t_n = 1\}$ of T, let us set for $x \in X$,

$$\ell_\sigma(x) := \sum_{i=1}^{n} d(x(t_{i-1}), x(t_i)),$$

and let $\ell(x)$ be the supremum of $\ell_\sigma(x)$ as σ varies in the set of finite subdivisions of T. The properties devised below yield that ℓ is lower semicontinuous when X is endowed with the topology of pointwise convergence (and a fortiori, when X is endowed with the metric of uniform convergence). However, ℓ is not continuous: one can increase ℓ by following a nearby curve that makes many small changes (a fact that every dog knows when tied with a leash). Details are given in Exercise 3 below.

The following characterizations are global ones.

Proposition 1.15. *For a function $f : X \to \overline{\mathbb{R}}$, the following assertions are equivalent:*

(a) *f is lower semicontinuous;*
(b) *The epigraph $E := E_f := \{(x, r) \in X \times \mathbb{R} : r \geq f(x)\}$ of f is closed;*
(c) *For all $r \in \mathbb{R}$ the sublevel set $S(r) := S_f(r) := \{x \in X : f(x) \leq r\}$ is closed.*

Proof. (a)⇒(b) It suffices to prove that $(X \times \mathbb{R}) \setminus E$ is open when f is lower semicontinuous. Given $(\bar{x}, \bar{r}) \in (X \times \mathbb{R}) \setminus E$, i.e., such that $\bar{r} < f(\bar{x})$, for every $r \in (\bar{r}, f(\bar{x}))$ one can find a neighborhood V of $\bar{x}$ such that $r < f(v)$ for all $v \in V$. Then $V \times (-\infty, r)$ is a neighborhood of $(\bar{x}, \bar{r})$ in $X \times \mathbb{R}$ that does not meet E. Hence $(X \times \mathbb{R}) \setminus E$ is open.

(b)⇒(c) Since for all $r \in \mathbb{R}$ one has $S(r) \times \{r\} = E \cap (X \times \{r\})$, $S(r)$ is closed when E is closed.

(c)⇒(a) Given $\bar{x} \in X$ and $r \in \mathbb{R}$ such that $r < f(\bar{x})$, one has $\bar{x} \in X \setminus S(r)$, which is open, and for all $v \in V := X \setminus S(r)$ one has $r < f(v)$. $\qquad\square$

The notion of lower semicontinuity is intimately tied to the concept of lower limit (denoted by liminf), which is a one-sided concept of limit that can be used even when there is no limit. In the following definition we suppose X is a subspace of a larger space W, $w \in \mathrm{cl}(X)$, a situation that will be encountered later, f.i., when $X = \mathbb{P} := (0, \infty)$, $W = \mathbb{R}$, and $w = 0$ or when $X = \mathbb{N}$, $W = \mathbb{R}_\infty$, and $w = +\infty$.

Definition 1.16. Given a topological space W, a subspace X of W, and a point $w \in \mathrm{cl}(X)$, the *lower limit* of a function $f : X \to \overline{\mathbb{R}}$ at w is the extended real number

$$\liminf_{x \to w} f(x) := \sup_{V \in \mathscr{N}(w)} \inf_{v \in V \cap X} f(v).$$

Here, as usual, $\mathscr{N}(w)$ denotes the family of neighborhoods of w in W.

Setting $m_V := \inf f(V \cap X)$, the supremum over $V \in \mathscr{N}(w)$ of the family $(m_V)_V$ can also be considered the limit of the net $(m_V)_V$. That explains the terminology. One can show that $\sup_V m_V$ is also the least cluster point of $f(x)$ as $x \to w$ in X. When W is metrizable, one can replace the family $\mathscr{N}(w)$ by the family of balls centered at w, so that $\liminf_{x \to w} f(x) = \sup_{r>0} m_r$, with $m_r := \inf f(B(w, r) \cap X)$, is the limit of a sequence. On the other hand, the lower limit of a net $(r_i)_{i \in I}$ of real numbers is a special case of the preceding definition, taking $w := \infty$ and $W := I \cup \{\infty\}$ with the topology described above.

Exercise. Deduce from what precedes that $\liminf_{x \to w} f(x)$ is the least of the limits of the convergent nets $(f(x_i))_{i \in I}$, where $(x_i)_{i \in I}$ is a net in X converging to w.

Let us give some useful calculus rules (with the convention $0r = 0$ for all $r \in \overline{\mathbb{R}}$).

Lemma 1.17. *For $f : X \to \overline{\mathbb{R}}$, $r \in \mathbb{R}_+$, one has $\liminf_{x \to w} rf(x) = r \liminf_{x \to w} f(x)$. If $f, g : X \to \overline{\mathbb{R}}$ are such that $\{\liminf_{x \to w} f(x), \liminf_{x \to w} g(x)\} \neq \{-\infty, +\infty\}$, then $\liminf_{x \to w}(f + g)(x) \geq \liminf_{x \to w} f(x) + \liminf_{x \to w} g(x)$.*

Proof. The first assertion being immediate, let us establish the second one. Let us set $\overline{f}(w) := \liminf_{x \to w} f(x)$ and $\overline{g}(w) := \liminf_{x \to w} g(x)$. If $\overline{f}(w) = -\infty$, or if $\overline{g}(w) = -\infty$, the result is obvious. Otherwise, given $r < \overline{f}(w)$, $s < \overline{g}(w)$, we can find $U, V \in \mathscr{N}(w)$ such that $\inf f(U) > r$, $\inf g(V) > s$, whence $\inf(f + g)(U \cap V) > r + s$. It follows that $\liminf_{x \to w}(f + g)(x) \geq r + s$. $\qquad\square$

Lower semicontinuity can be characterized with the notion of lower limit.

Lemma 1.18. *A function $f : X \to \overline{\mathbb{R}}$ on a topological space X is lower semicontinuous at some $w \in X$ if and only if one has $f(w) \le \liminf_{x \to w} f(x)$.*

Proof. Here $W = X$. Clearly, when f is lower semicontinuous at w, one has $f(w) \le \liminf_{x \to w} f(x)$. Conversely, when this inequality holds, for every $r < f(w)$, by the definition of the supremum over $\mathcal{N}(w)$, one can find $V \in \mathcal{N}(w)$ such that $r < \inf_{v \in V} f(v)$, so that f is lower semicontinuous at w. $\qquad\square$

One can also use nets for such a characterization.

Lemma 1.19. *A function $f : X \to \overline{\mathbb{R}}$ on a topological space X is lower semicontinuous at some $\overline{x} \in X$ if and only if for every net $(x_i)_{i \in I}$ in X converging to $\overline{x}$ one has $f(\overline{x}) \le \liminf_{i \in I} f(x_i)$.*

When $\overline{x}$ has a countable base of neighborhoods, one can replace nets by sequences in that characterization.

Proof. The condition is necessary: if f is lower semicontinuous at $\overline{x}$ and if a net $(x_i)_{i \in I}$ in X converges to $\overline{x}$, then for all $r < f(\overline{x})$ there exists some $V \in \mathcal{N}(\overline{x})$ such that $f(v) > r$ for all $v \in V$ and there exists some $h \in I$ such that $x_i \in V$ for $i \ge h$, so that $\inf_{i \ge h} f(x_i) \ge r$, whence $\liminf_{i \in I} f(x_i) \ge r$.

Conversely, suppose f is not lower semicontinuous at $\overline{x}$ and let $(V_i)_{i \in I}$ be a base of neighborhoods of $\overline{x}$: there exists some $r < f(\overline{x})$ such that for every $i \in I$ there exists some $x_i \in V_i$ such that $f(x_i) < r$. Ordering I by $j \ge i$ if $V_j \subset V_i$, we get a net $(x_i)_{i \in I}$ that converges to $\overline{x}$ and is such that $\liminf_{i \in I} f(x_i) \le r$. The second assertion follows from the fact that when $\overline{x}$ has a countable base of neighborhoods, one can take a decreasing sequence of neighborhoods for a base. $\qquad\square$

The family of lower semicontinuous functions enjoys stability properties.

Proposition 1.20. *If $(f_i)_{i \in I}$ is a family of functions that are lower semicontinuous at $\overline{x}$, then the function $f := \sup_{i \in I} f_i$ is lower semicontinuous at $\overline{x}$.*

For every $r \in \mathbb{R}_+$ and f, g that are lower semicontinuous at $\overline{x}$, the functions $\inf(f, g)$ and rf are lower semicontinuous at $\overline{x}$; $f + g$ is lower semicontinuous provided $\{f(\overline{x}), g(\overline{x})\} \ne \{-\infty, +\infty\}$.

If f and g are nonnegative and lower semicontinuous at $\overline{x}$, so is fg.

If $f : X \to \overline{\mathbb{R}}$ is lower semicontinuous at $\overline{x}$ and if $g : \overline{\mathbb{R}} \to \overline{\mathbb{R}}$ is nondecreasing and lower semicontinuous at $f(\overline{x})$, then $g \circ f$ is lower semicontinuous at $\overline{x}$.

One may observe that in the first assertion one cannot replace lower semicontinuity by continuity, as shown by the above example of arc length.

Proof. Let $r \in \mathbb{R}$, $r < f(\overline{x})$. There exists some $j \in I$ such that $r < f_j(\overline{x})$; hence one can find some $V \in \mathcal{N}(\overline{x})$ such that $r < f_j(v) \le f(v)$ for all $v \in V$. The proofs of the other assertions are also straightforward or follow from Lemma 1.18. $\qquad\square$

Proposition 1.21. *For every function $f : X \to \overline{\mathbb{R}}$ on a topological space X, the family of lower semicontinuous functions majorized by f has a greatest element $\overline{f}$ called the lower semicontinuous hull of f. Its epigraph is the closure of the epigraph of f. The function $\overline{f}$ is given by*

$$\bar{f}(x) = \liminf_{u \to x} f(u) = \sup_{U \in \mathcal{N}(x)} \inf_{u \in U} f(u).$$

Proof. The first assertion is a direct consequence of Proposition 1.20. The second one easily stems from the fact that the closure of the epigraph of f is the epigraph of a function. The proof of the explicit expression of $\bar{f}$ is left as an exercise. □

Results giving lower semicontinuity of the infimum of a family of functions require more exacting assumptions. In the next statements we change the notation, since in many applications, X is changed into a space of parameters W.

Proposition 1.22. *Let W, X be topological spaces, let $\bar{w} \in W$, and let $f : W \times X \to \overline{\mathbb{R}}$ be a function that is lower semicontinuous at $(\bar{w}, \bar{x})$ for every $\bar{x} \in X$. If the following compactness assumption is satisfied, then the performance function $p : W \to \overline{\mathbb{R}}$ given by $p(w) := \inf_{x \in X} f(w, x)$ is lower semicontinuous at $\bar{w}$:*
   (C) *for every net $(w_i)_{i \in I} \to \bar{w}$ there exist a subnet $(w_j)_{j \in J}$, a convergent net $(x_j)_{j \in J}$ in X, and $(\varepsilon_j)_{j \in J} \to 0_+$ such that $f(w_j, x_j) \le p(w_j) + \varepsilon_j$ for all $j \in J$.*

Proof. Given a net $(w_i)_{i \in I} \to \bar{w}$, one can find a subnet $(w_j)_{j \in J}$ such that $(p(w_j))_{j \in J}$ converges to $\liminf_{i \in I} p(w_i)$ and (taking a further subnet if necessary) such that for some $(\varepsilon_j) \to 0_+$ and some convergent net $(x_j)_{j \in J}$ one has $f(w_j, x_j) \le p(w_j) + \varepsilon_j$ for all $j \in J$. Then, if $\bar{x}$ is the limit of $(x_j)_{j \in J}$, one has

$$p(\bar{w}) \le f(\bar{w}, \bar{x}) \le \liminf_{j \in J} f(w_j, x_j) \le \liminf_{j \in J}(p(w_j) + \varepsilon_j) = \liminf_{i \in I} p(w_i).$$

These relations hold for every such net $(w_i)_{i \in I}$, so one has $p(\bar{w}) \le \liminf_{w \to \bar{w}} p(w)$. □

Corollary 1.23. *Let W and X be topological spaces, X being compact, and let $f : W \times X \to \mathbb{R}_\infty := \mathbb{R} \cup \{+\infty\}$ be lower semicontinuous at all points of $\{\bar{w}\} \times X$. Then the performance function p defined as above is lower semicontinuous at $\bar{w}$.*

When f is lower semicontinuous on $W \times X$, a simpler proof can be given using the Weierstrass theorem below.

Proof. Condition (C) is clearly satisfied when X is compact, since for every net $(w_i)_{i \in I} \to \bar{w}$ and for every sequence $(\alpha_n) \to 0_+$ one can take $H := I \times \mathbb{N}$, $w_h := w_i$, $\varepsilon_h := \alpha_n$ for $h := (i, n)$ and pick $x_h \in X$ satisfying $f(w_h, x_h) \le p(w_h) + \varepsilon_h$ and take a subnet $(x_j)_{j \in J}$ of $(x_h)_{h \in H}$ that converges in X. □

The following result is the main existence result in optimization theory.

Theorem 1.24 (Weierstrass). *Let $f : X \to \overline{\mathbb{R}}$ be a lower semicontinuous function on a compact topological space X. Then the set $M := \{w \in X : f(w) \le f(x) \ \forall x \in X\}$ of minimizers of f is nonempty.*

Proof. We may suppose $m := \inf f(X) < +\infty$, for otherwise, f is constant with value $+\infty$. Setting $S_f(r) := \{x \in X : f(x) \le r\}$, the family $\{S_f(r) : r > m\}$ is formed of nonempty closed subsets, and every finite subfamily has a nonempty intersection:

$\bigcap_{1\le i\le k} S_f(r_i) = S_f(r_j)$, where $r_j := \min_{1\le i\le k} r_i$. Therefore its intersection, which clearly coincides with M, is nonempty. $\square$

The compactness assumption in the preceding theorem can be relaxed by using a notion of coercivity. A function $f : X \to \mathbb{R}$ on a metric space X is said to be *coercive* if for all $r \in \mathbb{R}$ the sublevel set $S_f(r) := f^{-1}((-\infty, r])$ is bounded, or equivalently, if $f(x) \to +\infty$ as $d(x,x_0) \to +\infty$ (x_0 being an arbitrary point of X). This notion is essentially used in the case that X is a normed space and $f(x) \to +\infty$ as $\|x\| \to +\infty$. We will say that f is *inf-compact* if for all $r \in \mathbb{R}$ the sublevel set $S_f(r)$ is compact. When the closed balls of X are compact, this property coincides with coercivity. For such a function, the existence of minimizers is ensured by a well-known result and lower semicontinuity of f:

Corollary 1.25. *Let $f : X \to \mathbb{R}_\infty$ be an inf-compact function on a topological space X. Then f attains its minimum.*

In particular, if (X,d) is a metric space whose closed balls are compact for a topology weaker than the topology associated with d and if f is lower semicontinuous for this topology and coercive, then f attains its minimum.

Proof. The result being obvious when f takes the value $+\infty$ only, let $r \in \mathbb{R}$ be such that $S_f(r) := f^{-1}((-\infty, r])$ is nonempty. By assumption, $S_f(r)$ is compact. By Theorem 1.24, f attains its infimum on $S_f(r)$. Since $\inf f(X) = \inf f(S_f(r))$, every minimizer of the restriction $f \mid S_f(r)$ of f is also a minimizer of f. $\square$

Given a topological space X, one may try to weaken its topology in order to enlarge the family of compact subsets. Then a continuous function on X may not remain continuous. There are interesting cases, for instance making use of convexity assumptions, for which the function still remains lower semicontinuous, so that the preceding generalization of the classical existence of a minimizer is of interest.

Exercises

1. Using the relation $E = \bigcap_{i\in I} E_i$, where E_i is the epigraph of a function f_i and E is the epigraph of $f := \sup_{i\in I} f_i$, show that f is lower semicontinuous on X when each f_i is lower semicontinuous on X. Use a similar argument with sublevel sets.

2. Suppose X is a metric space. Show that f is lower semicontinuous at $\bar{x}$ iff f is *sequentially lower semicontinuous*, i.e., if for every sequence $(x_n) \to \bar{x}$ one has $f(\bar{x}) \le \liminf_n f(x_n)$.

3. Let (M,d) be a metric space, let $T := [0,1]$, and let $X := C(T,M)$ be the set of continuous maps from T to M. Given some $x \in X$ and some element s of the set S of nondecreasing sequences $s := (s_n)_{n\ge 0}$ satisfying $s_0 = 0$, $s_n = 1$ for n large, let

$$\ell_s(x) := \sum_{n\ge 0} d(x(s_n), x(s_{n+1}))$$

(observe that the preceding sum contains only a finite number of nonzero terms). Define the *length of a curve* $x \in X$ by $\ell(x) := \sup_{s \in S} \ell_s(x)$. Show that $\ell_s : X \to \mathbb{R}$ is continuous when X is endowed with the metric of uniform convergence (and even when X is provided with the topology of pointwise convergence). Conclude that the length ℓ is a lower semicontinuous function on X.

Show that ℓ is not continuous by taking $M := \mathbb{R}^2$, $\bar{x}$ given by $\bar{x}(t) := (t, 0)$, and by showing that there is some $x_n \in X$ such that $d(x_n, \bar{x}) \to 0$ and $\ell(x_n) \geq \sqrt{2}$. [Hint: For $n > 0$ define x_n by $x_n(t) = t - \frac{2k}{2n}$ for $t \in [\frac{2k}{2n}, \frac{2k+1}{2n}]$ and $x_n(t) = -t + \frac{k+1}{n}$ for $t \in [\frac{2k+1}{2n}, \frac{2k+2}{2n}]$, $k \in \mathbb{N}$, $k \leq n - 1$.]

4. Let (M, d) be a metric space whose closed balls are compact. Suppose X is arcwise connected. Show that every pair of points x_0, x_1 in X can be joined by a curve with least length (a so-called *geodesic*). Identify such a curve when M is $\mathbb{R}^d$, the unit sphere $\mathbb{S}^{d-1}$ of $\mathbb{R}^d$, and when M is the circular cylinder $C := \mathbb{S}^1 \times [0, h]$ in $\mathbb{R}^3$. Such curves prompted the development of differential geometry.

5. Show that the infimum of an infinite family of lower semicontinuous functions is not necessarily lower semicontinuous. [Hint: Every function f on a Hausdorff topological space X is the infimum of the family $(f_a)_{a \in X}$ given by $f_a(x) = f(a)$ if $x = a$, $+\infty$ otherwise.]

6. Let X be a closed subset of $\mathbb{R}^d$ and let $f : X \to \overline{\mathbb{R}}$ be *pseudo-coercive* in the sense that there exists some $x_0 \in X$ such that $f(x_0) < \liminf_{\|x\| \to \infty, x \in X} f(x)$ and lower semicontinuous. Show that f attains its infimum.

7. Let X be a closed subset of $\mathbb{R}^d$ and let $f : X \to \overline{\mathbb{R}}$. Assume that f is *finitely minimizable* in the sense that there exists $r \in \mathbb{R}_+$ such that for every $t > m :=$ $\inf f(X)$, there exists some $x \in X$ satisfying $\|x\| \leq r$, $f(x) < t$. Show that every pseudo-coercive function is finitely minimizable and that every finitely minimizable lower semicontinuous function on X attains its infimum at some point of $X \cap B[0, r]$, where r is the radius of essential minimization, i.e., the infimum of the real numbers r for which the above definition is satisfied.

8. Prove the Weierstrass theorem in the case that X is a compact metric space by using a minimizing sequence of f, i.e., a sequence (x_n) of X such that $(f(x_n)) \to \inf f(X)$.

9. Let $f : X \to \mathbb{R}_\infty$ be a lower semicontinuous function on a topological space X and let A be a nonempty subset of X whose closure is denoted by cl A. Show that $\sup f(A) = \sup f(\text{cl } A)$. Can one replace sup by inf?

10. Define a family of continuous functions whose supremum is not continuous.

11. Show that every lower semicontinuous function f on $[0, 1]$ (or on a second countable metric space X) is the supremum of the family of continuous functions $g \leq f$.

12. Prove Corollary 1.23 using open subsets.

13. Show that among all cylindrical barrels $B := B[0,r] \times [0,h]$ in $\mathbb{R}^3$ with a given area s there is one with greatest volume.

1.1.5 Baire Spaces and the Uniform Boundedness Theorem

In this subsection we review some facts from functional analysis that will be useful for our purposes. We will say that a subset G of some topological space T is *generic* if it contains the intersection of a countable family of open subsets of T (a so-called $\mathscr{G}_\delta$ set) that are dense in T; other terminologies are that G is *residual* or that the complement of G is *meager* or a *set of first category*. It is convenient to say that a property involving a point is generic if it holds on a generic subset. The main feature of this notion is that the intersection of a finite (or countable) family of generic subsets is still generic, a property that does not hold for dense subsets (consider the set of rational numbers and the set of irrational numbers in $\mathbb{R}$). The importance of this concept lies in the following result, valid in a large class of topological spaces called the class of Baire spaces (it also includes the class of locally compact spaces).

Lemma 1.26 (Baire's theorem). *In a complete metric space T, every generic subset is dense. Moreover, if T is the union of a countable family of closed subsets, then one of them has a nonempty interior.*

Proof. Let us show that every subset G of T containing the intersection $\bigcap_n T_n$ of a countable family of open dense subsets T_n of T is dense. Let (s_n) be a sequence of positive numbers with limit 0. Given a nonempty open subset U of T, the set $T_n \cap U$ is nonempty and open; in particular, $T_0 \cap U$ contains some closed ball $B[x_0, r_0]$ with $r_0 \in (0, s_0]$. Assume by induction that we have constructed open balls $B(x_k, r_k)$ with $r_k \le s_k$, $B[x_k, r_k] \subset B(x_{k-1}, r_{k-1}) \cap T_k$ for $k = 1, \dots, n$. Since $B(x_n, r_n)$ meets T_{n+1}, we can find a closed ball $B[x_{n+1}, r_{n+1}] \subset T_{n+1} \cap B(x_n, r_n)$ with $r_{n+1} \le s_{n+1}$. The sequence (x_n) obtained in this way is a Cauchy sequence (since $d(x_{n+p}, x_n) \le s_n$ for all n, p). Its limit belongs to $B[x_m, r_m] \subset T_m$ for each m and in particular to $B[x_0, r_0] \subset U$ and hence to $\bigcap_n T_n \cap U \subset G \cap U$: G is dense.

Now suppose $T = \bigcup_n F_n$, where each F_n is closed. Let $T_n := T \setminus F_n$; then T_n is open and if F_n has an empty interior then T_n is dense. If this happens for all $n \in \mathbb{N}$, then $\bigcap_n T_n$ is dense, an impossibility since $\bigcap_n T_n = \varnothing$. Thus, at least one F_n has a nonempty interior.   □

Theorem 1.27 (Banach–Steinhaus or uniform boundedness theorem). *Let X, Y be normed spaces, X being complete, and let F be a subset of the space $L(X,Y)$ of continuous linear maps from X to Y. If for all $x \in X$, the set $F(x) := \{f(x) : f \in F\}$ is bounded in Y, then F is bounded in $L(X,Y)$ for the usual norm.*

Proof. Denoting by B_Y unit ball of Y, for $n \in \mathbb{N}$ define the closed set

$$X_n := \{x \in X : \forall f \in F \ \|f(x)\| \le n\} = \bigcap_{f \in F} f^{-1}(nB_Y).$$

By assumption, X is the union of the family (X_n). By Baire's theorem above, there is some $k \in \mathbb{N}$ such that X_k has a nonempty interior. If $a \in \operatorname{int} X_k$, we also have $-a \in \operatorname{int} X_k$, since X_k is symmetric with respect to 0. It follows that $0 \in \operatorname{int} X_k$. Thus, if $r > 0$ is such that $rB_X \subset X_k$, we have $\|f\| \le r^{-1} k$ for all $f \in F$. $\qquad\square$

Corollary 1.28. *A weak* bounded subset of the dual X^* of a Banach space X is bounded. A weakly bounded subset of X is bounded. In particular, a weak* convergent sequence of X^* is bounded.*

Here a subset S of X (resp. X^*) is said to be *weakly* (resp. *weak**) *bounded* if for all $f \in X^*$ (resp. $x \in X$) the set $\{f(x) : x \in S\}$ (resp. $\{s^*(x) : s^* \in S\}$) is bounded.

Proof. The first assertion is the special case of the theorem corresponding to $Y := \mathbb{R}$. The second one stems from the fact that the embedding of X into X^{**} is isometric, as a consequence of the Hahn–Banach theorem, which we will prove below. $\qquad\square$

Exercises

1. Show that a *locally compact topological space* X (i.e., a space X such that for every $x \in X$ and every $U \in \mathcal{N}(x)$ there exists some $V \in \mathcal{N}(x)$ that is compact and contained in U) satisfies the *Baire property* that every generic subset is dense.

2. Show that the evaluation map $e : X \times X^* \to \mathbb{R}$ given by $e(x,x^*) := x^*(x)$ is sequentially continuous when X is endowed with the strong topology and X^* is provided with the weak* topology, but that it is never continuous when X is infinite-dimensional. [Hint: If there are neighborhoods U, V^* of the origins in X strong and X^* weak* respectively such that $e(x,x^*) \le 1$ for all $(x,x^*) \in U \times V^*$, a contradiction is obtained to the fact that V^* contains an infinite-dimensional linear subspace.]

3. Let X, Y be normed spaces, X being complete, and let F be a subset of $L(X,Y)$ such that for all $x \in X$ and all $y^* \in Y^*$ the set $F^{\mathsf{T}}(y^*)(x) := \{\langle y^*, f(x)\rangle : f \in F\}$ is bounded in $\mathbb{R}$. Show that F is bounded.

4. Let $X := C(T)$ be the space of continuous functions on $T := [0,1]$, endowed with the supremum norm $\|\cdot\|_\infty$ given by $\|f\|_\infty := \sup_{t \in T} |f(t)|$. For $n \in \mathbb{N} \setminus \{0\}$, let

$$F_n := \{f \in X : \exists r \in [0, 1 - 1/n], \ \forall s \in [0, 1/n] \ |f(r+s) - f(r)| \le ns\}.$$

Show that F_n is closed in X and that $G_n := X \setminus F_n$ is dense in X. Conclude that the set of bounded continuous functions that are nowhere differentiable on T is a generic subset of X. Such a conclusion reinforces the famous counterexample due to

Weierstrass of a continuous function that is nowhere differentiable. [Hint: Given $f \in X$ and $\varepsilon > 0$, in order to find some $g \in G_n \cap B(f, \varepsilon)$, pick a polynomial $p \in B(f, \varepsilon/2)$ and a piecewise affine function q such that $\|q\|_\infty < \varepsilon/2$ and $|q'(t)| > n + \|p'\|_\infty$ for all but finitely many $t \in [0,1]$ and set $g := p + q$.]

1.2 Set-Valued Mappings

Although poets keep celebrating the uniqueness of love, uniqueness is seldom met in human activities and real-world problems. For optimization problems, uniqueness seems to be far less important than stability of solutions. Two precise meanings can be given to the latter notion. The first means that given a solution $\bar{x}$, for all parameters w close enough to $\bar{w}$ one can find a solution x_w with limit $\bar{x}$ as w converges to $\bar{w}$ (persistence of solutions). The second means that if a sequence (or net) (x_n) of solutions corresponding to a sequence of parameters $(w_n) \to w_\infty$ has a limit x_∞, then this limit x_∞ belongs to the set of solutions for the limit parameter w_∞ (stability of solutions). Both concepts are important and useful in practice. In particular, in dealing with the parameterized minimization problem

$$(\mathscr{P}_w) \quad \text{minimize } f(w, \cdot) \text{ on } F(w) \subset X,$$

where W and X are topological spaces, f is an extended real-valued function on $W \times X$, and $F : W \to \mathscr{P}(X)$, the set of subsets of X, the two continuity properties can be applied to the feasible set $F(w)$ or to the solution set

$$S(w) := \{x \in F(w) : f(w, x) = \inf f(w, F(w))\}.$$

Another important example of the role of set-valued maps is provided by optimal control theory. There, a cost criterion is minimized over the set of solutions to an equation involving a parameter or control u at the disposal of the user (such as consumption of fuel or the like). In one of its simplest forms, the system is governed by a differential equation

$$\dot{x}(t) = f(t, x(t), u(t)), \quad t \in T,$$

where T is some interval of $\mathbb{R}$ and $f : T \times E \times U \to E$, E is a Banach space, U is a set, and $u(\cdot)$ is a map from T to U chosen in a certain class $\mathscr{U}$. It is often convenient to associate to such an equation the differential inclusion

$$\dot{x}(t) \in F(t, x(t)),$$

where $F(t, e) := \{f(t, e, u) : u \in U\}$ represents the set of potential velocities.

Mappings with sets as values appear in many practical problems such as image compression, image recovery, and evolution of oil reservoirs; they also appear in a number of mathematical problems, even if this is not always clearly recognized. For instance, they occur with every equivalence relation and with every order relation; they also appear as soon as one considers inverse images by a given mapping $g : X \rightarrow Y$, since $F(y) := g^{-1}(y) := \{x \in X : g(x) = y\}$ defines a set-valued map $F = g^{-1}$ from Y into X. Such set-valued mappings are often called multifunctions; other terms in use are multimappings, multis, correspondences, relations. We will use the terminology "multimaps," since it is concise and suggestive.

In this section, we present some generalities about such maps. Then we define limits of sets and we relate this notion to continuity properties for multimaps.

1.2.1 Generalities About Sets and Correspondences

A formal definition is in order.

Definition 1.29. A *multimap* (or correspondence, or set-valued mapping) $F : X \rightrightarrows Y$ from a set X into another set Y is a map from X into the set $\mathscr{P}(Y)$ of subsets of Y (also called the power set of Y and often denoted by 2^Y).

The notation $F : X \rightrightarrows Y$ suggests that F can be seen almost as a mapping from X into Y, the difference being that the image of a point of X is a subset of Y rather than a singleton. In this way, maps can be considered as special multimaps, a map $f : X \rightarrow Y$ being identified with the multimap F given by $F(x) = \{f(x)\}$. Any multimap $F : X \rightrightarrows Y$ induces a mapping from $\mathscr{P}(X)$ into $\mathscr{P}(Y)$ (still denoted by F) given by

$$F(A) := \bigcup_{a \in A} F(a),$$

so that $F(\{a\}) = F(a)$.

Multimaps can be composed: given $F : X \rightrightarrows Y$, $G : Y \rightrightarrows Z$, the *composition* of F and G is the multimap $G \circ F : X \rightrightarrows Z$ given by

$$(G \circ F)(x) := G(F(x)),$$

where $G(B)$, for $B := F(x)$, is defined as above. Then one has the associativity rule

$$H \circ (G \circ F) = (H \circ G) \circ F.$$

The *inverse* $F^{-1} : Y \rightrightarrows X$ of a multimap $F : X \rightrightarrows Y$ is the multimap given by

$$F^{-1}(y) := \{x \in X : y \in F(x)\}, \qquad y \in Y.$$

This definition is compatible with the familiar use of inverse images for maps. Let us note that here we do not leave the realm of multimaps, whereas the inverse of a map is in general a multimap, not a map (and this fact is a source of many mistakes made by beginners in mathematics). Let us also observe that the extension of F^{-1} to $\mathscr{P}(Y)$ is given by

$$F^{-1}(B) = \{x \in X : B \cap F(x) \neq \varnothing\}.$$

In particular, $F^{-1}(Y)$ is the *domain* dom F or $D(F)$ of F:

$$\operatorname{dom} F := D(F) := \{x \in X : F(x) \neq \varnothing\}.$$

It is also the *range* or image $R(F^{-1}) := \operatorname{Im} F^{-1} := F^{-1}(Y)$ of F^{-1}. Conversely, $\operatorname{dom} F^{-1}$ is the image of F: the roles of F and F^{-1} are fully symmetric. It is easy to check that when $F : X \rightrightarrows Y$, $G : Y \rightrightarrows Z$ are given multimaps, and when C is a subset of Z, one has

$$(G \circ F)^{-1}(C) = F^{-1}(G^{-1}(C)).$$

For all subsets A, A' of X (resp. B, B' of Y) one has $F(A \cup A') = F(A) \cup F(A')$ and $F^{-1}(B \cup B') = F^{-1}(B) \cup F^{-1}(B')$. Let us observe that in general, one has

$$F^{-1}(B \cap B') \neq F^{-1}(B) \cap F^{-1}(B'), \tag{1.1}$$

in contrast to what occurs for maps. Note that since $F = (F^{-1})^{-1}$, F^{-1} may be an arbitrary multimap from Y to X, and in fact, for a multimap $M : Y \rightrightarrows X$ one has $M(A \cap B) \neq M(A) \cap M(B)$; taking $F = M^{-1}$, so that $F^{-1} = M$, we obtain (1.1).

It is often convenient to associate to a multimap $F : X \rightrightarrows Y$ its *graph*

$$G(F) := \operatorname{gph}(F) := \{(x, y) \in X \times Y : y \in F(x)\}.$$

This subset of $X \times Y$ characterizes F, since $F(x) = \{y \in Y : (x, y) \in G(F)\}$. Conversely, to a given subset G of $X \times Y$, one can associate a multimap $F : X \rightrightarrows Y$ by

$$F(x) = \{y \in Y : (x, y) \in G\}, \qquad x \in X,$$

so that G is the graph of F. Moreover, when G is the graph $G(M)$ of some multimap M, one gets $F = M$ via this reverse process. Thus, there is a one-to-one correspondence between subsets of $X \times Y$ and multimaps from X into Y. This correspondence is simpler than the correspondence between maps and their graphs, since in the latter correspondence one has to consider only subsets G whose vertical slices $G \cap (\{x\} \times Y)$ (for $x \in X$) are singletons. In view of this one-to-one correspondence between a multimap and its graph, it is often convenient to identify a multimap with its graph and to say that a multimap has a property $\mathscr{P}$ if its graph has this property (such as closedness or convexity). This viewpoint is often fruitful and without any important risk of confusion; however, when X and Y are endowed with some operation $*$, one has to be aware that $F * F'$ usually denotes the multimap $x \mapsto F(x) * F'(x)$ and not the multimap whose graph is $G(F) * G(F')$ (see Exercise 5). Moreover, one has to be careful with the order of the terms in the product $X \times Y$, since it determines the direction of the multimap. In fact, the graph $G(F^{-1})$ of F^{-1} is the subset of $Y \times X$ that is symmetric to the graph $G(F)$ of F:

$$G(F^{-1}) = G(F)^{-1} := \{(y,x) : (x,y) \in G(F)\}.$$

When X is a product $X = X_1 \times X_2$, one has to be precise when one associates to a subset G of $X \times Y$ a multimap, since a partial multimap also can be defined in this way.

In order to get some practice, let us deal with the following "sum principle" (such an expression is a bit pompous for such a simple fact, but it is convenient).

Proposition 1.30. [26] *Let X and Y be two linear spaces (or additive groups), let c,d in Y, and let $A, B : X \rightrightarrows Y$ be two multimaps. Then the equations*

$$c + d \in A(x) + B(x), \tag{1.2}$$

$$0 \in A^{-1}(y + c) - B^{-1}(d - y), \tag{1.3}$$

$$y \in (A \circ B^{-1})(d - y) - c \tag{1.4}$$

are equivalent in the following sense: $x \in X$ is a solution to (1.2) if and only if $(A(x) - c) \cap (d - B(x))$ is nonempty and every y in this intersection is a solution to (1.3) and (1.4); conversely, y is a solution to (1.3) or (1.4) iff $A^{-1}(y+c) \cap B^{-1}(d - y)$ is nonempty and any x in this intersection is a solution to (1.2).

Proof. Clearly, x is a solution to (1.2) iff there exists $y \in A(x) - c$ such that $d - y \in B(x)$. This amounts to saying that $x \in A^{-1}(y + c) \cap B^{-1}(d - y)$ or that (1.3) holds. In other words, $y + c \in A(B^{-1}(d - y))$, i.e., (1.4) holds.   □

Although the preceding principle and the one in the exercises below are extremely simple, they give rise to interesting interpretations of various results.

Exercises

1. **(a)** Give an example to show that for a multimap $F : X \rightrightarrows Y$, the relations

$$F^{-1} \circ F = I_X, \quad F \circ F^{-1} = I_Y,$$

do not hold in general; here and elsewhere I_S denotes the *identity map* on a set S.
(b) Show that $I_X \subset F^{-1} \circ F$ (the inclusion being the inclusion of graphs or images) iff $D(F) = X$. Also show that $I_Y \subset F \circ F^{-1}$ iff $R(F) = Y$.
(c) Given multimaps $F : X \rightrightarrows Y$, $G : Y \rightrightarrows Z$ and $H := G \circ F$, give a sufficient condition in order to have $F \subset G^{-1} \circ H$. Show that this inclusion may not hold.

2. Give an example proving relation (1.1).

3. Given multimaps $F : X \rightrightarrows Y$, $G : Y \rightrightarrows Z$, show that

$$\mathrm{gph}(G \circ F) = (I_X \times G)(\mathrm{gph}F) = (F \times I_Z)^{-1}(\mathrm{gph}G).$$

4. Given a multimap $F : X \rightrightarrows Y$ and a subset B of Y one sometimes sets

$$F^+(B) := \{x \in X : F(x) \subset B\}.$$

(a) Observe that if F is a map, then $F^+(B) = F^{-1}(B)$ for every subset B of Y. Show conversely that if this relation holds for every subset B of Y, then F is single-valued.
(b) With the preceding notation, show that

$$F^+(Y \setminus B) = X \setminus F^{-1}(B), \qquad F^{-1}(Y \setminus B) = X \setminus F^+(B).$$

5. When the set Y is provided with a binary operation $\perp$ (say a sum), one can define a pointwise operation on the family of multimaps from X into Y by setting

$$(F \perp F')(x) := F(x) \perp F'(x).$$

If X is also provided with a binary operation $\perp$, one must note that the multimap $F \perp F'$ is not the multimap associated with $G(F) \perp G(F')$, where the operation $\perp$ on $X \times Y$ is defined componentwise by $(x, y) \perp (x', y') = (x \perp x', y \perp y')$ (and induces an operation on the set $\mathscr{P}(X \times Y)$ of subsets of $X \times Y$, as usual). Give examples showing that this operation on graphs does not correspond to the operation on multimaps via their values.

6. In Proposition 1.30, the roles of A and B (resp. c and d) are symmetric. One can also deduce nonsymmetric statements from that one. Let X and Y be two linear spaces, let $c \in Y$, and let $A, B : X \rightrightarrows Y$ be two multimaps. Then if $\widetilde{B}$ is the multimap given by $\widetilde{B}(x) := -B(-x)$, show that the equations

$$c \in A(x) + B(x), \tag{1.5}$$

$$0 \in A^{-1}(y + c) + (\widetilde{B})^{-1}(y), \tag{1.6}$$

$$y \in (A \circ B^{-1})(-y) - c \tag{1.7}$$

are equivalent in the following sense: x is a solution to (1.5) iff $(A(x) - c) \cap (-B(x))$ is nonempty and any y in this intersection is a solution to (1.6) and (1.7); conversely, y is a solution to (1.6) or (1.7) iff $A^{-1}(y + c) \cap B^{-1}(-y)$ is nonempty and every x in this intersection is a solution to (1.5).

7. Prove the composition principle of [868].

1.2.2 Continuity Properties of Multimaps

Let $F : T \rightrightarrows X$ be a multimap between two topological spaces. There are at least two reasons why the value $F(t)$ of the multimap at a point t close to a given point $t_0 \in T$ may be far from its value $F(t_0)$ at t_0: either $F(t)$ is much larger than $F(t_0)$, or else it is much smaller. Continuity of F means that these two events do not happen when t is close enough to t_0 and that $F(t)$ is not too far from $F(t_0)$. It is convenient to study separately these two regular behaviors. We start with a property ensuring that the values do not shrink abruptly.

Definition 1.31. A multimap $F : T \rightrightarrows X$ between two topological spaces is said to be *inward continuous* (or *inner continuous* or *lower semicontinuous*) at some point t_0 of T if for every open subset V of X such that $F(t_0) \cap V \neq \varnothing$ there exists some neighborhood U of t_0 in T such that $F(t) \cap V \neq \varnothing$ for all $t \in U$.

Example (e) below and a connection with limits of families of sets given in the next subsection explain the classical terminology "lower semicontinuity." However, since there is no order on X, we prefer the more intuitive terminology "inward continuity."

It is sometimes useful to deal with a pointwise version of the preceding concept: one says that F is *inward continuous* or *lower semicontinuous at some point* (t_0, x_0) of the graph of F if for every neighborhood V of x_0 there exists some neighborhood U of t_0 in T such that $F(t) \cap V \neq \varnothing$ for all $t \in U$. Thus, F is inward continuous at t_0 if and only if for all $x_0 \in F(t_0)$ the multimap F is inward continuous at (t_0, x_0). Let us note that F is inward continuous at (t_0, x_0) if and only if F^{-1} is *open at* (x_0, t_0) in the sense that for every neighborhood V of x_0, $F^{-1}(V)$ is a neighborhood of t_0. The multimap F is inward continuous on some subset S of T if it is inward continuous at all $s \in S$; for $S = T$, we just write that F is inward continuous.

Examples. (a) The multimap $D : \mathbb{R} \rightrightarrows \mathbb{R}^2$ given by $D(t) := \{(r \cos t, r \sin t) : r \in \mathbb{R}_+\}$ is inward continuous at every point of $\mathbb{R}$.

(b) The multimap $G : \mathbb{R} \rightrightarrows \mathbb{R}$ given by $G(0) := \{0\}$, $G(t) := [-1, 1]$ for $t \in \mathbb{R} \setminus \{0\}$ is everywhere inward continuous.

(c) The Heaviside multimap $H : \mathbb{R} \rightrightarrows \mathbb{R}$ given by $H(0) := [-1, 1]$, $H(t) := |t| t^{-1}$ for $t \in \mathbb{R} \setminus \{0\}$ is not inward continuous at 0.

(d) Let S be a subset of the set $C(T, X)$ of continuous maps from T to X. Then $F : T \rightrightarrows X$ given by $F(t) := \{f(t) : f \in S\}$ is inward continuous. In particular, if U is an arbitrary set, if $g : T \times U \to X$ is continuous in its first variable, then $F(\cdot) := g(\cdot, U)$ is inward continuous.

(e) Given $f : T \to \mathbb{R}$, its *hypograph multimap* $H_f : T \rightrightarrows \mathbb{R}$ given by $H_f(t) := (-\infty, f(t)]$ is inward continuous at $t_0 \in T$ iff f is lower semicontinuous at t_0. $\quad\square$

The proofs of the following properties are left as exercises.

Lemma 1.32. *(a) If F is the multimap $t \rightrightarrows \{f(t)\}$ associated with a map $f : T \to X$, F is inward continuous at t_0 if and only if f is continuous at t_0.*

(b) $F : T \rightrightarrows X$ is inward continuous if and only if for every open subset V of X, the set $F^{-1}(V)$ is open in T.

(c) If $F : T \rightrightarrows X$ is inward continuous at (t_0, x_0), if Y is another topological space, and if $G : X \rightrightarrows Y$ is inward continuous at (x_0, y_0), then $H := G \circ F : T \rightrightarrows Y$ is inward continuous at (t_0, y_0).

(d) If $F : S \rightrightarrows X$ and $G : T \rightrightarrows Y$ are two multimaps that are inward continuous at s_0, t_0 respectively, then their product H, given by $H(s, t) := F(s) \times G(t)$, is inward continuous at (s_0, t_0).

(e) If, in the preceding assertion, $S = T$ and $s_0 = t_0$, the multimap $(F, G) : T \rightrightarrows X \times Y$ given by $(F, G)(t) := F(t) \times G(t)$ is inward continuous at t_0.

(f) If F and G are two multimaps from T to X that are inward continuous at t_0, then their union H given by $H(t) := F(t) \cup G(t)$ is inward continuous at t_0.

The example of $F, G : \mathbb{R}_+ \rightrightarrows \mathbb{R}$ given by $F(t) := [t, 2t]$, $G(t) := [-2t, -t]$ shows that the intersection of two inward continuous multimaps may not be inward continuous.

Now we consider the other continuity property one may expect. It prevents the values to expand abruptly.

Definition 1.33. A multimap $F : T \rightrightarrows X$ between two topological spaces is said to be *Outward continuity* or, more classically, *upper semicontinuous* at some point t_0 of T if for every open subset V of X such that $F(t_0) \subset V$ there exists some neighborhood U of t_0 in T such that $F(t) \subset V$ for all $t \in U$. If F is outward continuous at each point s of some subset S of T, then F is said to be outward continuous on S.

Outward continuity is a stringent property that is seldom satisfied when the values of the multimap are noncompact. However, its mathematical content is simple and it occurs in some cases. The next properties are more likely to occur.

Definition 1.34. A multimap $F : T \rightrightarrows X$ between two topological spaces is said to be *compactly outward continuous* at some point t_0 of T if for every compact subset K of X the multimap $F_K : t \rightrightarrows F(t) \cap K$ is outward continuous at t_0.

A multimap $F : T \rightrightarrows X$ between two topological spaces is said to be *closed* at some point t_0 of T if for all $x \in X \setminus F(t_0)$ there exist neighborhoods U of t_0, V of x such that $F(t) \cap V = \varnothing$ for all $t \in U$.

Clearly outward continuity implies compact outward continuity. In fact, F is outward continuous (resp. compactly outward continuous) at t_0 if and only if for every closed (resp. compact) subset C of X contained in $X \setminus F(t_0)$, there exists a neighborhood U of t_0 such that for all $t \in U$, $F(t)$ and C are disjoint. The terminology "F is closed at t_0" is justified by its rephrasing in terms of closure: F is closed at t_0 if and only if

$$\mathrm{cl}(F) \cap (\{t_0\} \times X) = \{t_0\} \times F(t_0).$$

In fact, it is easy to see that F is closed at t_0 if and only if for every $x \in X$ and nets $(t_i)_{i \in I} \to t_0$, $(x_i)_{i \in I} \to x$, one has $x \in F(t_0)$ whenever $x_i \in F(t_i)$ for all $i \in I$. This property implies that $F(t_0)$ is closed in X, but is more demanding than closedness of $F(t_0)$ in general (Exercise 2). Also, one can check that F is closed at every point of T if and only if its graph is closed in $T \times X$. In many cases, outward continuity is more stringent than closedness.

Proposition 1.35. *(a)* If X is a regular (resp. Hausdorff) space, if $F : T \rightrightarrows X$ is outward continuous at some point $t_0 \in T$, and if $F(t_0)$ is closed (resp. compact), then F is closed at t_0.

(b) If F is closed at t_0 and if every open neighborhood W of $F(t_0)$ is such that $X \setminus W$ is compact, then $F : T \rightrightarrows X$ is outward continuous at t_0. In particular, if for some neighborhood U of t_0 the set $F(U)$ is contained in a compact subset Y of X, then F is closed at t_0 if and only if $F(t_0)$ is closed and F is outward continuous at t_0.

Proof. (a) When $F(t_0)$ is closed and X is regular, given $x \in X \setminus F(t_0)$ there exist neighborhoods V of x, W of $F(t_0)$ that are disjoint (take for V a closed neighborhood of x contained in $X \setminus F(t_0)$ and $W := X \setminus V$). If $U \in \mathcal{N}(t_0)$ is such that $F(t) \subset W$ for all $t \in U$, we get $F(t) \cap V = \varnothing$ for all $t \in U$. When X is just Hausdorff but $F(t_0)$ is compact, one can also find disjoint neighborhoods V, W of x and $F(t_0)$ respectively.

(b) Suppose F is closed at t_0 and for every open neighborhood W of $F(t_0)$, $X \setminus W$ is compact. If F is not outward continuous at t_0 one can find an open neighborhood W of $F(t_0)$ and a net $(t_i)_{i \in I} \to t_0$ such that for all $i \in I$, $F(t_i) \setminus W$ is nonempty. Since $X \setminus W$ is compact, taking $x_i \in F(t_i) \setminus W$ there exists a subnet $(x_j)_{j \in J}$ of $(x_i)_{i \in I}$ that converges. Its limit is in $X \setminus W$, hence in $X \setminus F(t_0)$, a contradiction to the closedness of F at t_0. The last assertion stems from the fact that one can replace T with U and X with Y. $\qquad\square$

Corollary 1.36. If $F : T \rightrightarrows X$ is closed at t_0, then F is compactly outward continuous at t_0 and $F(t_0)$ is closed. The converse holds when t_0 and the points of X have a countable base of neighborhoods.

Proof. If $F : T \rightrightarrows X$ is closed at t_0, then $F(t_0)$ is closed, and for every compact subset K of X the multimap $F_K : t \rightrightarrows F(t) \cap K$ is closed at t_0, hence is outward continuous at t_0 by Proposition 1.35 (b).

Suppose t_0 and the points of X have a countable base of neighborhoods, $F(t_0)$ is closed, and F is compactly outward continuous at t_0. Let (U_n) be a countable base of neighborhoods of t_0. If F is not closed at t_0, then there exist $x \in X \setminus F(t_0)$ and a countable base of neighborhoods (V_n) of x such that V_n meets $F(U_n)$ for all $n \in \mathbb{N}$. Let $t_n \in U_n$ and $x_n \in F(t_n) \cap V_n$. Then $K := \{x\} \cup \{x_n : n \in \mathbb{N}\}$ is compact. Since F_K is not closed, by Proposition 1.35 (a), it cannot be outward continuous. Thus we get a contradiction to the assumption that F is compactly outward continuous at t_0. $\quad\square$

Examples. (a) The multimap $D : \mathbb{R} \rightrightarrows \mathbb{R}^2$ of the preceding examples is closed at every point of $\mathbb{R}$ but is nowhere outward continuous.

(b) The multimap G of these examples is not outward continuous at 0.

(c) The multimap H of these examples is everywhere outward continuous.

(d) If U is a compact topological space, if $g : T \times U \to X$ is continuous, then $F(\cdot) := g(\cdot, U)$ is outward continuous.

(e) Given $f : T \to \mathbb{R}$, its hypograph multifunction $H_f : T \rightrightarrows \mathbb{R}$ is outward continuous at $t_0 \in T$ if and only if f is outward continuous at t_0. $\Box$

Again, the easy proofs of the following properties are left as exercises.

Lemma 1.37. *(a) If F is the multimap $t \rightrightarrows \{f(t)\}$ associated with a map $f : T \to X$, F is outward continuous at t_0 if and only if f is continuous at t_0.*

(b) $F : T \rightrightarrows X$ is outward continuous (resp. compactly outward continuous) if and only if for every closed (resp. compact) subset C of X, the set $F^{-1}(C)$ is closed in T.

(c) If $F : T \rightrightarrows X$ is outward continuous at t_0, if Y is another topological space, and if $G : X \rightrightarrows Y$ is outward continuous at every $x_0 \in F(t_0)$, then $H := G \circ F : T \rightrightarrows Y$ is outward continuous at t_0.

(d) If $F : S \rightrightarrows X$ and $G : T \rightrightarrows Y$ are two multimaps that are closed at s_0, t_0 respectively, then their product H given by $H(s,t) := F(s) \times G(t)$ is closed at (s_0, t_0). If $F(s_0)$ and $G(t_0)$ are compact, one can replace "closed" by "outward continuous" in this assertion.

(e) If F and G are two multimaps from T to X and Y respectively that are closed (resp. compactly outward continuous) at t_0, then the multimap (F,G) given by $(F,G)(t) := F(t) \times G(t)$ is closed (resp. compactly outward continuous) at t_0. If $F(t_0)$ and $G(t_0)$ are compact, one can replace "closed" by "outward continuous" in this assertion.

(f) If F and G are two multimaps from T to X that are outward continuous at t_0, then their union H given by $H(t) := F(t) \cup G(t)$ is outward continuous at t_0.

The following two results are easy extensions of known properties for continuous maps. The first one can be established by an easy covering argument. For the second one, we recall that a topological space X is said to be *connected* if it cannot be split into two nonempty disjoint open subsets.

Proposition 1.38. *If T is a compact space, if X is a Hausdorff space, and if $F : T \rightrightarrows X$ is outward continuous with compact values, then $F(T)$ is compact.*

Proposition 1.39. *If T is a connected topological space and $F : T \rightrightarrows X$ has connected values and is either inward continuous or outward continuous, then $F(T)$ is connected.*

Let us end this subsection with the quotation of a famous and useful result.

Theorem 1.40 (Michael [703]). *Let T be a topological space that is either metrizable or compact and let $F : T \rightrightarrows X$ be an inward continuous multimap with closed convex images in a Banach space X. Then F admits a continuous selection, i.e., a continuous map $f : T \to X$ such that $f(t) \in F(t)$ for all $t \in T$.*

Since the construction of f uses a partition of unity on T, it is valid in fact when T belongs to the class of paracompact spaces, a class encompassing both metrizable and compact spaces.

Exercises

1. Prove the assertions of this section that are not given proofs.

2. Check that if t_0 is not an *isolated point* of a topological space T (i.e., if t_0 is in the closure of $T \setminus \{t_0\}$) and if C is a closed proper subset of a topological space X, then $F : T \rightrightarrows X$ given by $F(t_0) := C$, $F(t) := X$ for $t \in T \setminus \{t_0\}$ is not closed at t_0.

3. Show that $F : T \rightrightarrows X$ is closed at every point of T if and only if its graph is closed in $T \times X$.

4. Let (X,d) be a metric space. Show that $F : T \rightrightarrows X$ is inward continuous at (t_0, x_0) if and only if $d(x_0, F(t)) \to 0$ as $t \to t_0$, if and only if there exists a map $f : T \to X$ that is continuous at t_0, with $f(t_0) = x_0$ and $f(t) \in \mathrm{cl}(F(t))$ for t close to t_0.

5. (**a**) Show that if F is the multimap $t \rightrightarrows \{f(t)\}$ associated with a map $f : T \to X$, then F is outward continuous at t_0 if and only if f is continuous at t_0.
(**b**) Check that for $f : \mathbb{R} \to \mathbb{R}$ given by $f(0) = 0$, $f(r) := 1/r$ otherwise, F is closed at 0 but not outward continuous at 0.

6. Prove that if T is a connected topological space, and if $F : T \rightrightarrows X$ has connected values and is such that for every $t \in T$ and every open subset V of X containing $F(t)$ the set $F^{-1}(V)$ is a neighborhood of t, then $F(T)$ is connected. Deduce Proposition 1.39 from that.

7. (**a**) Show that a multimap $F : T \rightrightarrows X$ from a topological space T to a metric space X is *Hausdorff outward continuous* at $t_0 \in T$ in the sense that $e_H(F(t), F(t_0)) := \sup_{x \in F(t)} d(x, F(t_0)) \to 0$ as $t \to t_0$ whenever F is outward continuous at t_0. Provide a counterexample to the converse.
(**b**) Show that F is inward continuous at t_0 whenever F is *Hausdorff inward continuous* at t_0 in the sense that $e_H(F(t_0), F(t)) := \sup_{x \in F(t_0)} d(x, F(t)) \to 0$ as $t \to t_0$.
(**c**) Consider the missing implications when one assumes that $F(t_0)$ is compact.

8. A multimap $M : X \rightrightarrows X^*$ between a Banach space and its dual is called a *monotone operator* if for all x, y in X and $x^* \in M(x)$, $y^* \in M(y)$ one has $\langle x^* - y^*, x - y \rangle \geq 0$. It is called *maximal monotone* if every monotone operator N whose graph contains the graph of M coincides with M. Prove that a maximal monotone operator is outward continuous from the strong topology to the weak* topology on the interior of its domain.

1.3   Limits of Sets and Functions

The notions of convergence for families of sets are important for their intrinsic interest, but also because they can serve to define convergences of families of functions that have better properties than traditional convergences.

1.3.1   Convergence of Sets

It is of common use to speak of limits of sets. For instance, in order to define a tangent to a curve C of a Euclidean space at some point a, one takes the limit of a secant to the curve C joining a to some point x of the curve when $x \to a$, with $x \in C$. As another example, let us consider the family $(E_t)_{t \in \mathbb{P}}$ of ellipses given by

$$E_t := \left\{ (r,s) \in \mathbb{R}^2 : tr^2 + s^2 = 1 \right\};$$

it is tempting to say that when $t \to 1$, i.e., when the eccentricity of the ellipse E_t tends to 1, E_t converges to the unit circle $E_1 := \mathbb{S}^1$. Many other examples can be given.

It is the purpose of the present subsection to explain how the intuitive idea of limit of a family of sets can be described in precise mathematical terms. Several approaches to a definition of limits of families of sets are possible. One can restrict one's attention to sequences of subsets of a topological space X, and for most purposes, such a framework is sufficient. In some cases, it may be useful to have at one's disposal a notion of convergence for nets of sets, i.e., for families $(F_i)_{i \in I}$ of subsets of X indexed by a directed set I. One may also adopt the framework of parameterized families. We choose the equivalent language of multimaps because it is closely connected with the notions of continuity of a multimap and it fits well the applications we have in view, in particular with generalized derivatives and tangent sets.

Let T be a (pointed) *parameter space*, i.e., a subspace T of a topological space P with which is associated a base point 0 of P in the closure $\mathrm{cl}(T)$ of T. We denote by $\mathscr{T}$ the trace $\mathscr{T} := \{N \cap T : N \in \mathcal{N}_P(0)\}$ on T of the family $\mathcal{N}_P(0)$ of neighborhoods of 0 in P. The usual case is the case $P := \mathbb{R}_+$, $T := \mathbb{P}$, the set of positive numbers, with 0 as base point. Another usual case is the case $T = \mathbb{N}$ in $P := \mathbb{R}_\infty$ and $+\infty$ is the base point; if one is interested only in limits of sequences of sets, one can restrict one's attention to this case. When one deals with a net $(S_i)_{i \in I}$ of subsets of a topological space X, one can set $T := I$ and endow the set $P := I_\infty := I \cup \{\infty\}$ with the topology defined in Sect. 1.1.2.

Given a topological space X and a multimap $F : T \rightrightarrows X$, it is not always the case that the limit of $F(t)$ as $t \to 0$ (in T) exists. However one can always consider two substitutes for the limit. They are described in the following definition.

Definition 1.41. The *limit inferior* (or inner limit) of a family $(F(t))_{t \in T}$ of subsets of a topological space $(X, \mathscr{O})$ is given by

$$\liminf_{t(\in T) \to 0} F(t) := \{x \in X : \forall V \in \mathscr{N}_X(x), \ \exists S \in \mathscr{T}, \ \forall s \in S, \ F(s) \cap V \neq \varnothing\}.$$

The *limit superior* (or outer limit) of the family $(F(t))_{t \in T}$ is given by

$$\limsup_{t(\in T) \to 0} F(t) := \{x \in X : \forall V \in \mathscr{N}_X(x), \ \forall S \in \mathscr{T}, \ F(S) \cap V \neq \varnothing\} = \bigcap_{S \in \mathscr{T}} \mathrm{cl}(F(S)).$$

If $\limsup_{t(\in T) \to 0} F(t) = \liminf_{t(\in T) \to 0} F(t)$ one says that $F(t)$ *converges* as $t \to 0$ in T and one denotes this set by $\lim_{t(\in T) \to 0} F(t)$ or $\lim_{t \to_T 0} F(t)$.

When the inclusion $t \in T$ can be implicitly assumed without great risk of ambiguity, we omit it in the notation for these one-sided limits and write $t \to 0$ instead of $t(\in T) \to 0$ or $t \to_T 0$. Clearly, one always has $\liminf_{t \to 0} F(t) \subset \limsup_{t \to 0} F(t)$.

There exist sequential versions of the preceding notions. One says that $x \in X$ is in the *sequential limit inferior* seq-$\liminf_{t \to 0} F(t)$ of $F(\cdot)$ as $t \to 0$ if for every sequence $(t_n) \to 0$ in T there exists a sequence $(x_n) \to x$ such that $x_n \in F(t_n)$ for all $n \in \mathbb{N}$ large enough. The *sequential limit superior* seq-$\limsup_{t \to 0} F(t)$ of $F(\cdot)$ as $t \to 0$ is defined as the set of $x \in X$ such that there exist sequences $(t_n) \to 0$ in T, $(x_n) \to x$ in X such that $x_n \in F(t_n)$ for all $n \in \mathbb{N}$.

Exercise. Show that $\liminf_{t \to 0} F(t) \subset$ seq-$\liminf_{t \to 0} F(t) \subset$ seq-$\limsup_{t \to 0} F(t) \subset \limsup_{t \to 0} F(t)$.

When X is metrizable and 0 has a countable base of neighborhoods, the sequential definition coincides with the original definition in view of the following characterization, whose simple proof is left to the reader.

Proposition 1.42. *A point x in a metric space (X, d) belongs to $\liminf_{t(\in T) \to 0} F(t)$ if and only if $\lim_{t(\in T) \to 0} d(x, F(t)) = 0$.*

It belongs to $\limsup_{t(\in T) \to 0} F(t)$ if and only if $\liminf_{t(\in T) \to 0} d(x, F(t)) = 0$.

Proposition 1.43. *If there exist $S \in \mathscr{T} := \{U \cap T : U \in \mathscr{N}_P(0)\}$ and a map $f : S \to X$ such that $\lim_{s(\in S) \to 0} f(s) = x$ and $f(s) \in F(s)$ for each $s \in S$, then $x \in \liminf_{t(\in T) \to 0} F(t)$.*

The converse holds when X is metrizable and either $F(t)$ is closed for all $t \in T$ or there exists a positive function on T with limit 0 as $t \to 0$.

Proof. The first assertion follows from the definitions.

Conversely, assume that X is metrizable and let d be a compatible metric. Let $x \in \liminf_{t(\in T) \to 0} F(t)$. If each $F(t)$ is nonempty (that occurs for t close to 0) and closed, we can find some $f(t) \in F(t)$ such that $d(x, f(t)) \leq 2d(x, F(t))$, since we can take $f(t) = x$ when $d(x, F(t)) = 0$. If there exists a positive function $\alpha(\cdot)$ on T with limit 0 as $t \to 0$, we can choose $f(t) \in F(t)$ in such a way that $d(x, f(t)) \leq d(x, F(t)) + \alpha(t)$. $\square$

Example. Let C, D be two closed subsets of a topological space X and let $F(n) := C$ for n an even integer, $F(n) := D$ for n an odd integer. Then $\limsup_{n \to \infty} F(n) = C \cup D$ and $\liminf_{n \to \infty} F(n) = C \cap D$.

Example. Let X be a normed space and let $(u_n) \to u$ be a convergent sequence in $X \setminus 0$. Then the sequence of half-lines $(\mathbb{R}_+ u_n)$ converges to $\mathbb{R}_+ u$.

Example. Let X_0 be a normed space and let $X = X_0 \times \mathbb{R}$ be endowed with the norm given by $\|(x,r)\| = (\|x\|^2 + r^2)^{1/2}$. For $t \in \mathbb{R}_+$ let S_t be the sphere $S_t := \{(x,r) \in X : \|x\|^2 + (r-t)^2 = t^2\}$. Then $(S_t) \to X_0 \times \{0\}$ as $t \to \infty$.

Example. For $f : T \to \mathbb{R}$ and $F(t) := (-\infty, f(t)] \subset X := \mathbb{R}$, one can check that

$$\liminf_{t(\in T) \to 0} F(t) = (-\infty, \liminf_{t(\in T) \to 0} f(t)], \qquad \limsup_{t(\in T) \to 0} F(t) = (-\infty, \limsup_{t(\in T) \to 0} f(t)],$$

where as usual, for $\mathscr{T} := \{N \cap T : N \in \mathscr{N}_P(0)\}$,

$$\liminf_{t(\in T) \to 0} f(t) := \sup_{S \in \mathscr{T}} \inf f(S), \qquad \limsup_{t(\in T) \to 0} f(t) := \inf_{S \in \mathscr{T}} \sup f(S).$$

Example. Let W, X be topological spaces, let $u \in W$, and let $g : W \times X \to \mathbb{R}^n$ be a mapping whose components g^i are lower semicontinuous at (u, x) for all $x \in X$. Let $F(w) := \{x \in X : g^i(w, x) \leq 0, i = 1, \ldots, n\}$. Then one can easily check that

$$\limsup_{w \to u} F(w) \subset F(u).$$

Such a multimap F appears as the feasible set in parameterized mathematical programming problems. The space $\mathbb{R}^n$ can be replaced with a general Banach space Z and the cone $\mathbb{R}^n_+$ with a closed convex cone C, provided one assumes that the epigraph $G := \{(x,z) \in X \times Z : z \in g(x) + C\}$ of g is closed.

Example. Let W, X, g, F be as in the preceding example. Suppose that for some $w_0 \in W$ and all $x_0 \in F^<(w_0) := \{x \in X : g^i(w_0, x) < 0, i = 1, \ldots, n\}$ the functions $w \mapsto g^i(w, x_0)$ are outward continuous at w_0. Then one has $g^i(w, x_0) < 0, i = 1, \ldots, n$, for w close to w_0, so that $x_0 \in \liminf_{w \to w_0} F(w)$. If $F(w_0)$ is contained in the closure of $F^<(w_0)$, then F is lower semicontinuous at w_0. Note that when X is a normed space, $g^i(w_0, \cdot)$ is convex for $i = 1, \ldots, n$, and $F^<(w_0)$ is nonempty, the condition $F(w_0) \subset \text{cl}(F^<(w_0))$ is satisfied.

We encourage the reader to devise calculus rules for compositions with continuous maps and usual operations such as unions, intersections, products, inverse images (or sums when X is endowed with an addition; any other operation can also be considered).

Links with continuity of multimaps are presented in the following statements, whose simple proofs are left to the reader.

Proposition 1.44. *One has* $\limsup_{t(\in T)\to 0} F(t) \subset E$ *if and only if the multimap* G *obtained by extending* F *to* $T \cup \{0\}$ *by setting* $G(0) := E$, $G(t) := F(t)$ *for* $t \in T$ *is closed at* 0.

One has $E \subset \liminf_{t(\in T)\to 0} F(t)$ *if and only if the multimap* G *obtained by extending* F *to* $T \cup \{0\}$ *by setting* $G(0) := E$ *is inward continuous at* 0.

Corollary 1.45. *Given a multimap* $F : X \rightrightarrows Y$ *between two topological spaces* X *and* Y *and* $\bar{x} \in X$ *one has* $\limsup_{x\to\bar{x}} F(x) \subset F(\bar{x})$ *if and only if* F *is closed at* $\bar{x}$.

One has $F(\bar{x}) \subset \liminf_{x(\neq\bar{x})\to\bar{x}} F(x)$ *if and only if* F *is inward continuous at* $\bar{x}$.

Since one always has $F(\bar{x}) \subset \limsup_{x\to\bar{x}} F(x)$, the first assertion can be rephrased as follows: F is closed at $\bar{x}$ if and only if $\limsup_{x\to\bar{x}} F(x) = F(\bar{x})$. Since one always has $\liminf_{x\to\bar{x}} F(x) \subset \mathrm{cl}(F(\bar{x}))$, when $F(\bar{x})$ is closed, the last assertion can be reformulated in the following manner: F is inward continuous at $\bar{x}$ if and only if $\liminf_{x\to\bar{x}} F(x) = F(\bar{x})$. This last observation justifies the use of the expression lower semicontinuity instead of inward continuity.

The next result shows the interest of the notions of limits of sets for optimization. We formulate it in terms of maximization because the statement is easier to memorize and is convenient for economical questions; however, mathematicians often prefer the minimization version.

Proposition 1.46. *Let* $(F(t))_{t\in T}$ *be a parameterized family of subsets of* X *and let* $F_0 \subset \liminf_{t(\in T)\to 0} F(t)$. *Given a lower semicontinuous function* $h : X \to \overline{\mathbb{R}}$, *let* $m_0 := \sup h(F_0)$, $m(t) := \sup h(F(t))$ *for* $t \in T$. *Then one has* $m_0 \leq \liminf_{t\to_T 0} m(t)$.

Proof. The result is obvious when $m_0 = -\infty$. Assume $m_0 > -\infty$ and take $r \in \mathbb{R}$, $r < m_0$. Since h is lower semicontinuous, the set $V := h^{-1}((r, +\infty])$ is open in X and it meets F_0. Since $F_0 \subset \liminf_{t(\in T)\to 0} F(t)$, there exists $N \in \mathcal{N}_P(0)$ such that $F(t) \cap V \neq \varnothing$ for all $t \in N \cap T$. Picking $x_t \in F(t) \cap V$, we get $m(t) \geq h(x_t) > r$ for all $t \in N \cap T$; hence $\liminf_{t(\in T)\to 0} m(t) \geq r$. Since r is arbitrarily close to m_0, the announced inequality holds. $\qquad\square$

Corollary 1.47. *Let* $F : T \rightrightarrows X$ *be a parameterized family of sets as above and let* $F_0 \subset \liminf_{t(\in T)\to 0} F(t)$. *Given an outward continuous function* $j : X \to \overline{\mathbb{R}}$, *let* $p_0 := \inf j(F_0)$, $p(t) := \inf j(F(t))$ *for* $t \in T$. *Then one has* $p_0 \geq \limsup_{t(\in T)\to 0} p(t)$.

Proof. It suffices to note that for $h := -j$ one has $p(t) = -m(t)$, $p_0 = -m_0$, where m is defined as in the preceding proposition. $\qquad\square$

In the next statement we give an answer to the question, if x_0 is a cluster point of a family $(x_t)_{t\in T}$, where x_t is a solution to the problem $(\mathcal{P}_t)$ of maximizing h over $F(t)$, under what conditions can one assert that x_0 is a solution to the limiting problem? We even consider the more general (and more realistic) case of approximate solutions. Here, given $t \in T$, $\alpha > 0$, we define the set of α-approximate solutions to $(\mathcal{P}_t)$ as the set

$$S(t, \alpha) := \{x \in F(t) : h(x) \geq m(t) - \alpha\},$$

where the subtraction in $\mathbb{R}$ has been extended to $\overline{\mathbb{R}} \times \mathbb{R}$ by setting $r - s = +\infty$ if $r = +\infty$, $s \in \mathbb{R}_-$, $r - s = 1/s$ if $r = +\infty$, $s \in \mathbb{P} := (0, +\infty)$, $r - s = -\infty$ if $r = -\infty$, $s \in \mathbb{R}_+$, $r - s = 1/s$ if $r = -\infty$, $s < 0$. Let $S(0) := \{x \in F_0 : h(x) = m(0)\}$.

Proposition 1.48. *Let $F : T \rightrightarrows X$ be a parameterized family of sets as above, with $T \subset P$ and $0 \in \mathrm{cl}(T)$, and let $h : X \to \overline{\mathbb{R}}$ be outward continuous at some $x_0 \in F_0 \subset X$. Assume that $m_0 \le \liminf_{t(\in T) \to 0} m(t)$, with $m_0 := \sup h(F_0)$, $m(t) := \sup h(F(t))$ for $t \in T$. Given mappings $t \mapsto \varepsilon(t)$, $t \mapsto x_t$ from T to $\mathbb{R}_+$ and X respectively such that $\varepsilon(t) \to 0$, $x_t \to x_0$ as $t \to 0$ in T, $x_t \in S(t, \varepsilon(t))$, one has $x_0 \in S(0)$.*

Proof. Suppose, to the contrary, that $h(x_0) < m_0$. Let $r, r' \in (h(x_0), m_0)$ with $r < r'$. Since h is outward continuous at x_0 and $x_t \to x_0$ as $t \to 0$ in T, one can find $N \in \mathcal{N}_P(0)$ such that $h(x_t) < r$ for all $t \in N \cap T$. Since $m_0 \le \liminf_{t(\in T) \to 0} m(t)$, shrinking N if necessary, we may assume that $r' < m(t)$ for all $t \in T \cap N$. Since $\varepsilon(t) \to 0$ as $t \to 0$ in T, we have $\min(m(t) - \varepsilon(t), 1/\varepsilon(t)) > r$ for $t \in N \cap T$, shrinking N again if necessary. Then we obtain a contradiction to $x_t \in S(t, \varepsilon(t))$ for $t \in N \cap T$. $\qquad\square$

1.3.2 Supplement: Variational Convergences

Simple examples show that pointwise convergence of a family $(f_t)_{t \in T}$ of functions is too weak a property to bring some usefulness in optimization: one may have $(f_t) \to f$ pointwise while $\inf f_t$ does not converge to $\inf f$. On the other hand, uniform convergence is too stringent a property to be satisfied in most practical problems. Such a situation led mathematicians to consider variational convergences, in particular epiconvergence. These convergences are of interest because one can define one-sided epi-limits even when (f_t) does not converge in any sensible sense. Moreover, they have a bearing on nonsmooth analysis, because given a function $f : X \to \overline{\mathbb{R}}$ and $\bar{x} \in X$ at which f is finite, one may wish to study the convergence of the differential quotients $f_t : v \mapsto (1/t)[f(\bar{x} + tv) - f(\bar{x})]$ as $t \to 0_+$.

In the sequel, T is a subset of a pointed topological space $(P, 0)$ with $0 \in \mathrm{cl}(T)$, X is a topological space, and $(f_t)_{t \in T}$ is a family of functions from X to $\overline{\mathbb{R}}$ parameterized by T. We denote by E_t the epigraph of f_t: $E_t := \{(x, r) \in X \times \mathbb{R} : f_t(x) \le r\}$.

Definition 1.49. The function e-$\liminf_t f_t$ (called the *lower epilimit*) is the function f whose epigraph E is $\limsup_t E_t$. The function e-$\limsup_t f_t$ (called the *upper epilimit*) is the function whose epigraph is $\liminf_t E_t$. One says that (f_t) epi-converges to some function f if (E_t) converges to the epigraph of f.

This definition is justified by the fact that the sets $\limsup_t E_t$ and $\liminf_t E_t$ are closed and are stable by addition of vectors of the form $(0, r)$ with $r \in \mathbb{R}_+$, and hence so are epigraphs. Because the study of variational convergences is outside the scope of this book, we limit our glimpse to a single result.

Proposition 1.50. *Let $f_0 \in \overline{\mathbb{R}}^X$, $(f_t)_{t \in T}$ be such that e-$\lim\sup_t f_t \leq f_0$. Then*

$$\limsup_{t(\in T) \to 0} \inf f_t(X) \leq \inf f_0(X).$$

Proof. It suffices to apply Corollary 1.47 to the epigraph $F(t)$ of f_t (resp. F_0 of f_0), taking for j the linear functional $(x, r) \mapsto r$.   □

Exercises

1. Give the proofs of the statements presented without proofs in this section.

2. Given a parameterized family $(F_t)_{t \in T}$ of subsets of a topological space X, indexed by $T \subset P$, with $0 \in \mathrm{cl}(T)$, $0 \notin T$, show that $\limsup_{t(\in T) \to 0} F(t)$ is the smallest subset C of X such that the multimap $G : T \cup \{0\} \rightrightarrows X$ given by $G(0) := C$, $G(t) = F(t)$ for $t \in T$ is closed at 0.

3. Given a parameterized family $(F_t)_{t \in T}$ of subsets of X as in Exercise 2, show that $\liminf_{t(\in T) \to 0} F(t)$ is the greatest subset C of X such that the multimap $G : T \cup \{0\} \rightrightarrows X$ given by $G(0) := C$, $G(t) = F(t)$ for $t \in T$ is inward continuous at 0.

4. Define the *outward limit* of a parameterized family $F(t)$ as $t(\in T) \to 0$ as the set $\mathrm{outlim}_{t(\in T) \to 0} F(t)$ of $x \in X$ for which there exists a compact subset K of X such that $x \in \limsup_{t(\in T) \to 0} F(t) \cap K$.
(a) Show that $\mathrm{outlim}_{t(\in T) \to 0} F(t) = \limsup_{t(\in T) \to 0} F(t)$ if X and P are metrizable.
(b) Prove that in general one has

$$\text{seq-}\limsup_{t(\in T) \to 0} F(t) \subset \mathrm{outlim}_{t(\in T) \to 0} F(t) \subset \limsup_{t(\in T) \to 0} F(t).$$

(c) Find a link between the notion of outward limit and the concept of compact outward continuity of a multimap.
(d) Suppose the parameter space P is metrizable and X is the dual of a Banach space X_* endowed with its weak* topology. Show that $x \in \mathrm{outlim}_{t(\in T) \to 0} F(t)$ if x is a weak* cluster point of a bounded sequence (x_n) satisfying $x_n \in F(t_n)$ for some sequence $(t_n) \to 0$ in T.

5. Given an increasing sequence (F_n) of subsets of X (for the order defined by inclusion), show that $\lim_n F_n = \mathrm{cl}(F)$, where F is the union of the F_n's.

6. Given a decreasing sequence (F_n) of subsets of X (for the order defined by inclusion), show that $\lim_n F_n = \mathrm{cl}(S)$, where S is the intersection of the F_n's.

7. Given a parameterized family $(F_t)_{t \in T}$ of subsets of a metric space X, show that $C \subset \liminf_{t(\in T) \to 0} F_t$ if and only if $d(x, C) \geq \limsup_{t(\in T) \to 0} d(x, F_t)$ for all $x \in X$.

Show that if $d(x,C) \leq \liminf_{t(\in T) \to 0} d(x,F_t)$ for all $x \in X$ and if C is closed, then $\limsup_{t(\in T) \to 0} F_t \subset C$. Prove the converse when the closed balls of X are compact.

1.4 Convexity and Separation Properties

Separation properties are among the pillars of functional analysis. They will be used throughout the book, in particular in Chap. 3. First, we need to review some properties of convex sets and functions.

1.4.1 Convex Sets and Convex Functions

Let us recall that a subset C of a linear space X is said to be *convex* if a segment whose extremities are in C is entirely contained in C: for all $x_0, x_1 \in C$, $t \in [0,1]$, one has $x_t := (1-t)x_0 + tx_1 \in C$. Among convex subsets, the simplest ones are *affine subspaces* obtained by translating linear subspaces, *half-spaces* (subsets D such that there exist a linear form ℓ on X and $r \in \mathbb{R}$ for which $D = \ell^{-1}((-\infty,r))$ or $D = \ell^{-1}(-\infty,r])$) and *convex cones*. The latter are the subsets that are stable by positive homotheties $h_r : x \mapsto rx$ (with $r \in \mathbb{P} := (0,+\infty)$ fixed) and addition, as easily checked. From antiquity to the present, *polyhedral subsets*, i.e., finite intersections of closed half-spaces, have played a special role among convex subsets, since they enjoy particular properties not shared by all convex sets.

A function f from a linear space X to $\overline{\mathbb{R}} := \mathbb{R} \cup \{-\infty, \infty\}$ is said to be *convex* if its *epigraph*

$$E_f := \mathrm{epi}(f) := \{(x,r) \in X \times \mathbb{R} : r \geq f(x)\}$$

is convex, or equivalently, if for every $t \in [0,1]$, $x_0, x_1 \in X$,

$$f((1-t)x_0 + tx_1) \leq (1-t)f(x_0) + tf(x_1)$$

(with the convention that $(-\infty) + (+\infty) = +\infty$ and $0 \cdot (+\infty) = +\infty$, $0 \cdot (-\infty) = -\infty$, which we adopt in the sequel). It is easy to show that f is convex if and only if its *strict epigraph*

$$E'_f := \mathrm{epi}_s(f) := \{(x,r) \in X \times \mathbb{R} : r > f(x)\}$$

is convex. A function f is *concave* if $-f$ is convex. A function $s : X \to \overline{\mathbb{R}}$ is said to be *sublinear* if its epigraph is a convex cone, i.e., if it is *subadditive* ($s(x+x') \leq s(x) + s(x')$ for all $x, x' \in X$) and *positively homogeneous* ($s(tx) = ts(x)$ for all $t \in \mathbb{P}$, $x \in X$). A sublinear function p with nonnegative values is called a *gauge*; if moreover p is finite and *even*, i.e., if $p(-x) = p(x)$ for every $x \in X$, then p is a *seminorm*.

Convex functions taking the value $-\infty$ are very special (for instance, they do not take any finite value if they are lower semicontinuous); therefore we will usually discard them and consider only functions with values in $\mathbb{R}_\infty := \mathbb{R} \cup \{+\infty\}$. In contrast, it is useful to admit functions taking the value $+\infty$; among them is the *indicator function* ι_C of a subset C of X: let us recall that it is given by $\iota_C(x) = 0$ for $x \in C$, $\iota_C(x) = +\infty$ for $x \in X \setminus C$. For instance, one can take a constraint C into account by replacing an objective function f by $f_C := f + \iota_C$. One calls *proper* a function that does not take the value $-\infty$ and takes at least one finite value. The expression nonimproper would be less ambiguous, but the risk of confusion with the topological concept is limited, so that we keep the usual terminology. Moreover, the epigraph of a function $f : X \to \mathbb{R}_\infty$ is a proper subset (nonempty and not the whole space) of $X \times \mathbb{R}$ if and only if f is proper. We denote by D_f or $\mathrm{dom}\, f$ the *domain* of f, i.e., the projection on X of $E_f := \mathrm{epi}\, f$:

$$D_f := \mathrm{dom}\, f := \{x \in X : f(x) < +\infty\}.$$

The following statement will be used repeatedly; it relies on the obvious fact that the image of a convex set under a linear map is convex.

Lemma 1.51. *Let W and X be linear spaces and let $f : W \times X \to \overline{\mathbb{R}}$ be convex. Then the performance function $p : W \to \overline{\mathbb{R}}$ defined as follows is convex:*

$$p(w) := \inf_{x \in X} f(w, x).$$

Proof. The result follows from the fact that the strict epigraph of p is the projection on $W \times \mathbb{R}$ of the strict epigraph of f.   □

Let us add that if f is positively homogeneous in the variable w, then so is p.

Example. If C is a convex subset (resp. a convex cone) of a normed space, then the associated *distance function* $d_C : w \mapsto \inf_{x \in C} \|w - x\|$ is convex (resp. sublinear).

Example. Given $f, g : X \to \overline{\mathbb{R}}$, their *infimal convolution* $f \square g : X \to \overline{\mathbb{R}}$ defined by

$$(f \square g)(w) := \inf\{f(u) + g(v) : u, v \in X, \ u + v = w\} = \inf\{f(w - x) + g(x) : x \in X\}$$

is convex whenever f and g are convex. If f and g are sublinear then $f \square g$ is sublinear. The preceding example is a special case corresponding to $f := \|\cdot\|$, $g := \iota_C$.

Besides the indicator function and the distance function, two other functions associated with a convex set play a noteworthy role. If C is a subset of X containing the origin, the *gauge function* (or Minkowski gauge) μ_C of C is defined by

$$\mu_C(x) := \inf\{r \in \mathbb{R}_+ : x \in rC\}, \qquad\qquad x \in X.$$

Clearly, μ_C is positively homogeneous and one has $C \subset \mu_C^{-1}([0,1])$. If C is *star-shaped*, i.e., if for all $x \in C$, $t \in [0,1]$ one has $tx \in C$, then $\mu_C^{-1}([0,1)) \subset C$. If moreover C is algebraically closed in the sense that its intersection with every ray

$L_u := \mathbb{R}_+ u$, $u \in X \setminus \{0\}$, is closed in L_u, then $C = \mu_C^{-1}([0,1])$. In particular, the gauge function of the closed unit ball B_X of a normed space $(X, \|\cdot\|)$ is just $\|\cdot\|$. We leave the proof of the next two lemmas as exercises. Hereinafter, a subset C of a linear space X is said to be *absorbing* if for all $x \in X$ there exists some $r > 0$ such that $x \in rC$.

Lemma 1.52. *A subset C of X is absorbing if and only if its gauge μ_C is finitely valued. If C is a convex subset of X, then μ_C is sublinear.*

Another function one can associate to a subset C of a normed space X is its *support function* σ_C or $h_C : X^* \to \overline{\mathbb{R}}$ given by

$$\sigma_C(x^*) := h_C(x^*) := \sup\{\langle x^*, x \rangle : x \in C\}, \qquad x^* \in X^*. \qquad (1.8)$$

Lemma 1.53. *If C is a nonempty subset of X, its support function $\sigma_C := h_C$ is a lower semicontinuous sublinear function on the (topological) dual X^* of X.*

Since the intersection of a family of convex subsets is convex, any nonempty subset A of a linear space X is contained in a convex set C that is the smallest in the family $\mathscr{C}_A$ of convex sets containing A. It is denoted by $\mathrm{co}(A)$ and called the convex set generated by A or the *convex hull* of A. It is obtained as the intersection of the family $\mathscr{C}_A$. It is easy to check that $\mathrm{co}(A)$ is the set of convex combinations of elements of A, i.e., $\mathrm{co}(A)$ is the set of $x \in X$ that can be written as

$$t_1 a_1 + \cdots + t_n a_n$$

with $n \in \mathbb{N} \setminus \{0\}$, $a_i \in A$, $t := (t_1, \ldots, t_n)$ being an element of the *canonical simplex* Δ_n, i.e., the set of $t := (t_1, \ldots, t_n) \in \mathbb{R}^n_+$ satisfying $t_1 + \cdots + t_n = 1$. The *convex hull* $\mathrm{co}(h)$ of a function $h : X \to \mathbb{R}_\infty$ is the greatest convex function g bounded above by h. Its epigraph is almost the convex hull of the epigraph E_h of h. In fact, it is the vertical closure of $\mathrm{co}(E_h)$ in the sense that one has $\mathrm{epi}_s g \subset \mathrm{co}(E_h) \subset \mathrm{epi}\, g$. Thus

$$g(x) := \inf\left\{ \sum_{i=1}^m t_i h(x_i) : m \geq 1, \, t := (t_1, \ldots, t_m) \in \Delta_m, \, x_i \in X, \, t_1 x_1 + \cdots + t_m x_m = x \right\}.$$

Exercise. Show that for $g := \mathrm{co}(h)$ the inclusions $\mathrm{epi}_s g \subset \mathrm{co}(\mathrm{epi}\, h) \subset \mathrm{epi}\, g$ may be strict. [Hint: Consider $h : \mathbb{R} \to \mathbb{R}$ given by $h(0) := 1$, $h(x) := |x|$ for $x \in \mathbb{R} \setminus \{0\}$.]

Note that in general, the union of a family (C_p) of convex subsets is no longer convex; but when (C_p) is an increasing sequence (with respect to inclusion), the union is convex. Similarly, the infimum of a countable family (k_p) of convex functions is convex when the sequence (k_p) is decreasing; but that is not the case if the sequence (k_p) does not satisfy this property. The following lemma describes the convex hull of the infimum of an arbitrary sequence of functions. It can be taken as an exercise, but it will be used in Chap. 4.

Lemma 1.54. *(a) Given a sequence $(E_n)_{n \geq 1}$ of nonempty subsets of a linear space Z, the convex hull C of the union E of the E_n's is the union over $p \in \mathbb{N} \setminus \{0\}$ of the convex hulls C_p of $E_1 \cup \cdots \cup E_p$:*

$$C := \mathrm{co}(E) = \bigcup_p C_p, \quad \text{where } C_p := \mathrm{co}\left(\bigcup_{1 \leq n \leq p} E_n\right). \tag{1.9}$$

For $m, p \in \mathbb{N} \setminus \{0\}$, setting $\mathbb{N}_m := \{1, \ldots, m\}$ and denoting by $J_{m,p}$ the set of maps $j : \mathbb{N}_m \to \mathbb{N}_p$, the set C_p is given by

$$C_p := \bigcup_{m \geq 1} \bigcup_{j \in J_{m,p}} \left\{ \sum_{i=1}^m t_i x_i : t := (t_1, \ldots, t_m) \in \Delta_m, \ x_i \in E_{j(i)} \right\}. \tag{1.10}$$

(b) Given a sequence (h_n) of functions on a linear space X, the convex hull k of the function $h := \inf_n h_n$ is the infimum over $p \in \mathbb{N} \setminus \{0\}$ of the convex hulls $k_p := \mathrm{co}\,(h_1, \ldots, h_p)$ of the functions $h_1, \ldots, h_p$. The function k_p is given by

$$k_p(x) = \inf_{m \geq 1} \inf_{j \in J_{m,p}} \inf\left\{ \sum_{i=1}^m t_i h_{j(i)}(x_i) : (t_1, \ldots, t_m) \in \Delta_m, \ x_i \in X, \ \sum_{i=1}^m t_i x_i = x \right\}. \tag{1.11}$$

Proof. (a) In fact, every element of C can be written as a convex combination of a finite family of elements of E, hence is an element of C_p for some p. The reverse containment is obvious since $C_p \subset C$ for all p.

The right-hand side of (1.10) is clearly contained in C_p. Using the fact that the concatenation of an element j of $J_{m,p}$ with an element j' of $J_{n,p}$ is an element of $J_{m+n,p}$, it is easily seen that this set is convex and contains all E_n's for $n \in \mathbb{N}_p$, so that it coincides with C_p.

(b) Now, when E_n is the epigraph of a function h_n, the vertical closure of C_p is the epigraph of $k_p := \mathrm{co}\,(h_1, \ldots, h_p)$, the greatest convex function majorized by $h_1, \ldots, h_p$.

The right-hand side of (1.11) defines a function that is clearly minorized by k_p. Since it is easily seen that it is convex and bounded above by h_n for all $n \in \mathbb{N}_p$, it coincides with k_p. One can also derive this formula from (1.10) using epigraphs. $\square$

When X is a normed space, every subset S is contained in a smallest closed convex subset, its *closed convex hull* $\overline{\mathrm{co}}(S)$. It is easy to check using the following elementary result that this set is just the closure of $\mathrm{co}(S)$. In fact, the lemma and the preceding assertion are valid in any topological linear space. In the sequel, a number of results given for normed spaces are valid for topological linear spaces. We leave the proofs of the next two results as exercises.

Lemma 1.55. *The closure $\mathrm{cl}(C)$ and the interior $\mathrm{int}(C)$ of a convex subset C of a normed space are convex.*

Lemma 1.56. *If the interior of a convex subset C of a normed space is nonempty, then one has* $\mathrm{cl}(C) = \mathrm{cl}(\mathrm{int}(C))$ *and* $\mathrm{int}(\mathrm{cl}(C)) = \mathrm{int}(C)$.

Lemma 1.57. *If C is a nonempty convex subset of a finite-dimensional space, then C has a nonempty interior (called the* relative interior *and denoted by* $\mathrm{ri}(C)$*) in the affine subspace A it generates.*

Proof. By definition, A is the smallest affine subspace containing C. Using a translation, we may suppose $0 \in C$, so that A is the linear subspace generated by C. Let n be the dimension of A and let m be the greatest integer k such that there exists a linearly independent family $\{e_1, \ldots, e_k\}$ in C satisfying

$$\mathrm{co}\{e_1, \ldots, e_k\} := \{t_1 e_1 + \cdots + t_k e_k : (t_1, \ldots, t_k) \in \Delta_k\} \subset C.$$

Let $\{e_1, \ldots, e_m\}$ be such a family and let L be the linear space it generates. Then C is contained in L: otherwise, we could find some $e \in C \setminus L$ and the family $\{e_1, \ldots, e_m, e\}$ would satisfy the above conditions and be strictly larger than $\{e_1, \ldots, e_m\}$. Thus $L = A$ and the set $\mathrm{co}\{e_1, \ldots, e_k\}$ has nonempty interior in A for the unique Hausdorff linear topology on A obtained by transporting the topology of $\mathbb{R}^m$ by the isomorphism defined by the base $\{e_1, \ldots, e_m\}$. □

For the next result we need the notion of *core* of a subset C of a linear space X:

$$\mathrm{core}\, C := \{u \in X : \forall v \in X\ \exists \varepsilon > 0,\ u + [-\varepsilon, \varepsilon]v \subset C\} \tag{1.12}$$

$$= \{u \in C : \forall v \in X \setminus \{0\}\ \exists \alpha > 0,\ u + [0, \alpha]v \subset C\}. \tag{1.13}$$

The elements of the core of C are also said to be *internal* elements of C and the core of C is also called the *algebraic interior* of C. For a convex subset one has the following characterizations.

Lemma 1.58. *For a nonempty convex subset C of a linear space X and $u \in X$, the following assertions are equivalent:*

(a) $u \in \mathrm{core}\, C$;
(b) $C - u$ is absorbing: for every $x \in X$ there exists $t > 0$ such that $tx \in C - u$;
(c) $X = \mathbb{R}_+ (C - u) := \{r(c - u) : r \in \mathbb{R}_+,\ c \in C\}$.

Proof. The implications (a)⟹(b)⟹(c) are obvious. Now (c) implies that $0 \in C - u$: this is obvious if $X = \{0\}$, and otherwise, taking $v \neq 0$ in X, we can write $v = r(c - u)$, $-v = r'(c' - u)$, for some $c, c' \in C$ and $r, r' > 0$ (since $v \neq 0$), so that $0 = r(r + r')^{-1} c + r'(r + r')^{-1} c' - u \in C - u$ by convexity; hence $u \in C$. Moreover, if $v \in X \setminus \{0\}$, we can find $r \in \mathbb{P}$, $c \in C$ such that $v = r(c - u)$; then for $s \in [0, r^{-1}]$ we have $u + sv = (1 - sr)u + src \in C$, so that by (1.13), we have $u \in \mathrm{core}\, C$. □

Let us compare the core and the interior of a convex subset.

Proposition 1.59. *The core of a convex subset C of a normed space X coincides with its interior* $\mathrm{int}C$ *whenever one of the following conditions is satisfied:*

(a) $\mathrm{int}C \neq \varnothing$;
(b) X *is finite-dimensional;*
(c) X *is a Banach space and C is closed.*

Proof. The interior $\mathrm{int}C$ of a convex subset C of a normed space X is always contained in its core, since for every $u \in \mathrm{int}C$ and every $v \in X$ the map $f : t \mapsto u + tv$ is continuous and $f(0) \in \mathrm{int}C$, so that $f(t) \in C$ for $t > 0$ small enough.

(a) In order to prove the equality $\mathrm{core}C = \mathrm{int}C$ when $\mathrm{int}C$ is nonempty, let $u \in \mathrm{int}C$ and let $\bar{x} \in \mathrm{core}C$. The definition of $\mathrm{core}C$ yields some $\varepsilon > 0$ such that $z := \bar{x} + \varepsilon(\bar{x} - u) \in C$. Then the mapping h given by $h(x) = z + \varepsilon(1 + \varepsilon)^{-1}(x - z)$ is a homeomorphism of X onto X satisfying $h(u) = \bar{x}$, $h(C) \subset C$. Then $h(\mathrm{int}C)$ is a neighborhood of $\bar{x}$ that is contained in C, so that $\bar{x} \in \mathrm{int}C$.

Assertion (b) follows from assertion (a), taking into account Lemma 1.57, which asserts that C has a nonempty interior in the affine subspace it generates, which is the whole space if $\mathrm{core}C$ is nonempty.

(c) We may suppose $\mathrm{core}C \neq \varnothing$ and, using a translation if necessary, $0 \in \mathrm{core}C$. Then X is the union of the closed subsets nC for $n \in \mathbb{N} \setminus \{0\}$. Since X is a Baire space, one of these sets has nonempty interior. Thus $\mathrm{int}C$ is nonempty and (a) applies. $\square$

In order to obtain important, but more advanced, interiority results, we need to introduce a special class of convex sets that has remarkable preservation properties. Let us say that a subset C of a normed (or topological) linear space X is *ideally convex* if for every bounded sequence (x_n) of C and element (t_n) of $\Delta_\infty := \{(t_n) \in \mathbb{R}_+^{\mathbb{N}} : \sum_{n \geq 0} t_n = 1\}$, the series $\sum_{n \geq 0} t_n x_n$ converges in C whenever it converges (which means that for $s_n := t_0 x_0 + \cdots + t_n x_n$ the sequence (s_n) converges). We leave as an exercise the proof of the next lemma, giving some examples and some easy properties; see [507, 984].

Lemma 1.60. *Let X be a normed space (or a topological linear space).*

(a) Every closed convex subset of X is ideally convex.
(b) Every open convex subset of X is ideally convex.
(c) Every convex subset of X is ideally convex if X is finite-dimensional.
(d) The intersection of a family of ideally convex subsets is again ideally convex.
(e) If X, Y are normed spaces, if $A \in L(X,Y)$, and if D is an ideally convex subset of Y, then $C := A^{-1}(D)$ is ideally convex in X.

Another permanence property will play an important role.

Lemma 1.61. *Let W and X be Banach spaces, and let C be the projection $p_X(F)$ on X of a closed convex subset F of $W \times X$. If the projection $p_W(F)$ of F on W is bounded, then C is ideally convex.*

Proof. Let x be the sum of a series with general term $t_n x_n$, where $(t_n) \in \Delta_\infty$, (x_n) is bounded and $x_n \in C$ for all $n \in \mathbb{N}$. For all $n \in \mathbb{N}$, there exists some $w_n \in W$ such that $(w_n, x_n) \in F$. Since $p_W(F)$ is bounded, (w_n) is bounded. Thus $((w_n, x_n))$ is bounded and the series $\Sigma t_n(w_n, x_n)$ satisfies the Cauchy criterion. Its sum $(\overline{w}, \overline{x})$ is in F since F is closed and convex (or ideally convex). Then $x = \overline{x} \in C$, so that C is ideally convex. $\qquad\qquad\qquad\qquad\qquad\qquad\qquad\qquad\qquad\qquad\qquad\qquad\qquad\qquad\qquad\qquad\Box$

For an ideally convex set C, proving that a point x belongs to $\mathrm{int}(C)$ is reduced to checking the algebraic equality $X = \mathbb{R}_+(C - x)$, in view of the next lemma.

Lemma 1.62. *Let C be an ideally convex subset of a Banach space X. Then*

$$\mathrm{int}(C) = \mathrm{core}(C) = \mathrm{core}(\mathrm{cl}(C)) = \mathrm{int}(\mathrm{cl}(C)).$$

Proof. By Proposition 1.59 (c), we already know that $\mathrm{core}(\mathrm{cl}(C)) = \mathrm{int}(\mathrm{cl}(C))$. Given $\overline{x} \in \mathrm{int}(\mathrm{cl}(C))$, let us show that $\overline{x} \in \mathrm{int}(C)$. That will prove the chain of equalities. Performing a translation, we may suppose $\overline{x} = 0$. Then there exists some $r > 0$ such that $B := rB_X \subset \mathrm{cl}C$. Then for all $b \in \mathrm{int}(B)$ and for all V in the family $\mathcal{N}(b)$ of neighborhoods of b, one has $V \cap B \in \mathcal{N}(b)$, hence $V \cap B \cap C \neq \varnothing$, so that $\mathrm{int}(B) \subset \mathrm{cl}(B \cap C)$. Thus, for $q \in (0, 1/2)$ we have

$$B = \mathrm{cl}(\mathrm{int}(B)) \subset \mathrm{cl}(B \cap C) \subset B \cap C + qB. \qquad (1.14)$$

Given $u_0 \in B$, let us inductively construct sequences $(x_n)_{n \geq 1}$ in $B \cap C$, $(u_n)_{n \geq 1}$ in B such that

$$q^n x_n = q^n u_{n-1} - q^{n+1} u_n. \qquad (1.15)$$

We obtain $x_1 \in B \cap C$ and $u_1 \in B$ by writing $u_0 = x_1 + q u_1$ according to (1.14). Suppose x_k and u_k have been obtained for $k = 1, \ldots, n-1$. Then inclusion (1.14) yields some $x_n \in B \cap C$ and $u_n \in B$ such that $u_{n-1} = x_n + q u_n$, so that (1.15) holds. Then for $p \in \mathbb{N} \setminus \{0\}$, we get

$$\left\| \sum_{n=1}^{p} q^n x_n - q u_0 \right\| = \left\| q^{p+1} u_p \right\| \leq q^{p+1} r.$$

Thus $q u_0$ is the sum of the series with general term $q^n x_n$ $(n \geq 1)$, and since the sum of the series with general term q^n is $q(1-q)^{-1} \leq 1$, by the ideal convexity of C, we get $(1-q)u_0 \in C$, and since $q u_0 \in C$, by convexity, $u_0 \in C$. Thus B is a neighborhood of $\overline{x}$ contained in C, and hence $\overline{x} \in \mathrm{int}(C)$. $\qquad\qquad\qquad\qquad\qquad\qquad\Box$

The preceding results can be used to obtain an open mapping theorem for multimaps with closed convex graphs.

Theorem 1.63 (Robinson–Ursescu). *Let W, X be Banach spaces, let $F : W \rightrightarrows X$ be a multimap with closed convex graph. Then for every $(\overline{w}, \overline{x})$ in the graph of F such that $\overline{x} \in \mathrm{core}\, F(W)$, the multimap F is open at $(\overline{w}, \overline{x})$. In fact, F is open at $(\overline{w}, \overline{x})$ with a linear rate in the sense that there exists some $c > 0$ such that*

$$\forall t \in (0, 1], \qquad B(\overline{x}, tc) \subset F(B(\overline{w}, t)).$$

Proof. Without loss of generality, we may suppose $(\overline{w}, \overline{x}) = (0,0)$, so that $F(W)$ is absorbing. Let B be the closed ball with center $\overline{w} = 0$ and radius r in W and let $C := F(B) = p_X((B \times X) \cap \mathrm{gph} F)$. Let us check that C is absorbing. That will prove that $0 \in \mathrm{int} C$ by Lemmas 1.61, 1.62. Let x be an arbitrary point of X. Since $F(W)$ is absorbing, there exist some $s > 0$ and $w \in W$ such that $sx \in F(w)$. If $w \in B$, then $sx \in C$. If $w \in W \setminus B$, let $t := r \|w\|^{-1}$, so that $tw \in B$. The convexity of F yields

$$stx = tsx + (1-t)0 \in tF(w) + (1-t)F(0) \subset F(tw + (1-t)0) = F(tw) \subset F(B) = C,$$

since $t \in (0,1)$. Thus, C is absorbing, and every neighborhood of $\overline{w}$ is mapped by F onto a neighborhood of $\overline{x}$: F is open at $(\overline{w}, \overline{x})$.

The last assertion stems from the convexity of F: if $B(\overline{x}, c)$ is contained in $F(B(\overline{w}, 1))$, then for $t \in (0,1]$, one has

$$B(\overline{x}, tc) = tB(\overline{x}, c) + (1-t)\overline{x} \subset tF(B(\overline{w}, 1)) + (1-t)F(\overline{w})$$

$$\subset F(tB(\overline{w}, 1) + (1-t)\overline{w}) = F(B(\overline{w}, t)).$$

Exercises

1. Let $f : X \to \overline{\mathbb{R}}$ be a convex function on a linear space. Show that if for some $\overline{x} \in X$ one has $f(\overline{x}) = -\infty$, then for all $v \in X \setminus \{0\}$ there is at most one $t \in \mathbb{R}$ such that $f(\overline{x} + tv)$ is finite. If f is sublinear and if $f(0) = -\infty$, then either for all $x \in X$ one has $f(x) = -\infty$, or else $f(x) = +\infty$ for all $x \in X \setminus \{0\}$.

If X is a normed space and if f is lower semicontinuous with $f(\overline{x}) = -\infty$, show that f cannot take a finite value and that f has a closed domain. Give an example of such a function. [Hint: Take the *valley function* υ_C associated with a nonempty closed convex subset C, given by $\upsilon_C(x) = -\infty$ for $x \in C$, $\upsilon_C(x) = +\infty$ for $x \in X \setminus C$.]

2. Show that if a function $f : X \to \overline{\mathbb{R}}$ takes at least one finite value on a subset C of a set X, then the problem

$$(\mathscr{C}) \qquad \text{minimize } f(x) \quad \text{for } x \in C$$

is equivalent (in the sense that it has the same value and the same set of solutions) to the unconstrained problem of minimizing $f + \iota_C$ on X, where ι_C is the indicator function of C.

3. Show that a lower semicontinuous function $f : X \to \mathbb{R}_\infty$ is convex whenever it is *midpoint convex* in the sense that for all $x, y \in X$ one has $f((1/2)(x+y)) \le (1/2)f(x) + (1/2)f(y)$. Prove that the lower semicontinuity assumption cannot be dropped.

4. Let $(f_i)_{i \in I} \to f$ pointwise, I being a directed set. Show that f is convex if for all $i \in I$, f_i is convex.

5. Let $(f_i)_{i \in I}$ be a net of convex functions on X. Show that the function f given by $f(x) := \limsup_i f_i(x)$ is convex.

6. Let (x_n) be a bounded sequence of a normed space X and let $p \in [1, +\infty)$. Check that the set C of minimizers of the function f given by $f(x) := \liminf_n \|x - x_n\|^p$ is convex. Give conditions ensuring that C is a singleton (called the *asymptotic center* of (x_n)).

7. Let X be a linear space, $g : X \to \mathbb{R}_\infty$ convex, $h : \mathbb{R}_\infty \to \mathbb{R}_\infty$ nondecreasing and such that $h \mid \mathbb{R}$ is convex and $h(+\infty) = +\infty$. Check that $f := h \circ g$ is convex. Deduce from that for every norm $\|\cdot\|$ on X and nondecreasing convex function $h : \mathbb{R} \to \mathbb{R}_\infty$, the function $x \mapsto h(\|x\|)$ is convex.

8. Prove that for a convex subset C of a finite-dimensional space, the set C is a linear subspace iff cl$C$ is a linear subspace.

9. (*Homogenization*) Let C be a convex subset of a linear space W and let $Q := \mathbb{R}_+(C \times \{1\}) \subset W \times \mathbb{R}$. Check that Q is a convex cone in $Z := W \times \mathbb{R}$. Suppose that $W := X \times \mathbb{R}$ and that C is the epigraph of a convex function f on X. Show that Q is the epigraph of a sublinear function $s : X \times \mathbb{R} \to \overline{\mathbb{R}}$ and give an analytical expression for s. Such a construction makes it possible to reduce a question about convex functions to a question about sublinear functions.

10. Recall that the *relative interior* ri$C$ of a convex subset of a normed space X is the set of points of C that are interior to C when C is considered as a subset of its affine hull aff$C$, the smallest affine subspace of X containing C.

(a) Show that if $x \in$ ri$C$, $y \in$ cl$C \cap$ aff$C$ then for all $t \in [0, 1)$ one has $(1 - t)x + ty \in C$.
(b) Prove that if X is finite-dimensional and if C is nonempty, then ri$C$ is nonempty and $u \in$ ri$C$ iff $\mathbb{R}_+(C - u)$ is a linear subspace.
(c) Prove that if X is finite-dimensional, then C, ri$C$, and cl$C$ have the same affine hull.
(d) Deduce from what precedes that in a finite-dimensional space X one has cl(ri$C$) = cl$C$ and ri(cl$C$) = ri$C$.
(e) Prove that for a nonempty convex subset of a finite-dimensional space one has $u \in$ ri$C$ iff for every $x \in C$ there exists $t > 1$ such that $(1 - t)x + tu \in C$.
(f) Prove that if $(C_i)_{i \in I}$ is a family of convex subsets of a finite-dimensional space X and if the family $(\mathrm{ri}C_i)_{i \in I}$ has a nonempty intersection, then

$$\mathrm{cl}\left(\bigcap_{i \in I} C_i\right) = \bigcap_{i \in I} \mathrm{cl}C_i,$$

and if the family is finite then

$$\mathrm{ri}\left(\bigcap_{i \in I} C_i\right) = \bigcap_{i \in I} \mathrm{ri}C_i$$

(see [871]).

11. The *quasirelative interior* of a convex subset C of a normed space X is the set qri$C$ of $u \in C$ such that $T(C,u) := \mathrm{cl}(\mathbb{R}_+(C-u))$ is a linear subspace.

(a) Show that if C is contained in a finite-dimensional subspace, then qri$C =$ ri$C$.
(b) Show that $u \in$ qri$C$ iff the *normal cone* to C at u defined by

$$N(C,u) := \{x^* \in X^* : \langle x^*, x-u \rangle \leq 0 \; \forall x \in C\}$$

is a linear subspace of X^*.
(c) Prove that if int$C$ is nonempty then qri$C =$ int$C$.
(d) Prove that if int$C$ is nonempty then $\mathrm{cl}($qri$C) =$ cl$C$.
(e) Let $A : X \to Y$ be a continuous linear map and let C be a convex subset of X. Show that $A($qri$C) \subset$ qri$A(C)$ and in particular, when Y is finite-dimensional, $A($qri$C) \subset$ ri$A(C)$.
(f) Suppose C is a convex cone of X such that $X = \mathrm{cl}(C-C)$ and $C \cap (-C) = \{0\}$, $C^+ \cap (-C^+) = \{0\}$, where $C^+ = -C^0 := \{x^* \in X^* : \langle x^*, x \rangle \geq 0 \; \forall x \in C\}$. Show that qri$C$ is the set of $x \in C$ that are strictly positive (i.e., $\langle x^*, x \rangle > 0$ for all $x^* \in C^+ \setminus \{0\}$).
(g) Find qri$C$ when $C = \{x \in \ell_2 : \|x\|_1 \leq 1\}$ (show that $x \in$ qri$C$ iff $x \in C$ and the set of n such that $x_n \neq 0$ is infinite) (see [126]).

12. Find the interior of the set of positive semidefinite symmetric matrices with n rows and n columns.

13. A function $q : Z \to \overline{\mathbb{R}}$ on a linear space Z is said to be *quasiconvex* if $q((1-t)z_0 + tz_1) \leq \max(q(z_0), q(z_1))$ for all $z_0, z_1 \in Z$ and $t \in [0,1]$. Show that the performance function $p : W \to \overline{\mathbb{R}}$ associated with a quasiconvex function $f : W \times X \to \overline{\mathbb{R}}$ by $p(w) := \inf_{x \in X} f(w,x)$ is quasiconvex.

14. Check that the performance function p of the preceding exercise is convex whenever f is *convexlike* in the sense for every $(w_i, x_i, r_i) \in$ epi$f$ $(i = 0,1)$ and $t \in (0,1)$, $r \in \mathbb{R}$ with $r > (1-t)r_0 + tr_1$ there exists some $x \in X$ satisfying $f((1-t)w_0 + tw_1, x) < r$.

15. (*Locally convex functions* that are not convex) Let $W := \{(x,y) \in \mathbb{R}^2 : y < 4x^2\}$ and let $f : W \to \mathbb{R}$ be given by $f(x,y) := x^4 + (y-x^2)^2$. Show that for all $w \in W$ the function f is convex on some neighborhood of w, but that one cannot find a convex function on $\mathbb{R}^2$ whose restriction to W is f.

16. Let A be a nonempty subset of a normed space X and let $f : X \to \mathbb{R}$ be a convex function. Let C be the convex hull of A. Check that $\sup f(C) = \sup f(A)$.

17. Let A, B be nonempty subsets of X and let C, D be their respective convex hulls. Check that $f := d(\cdot, D)$ is convex and majorized by $d(\cdot, B)$. Deduce from this and from the preceding exercise that $e_H(C,D) \leq e_H(A,B)$, where the *excess* e_H is defined by $e_H(A,B) := \sup\{d(a,B) : a \in A\}$, with the usual convention when one of these sets is empty ($e_H(\varnothing, B) = 0$ for all $B \subset X$, $e_H(A, \varnothing) = +\infty$ when A is nonempty).

Conclude that the *Pompeiu–Hausdorff metric* d_H given by

$$d_H(A,B) := \max(e_H(A,B), e_H(B,A))$$

satisfies $d_H(C,D) \leq d_H(A,B)$.

18. The *recession cone* of a nonempty subset C of a topological linear space X is the set 0^+C of $v \in X$ such that $C + \mathbb{R}_+ v \subset C$.

(a) Check that if C is the nonempty intersection of a family $(C_i)_{i \in I}$, then 0^+C is the intersection of the family $(0^+C_i)_{i \in I}$. Is a similar property valid for the union?
(b) For $C := A \times B$ with $A \subset Y$, $B \subset Z$, check that $0^+C = 0^+A \times 0^+B$.
(c) Prove that if C is closed convex then 0^+C is a closed convex cone and that one has $v \in 0^+C$ iff $C + v \subset C$. Show that this equivalence may fail if C is not closed.
(d) Prove that if C is closed convex then 0^+C coincides with the *asymptotic cone* C_∞ or $T_\infty(C)$ of C defined as the set $\limsup_{t \to +\infty} (1/t)C$.
(e) Prove that if X is finite-dimensional then C is bounded if and only if $C_\infty = \{0\}$.

19. Let $A : X \to Y$ be a linear map between two normed spaces and let $C \subset X$ be a closed convex set.

(a) Show that $A(0^+C) \subset 0^+(A(C))$, 0^+C denoting the recession cone of C defined in Exercise 18.
(b) Prove that when X is finite-dimensional and $\ker A \cap 0^+C$ is a linear subspace, then $A(C)$ is closed.
(c) Show that if D is another nonempty closed convex subset of X, if X is finite-dimensional, and if $0^+C \cap 0^+D$ is a linear subspace, then $C - D$ is closed. [Hint: Reduce the question to the preceding one.]

20. Study the passage from the convergence of a family $(C_t)_{t \in T}$ of closed convex subsets of a normed space X to the convergence of $(0^+C_t)_{t \in T}$ (see [681, 883]).

21. Introduce a notion of asymptotically compact set in a normed space and use it to consider generalizations of the properties displayed in the preceding two exercises (see [787, 830]).

22. (Carathéodory's theorem) Let $A := \{a_i : i \in I\}$ be a finite subset of $\mathbb{R}^d$. For $J \subset I$ let $Q_J := \{\sum_{j \in J} r_j a_j : r_j \in \mathbb{R}_+\}$.

(a) Show that Q_J is closed if the family $(a_j)_{j \in J}$ is linearly independent.
(b) Check that the cone $Q := Q_I$ generated by A is the union of the family of cones Q_J for $J \subset I$ such that $(a_j)_{j \in J}$ is linearly independent. Conclude that Q is closed.
(c) Let $C := \mathrm{co}(A)$ be the convex hull of A. Given $x \in C$, prove that there exists a subset $A_J := \{a_j : j \in J\}$ of A of at most $d + 1$ elements such that $x \in \mathrm{co}(A_J)$. [Hint: Apply part (a) to the family of vectors $b_i := (a_i, 1)$ in $\mathbb{R}^{d+1}$.]

23. Deduce from Exercise 22 that the convex hull of a compact subset of $\mathbb{R}^d$ is compact.

1.4.2 Separation and Extension Theorems

Again in this subsection we will experience that like Janus, convex analysis has two faces. Looking at both the analytical face and the geometrical face is fruitful. In fact, the following extension and separation theorems are closely intertwined. We start with a finite-dimensional version of the separation theorem.

Theorem 1.64 (Finite-dimensional geometric form of the Hahn–Banach theorem). *Let C be a nonempty convex subset of a finite-dimensional linear space X and let $a \in X \setminus C$. Then there exists some $f \in X^* \setminus \{0\}$ such that $f(a) \geq \sup_C f$.*
If, moreover, C is closed, one can require that $f(a) > \sup_C f$.

Proof. (a) Let us first consider the case that C is closed. Since X is finite-dimensional, we may endow X with the norm associated with a scalar product $(\cdot \mid \cdot)$. Then the point a has a best approximation p in C characterized by

$$\forall z \in C, \qquad (z - p \mid a - p) \leq 0.$$

Let $f \in X^*$ be defined by $f(x) := (x \mid a - p)$. For all $z \in C$ we have $f(p) \geq f(z)$, and the second conclusion is established, because $f(a) - f(p) = \|a - p\|^2 > 0$, since $a \notin C$.

(b) Now let us consider the general case in which C is not assumed to be closed. Let S_{X^*} be the unit sphere of X^* and for $x \in C$ let

$$S_x := \{u^* \in S_{X^*} : \langle u^*, x \rangle \leq \langle u^*, a \rangle\},$$

so that $f \in S_{X^*}$ is such that $f(a) \geq \sup_C f$ if and only if $f \in \cap_{x \in C} S_x$. Since X is finite-dimensional, S_{X^*} is compact; hence this intersection is nonempty, provided the family of closed subsets $(S_x)_{x \in C}$ has the finite intersection property. Thus, we have to show that for every finite subset $F := \{x_1, \ldots, x_n\}$ of C one has $S_{x_1} \cap \cdots \cap S_{x_n} \neq \emptyset$. Let $E := \text{co}(F)$. Since F is finite, E is the image of the *canonical simplex*

$$\Delta_n := \{(t_1, \ldots, t_n) \in \mathbb{R}_+^n : t_1 + \cdots + t_n = 1\}$$

by the map $(t_1, \ldots, t_n) \mapsto t_1 x_1 + \cdots + t_n x_n$, so that E is compact, hence closed, and contained in C. Part (a) of the proof yields some f in $X^* \setminus \{0\}$ satisfying $f(a) > f(z)$ for all $z \in E$. Without loss of generality we may suppose $\|f\| = 1$. Thus $f \in S_x$ for all $x \in F$. $\qquad\square$

Corollary 1.65. *Let A and B be two disjoint nonempty convex subsets of a finite-dimensional space X. Then there exists some $f \in X^* \setminus \{0\}$ such that $f(a) \geq f(b)$ for all $a \in A$ and all $b \in B$.*

Proof. Since $C := A - B$ is convex and since A and B are disjoint, one has $0 \notin C$ and it suffices to take the linear form f provided by the preceding statement. $\qquad\square$

We now deal with the possibly infinite-dimensional case, for which one has to use the axiom of choice under the form of Zorn's lemma. The analytical versions are intimately linked with the geometrical versions. In the latter cases, one is led to detect the special place of half-spaces among convex subsets; in the analytical versions, it is the special place of linear forms among sublinear forms we put under full light.

Proposition 1.66. *The space $S(X)$ of finitely valued sublinear functions on a linear space X, ordered by the pointwise order, is lower inductive, hence has minimal elements. Each such element is a linear form.*

Proof. We have to show that every totally ordered subset C of $S(X)$ has a lower bound. Let s_0 be a fixed element of C. For every $s \in C, x \in X$, we have

$$s(x) \geq \inf(s_0(x), -s_0(-x)),$$

since we have either $s \geq s_0$, hence $s(x) \geq s_0(x)$, or $s \leq s_0$, hence $s(x) \geq -s(-x) \geq -s_0(-x)$. It follows that p given by $p(x) := \inf\{s(x) : s \in C\}$ is finitely valued and, as easily checked, sublinear. Thus, by Zorn's lemma, $S(X)$ has minimal elements.

The second assertion is a consequence of the next lemma and of the observation that if $s \in S(X)$ is odd, then it is linear: for every $x, y \in X, r \in \mathbb{R}, r < 0$, one has

$$s(x+y) \leq s(x) + s(y) = -s(-x) - s(-y) \leq -s(-x-y) = s(x+y),$$
$$s(rx) = s(-|r|x) = -|r|s(x) = rs(x).$$

Lemma 1.67. *Let $s \in S(X)$ and let $u \in X$. Then the function s_u given by*

$$s_u(x) := \inf\{s(x - tu) - s(-tu) : t \in \mathbb{R}_+\} \tag{1.16}$$

is sublinear and such that $s_u \leq s$, $s_u(u) = -s_u(-u)$.

Thus, when s is minimal in $S(X)$, one has $s_u = s$ and $s(u) = -s(-u)$ for all $u \in X$, so that the proof of the proposition is complete.

Proof. For all $x \in X$, the inequality $s_u(x) \leq s(x)$ stems from the choice $t = 0$ in relation (1.16). Now, since for $t \in \mathbb{R}_+$ we have $s(-tu) \leq s(x - tu) + s(-x)$, we get

$$\forall t \in \mathbb{R}_+, \qquad -s(-x) \leq s(x - tu) - s(-tu),$$

so that the infimum in relation (1.16) is finite. It is easy to check that s_u is sublinear. Taking $t = 1, x = u$ in (1.16), we get $s_u(u) \leq -s(-u)$. But since $0 \leq s_u(u) + s_u(-u)$ and $s_u \leq s$, we obtain

$$-s(-u) \leq -s_u(-u) \leq s_u(u) \leq -s(-u),$$

hence $-s_u(-u) = s_u(u)$.   □

Corollary 1.68. *For every $s \in S(X)$ there exists some linear form ℓ on X such that $\ell \leq s$.*

Proof. Let $S_s(X) := \{s' \in S(X) : s' \leq s\}$. The induced order on $S_s(X)$ by $S(X)$ is inductive, any chain C in $S_s(X)$ being a chain in $S(X)$ and any lower bound of C in $S(X)$ being a lower bound of C in $S_s(X)$. Thus $S_s(X)$ has a minimal element, which is clearly minimal in $S(X)$, hence is linear. □

Corollary 1.69. *Let X be a topological linear space and let h be a continuous sublinear function. Then there exists a continuous linear form ℓ on X such that $\ell \leq h$.*

Proof. It suffices to prove that if ℓ is a linear form bounded above by h, then ℓ is continuous. Given $\varepsilon > 0$ we can find a symmetric neighborhood V of 0 such that $h(v) \leq \varepsilon$ for every $v \in V$. Then for $v \in V$, we have $\ell(v) \leq h(v) \leq \varepsilon$ and $\ell(-v) \leq h(-v) \leq \varepsilon$, so that $|\ell(v)| \leq \varepsilon$. Thus ℓ is continuous. □

The following lemma is a prototype of so-called sandwich theorems.

Lemma 1.70. *Given a linear space X, $h : X \to \mathbb{R}$, $k : X \to \mathbb{R}_\infty$ both sublinear and such that $-k \leq h$, there exists some linear form ℓ on X such that $-k \leq \ell \leq h$.*

Proof. Let us introduce $s : X \to \overline{\mathbb{R}}$ by

$$s(x) := \inf\{h(x+y) + k(y) : y \in X\}.$$

Since $h(y) \leq h(x+y) + h(-x)$ and since $h(y) \geq -k(y)$ for all $y \in X$, we have

$$h(x+y) + k(y) \geq h(y) - h(-x) + k(y) \geq -h(-x),$$

so that $s(x) > -\infty$ for all $x \in X$, and of course, $s(x) \leq h(x) < +\infty$. We easily check that s is sublinear (in fact, s is the infimal convolution of h and $g : X \to \mathbb{R}_\infty$ given by $g(x) = k(-x)$ for $x \in X$) and that $s \leq h$, $s \leq g$. Thus, taking a linear form $\ell \leq s$, as given by Corollary 1.68, we have $\ell \leq h$, $\ell \leq g$, whence for $x \in X$, $\ell(x) = -\ell(-x) \geq -g(-x) = -k(x)$. □

Theorem 1.71 (Hahn–Banach). *Let X_0 be a linear subspace of a real linear space X, let ℓ_0 be a linear form on X_0, and let $h : X \to \mathbb{R}$ be a sublinear functional such that $\ell_0(x) \leq h(x)$ for all $x \in X_0$. Then there exists a linear form ℓ on X extending ℓ_0 such that $\ell \leq h$.*

Proof. Defining k by $k(x) := -\ell_0(x)$ for $x \in X_0$, $k(x) := +\infty$ otherwise, so that $-k \leq h$, Lemma 1.70 yields a linear form ℓ satisfying $-k \leq \ell \leq h$. Then we get $\ell|X_0 - \ell_0 = 0$, the unique linear form on X_0 that is nonnegative being the null functional. □

Corollary 1.72. (a) *Let X be a normed space and let $\overline{x} \in X$. Then there exists a continuous linear form ℓ on X such that $\|\ell\| := \sup \ell(B_X) \leq 1$ and $\ell(\overline{x}) = \|\overline{x}\|$.*
(b) *The map $j : X \to X^{**}$ defined by $j(x)(x^*) := x^*(x)$ is a monometry, i.e., $\|j(x)\| = \|x\|$ for all $x \in X$.*

Proof. (a) Let $X_0 := \mathbb{R}\bar{x}$ and let ℓ_0 be the linear form on X_0 given by $\ell_0(r\bar{x}) = r\|\bar{x}\|$ for $r \in \mathbb{R}$. Thus, for every $x \in X_0$, one has $\ell_0(x) \leq h(x) := \|x\|$. Theorem 1.71 yields some linear form ℓ on X extending ℓ_0 such that $\ell \leq h$. Then one has $\|\ell\| \leq 1$ and $\ell(\bar{x}) = \|\bar{x}\|$.

(b) For all $x \in X$, the inequality $\|j(x)\| \leq \|x\|$ follows from the definition of a dual norm. The reverse inequality stems from (a). □

The preceding corollary is a special case of the next one.

Corollary 1.73. *Let X be a normed space and let Y be a linear subspace of X. Then every continuous linear form y^* on Y has a linear continuous extension x^* to X such that $\|x^*\| = \|y^*\|$.*

Proof. Let $c := \|y^*\|$. Theorem 1.71 provides some linear form ℓ on X that extends y^* and is bounded above by $c\|\cdot\|$, hence is continuous. Clearly, $\|\ell\| = \|y^*\|$. □

Corollary 1.74. *Let Y be a closed linear subspace of a normed space X. If $Y \neq X$, there exists a nonnull continuous linear form f on X that is null on Y.*

Proof. Let $p : X \to X/Y$ be the quotient map. Since $Y \neq X$, one can find some nonnull $z \in X/Y$. Then Corollary 1.72 yields some ℓ in the dual of X/Y such that $\ell(z) \neq 0$. Then $f = \ell \circ p$ is nonnull and null on Y. □

Corollary 1.75. *Let Y be a closed linear subspace of a Banach space X. Then Y^* is isometric to $X^*/Y^\perp$, where $Y^\perp := \{x^* \in X^* : x^*(y) = 0 \;\forall y \in Y\}$.*

Proof. Let $r : X^* \to Y^*$ be the restriction map given by $r(x^*) := x^*\,|_Y$. Corollary 1.73 ensures that r is onto. The kernel of r being precisely $Y^\perp$, one can factorize r as $r = q \circ p$, where $p : X^* \to X^*/Y^\perp$ is the canonical projection and $q : X^*/Y^\perp \to Y^*$ is an isomorphism. If we give to $X^*/Y^\perp$ the quotient norm defined by $\|z\| := \inf\{\|x^*\| : x^* \in p^{-1}(z)\}$, Corollary 1.73 can serve to prove that q is isometric. □

The following statement is suggestive. Its proof is left as an exercise using later results (the sum rule in convex analysis).

Corollary 1.76 (Sandwich theorem). *Let X be a normed space, let $f : X \to \mathbb{R}_\infty$ be a convex function, and let g be a concave function such that $f \geq g$. If g is continuous and finite at some point of the domain of f, then there exists a continuous affine function h on X such that*

$$f \geq h \geq g.$$

Now let us turn to geometric forms of the Hahn–Banach theorem. In the sequel we say that a subset H of a linear space X is a *hyperplane* if there exist $c \in \mathbb{R}$ and a linear form $h \neq 0$ on X such that $H = h^{-1}(c)$. We first consider an algebraic version.

Proposition 1.77. *Let C be an absorbing convex subset of a linear space X and let $e \in X \setminus \text{core}\, C$. Then there exists a hyperplane H of X such that $e \in H$, $H \cap \text{core}\, C = \varnothing$. Moreover, C is contained in one of the open half-spaces determined by H.*

Proof. Let $j := \mu_C$ be the Minkowski gauge of C:

$$j(x) := \inf\{t > 0 : x \in tC\}.$$

Since C is absorbing and convex, j is finite on X and sublinear. For all $x \in \mathrm{core}\,C$ one has $j(x) < 1$, since there exists some $r > 0$ such that $rx \in C - x$, hence $j(x) \le (1 + r)^{-1}$. Conversely, if $j(x) < 1$, then $x \in \mathrm{core}\,C$, since for all $u \in X$ and for $\varepsilon > 0$ such that $\varepsilon j(u) < 1 - j(x)$ one has, for all $r \in [0, \varepsilon]$, $j(x + ru) \le j(x) + j(ru) < 1$, hence $x + ru \in tC \subset C$ for some $t \in (0, 1)$. Since $e \in X \setminus \mathrm{core}\,C$, we have $j(e) \ge 1$. Let $X_0 := \mathbb{R}e$, and let $\ell_0 : X_0 \to \mathbb{R}$ be given by $\ell_0(re) := rj(e)$. Then, since $rj(e) \le 0 \le j(re)$ for $r \le 0$, we have $\ell_0 \le j \,|\, X_0$, so that there exists some linear form h on X extending ℓ_0 with $h \le j$. Let $H := \{x \in X : h(x) = j(e)\}$. Then $e \in H$ and for $x \in \mathrm{core}\,C$ we have $h(x) \le j(x) < 1 \le j(e)$, and hence $x \notin H$ and $\mathrm{core}\,C \subset h^{-1}((-\infty, j(e)))$. $\square$

Theorem 1.78 (Eidelheit). *Let A and B be two disjoint nonempty convex subsets of a topological linear space X. If A is open, then there exist some $f \in X^* \setminus \{0\}$, $r \in \mathbb{R}$ such that*

$$\forall a \in A, \ \forall b \in B, \qquad f(a) > r \ge f(b).$$

Proof. Let $D := A - B := \{a - b : a \in A, \ b \in B\}$. It is a convex subset of X that is open as the union over $b \in B$ of the translated sets $A - b$, and $0 \notin D$. Taking $e \in D$ and setting $C := e - D$, we see that $e \notin C$, $0 \in C$, and C is absorbing. Thus, there exist some $s > 0$ and some linear form f on X such that $f(e) = s$ and $f(x) < s$ for all $x \in C$. Since f is bounded above on the neighborhood C of 0, f is continuous. Moreover, for $a \in A, b \in B$, one has $f(e - a + b) < s = f(e)$, hence $f(a) \ge \sup f(B)$. In fact, since A is open and $f \ne 0$, one must have $f(a) > r := \sup f(B)$. $\square$

Theorem 1.79 (Hahn–Banach strong separation theorem). *Let A and B be two disjoint nonempty convex subsets of a normed space (or a locally convex topological linear space) X. If A is compact and B is closed, then there exist some $f \in X^* \setminus \{0\}$ and some $r \in \mathbb{R}$, $\delta > 0$ such that*

$$\forall a \in A, \ \forall b \in B, \qquad f(a) > r + \delta > r > f(b).$$

Proof. For every $a \in A$ there exists a symmetric open convex neighborhood V_a of 0 in X such that $(a + 2V_a) \cap B = \varnothing$. Let F be a finite subset of A such that the family $(a + V_a)_{a \in F}$ forms a finite covering of A. Then if V is the intersection of the family $(V_a)_{a \in F}$, V is an open neighborhood of 0 and $A' \cap B = \varnothing$ for $A' := A + V$. The Edelheit theorem yields $f \in X^* \setminus \{0\}$ and $s \in \mathbb{R}$ such that $f(a) > s \ge f(b)$ for all $a \in A', b \in B$. The compactness of A ensures that there exists $\delta > 0$ such that $f(a) > s + 2\delta$ for all $a \in A$. Setting $r := s + \delta$, we get the result. $\square$

Example. The compactness assumption on A cannot be omitted, as shown by the example of $X = \mathbb{R}^2$, $A := \{(r, s) \in \mathbb{R}_+^2 : rs \ge 1\}$, $B := \mathbb{R} \times (-\infty, 0]$.

Corollary 1.80 (Mazur). *Every closed convex subset C of a normed space X is weakly closed.*

Proof. It suffices to show that if C is a nonempty closed convex subset of X and if $C \neq X$, then C coincides with the intersection of the family $\mathscr{D}$ of closed half-spaces containing C. That amounts to showing that for $a \in X \setminus C$ there exists some $D \in \mathscr{D}$ such that $a \notin D$. Taking $A := \{a\}$ and $B := C$ in the preceding separation theorem, we get some $f \in X^* \setminus \{0\}$ and some $r \in \mathbb{R}$ such that $f(a) > r > f(b)$ for all $b \in C$. Thus $D := f^{-1}((-\infty, r])$ belongs to $\mathscr{D}$ but $a \notin D$. $\qquad\square$

A special case of the Fenchel transform we will study later on is the passage from closed convex subsets (or their indicator functions) to their support functions. The *support function* h_C or σ_C of a subset C of a normed space X has been defined in (1.8) as the function $h_C : X^* \to \overline{\mathbb{R}}$ given by

$$h_C(x^*) := \sigma_C(x^*) := \sup\{\langle x^*, x \rangle : x \in C\}.$$

Corollary 1.81 (Hörmander). *The map $h : C \mapsto h_C$ is an injective map from the set $\mathscr{C}(X)$ of nonempty closed convex subsets of the normed space X into the space $\mathscr{H}(X)$ of positively homogeneous functions on X null at 0. Moreover, $h_{\lambda C} = \lambda h_C$ for all $\lambda \in \mathbb{R}_+$, $C \in \mathscr{C}(X)$ and $h_{\mathrm{cl}(A+B)} = h_A + h_B$, $h_{\overline{\mathrm{co}}(A \cup B)} = \max(h_A, h_B)$ for all $A, B \in \mathscr{C}(X)$.*

Its restriction to the space $\mathscr{C}_b(X)$ of nonempty closed bounded convex subsets of X is an isometry onto the set $S_c(X)$ of continuous sublinear functions on X when $\mathscr{C}_b(X)$ is endowed with the Pompeiu–Hausdorff metric and $S_c(X)$ is provided with the norm given by $\|s\| := \sup\{|s(x)| : x \in B_X\}$.

Proof. We just prove the injectivity of h, leaving the other assertions as exercises (see [198, 591]). It suffices to prove that for $C, D \in \mathscr{C}(X)$ satisfying $C \setminus D \neq \varnothing$ one has $h_C \neq h_D$. Given $b \in C \setminus D$ we can find $x^* \in X$ such that $\langle x^*, b \rangle > \sup_{x \in D}\langle x^*, x \rangle$. Then we have $h_C(x^*) \geq \langle x^*, b \rangle > \sup_{x \in D}\langle x^*, x \rangle = h_D(x^*)$. $\qquad\square$

Let us give a short account of polarity, a passage from a subset of a normed space X to a subset of the dual X^* of X (or the reverse, or, more generally, from a subset of X to a subset of a space Y paired with X by a bilinear coupling function). This correspondence is a geometric analogue of a correspondence for functions, the Fenchel conjugacy, that we will study in Chap. 3. This correspondence assigns to a subset S of X its *polar* set defined by

$$S^0 := \{x^* \in X^* : \forall x \in S \ \langle x^*, x \rangle \leq 1\}. \tag{1.17}$$

It is obvious that S^0 is a weak* closed convex subset of X^* containing 0. If S is a cone, then S^0 is a convex cone and $S^0 := \{x^* \in X^* : \forall x \in S \ \langle x^*, x \rangle \leq 0\}$; if S is a linear subspace, then S^0 is the linear subspace $S^\perp := \{x^* \in X^* : \forall x \in S \ \langle x^*, x \rangle = 0\}$, also called the *orthogonal* of S. It is also easy to show that

$$(S \cup T)^0 = S^0 \cap T^0.$$

A base of neighborhoods of 0 for the weak* topology is formed by the polar sets of finite subsets. On the other hand, one has the following classical theorem.

Theorem 1.82 (Alaoglu–Bourbaki). *Let X be a normed space and let S be a neighborhood of 0 in X. Then S^0 is weak* compact.*

Proof. Since $S^0 \subset T^0$ when $T \subset S$ and since S^0 is weak* closed, it suffices to prove the result when S is a ball centered at 0. Since $(rS)^0 = r^{-1}S^0$ for $r > 0$, we may suppose $S = B_X$. Then $S^0 = B_{X^*}$ and the result has been shown in Theorem 1.5 in that case. $\qquad\square$

The polar T^0 of a subset T of X^* is defined similarly by $T^0 := \{x \in X : \forall x^* \in T \ \langle x^*, x \rangle \le 1\}$; if S is a subset of X, then its *bipolar* is the set $S^{00} := (S^0)^0$.

Corollary 1.83 (Bipolar theorem). *For every nonempty subset S of a normed space X, its bipolar is the closed convex hull of $S \cup \{0\}$: $S^{00} := \overline{co}(S \cup \{0\})$.*

Proof. Let $C := \overline{co}(S \cup \{0\})$. Since one has $S \subset S^{00}$, and since S^{00} is closed convex and contains 0, one has $C \subset S^{00}$. Given $a \in X \setminus C$, Theorem 1.79 yields $x^* \in X^*$ and $r \in \mathbb{R}$ such that $\langle x^*, a \rangle > r > \langle x^*, c \rangle$ for all $c \in C$. Since $0 \in C$, one has $r > 0$ and $r^{-1}x^* \in C^0 \subset S^0$, hence $a \notin S^{00}$. Therefore $S^{00} = C$. $\qquad\square$

The next result answers a natural question about taking limits of polar sets. The notion of outward limit introduced in Exercise 4 of Sect. 1.3 is recalled in the beginning of the proof.

Theorem 1.84. *Let $(F(t))_{t \in T}$ be a parameterized family of subsets of a normed space X. Then, X^* being endowed with the weak* topology, the following inclusions hold. When for all $t \in T$ the set $F(t)$ is a closed convex cone, the second one is an equality. If the unit ball of X^* is sequentially compact for the weak* topology, one can replace the outward limit with the sequential $\limsup$:*

$$\text{outlim}_{t(\in T)\to 0}F(t)^0 \subset \left(\liminf_{t(\in T)\to 0} F(t)\right)^0, \quad \liminf_{t(\in T)\to 0} F(t) \subset \left(\text{outlim}_{t(\in T)\to 0}F(t)^0\right)^0.$$

$$(1.18)$$

Proof. We encourage the reader to devise a proof in the sequential case by simplifying the proof for the general case that follows. Let $\bar{x}^* \in \text{outlim}_{t(\in T)\to 0}F(t)^0$: there exist nets $(t_i)_{i \in I} \to 0$, $(x_i^*)_{i \in I} \to \bar{x}^*$ for the weak* topology and a compact subset K of X^* such that $x_i^* \in F(t_i)^0 \cap K$ for all $i \in I$. Let $m > 0$ be such that $K \subset mB_{X^*}$. For all $x \in \liminf_{t(\in T)\to 0} F(t)$ we can find a cofinal subset J of I and a net $(x_j)_{j \in J} \to x$ such that $x_j \in F(t_j)$ for all $j \in J$ (we can take $J = I$ if $F(t)$ is nonempty for all $t \in T$, since $d(x, F(t)) \to 0$ as $t(\in T) \to 0$). Since for all $j \in J$,

$$\left|\langle x_j^*, x_j \rangle - \langle \bar{x}^*, x \rangle\right| \le \left|\langle x_j^*, x_j - x \rangle\right| + \left|\langle x_j^* - \bar{x}^*, x \rangle\right| \le m\|x_j - x\| + \left|\langle x_j^* - \bar{x}^*, x \rangle\right|,$$

we get that $\langle \bar{x}^*, x \rangle = \lim_j \langle x_j^*, x_j \rangle \le 1$. Since x is arbitrary in $\liminf_{t(\in T)\to 0} F(t)$, the two equivalent inclusions of (1.18) are established.

Suppose now that for all $t \in T$, the set $F(t)$ is a closed convex cone and let $\bar{x} \in X \setminus \liminf_{t(\in T) \to 0} F(t)$: there exist $r > 0$ and a subset S of T such that $0 \in \mathrm{cl}(S)$ and $d(\bar{x}, F(t)) \geq r$ for all $t \in S$. For all $t \in S$, the Eidelheit separation theorem yields some $x_t^* \in S_{X^*}$ and $c_t \in \mathbb{R}$ such that

$$\sup\{\langle x_t^*, x \rangle : x \in F(t)\} \leq c_t \leq \inf\{\langle x_t^*, y \rangle : y \in B(\bar{x}, r)\} = \langle x_t^*, \bar{x} \rangle - r.$$

Since $0 \in F(t)$, we have $c_t \geq 0$ for all $t \in S$. Since $F(t)$ is a cone, we must have $\langle x_t^*, x \rangle \leq 0$ for all $t \in S$, $x \in F(t)$, i.e., $x_t^* \in F(t)^0 \cap S_{X^*}$. Since B_{X^*} is weak* compact, (x_t^*) has a weak* cluster point $\bar{x}^* \in \mathrm{outlim}_{t(\in T) \to 0} F(t)^0$. Passing to the limit in the preceding inequality, we get $\langle \bar{x}^*, \bar{x} \rangle \geq r > 0$, so that $\bar{x} \notin (\mathrm{outlim}_{t(\in T) \to 0} F(t)^0)^0$.   □

The following lemma will be used to establish a minimax theorem. In it, we denote by $\vee$ and $\wedge$ the operations on $\mathbb{R}$ given by $r_1 \vee \cdots \vee r_k = \max(r_1, \ldots, r_k)$ and $r_1 \wedge \cdots \wedge r_k = \min(r_1, \ldots, r_k)$ for $r_i \in \mathbb{R}$, $i \in \mathbb{N}_k$, and as above, Δ_k stands for the canonical simplex of $\mathbb{R}^k$: $\Delta_k := \{(s_1, \ldots, s_k) \in \mathbb{R}_+^k : s_1 + \cdots + s_k = 1\}$. As usual, max (resp. min) means that one has attainment of the supremum (resp. infimum) when it is finite.

Lemma 1.85. *Let $f_1, \ldots, f_k$ be convex functions on a convex subset C of a linear space X. Then*

$$\inf_C (f_1 \vee \cdots \vee f_k) = \max\{\inf_C (s_1 f_1 + \cdots + s_k f_k) : s := (s_1, \ldots, s_k) \in \Delta_k\}.$$

If $g_1, \ldots, g_k$ are concave functions on C then

$$\sup_C (g_1 \wedge \cdots \wedge g_k) = \min\{\sup_C (s_1 g_1 + \cdots + s_k g_k) : s := (s_1, \ldots, s_k) \in \Delta_k\}.$$

Proof. For each $s := (s_1, \ldots, s_k) \in \Delta_k$, we obviously have $h := f_1 \vee \cdots \vee f_k \geq h_s := s_1 f_1 + \cdots + s_k f_k$, hence $\inf_C h \geq m_C(s) := \inf h_s(C) := \inf_C(s_1 f_1 + \cdots + s_k f_k)$ and $\inf_C h \geq \sup\{m_C(s) : s := (s_1, \ldots, s_k) \in \Delta_k\}$, with equality if $\inf_C h = -\infty$. Now let $t \in \mathbb{R}$ be such that $t \leq \inf_C h$ and let

$$A := \{r = (r_1, \ldots, r_k) \in \mathbb{R}^k : \exists x \in C, \ r_i > f_i(x) \ i = 1, \ldots, k\},$$

a convex subset of $\mathbb{R}^k$. The choice of t shows that $b := (t, \ldots, t) \notin A$. The finite-dimensional separation theorem yields some $\bar{s} = (\bar{s}_1, \ldots, \bar{s}_k) \in \mathbb{R}^k \setminus \{0\}$ such that

$$\bar{s}_1 r_1 + \cdots + \bar{s}_k r_k \geq \bar{s}_1 t + \cdots + \bar{s}_k t \qquad \forall r = (r_1, \ldots, r_k) \in A.$$

We have $\bar{s}_i \geq 0$ for $i = 1, \ldots, k$, since r_i can be arbitrarily large. Since $\bar{s} \neq 0$, by homogeneity, we may suppose $\bar{s}_1 + \cdots + \bar{s}_k = 1$, i.e., $\bar{s} \in \Delta_k$. Then for each $x \in C$, since r_i can be arbitrarily close to $f_i(x)$, we get

$$h_{\bar{s}}(x) := \bar{s}_1 f_1(x) + \cdots + \bar{s}_k f_k(x) \geq t.$$

Therefore $m_C(\bar{s}) \geq t$, and we get $\sup_{s \in \Delta_k} m_C(s) \geq \inf_C h$, so that equality holds. When $\inf_C h$ is finite we can take $t = \inf_C h$, and the inequality $\inf_C(\bar{s}_1 f_1 + \cdots + \bar{s}_k f_k) \geq t$ shows that we have attainment for this $\bar{s} \in \Delta_k$. The second assertion is obtained by setting $f_i := -g_i$. □

Theorem 1.86 (Infmax theorem). *Let A and B be nonempty convex subsets of linear spaces X and Y respectively, and let $\ell : A \times B \to \mathbb{R}$ be a function that is convex in its first variable and concave in its second variable. Then if B is compact for some topology and if ℓ is outward continuous in its second variable, one has*

$$\inf_{x \in A} \max_{y \in B} \ell(x,y) = \max_{y \in B} \inf_{x \in A} \ell(x,y).$$

Proof. The inequality $\alpha := \inf_{x \in A} \sup_{y \in B} \ell(x,y) \geq \beta := \sup_{y \in B} \inf_{x \in A} \ell(x,y)$ is valid without any assumption. Here we can write max instead of sup, since B is compact and $\ell(x, \cdot)$ is outward continuous for all $x \in A$, as is $\inf_{x \in A} \ell(x, \cdot)$. Given $k \in \mathbb{N} \setminus \{0\}$ and $a_1, \ldots, a_k \in A$, applying the preceding lemma with $C = B$, $g_i = \ell(a_i, \cdot)$, we can find $s \in \Delta_k$ such that

$$\sup_{b \in B}(\ell(a_1, b) \wedge \cdots \wedge \ell(a_k, b)) = \sup_{b \in B}(s_1 \ell(a_1, b) + \cdots + s_k \ell(a_k, b)).$$

Since $\ell(\cdot, b)$ is convex for all $b \in B$, we get

$$\sup_{b \in B}(\ell(a_1, b) \wedge \cdots \wedge \ell(a_k, b)) \geq \sup_{b \in B}(\ell(s_1 a_1 + \cdots + s_k a_k, b)) \geq \alpha.$$

Introducing for $a \in A$ the closed subset $B_a := \{b \in B : \ell(a, b) \geq \alpha\}$, which is nonempty by the Weierstrass theorem, we deduce from these inequalities that $B_{a_1} \cap \cdots \cap B_{a_k}$ is nonempty. The finite intersection property of the compact space B ensures that $\bigcap_{a \in A} B_a$ is nonempty. That means that there exists some $\bar{b} \in B$ such that $\inf_{a \in A} \ell(a, \bar{b}) \geq \alpha$. Thus $\beta \geq \alpha$ and equality holds. □

Exercises

1. Prove the **Mazur–Orlicz theorem**: Let $h : X \to \mathbb{R}$ be a sublinear functional on some linear space X and let C be a nonempty convex subset of X. Then there exists a linear form ℓ on X such that $\ell \leq h$ and $\inf \ell(C) = \inf h(C)$. [See [893, p. 13].]

2. Prove the **Mazur–Bourgin theorem**: Let C be a convex subset with nonempty interior in a topological linear space X and let A be an affine subspace of X such that $A \cap \operatorname{int} C = \varnothing$. Prove that there exists a closed hyperplane H of X containing A that does not meet $\operatorname{int} C$. [See [506, p. 5].]

3. Prove **Mazur's theorem**: Let (x_n) be a sequence of a normed space X that weakly converges to some $x \in X$. Then there exists a sequence (y_n) strongly converging to x such that for all $k \in \mathbb{N}$, y_k is a convex combination of the x_n's. [Hint: Consider the closed convex hull of $\{x_n : n \in \mathbb{N}\}$.]

4. Prove the Sandwich theorem using the Eidelheit theorem.

5. Prove **Stone's theorem**: Let A and B be disjoint convex subsets of a normed space X. Show that there exists a pair (C, D) of disjoint convex subsets satisfying $A \subset C$, $B \subset D$ that is maximal for the order induced by inclusion. Show that when A is open one can take for C and D opposite half-spaces, C being open.

6. Let $j : Y \to X$ be the canonical injection of a linear subspace Y of a normed space X into X and let $j^{\mathsf{T}} : X^* \to Y^*$ be its transpose map given by $j^{\mathsf{T}}(x^*) := x^* \circ j$ for $x^* \in X^*$. Rephrase Corollary 1.73 thus: j^{T} is surjective. Show that the kernel of j^{T} is the polar set Y^0 of Y and that Y^* can be isometrically identified with X^*/Y^0.

7. Prove Theorem 1.86 under the assumption that f is convex–concave-like in the following sense: for all $t \in [0, 1]$ and all $x_1, x_2 \in A$, $y_1, y_2 \in B$ there exist some $x_3 \in A$, $y_3 \in B$ such that

$$\ell(x_3, y) \le (1 - t)\ell(x_1, y) + t\ell(x_2, y) \quad \forall y \in B,$$

$$\ell(x, y_3) \ge (1 - t)\ell(x, y_1) + t\ell(x, y_2) \quad \forall x \in A.$$

[Hint: Adapt the proof above or see [140].]

8. Let $A : W \to X$ be a continuous linear operator with transpose map $A^{\mathsf{T}} : X^* \to W^*$ given by $A^{\mathsf{T}}(x^*) = x^* \circ A$. Show that for $D := A(C)$ one has $D^0 = (A^{\mathsf{T}})^{-1}(C^0)$.

1.5 Variational Principles

It is well known that not all lower semicontinuous functions on a noncompact metric space that are bounded below attain their infima; as a classical example, one can take the exponential function on $\mathbb{R}$. However, one can show that a simple and small perturbation of the given function does attains its infimum. This is the content of the Ekeland variational principle. It has important and numerous consequences, in particular for existence results without compactness assumptions. Thus it is a fundamental tool of nonlinear analysis and of nonsmooth analysis.

In this section we present such minimization results (also called variational principles). The main one is the Ekeland variational principle. As a preview of this result, we distill a simple version of it under some restrictive assumptions. We will prove the general version just after. In a supplement, some detours of independent interest are made. Algorithmic and dynamical interpretations are proposed in the exercises. A number of supplementary readings are suggested. They deal with fixed-point results, metric convexity, the Banach open mapping theorem, and the Palais–Smale condition.

1.5.1   The Ekeland Variational Principle

We first state and prove the promised rudimentary version. It relies on a compactness property rather than on a completeness argument. For this reason, this version is limited to metric spaces in which closed balls are compact (sometimes called *Heine spaces*).

Claim. *Let* $f : X \to \mathbb{R}$ *be a lower semicontinuous function on some finite-dimensional normed space that is bounded below. Then for every* $\gamma > 0$ *one can find* $u \in X$ *such that* $f + \gamma d(u, \cdot)$ *attains its minimum at* u:

$$\forall x \in X, \qquad f(u) \le f(x) + \gamma \|x - u\|. \qquad (1.19)$$

Moreover, for every $\varepsilon > 0$, *one can require* $f(u) < \inf f(X) + \varepsilon$.

Proof. [487] Given $\gamma, \varepsilon > 0$ and $v \in X$ with $f(v) < \inf f(X) + \varepsilon$, the function g defined by $g(x) := f(x) + \gamma \|x - v\|$ is lower semicontinuous and coercive, since f is lower semicontinuous and bounded below. Therefore g attains its infimum at some point u:

$$\forall x \in X, \qquad f(u) + \gamma \|u - v\| \le f(x) + \gamma \|x - v\|, \qquad (1.20)$$

whence by the triangle inequality,

$$\forall x \in X, \qquad f(u) + \gamma \|u - v\| \le f(x) + \gamma \|x - u\| + \gamma \|u - v\|.$$

Subtracting $\gamma \|u - v\|$ from both sides of this relation, we get (1.19). Taking $x = v$ in (1.20), we get $f(u) \le f(v) \le \inf f(X) + \varepsilon$. $\square$

The preceding proof shows that the result holds in any dual space, in particular in any reflexive space, whenever f is lower semicontinuous for the weak* topology. In fact, we will show that it holds in any Banach space, and even in any complete metric space. Some more information will also be provided in the full version that follows. It asserts that every bounded-below lower semicontinuous function on a complete metric space can be approximated by a function attaining its infimum. Moreover, the approximation can be made uniform (by changing the metric to a uniformly equivalent one such as $\max(d, 1)$ or $d/(d+1)$) and is of a simple nature. Furthermore, some localization of a minimizer can be obtained.

Theorem 1.87 (Ekeland variational principle). *Let* (X, d) *be a complete metric space and let* $f : X \to \mathbb{R}_\infty := \mathbb{R} \cup \{+\infty\}$ *be a bounded-below lower semicontinuous function with nonempty domain. Then for every* $\gamma > 0$ *one can find* $u \in X$ *such that* $f + \gamma d(u, \cdot)$ *attains a strict minimum at* u:

$$f(u) < f(x) + \gamma d(u, x) \quad \text{for all } x \in X \setminus \{u\}. \qquad (1.21)$$

Moreover, given $\bar{x} \in X$, *one can require that* $f(u) + \gamma d(u, \bar{x}) \le f(\bar{x})$.

One can give a slightly more precise statement. Hereinafter, we say that $\bar{x} \in X$ is an ε-*minimizer* of f if $f(\bar{x}) \leq \inf f(X) + \varepsilon$.

Theorem 1.88 (Ekeland variational principle and approximate minimization).
Let (X,d) be a complete metric space and let $f : X \to \mathbb{R}_\infty$ be a bounded-below lower semicontinuous proper function. Given an ε-minimizer $\bar{x}$ of f and given $\gamma, \rho > 0$ satisfying $\gamma\rho \geq \varepsilon$, one can find $u \in X$ such that the following inequalities hold:

(a) $d(u,\bar{x}) \leq \rho$,
(b) $f(u) + \gamma d(u,\bar{x}) \leq f(\bar{x})$,
(c) $f(u) < f(x) + \gamma d(u,x)$ *for all* $x \in X \setminus \{u\}$.

Proof. Changing the metric d to γd would reduce the proof to the case $\gamma = 1$, but it would not be really simpler. It consists in associating to f an order on X by

$$A(x) := \{y \in X : f(y) + \gamma d(x,y) \leq f(x)\}, \qquad\qquad x \in X.$$

We have $x \in A(x)$ for all $x \in X$, and the relations $y \in A(x)$, $x \in A(y)$ imply $x = y$. Let us check that the relation A satisfies the transitivity property $A(y) \subset A(x)$ for all $x \in X$, $y \in A(x)$. We may assume $x \in \mathrm{dom} f$, so that $f(y) < +\infty$. Then for all $z \in A(y)$ we also have $f(z) < +\infty$ and $\gamma d(y,z) \leq f(y) - f(z)$. Since $y \in A(x)$, we also have $\gamma d(x,y) \leq f(x) - f(y)$. Adding the respective sides of these two inequalities and using the triangle inequality, we get $\gamma d(x,z) \leq f(x) - f(z)$, or $z \in A(x)$. Thus A defines an order; we shall construct a minimal element.

Given $\bar{x} \in \mathrm{dom} f$, we can define inductively a sequence (x_n) starting from $x_0 := \bar{x}$ by picking $x_{n+1} \in A(x_n)$ satisfying

$$f(x_{n+1}) \leq \frac{1}{2}f(x_n) + \frac{1}{2}\inf f(A(x_n)). \tag{1.22}$$

Such a choice is possible: it suffices to use the definition of an infimum when $\inf f(A(x_n)) < f(x_n)$ and to take $x_{n+1} = x_n$ when $\inf f(A(x_n)) = f(x_n)$. Since $x_n \in A(x_n)$, (1.22) ensures that the sequence $(f(x_n))$ is nonincreasing, hence is convergent, since f is bounded below. Let $\lambda := \lim_n f(x_n)$.

Since $x_{n+1} \in A(x_n)$, we have $\gamma d(x_n, x_{n+1}) \leq f(x_n) - f(x_{n+1})$, and by induction,

$$\gamma d(x_n, x_{n+p}) \leq f(x_n) - f(x_{n+p}) \tag{1.23}$$

for all $n, p \geq 0$. Thus (x_n) is a Cauchy sequence, hence has a limit, that we denote by u.

Because f is lower semicontinuous, $A(x_n)$ is closed for each n. Since relation (1.23) says that $x_{n+p} \in A(x_n)$ for all $p \geq 0$, we get $u \in A(x_n)$. In particular, taking $n = 0$ and remembering that $x_0 = \bar{x}$, we get

$$f(u) + \gamma d(\bar{x}, u) \leq f(\bar{x}).$$

Moreover, by the transitivity property of relation A, for all $v \in A(u)$ and all $n \in \mathbb{N}$, we have $v \in A(x_n)$. Thus $\inf f(A(x_n)) + \gamma d(x_n, v) \leq f(v) + \gamma d(x_n, v) \leq f(x_n)$ and relation (1.22) yields

$$\gamma d(x_n, v) \leq f(x_n) - \inf f(A(x_n)) \leq 2(f(x_n) - f(x_{n+1})) \to 0,$$

hence $d(v, u) = 0$. It follows that $A(u) = \{u\}$. Hence item (c), and equivalently (1.21), is satisfied.

If $\bar{x}$ is such that $f(\bar{x}) \leq \inf f(X) + \varepsilon$, and $\varepsilon \leq \gamma \rho$, we have

$$\inf f(X) + \gamma d(u, \bar{x}) \leq f(u) + \gamma d(u, \bar{x}) \leq f(\bar{x}) \leq \inf f(X) + \gamma \rho,$$

so that $d(u, \bar{x}) \leq \rho$. $\qquad \square$

The assertions of the preceding statement can be interpreted in the following way. Given an approximate minimizer $\bar{x}$ of a lower semicontinuous function f on a complete metric space, one can find nearby $\bar{x}$ a genuine minimizer u of a modified function $f_{\gamma, u} := f + \gamma d(u, \cdot)$ that can be considered as being not too far from the original one. However, there is a tradeoff between the accuracies of the two approximating elements u, $f_{\gamma, u}$: one cannot expect to get arbitrarily good approximations of f and of $\bar{x}$ at the same time.

As mentioned above, replacing the metric d by an equivalent bounded metric and γ by the general term of a sequence $(\gamma_n) \to 0_+$, we obtain an approximation result.

Corollary 1.89. *Let $f : X \to \mathbb{R}_\infty$ be a bounded-below lower semicontinuous proper function on a complete metric space (X, d). Then there exists a sequence (f_n) of lower semicontinuous functions attaining their infima that converges uniformly to f. More precisely, one may require that for every ε, one have $f \leq f_n \leq f + \varepsilon$ for n large enough.*

Another approximation result involving functions of two variables will be useful. In it, the perturbation bears on the first variable only.

Corollary 1.90 (Partial Ekeland theorem). *Let (W, d) be a complete metric space, let X be a topological space, and let $f : W \times X \to \mathbb{R}_\infty$ be a lower semicontinuous function with nonempty domain that is bounded below. Suppose that for every $w \in W$ there exist a neighborhood V of w and a compact subset K of X such that for all $v \in V$ one has $\inf\{f(v, x) : x \in K\} = \inf\{f(v, x) : x \in X\}$. Then given a sequence $(\varepsilon_n) \to 0_+$, there exists a sequence $((w_n, x_n))$ in $W \times X$ such that for all $n \in \mathbb{N}$ and all $(w, x) \in W \times X$,*

$$f(w_n, x_n) \leq f(w, x) + \varepsilon_n d(w, w_n).$$

The essence of this statement can be more easily grasped in the case that X is compact; then one can take $V = W$ and $K = X$.

Proof. Let $p : W \to \mathbb{R}_\infty$ be the function given by

$$p(w) := \inf_{x \in X} f(w,x).$$

It is bounded below, proper, and lower semicontinuous by Corollary 1.23. Let (v_n) be a sequence of W such that $p(v_n) < \inf\{p(w) : w \in W\} + \varepsilon_n$. Then for each $n \in \mathbb{N}$, Theorem 1.88 yields some $w_n \in B[v_n, 1]$ such that $p(w_n) \leq p(w) + \varepsilon_n d(w, w_n)$. Taking a minimizer x_n of $f(w_n, \cdot)$, we get the result. $\qquad\square$

Theorem 1.88 can be rephrased in such a way that it appears as an existence result.

Corollary 1.91. [906] *Let (X,d) be a complete metric space and let $f : X \to \mathbb{R}_\infty$ be a bounded-below lower semicontinuous function. Suppose that for some $\gamma > 0$ and for all $w \in X$ such that $f(w) > \inf f(X)$ there exists some $x \in X \setminus \{w\}$ such that $f(x) \leq f(w) - \gamma d(w,x)$. Then there exists $u \in X$ such that $f(u) = \inf f(X)$.*

Proof. We may assume $\inf f(X) < +\infty$. Let $u \in X$ be given by Theorem 1.87. Assuming $f(u) > \inf f(X)$ and taking $w = u$ in our assumption, we get a contradiction to relation (1.21). Thus $f(u) = \inf f(X)$. $\qquad\square$

Exercises

1. Show that a metric space (X,d) satisfying the Ekeland variational principle in the form of Theorem 1.87 is complete. [Hint: Given a Cauchy sequence (x_n) of X, let $f : X \to \mathbb{R}_+$ be given by $f(x) = \lim_n d(x, x_n)$. Show that f is well defined, is Lipschitzian, and that if $\gamma \in (0,1)$ and if $u \in X$ satisfies $f(u) \leq f(x) + \gamma d(x,u)$ for every $x \in X$, then $f(u) = 0$ and $(x_n) \to u$; see [961].]

2. Give to the proof of the Ekeland principle the following interpretations.

(a) Consider $A(x_n)$ as a set of possible choices for passing from an iteration n to the following one in some algorithm. The fact that the next iterate x_{n+1} is not uniquely determined is not unusual in theoretical studies of algorithms. Of course, for a practical use, a more specific rule must be given to process each iteration. Note that the class of algorithms described by the proof of Theorem 1.88 belongs to the family of descent algorithms.

(b) The iterative process given by $x_{n+1} \in A(x_n)$ can be seen as a discrete dynamical system. The sequence (x_n) then appears as an orbit or a trajectory of the system. Its limit point u is a *rest point* or *stationarity point* of the system in the sense that $A(u) = \{u\}$. See [37, 892].

(c) More precisely, prove the following. Every multimap $A : X \rightrightarrows X$ from a complete metric space into itself with closed nonempty values has a rest point whenever the following conditions are satisfied: (i) for all $x \in X$, $y \in A(x)$ one has $A(y) \subset A(x)$; (ii) if (x_n) is a trajectory of A, then $\sum_n d(x_n, x_{n+1}) < +\infty$.

3. Prove that Corollary 1.91 implies Theorem 1.88.

4. Deduce Theorem 1.88 from its more rudimentary forms in which

(a) f is continuous.
(b) X is a bounded closed subset of a Banach space. [Hint: First show that there is no loss of generality in assuming that X is bounded, replacing X itself with its subset $A(\bar{x})$ if necessary, and then use an embedding of X into the space of bounded continuous functions on X.]

5. Check that the proof of the Ekeland principle has not used the symmetry of the function d. For further generalizations of the Ekeland principle, see [137, Theorem 2.5.2] and [655].

1.5.2 Supplement: Some Consequences of the Ekeland Principle

Under additional assumptions, the conclusion of Theorem 1.88 can be transformed into interesting consequences. The first one requires the knowledge of Gâteaux differentiability described in Chap. 2; the second one, in Corollary 1.93, requires the concept of subdifferential for a convex function introduced in Chap. 3.

Corollary 1.92. *Let $f : B \to \mathbb{R}$ be a lower semicontinuous function on an open ball $B := B(\bar{x}, r)$ of a Banach space E. Suppose f is bounded below and Gâteaux differentiable on B. Then given $\alpha > f(\bar{x}) - \inf f(B)$, there exists $u \in B$ such that $\|Df(u)\| < \alpha/r$.*

Proof. Let $\varepsilon \in (0, \alpha)$ be such that $\varepsilon > f(\bar{x}) - \inf f(B)$ and let $\rho < \sigma < r$ be such that $\rho > \varepsilon \alpha^{-1} r$. Let us set $X := B[\bar{x}, \sigma]$. Theorem 1.88 yields some $u \in B[\bar{x}, \rho]$ that is a minimizer of $g : x \mapsto f(x) + \varepsilon \rho^{-1} \|x - u\|$ on X. Since the function $h : x \mapsto \|x - u\|$ has a directional derivative at $x = u$ given by $h'(u, v) = \|v\|$ and since u is interior to $B[\bar{x}, \sigma]$, we have $g'(u, v) \geq 0$ for all $v \in E$, and we obtain

$$Df(u)(v) + \varepsilon \rho^{-1} \|v\| \geq 0 \quad \forall v \in E.$$

This inequality shows that $\|Df(u)\| \leq \varepsilon/\rho < \alpha/r$. $\square$

The preceding result can be rephrased in the following way: if for some $c > 0$ one has $\|Df(u)\| \geq c$ for every $u \in B$, then $\inf_{u \in B} f(u) \leq f(\bar{x}) - cr$: the function f has a significant decrease on the ball B. For a convex function one can get rid of differentiability assumptions. A generalization encompassing both cases will be given later on. Here $\partial f(u) := \{u^* \in X^* : f \geq u^* + f(u) - u^*(u)\}$.

Corollary 1.93. *Let E be a Banach space and let $f : X \to \mathbb{R}_\infty$ be a bounded-below lower semicontinuous convex function on a ball $B := B(\bar{x}, r)$ and finite at $\bar{x}$. Let $\alpha > f(\bar{x}) - \inf f(B)$. Then there exist $u \in B$ and $u^* \in \partial f(u)$ such that $\|u^*\| < \alpha/r$.*

Proof. Let ε, ρ, σ, g be as in the preceding proof. Again we get a minimizer u of the function g belonging to the interior of $B[\bar{x}, \sigma]$, so that we have

$$f'(u, v) + \varepsilon \rho^{-1} \|v\| \geq 0 \quad \forall v \in E.$$

The sandwich theorem (or the sum rule for subdifferentials) yields some $u^* \in \partial f(u)$ such that $\|u^*\| \leq \varepsilon / \rho < \alpha / r$.   □

Exercise. Prove the following variant of Corollary 1.92 that makes it more precise. With the same assumptions, setting $\beta = f(\bar{x}) - \inf f(B)$, for every $t \in (0, 1)$ there exists $u \in B[\bar{x}, tr]$ such that $\|f'(u)\| \leq \beta / tr$.

1.5.3 Supplement: Fixed-Point Theorems via Variational Principles

Fixed-point theorems are important tools for solving equations and proving existence results. The most important ones are the Brouwer theorem, its extensions by Schauder and Tikhonov and the contraction theorem. We quote the first one, referring to [126] for an elegant proof, and we show that the Ekeland variational principle is powerful enough to imply the contraction theorem. It even yields an extension to multimaps.

Theorem 1.94 (Brouwer). *If C is a convex compact subset of a finite-dimensional normed space and if $f : C \to C$ is continuous, then f has at least one fixed point u, i.e., a point $u \in C$ such that $f(u) = u$.*

Given a multimap $F : X \rightrightarrows X$ with nonempty closed values from a metric space into itself, one says that F is a *contraction* if there exists $c \in [0, 1)$ such that for every $x, x' \in X$ one has $d_H(F(x), F(x')) \leq c d(x, x')$, where d_H is the *Pompeiu–Hausdorff distance* given for two nonempty closed subsets A, A' of X by

$$d_H(A, A') := \max(e_H(A, A'), e_H(A', A)) \quad \text{with } e_H(A, A') := \sup_{a \in A} d(a, A').$$

Theorem 1.95 (Picard, Banach, Nadler [739]). *Let (X, d) be a complete metric space and let $F : X \rightrightarrows X$ be a multivalued contraction with nonempty closed values. Then F has a fixed point: there exists $u \in X$ such that $u \in F(u)$.*

Moreover, if c is the Lipschitz rate of F, then for every $\bar{x} \in X$ and for every $r > (1 - c)^{-1} d(\bar{x}, F(\bar{x}))$, one can find $u \in B(\bar{x}, r)$ such that $u \in F(u)$.

Note that when F is single-valued, uniqueness of the fixed point follows from the contraction property, since for any pair of fixed points u, u' of F one has

$$d(u,u') = d(F(u),F(u')) \leq cd(u,u'),$$

whence $d(u,u') = 0$. When F is multivalued, uniqueness is lost (consider the case $F(x) = X$ for all $x \in X$, for instance).

Proof. Let us define f on X by $f(x) = d(x,F(x)) := \inf\{d(x,y) : y \in F(x)\}$. Since

$$d(x,F(x)) \leq d(x,x') + d(x',F(x')) + d_H(F(x'),F(x)),$$

hence $|f(x) - f(x')| \leq (c+1)d(x,x')$, we see that f is continuous. Since f is nonnegative, given $\bar{x} \in X$, setting $\varepsilon := d(\bar{x},F(\bar{x}))$, and choosing $\gamma \in (0,1-c)$ so close to $1-c$ that $\rho := \varepsilon/\gamma < r$, Theorem 1.88 yields some $u \in B[\bar{x},\rho]$ such that

$$\forall x \in X, \quad d(u,F(u)) \leq d(x,F(x)) + \gamma d(u,x).$$

Taking $x \in F(u)$ in this relation and noting that $d(x,F(x)) \leq d_H(F(u),F(x))$, we get

$$d(u,F(u)) \leq d_H(F(u),F(x)) + \gamma d(u,x) \leq (c+\gamma)d(u,x).$$

Passing to the infimum over $x \in F(u)$, we obtain $d(u,F(u)) \leq (c+\gamma)d(u,F(u))$. Since $c+\gamma < 1$, we must have $d(u,F(u)) = 0$, and hence $u \in F(u)$ as $F(u)$ is closed. $\square$

Another fixed-point theorem can be deduced from the Ekeland principle.

Theorem 1.96 (Caristi, Kirk). *Let X be a complete metric space and let $F : X \rightrightarrows X$ be a multimap with nonempty values. Suppose there exists some lower semicontinuous function $h : X \to \overline{\mathbb{R}}_+ := \mathbb{R}_+ \cup \{+\infty\}$ such that for every $x \in X$, $y \in F(x)$ one has*

$$h(y) \leq h(x) - d(x,y). \tag{1.24}$$

Then there exists some $u \in X$ such that $F(u) = \{u\}$.

Proof. Applying Theorem 1.87 with $f := h$, $\gamma := 1$, we get some $u \in X$ such that $h(u) < h(x) + d(x,u)$ for all $x \neq u$. If we could find $v \in F(u)$ with $v \neq u$, taking $x := v$ and using (1.24), we would have $h(u) < h(v) + d(v,u) \leq h(u)$, an impossibility. $\square$

Exercises

1. Show that the Caristi–Kirk theorem is equivalent to the Ekeland principle.

2. Define a function $f : X \to \mathbb{R} \cup \{+\infty\}$ on a metric space X to be decreasingly lower semicontinuous if for every convergent sequence $(x_n) \to x$ such that $f(x_{n+1}) < f(x_n)$ one has $f(x) \leq \lim_n f(x_n)$. Show that the Caristi–Kirk fixed-point theorem and the Ekeland variational principle are still valid under the assumption that the function is decreasingly lower semicontinuous

3. Let D be a nonempty subset of a complete metric space and let $F : D \rightrightarrows X$ be a multimap with closed graph in $X \times X$. Let us say that F satisfies the (A-b-C) property if for $c \in [0,1)$, $b > 0$ such that $b < (1-c)/(1+c)$ and all (x,y) in the graph of F with $x \neq y$ there exist $u \in X$, $z \in D$ such that

$$d(x,y) = d(x,u) + d(u,y), \quad d(u,z) < bd(u,x), \quad d(y, F(z)) \leq cd(x,z).$$

We say that F satisfies the (A–C) property if there exists some $c \in [0,1)$ such that the preceding property holds for all $b \in (0, (1-c)/(1+c))$.

The multimap F is said to be a *directional contraction* if its graph is closed and for some $c \in (0,1)$ and all $(x,y) \in F$ with $x \neq y$ there exists some $z \in X \setminus \{x\}$ satisfying

$$d(x,y) = d(x,z) + d(z,y), \quad d(y, F(z)) \leq cd(x,z).$$

(a) Check that a closed-valued contraction is a directional contraction.
(b) Check that a directional contraction satisfies the (A–C) property.
(c) Using the Ekeland variational principle, show that if F satisfies the (A–b–C) property, then F has a fixed point u and for all $x \in D$ and all $a > (1-b)/(1+b) - c$ one can find a fixed point u of F such that $d(x,u) \leq (1/a)d(x,F(x))$.
(d) Deduce from the preceding question that if F satisfies the (A–C) property, then F has a fixed point u and, denoting by S the set of fixed points of F, for all $x \in D$, one has that $d(x,S) \leq (1/(1-c))d(x,F(x))$. [Hint: See [56, 214].]

1.5.4 Supplement: Metric Convexity

As an application of the preceding theorem, let us give a simple proof of Menger's theorem. It deals with a notion of metric convexity (in spaces without linearity structure). Let us give precise definitions. A metric space (M,d) is said to be *metrically convex* (or for short, convex) if for every two distinct points a,b of M there exists some $c \in M \setminus \{a,b\}$ between a and b in the sense that

$$d(a,c) + d(c,b) = d(a,b).$$

We write acb when this relation holds; more generally, for a finite sequence $a_0, a_1, \ldots, a_n$ of points of M, we write $a_0 a_1 \ldots a_n$ if $d(a_0, a_1) + \cdots + d(a_{n-1}, a_n) = d(a_0, a_n)$. The space is said to be a *metric segment space* if for every two points a, b in M there exists a *geodesic* joining them, i.e., an isometric mapping $g : [0, \ell] \to g([0, \ell]) \subset M$, where $\ell := d(a,b)$, such that $g(0) = a$, $g(\ell) = b$. Let us call it a *metric midpoint space* if for every two points a, b in M there exists $c \in M$ such that $d(a,c) = d(c,b) = \frac{1}{2}d(a,b)$.

Theorem 1.97 (Menger). *A complete metric space is a metric segment space iff it is a metric midpoint space iff it is a metrically convex space.*

Proof. Clearly, a metric segment space (M,d) is a metric midpoint space. Let us first prove the converse when (M,d) is complete. Let (M,d) be a complete metric midpoint space. Let $a,b \in M$ with $\ell := d(a,b) > 0$ (the case $a = b$ is obvious). Let D be the set of real numbers of the form $2^{-n}k\ell$ with $n \in \mathbb{N} \setminus \{0\}$, $k \in \mathbb{N} \cap [0,2^n]$. Using induction on n we can find an isometry i from D onto a subset D' of M by defining it on the set

$$D_n := \{2^{-n}k\ell : k \in \mathbb{N} \cap [0,2^n]\}$$

using the midpoint property. Since D is dense in $[0,\ell]$, since M is complete and since i is uniformly continuous, we can extend i to a uniformly continuous mapping g on $[0,\ell]$ with the same Lipschitz rate 1. In fact, g is isometric, and it is the expected geodesic.

Obviously, a metric midpoint space (M,d) is a metrically convex space. Let us prove the converse when (M,d) is complete. Let $a,b \in M$ with $\ell := d(a,b) > 0$ (again the case $a = b$ is obvious). Let us endow

$$X := \left\{ (u,v) \in M \times M : auvb, \ d(a,u) \leq \frac{1}{2}\ell, \ d(v,b) \leq \frac{1}{2}\ell \right\}$$

with the metric given by $d((u,v),(u',v')) := d(u,u') + d(v,v')$, making it a nonempty complete metric space (it contains (a,b) and is closed in $M \times M$). Let $F : X \rightrightarrows X$ be given by

$$F(u,v) = \{(u',v') \in X : auu'v'vb\}.$$

Clearly, F has a nonempty set of values. Let $h : X \to \mathbb{R}$ be given by $h(u,v) := d(u,v)$. Since for every $(u,v) \in X$ and any $(u',v') \in F(u,v)$ we have

$$d(u,u') + d(u',v') + d(v',v) = d(u,v),$$

we see that $h(u',v') = h(u,v) - d((u,v),(u',v'))$. Moreover, h is continuous.

Theorem 1.96 yields some $(u,v) \in X$ such that $F(u,v) = \{(u,v)\}$. If we had $u \neq v$ we could find $w \in M$ such that uwv and $w \neq u$, $w \neq v$. Then we would have awb and either $d(a,w) \leq \frac{1}{2}\ell$ or $d(w,b) \leq \frac{1}{2}\ell$. In the first case we would have $(w,v) \in F(u,v)$, and in the second case we would have $(u,w) \in F(u,v)$; in both cases we would have $F(u,v) \neq \{(u,v)\}$, a contradiction. $\square$

Exercise. [92, 944] A metric space (M,d) is said to be *approximately metrically convex* if for every $x,z \in M$, $t \in (0,1)$, $\varepsilon > 0$ there exists some $y \in M$ such that $d(x,y) < td(x,z)$, $d(y,z) < (1-t)d(x,z) + \varepsilon$. For two subsets C, D of (M,d) and $r \in \mathbb{R}_+$, set $B(C,r) := \{x \in M : d_C(x) < r\}$, $B[C,r] := \{x \in M : d_C(x) \leq r\}$ and $\mathrm{gap}(C,D) := \inf\{d(x,y) : x \in C, y \in D\}$. Show that each of the following properties characterizes approximately metrically convex metric spaces:

(a) For all $x \in X$, $r,s > 0$ one has $B(B(x,r),s) = B(x,r+s)$;
(b) For all $x \in X$, $r,s > 0$ one has $B[B[x,r],s] = B[x,r+s]$;
(c) For all $w,x \in X$ and $r \in [0,d(w,x)]$ one has $d(w,B(x,r)) = d(w,x) - r$;

(d) For all $w, x \in X$ and $r \in [0, d(w,x)]$ one has $d(w, B[x,r]) = d(w,x) - r$;

(e) For all $x \in X$, $r, s \geq 0$ one has $d_H(B[x,r], B[x,s]) = |r - s|$;

(f) For all $w, x \in X$, $r, s \geq 0$ one has $\mathrm{gap}(B[w,r], B[x,s]) \leq \max(d(w,x) - r - s, 0)$.

Exercise. Show that the space $\mathscr{C}_b$ of nonempty, bounded, closed convex subsets of a Banach space X is metrically convex and complete when endowed with the Pompeiu–Hausdorff distance. [Hint: Given C_0, C_1 in $\mathscr{C}_b$, $t \in [0,1]$, using Hörmander's isometry $h : C \mapsto h_C$ from $\mathscr{C}_b$ into the space $BC(S)$ of bounded continuous functions on the unit sphere S of the dual X^* of X given by $h_C(x^*) :=$ $\sup\{\langle x^*, x \rangle : x \in C\}$, show that for $C_t := (1-t)C_0 + tC_1$ one has $d_H(C_t, C_0) = t d_H(C_0, C_1)$, $d_H(C_t, C_1) = (1-t) d_H(C_0, C_1)$. Here $BC(S)$ is endowed with the sup norm.]

1.5.5 Supplement: Geometric Principles

The Ekeland principle is equivalent to geometric results of interest. Here we only show how one can deduce the Drop theorem from Theorem 1.88; we refer to the exercises for the reverse direction. In the sequel, the *drop* $D(a, B)$ generated by a point $a \in X$ and a subset B of a normed space X, also denoted by $[a, B]$ (because it reduces to a segment when B is a singleton), is the set

$$D(a, B) := \mathrm{co}(\{a\} \cup B) = \{(1-t)a + tb : t \in [0,1],\ b \in B\}.$$

Recall that the *gap* between two subsets A, B of X is defined by

$$\mathrm{gap}(A, B) := \inf_{(a,b) \in A \times B} d(a, b).$$

Theorem 1.98 (Drop theorem [253]). *Let E be a nonempty complete subset of a normed space X and let B be a nonempty, bounded, closed, convex subset of X such that $\delta := \mathrm{gap}(B, E) > 0$. Then there exists some $e \in E$ such that $D(e, B) \cap E = \{e\}$. Moreover, given $\bar{e} \in E$ one can choose $e \in E \cap D(\bar{e}, B)$.*

Proof. Let $\beta := \mathrm{diam}(B) := \sup\{d(w, x) : w, x \in B\}$ be the *diameter* of B and let $\gamma > 0$ be such that $\gamma(1 + \beta/\delta) < 1$. Let us apply Theorem 1.88 to the function $f := d(\cdot, B)$ that is continuous and bounded below. Given $\bar{e} \in E$, we can find $e \in A(\bar{e}) := \{x \in E : f(x) + \gamma d(x, \bar{e}) \leq f(\bar{e})\}$ such that

$$d(e, B) < d(x, B) + \gamma \|x - e\| \quad \forall x \in E \setminus \{e\}. \tag{1.25}$$

Let us show that $D(e, B) \cap E = \{e\}$. Suppose, to the contrary, that there exists some $y \in D(e, B) \cap E \setminus \{e\}$: there exist $t \in [0, 1)$ and $z \in B$ such that $y = te + (1-t)z$. For any $\varepsilon > 0$, $b \in B$ such that $\|e - b\| < (1 + \varepsilon)d(e, B)$ we have

$$d(y, B) \leq \|tb + (1-t)z - y\| = \|tb - te\| \leq t(1 + \varepsilon)d(e, B),$$

and since $\varepsilon > 0$ can be taken arbitrarily small, $d(y,B) \leq td(e,B)$. Now $\|y - e\| = (1-t)\|z - e\|$ and $\|z - e\| \leq d(e,B) + \beta \leq (1 + \beta/\delta)d(e,B)$, so that gathering the preceding estimates with the inequality $d(e,B) \leq d(y,B) + \gamma\|y - e\|$ obtained by taking $x := y$ in (1.25), we get

$$d(e,B) < td(e,B) + \gamma(1-t)\|z - e\| \leq td(e,B) + \gamma(1-t)(1 + \beta/\delta)d(e,B),$$

a contradiction to $\gamma(1 + \beta/\delta) < 1$ and $d(e,B) \geq \delta > 0$. $\square$

The next consequence involving a truncated cone rather than a drop will be useful.

Lemma 1.99. *Let E be a nonempty complete subset of a normed space X, let $\overline{w} \in X \setminus E$, $\overline{e} \in E$, $s > r > 0$ such that $B[\overline{w},s] \cap E = \varnothing$, and let $B = B[\overline{w},r]$, $C := \mathbb{R}_+(B - \overline{e})$. Then there exists some $e \in E \cap D(\overline{e},B)$ such that $(e + C) \cap E \cap B(e,s-r) = \{e\}$.*

Proof. Theorem 1.98 yields some $e \in E \cap D(\overline{e},B)$ such that $D(e,B) \cap E = \{e\}$. Let $b \in B$, $t \in (0,1]$ be such that $e = (1-t)b + t\overline{e}$. Then since $tB + (1-t)b \subset B$, we have

$$B - \overline{e} = B + t^{-1}(1-t)b - t^{-1}e \subset t^{-1}(B - e),$$

so that $C := \mathbb{R}_+(B - \overline{e}) \subset \mathbb{R}_+(B - e)$. Thus, for $x \in E \cap (e + C) \cap B(e,s-r)$ we can find $q \in \mathbb{R}_+$ and $b' \in B$ such that

$$x - e = q(b' - e).$$

Since $\|x - e\| < s - r$ and $\|b' - e\| \geq s - r$, we have $q = \|b' - e\|^{-1}\|x - e\| < 1$. Thus $x = e + q(b' - e) \in D(e,B) \cap E = \{e\}$, and $x = e$. $\square$

Let us apply the preceding lemma to density properties. We need the following terminology. A point e of a subset E of a normed space X is said to be a *support point* of E if there exists a closed convex proper cone C with nonempty interior and some $\varepsilon > 0$ such that $(e + C) \cap E \cap B(e,\varepsilon) = \{e\}$. When E is convex, e is a support point of E iff it is an *exposed point* of E, i.e., a point $e \in E$ such that there exists some $f \in X^* \setminus \{0\}$ satisfying $f(e) > f(x)$ for all $x \in E \setminus \{e\}$. The condition is obviously sufficient (take $C := \{w \in X : f(w) \geq (1/2)\|f\| \cdot \|w\|\}$); it is necessary by the Hahn–Banach theorem. The preceding lemma provides plenty of support points, even in the nonconvex case.

Corollary 1.100. *For every nonempty complete subset E of a normed space X, the set S of support points of E is dense in the boundary $\mathrm{bdry}(E)$ of E.*

Proof. The conclusion is obvious if the boundary of E is empty. Let $\overline{e}$ be a boundary point of E and let $\alpha > 0$ be given. There exists some $\overline{w} \in B(\overline{e},\alpha/2) \setminus E$. Let $s > 0$ be the radius of a closed ball with center $\overline{w}$ contained in $X \setminus E$, so that $s \leq d(\overline{w},E) \leq d(\overline{w},\overline{e}) < \alpha/2$, and let $r \in (0,s)$. The preceding lemma provides some $e \in E \cap D(\overline{e}, B[\overline{w}, r])$ such that $(e+C) \cap E \cap B(e, s-r) = \{e\}: e \in S$. Moreover, $d(\overline{e}, e) \le \mathrm{diam} D(\overline{e}, B[\overline{w}, r]) < \alpha$. That shows that S is dense in $\mathrm{bdry}(E)$.   □

The preceding observations entail a result whose discovery is at the origins of the results of the present section.

Theorem 1.101 (Bishop–Phelps [104]). *For any closed convex subset C of a Banach space, the set of support points of C is dense in the boundary of C.*

The next theorem reveals another density result that can be obtained by the method of the present section. A nonconvex version will be given later on (Theorem 1.153).

Theorem 1.102 (Bishop–Phelps). *For every bounded, closed, convex subset C of a Banach space X, the set of continuous linear forms that attain their maximum on C is dense in X^*.*

Proof. Let $h \in X^*$ and let $\varepsilon > 0$ be given. Let g be the indicator function of C and let $f := g - h$. Let $\overline{x} \in C$ be such that $f(\overline{x}) < \inf f(C) + \alpha$ with $\alpha := 1$. Taking $r > 1/\varepsilon$ large enough, the ball $B := B(\overline{x}, r)$ contains C and $\inf f(B) = \inf f(C)$. Corollary 1.93 yields some $e \in \mathrm{dom} f = C$ and $e^* \in \partial f(e)$ such that $\|e^*\| < \alpha/r < \varepsilon$. Then $e^* + h \in \partial g(e)$. This means that $\langle h + e^*, x - e \rangle \le 0$ for all $x \in C$.   □

Taking for C the closed unit ball of X, we get that the *duality mapping* $J : X \rightrightarrows X^*$ defined by $J(x) := \{x^* \in X^* : \|x^*\| = \|x\|, \langle x^*, x \rangle = \|x\|^2\}$ has a dense image in X^*.

Let us quote the following result, for it shows the limitations of what one can expect. See [374, 507], for instance, for the proof.

Theorem 1.103 (James). *A bounded, weakly closed subset E of a Banach space X is weakly compact if and only if every continuous linear form on X attains its maximum on E. In particular, X is reflexive if and only if every $x^* \in X^*$ attains its supremum on the unit ball of X.*

1.5.6   Supplement: The Banach–Schauder Open Mapping Theorem

Let us also show that the Ekeland variational principle yields the Banach–Schauder open mapping theorem, one of the cornerstones of linear functional analysis. It is a special case of the Robinson–Ursescu theorem, but its importance justifies a new proof. The one we present uses some results from convex analysis (Chap. 3).

Theorem 1.104 (Banach–Schauder open mapping theorem). *Let X and Y be Banach spaces and let $A : X \to Y$ be a continuous linear mapping such that $A(X) = Y$. Then A is open i.e., for every neighborhood U of a point $x \in X$, the image $V := A(U)$ is a neighborhood of $A(x)$.*

Theorem 1.105 (Banach isomorphism theorem). *If A is a linear continuous bijection between two Banach spaces, then A is an isomorphism.*

We can even give a quantitative form to this result by introducing the *openness rate* of A as

$$\omega(A) := \sup\{r \in \mathbb{R}_+ : rU_Y \subset A(U_X)\},$$

where U_X (resp. U_Y) is the open unit ball of X (resp. Y). It would not change if the closed unit balls B_X (resp. B_Y) were substituted for the open balls; here we make the usual convention that the supremum of the empty subset of $\mathbb{R}_+$ is 0.

We will relate $\omega(A)$ to the injectivity constant of the *transpose map* $A^{\mathsf{T}} : Y^* \to X^*$ of A. The *injectivity constant* $\alpha(B)$ of a linear mapping $B : V \to W$ between two normed spaces is defined by

$$\alpha(B) := \inf\{\|B(v)\| : v \in S_V\} = \max\{c \in \mathbb{R}_+ : \|B(v)\| \geq c\|v\| \quad \forall v \in V\}.$$

Clearly, B is injective when $\alpha(B)$ is positive, but the converse is not true.

Exercise. Define an injective continuous linear map B from a separable Hilbert space into itself such that $\alpha(B) = 0$.

Exercise. Let b be a bounded positive function on a set S that is not bounded away from 0. Check that the map $B : f \mapsto bf$ on the space $V = W$ of bounded functions on S is injective but $\alpha(B) = 0$.

When B is an isomorphism, the constant $\alpha(B)$ is related to the norm of the inverse B^{-1} of B by the relation $\alpha(B) = \|B^{-1}\|^{-1}$ (with the usual conventions).

We define the *Banach constant* $\beta(A)$ of $A : X \to Y$ as the injectivity constant of the transpose A^{T} of A: $\beta(A) := \alpha(A^{\mathsf{T}})$. Then we observe in the following proposition that $\beta(A) > 0$ whenever A is open.

Proposition 1.106. *For any continuous linear operator $A : X \to Y$ between two normed spaces one has $\beta(A) := \alpha(A^{\mathsf{T}}) \geq \omega(A)$. In particular, when A is open, one has $\alpha(A^{\mathsf{T}}) > 0$ and A^{T} is injective. If X is complete, one has $\beta(A) = \omega(A)$.*

Proof. Let $r > 0$ be such that $rU_Y \subset \mathrm{cl}(A(B_X))$. By definition of the dual norm on X^* and the fact that $\|y\| = \sup\{\langle y^*, y \rangle : y^* \in S_{Y^*}\}$ (by Corollary 1.72), we have

$$\alpha(A^{\mathsf{T}}) = \inf\{\sup\{\langle A^{\mathsf{T}}(y^*), x \rangle : x \in B_X\} : y^* \in S_{Y^*}\}$$
$$= \inf\{\sup\{\langle y^*, A(x) \rangle : x \in B_X\} : y^* \in S_{Y^*}\}$$
$$= \inf\{\sup\{\langle y^*, y \rangle : y \in \mathrm{cl}(A(B_X))\} : y^* \in S_{Y^*}\}$$
$$\geq \inf\{\sup\{\langle y^*, y \rangle : y \in rU_Y\} : y^* \in S_{Y^*}\} = r.$$

It follows that $\alpha(A^{\mathsf{T}}) \geq \varpi(A) := \sup\{r \in \mathbb{R}_+ : rU_Y \subset \mathrm{cl}(A(B_X))\} \geq \omega(A)$.

Let us prove the equality $\alpha(A^\mathsf{T}) = \omega(A)$ when X is complete. Since this equality holds when $\alpha(A^\mathsf{T}) = 0$, it suffices to show that when $\alpha(A^\mathsf{T}) > 0$, for all $r \in (0, \alpha(A^\mathsf{T}))$ one has $rU_Y \subset A(B_X)$, and hence $rU_Y \subset A(sU_X)$ for all $s > 1$ and $r \leq \omega(A)$.

Given $y \in Y \setminus A(B_X)$, we wish to show that $y \notin rB_Y$. Let us set $f(x) = \|A(x) - y\|$ for $x \in X$. Then $f(0) = \|y\| \leq \inf f(X) + \|y\|$. Let us apply Theorem 1.88 with $\bar{x} = 0$, $\varepsilon = \|y\|$, $\rho = 1$ on X. It yields some $u \in B_X$ that is a minimizer on X of the function $g : x \mapsto f(x) + \varepsilon \|x - u\|$. Since g is convex and u is a minimizer of g on X, we have $0 \in \partial g(u)$. By the classical rules of convex analysis (Theorems 3.39 and 3.40), we get some $u^* \in \partial f(u)$ such that $\|u^*\| \leq \varepsilon$ and some $v^* \in \partial \|\cdot\| (A(u) - y)$ such that $u^* = A^\mathsf{T}(v^*)$. Since $y \notin A(B_X)$, we have $v := A(u) - y \neq 0$, and hence $\|v^*\| = 1$ (one can easily check that $v^* \in \partial \|\cdot\| (v)$ iff $v^* \in B_{X^*}$ and $\langle v^*, v \rangle = \|v\|$). Thus, by definition of $\alpha(A^\mathsf{T})$,

$$\alpha(A^\mathsf{T}) \leq \|u^*\| \leq \varepsilon = \|y\|$$

and $r < \alpha(A^\mathsf{T}) \leq \|y\|: y \notin rB_Y$. Therefore $rB_Y \subset A(B_X)$ and $r \leq \omega(A)$. □

Exercise. Deduce from the Hahn–Banach theorem that $\alpha(A^\mathsf{T}) = \varpi(A)$.

The open mapping theorem easily follows from the preceding proposition. In fact, when A is surjective, we have

$$Y = \bigcup_{n=1}^{\infty} A(nB_X) = \bigcup_{n=1}^{\infty} \mathrm{cl} A(nB_X).$$

The Baire category theorem asserts that for some $n \geq 1$ the set $\mathrm{cl} A(nB_X)$ has nonempty interior. Thus $\mathrm{cl} A(B_X)$ has nonempty interior. Since this set is convex and symmetric with respect to 0, we get that 0 belongs to the interior of $\mathrm{cl} A(B_X)$: there exists $r > 0$ such that $rB_Y \subset \mathrm{cl} A(B_X)$. Then $\varpi(A) > 0$ and thus $\omega(A) > 0$ by the proof of the preceding proposition: A is open. □

Corollary 1.107 (Closed graph theorem). *Every linear map with closed graph between two Banach spaces is continuous.*

Proof. Let $B : Y \to Z$ be such a map. The graph X of B, being a closed linear subspace of $Y \times Z$, is a Banach space, and $A : (y, By) \mapsto y$ is a continuous bijection from X onto Y. Its inverse $y \mapsto (y, By)$ being continuous, B is continuous. □

Exercise. Let X and Y be Banach spaces and let $A : X \to Y$ be a linear map such that for every $y^* \in Y^*$ the linear form $y^* \circ A$ is continuous on X. Show that A is continuous. [Hint: Prove that the graph G of A is such that $G = \{(x,y) \in X \times Y : \forall y^* \in Y^*, y^*(A(x)) = y^*(y)\}$, hence is closed.]

A classical factorization result will be helpful.

Lemma 1.108. *Let X, Y be Banach spaces, let $A : X \to Y$ be a surjective continuous linear map, and let $\ell \in X^*$ be such that $\ell(x) = 0$ for all $x \in N := \ker A$. Then there exists some y^* in the dual Y^* of Y such that $\ell = y^* \circ A$.*

Proof. Since A is surjective and since for every $x, x' \in X$ satisfying $A(x) = A(x')$ one has $\ell(x) = \ell(x')$, there exists a map $k : Y \to \mathbb{R}$ such that $\ell = k \circ A$. It is easy to check that k is linear. Now, the Banach open mapping theorem asserts that there exists some $c > 0$ such that for all $y \in Y$ one can find some $x \in A^{-1}(y)$ satisfying $\|x\| \le c\|y\|$. It follows that for all $y \in Y$ one has $k(y) = k(A(x)) = \ell(x) \le \|\ell\|c\|y\|$. Thus k is continuous and we can take $y^* = k$. $\square$

Remarks. Instead of using the Banach open mapping theorem, one can introduce the canonical projection $p : X \to X/N$, observe that ℓ can be factorized as $\ell = \bar{\ell} \circ p$ for some $\bar{\ell}$ in the dual of X/N, and use the Banach isomorphism theorem to get that the map $\bar{A} : X/N \to Y$ induced by A is an isomorphism, so that one has $\ell = y^* \circ A$ with $y^* := \bar{\ell} \circ \bar{A}^{-1}$. $\square$

1.6 Decrease Principle, Error Bounds, and Metric Estimates

The framework of metric spaces allows one to give a quantitative approach as well as a qualitative approach to a number of questions. Moreover, it is such a bare framework that the main ideas are not hidden by secondary facts or additional structures. The present section takes advantage of these favorable features. We first experience them when dealing with decrease principles and estimates bearing on the conditioning of a function $f : X \to \overline{\mathbb{R}}_+ := \mathbb{R} \cup \{+\infty\}$ on a metric space (X, d) of the form

$$\forall x \in X, \qquad d(x, S) \le cf(x),$$

where $S := f^{-1}(\{0\})$, $c > 0$, and for a subset S of X, $d_S(\cdot) := d(\cdot, S)$ is the *distance function* given by $d(x, S) := \inf\{d(x, y) : y \in S\}$, with the convention $d(x, S) := +\infty$ if S is empty. More generally, if F is a map or a multimap from X into another metric space Y and if b is a given element of Y, it is of interest to estimate the distance to the solution set $S := F^{-1}(\{b\})$ of the equation $F(x) = b$ or the inclusion $b \in F(x)$ by the computable value $d(b, F(x))$. The possibility of getting such an estimate is obtained by the study of what is now known as the study of calmness, of error bounds, and of metric regularity; it has given rise to a vast literature. Estimates of the form

$$d(x, F^{-1}(b)) \le c\, d(b, F(x))$$

are particularly useful in nonsmooth analysis. They can also be used in connection with penalization techniques for optimization problems. We devote the third subsection below to such a motivation. Its simple penalization lemma is of great use. The last subsections deal with a systematic study of relations similar to the preceding one. We also connect this subject with the notion of open map or multimap and with Lipschitzian properties of multimaps appearing in a number of problems.

1.6.1 Decrease Principle and Error Bounds

The following question arises frequently: given a function $f : X \to \overline{\mathbb{R}}_+ := [0, \infty]$ on a metric space (X, d), if the value of f at $x \in X$ is small, is x close to the zero set $S := f^{-1}(0)$ of f? Such a question is of importance for algorithms, but its application is much larger.

The above question can be given a precise quantitative form. The following elementary lemma answers such a need in describing a desirable behavior of f that secures a positive answer to the question. In its statement, we say that $\gamma : \mathbb{R}_+ \to \mathbb{R}_+$ is a *gauge* or a *growth function* if γ is nondecreasing and $\gamma(t) > 0$ for $t > 0$, and we recall that $\mu : \mathbb{R}_+ \to \overline{\mathbb{R}}_+$ is a *modulus* if μ is nondecreasing and $\mu(t) \to 0$ as $t \to 0$. The proof is left as an exercise.

Lemma 1.109. [808] *The following assertions about f are equivalent:*

(a) f is well set: for every sequence (x_n) of X, $(f(x_n)) \to 0 \Longrightarrow (d_S(x_n)) \to 0$;
(b) There exists a modulus μ such that $d_S(\cdot) \leq \mu(f(\cdot))$;
(c) There exists a gauge γ such that $\gamma(d_S(\cdot)) \leq f(\cdot)$.

These quantitative notions can be useful for the study of the speed of convergence of algorithms. However, we will restrict our attention to the case that the function f is *well conditioned* or *linearly conditioned* in the sense there exists a constant $c > 0$ such that $\gamma(r) := cr$ for $r > 0$ small enough. Then the *conditioning rate* $\gamma_f(\overline{x})$ of f at $\overline{x} \in S$ is the supremum of the constants c such that $cd_S(\cdot) \leq f(\cdot)$ near $\overline{x}$:

$$\gamma_f(\overline{x}) := \liminf_{x \to \overline{x}, \ x \in X \setminus S} \frac{f(x)}{d(x, S)}.$$

The terminology is justified by the following example, showing a relationship with the notion of conditioning of a matrix, the ratio between its smallest eigenvalue and its largest one.

Example. Let A be a positive definite symmetric operator on a Euclidean space X such that $\|A\| = 1$. Let q be the quadratic form associated with A by $q(x) := (1/2)(Ax \mid x)$, where $(\cdot \mid \cdot)$ is the scalar product of X, and let $f := \sqrt{q}$. Then $\gamma_f(r) = \sqrt{\alpha/2}r$, where α is the least eigenvalue of A, whereas q is nonlinearly conditioned: the greatest gauge γ satisfying $\gamma(d_S(\cdot)) \leq q(\cdot)$ is $\gamma_q(r) = (1/2)\alpha r^2$. Since the greatest eigenvalue of A is 1, the conditioning of A is α. □

The Ekeland variational principle can be used to obtain a useful estimate about the distance to the set of solutions of an equation or to the set of minimizers of a function. The proof of this estimate is similar to the proof of Corollary 1.91, but it has a more local character.

Proposition 1.110 (Error bound property). *Let $f : X \to \overline{\mathbb{R}}_+$ be a nonnegative lower semicontinuous function on a complete metric space (X, d) and let $S := f^{-1}(\{0\})$. Let $x \in \mathrm{dom} f$, $c > 0$, and $\rho > f(x)/c$ be such that for every $\gamma \in (0, c)$,*

$u \in B(x,\rho) \setminus S$ *there exists* $v \in X \setminus \{u\}$ *satisfying* $f(v) + \gamma d(u,v) \le f(u)$. *Then* S *is nonempty and*

$$d(x,S) \le c^{-1} f(x).$$

Proof. In Theorem 1.88 take $\bar{x} := x$, $\gamma \in (0,c)$ with $f(x)/\gamma < \rho$. Let $u \in X$ be given by the conclusion of Theorem 1.88, i.e., such that

$$f(u) + \gamma d(u,x) \le f(x), \quad f(u) < f(x') + \gamma d(u,x') \quad \text{for all } x' \in X \setminus \{u\}.$$

Then one has $d(u,x) \le \gamma^{-1} f(x) < \rho$. The second inequality and our assumption ensure that we cannot have $u \in B(x,\rho) \setminus S$. Thus $u \in S$ and $d(x,S) \le d(x,u) \le \gamma^{-1} f(x)$. Since γ is arbitrarily close to c, we get the announced estimate. $\qquad\square$

The lower semicontinuity condition in such a result can be relaxed. Here we say that a function $f : X \to \overline{\mathbb{R}}_+$ is such that f^{-1} is *closed at* 0 if for every convergent sequence (x_n) such that $(f(x_n)) \to 0$, one has $\lim_n x_n \in S := f^{-1}(\{0\})$. This concept coincides with the notion of closed multimap for $F := f^{-1}$. Given $s \in (0,1)$, let us say that a function f is s-*steep* at $x \in \mathrm{dom}\, f$ if for all $w \in X \setminus S$ satisfying $d(w,x) + f(w) < f(x)/s$, one has $\inf\{f(v) : v \in B[w, f(w)]\} < (1-s) f(w)$.

Proposition 1.111 (Steepness principle [225]). *Let* (X,d) *be a complete metric space and let* $f : X \to \overline{\mathbb{R}}_+$, $S := f^{-1}(\{0\})$. *Suppose that* f^{-1} *is closed at* 0 *and that for some* $s \in (0,1)$ *the function* f *is* s-*steep at* $x \in \mathrm{dom}\, f$. *Then* S *is nonempty and* $d(x,S) \le f(x)/s$. *In fact, there is some* $u \in S$ *such that* $d(u,x) \le f(x)/s$.

Proof. We may assume that $f(x) > 0$ and $B(x, f(x)/s) \cap S = \varnothing$ since otherwise the inequality is trivial. Let us show that $B[x, f(x)/s] \cap S \ne \varnothing$. Starting with $u_0 := x$, we construct inductively a sequence (u_n) of $X \setminus S$ satisfying $d(u_{n+1}, u_n) \le f(u_n)$, $f(u_{n+1}) \le t f(u_n)$ for $t := 1 - s$. Such a construction is possible since $d(u_0,x) + f(u_0) < f(x)/s$, so that there exists some $u_1 \in B[u_0, f(u_0)]$ satisfying $f(u_1) < t f(x)$, and assuming that $u_0, \ldots, u_n$ have been obtained, we have

$$d(u_n,x) + f(u_n) \le \sum_{k=0}^{n-1} d(u_{k+1}, u_k) + f(u_n) \le \sum_{k=0}^{n} f(u_k) \le \sum_{k=0}^{n} t^k f(x) < \frac{1}{s} f(x),$$

hence $u_n \in B(x, f(x)/s) \subset X \setminus S$, so that we can pick $u_{n+1} \in B[u_n, f(u_n)]$ satisfying $f(u_{n+1}) \le t f(u_n)$. Since $d(u_{n+1}, u_n) \le f(u_n) \le t^n f(x)$, the sequence (u_n) is a Cauchy sequence, hence has a limit u in the closed ball $B[x, f(x)/s]$. Our closure condition ensures that $f(u) = 0$. Thus $u \in B[x, f(x)/s] \cap S$. $\qquad\square$

The following concept will be convenient for using the preceding result in a systematic way and for giving it an infinitesimal character. The terminology we adopt reflects its role.

Definition 1.112. Given a metric space (X,d) and $f : X \to \mathbb{R}_\infty$, a function $\delta_f : X \to \mathbb{R}_+$ is said to be a *decrease index* for f if for every $x \in X$, $r, c > 0$, we have

$$\inf_{u \in B(x,r)} \delta_f(u) \geq c \quad \Rightarrow \quad \inf_{u \in B(x,r)} f(u) \leq f(x) - cr, \tag{1.26}$$

or equivalently,

$$f(x) < \inf_{u \in B(x,r)} f(u) + cr \quad \Rightarrow \quad \exists u \in B(x,r) : \delta_f(u) < c. \tag{1.27}$$

The following result is an easy but useful consequence of the definition of a decrease index. Observe that the assumption is weaker than the one in implication (1.26), since it bears on $B(\bar{x}, \rho) \setminus S$ and that the conclusion is somewhat stronger, since it yields a minimizer. Here $B(\bar{x}, \rho) = X$ if $\rho = +\infty$.

Theorem 1.113 (Decrease principle). *Let X be a complete metric space, let $f : X \to \mathbb{R}_+$ be a function such that f^{-1} is closed at 0, and let $S := f^{-1}(\{0\})$. Let $\delta_f : X \to \mathbb{R}_+$ be a decrease index for f. Suppose there are $x \in \operatorname{dom} f$, $c > 0$, and $\rho \in (c^{-1} f(x), +\infty]$ such that $\delta_f(u) \geq c$ for all $u \in B(x, \rho) \setminus S$. Then S is nonempty and*

$$d(x, S) \leq c^{-1} f(x). \tag{1.28}$$

In particular, if for some positive number c one has $\delta_f(u) \geq c$ for all $u \in X \setminus S$, and if $\operatorname{dom} f$ is nonempty, then S is nonempty and for all $x \in X$ inequality (1.28) holds.

Proof. If for all $s \in (0, c)$ one has $B(x, f(x)/s) \cap S \neq \varnothing$, then the result holds. Thus, we consider the case that there exists some $\bar{s} \in (0, c)$ such that $B(x, f(x)/\bar{s}) \cap S = \varnothing$, and using the fact that $f(x)/c < \rho$, we take $s \in [\bar{s}, c)$ satisfying $f(x)/s < \rho$. Let us prove that f is s-steep at x. Given $w \in X \setminus S$ satisfying $d(w, x) + f(w) < f(x)/s$, for all $u \in B(w, f(w))$ we have

$$d(u, x) \leq d(u, w) + d(w, x) < f(w) + d(w, x) < f(x)/s < \min(f(x)/\bar{s}, \rho),$$

hence $u \in B(x, \rho) \setminus S$ and $\delta_f(u) \geq c$, so that by definition of a decrease index, taking $r := f(w)$, we get $\inf f(B(w, f(w))) \leq f(w) - cf(w) < (1 - s) f(w)$ and f is s-steep at x. By Proposition 1.111, $d(x, S) \leq f(x)/s$, and the case $B(x, f(x)/\bar{s}) \cap S = \varnothing$ for some $\bar{s} \in (0, c)$ is excluded. $\qquad\square$

The preceding theorem is an existence result for the zero set of f. A variant assumes existence but yields a useful localization property.

Theorem 1.114 (Local decrease principle). *Let X be a complete metric space, let $f : X \to \mathbb{R}_+$ be such that f^{-1} is closed at 0 on X, and let $\bar{x} \in S := f^{-1}(\{0\})$, $c > 0$, and $\rho \in (0, +\infty]$. Let $\delta_f : X \to \mathbb{R}_+$ be a decrease index for f such that $\delta_f(u) \geq c$ for all $u \in B(\bar{x}, 2\rho) \setminus S$. Then for all $x \in B(\bar{x}, \rho)$, relation (1.28) holds.*

Proof. Given $x \in B(\bar{x}, \rho)$ satisfying $f(x) \geq c\rho$, we obviously have $d(x, S) \leq c^{-1} f(x)$, since $d(x, S) \leq d(x, \bar{x}) < \rho$. Now, for $x \in B(\bar{x}, \rho)$ satisfying $f(x) < c\rho$, for $u \in B(x, \rho) \setminus S$ we have $u \in B(\bar{x}, 2\rho) \setminus S$, hence $\delta_f(u) \geq c$, and Theorem 1.113 yields $d(x, S) \leq c^{-1} f(x)$. □

The following examples borrow concepts that will be explained later on; thus, they should be skipped in a first reading. We display them here just to show that the notion of decrease index is versatile.

Example. If X is a Banach space and if $f : X \to \mathbb{R}$ is Gâteaux differentiable, setting $\delta_f(x) := \|Df(x)\|$ gives a decrease index in view of Corollary 1.92.

Example. If X is a Banach space and if $f : X \to \mathbb{R}_\infty$ is convex lower semicontinuous, setting $\delta_f(x) := d(0, \partial f(x))$ with the convention $\inf \varnothing = +\infty$, ∂f being the Moreau–Rockafellar subdifferential of f studied in Chap. 3, gives a decrease index in view of Corollary 1.93.

Example. If X is an open convex subset of a normed space and if $f : X \to \mathbb{R}$ is concave (and extended by $-\infty$ outside X), setting $\delta_f(x) := \sup\{\|x^*\| : x^* \in \partial(-f)(x)\}$ yields a decrease index on X. In fact, given $x \in X$, r, $c > 0$ such that $f(x) < \inf f(B(x, r)) + cr$, taking $c' < c$ satisfying $f(x) < \inf f(B(x, r)) + c'r$, for all $x^* \in \partial(-f)(x)$ one has

$$c'r > \sup_{w \in B(x,r)} (-f)(w) - (-f)(x) \geq \sup_{w \in B(x,r)} \langle x^*, w - x \rangle = r\|x^*\|,$$

hence $\|x^*\| \leq c'$ and $\delta_f(x) \leq c' < c$. □

We will see later that subdifferentials yield decrease indexes. In the general framework of complete metric spaces, the concept of slope introduced by De Giorgi et al. [263, 265] is the main example of decrease index. We now present it.

Definition 1.115. For a function $f : X \to \mathbb{R}_\infty$ finite at $x \in X$, the *slope* (or strong slope or calmness rate) of f at x is the function $|\nabla|(f) : X \to \mathbb{R}_\infty$ given by

$$|\nabla|(f)(x) := \limsup_{v \to x,\, v \neq x} \frac{(f(x) - f(v))^+}{d(x, v)} := \inf_{\varepsilon > 0} \sup_{v \in B(x,\varepsilon) \setminus \{x\}} \frac{(f(x) - f(v))^+}{d(x, v)},$$

where the positive part r^+ of an extended real number r is $\max(r, 0)$.

The terminology "calmness rate" is justified by the following observation. If f is *calm* at $x \in X$ in the sense that $f(x) < +\infty$ and for some $r, c > 0$ one has

$$\forall v \in B(x, r), \qquad f(v) \geq f(x) - cd(v, x),$$

then one has $|\nabla|(f)(x) \leq c$. In fact, $|\nabla|(f)(x)$ is the infimum of the constants $c > 0$ such that the preceding inequality holds for some $r > 0$. We observe that the terms "calmness rate" or "downward slope" is more justified than the classical term "slope," since the behavior of f on the superlevel set $\{u \in X : f(u) > f(x)\}$ is not involved in the definition. Moreover, with the convention $0/0 = 0$, one has

$$|\nabla|(f)(x) = \max\left(\inf_{\varepsilon>0}\sup_{v\in B(x,\varepsilon)\setminus\{x\}}\frac{f(x)-f(v)}{d(x,v)},0\right) = \limsup_{v\to x}\frac{f(x)-f(v)}{d(x,v)}.$$

We also note that $|\nabla|(f)(x) = 0$ when x is a local minimizer of f, but the converse does not hold, as the example of $f : \mathbb{R} \to \mathbb{R}$ given by $f(t) = -t^2$ shows for $x = 0$.

The notation we use (that is a slight variant of the original notation) evokes the following example and takes into account the fact that in general one cannot speak of the gradient of f or of the derivative of f. Here we use notions from Chap. 2.

Example–Exercise. If f and g are finite at x and tangent at x in the sense that $g(v) = f(v) + o(d(v,x))$, then $|\nabla|(g)(x) = |\nabla|(f)(x)$. Deduce from this property that if X is a normed space and if f is Fréchet differentiable at $x \in X$ then $|\nabla|(f)(x) = \|Df(x)\|$.

Remark. Let δ_f be a decrease index for f and let $\delta_f' : X \to \mathbb{R}_+$ be such that $\delta_f' \le \delta_f$. Then δ_f' is a decrease index for f.

This simple observation shows the versatility of the notion of decrease index.

Remark–Exercise. A function δ_f on X is a decrease index for f if and only if its lower semicontinuous hull $\overline{\delta_f}$ given by $\overline{\delta_f}(x) := \liminf_{v\to x}\delta_f(v)$ is a decrease index for f. [Hint: Use the fact that the infimum of the lower semicontinuous hull $\overline{\varphi}$ of a function φ on an open subset B of X coincides with the infimum of φ on B.]

Taking into account the preceding two remarks, the next result states in essence that the slope is somewhat the best decrease index: it is almost the largest one.

Proposition 1.116. *For every bounded-below lower semicontinuous function f on a complete metric space X, the slope of f is a decrease index. Moreover, for every decrease index δ_f for f, the lower semicontinuous hull $\overline{\delta_f}$ of δ_f satisfies $\overline{\delta_f} \le |\nabla|(f)$.*

This result is a consequence of a lemma that brings some additional information.

Lemma 1.117. *Let X be a complete metric space and let f be a lower semicontinuous function on the open ball $B := B(x,r)$. Suppose $\inf f(B) > -\infty$ and let $\beta := f(x) - \inf f(B)$. Then for all $t \in (0,1)$ there exists $u \in B[x,rt]$ such that $|\nabla|(f)(u) \le \beta/rt$, $f(u) \le f(x)$.*

Proof. We may suppose $\beta < +\infty$. We apply Theorem 1.88 to the restriction of f to the closed ball $B[x,rs]$, where $s \in (t,1)$. Then there exists $u \in B[x,rt]$ satisfying $f(u) \le f(x)$ that is a minimizer of $f(\cdot) + (\beta/rt)d(\cdot,u)$ on $B[x,rs]$, hence on $B[u,\sigma]$ for $\sigma := r(s-t)$, so that

$$|\nabla|(f)(u) = \inf_{\rho>0}\sup_{w\in B(u,\rho)\setminus\{u\}}\frac{(f(u)-f(w))^+}{d(u,w)} \le \sup_{w\in B(u,\sigma)\setminus\{u\}}\frac{(f(u)-f(w))^+}{d(u,w)} \le \frac{\beta}{rt}.$$

Proof of Proposition 1.116. Given $x \in X$, $r, c > 0$ such that $cr > \beta := f(x) - \inf f(B(x,r))$, one picks $t \in (0,1)$ such that $crt > \beta$. Then the lemma yields some $u \in B[x, rt] \subset B(x,r)$ such that $|\nabla|(f)(u) \le \beta/rt < c$, so that (1.27) is satisfied and $|\nabla|(f)$ is a decrease index.

Now let us prove that for every decrease index δ_f of f and $x \in X$ one has $\overline{\delta_f}(x) \le |\nabla|(f)(x)$. Let $c > |\nabla|(f)(x)$. Then for every $b \in (|\nabla|(f)(x), c)$, there exists some $s > 0$ such that

$$\sup_{u \in B(x,s) \setminus \{x\}} \frac{(f(x) - f(u))^+}{d(u,x)} < b;$$

hence for all $r \in (0,s)$ and all $u \in B(x,r)$, one has $f(x) \le f(u) + bd(u,x) \le f(u) + br$ and $f(x) \le \inf f(B(x,r)) + br < \inf f(B(x,r)) + cr$. Thus, by (1.27) there exists some $u \in B(x,r)$ such that $\delta_f(u) < c$. Thus $\overline{\delta_f}(x) = \sup_{r \in (0,s)} \inf_{u \in B(x,r)} \delta_f(u) \le c$. Since c is arbitrarily close to $|\nabla|(f)(x)$, we get $\overline{\delta_f}(x) \le |\nabla|(f)(x)$.   □

It will be useful to dispose of a parameterized version of the decrease principle. The novelty here lies in the (inward) continuous dependence of the solution set on the parameter w.

Theorem 1.118 (Parameterized decrease principle). *Let W be a topological space and let X be a complete metric space. Let $f : W \times X \to \overline{\mathbb{R}}_+$ and let $(\overline{w}, \overline{x}) \in S := \{(w,x) \in W \times X : f(w,x) = 0\}$. For each $w \in W$ let $\delta_w : X \to \overline{\mathbb{R}}_+$ be a decrease index for $f_w := f(w, \cdot)$ and let $S(w) := f_w^{-1}(\{0\})$. Suppose there exist $c, r > 0$ and a neighborhood U of $\overline{w}$ such that*

(a) $\delta_w(x) \ge c$ for all $(w,x) \in (U \times B(\overline{x}, r)) \setminus S$;
(b) The multimap $w \rightrightarrows \mathrm{epi} f_w$ is inward continuous at $(\overline{w}, (\overline{x}, 0))$;
(c) For all $w \in U$ the function f_w is such that f_w^{-1} is closed at 0.

Then the multimap $S(\cdot)$ is inward continuous at $(\overline{w}, \overline{x})$. Moreover, for all $s \in (0, r/2)$ there exists a neighborhood V of $\overline{w}$ such that for all $w \in V$ the set $S(w)$ is nonempty and

$$\forall x \in B(\overline{x}, s), \qquad d(x, S(w)) \le c^{-1} f(w,x). \tag{1.29}$$

Assumption (b) can be phrased in terms of upper epi-limits. It is rather mild: it is satisfied whenever $f(\cdot, \overline{x})$ is outward continuous at $\overline{w}$.

Proof. Let U, c, r be as in the assumptions. Assumption (b) means that $q_w := 2(c+1)d((\overline{x}, 0), \mathrm{epi} f_w)$ has limit 0 as $w \to 0$. Assumption (c) implies that $\overline{x} \in S(w)$ for all $w \in U \cap W_0$ with $W_0 := \{w \in W : q_w = 0\}$. Thus, to prove that $S(\cdot)$ is inward continuous at $(\overline{w}, \overline{x})$, it suffices to find some neighborhood $V \subset U$ of $\overline{w}$ such that $d(\overline{x}, S(w)) \le 2q_w$ for all $w \in V \cap W_+$, where $W_+ := W \setminus W_0$. Given $s \in (0, r/2)$, there is a neighborhood V of $\overline{w}$ such that $q_w < \min(s, r/2 - s)$ for all $w \in V$. Since $q_w > 0$ for $w \in W_+$, one can pick $x_w \in B(\overline{x}, s)$ and $r_w \in (0, q_w)$ such that $d(\overline{x}, x_w) < r_w$, $f(w, x_w) < cr_w$. For every $w \in V \cap W_+$ one has $B(x_w, s) \subset B(\overline{x}, r)$, so that assumption (a) ensures that one has $\delta_w(u) \ge c$ for all $u \in B(x_w, r_w) \setminus S(w)$. It follows from

Theorem 1.113, in which we replace f by f_w and set $x := x_w$, $\rho = s$, that $S(w)$ is nonempty and

$$d(x_w, S(w)) \leq c^{-1} f(w, x_w) < r_w.$$

That proves that $d(\bar{x}, S(w)) < 2r_w \leq 2q_w$ and that $S(\cdot)$ is inward continuous at $(\bar{w}, \bar{x})$.

Now let us prove (1.29). Assume, in view of a contradiction, that there exist $w \in V$, $x \in B(\bar{x}, s)$ such that $f(w, x) < cd(x, S(w))$. Setting $t := d(x, S(w)) > 0$, we have $t \leq d(x, \bar{x}) + d(\bar{x}, S(w)) < s + 2r_w$, hence $d(x, \bar{x}) + t < 2s + 2r_w \leq r$; thus $B(x, t) \subset B(\bar{x}, r)$ and $\delta_w(u) \geq c$ for all $u \in B(x, t) \setminus S(w)$. Since $f_w(x) < ct$, applying again Theorem 1.113, we get $d(x, S(w)) \leq c^{-1} f_w(x) < t$, a contradiction. □

Exercises

1. (a) Prove Lemma 1.109.
(b) Detect relationships between the best γ and μ of that lemma.

2. Prove the observation following Definition 1.115 above.

3. Deduce from Proposition 1.110 the following statement [53, Theorem 1.3]. Let (X, d) be a complete metric space, let $\alpha, \beta \in \mathbb{R}$, and let $f : X \to \mathbb{R}_\infty$ be lower semicontinuous such that $f^{-1}((-\infty, \beta]) \neq \varnothing$ and such that for all $x \in f^{-1}((\alpha, \beta))$ there exists $y \in f^{-1}([\alpha, +\infty)) \setminus \{x\}$ satisfying $f(y) + d(x, y) \leq f(x)$. Then $f^{-1}((-\infty, \alpha]) \neq \varnothing$ and $d(x, f^{-1}((-\infty, \alpha])) \leq (f(x) - \alpha)^+$ for all $x \in f^{-1}((-\infty, \beta))$. [Hint: Given $x \in f^{-1}((-\infty, \beta))$, pick $\gamma \in (f(x), \beta)$ and apply Proposition 1.110 to the function g given by $g(u) := (f(u) - \alpha)^+$ for $u \in f^{-1}((-\infty, \gamma])$, $g(u) := +\infty$ for $u \in f^{-1}((\gamma, +\infty))$.]

4. The definition of a decrease index does not lead to good calculus rules for sums or suprema. Prove, however, that if δ_f and δ_g are decrease indexes for f and g respectively, then $\inf(\delta_f, \delta_g)$ is a decrease index for $h := \min(f, g)$.

5. Let $h : X \to Y$ be a surjective map between two metric spaces such that for all $r > 0$, $y \in Y$ one has $B(y, r) = \bigcup_{x \in h^{-1}(y)} h(B(x, r))$. Let $g : Y \to \mathbb{R}_\infty$ and $f := g \circ h$.

(a) Show that if δ_f is a decrease index for f, then δ_g given by $\delta_g(y) := \inf\{\delta_f(x) : x \in h^{-1}(y)\}$ is a decrease index for g.
(b) Prove that the condition on h is satisfied whenever for every $x, x' \in X$, $y \in Y$ with $h(x) = h(x')$ one has $d(x, h^{-1}(y)) = d(x', h^{-1}(y))$. Show that the latter condition holds whenever Y is the quotient of X by the action of a group of isometries on X, in particular, when Y is the quotient of a normed space by a closed linear subspace.

6. Let δ_f be a decrease index for $f : X \to \mathbb{R}_\infty$. Check that every function $\delta'_f : X \to \overline{\mathbb{R}}_+$ minorizing δ_f is a decrease index for f. In particular, $\overline{\delta}_f$ given by $\overline{\delta}_f(x) := \liminf_{v \to x} \delta_f(v)$ is a decrease index for f.

7. Show by modifying the function $f_0 : \mathbb{R} \to \mathbb{R}$ given by $f_0(x) = -|x|$ in such a way that there exists a sequence of local minimizers of the new function f that converges to 0, that the slope of f is not necessarily a lower semicontinuous decrease index.

8. Deduce Corollaries 1.92 and 1.93 from Proposition 1.116. For this purpose, show that if f is Gâteaux differentiable at u, then $\|f'(u)\| \leq |\nabla|(f)(u)$, whereas if f is convex and continuous at u, then $\inf\{\|u^*\| : u^* \in \partial f(u)\} \leq |\nabla|(f)(u)$. [Hint: Use the Hahn–Banach theorem.]

9. Given a complete metric space (X,d) and a lower semicontinuous function $f : X \to \overline{\mathbb{R}}_+$, let $S = f^{-1}(0)$, let $\ell : X \to \overline{\mathbb{R}}$ be given by $\ell(x) := \sup\{(f(x) - f(w))/d(w,x) : w \in X \setminus \{x\}\}$, and $c(x) := \inf\{\ell(x') : x' \in [f \leq f(x)] \cap B(x, d(x,S))\}$. Use the Ekeland principle to show that $c(x)d(x,S) \leq f(x)$ for all $x \in X$ (with the convention $0.(+\infty) := 0$). (See [752].)

1.6.2 Supplement: A Palais–Smale Condition

Adding a convenient compactness condition to the use of minimization principles, one gets an existence theorem. This condition is a variant of the original Palais–Smale condition that is fulfilled in a number of concrete problems. It is as follows.

Definition 1.119. A function $f : X \to \mathbb{R}_\infty$ on a metric space X is said to satisfy the *Palais–Smale condition* at level $\ell \in \mathbb{R}$, denoted by $(PS)_\ell$, if every sequence (x_n) in X such that $(f(x_n)) \to \ell$, $(|\nabla|(f)(x_n)) \to 0$ has a convergent subsequence.

When X is a normed space and f is differentiable, the assumption $(|\nabla|(f)(x_n)) \to 0$ can be replaced with the assumption $(f'(x_n)) \to 0$; when f is convex, it can be replaced with the assumption $(d(0, \partial f(x_n))) \to 0$. This condition can also be given for functions on Riemannian or Finslerian manifolds. It can be used for critical values of f that are not necessarily the infimum of f.

Proposition 1.120. *Let $f : X \to \mathbb{R}_\infty$ be a bounded-below lower semicontinuous function on a complete metric space X and let $\ell := \inf f(X) \in \mathbb{R}$. If f satisfies the Palais–Smale condition (PS_ℓ) at level ℓ, then f attains its infimum.*

Proof. Let (w_n) be a minimizing sequence of f: setting $\varepsilon_n := f(w_n) - \ell$ one has $(\varepsilon_n) \to 0$. Taking $r = 2, t = 1/2, x = w_n$ in Lemma 1.117, we can find $u_n \in X$ such that $|\nabla|(f)(u_n) \leq \varepsilon_n, f(u_n) \leq f(w_n)$. Then $(f(u_n)) \to \ell$ and condition (PS_ℓ) yields a limit point u of (u_n). Since f is lower semicontinuous, we get $f(u) \leq \liminf f(u_n) = \ell$, hence $f(u) = \ell$. □

Exercise. Observe that instead of assuming that f is lower semicontinuous, one may suppose that $g := f - \ell$ is such that g^{-1} is closed at 0.

1.6.3 Penalization Methods

The question of dealing with constraints is a delicate one in tackling minimization problems (and with real-life problems, too!). Several possibilities exist to eliminate constraints. The first one is a theoretical means. It consists in replacing the objective function $f : X \to \mathbb{R}_{\infty} := \mathbb{R} \cup \{+\infty\}$ in the problem with constraint

$$(\mathscr{P}) \quad \text{minimize } f(x), \ x \in A,$$

where A is some *admissible* or *feasible set*, by the modified function $f_A := f + \iota_A$ given by

$$f_A(x) := f(x) \text{ if } x \in A, \qquad f_A := +\infty \text{ if } x \in X \setminus A,$$

that takes into account the admissible set A through the *indicator function* ι_A of A given by $\iota_A(x) := 0$ if $x \in A$, $\iota_A(x) := +\infty$ if $x \in X \setminus A$. Although this trick is of mathematical interest, it is not of great practical use, since the function f_A is wildly nonsmooth in general and difficult to handle (see the following chapters, however). When the admissible set is defined by implicit constraints such as

$$A := \{x \in X : \ g_e(x) = 0, \ g_i(x) \leq 0, \ e \in E, \ i \in I\}, \tag{1.30}$$

where I and E are finite sets and $g_j : X \to \mathbb{R}$ for $j \in I \cup E$, the use of multipliers has proved to be a tool of prominent importance since the work of Lagrange in the eighteenth century. Then the objective is changed into a combination of f and the constraint functions:

$$\ell_{\lambda,\mu} := f + \sum_{e \in E} \mu_e g_e + \sum_{i \in I} \lambda_i g_i,$$

where $\lambda_i \in \mathbb{R}_+$, $\mu_e \in \mathbb{R}$. More generally, when A is the inverse image under some map $g : X \to Z$ of some subset C (for instance a closed convex cone) of a normed space Z,

$$A := \{x \in X : g(x) \in C\}, \tag{1.31}$$

the function $\ell_y := f + y \circ g$, where $y \in Z^*$, known as the *Lagrangian function* of the problem, plays a key role both for optimality conditions and for algorithms.

A third idea (bearing connections with the other two devices) consists in introducing some penalty terms, replacing the objective f by the penalized objective

$$p_r := f + r \sum_{e \in E} |g_e| + r \sum_{i \in I} g_i^+,$$

where $t^+ := \max(t, 0)$ and $r \in \mathbb{R}_+$ is some well-chosen rate of penalization, or

$$p_r := f + r d_C(g(\cdot))$$

in the case $A = g^{-1}(C)$, where $d_C(z) := d(z,C) := \inf\{d(z,w) : w \in C\}$ for $z \in Z$. One may expect that by taking r large enough, the value of the infimum of this penalized objective on the whole space will converge to the value of the constrained problem $(\mathscr{P})$. In general, one has to replace r by an infinite sequence $(r_n) \to \infty$, so that one has to solve a sequence of unconstrained problems. In some favorable cases a single penalized problem suffices, provided the penalization constant r is large enough. A simple case is presented in the following result that is of frequent use in optimization theory and in nonsmooth analysis.

Proposition 1.121 (Penalization lemma). *Let A be a nonempty subset of a metric space X and let $f : X \to \mathbb{R}$ be a Lipschitzian function with rate r. Then for every $s \geq r$,*

$$\inf_{x \in A} f(x) = \inf_{x \in X} (f(x) + s d_A(x)). \tag{1.32}$$

Moreover, $\bar{x} \in A$ is a minimizer of f on A if and only if $\bar{x}$ is a minimizer of $f_s := f + s d_A$ on X. If A is closed and if $s > r$, every minimizer z of f_s belongs to A.

Replacing X by a neighborhood of $\bar{x}$, we get a similar statement in terms of local minimizers.

Proof. Since $f_s = f$ on A, we have $m := \inf f(A) \geq \inf f_s(X)$. If we had strict inequality we could find $x \in X$ such that $f_s(x) < m$. Then we would have $s d_A(x) < m - f(x)$, so that we could pick $x' \in A$ such that $s d(x,x') < m - f(x)$. Since f is Lipschitzian with rate $r \leq s$, we would get $f(x') \leq f(x) + s d(x,x') < m$, a contradiction. The second assertion follows immediately.

Suppose now that A is closed and that a minimizer z of f_s belongs to $X \setminus A$ for some $s > r$. Then $d_A(z)$ is positive, so that by the relations $\inf f(A) = \inf f_s(X) = f(z) + s d_A(z)$,

$$r d_A(z) < s d_A(z) = \inf f(A) - f(z),$$

one can find $a \in A$ such that $r d(a,z) < \inf f(A) - f(z)$, a contradiction to the relations $\inf f(A) = \inf f_r(X) \leq f(z) + r d(a,z)$. $\square$

When X is a normed space, we observe that the new function f_s is nonsmooth in general. This example is one of the main incentives to the study of nonsmooth analysis. We also note that when the admissible subset A is defined by equations or inequalities, the function d_A is not explicitly given and may be difficult to compute. For instance, if A is given by (1.30), it would be preferable to substitute for f_s the function

$$p_s : x \mapsto f(x) + s \sum_{e \in E} |g_e(x)| + s \sum_{i \in I} g_i(x)^+$$

mentioned above. This function is still nonsmooth, but it is explicitly determined, and its nonsmoothness is "reasonable" if the g_j's are smooth. More generally, if $A := g^{-1}(C)$, where $g : X \to Z$ is some mapping with values in a metric space Z and C is a closed nonempty subset of Z, as in (1.31), one would like to substitute for f_s the function

$$x \mapsto f(x) + s d_C(g(x))$$

for some appropriate constant $s > 0$. When C is simple enough, for instance when C is the negative (resp. positive) cone of some Euclidean space or some Banach lattice, d_C may be rather simple: under appropriate assumptions one has $d_C(z) = \|z^+\|$ (resp. $d_C(z) = \|z^-\|$), where $z^+ := \max(z,0)$, $z^- := (-z)^+$. Quite often z^+ is easy to compute (see the exercises below).

In the following section we will consider the case that the mapping g is metrically regular with respect to C around some point $\bar{x}$ in the sense that it satisfies an inequality of the form

$$d(x, g^{-1}(C)) \le c\, d(g(x), C)$$

on some neighborhood U of $\bar{x} \in X$. Such an inequality ensures that we can pass from $f + s d_A$ to $f + c s d(g(\cdot), C)$ as expected. Other reasons justify the interest of such an estimate (calculus of tangent and normal cones, optimality conditions, etc.).

Exercises

1. Characterize the norms on $\mathbb{R}^n$ for which $d_C(z) = \|z^+\|$, C being the negative orthant of $\mathbb{R}^n$ and $z^+ := (z_i^+)$ for $z := (z_i)$.

2. Show that if E and I are finite sets and if $C := \{0\} \times \mathbb{R}_-^I \subset Z := \mathbb{R}^E \times \mathbb{R}^I$, then for the usual norms on Z, one has

$$d_C(z) = \|(z_E, z_I^+)\| \quad \text{for } z = (z_E, z_I) \in \mathbb{R}^E \times \mathbb{R}^I,$$

so that $f + r d_C \circ g = p_r := f + r\sum_{i \in I} g_i^+ + r\sum_{e \in E} |g_e|$.

3. **(a)** Let T be a compact topological space and let C be the negative cone of $Z := C(T)$. Show that $d_C(z) = \|z^+\|$ with $z^+(t) := (z(t))^+$.
   (b) Prove the same result when T is a measure space and $Z = L_p(T)$ for some $p \in [1, \infty]$ endowed with its usual norm.

4. **(Penalization algorithms)** Consider the problem

$$\text{Minimize } f(x) \text{ subject to } x \in A,$$

where $f : X \to \mathbb{R}$ is lower semicontinuous and A is a closed subset of X. Suppose there exists a continuous function $q : X \to \mathbb{R}_+$ such that $A = q^{-1}(0)$. Given an increasing sequence (r_n) of positive numbers, let x_n be a minimizer of the penalized function $p_n := f + r_n q$. Show that $p_n(x_n) \le p_{n+1}(x_{n+1})$, $q(x_{n+1}) \le q(x_n)$, $f(x_n) \le f(x_{n+1})$ and that $f(x_n) \le p_n(x_n) \le \inf f(A)$ for all n. Prove that every cluster point of the sequence (x_n) is a solution of the given problem whenever $(r_n) \to \infty$.

5. **(Debreu's lemma)** Let $A : X \to Y$ be a linear map between two Euclidean spaces and let $Q : X \to X^*$ be such that $\langle Qx, x \rangle > 0$ for all $x \in \text{Ker}A \setminus \{0\}$. Prove that there exists $c > 0$ such that $\langle Qx, x \rangle + c\|Ax\|^2 > 0$ for all $x \in X \setminus \{0\}$.

1.6.4 Robust and Stabilized Infima

When one minimizes a composite function of the form $h \circ g$, where $g : X \to Y$ is a map between two metric spaces X, Y and $h : Y \to \mathbb{R}_\infty$, one has to admit that some inaccuracies may occur in computing the value of h at $g(x)$. Similarly, if Z is a subset of Y that is not easy to determine, the computation of $\inf h(Z)$ may be eased by replacing the inclusion $y \in Z$ by the inclusion $y \in Z_\delta := \{y \in Y : d(y,Z) \le \delta\}$ or $y \in B(Z,\delta) := \{y \in Y : d(y,Z) < \delta\}$ for some small $\delta > 0$. Thus, one is led to the following concept, in which $f : X \to \mathbb{R}$ and h, g are as above.

Definition 1.122. The *stabilized infimum* of $h : Y \to \overline{\mathbb{R}}$ on the subset Z of Y is

$$\wedge_Z h := \lim_{\delta \to 0+} \inf h(Z_\delta) = \sup_{\delta > 0} \inf \{h(y) : y \in Y, \ d(y,Z) < \delta\}.$$

The *stabilized infimum* of the composed function $h \circ g$, with g and h as above, is

$$\wedge_g h := \wedge_{g(X)} h := \sup_{\delta > 0} \inf \{h(y) : d(y, g(X)) < \delta\}.$$

The *stabilized infimum* of the sum $f_1 + \cdots + f_k$ of a finite family $(f_1, \dots, f_k)$ of functions on X is

$$\wedge(f_1, \dots, f_k) := \sup_{\delta > 0} \inf \{f_1(x_1) + \cdots + f_k(x_k) : \operatorname{diam}(x_1, \dots, x_k) < \delta\}.$$

The *stabilized infimum* of the mixed function $f + h \circ g$ is

$$\wedge_g(f, h) := \sup_{\delta > 0} \inf \{f(x) + h(y) : w, x \in X, \ y \in Y, \ d(w,x) < \delta, \ d(y, g(w)) < \delta\}.$$

The infimum $\inf h(Z)$ will be called *robust* if it is equal to the stabilized infimum $\wedge_Z h$. Then a minimizer of h on Z will be called a *robust minimizer*. A similar terminology will be used in the other cases. The concept for sums can be given for every operation on a family of k functions.

Clearly, one has the inequalities

$$\wedge_Z h \le \inf h(Z), \qquad \wedge_g h \le \inf h(g(X)), \qquad \wedge(f_1, \dots, f_k) \le \inf_X(f_1 + \cdots + f_k).$$

These inequalities may be strict, as simple examples show.

Example. Let $X := \mathbb{R}$, $Y := \mathbb{R}^2$, h, g being given by $h(r,s) = rs$, $g(x) = (x,0)$. Then $\wedge_g h = -\infty$, while $\inf h \circ g = 0$.

We shall soon present criteria for equality. Before that, let us relate the different concepts of the preceding definition. What follows shows that the last concept encompasses the other ones but can be reduced to each of them. Of course,

$\wedge_g h = \wedge_g(f,h)$ with $f := 0$. Moreover, as is easily seen, $\wedge_g(f,h) = \wedge(f \circ p_X, h \circ p_Y, \iota_G)$, where $p_X : X \times Y \to X$, $p_Y : X \times Y \to Y$ are the canonical projections and ι_G is the indicator function of the graph $G \subset X \times Y$ of g. In particular, $\wedge_g h = \wedge(h \circ p_Y, \iota_G) = \wedge_G j$, with $j := h \circ p_Y$. Also, $\wedge_g(f,h) = \wedge_{(I_X,g)}(f \times h)$.

We have already observed that $\wedge_g h := \wedge_{g(X)} h$. Conversely, if Z is a subset of Y, denoting by $g : Z \to Y$ the canonical injection, we have $\wedge_Z h = \wedge_g h$. Also, we obviously have $\wedge_Z h = \wedge(h, \iota_Z)$.

Given a family $(f_1, \ldots, f_k)$ of functions on X, denoting by $g : X \to Y := X^k$ the diagonal map defined by $g(x) := (x, \ldots, x)$, by $h : Y \to \mathbb{R}_\infty$ the function given by $h(x_1, \ldots, x_k) := f_1(x_1) + \cdots + f_k(x_k)$, setting $Z := \Delta := g(X)$, and endowing X^k with the metric $d := d_\infty$, for all $y := (x_1, \ldots, x_k) \in X^k$ let us check the inequalities

$$\frac{1}{2}\mathrm{diam}(x_1, \ldots, x_k) \le d(y, g(X)) \le \mathrm{diam}(x_1, \ldots, x_k) := \max_{1 \le i, j \le k} d(x_i, x_j).$$

The second one is obvious, and for all $r > d(y, g(X))$ there exists some $x \in X$ such that $d(x, x_i) < r$ for $i \in \mathbb{N}_k := \{1, \ldots, k\}$, so that $d(x_i, x_j) < 2r$ for all $i, j \in \mathbb{N}_k$. Thus,

$$\wedge(f_1, \ldots, f_k) = \wedge_g h = \wedge_Z h. \tag{1.33}$$

When Z is a singleton $\{z\}$, $\wedge_Z h$ is the value $\overline{h}(z) := \liminf_{y \to z} h(y)$ at z of the lower semicontinuous hull $\overline{h}$ of h. This simple observation leads us to consider criteria involving semicontinuity concepts.

In the following statement we say that $h : Y \to \mathbb{R}_\infty$ is *uniformly lower semicontinuous around a subset* Z of Y if for every $\varepsilon > 0$, there exists some $\delta > 0$ such that for all $y \in Y \setminus Z$, $z \in Z$ satisfying $d(y,z) < \delta$ one has $h(y) > h(z) \widehat{-} \varepsilon$, where $r \widehat{-} \varepsilon := r - \varepsilon$ for $r \in \mathbb{R}$, $r \widehat{-} \varepsilon := 1/\varepsilon$ for $r = +\infty$. In the sequel, we simply write $r - \varepsilon$.

Lemma 1.123. (a) *If h is uniformly lower semicontinuous around Z, then $\wedge_Z h = \inf h(Z)$.*
(b) *If h is lower semicontinuous at each point of Z and if Z is compact, then $\wedge_Z h = \inf h(Z)$.*
(c) *If $(f_1, \ldots, f_k)$ is a finite family of lower semicontinuous functions that are bounded below and if f_1 is inf-compact in the sense that its sublevel sets are compact, then $\wedge(f_1, \ldots, f_k) = \inf_{x \in X}(f_1(x) + \cdots + f_k(x))$.*

Proof. (a) The relation $\wedge_Z h = \inf h(Z)$ is obvious if $\wedge_Z h = +\infty$. Thus, it suffices to prove that for every $s > r > \wedge_Z h$ one has $s > \inf h(Z)$. Let $\varepsilon \in (0, s - r)$ with $1/\varepsilon > r$ and let $\delta > 0$ be such that $h(y) > h(z) - \varepsilon$ whenever $y \in Y \setminus Z$, $z \in Z \cap B(y, \delta)$. By definition of $\wedge_Z h$ one has $\inf h(Z_\delta) < r$, so that there exist some $y \in Y$, $z \in Z$ satisfying $h(y) < r$, $d(y,z) < \delta$. If $y \in Z$, one has $\inf h(Z) \le h(y) < s$. If $y \in Y \setminus Z$, one cannot have $h(z) = +\infty$, since $h(y) < r < 1/\varepsilon$; thus $h(z) < h(y) + \varepsilon < h(y) + s - r < s$, whence $\inf h(Z) < s$.

(b) We may suppose $\inf h(Z) > -\infty$. Let us show that for all $r < \inf h(Z)$ we can find some $\delta > 0$ such that $r \le \inf h(Z_\delta)$. For all $z \in Z$ there exists an open neighborhood V_z of z in Y such that $h(y) > r$ for all $y \in V_z$. Since the union V of the family $(V_z)_{z \in Z}$ is an open neighborhood of Z, one can find some $\delta > 0$ such that $Z_\delta \subset V$. It is the required δ.

(c) We may suppose $\wedge(f_1, \ldots, f_k) < +\infty$, the result being obvious otherwise. For $i \in \mathbb{N}_k$ let $(x_{i,n})_n$ be a sequence of X such that $(\mathrm{diam}(x_{1,n}, \ldots, x_{k,n})) \to 0$ and $(f_1(x_{1,n}) + \cdots + f_k(x_{k,n})) \to \wedge(f_1, \ldots, f_k)$. Since $f_2, \ldots, f_k$ are bounded below, $(x_{1,n})_n$ is contained in some sublevel set of f_1. This sequence has a convergent subsequence $(x_{1,p(n)})_n$. The sequences $(x_{i,p(n)})_n$ have the same limit $\bar{x}$. By lower semicontinuity, we get $f_1(\bar{x}) + \cdots + f_k(\bar{x}) = \wedge(f_1, \ldots, f_k)$. $\qquad \square$

Given a family $(f_1, \ldots, f_k)$ of functions on X with values in $\mathbb{R}_\infty$, the equality $\wedge(f_1, \ldots, f_k) = \inf_X (f_1 + \cdots + f_k)$ means that the behaviors of the functions $f_1, \ldots, f_k$ are not too antagonistic, at least for what concerns minimization. In the next obvious lemma we consider a collective behavior that is not bound to minimization.

Lemma 1.124. *Let $(f_1, \ldots, f_k)$ be a* (lower) coherent *family of functions on X in the sense that for every sequences $(x_{1,n})_n, \ldots, (x_{k,n})_n$ satisfying $(d(x_{i,n}, x_{j,n}))_n \to 0$ for $i, j \in \mathbb{N}_k$ there exist sequences $(\varepsilon_n)_n \to 0$, $(x_n)_n$ such that $(d(x_n, x_{i,n}))_n \to 0$ for $i \in \mathbb{N}_k$ and for all $n \in \mathbb{N}$,*

$$f_1(x_n) + \cdots + f_k(x_n) - \varepsilon_n \le f_1(x_{1,n}) + \cdots + f_k(x_{k,n}). \qquad (1.34)$$

Then $\wedge(f_1, \ldots, f_k) = \inf\{f_1(x) + \cdots + f_k(x) : x \in X\}$.

This equality also holds if $(f_1, \ldots, f_k)$ is quasicoherent *in the sense that for all sequences $(x_{1,n})_n, \ldots, (x_{k,n})_n$ satisfying $(\mathrm{diam}\{x_{i,n} : i \in \mathbb{N}_k\})_n \to 0$ there exist an infinite subset N of $\mathbb{N}$ and sequences $(\varepsilon_n) \to 0$, (x_n) such that (1.34) holds for all $n \in N$.*

Proof. Let $f := f_1 + \cdots + f_k$ and let $h : X^k \to \mathbb{R}_\infty$ be defined for $y := (y_1, \ldots, y_k)$ by $h(y) := f_1(y_1) + \cdots + f_k(y_k)$ as in relation (1.33). Let Δ be the diagonal of X^k. We may suppose $\wedge(f_1, \ldots, f_k) < +\infty$, so that $\inf h(B(\Delta, \delta)) < +\infty$ for all $\delta > 0$. Let us first suppose there exists some $\delta > 0$ such that $\inf h(B(\Delta, \delta)) > -\infty$. Given a sequence $(\delta_n) \to 0_+$ in $(0, \delta)$, let $y_n \in B(\Delta, \delta_n)$ be such that $h(y_n) < \inf h(B(\Delta, \delta_n)) + \delta_n$. The quasicoherence condition yields sequences (x_n), $(\varepsilon_n) \to 0_+$ and an infinite subset N of $\mathbb{N}$ such that $h(y_n) \ge f(x_n) - \varepsilon_n$ for all $n \in N$. It follows that $\wedge(f_1, \ldots, f_k) = \sup_{n \in N} \inf h(B(\Delta, \delta_n)) \ge \inf_{n \in N}(f(x_n) - \varepsilon_n - \delta_n) \ge \inf f(X)$. When $\inf h(B(\Delta, \delta)) = -\infty$ for all $\delta > 0$, given a sequence $(\delta_n) \to 0_+$, we pick $y_n \in B(\Delta, \delta_n)$ such that $h(y_n) < -n$. Using the quasicoherence condition as above, we get sequences (x_n), $(\varepsilon_n) \to 0_+$ and an infinite subset N of $\mathbb{N}$ such that $h(y_n) \ge f(x_n) - \varepsilon_n$ for all $n \in N$. Then we get $\inf f(X) \le \inf_n f(x_n) = -\infty = \wedge(f_1, \ldots, f_k)$, and equality holds again. $\qquad \square$

Coherence carries a localization condition for the sequence (x_n) of (1.34), since it requires $(d(x_n,x_{i,n})) \to 0$ for $i \in \mathbb{N}_k$; this requirement is not used in the preceding proof but will play a role later on.

Lemma 1.125. *Let $(f_1,\ldots,f_k)$ be a coherent family of functions on X and let f_{k+1} be uniformly continuous. Then $(f_1,\ldots,f_{k+1})$ is a coherent family.*

Proof. Given sequences $(x_{1,n}),\ldots,(x_{k+1,n})$ satisfying $(d(x_{i,n},x_{j,n})) \to 0$ for $i,j \in \mathbb{N}_{k+1}$, picking $(\varepsilon_n) \to 0$, (x_n) such that $(d(x_n,x_{i,n})) \to 0$ for $i \in \mathbb{N}_k$ and (1.34) holds, we note that $d(x_n,x_{k+1,n}) \le d(x_n,x_{k,n}) + d(x_{k,n},x_{k+1,n}) \to 0$, so that $(f_{k+1}(x_{k+1,n}) - f_{k+1}(x_n)) \to 0$, and relation (1.34) can be extended to $(f_1,\ldots,f_{k+1})$. $\square$

The preceding lemmas and an induction entail the following proposition.

Proposition 1.126. *If $(f_0,\ldots,f_k)$ is a finite family of functions on X that are uniformly continuous except for f_0, then $(f_0,\ldots,f_k)$ is coherent.*

The notion of coherence can be localized and then characterized. We say that the family $(f_1,\ldots,f_k)$ is *coherent around some* $\bar{x} \in X$ if there exists some neighborhood V of $\bar{x}$ such that the restriction of $(f_1,\ldots,f_k)$ to V is coherent.

Proposition 1.127 [526, Prop. 2.3]. *A family $(f_1,\ldots,f_k)$ of lower semicontinuous functions with values in $\mathbb{R}_\infty$ is coherent around some $\bar{x} \in X$ at which these functions are finite if and only if there exist a neighborhood V of $\bar{x}$ and a modulus μ such that for all $x \in V$ and $(t_1,\ldots,t_k) \in \mathbb{R}^k$, for $f := f_1 + \cdots + f_k$, one has*

$$d((x,t_1 + \cdots + t_k),\mathrm{epi}\,f) \le \mu\left(d((x,t_1),\mathrm{epi}\,f_1) + \cdots + d((x,t_k),\mathrm{epi}\,f_k)\right). \quad (1.35)$$

Proof. Suppose that for some neighborhood V of $\bar{x}$ and some modulus μ, (1.35) holds for every $x \in V$, $t_i \in \mathbb{R}$. Let $(x_{1,n}),\ldots,(x_{k,n})$ be sequences of V satisfying $(d(x_{i,n},x_{j,n})) \to 0$ for $i,j \in \mathbb{N}_k$. Let $M := \{n \in \mathbb{N} : \exists i \in \mathbb{N}_k\ f_i(x_{i,n}) = +\infty\}$. For any sequence $(\varepsilon_n) \to 0_+$ and for every $n \in M$ we can take $x_n = \bar{x}$ in (1.34). Thus, dropping M, we may assume that $t_{i,n} := f_i(x_{i,n}) < +\infty$ for all i,n. Let $w_n := x_{1,n}$. Then $(d(w_n,x_{i,n})) \to 0$ and $d((w_n,t_{i,n}),\mathrm{epi}\,f_i) \le d(w_n,x_{i,n})$ for $i \in \mathbb{N}_k$. Thus, relation (1.35) shows that $(d((w_n,t_{1,n} + \cdots + t_{k,n}),\mathrm{epi}\,f)) \to 0$. Thus, there exists some $(x_n,t_n) \in \mathrm{epi}\,f$ such that $(d(x_n,w_n)) \to 0$, $(\varepsilon_n) := (|t_{1,n} + \cdots + t_{k,n} - t_n|) \to 0$. Then

$$f_1(x_{1,n}) + \cdots + f_k(x_{k,n}) = t_{1,n} + \cdots + t_{k,n} \ge t_n - \varepsilon_n \ge f(x_n) - \varepsilon_n,$$

and the family $(f_1,\ldots,f_k)$ is coherent on V.

Conversely, let $(f_1,\ldots,f_k)$ be a family that is coherent on some neighborhood V of $\bar{x}$. Let $r > 0$ be such that $B(\bar{x},2r) \subset V$. For $s \in \mathbb{P}$ let

$$\mu(s) := \sup\left\{d\left(\left(x,\sum_{i=1}^{k} t_i\right),\mathrm{epi}\,f\right) : x \in B(\bar{x},r),\ t_1,\ldots,t_k \in \mathbb{R},\ \sum_{i=1}^{k} d((x,t_i),\mathrm{epi}\,f_i) \le s\right\}$$

and $\mu(0) = 0$. Since μ is nondecreasing, it remains to check that μ is continuous at 0. Let $(s_n) \to 0$ in $(0, r)$ with $\mu(s_n) > 0$ for all n and let $u_n \in B(\bar{x}, r)$, $t_{1,n}, \dots t_{k,n} \in \mathbb{R}$ satisfying $d((u_n, t_{1,n}), \operatorname{epi} f_1) + \cdots + d((u_n, t_{k,n}), \operatorname{epi} f_k) \le s_n$ and

$$d((u_n, t_{1,n} + \cdots + t_{k,n}), \operatorname{epi} f) \ge q_n := \min(n, \mu(s_n)/2)$$

(q_n is chosen to take into account the case $\mu(s_n) = +\infty$). Then for $i \in \mathbb{N}_k$, there exist $(x_{i,n}, t'_{i,n}) \in \operatorname{epi} f_i$ such that $d(x_{i,n}, u_n) < s_n$, $|t'_{i,n} - t_{i,n}| < s_n$. Since $x_{i,n} \in B(\bar{x}, 2r) \subset V$ and $d(x_{i,n}, x_{j,n}) \le 2s_n$, the coherence of the family $(f_1, \dots, f_k)$ yields some sequences $(\varepsilon_n) \to 0$, (x_n) in $B(\bar{x}, r)$ such that $(d(x_n, x_{i,n})) \to 0$ for $i \in \mathbb{N}_k$ and

$$f_1(x_{1,n}) + \cdots + f_k(x_{k,n}) \ge f_1(x_n) + \cdots + f_k(x_n) - \varepsilon_n.$$

Since $t_{i,n} > t'_{i,n} - s_n \ge f_i(x_{i,n}) - s_n$, we get

$$t_{1,n} + \cdots + t_{k,n} \ge f(x_n) - \varepsilon_n - k s_n.$$

Since $d(u_n, x_n) \le d(u_n, x_{i,n}) + d(x_{i,n}, x_n)$, which has limit 0, our choice of u_n, $t_{i,n}$ yields

$$q_n \le d((u_n, t_{1,n} + \cdots + t_{k,n}), \operatorname{epi} f) \le d(u_n, x_n) + \left(f(x_n) - (t_{1,n} + \cdots + t_{k,n})\right)^+$$
$$\le d(u_n, x_n) + \varepsilon_n + k s_n.$$

Thus $q_n = \mu(s_n)/2$ for n large, $(\mu(s_n)) \to 0$, and μ is a modulus. $\square$

The preceding characterization incites to introduce a case of particular interest.

Definition 1.128 (Ioffe). A family $(f_1, \dots, f_k)$ of lower semicontinuous functions on X with sum f is said to be *linearly coherent* around some $\bar{x} \in X$, or to satisfy the linear metric qualification condition around $\bar{x}$, if there exist $c > 0$, $\rho > 0$ such that for all $x \in B(\bar{x}, \rho)$, $(t_1, \dots, t_k) \in \mathbb{R}^k$ relation (1.35) holds with $\mu(r) := cr$ for $r \in \mathbb{R}_+$, i.e., one has

$$d((x, t_1 + \cdots + t_k), \operatorname{epi} f) \le cd((x, t_1), \operatorname{epi} f_1) + \cdots + cd((x, t_k), \operatorname{epi} f_k). \qquad (1.36)$$

An analogue of Proposition 1.126 can be given.

Proposition 1.129. (a) *Every family $(f_1, \dots, f_k)$ of lower semicontinuous functions on X all but one of which are locally Lipschitzian around $\bar{x}$ is linearly coherent.*

(b) *If $(f_1, \dots, f_k)$ is a family of lower semicontinuous functions that is linearly coherent around $\bar{x} \in X$ and if f_{k+1} is Lipschitzian around $\bar{x}$, then $(f_1, \dots, f_{k+1})$ is linearly coherent around $\bar{x}$.*

Proof. It suffices to prove that if f is lower semicontinuous and if g is Lipschitzian around $\bar{x}$, then (f, g) is linearly coherent around $\bar{x}$. Then assertions (a) and (b) follow by induction on k. Since (1.36) is preserved when one changes the norm in $X \times \mathbb{R}$ to an equivalent one, we may suppose the Lipschitz rate of g is 1 on some ball $B(\bar{x}, \rho)$. Let F, G, H be the epigraphs of f, g, and $h := f + g$ respectively and let $x \in B(\bar{x}, \rho)$, $s, t \in \mathbb{R}$. Given $\varepsilon > 0$, let $(u, r) \in F$ satisfying $\|u - x\| + |r - s| \le d_F(x, s) + \varepsilon$. When $g(x) \ge t$, we have $|g(x) - t| = (g(x) - t)^+ = d_G(x, t)$, and since $(u, r + g(u)) \in H$ and

$$\|u - x\| + |(r + g(u)) - (s + t)| \le \|u - x\| + |r - s| + |g(u) - g(x)| + |g(x) - t|$$
$$\le 2\|u - x\| + |r - s| + |g(x) - t|,$$

we get $d_H(x, s + t) \le 2d_F(x, s) + 2\varepsilon + d_G(x, t)$. When $g(x) < t$ we have $(u, r + g(u) + t - g(x)) \in H$ and

$$\|u - x\| + |(r + g(u) + t - g(x)) - (s + t)| \le \|u - x\| + |r - s| + |g(u) - g(x)|$$
$$\le 2\|u - x\| + |r - s| \le 2d_F(x, s) + 2\varepsilon.$$

Thus, in both cases we have $d_H(x, s + t) \le 2d_F(x, s) + 2\varepsilon + d_G(x, t)$. Since $\varepsilon > 0$ is arbitrary, we get $d_H(x, s + t) \le 2d_F(x, s) + d_G(x, t)$. $\square$

1.6.5 Links Between Penalization and Robust Infima

Now let us point out the links of the preceding concepts with penalization. This can be done for each of the various cases of stabilized infimum. In view of the passages described above, we limit our study to the case of a composition $h \circ g$, where $g : X \to Y$ and $h : Y \to \mathbb{R}_\infty$. In order to get some flexibility, we make use of a function $k : Y \times Y \to \overline{\mathbb{R}}_+ := [0, +\infty]$ such that

$$k(y, y') \to 0 \iff d(y, y') \to 0. \qquad (1.37)$$

We call such a function a *forcing bifunction*. For instance, one may choose $k := d^p$ with $p > 0$ or, more generally, $k := \mu \circ d$, where $\mu : \mathbb{R}_+ \to \overline{\mathbb{R}}_+$ is continuous at 0 with $\mu(0) = 0$ and *firm* (i.e., $(t_n) \to 0$ whenever $(\mu(t_n)) \to 0$). Given $c > 0$, we define the *penalized infimum* of $h \circ g$ by

$$m_c := \inf\{h(y) + ck(g(x), y) : (x, y) \in X \times Y\}, \qquad m := \sup_{c > 0} m_c.$$

Then $c \mapsto m_c$ is clearly nondecreasing. One may have $m_c = -\infty$ for all $c > 0$ while $\wedge_g h$ is finite. This fact occurs for $X := \{0\} \subset Y := \mathbb{R}$, $g(0) := 0$, $h(y) := -y^2$, $k(y, y') := |y - y'|$; note that it does not occur when k is given by $k(y, y') = (y - y')^2$, so that it is of interest to choose k appropriately. When $m_c > -\infty$ for at least one $c > 0$, one has a remarkable relationship between $m := \sup_{c > 0} m_c$ and $\wedge_g h$. It shows that m does not depend on the choice of k among those ensuring $m > -\infty$.

Proposition 1.130. *One always has $m \leq \wedge_g h$. If $m > -\infty$, equality holds.*

Proof. Let us first prove that for all $c > 0$ we have $m_c \leq \wedge_g h$. We may suppose that $\wedge_g h < +\infty$. Let $s > r > \wedge_g h$. By definition of $\wedge_g h$, for all $\delta > 0$, there exist some $x \in X$, $y \in Y$ satisfying $d(g(x), y) < \delta$ and $h(y) < r$. Taking $\delta > 0$ such that $k(y', y'') < \eta := c^{-1}(s-r)$ when $d(y', y'') < \delta$, we get some $(x, y) \in X \times Y$ such that $h(y) + ck(y, g(x)) < r + c\eta = s$, hence $m_c < s$ and, s being arbitrarily close to $\wedge_g h$, $m_c \leq \wedge_g h$. Thus $m := \sup_{c>0} m_c \leq \wedge_g h$.

Now let us show that $\wedge_g h \leq m$ when $m > -\infty$. We may suppose that $m < +\infty$. Let $b > 0$ be such that $m_b > -\infty$ and let $r > m$, $\delta > 0$ be given. Let $\alpha > 0$ be such that $d(y, y') < \delta$ whenever $y, y' \in Y$ satisfy $k(y, y') < \alpha$. Now we pick $c > b$ large enough that $(c-b)\alpha \geq r - m_b$. Since $r > m \geq m_c$, we can find $(x, y) \in X \times Y$ such that $h(y) + ck(g(x), y) < r$. Since $m_b \leq h(y) + bk(g(x), y)$, we have $(c-b)k(g(x), y) < r - m_b$, hence $k(g(x), y) < \alpha$ and $d(g(x), y) < \delta$. Thus

$$\inf\{h(y') : y' \in Y, \, d(y', g(X)) < \delta\} \leq h(y) < r.$$

Taking the supremum over $\delta > 0$, we get $\wedge_g h \leq r$, hence $\wedge_g h \leq m$. $\qquad\square$

A similar result holds for a sum. We leave the proof as an exercise. This time, given a family $(f_1, \ldots, f_k)$ of functions on X and a forcing bifunction $k_X : X \times X \to \overline{\mathbb{R}}_+$, we set $m := \sup_{c>0} m_c$ with

$$m_c := \inf\left\{ f_1(x_1) + \cdots + f_k(x_k) + c \sum_{i,j=1}^{k} k_X(x_i, x_j) : (x_1, \ldots, x_k) \in X^k \right\}.$$

Proposition 1.131. *One always has $m \leq \wedge(f_1, \ldots, f_k)$. If the functions f_i are bounded below, or, more generally, if $m > -\infty$, equality holds.*

Penalization methods are not limited to convergence of values. They also bear on convergence of approximate minimizers, as we are going to show for the minimization of a sum of functions and then for the minimization of a composite function. For the sake of simplicity, we slightly change the notation of Proposition 1.131, considering two functions $f, g : X \to \mathbb{R}_\infty$ and setting, for $c \in \mathbb{R}_+$,

$$p_c(x, y) := f(x) + g(y) + ck_X(x, y).$$

Proposition 1.132. *Let f, g be bounded below, or more generally, let them be such that $m_b := \inf p_b(X \times X) > -\infty$ for some $b > 0$. Suppose $\wedge(f, g) < \infty$. Then given a sequence $(\varepsilon_n) \to 0_+$, if (x_n, y_n) is an ε_n-minimizer of p_n, one has $(d(x_n, y_n)) \to 0$ as $n \to +\infty$.*

Proof. Since k_X is a forcing function, the result stems from the inequalities

$$\wedge(f, g) + \varepsilon_n \geq \inf p_n + \varepsilon_n \geq f(x_n) + g(y_n) + nk_X(x_n, y_n) \geq m_b + (n-b)k_X(x_n, y_n).$$
$$\square$$

Similar results hold for the minimization of the function $f + h \circ g$, where $f : X \to \mathbb{R}_\infty$, $g : X \to Y$ is continuous, $h : Y \to \mathbb{R}_\infty$, X, Y being two metric spaces. Moreover, such results can be localized. Given forcing bifunctions k_X and k_Y on X and Y respectively and a robust minimizer $\bar{x}$ of $f + h \circ g$, let us set, for $t \in \mathbb{R}_+$,

$$p_t(x,y) := f(x) + h(y) + t k_Y(g(x), y) + k_X(x, \bar{x}), \qquad (x,y) \in X \times Y.$$

Theorem 1.133. *Let f, h, g, p_t be as above and let $\bar{x}$ be a robust minimizer of $f + h \circ g$ with $(f + h \circ g)(\bar{x})$ finite. Suppose g is continuous at $\bar{x}$ and, for some $b > 0$, $m_b := \inf p_b(X \times Y) > -\infty$. Given a sequence $(\varepsilon_n) \to 0_+$, every sequence $((x_n, y_n))_n$ such that (x_n, y_n) is an ε_n-minimizer of p_n converges to $(\bar{x}, \bar{y}) := (\bar{x}, g(\bar{x}))$.*

Proof. Let $m := f(\bar{x}) + h(\bar{y})$. Since (x_n, y_n) is an ε_n-minimizer of p_n, we have

$$m + \varepsilon_n \geq m_n + \varepsilon_n \geq p_n(x_n, y_n) \geq m_b + (n - b) k_Y(g(x_n), y_n),$$

so that $(r_n) := (d(g(x_n), y_n)) \to 0$. For $r \in \mathbb{R}_+$, let

$$\mu(r) := \inf\{f(x) + h(y) : d(g(x), y) \leq r\}.$$

Since $\bar{x}$ is a robust minimizer, we have $(\mu(r_n)) \to m := f(\bar{x}) + h(\bar{y})$,

$$m + \varepsilon_n \geq m_n + \varepsilon_n \geq p_n(x_n, y_n) \geq f(x_n) + h(y_n) + k_X(x_n, \bar{x}) \geq \mu(r_n) + k_X(x_n, \bar{x}).$$

Thus $(k_X(x_n, \bar{x})) \to 0$, so that $(x_n) \to \bar{x}$, $(g(x_n)) \to g(\bar{x})$ and $(y_n) \to g(\bar{x})$. $\qquad \square$

Exercises

1. Given $h : Y \to \mathbb{R}_\infty$ and $Z \subset Y$, show that $\wedge_Z h = \inf h(Z)$ if for every $\varepsilon > 0$, there exists some $\delta > 0$ such that for all $y \in Y$ satisfying $d(y, Z) < \delta$, there exists some $z \in Z$ satisfying $h(z) < h(y) + \varepsilon$.

2. Using the preceding exercise, show that if Z is compact and $h : Y \to \mathbb{R}_\infty$ is lower semicontinuous and finite at each point of Z, then $\wedge_Z h = \inf h(Z)$. [Hint: Given $\varepsilon > 0$, for $z \in Z$ let $O_z := \{y \in Y : h(y) > h(z) - \varepsilon\}$; then using the Lebesgue lemma, find $\delta > 0$ such that for every $w \in Z$ there exists some $z \in Z$ such that $B(w, \delta) \subset O_z$.]

3. Show that if Z is compact, if h is lower semicontinuous, and if the restriction of h to Z is finite and continuous, then h is uniformly lower semicontinuous around Z. [Hint: Use the Lebesgue lemma as in the preceding exercise and use the fact that the restriction of h to Z is uniformly continuous.]

4. Prove Proposition 1.131 or deduce it from Proposition 1.130.

5. Give a direct proof of the fact that if $\{f_1, \ldots, f_{k-1}\}$ is a finite family of functions on X that are uniformly continuous, then for every lower semicontinuous function f_k, the family $(f_1, \ldots, f_k)$ is coherent, so that $\wedge(f_1, \ldots, f_k) = \inf\{f_1(x) + \cdots + f_k(x) : x \in X\}$.

6. Show that if μ is a modulus, then ν given by $\nu(s) := \sup\{\mu(r) : r \in [0, s]\}$ is a nondecreasing modulus.

7. Show that if $f := h \circ g$ and if $\wedge_g h$ is finite, then there exists a modulus μ such that for $k := \mu \circ d$, one has $m := \sup_{c>0} m_c = \wedge_g h$. [Hint: Take $\mu(r) = r$ for $r \in [0, \delta]$, $\mu(r) := +\infty$ for $r > \delta$, where $\delta > 0$ is such that $\inf\{h(y) : d(y, g(X)) < \delta\} > -\infty$.]

1.6.6 Metric Regularity, Lipschitz Behavior, and Openness

The notion of open mapping is so simple and so natural that it is often unduly taken in place of continuity by beginners in topology. A classical use of the notion of open map is the Banach–Schauder open mapping theorem, which we saw in Sect. 1.5.6. In the Robinson–Ursescu theorem, it made another appearance for multimaps with closed convex graphs. However, the notion of open multimap is not limited to the convex case.

Definition 1.134. A multimap $F : X \rightrightarrows Y$ between two topological spaces is said to be *open at* $(\bar{x}, \bar{y}) \in F$ (identified with $\mathrm{gph}(F)$) if for every neighborhood U of $\bar{x}$, $F(U)$ is a neighborhood of $\bar{y}$: $U \in \mathcal{N}(\bar{x}) \Rightarrow F(U) \in \mathcal{N}(\bar{y})$.

Clearly, $F : X \rightrightarrows Y$ is open at $(\bar{x}, \bar{y}) \in F$ if and only if $F^{-1} : Y \rightrightarrows X$ is lower semicontinuous at $(\bar{y}, \bar{x})$. When X and Y are metric spaces, a quantitative notion can be related to the preceding one. In order to get a versatile definition, we present it with the use of a subset P of $X \times Y$. The reader is advised to drop its occurrences in a first reading, i.e., to take $P = X \times Y$. Recall that if P is a subset of $X \times Y$, given $x \in X$, $P(x)$ stands for $\{y \in Y : (x, y) \in P\}$ and $P^{-1} := \{(y, x) \in Y \times X : (x, y) \in P\}$.

Definition 1.135. A multimap $F : X \rightrightarrows Y$ between two metric spaces is said to be *open on a subset* P of $X \times Y$ at rate $a > 0$ if

$$\forall r > 0, \ \forall (x, y) \in F, \qquad B(y, ar) \cap P(x) \subset F(B(x, r)),$$

or in other words,

$$r > 0, \ (x, y) \in F, \ (x, y') \in P, \ d(y, y') < ar \implies \exists x' \in B(x, r) : y' \in F(x'). \quad (1.38)$$

Thus, when $P = X \times Y$, the rate a measures the ratio between the radius of an open ball with center in $F(x)$ that can be guaranteed to be contained in the image of the open ball $B(x, r)$ and r itself. It is shown in a supplement that the case of multimaps can be reduced to the case of maps. Introducing P allows some versatility: besides the case $P := X \times Y$, or $P \in \mathcal{N}(\bar{x}, \bar{y})$ (for which one says that F is *open at a linear rate around* $(\bar{x}, \bar{y})$), one can take $P := \{\bar{x}\} \times Z$ with $Z \subset Y$ (which corresponds to linear openness at $\bar{x}$ when $Z \in \mathcal{N}(\bar{y})$) and intermediate cases. The *exact rate of openness* of F around $(\bar{x}, \bar{y})$ is the supremum of the constants a such that (1.38) holds for some $P \in \mathcal{N}(\bar{x}, \bar{y})$.

The concept of metric regularity we introduce now is an extremely useful notion for handling estimates about solutions of systems of equalities and inequalities. As we have seen in the preceding section, it is also a key ingredient for using penalization techniques. In fact, this concept enables us to treat general correspondences. Its interest lies in the fact that dealing with the inverse image $F^{-1}(y)$ of some point $y \in Y$ (or $F^{-1}(C)$ for some subset $C \subset Y$) by some map or multimap F is often a delicate matter. In particular, one would like to replace the distance function $d(\cdot, F^{-1}(y))$ to $F^{-1}(y)$ by a more tractable function such as $d(y, F(\cdot))$. When $F(\cdot) := g(\cdot) + C$, for some convex cone C of a normed space Y, one has $d(y, F(x)) = d(y - g(x), C)$. In particular, when $Y = \mathbb{R}^n$ with the sum norm, $C = \mathbb{R}^n_+$, then $d(y, F(x)) = (\sum_{i=1}^n (g_i(x) - y_i)^+)$.

Definition 1.136. Given a positive number c, a multimap $F : X \rightrightarrows Y$ between two metric spaces is said to be *metrically regular with rate c* (or c-*regular*) on a subset P of $X \times Y$ if

$$\forall (x, v) \in P, \quad d\left(x, F^{-1}(v)\right) \le c\, d\left(v, F(x)\right). \tag{1.39}$$

It is c-*regular around* $(\bar{x}, \bar{y}) \in F$ if it is c-regular on some $P \in \mathcal{N}(\bar{x}, \bar{y})$.

It is c-*subregular* at some $(\bar{x}, \bar{y}) \in F$ if it is c-regular on $P := U \times \{\bar{y}\}$ for some $U \in \mathcal{N}(\bar{x})$, i.e., if $d(x, F^{-1}(\bar{y})) \le cd(\bar{y}, F(x))$ for x near $\bar{x}$.

A multimap $F : X \rightrightarrows Y$ is said to be c-regular on P with respect to some subset S of X if the multimap F_S whose graph is $\text{gph}(F) \cap (S \times Y)$ is c-regular on P.

It is useful to observe that F is c-regular on $P \subset X \times Y$ if and only if one has

$$(x, y) \in F, \ (x, y') \in P \Longrightarrow d\left(x, F^{-1}(y')\right) \le c\, d(y, y'). \tag{1.40}$$

The *regularity rate* $\text{reg}_P(F)$ of F on P is the infimum of the set of positive numbers c such that F is c-regular on P. Similarly, the *regularity rate of F around* $(\bar{x}, \bar{y})$ is the infimum $\text{reg}(F, (\bar{x}, \bar{y})) := \inf\{\text{reg}_P(F) : P \in \mathcal{N}(\bar{x}, \bar{y})\}$ of the regularity rates of F over the family $\mathcal{N}(\bar{x}, \bar{y})$ of neighborhoods of $(\bar{x}, \bar{y})$. Since F is metrically regular with rate c around $(\bar{x}, \bar{y})$ iff there exists some $r > 0$ such that (1.39) holds with $P = B(\bar{x}, r) \times B(\bar{y}, r)$, using the convention $\frac{0}{0} = 0$, $\frac{t}{0} = \infty$ for $t > 0$, the regularity and subregularity rates of F around $(\bar{x}, \bar{y})$ are respectively

$$\text{reg}(F, (\bar{x}, \bar{y})) = \limsup_{(x,v) \to (\bar{x}, \bar{y})} \frac{d(x, F^{-1}(v))}{d(v, F(x))}$$

and

$$\text{subreg}(F, (\bar{x}, \bar{y})) = \limsup_{x \to \bar{x}} \frac{d(x, F^{-1}(\bar{y}))}{d(\bar{y}, F(x))}.$$

Clearly, c-subregularity of F at $(\bar{x}, \bar{y})$ is a weaker property than c-regularity of F around $(\bar{x}, \bar{y})$. Moreover, c-subregularity of F at $(\bar{x}, \bar{y}) \in F$ is equivalent to the existence of some $U \in \mathcal{N}(\bar{x})$, $V \in \mathcal{N}(\bar{y})$ such that $d\left(x, F^{-1}(\bar{y})\right) \leq c\, d(y, \bar{y})$ for all $x \in U$, $y \in F(x) \cap V$. In fact, if this property holds and if $q, r, s > 0$ are such that $B(\bar{x}, r) \subset U$, $B(\bar{y}, s) \subset V$, $q \leq \min(r, cs)$, one also has $d\left(x, F^{-1}(\bar{y})\right) \leq c\, d(y, \bar{y})$ for all $x \in B(\bar{x}, q)$, $y \in F(x) \setminus V$, since $d\left(x, F^{-1}(\bar{y})\right) \leq d(x, \bar{x}) < q \leq cs \leq c\, d(y, \bar{y})$.

It is often convenient to restrict the verification of (1.39) to pairs (x, v) such that $d(v, F(x))$ is small enough. The following proposition allows this easing of the task.

Proposition 1.137. *A multimap $F : X \rightrightarrows Y$ is c-regular around $\bar{z} := (\bar{x}, \bar{y}) \in F$ if and only if there exists some $\varepsilon > 0$ such that F is metrically regular on the set*

$$P := P_\varepsilon := \{(x, v) \in B(\bar{x}, \varepsilon) \times B(\bar{y}, \varepsilon) : d(v, F(x)) < \varepsilon\},$$

if and only if there exists some $\varepsilon > 0$ such that

$$(x, v) \in B(\bar{x}, \varepsilon) \times B(\bar{y}, \varepsilon),\ y \in F(x) \cap B(v, \varepsilon) \Rightarrow d\left(x, F^{-1}(v)\right) \leq c\, d(v, y). \quad (1.41)$$

Proof. Regularity on P_ε being obviously necessary, let us prove that it is sufficient. Suppose F is metrically regular on the set P_ε for some $\varepsilon > 0$. Take $\delta > 0$ such that $(c+1)\delta < c\varepsilon$ and let $(x, v) \in B(\bar{x}, \delta) \times B(\bar{y}, \delta)$. By assumption, relation (1.39) holds when $d(v, F(x)) < \varepsilon$. When $d(v, F(x)) \geq \varepsilon$, using the fact that $d(v, F(\bar{x})) \leq d(v, \bar{y}) < \delta < \varepsilon$, we see that relation (1.39) also holds, since we have

$$d\left(x, F^{-1}(v)\right) \leq d(x, \bar{x}) + d\left(\bar{x}, F^{-1}(v)\right)$$
$$\leq d(x, \bar{x}) + c\, d(v, F(\bar{x})) \leq d(x, \bar{x}) + c\, d(v, \bar{y}) < c\varepsilon \leq c\, d(v, F(x)).$$

In order to prove the last assertion, we note that when F is metrically regular on P_ε, implication (1.41) holds, since for all $y \in F(x) \cap B(v, \varepsilon)$ we have $d(v, F(x)) \leq d(v, y)$. To prove the converse, we observe that for every $(x, v) \in B(\bar{x}, \varepsilon) \times B(\bar{y}, \varepsilon)$ such that $d(v, F(x)) < \varepsilon$, the set $F(x) \cap B(v, \varepsilon)$ is nonempty and one has $d(v, F(x)) = d(v, F(x) \cap B(v, \varepsilon))$, so that taking the infimum over $y \in F(x) \cap B(v, \varepsilon)$, we get relation (1.39). $\qquad\square$

It is useful to compare the notions of metric regularity and of openness at a linear rate with a Lipschitzian property.

Definition 1.138 ([33]). A multimap $M : Y \rightrightarrows Z$ between two metric spaces is said to be *pseudo-Lipschitzian* (or Lipschitz-like or satisfying the Aubin property) with rate c on $Q \subset Y \times Z$ if

$$\forall y, y' \in Y, \qquad e_H(M(y) \cap Q(y'), M(y')) \leq c d(y, y'). \quad (1.42)$$

Instead of using the Pompeiu–Hausdorff excess e_H, one can reformulate this requirement (which can be restricted to $y' \in p_Y(Q)$, since $e_H(\varnothing, S) = 0$ for every subset S) as follows:

$$\forall (y,z) \in M, \ (y',z) \in Q, \quad d(z,M(y')) \le cd(y,y'). \tag{1.43}$$

When $Q = V \times W$, it takes the simpler form

$$\forall y \in Y, \ y' \in V, \ z \in W \cap M(y), \quad d(z,M(y')) \le cd(y,y'). \tag{1.44}$$

One says that $M : Y \rightrightarrows Z$ is *pseudo-Lipschitzian* (or *Lipschitz-like*) *with rate c around* $(\bar{y},\bar{z}) \in M$ if (1.44) holds for some $V \in \mathcal{N}(\bar{y})$, $W \in \mathcal{N}(\bar{z})$. If in (1.44), $W = Z$, this last property implies Lipschitz behavior (with respect to the *Pompeiu–Hausdorff metric d_H*) with rate c on V. However, this last property is seldom met in applications, whereas pseudo-Lipschitz behavior frequently occurs.

When $Q = \{\bar{y}\} \times W$, where $W \in \mathcal{N}(\bar{z})$, i.e., when one has

$$\forall y \in Y, \ w \in W \cap M(y), \qquad d(w,M(\bar{y})) \le cd(y,\bar{y}), \tag{1.45}$$

one says that M is *c-calm* at $(\bar{y},\bar{z})$. Let us note that this property is satisfied whenever for some $V \in \mathcal{N}(\bar{y})$, $W \in \mathcal{N}(\bar{z})$ one has $d(w,M(\bar{y})) \le cd(y,\bar{y})$ for all $y \in V$, $w \in W \cap M(y)$, since for $q,r,s > 0$ such that $B(\bar{y},q) \subset V$, $B(\bar{z},r) \subset W$, $s \le \min(r,cq)$, for $y \in Y \setminus V$, $w \in B(\bar{z},s) \cap M(y)$, one has $d(w,M(\bar{y})) \le d(w,\bar{z}) < s \le cq \le cd(y,\bar{y})$.

Example: For $f : Y \to \mathbb{R}_\infty$, $M(y) := [f(y),+\infty)$, M is calm at $(\bar{y},f(\bar{y})) \in X \times \mathbb{R}$ iff f is *calm* at $\bar{y} \in \mathrm{dom} f$ in the sense that there exist $c,r > 0$ such that $f(y) \ge f(\bar{y}) - cd(y,\bar{y})$ for all $y \in B(\bar{y},r)$. The example of $x \mapsto \sqrt{|x|}$ on $\mathbb{R}$ shows that calmness is less demanding than the pseudo-Lipschitz property.

Similarly, the pseudo-Lipschitz property is a purely local property: one can ensure that M is pseudo-Lipschitzian around $(\bar{y},\bar{z})$ with rate c provided there exist some $U \in \mathcal{N}(\bar{y})$, $W \in \mathcal{N}(\bar{z})$ such that

$$\forall y, \ y' \in U, \ w \in W \cap M(y) \quad d(w,M(y')) \le cd(y,y'). \tag{1.46}$$

In fact, assuming $U = B(\bar{y},\alpha)$, $W = B(\bar{z},\beta)$ for some $\alpha,\beta > 0$ as we may, taking $\gamma \in (0,\alpha/2)$, $\delta > 0$, $\delta \le c(\alpha - 2\gamma)$, $V := B(\bar{y},\gamma)$, $W' := B(\bar{z},\delta)$, for $y \in Y \setminus U$, $y' \in V$, $w \in W' \cap M(y)$, we have $d(y,y') \ge \alpha - \gamma$ and $d(w,M(y')) < \delta + d(\bar{z},M(y')) \le \delta + c\gamma \le c(\alpha - \gamma)$; thus (1.44) holds with W changed into W'.

The *exact pseudo-Lipschitz* (resp. *calmness*) *rate* of M around $(\bar{x},\bar{y})$ is the infimum of the constants c such that (1.43) (resp. (1.45)) holds for some $Q \in \mathcal{N}(\bar{x},\bar{y})$ (resp. $W \in \mathcal{N}(\bar{z})$).

The preceding notions are closely related, as the following theorem shows. Let us note that it contains equalities for the rates we defined.

Theorem 1.139. *For a multimap $F : X \rightrightarrows Y$ between two metric spaces, a subset P of $X \times Y$, and a positive number c, the following assertions are equivalent:*

(a) F is open at the linear rate $a := c^{-1}$ on P;
(b) $F^{-1} : Y \rightrightarrows X$ is pseudo-Lipschitzian on $Q := P^{-1}$ with rate c;
(c) F is metrically regular on P with rate c.

Proof. The equivalence (b)⇔(c) consists in taking $Z := X$, $M := F^{-1}$, $z = x$ to pass from (1.40) to (1.43) and vice versa.

(c)⇒(a) Suppose F is c-metrically regular on P. Given $r > 0$, $(x,y) \in F$, $(x,y') \in P$ with $d(y,y') < c^{-1}r$, relation (1.40) implies that $d(x, F^{-1}(y')) < r$, so that there exists $x' \in F^{-1}(y')$ such that $d(x,x') < r$ and $y' \in F(x')$: relation (1.38) holds.

(a)⇒(c) Suppose F is open at a linear rate a on P. Let $c := a^{-1}$. Given $(x,y) \in F$, $y' \in P(x)$, $s > d(y',y)$, setting $r := cs$, so that $y' \in P(x) \cap B(y,ar)$, relation (1.38) ensures that there exists $x' \in B(x,r)$ such that $y' \in F(x')$ or $x' \in F^{-1}(y')$, and hence $d(x, F^{-1}(y')) \le d(x,x') < r = cs$. Since s is arbitrarily close to $d(y',y)$, we get $d(x, F^{-1}(y')) \le c\, d(y',y)$, and (1.40) holds. □

Taking $P = U \times V$ for some $U \in \mathcal{N}(\bar{x})$, $V \in \mathcal{N}(\bar{y})$, we get a local version.

Corollary 1.140. *A multimap $F : X \rightrightarrows Y$ between two metric spaces is open at a linear rate a around $(\bar{x}, \bar{y}) \in F$ if and only if it is metrically regular around $(\bar{x}, \bar{y})$ with rate $c = a^{-1}$ if and only if F^{-1} is pseudo-Lipschitzian around $(\bar{y}, \bar{x})$ with rate c.*

Taking $P := U \times \{\bar{y}\}$ in Theorem 1.139, with $U \in \mathcal{N}(\bar{x})$, we get a characterization of calmness.

Corollary 1.141. *A multimap $F : X \rightrightarrows Y$ between two metric spaces is c-subregular at $(\bar{x}, \bar{y}) \in F$ if and only if F^{-1} is c-calm at $(\bar{y}, \bar{x})$.*

1.6.7 Characterizations of the Pseudo-Lipschitz Property

One can give a characterization of pseudo-Lipschitz behavior of a multimap F in terms of the function $(x,y) \mapsto d(y, F(x))$ or in terms of the distance function d_F to the graph of the multimap F given by

$$d_F(u,v) := \inf\{d((u,v),(x,y)) : (x,y) \in F\}.$$

In the sequel, we change the metric on $X \times Y$ in order to be reduced to the simpler case of rates one: given $c > 0$, we set

$$d_c((x,y),(x',y')) = \max\left(c\, d(x,x'), d(y,y')\right). \tag{1.47}$$

Theorem 1.142. *Given $c > 0$ and two metric spaces X, Y whose product is endowed with the metric d_c, the following assertions about a multimap $F : X \rightrightarrows Y$ and $(\bar{x}, \bar{y}) \in F$ are equivalent:*

(a) *For some $N \in \mathcal{N}(\bar{x}, \bar{y})$ and all $(u,v) \in N$ one has $d(v, F(u)) \le d_c((u,v), F)$;*
(b) *For some $N \in \mathcal{N}(\bar{x}, \bar{y})$ the function $(x,y) \mapsto d(y, F(x))$ is 1-Lipschitzian on N;*
(c) *F is pseudo-Lipschitzian with rate c around $(\bar{x}, \bar{y})$.*

Proof. We use the fact that for a subset F of a metric space Z and for a given neighborhood N of a point $\bar{z} \in F$ there exists a neighborhood P of $\bar{z}$ such that for all

$w \in P$ one has $d(w,F) = d(w,F \cap N)$. In fact, taking $P := B(\bar{z},r)$, where $r > 0$ is such that $B(\bar{z},2r) \subset N$, for all $w \in P$, $z \in F \setminus N$ one has $d(w,z) \geq d(z,\bar{z}) - d(w,\bar{z}) \geq 2r - r$, while $d(w,F) \leq d(w,\bar{z}) < r$.

(a)$\Rightarrow$(b) When (a) holds, for all $(x,y),(x',y') \in N$ one has $d(y,F(x)) \leq d_c((x,y),F)$,

$$d(y,F(x)) - d(y',F(x')) \leq \inf\{\max(cd(x,x'),d(y,z)) : z \in F(x')\} - d(y',F(x'))$$
$$\leq \max\big(cd(x,x'),d(y,F(x'))\big) - d(y',F(x'))$$
$$\leq \max(cd(x,x'),d(y,y')).$$

(b)$\Rightarrow$(c) Assume (b) holds and take a neighborhood $P := U \times V$ of $\bar{z} := (\bar{x},\bar{y})$ associated to N as in the preliminary part of the proof, so that for every $w := (u,v) \in P$ one has $d(w,F) = d(w,F \cap N)$. Then for all $u,x \in U$, $v \in F(x) \cap V$ one has

$$d(v,F(u)) \leq d(v,F(x)) + d_c((u,v),(x,v)) = cd(u,x).$$

Thus, by (1.46), F is pseudo-Lipschitzian with rate c around $(\bar{x},\bar{y})$.

(c)$\Rightarrow$(a) Let $U \in \mathcal{N}(\bar{x})$, $V \in \mathcal{N}(\bar{y})$ be such that for every $u,x \in U$, $v \in F(x) \cap V$, one has $d(v,F(u)) \leq cd(u,x)$. Let $U' \in \mathcal{N}(\bar{x})$, $V' \in \mathcal{N}(\bar{y})$ be such that for all $(u,v) \in U' \times V'$ one has $d_c((u,v),F) = d_c((u,v),F \cap (U \times V))$. Since for all $y \in F(x)$ we have $cd(x,u) \leq d_c((u,v),(x,y))$, taking the infimum over $(x,y) \in F \cap (U \times V)$, we get $d(v,F(u)) \leq d_c((u,v),F)$. $\qquad\square$

1.6.8 Supplement: Convex-Valued Pseudo-Lipschitzian Multimaps

Pseudo-Lipschitzian multimaps with convex values in a normed space can be characterized in a simple way (the statement below is a characterization because any Lipschitzian multimap is obviously pseudo-Lipschitzian).

Proposition 1.143. *Let X be a metric space, let Y be a normed space, let $F : X \rightrightarrows Y$ be a multimap with convex values, and let $\bar{x} \in X$, $\bar{y} \in F(\bar{x})$. If F is pseudo-Lipschitzian around $(\bar{x},\bar{y})$, then for some ball B with center $\bar{y}$ the multimap G given by $G(x) := F(x) \cap B$ is Lipschitzian. More precisely, if for some $q,r,\ell \in \mathbb{P} := (0,+\infty)$, one has*

$$e_H(F(x) \cap B[\bar{y},r],F(x')) \leq \ell d(x,x') \quad \forall x,x' \in B[\bar{x},q],$$

then for $B := B[\bar{y},r]$, $G(\cdot) := F(\cdot) \cap B$ and $p < \min(\ell^{-1}r,q)$, one has

$$d_H(G(x),G(x')) \leq 2r(r - \ell p)^{-1}\ell d(x,x') \quad \forall x,x' \in B[\bar{x},p].$$

Proof. Without loss of generality we may assume that $\bar{y} = 0$. Taking $x := \bar{x}$, and observing that $\bar{y} \in F(\bar{x}) \cap B[\bar{y},r]$, we get that for all $x' \in B[\bar{x},q]$, $F(x')$ is nonempty.

Let $k \in (\ell, p^{-1}r)$. Given $x, x' \in B[\bar{x}, p]$, let us prove that for all $y' \in G(x')$ we have $d(y', G(x)) \leq 2r(r - kp)^{-1}kd(x, x')$, which will ensure that

$$e_H(G(x'), G(x)) \leq 2r(r - kp)^{-1}kd(x, x'). \tag{1.48}$$

The result will follow from the symmetry of the roles of x and x' and by taking the infimum over $k \in (\ell, p^{-1}r)$. Given $y' \in G(x')$, we can pick $w, z \in F(x)$ such that $\|w\| \leq kd(x, \bar{x}) \leq kp$ and $\|z - y'\| \leq k\delta$ for $\delta := d(x, x')$. If $z \in B[0, r]$, the expected inequality is satisfied, since $d(y', G(x)) \leq \|z - y'\| \leq kd(x, x')$ and $2r(r - kp)^{-1} \geq 1$. Suppose $s := \|z\| > r$. Let $y := tw + (1 - t)z$, with $t := (s - r)(s - kp)^{-1}$. Then $t \in [0, 1]$, $y \in F(x)$, and since $s - r := \|z\| - r \leq \|z - y'\| + \|y'\| - r \leq k\delta$, $\|w\| \leq kp$, $\|z\| = s$, we have

$$\|y\| \leq (s - r)(s - kp)^{-1}kp + (r - kp)(s - kp)^{-1}s \leq r$$

and

$$\|z - y\| = t\|z - w\| \leq (s - r)(s - kp)^{-1}(s + kp) \leq (s - kp)^{-1}(s + kp)k\delta;$$

hence since $\|z - y'\| \leq k\delta$, $(s - kp)^{-1}(s + kp)k\delta + k\delta = 2s(s - kp)^{-1}k\delta$ and since $u \mapsto u(u - kp)^{-1}$ is nonincreasing,

$$\|y - y'\| \leq \|y - z\| + \|z - y'\| \leq 2s(s - kp)^{-1}k\delta \leq 2r(r - kp)^{-1}k\delta.$$

1.6.9 Calmness and Metric Regularity Criteria

Now we devise criteria for calmness and metric regularity. Since these concepts do not require any linear structure, it is appropriate first to present criteria in terms of metric structures. Later on, we will devise criteria in terms of concepts from nonsmooth analysis. The following obvious statement explains why the decrease principle may be useful for subregularity and calmness.

Proposition 1.144. *Let $F : X \rightrightarrows Y$ be a multimap with closed values between two metric spaces and let $\bar{x} \in X$, $\bar{y} \in F(\bar{x})$. Then F is subregular at $(\bar{x}, \bar{y})$ (and F^{-1} is calm at $(\bar{y}, \bar{x})$) if and only if the function $f : X \to \mathbb{R}$ given by $f(x) := d(\bar{y}, F(x))$ is linearly conditioned at $\bar{x}$ in the sense that there exist $c > 0$ and $U \in \mathcal{N}(\bar{x})$ such that $d(x, f^{-1}(\{0\})) \leq cf(x)$ for all $x \in U$.*

Note that since $f^{-1}(\{0\}) = F^{-1}(\bar{y})$, the assumption made in Theorem 1.114 that f^{-1} is closed at 0 means that F^{-1} is closed at $\bar{y}$, i.e., that $\mathrm{cl}(\mathrm{gph}F^{-1}) \cap (\{\bar{y}\} \times X) = \mathrm{gph}F^{-1} \cap (\{\bar{y}\} \times X)$ or that for every sequence $((x_n, y_n)) \to_F (x, \bar{y})$ one has $x \in F^{-1}(\bar{y})$.

Corollary 1.145. *Let $F : X \rightrightarrows Y$ be a multimap with closed values between two metric spaces and let $\bar{x} \in X$, $\bar{y} \in F(\bar{x})$. Suppose F^{-1} is closed at $\bar{y}$ and there exist $b > 0$, a neighborhood U of $\bar{x}$, and a decrease index δ_f of the function $f : X \to \overline{\mathbb{R}}_+$ given by $f(\cdot) := d(\bar{y}, F(\cdot))$ such that $\delta_f(x) \geq b$ for $x \in U \setminus F^{-1}(\bar{y})$.*
Then F is subregular at $(\bar{x}, \bar{y})$ (and F^{-1} is calm at $(\bar{y}, \bar{x})$) and

$$\forall x \in U \quad d(x, F^{-1}(\bar{y})) \leq \frac{1}{b} d(\bar{y}, F(x)).$$

For metric regularity, we may use the parameterized decrease principle. We endow $X \times Y$ with the metric d_c given by $d_c((x,y),(x',y')) = cd(x,x') \vee d(y,y')$ and we recall that ι_F denotes the indicator function of a subset F of $X \times Y$.

Theorem 1.146. *Let $F : X \rightrightarrows Y$ be a multimap with closed graph between two complete metric spaces and let b be a positive number. Endow $X \times Y$ with the metric d_c with $c := 1/b$. Let $f : Y \times X \to \mathbb{R}_\infty$ be given by $f(w,x) := d(w, F(x))$. Let $(\bar{x}, \bar{y}) \in F := \mathrm{gph}(F)$ and let $W \times U$ be a neighborhood of $(\bar{y}, \bar{x})$. Suppose that for some decrease index δ_w of $f_w := f(w, \cdot)$ one has*

$$\delta_w(x) \geq b \quad \forall (w,x) \in W \times U, \ w \notin F(x). \tag{1.49}$$

Then F is metrically regular around $(\bar{x}, \bar{y})$ with rate $c := 1/b$.

Proof. Let $S := f^{-1}(\{0\})$ and for $w \in W$, $S(w) := \{x \in X : (w,x) \in S\}$. Since for $x \in X$, $F(x)$ is closed, we have $x \in S(w)$ if and only if $w \in F(x)$, so that $S(w) = F^{-1}(w)$ for all $w \in W$. Since F has a closed (or locally closed) graph, f^{-1} is closed at 0. Moreover, the multimap $w \rightrightarrows \mathrm{epi} f$ is inward continuous at $(\bar{y}, (\bar{x}, 0))$, since $(\bar{x}, f(w, \bar{x})) \to (\bar{x}, 0)$ as $w \to \bar{y}$. Then the parameterized decrease principle ensures that there exists some $\varepsilon > 0$ such that

$$\forall (w,x) \in B((\bar{y}, \bar{x}), \varepsilon), \qquad d(x, F^{-1}(w)) \leq cf(w,x). \qquad \square$$

Exercises

1. Let f be a bounded map between two metric spaces X, Y. Suppose there exist $\bar{x} \in X$ and $U \in \mathcal{N}(\bar{x})$ such that f is Lipschitzian with rate k on U. Prove that there exist $k' > 0$ and $U' \in \mathcal{N}(\bar{x})$ such that $d(f(x), f(u)) \leq k'd(x,u)$ for all $u \in U'$, $x \in X$.

2. Let F be a multimap between two metric spaces X, Y. Prove that the following two assertions about a point (x_0, y_0) of the graph of F and $k > 0$ are equivalent:

(a) There exist neighborhoods U, V of x_0, y_0 respectively such that

$$e_H(F(u) \cap V, F(u')) \leq kd(u,u') \ \forall u,u' \in U.$$

(b) There exist neighborhoods U, V of x_0, y_0 respectively such that

$$e_H(F(u) \cap V, F(x)) \leq k d(u,x) \quad \forall u \in U \; \forall x \in X.$$

(This characterization of the pseudo-Lipschitz property is due to Henrion.)

3. (a) Let C be a nonempty convex subset of a normed space X and let $h : X \to \mathbb{R}$ be a convex function null on C. Let $g := d(\cdot, C)$. Suppose that for some $\delta > 0$ one has $g(x) \leq h(x)$ whenever $g(x) < \delta$. Show that $d(x,C) \leq h(x)$ for all $x \in X$.

(b) Let $F : X \rightrightarrows Y$ be a multimap with convex graph between two normed spaces. Let $c, \delta > 0$, $y \in Y$ be such that for all $x \in X$ satisfying $d(x, F^{-1}(y)) < r$ one has

$$d(x, F^{-1}(y)) \leq c \, d(y, F(x)).$$

Prove that the preceding inequality holds for every $x \in X$. (See [656].)

4. Let $F: \mathbb{R} \rightrightarrows \mathbb{R}^2$ be the multimap with domain $\mathbb{R}_+$ given by

$$F(x) := \{(y, z) \in \mathbb{R} \times \mathbb{R}_+ : y^2 \leq 2xz\}.$$

(a) Show that F has a closed convex graph that can be interpreted as the set of vectors in $\mathbb{R}^3$ that make an angle no more than $\pi/4$ with the vector $(1, 0, 1)$.

(b) Show that $(0,0) \in F(0)$ but that $(0,0)$ is not an internal point of $F(X)$ and that for every neighborhood V of $(0,0)$, the point $(\varepsilon, \varepsilon^3)$ is in V for $\varepsilon > 0$ small enough but $x \in F^{-1}(\varepsilon, \varepsilon^3)$ iff $x \geq (2\varepsilon)^{-1}$.

5. (a) Prove that a multimap $F : X \rightrightarrows Y$ between two metric spaces that is open at a linear rate a on some subset P of $X \times Y$ is metrically regular with rate $c = a^{-1}$ on every subset Z of $X \times Y$ such that for some $\sigma > 0$ one has $\{x\} \times (B(v, \sigma) \cap F(x)) \subset P$ whenever $(x, v) \in Z$.

(b) Deduce from this result that if F is open with a linear rate a on some neighborhood $U \times V$ of $(\overline{x}, \overline{y}) \in F$ then it is metrically regular around $(\overline{x}, \overline{y})$ with rate $c = a^{-1}$.

6. A multimap $F : X \rightrightarrows Y$ between two metric spaces is said to be *globally open* at a linear rate a around $\overline{x} \in X$ if there exist some $\rho > 0$ and some $U \in \mathcal{N}(\overline{x})$ such that

$$B(F(x), ar) \subset F(B(x, r)) \quad \text{whenever } r \in (0, \rho) \text{ and } x \in U,$$

where, for a subset S of Y, we set $B(S, r) := \{y \in Y : d(y, S) < r\}$.

A multimap $F : X \rightrightarrows Y$ between two metric spaces is *globally regular* around $\overline{x} \in X$ with rate c if there exist $\delta > 0$ and a neighborhood U of $\overline{x}$ such that

$$d(u, F^{-1}(y)) \leq c \, d(y, F(u)) \quad \text{whenever } u \in U, \; y \in Y \text{ and } d(y, F(u)) < \delta.$$

Prove that $F : X \rightrightarrows Y$ is globally open at a linear rate a around $\bar{x} \in X$ iff F is globally regular with rate $c := a^{-1}$ around $\bar{x} \in X$. (See [721].)

7. Reduction to mappings.

(a) Show that a multimap $F : X \rightrightarrows Y$ is open on $P \subset X \times Y$ with rate a iff the restriction $p := p_Y \mid F$ of the canonical projection $p_Y : X \times Y \to Y$ to F is open on $P' := \{(x, y, y') \in X \times Y \times Y : (x, y') \in P\}$ with rate a, $X \times Y$ being endowed with the distance d_c with $c := a^{-1}$.

(b) Show that a multimap $F : X \rightrightarrows Y$ is c-regular on $W \subset X \times Y$ iff the restriction $p = p_Y \mid F$ of the canonical projection $p_Y : X \times Y \to Y$ to (the graph of) F is c-regular on $W' := \{(u, y, v) \in X \times Y \times Y : (u, v) \in W\}$.

1.7 Well-Posedness and Variational Principles

In this section we show that three important topics in nonlinear analysis and optimization are intimately related: variational principles, the theory of perturbations, and the notion of well-posedness. The concept of genericity plays a key role in these connections. Our route takes a simple and versatile approach to these "smooth" variational principles, which will play an important role in the sequel.

We recall that a subset G of some topological space T is *generic* if it contains the intersection of a countable family (G_n) of open subsets of T (a so-called $\mathscr{G}_\delta$ set) that are dense in T (i.e., $\mathrm{cl} G_n = T$). We also recall that Baire's theorem ensures that in a complete metric space every generic subset is dense.

We will use the classical concept of well-posed minimization problem. Given a function $f : X \to \mathbb{R}_\infty := \mathbb{R} \cup \{+\infty\}$ on a complete metric space X, the minimization problem of f is said to be well-posed (in the sense of Tykhonov) if every minimizing sequence (x_n) of f converges; here (x_n) is said to be *minimizing* if $(f(x_n)) \to \inf_X f$. We will say in brief that f is well-posed. This property entails uniqueness of minimizers. Its main interest concerns the case in which f belongs to the set $\mathrm{BLS}(X)$ of lower semicontinuous (lower semicontinuous) functions from X into $\mathbb{R}_\infty$ that are bounded below. Then the well-posedness of f entails existence (and uniqueness) of a minimizer.

While common experience shows that one cannot expect that any minimization problem is well-posed, it can be proved under appropriate assumptions that most problems are, i.e., that well-posedness is a generic property. In order to give a precise meaning to the preceding assertion, we will use the formalism of parameterized minimization problems. It is a framework that has proved to be efficient and versatile for optimization problems and duality theory.

In this section, unless otherwise stated, the parameter space W is a topological space and the decision space X is a complete metric space with metric d. In several instances W will be a space of functions on X. If $F : W \times X \to \mathbb{R}_\infty$ is a function, the partial functions $F_w : X \to \mathbb{R}_\infty$ and $F_x : W \to \mathbb{R}_\infty$ are defined by $F_w(\cdot) = F(w, \cdot)$ and $F_x(\cdot) = F(\cdot, x)$. We say that F is a *perturbation* of a given function $f : X \to \mathbb{R}_\infty$ if for some given base point 0 of W we have $F_0 = f$. In such a case, the study of the *performance function p* given by

$$p(w) := \inf_{x \in X} F(w, x)$$

gives precious information about the minimization of f.

The following easy characterization of well-posedness justifies the use of the set of approximate solutions to the problem of minimizing the given function $f : X \to \mathbb{R}_\infty$. We assume that f is bounded below and, for $\varepsilon > 0$, $C \subset X$, we set

$$S(f, C, \varepsilon) := \{x \in C : f(x) < \inf_C f + \varepsilon\}, \quad S_f(\varepsilon) := S(f, \varepsilon) := S(f, X, \varepsilon). \quad (1.50)$$

As above, $\mathscr{N}(w)$ denotes the family of neighborhoods of w in W and the *diameter* of a subset Y of X is denoted by diam$Y$: diam$Y := \sup \{d(y, z) : y, z \in Y\}$.

Lemma 1.147. *A function $f : X \to \mathbb{R}_\infty$ that is bounded below is well-posed if and only if* diam $\left(S_f(\varepsilon)\right) \to 0$ *as $\varepsilon \to 0_+$.*

Proof. The condition is sufficient, as it implies that any minimizing sequence is a Cauchy sequence. It is also necessary: if there were $\delta > 0$ and a sequence $(\varepsilon_n) \to 0_+$ such that diam $\left(S_f(\varepsilon_n)\right) > \delta$ for all $n \in \mathbb{N}$, we could find two sequences $(x'_n), (x''_n)$ in X such that $x'_n, x''_n \in S_f(\varepsilon_n)$ and $d(x'_n, x''_n) > \delta$ for all $n \in \mathbb{N}$. Then the sequence (x_n) given by $x_n = x'_p$ if $n = 2p$, $x_n = x''_p$ if $n = 2p + 1$ would be minimizing but it could not converge. $\qquad \square$

Given a perturbation $F : W \times X \to \mathbb{R}_\infty$, our genericity criterion involves the sets

$$W(r) := \{w \in W : \exists \varepsilon > 0,\ \exists a \in X,\ S(F_w, \varepsilon) \subset B(a, r)\} \qquad r > 0.$$

Theorem 1.148. *Let W be a topological space, let X be a complete metric space and let $F : W \times X \to \mathbb{R}_\infty$ be such that for all $w \in W$ the function F_w is bounded below on X and has a nonempty domain. If the following two conditions are satisfied, then there exists a generic subset G of W such that for all $w \in G$ the minimization of F_w on X is a well-posed problem:*

(a) For every $r > 0$ the set $W(r)$ is open in W;
(b) For every $r > 0$ the set $W(r)$ is dense in W.

In particular, if W is a metrizable Baire space and if F is a perturbation of f, there exists a sequence $(w_n) \to 0$ such that $F(w_n, \cdot)$ is well-posed for all $n \in \mathbb{N}$.

Proof. Given a sequence (r_n) in $\mathbb{P} := (0, +\infty)$ with limit 0, we set

$$G = \bigcap_{n \in \mathbb{N}} W(r_n),$$

so that by our assumptions, G is a generic subset of W. Let us show that for every $w \in G$, any minimizing sequence (x_n) of F_w is convergent. It suffices to show that (x_n) is a Cauchy sequence. Given $\alpha > 0$, let $k \in \mathbb{N}$ be such that $r_k < \alpha/2$. Since $w \in W(r_k)$, we can find $a \in X$ and $\varepsilon > 0$ such that $S(F_w, \varepsilon) \subset B(a, r_k)$. Since (x_n) is a minimizing sequence of F_w, we can find $m \in \mathbb{N}$ such that $x_n \in S(F_w, \varepsilon)$ for $n \geq m$. Thus diam $\{x_n : n \geq m\} \leq 2r_k < \alpha$. $\square$

Let us give criteria ensuring conditions (a) and (b). We start with condition (a). The assumption that for all $w \in W$ the function F_w is bounded below is still in force. Let us recall that the topology of uniform convergence on the space $(\mathbb{R}_\infty)^X$ of functions from X to $\mathbb{R}_\infty$ is the topology induced by the metric d_∞ given by

$$d_\infty(f, g) := \sup_{x \in X} \left| \frac{f(x)}{1 + |f(x)|} - \frac{g(x)}{1 + |g(x)|} \right|.$$

This topology is generated by the sets

$$U(f, \alpha) := \{g : f(x) - \alpha < g(x) < f(x) + \alpha\} \qquad \alpha \in (0, 1),\ f \in (\mathbb{R}_\infty)^X,$$

where, by convention, $+\infty + \alpha = +\infty$, $+\infty - \alpha = 1/\alpha$, $+\infty/+\infty = 1$.

Lemma 1.149. *Let W be a topological space, let X be a metric space and let $F : W \times X \to \mathbb{R}_\infty$ be such that for all $w \in W$ the function F_w is bounded below on X and has a nonempty domain. If the mapping $w \longmapsto F_w$ is continuous for the topology of uniform convergence on $(\mathbb{R}_\infty)^X$, then for all $r > 0$ the set $W(r)$ is open.*

Proof. Let $r > 0$ and let $w \in W(r)$. There exist $\varepsilon > 0$ and $a \in X$ such that $S(F_w, \varepsilon) \subset B(a, r)$ and there exists a neighborhood V of w such that $F_v \in U(F_w, \varepsilon/3)$ for all $v \in V$. Then for all $x \in S(F_v, \varepsilon/3)$, one has $x \in S(F_w, \varepsilon) \subset B(a, r)$, since $F_w(x) < F_v(x) + \varepsilon/3 < \inf F_v(X) + 2\varepsilon/3 < \inf F_w(X) + \varepsilon$. Thus V is contained in $W(r)$. $\square$

Now our task consists in giving verifiable conditions ensuring condition (b) of Theorem 1.148. The next criterion uses the possibility of deforming F_w in a sufficiently steep manner around an approximate solution x as if its graph were given a blow. We will obtain this possibility by introducing "bumps," i.e., nonnull functions that are zero outside of a bounded subset.

Lemma 1.150. *The density assumption (b) of Theorem 1.148 is satisfied whenever the following condition holds: for all $w \in W$, $V \in \mathcal{N}(w)$, $r > 0$ there exist $\varepsilon > \eta > 0$, $a \in S(F_w, \eta)$, $v \in V$ such that*

$$F_v(a) \leq F_w(a) - \varepsilon, \tag{1.51}$$

$$F_v \geq F_w \qquad on\ X \setminus B(a, r). \tag{1.52}$$

Proof. In order to prove that for every $r > 0$ the set $W(r)$ is dense in W, let us show that for all $w \in W$, $V \in \mathcal{N}(w)$, the set $W(r) \cap V$ is nonempty. Taking $\varepsilon > \eta > 0$, $v \in V$, $a \in S(F_w, \eta)$ as in the assumption, by (1.51), for all $x \in S(F_v, \varepsilon - \eta)$ we have

$$F_v(x) < p(v) + \varepsilon - \eta \le F_v(a) + \varepsilon - \eta \le F_w(a) - \eta \le p(w) \le F_w(x),$$

so that $x \in B(a, r)$ by (1.52). Thus $S(F_v, \varepsilon - \eta) \subset B(a, r)$ and $v \in W(r) \cap V$. $\square$

The criterion of Lemma 1.150 is satisfied whenever the perturbation is rich enough, as the following examples show.

Example. Let X be an arbitrary metric space and let W be the space $\mathrm{BLS}(X)$ of bounded-below lower semicontinuous functions on X endowed with the topology of uniform convergence. Let us show that the evaluation $F : W \times X \to \mathbb{R}_\infty$ given by $F(w, x) = w(x)$ satisfies the criterion of Lemma 1.150. Given $w \in W$, $V \in \mathcal{N}(w)$, $r > 0$, we pick $\eta > 0$ such that $v \in V$ whenever v satisfies $w - \varepsilon \le v \le w + \varepsilon$ with $\varepsilon := 2\eta$, and taking $a \in S(F_w, \eta)$, we define v by $v(a) = w(a) - \varepsilon$, $v(x) = w(x)$ for $x \in X \setminus \{a\}$ and we see that $v \in \mathrm{BLS}(X)$ satisfies conditions (1.51) and (1.52). $\square$

Example. Let X be an arbitrary metric space and let W be the space $\mathrm{BC}(X)$ of bounded continuous functions on X endowed with the topology of uniform convergence. Let us show that the evaluation $F : W \times X \to \mathbb{R}_\infty$ also satisfies the criterion of Lemma 1.150. Given $w \in W$, $V \in \mathcal{N}(w)$, $r > 0$, we pick $\eta > 0$ such that $v \in V$ whenever v satisfies $w - 3\eta \le v \le w + 3\eta$, and taking $\varepsilon := 2\eta$, $a \in S(F_w, \eta)$ and $\delta \in (0, r)$ such that $w(B[a, \delta]) \subset [w(a) - \eta, w(a) + \eta]$, using Urysohn's theorem on the closed ball $B[a, \delta]$, we can find $v \in \mathrm{BC}(X)$ such that $v(a) = w(a) - \varepsilon$, $v(x) = w(x)$ for $x \in X \setminus B(a, \delta)$,

$$v(x) \in [w(a) - 2\eta, w(a) + 2\eta] \subset [w(x) - 3\eta, w(x) + 3\eta] \quad \text{for } x \in B(a, \delta),$$

so that $v \in V$. $\square$

These examples suggest that we consider the case that W is a set of functions from X to $\mathbb{R}_\infty$, f is a given function on X, and F is the perturbation $(w, x) \mapsto f(x) + w(x)$. In particular, we say that a normed space $(W, \|\cdot\|)$ of bounded functions from X to $\mathbb{R}$ is *bumpable* if there exists a subset B of W satisfying the following condition: for every $\alpha, r > 0$ one can find $\varepsilon > 0$ such that for all $a \in X$ there exists $b \in B$ such that

$$\|b\| < \alpha, \quad b(a) > \varepsilon, \quad b(x) \le 0 \quad \text{for } x \in X \setminus B(a, r). \tag{1.53}$$

Corollary 1.151 (Metric variational principle). *Let $(W, \|\cdot\|)$ be a bumpable space of bounded functions on X. Then given a bounded-below, proper, lower semicontinuous function $f : X \to \mathbb{R}_\infty$ the set of $g \in W$ such that $f + g$ is well-posed is generic in W.*

Proof. Again, let $F : W \times X \to \mathbb{R}_\infty$ be given by $F(w, x) = f(x) + w(x)$. It suffices to check that assumption (1.53) implies the conditions of Lemma 1.150. Let $w \in W$,

$V \in \mathcal{N}(w)$, $r > 0$ be given. We pick $\alpha > 0$ such that $B(w, \alpha) \subset V$ and we associate to α, r some $\varepsilon > 0$ as in the definition of a bumpable space. Then for $\eta := \varepsilon/2$ we pick $a \in S(F_w, \eta)$ and $b \in B$ satisfying (1.53). Then for $v := w - b \in V$, conditions (1.51) and (1.52) are satisfied.   □

When X is a normed space, it is natural to take for B a family of bumps deduced by translations and dilations from a single bump located near 0. Recall that the support of a function f on X is the closure of the set of points at which f is non null.

Theorem 1.152 (Deville–Godefroy–Zizler variational principle). *Let X be a normed space and let W be a linear subspace of $\mathrm{BC}(X)$ endowed with a norm $\|\cdot\|$ stronger than the norm $\|\cdot\|_\infty$ of uniform convergence, for which it is complete. Suppose*

(a) There exists some $\overline{b} \in W$ with bounded support such that $\overline{b}(0) > 0$
(b) For all $w \in W$, $a \in X$, the function $w_a : x \mapsto w(x+a)$ is in W and $\|w_a\| = \|w\|$
(c) For all $w \in W$ and all $t > 0$ the function $w(t\cdot) : x \mapsto w(tx)$ belongs to W

Then given a bounded-below, lower semicontinuous function $f : X \to \mathbb{R}_\infty$, the set G of $g \in W$ such that $f + g$ is well-posed is generic in W.

Moreover, there exists some function $\alpha : \mathbb{P} \to \mathbb{P}$ depending only on $(W, \|\cdot\|)$ such that for every $\varepsilon > 0$ and $y \in X$ satisfying $f(y) < \inf f(X) + \alpha(\varepsilon)$ one can find some $g \in G$ with $\|g\| \leq \varepsilon$ such that the minimizer z of $f + g$ belongs to $B(y, \varepsilon)$. If for all $w \in W$, one has $\sup_{t \geq 1} \|w(t\cdot)\| < +\infty$ (resp. $\sup_{t \geq 1} t^{-1} \|w(t\cdot)\| < +\infty$), then for some $c > 0$, one can take $\alpha(\varepsilon) = c\varepsilon$ (resp. $\alpha(\varepsilon) = c\varepsilon^2$) for $\varepsilon \in (0, 1)$.

Proof. It suffices to check that the family of functions

$$B = \left\{ s\overline{b}(a + t\cdot) : s \in \mathbb{R},\ t > 0,\ a \in X \right\}$$

makes $(W, \|\cdot\|)$ bumpable. Let $\sigma > 0$ be such that the support of $\overline{b}$ is contained in $B(0, \sigma)$. Given $\alpha, r > 0$ and $a \in X$, we take $t := r^{-1}\sigma$, $s > 0$ such that $s \|\overline{b}(t\cdot)\| < \alpha$, $\varepsilon \in (0, s\overline{b}(0))$ and we set $b := s\overline{b}(t \cdot -ta)$. Then $\|b\| < \alpha$, $b(a) > \varepsilon$, and $b = 0$ on $X \setminus B(a, r)$.

In order to prove the final assertions, we pick $b_1 \in B$ such that $b_1(0) = 1$ and the support of b_1 is contained in $B(0, 1)$. For $\varepsilon > 0$ we define $b_\varepsilon \in W$ and $\alpha(\varepsilon)$ by

$$b_\varepsilon(x) := b_1(x/\varepsilon), \qquad \alpha(\varepsilon) := \varepsilon/(4\|b_\varepsilon\|).$$

Given $y \in X$ satisfying $f(y) < \inf f(X) + \alpha(\varepsilon)$, let us set

$$h(x) := -3\alpha(\varepsilon)b_\varepsilon(x - y).$$

The first part of the theorem yields some $k \in W$ and some $z \in X$ such that $\|k\| < \alpha(\varepsilon)$ and $(f + h) + k$ attains its minimum at $z \in X$. Since $\|b_\varepsilon\| \geq \|b_\varepsilon\|_\infty \geq 1$, one has $\|k\|_\infty \leq \|k\| \leq \alpha(\varepsilon) \leq \varepsilon/4$. Then $g := h + k$ satisfies $\|g\| \leq \|h\| + \|k\| \leq \varepsilon$ and

$$(f+g)(z) \le f(y) + g(y) \le f(y) - 3\alpha(\varepsilon) + k(y) < \inf f(X) - \alpha(\varepsilon), \qquad (1.54)$$

$$\forall x \in X \setminus B(y,\varepsilon) \quad (f+g)(x) = f(x) + k(x) \ge \inf f(X) - \alpha(\varepsilon), \qquad (1.55)$$

so that $z \in B(y,\varepsilon)$. Finally, when for some $c > 0$ and all $\varepsilon \in (0,1)$ one has $\|b_\varepsilon\| \le 1/4c$ (resp. $\varepsilon \|b_\varepsilon\| \le 1/4c$), one can take $\alpha(\varepsilon) = c\varepsilon$ (resp. $\alpha(\varepsilon) = c\varepsilon^2$). $\qquad\square$

In turn, the Deville–Godefroy–Zizler variational principle encompasses a smooth variational principle, the Borwein–Preiss variational principle, which will be displayed in the next chapter, since it requires some differentiability notions. It will play an important role in the sequel.

Exercises

1. Use Lemma 1.147 and the proof of Theorem 1.148 to show that for every sequence $(r_n) \to 0_+$, the set W_P of $w \in W$ such that F_w is well-posed is exactly $G := \bigcap_{r>0} W(r) = \bigcap_n W(r_n)$.

2. (a) Show that $W(r)$ is open whenever the performance function p is outward continuous and the perturbation F is *upper epi-hemicontinuous* in the sense that for all $u \in W$ such that $\mathrm{dom}(F_u) \ne \varnothing$, one has $e_H(\mathrm{epi}(F_v), \mathrm{epi}(F_u)) \to 0$ as $v \to u$, where e_H is the Hausdorff–Pompeiu excess.

(b) Show that when $w \longmapsto F_w$ is continuous for the topology of uniform convergence on $(\mathbb{R}_\infty)^X$, then the performance function p is outward continuous and the perturbation F is upper epi-hemicontinuous.

(c) Observe that the assumptions of **(a)** are strictly more general than the assumption of Lemma 1.149. [Hint: Consider the example $W = \mathbb{R}_+$, $X = \mathbb{R}$, F given by $F(w,x) = \max(1 - w^{-1}|x|, 0)$ for $w \ne 0$, $F(0,x) = 0$.]

3. Let us say that the perturbation F is *boundedly epi-hemicontinuous* if for every bounded subset B of $X \times \mathbb{R}$ and for all $u \in W$ one has $e_H(\mathrm{epi}(F_v) \cap B, \mathrm{epi}(F_u)) \to 0$ and $e_H(\mathrm{epi}(F_u) \cap B, \mathrm{epi}(F_v)) \to 0$ as $v \to u$. Show that if F is boundedly epi-hemicontinuous and if for each $w \in W$ the function F_w has connected sublevel sets, then for all $r > 0$ the set $W(r)$ is open.

4. Let X be a metric space, let W be a linear subspace of the space $\mathrm{BC}(X)$ endowed with a norm $\|\cdot\|$ stronger than the norm of uniform convergence. Show that $(W, \|\cdot\|)$ is bumpable when the following two conditions involving some point $\bar{x} \in X$, some subset B_0 of W, and some family H of isometries of X are satisfied:

(a) For all $r > 0$ there exists $b \in B_0$ such that $b(\bar{x}) > 0$, $b \mid (X \setminus B(\bar{x}, r)) \le 0$, $\|b\| \le r$;

(b) For all $x \in X$, $b \in B_0$ there exists $h \in H$ such that $h(x) = \bar{x}$, $b \circ h \in W$, $\|b \circ h\| \le \|b\|$.

Assuming that X is a normed space, taking $\bar{x} = 0$ and for H the family of translations of X, deduce Theorem 1.152. (See [814].)

5. Let E be a normed space and let X be the *projective space* associated with E: X is the quotient of $E \setminus \{0\}$ for the equivalence relation $e \sim e'$ iff there exists $\lambda \in \mathbb{R}$ such that $e' = \lambda e$. Let p be the canonical projection $p : E \setminus \{0\} \to X$. Let W be the space of functions f on X such that $f \circ p$ is of class C^1 on $E \setminus \{0\}$. Suppose there exists on E a bump function (i.e., a function b with bounded nonempty support) that is of class C^1 and Lipschitzian. Show that W is bumpable for the topology of uniform convergence.

6. Let X be a complete metric space and let $k : X \times X \to \overline{\mathbb{R}}_+ := [0,+\infty]$ be a forcing bifunction, or, more generally, a function $k : X \times X \to \overline{\mathbb{R}}_+$ such that $k(x,x) = 0$ for all $x \in X$ and $d \leq \mu \circ k$ for some modulus $\mu : \mathbb{R}_+ \to \mathbb{R}_+$. Prove the *Borwein–Preiss variational principle* [128] in the version of [655] and [137, Theorem 2.5.2]: Given a bounded-below lower semicontinuous function $f : X \to \mathbb{R}_\infty$, a sequence (c_n) of positive numbers, and for $\varepsilon > 0$ some ε-minimizer $\bar{x}$ of f, there exist $u \in X$ and a sequence (u_n) of X such that for $g := \sum_n c_n k(\cdot, u_n)$, one has

(a) $k(\bar{x}, u) \leq \varepsilon/c_0$, $k(x_n, u) \leq 2^{-n}\varepsilon/c_0$;
(b) $f(u) + g(u) \leq f(\bar{x})$;
(c) $f(u) + g(u) < f(x) + g(x)$ for all $x \in X \setminus \{u\}$.

Observe that when X is a Banach space with a smooth norm $\|\cdot\|$ and for some $p > 1$ one has $k(x,x') = \|x - x'\|^p$, then g is smooth.

7. Show that the Ekeland variational principle follows from Corollary 1.151 or Theorem 1.152 and one can even get a reinforced assertion as follows.

Given a bounded-below proper lower semicontinuous function $f : X \to \mathbb{R}_\infty$ on the complete metric space X and given $\varepsilon > 0$, there exists $x_\varepsilon \in X$ such that $f + \varepsilon d(x_\varepsilon, .)$ is well-posed. [Hint: Take for W the space of Lipschitz functions on X with an appropriate norm involving the Lipschitz rate; see [430].]

1.7.1   Supplement: Stegall's Principle

A subset Y of a Banach space X is said to have the *Radon–Nikodým property* (RNP) if every nonempty bounded subset Z of Y is *dentable* in the following sense: for every $\delta > 0$ there exist $\varepsilon > 0$ and $x^* \in X^* \setminus \{0\}$ such that diam $S(x^*, Z, \varepsilon) \leq \delta$, where $S(x^*, Z, \varepsilon) := \{z \in Z : x^*(z) \leq \inf x^*(Z) + \varepsilon\}$ is the *slice* defined in (1.50). When X is a dual space, Z is said to be *weak* dentable* when it has weak* slices (i.e., slices defined by elements of the predual of X) of arbitrarily small diameter.

Theorem 1.153 (Stegall).   *Let Y be a nonempty closed and bounded subset of a Banach space X with the RNP and let f be a bounded-below lower semicontinuous function on Y. Then the set of $x^* \in X^*$ such that $f + x^*$ is well-posed on Y is a generic subset of X^*.*

Proof. Let $W = X^*$ and let $F : W \times Y \to \mathbb{R}_\infty$ be given by $F(w, y) = f(y) + \langle w, y \rangle$. Let us endow W with the dual norm. Then F is lower semicontinuous and $w \mapsto F_w$ is continuous for the topology of uniform convergence, so that assumption (a) of Theorem 1.148 is satisfied by Lemma 1.149. For the proof that assumption (b) is satisfied, we refer to [832, pp. 85–87] or [137, pp. 267–271]. $\square$

In the following variant, the boundedness assumption on Y is replaced with a coercivity assumption: f is said to be *super-coercive* if $\liminf_{\|x\| \to \infty} f(x)/\|x\| > 0$.

Corollary 1.154 (Fabian). *Let X be a Banach space with the RNP and let f be a lower semicontinuous, bounded-below, super-coercive function on X. Then the set of $x^* \in X^*$ such that $f + x^*$ is well-posed on X is a generic subset of X^*.*

Proof. Let $a > 0$ with $\liminf_{\|x\| \to \infty} f(x)/\|x\| > a$, so that there exists $r > 0$ such that $f(x) \geq a \|x\|$ for $x \in X \setminus rB_X$. Since f is bounded below, adding a constant to f if necessary, we may suppose $f(x) \geq a \|x\|$ for all $x \in X$. Given $s \in (0, a)$, for all $x^* \in (a - s)B_{X^*}, x \in X$ we have

$$f(x) + \langle x^*, x \rangle \geq a \|x\| - (a - s) \|x\| = s \|x\|. \tag{1.56}$$

Let $t > f(0)/s$, $\alpha := st - f(0)$. Applying Stegall's principle to the ball tB_X and taking $\varepsilon := a - s$, we can find some $x^* \in X^*$ with $\|x^*\| < a - s$ such that $f + x^*$ is well-posed on tB_X. To get the conclusion, it suffices to show that $f(x) + \langle x^*, x \rangle \geq \inf(f + x^*)(tB_X) + \alpha$ for all $x \in X \setminus tB_X$. If, on the contrary, there exists some $x \in X \setminus tB_X$ such that $f(x) + \langle x^*, x \rangle < \inf(f + x^*)(tB_X) + \alpha$, then by (1.56), we have $s \|x\| < (f + x^*)(0) + \alpha = st$, hence $\|x\| < t$, a contradiction. $\square$

Stegall's principle can be used to get a representation of Radon–Nikodým sets in terms of exposed points. Given a bounded, closed, convex subset C of a Banach space X, a point $\bar{x} \in C$ is said to be *firmly (or strongly) exposed* by some $x^* \in X^*$ if $-x^*$ is well-posed on C and $\bar{x}$ is a minimizer of $-x^*$, or equivalently, if every sequence (x_n) of C such that $(\langle x^*, x_n \rangle) \to \sigma_C(x^*) := \sup x^*(C)$ converges to $\bar{x}$.

Theorem 1.155. *Let X be a Banach space and let C be a bounded, closed, convex subset of X with the RNP. Then the set G of continuous linear forms on X that firmly expose C is a dense $\mathscr{G}_\delta$ subset of X^*. Moreover, C is the closed convex hull of the set E of firmly exposed points of C.*

Proof. Let $W := X^*$, let $F : W \times X \to \mathbb{R}$ be the evaluation, and for $r > 0$, let

$$W(r) := \{x^* \in X^* : \exists \varepsilon > 0, \exists a \in C, \ S(x^*, C, \varepsilon) \subset B(a, r)\}.$$

Then given a sequence $(r_n) \to 0_+$, as in the proof of Theorem 1.148, G is seen to be the intersection of the family $(W(r_n))_n$. Since C has the RNP, all $W(r_n)$ are dense in X^*. Now, for all $r > 0$, $W(r)$ is open, since for $x^* \in W(r)$, if $\varepsilon > 0$, $a \in C$ are such that $S(x^*, C, \varepsilon) \subset B(a, r)$ and if $b > 0$ is such that $C \subset bB_X$, then for $\eta \in (0, \varepsilon)$, $\beta > 0$ with $\eta + 2\beta b \leq \varepsilon$ and for $y^* \in B(x^*, \beta)$ one has $S(y^*, C, \eta) \subset S(x^*, C, \varepsilon)$, as is easily seen, so that $y^* \in W(r)$. For the proof of the last assertion, we refer to [376] or [832]. A nice joint approach to Stegall's principle, the Asplund–Namioka–Phelps theorem, and Collier's theorem is given in [616].   □

1.8 Notes and Remarks

The notions of metric space and topology that seem so natural today were once undecipherable. Many other notions in mathematics became obvious, even if they were once mysterious. It is likely that nets remain obscure objects for many readers. Thus, we give elements to master this notion of generalized sequences. Another example is prevalence, a vague idea. It can be made precise in different ways. The one we consider, genericity, has a great importance in mathematics for various purposes: analysis [37, 219, 289, 341, 902, 942], convergence [260, 859], differentiability [20, 114, 259, 437, 623], geometry and differential topology [1], mathematical programming [544, 545, 552], optimal control theory, optimization [71, 72, 261, 323, 352, 860–863], partial differential equations [887, 888], to name a few.

The weak* topology on a dual space plays an important role in the sequel. We present some important results without proofs just to set the stage. They also serve to show the differences between the properties of the weak topology and those of the weak* topology, in particular for what concerns sequential compactness. We give a complete proof of Theorem 1.13 inspired by [507, lemma p. 151] because we feel that the use in nonsmooth analysis of the bounded weak* topology should be promoted.

Convergences and topologies can be introduced on the power set $\mathscr{P}(X)$ of a topological space X, i.e., the set of subsets of X. Although this point of view may be illuminating, we have refrained from adopting it; we refer the reader to [86, 209, 591, 607, 900]. In spite of its deficiencies, we have kept most of the traditional terminology for limits of sets and continuity of multimaps. However, we have introduced some notions that may be useful but are not classical. More drastic changes of terminology are to be found in the monograph [883]. Variational convergences are studied there in more detail. See also [24, 29, 39, 86]. Note that the definitions of epi-limits would be more natural if one were to consider hypoconvergence (i.e., convergence of hypographs) instead of convergence of epigraphs, but the great importance of convexity incites us to keep epigraphs.

The preliminary study of convexity we introduce in this first chapter is justified by the use of separation properties. More will be presented in Chap. 3. The notion of ideally convex set is due to Lifshits; see [507, pp. 138, 201] and [7, 694] for variants. The approach to the minimax theorem presented here is taken from [893].

The Ekeland variational principle is a tool of utmost importance for many questions in nonlinear analysis and nonsmooth analysis. See [37, 262, 341–348, 350–353] and for equivalent forms [756, 792, 906]. In particular, a geometric equivalent form is given in [253, 792] and a reinforced version appears in [430]. Related geometric statements of Brønsted [174] and Bishop–Phelps [104] paved the way to the use of order methods in nonlinear analysis. The relatives of the drop theorem given in [792] were attempts to obtain a smooth version of the variational principle. This aim has been reached by Borwein and Preiss [128] and generalized by Deville–Godefroy–Zizler [289]. The presentation of the latter here follows [814].

We have offered the reader some detours and complements to variational principles in order to stress their usefulness. Several circular tours exist [253, 262, 465, 792, 906]. The fact that the Ekeland variational principle characterizes completeness of a metric space [961] is a testimony of the generality of that result. Thus, it has attracted much attention. Some geometrical forms have been given to it [792] and some extensions to vector-valued functions, more general spaces [386] or other perturbations [128], [137, Sect. 2.5], [655, 674] have been given; they are outside the scope of this book. The quantitative form of the Banach open mapping theorem is adapted from [529].

The Palais–Smale theory is an important tool in nonlinear functional analysis; see for instance [37, 338, 765, 767, 988].

Section 1.6 benefited from the paper [216], from the book [218], and from a mutual influence from the works of Azé–Corvellec [55], Azé [53, 54], and Ioffe [531]. The decrease principle of Theorem 1.114 appeared in [825].

Given a function $f : X \to \overline{\mathbb{R}}_+ := [0, \infty]$ on a metric space X, the following question arises: if the value of f at x is small, is x close to the zero set $S := f^{-1}(0)$ of X? Such a question is of importance for algorithms, but its bearing is much larger. Numerous authors have tackled it, among whom are Hoffman, Burke–Ferris, Cominetti, Lemaire, Zhang–Treiman, Cornejo–Jourani–Zalinescu, Pang, Penot, Ioffe, Ng–Zheng, Ngai–Théra, Azé–Corvellec, Azé, Henrion–Jourani–Outrata, Dontchev–Rockafellar, Henrion–Outrata, Kummer, Łojasiewicz, Bolte–Daniilidis–Lewis, Wu–Ye, and many more. The convex case is specially rich. It has been treated by Auslender, Cominetti, Crouzeix, Klatte, Lewis, Li, Mangasarian, Pang, Robinson, Song, Zalinescu among others.

The first study of error bounds was the one by Hoffmann [505] in the framework of polyhedral functions. The use of subdifferentials for the study of error bounds began in [990] and has been followed by [229, 808]. The language in these papers was related to *nonlinear conditioning*, with the same objective of evaluating the distance to a set of solutions by the value of the objective. The route we take for reaching such an aim avoids the lower semicontinuity assumption of the Ekeland principle by using a variant of a result of Cominetti [225], Proposition 1.111. Here, for the sake of simplicity, we neglect nonlinear estimates obtained in [64, 229], [531, Sect. 3.5], and [808]. The notion of (strong) slope due to De Giorgi et al. [263] brought noteworthy improvements for decrease principles. See [46] and the comprehensive surveys [53, 54, 531].

The method of penalization is a widely used process (see [393, 692], for example). It was introduced in nonsmooth analysis by Ioffe [512, 515, 516] and Fabian [363] in order to get decoupling conditions for sums. A similar method was used for the study of Hamilton–Jacobi equations by Crandall and Lions [244]. The notions of stabilized infimum and robust minimizer were detected by Aussel, Lassonde, and Corvellec in [46] under another terminology and used for nonsmooth analysis; see also [137, 582, 615]. The presentation given here is new. See also [511, 530, 531], in which a general condition is introduced and Proposition 1.127 is proved. The concept of coherence for a family of functions was introduced by Ioffe [530]. Its variant, called quasicoherence, is new. The terminology for these new notions has varied, but they seem to be basic enough to take place here. They are used in Chap. 4.

Metric regularity was initiated by Liusternik [693, 694] and developed in [140, 303, 305, 514, 794]. The relationships between openness at a linear rate, metric regularity, and pseudo-Lipschitz behavior partly detected in [305] were fully disclosed in [794] and were completed with perturbation properties in [63, 140]. Other works have been revealed to the author in [531]; see also [319]. The use of the Ekeland principle and the tools from nonsmooth analysis for metric regularity was initiated in [511] and developed for the computation of tangent cones and for optimality conditions in [111, 785, 789]; the terminology "metric regularity" appeared in [788]. Openness at a linear rate is called "covering property with bounded modulus" in the Russian literature. We do not adopt the latter terminology because the term "covering" already has two different meanings in topology. Moreover, we save the term "modulus" for nonnegative functions on $\mathbb{R}_+$ that are continuous at 0 and null at 0. The novelty and importance of the notion of Lipschitz-like behavior justify the terminology "Aubin property" introduced by Rockafellar. However, we keep the traditional one.

The theory of perturbations we adopt here as a convenient framework has proved to be very fruitful, especially in the convex case (see Chap. 3 and [323, 353, 822, 872], for example).

In [551], Ioffe and Zaslavski address a stronger form of well-posedness, close to those considered in [999, 1000], with uniqueness and continuity of the performance function. In their generic variational principle the decision space X is not supposed to be complete, but the parameter space W is metrizable. Their method assumes existence of an appropriate dense subset of the parameter space instead of relying on the intrinsic sets $W(r)$ we used in [808] and here. In [543–545, 552, 553] the principle of [551] is applied to various constrained optimization problems, taking into account the specific structure of each problem. The case of an explicitly constrained convex program is considered in [863] with respect to the bounded-Hausdorff topology; the results of [863] deal with specific notions of well-posedness that cannot be derived from Tykhonov well-posedness, since they involve Levitin–Polyak minimizing sequences that are not minimizing sequences when one adds to the objective function the indicator function of the constraint set. The terminology "strong minimizer" (resp. "strongly exposed") is commonly used instead of "firm minimizer" (resp. "firmly exposed") but it may be misleading when a strong topology and a weak topology are present.

In [318] a general approach to generic well-posedness is considered. Genericity is reinforced, since the complement of the set of well-posed functions is shown to be σ-porous; in a finite-dimensional space, this notion implies that the set is not only meager but also negligible with respect to the Lebesgue measure (see [981]). In [999, 1000] strong forms of well-posedness are related to differentiability of the performance function in a general framework, and applications to the calculus of variations are given. It is also shown there that such relationships have their origins in classical papers of Šmulian [898] and Asplund–Rockafellar [23].

The original proof of the Borwein–Preiss variational principle is more constructive than the one we present here and in Chap. 2; see [128, 137, 890]. A variant due to Loewen and Wang [674] is presented in [890]. The Borwein–Preiss variational principle brings slightly more precise information. However, the statement we present is sufficient for the applications we have in view and provides localization information. Its smoothness content, in particular the C^1 and D^1 smooth cases, will be expounded in the next chapter.

Chapter 2
Elements of Differential Calculus

If I have seen further, it is by standing on the shoulders of giants.

—Isaac Newton, letter to Robert Hooke, February 5, 1675

Differential calculus is at the core of several sciences and techniques. Our world would not be the same without it: astronomy, electromagnetism, mechanics, optimization, thermodynamics, among others, use it as a fundamental tool.

The birth of differential calculus is usually attributed to Isaac Newton and Gottfried Wilhelm Leibniz in the latter part of the seventeenth century, with several other contributions. The pioneer work of Pierre de Fermat is seldom recognized, although he introduced the idea of approximation that is the backbone of differential calculus and that enabled him (and others) to treat many applications. During the eighteenth century, the topic reached maturity, and its achievements led to the principle of determinism in the beginning of the nineteenth century. But it is only with the appearance of functional analysis that it took its modern form.

Several notions of differentiability exist; they correspond to different needs or different situations. The most usual one is the notion of Fréchet differentiability, that is presented in Sect. 2.4. However, a weaker notion of directional differentiability due to Hadamard has some interest. We present it in Sect. 2.3 as a passage from the case of one-variable maps to the case of maps defined on open subsets of normed spaces. Its study has an interest of its own and forms a basis for a notion of subdifferential that will come to the fore in Chap. 4 along with a notion corresponding to the Fréchet derivative. For some results, differentiability does not suffice and one needs some continuity property of the derivative. Besides classical continuity, we consider a weaker continuity condition. The latter is seldom given attention. Still, it will serve as preparation for the limiting processes considered in Chap. 6.

The main questions we treat are the invertibility of nonlinear maps, its applications to geometrical notions, and its uses for optimization problems. The notions of normal cone and tangent cones appearing for optimality conditions in fact belong

J.-P. Penot, *Calculus Without Derivatives*, Graduate Texts in Mathematics 266, DOI 10.1007/978-1-4614-4538-8_2, © Springer Science+Business Media New York 2013

to the realm of nonsmooth analysis. Many practitioners are unaware of this—rather like Molière's Monsieur Jourdain, who had been speaking prose all his life without knowing it. We end the chapter with an introduction to the calculus of variations, that has been a strong incentive for the development of differential calculus since the end of the seventeenth century. Differentiability questions for convex functions will be considered in the next chapter.

2.1 Derivatives of One-Variable Functions

The differentiation of one-variable vector-valued functions is not very different from the differentiation of one-variable real-valued functions. In both cases, the calculus relies on rules for limits. The aims are similar too. In both cases, the purpose consists in drawing some information about the behavior of the function from some knowledge concerning the derivative. In the vector-valued case, the direction of the derivative takes as great importance as its magnitude.

2.1.1 Differentiation of One-Variable Functions

In this section unless otherwise mentioned, T is an open interval of $\mathbb{R}$ and $f : T \to X$ is a map with values in a normed space X.

Definition 2.1. A map f is said to be *right-differentiable* (resp. *left-differentiable*) at $t \in T$ if the quotient $(f(t+s) - f(t))/s$ has a limit as $s \to 0_+$, i.e., $s \to 0$ with $s > 0$ (resp. as $s \to 0_-$, i.e., $s \to 0$ with $s < 0$). These limits, denoted by $f'_+(t)$ and $f'_-(t)$ respectively, are called the right and the left *derivatives* of f at t.

When these limits coincide, f is said to be *differentiable* at t, and their common value $f'(t)$ is called the *derivative* of f at t.

Thus f is differentiable at t if and only if the quotient $(f(t+s) - f(t))/s$ has a limit as $s \to 0$, with $s \neq 0$, or equivalently, if there exist some vector $v(= f'(t)) \in X$ and some function $r : T' := T - t \to X$ called a *remainder* such that $r(s)/s \to 0$ as $s \to 0$, for which one has the expansion

$$f(t') = f(t) + (t' - t)v + r(t' - t), \qquad (2.1)$$

as can be seen by setting $s = t' - t$, $r(0) = 0$, $r(s) = s^{-1}(f(t+s) - f(t)) - v$ for $s \in T' \setminus \{0\}$.

The following rules are immediate consequences of the rules for limits.

Proposition 2.2. *If $f,g : T \to X$ are differentiable at $t \in T$ and $\lambda, \mu \in \mathbb{R}$, then $h := \lambda f + \mu g$ is differentiable at t and its derivative at t is $h'(t) = \lambda f'(t) + \mu g'(t)$.*

Proposition 2.3. *If $f : T \to X$ is differentiable at $t \in T$, if Y is another normed space and if $A : X \to Y$ is linear and continuous, then $g := A \circ f$ is differentiable at $t \in T$ and $g'(t) = A(f'(t))$.*

Similar rules hold for right derivatives and left derivatives. We will see later a more general composition rule (or chain rule). The following composition rule can be proved using quotients as for scalar functions. We prefer using expansions as in (2.1) because such expansions give the true flavor of differential calculus, i.e., approximations by continuous affine functions. Moreover, one does not need to take care of denominators taking the value 0.

Proposition 2.4. *If T, U are open intervals of $\mathbb{R}$, if $g : T \to U$ is differentiable at $\bar{t} \in T$, and if $h : U \to X$ is differentiable at $\bar{u} := g(\bar{t})$, then $f := h \circ g$ is differentiable at $\bar{t}$ and $f'(\bar{t}) = g'(\bar{t})h'(\bar{u})$.*

Proof. Let $v := h'(\bar{u})$ and let $\alpha : T \to \mathbb{R}$, $\beta : U \to X$ be such that $\alpha(t) \to 0$ as $t \to \bar{t}$, $\beta(u) \to 0$ as $u \to \bar{u}$ with $g(t) - g(\bar{t}) = (t - \bar{t})g'(\bar{t}) + (t - \bar{t})\alpha(t)$, $h(u) - h(\bar{u}) = (u - \bar{u})v + (u - \bar{u})\beta(u)$. Then one has

$$f(t) - f(\bar{t}) = h(g(t)) - h(\bar{u}) = (g(t) - \bar{u})v + (g(t) - \bar{u})\beta(g(t))$$
$$= (t - \bar{t})g'(\bar{t})v + (t - \bar{t})\alpha(t)v + (t - \bar{t})(g'(\bar{t}) + \alpha(t))\beta(g(t)).$$

Since $g(t) \to \bar{u}$ as $t \to \bar{t}$, one sees that $\alpha(t)v + (g'(\bar{t}) + \alpha(t))\beta(g(t)) \to 0$ as $t \to \bar{t}$, so that f is differentiable at $\bar{t}$ and $f'(\bar{t}) = g'(\bar{t})v = g'(\bar{t})h'(\bar{u})$.   □

Now let us devise a rule for the derivative of a product. It can be generalized to a finite number of factors.

Proposition 2.5 (Leibniz rule). *Let X, Y, Z be normed spaces and let $b : X \times Y \to Z$ be a continuous bilinear map. Given functions $f : T \to X$, $g : T \to Y$ that are differentiable at t, the function $h : r \mapsto b(f(r), g(r))$ is differentiable at t and*

$$h'(t) = b(f'(t), g(t)) + b(f(t), g'(t)).$$

Proof. By assumption, there exist some $\alpha : (T - t) \to X$, $\beta : (T - t) \to Y$ satisfying $\alpha(s) \to 0$, $\beta(s) \to 0$ as $s \to 0$ such that

$$f(t + s) = f(t) + sf'(t) + s\alpha(s), \qquad g(t + s) = g(t) + sg'(t) + s\beta(s).$$

Plugging these expansions into b and setting $\gamma(s) := b(\alpha(s), g(t)) + b(f(t), \beta(s)) + sb(\alpha(s), \beta(s))$, so that $\gamma(s) \to 0$ as $s \to 0$, we get

$$h(t + s) - h(t) = sb(f'(t), g(t)) + sb(f(t), g'(t)) + s\gamma(s)$$

and $s^{-1}(h(t + s) - h(t)) \to b(f'(t), g(t)) + b(f(t), g'(t))$.   □

2.1.2   The Mean Value Theorem

The mean value theorem is a precious tool for devising estimates. For this reason, it is a cornerstone of differential calculus. Let us note that the elementary version recalled in the following lemma is not valid when the function takes its values in a linear space of dimension greater than one.

Lemma 2.6. *Let $f : T \to \mathbb{R}$ be a continuous function on some interval $T := [a,b]$ of $\mathbb{R}$, with $a < b$. If f is differentiable on (a,b) then there exists some $c \in (a,b)$ such that*

$$f(b) - f(a) = f'(c)(b-a).$$

Example. Let $f : [0,1] \to \mathbb{R}^2$ be given by $f(t) := (t^2, t^3)$ for $t \in T := [0,1]$. Then one cannot find every $c \in \mathrm{int}\,T$ satisfying the preceding relation, since the system $2c = 1$, $3c^2 = 1$ has no solution.   □

Instead, a statement under the form of an inequality is valid.

Theorem 2.7. *Let X be a normed space, $T := [a,b]$ a compact interval of $\mathbb{R}$, and $f : T \to X$, $g : T \to \mathbb{R}$ continuous on T with right derivatives on (a,b) such that $\left\| f'_+(t) \right\| \le g'_+(t)$ for every $t \in (a,b)$. Then*

$$\|f(b) - f(a)\| \le g(b) - g(a). \tag{2.2}$$

Proof. It suffices to prove that for every given $\varepsilon > 0$, b belongs to the set

$$T_\varepsilon := \{t \in T : \|f(t) - f(a)\| \le g(t) - g(a) + \varepsilon(t - a)\}.$$

This set is nonempty, since $a \in T_\varepsilon$, and closed, being defined by an inequality whose sides are continuous. Let $s := \sup T_\varepsilon \le b$. Then $s \in T_\varepsilon$.

We first suppose f and g have right derivatives on $[a,b)$ and we show that assuming $s < b$ leads to a contradiction. The existence of the right derivatives of f and g at s yields some $\delta \in (0, b-s)$ such that

$$r \in (0,\delta] \Rightarrow \left\| \frac{f(s+r) - f(s)}{r} - f'_+(s) \right\| \le \frac{\varepsilon}{2}, \quad \left| \frac{g(s+r) - g(s)}{r} - g'_+(s) \right| \le \frac{\varepsilon}{2}.$$

It follows that for $r \in (0,\delta]$ one has

$$\|f(s+r) - f(s)\| \le r \left\| f'_+(s) \right\| + r\varepsilon/2, \quad g(s+r) - g(s) \ge rg'_+(s) - r\varepsilon/2.$$

Therefore, since $s \in T_\varepsilon$ and $\left\| f'_+(s) \right\| \le g'_+(s)$,

$$\|f(s+r) - f(a)\| \le \|f(s+r) - f(s)\| + \|f(s) - f(a)\|$$
$$\le rg'_+(s) + r\varepsilon/2 + g(s) - g(a) + \varepsilon(s-a)$$
$$\le g(s+r) - g(s) + r\varepsilon + g(s) - g(a) + \varepsilon(s-a)$$
$$\le g(s+r) - g(a) + \varepsilon(s+r-a).$$

This string of inequalities shows that $s+r \in T_\varepsilon$, a contradiction to the definition of s. Thus $b \in T_\varepsilon$ and the result is established under the additional assumption that the right derivatives of f and g exist at a (note that we may have $s = a$ in what precedes).

When this additional assumption is not made, we take $a' \in (a,b]$ and we apply the preceding case to the interval $[a',b]$:

$$\left\| f(b) - f(a') \right\| \le g(b) - g(a').$$

Then passing to the limit as $a' \to a_+$, we get the announced inequality.   □

Remark. Since we allow the possibility that the right derivatives do not exist at the extremities of the interval, we may assume that the derivatives do not exist (or do not satisfy the assumed inequality) at a finite number of points of T. To prove this, it suffices to subdivide the interval into subintervals and to gather the obtained inequalities using the triangular inequality. In fact, one can exclude a countable set of points of T, but the proof is more involved; see [197], [294, p.153].

Theorem 2.8. *With the notation of Theorem 2.7, the estimate (2.2) holds when f and g are continuous on T and have right derivatives on $T \setminus D$, where D is countable, such that $\left\| f'_+(t) \right\| \le g'_+(t)$ for every $t \in T \setminus D$.*

The most usual application is given in the following corollary, in which we take $g(t) = mt$ for some $m \in \mathbb{R}_+$ and $t \in T$. The Lipschitz property is obtained on substituting an arbitrary pair t,t' (with $t \le t'$) for a, b.

Corollary 2.9. *Let $f : T \to X$ be continuous on $T := [a,b]$, let $m \in \mathbb{R}_+$, and let D be a countable subset of T. Suppose that for all $t \in (a,b) \setminus D$, f has a right derivative at t such that $\left\| f'_+(t) \right\| \le m$. Then f is Lipschitzian with rate m on T, and in particular,*

$$\|f(b) - f(a)\| \le m(b-a).$$

The case $m = 0$ yields the following noteworthy consequence.

Corollary 2.10. *Let $f : [a,b] \to X$ be continuous and such that f has a right derivative f'_+ on $(a,b) \setminus D$ that is null, D being countable. Then f is constant on $[a,b]$.*

The purpose of obtaining estimates often requires the introduction of auxiliary functions, as in the proof of the following useful corollary.

Corollary 2.11. *Let* $f : T \to X$ *be continuous on* $T := [a,b]$, *let* $v \in X$, $r \in \mathbb{R}_+$, *and let D be a countable subset of T. Suppose f has a right derivative on* $(a,b) \setminus D$ *such that* $f'_+(t) \in v + rB_X$ *for every* $t \in (a,b) \setminus D$. *Then*

$$f(b) \in f(a) + (b-a)v + (b-a)rB_X.$$

Proof. Define $h : T \to X$ by $h(t) := f(t) - tv$. Then h is continuous and for $t \in (a,b) \setminus D$ one has $\|h'_+(t)\| = \|f'_+(t) - v\| \le r$. Then Corollary 2.10 entails that

$$\|f(b) - f(a) - (b-a)v\| = \|h(b) - h(a)\| \le (b-a)r,$$

an estimate equivalent to the inclusion of the statement.　　　　　　　　□

Remark. The terminology for the theorem stems from the fact that the mean value $\bar{v} := (b-a)^{-1}(f(b) - f(a))$ is estimated by the approximate speed v, with an error r that is exactly the magnitude of the uncertainty of the estimate of the instantaneous speed $f'_+(t)$. Note that the shorter the lapse of time $(b-a)$, the more precise the localization of $f(b)$ by $f(a) + (b-a)v$. Thus, if you lose your dog, be sure to have a rather precise idea of his speed and direction and do not lose time in pursuing him.

2.2　Primitives and Integrals

The aim of this subsection is to present an inverse of the differentiation operator. In fact, as revealed by the Darboux property (Exercise 1), not all functions from some interval T of $\mathbb{R}$ to a real Banach space X are derivatives. Therefore, we will get a primitive g of a function f on T only if f is regular enough. Here we use the following terminology.

Definition 2.12. A function $g : T \to X$ is said to be a *primitive* of $f : T \to X$ if g is continuous and if there exists a countable subset D of T such that for all $t \in T \setminus D$, g is differentiable at t and $g'(t) = f(t)$.

Corollary 2.10 ensures uniqueness of g.

Proposition 2.13. *If* g_1 *and* g_2 *are two primitives of an arbitrary function* $f : T \to X$, *then* $g_1 - g_2$ *is constant.*

Proof. If g_1 and g_2 are two primitives of f, then there exist countable subsets D_1 and D_2 of T such that g_i is differentiable on $T \setminus D_i$ and $g'_i(t) = f(t)$ for all $t \in T \setminus D_i$ $(i = 1,2)$. Then for the countable set $D := D_1 \cup D_2$, the continuous function $g_1 - g_2$ is differentiable on $T \setminus D$ and its derivative is 0 there; thus $g_1 - g_2$ is constant.　□

In order to construct g from f, we use an integration process. Such a process is useful for many other purposes and is well known when $X = \mathbb{R}$. Since we focus on vector-valued functions, we are not too exacting about regularity assumptions, so that we choose a construction that is simpler than the Lebesgue–Bochner integration theory. For most purposes, integrating continuous functions would suffice. However, admitting simple discontinuities may be useful. The class we select is described in the next definition.

Definition 2.14. A function $f : T \to X$ from a compact interval $T := [a,b]$ of $\mathbb{R}$ to a real Banach space X is said to be *regulated* if for all $t \in [a,b)$ (resp. $t \in (a,b]$), f has a limit on the right $f(t_+) := \lim_{r \to t, r > t} f(r)$ (resp. on the left $f(t_-) := \lim_{s \to t, s < t} f(s)$).

The function f is said to be a (right-) *normalized regulated function* if it is regulated, if $f(b) = f(b_-)$, and if for all $t \in [a,b)$ one has $f(t_+) = f(t)$.

Real-valued monotone functions, vector-valued continuous functions, and step functions are regulated functions. Recall that $f : T \to X$ is a *step function* if there is a finite sequence $\sigma := (s_0, s_1, \ldots, s_k)$ with $s_0 = a < s_1 < \cdots < s_k = b$, called a *subdivision* of T, such that f is constant on each open interval (s_{i-1}, s_i) for $i = 1, \ldots, k$. The step function f is said to be a (right-) normalized step function if f is constant on $[s_{i-1}, s_i)$ for $i = 1, \ldots, k-1$ and on $[s_{k-1}, b]$. We leave the proofs of the following results as exercises (see [294]).

Proposition 2.15. *Let X be a Banach space and let T be a compact interval of $\mathbb{R}$. A function $f : T \to X$ is regulated (resp. normalized regulated) if and only if it is the uniform limit of a sequence (f_n) of step functions (resp. normalized step functions).*

It follows that a regulated function on T is bounded. Moreover:

Proposition 2.16. *For every regulated function $f : T \to X$, the set $f(T)$ is relatively compact in X (i.e., $\mathrm{cl}(f(T))$ is compact). Moreover, the set of discontinuities of f is at most countable.*

The next statement can be either derived from Proposition 2.15 or proved directly (see [294]).

Proposition 2.17. *The space $R(T,X)$ (resp. $R_n(T,X)$) of regulated functions (resp. normalized regulated functions) from a compact interval T to a Banach space X endowed with the norm $\|\cdot\|_\infty$ given by $\|f\|_\infty := \sup_{t \in T} \|f(t)\|$ is a Banach space.*

The *integral* of a step function f can be defined unambiguously as follows: if $s_0 = a < s_1 < \cdots < s_k = b$ is such that $f(t) = c_i$ for $t \in (s_{i-1}, s_i)$, $i \in \mathbb{N}_k$, then

$$\int_T f := \int_a^b f(t)dt := \sum_{i=1}^k (s_{i+1} - s_i)c_i.$$

It is easy to show that this element of X does not depend on the subdivision of T. Moreover, for every step function f from T to X, the triangle inequality ensures that

$$\left\| \int_T f \right\| \leq (b-a)\|f\|_\infty. \tag{2.3}$$

Since the space $S(T,X)$ of step functions is dense in the space $R(T,X)$, the map $f \mapsto \int_T f$ can be extended by continuity from $S(T,X)$ to $R(T,X)$:

$$\int_T f = \lim_n \int_T f_n \quad \text{if} \ f = \lim_n f_n, \quad f_n \in S(T,X).$$

This extension is linear, continuous, and with norm $b - a$, since (2.3) remains valid for $f \in R(T,X)$. Moreover, given $a \le b \le c$ in $\mathbb{R}$, for all $f \in R([a,c],X)$ one has *Chasles's relation*

$$\int_a^c f = \int_a^b f + \int_b^c f.$$

It easily follows from the case of step functions by a passage to the limit.

The following composition property is crucial: using continuous linear forms x^* on X, it enables one to uniquely determine the integral of a regulated function $f \in R(T,X)$ with the help of the integrals of the real-valued functions $x^* \circ f$.

Proposition 2.18. *Given Banach spaces X, Y and $A \in L(X,Y)$, for every $f \in R(T,X)$ one has $A \circ f \in R(T,Y)$ and $\int_T A \circ f = A(\int_T f)$.*

Proof. The first assertion is a direct consequence of the definition. It can also be checked by taking a sequence (f_n) in $S(T,X)$ that converges uniformly to f. Since the relation $\int_T A \circ f_n = A(\int_T f_n)$ is immediate, the second assertion follows from the definition of the integral of $A \circ f$ as $\lim \int_T A \circ f_n$, since $(A(\int_T f_n)) \to A(\int_T f)$, A being continuous and $(\int_T f_n)$ converging to $\int_T f$. $\qquad\square$

The next result gives a partial inverse of the differentiation operator.

Theorem 2.19. *For $f \in R(T,X)$, the map $g : t \mapsto \int_a^t f(s)\mathrm{d}s$ is a primitive of f on T.*

Proof. Given $t \in [a,b]$, $\varepsilon > 0$, let $\delta \in (0, b-t)$ be such that $\|f(t+r) - f(t_+)\| \le \varepsilon$ for every $r \in (0,\delta]$. Since for $c := f(t_+)$ one has $\int_t^{t+r} c = rc$, it follows from Chasles's relation and (2.3) that

$$\left\| \int_a^{t+r} f - \int_a^t f - rc \right\| = \left\| \int_t^{t+r} (f - c) \right\| \le r\varepsilon.$$

This relation shows that $g : t \mapsto \int_a^t f(s)\mathrm{d}s$ has a right derivative at t whose value is c. Similarly, if $t \in (a,b]$, then g has $f(t_-)$ as a left derivative at t. Therefore, if f is continuous at $t \in (a,b)$, then g is differentiable at t and $g'(t) = f(t)$. Since the set D of discontinuities of f is countable, we get that g is differentiable on $T \setminus D$ with derivative f. Moreover, g is continuous on T in view of Chasles's relation and (2.3). $\qquad\square$

Corollary 2.20. *If $f : T \to X$ is continuous, then $g : t \mapsto \int_a^t f(s)\mathrm{d}s$ is of class C^1 (i.e., differentiable with a continuous derivative) and its derivative is f.*

Let us give two rules that are useful for the computation of primitives.

Proposition 2.21 (Change of variables). *Let $h : S = [\alpha, \beta] \to \mathbb{R}$ be the primitive of a regulated function h' such that $h(S) \subset T$ and let $f \in R(T,X)$. If either f is continuous or h is strictly monotone, then $s \mapsto h'(s)f(h(s))$ is regulated and for all $r \in [\alpha, \beta]$ one has*

$$\int_{\alpha}^{r} h'(s)f(h(s))\mathrm{d}s = \int_{h(\alpha)}^{h(r)} f(t)\mathrm{d}t. \tag{2.4}$$

Proof. When f is continuous, since h is continuous, $f \circ h$ is continuous and then $k : s \mapsto h'(s)f(h(s))$ is regulated; the same is true when h is either increasing or decreasing. Then the left-hand side of equality (2.4) is the value at r of the primitive j of k satisfying $j(\alpha) = 0$. The right-hand side is $g(h(r))$, where g is the primitive of f satisfying $g(h(\alpha)) = 0$. Under each of our assumptions, for a countable subset D of S, the derivative of $g \circ h$ at $r \in S \setminus D$ exists and is $h'(r)g'(h(r)) = h'(r)f(h(r))$. The uniqueness of the primitive of k null at α gives the equality. □

Proposition 2.22 (Integration by parts). *Let X, Y, and Z be Banach spaces, let $(x,y) \mapsto x * y$ be a continuous bilinear map from $X \times Y$ into Z, and let $f : T \to X$, $g : T \to Y$ be primitives of regulated functions, with $T := [a,b]$. Then*

$$\int_{a}^{b} f(t) * g'(t)\mathrm{d}t = f(b) * g(b) - f(a) * g(a) - \int_{a}^{b} f'(t) * g(t)\mathrm{d}t.$$

Proof. The functions $t \mapsto f(t) * g'(t)$ and $t \mapsto f'(t) * g(t)$ clearly have one-sided limits at all points of $T := [a,b]$. Moreover, their sum is the derivative of $h : t \mapsto f(t) * g(t)$ on $T \setminus D$, where D is the countable set of nondifferentiability of f or g. Thus the result amounts to the equality $\int_{a}^{b} h'(t)\mathrm{d}t = h(b) - h(a)$, which stems from the uniqueness of the primitive of h' that takes the value 0 at a. □

Exercises

1. **(Darboux property)** Show that the derivative f of a differentiable function $g : T \to \mathbb{R}$ satisfies the intermediate value property: given $a, b \in T$ with $f(a) < f(b)$ and $r \in (f(a), f(b))$, there exists some c between a and b such that $f(c) = r$.

2. Show that there exist a continuous function $f : \mathbb{R} \to \mathbb{R}$ and two continuous functions g_1, g_2 whose difference is not constant and are such that g_1 and g_2 are differentiable on $\mathbb{R} \setminus N$, where N is a set of measure zero, with $g_1'(t) = g_2'(t) = f(t)$ for all $t \in \mathbb{R} \setminus N$. [Hint: Take $f = 0$, $g_1 = 0$ and for g_2 take an increasing function whose derivative is 0 a.e.]

3. Prove Theorem 2.8. [See [197, 294].]

4. Prove Proposition 2.15. [See [294, 7.6.1].]

5. Show that every (right-) normalized step function on $T := [a,b]$ can be written as a linear combination of the functions $(e_t)_{t \in T}$ given by $e_b = 1$, and for $t \in [a,b)$, $e_t(r) = 1$ for $r \in [a,t)$, $e_t(r) = 0$ for $r \in [t,b]$. Give a generalization to the case of step functions taking their values in a normed space.

6. A function $v : T \to X$ from an interval $T := [a,b]$ of $\mathbb{R}$ to a normed space X is said to be of *bounded variation* if there exists some $c \in \mathbb{R}_+$ such that for every subdivision $\sigma := (s_0, s_1, \ldots, s_k)$ of T one has $\Sigma_{1 \le i \le k} \|v(s_i) - v(s_{i-1})\| \le c$. The infimum of such constants c is denoted by $V_a^b(v)$ and called the variation of v on $[a,b]$.

(a) Prove that the space $BV(T,X)$ of functions of bounded variation on T forms a normed space for the norm $v \mapsto \|v\|_{BV(T,X)} := \|v(a)\| + V_a^b(v)$.
(b) Show that a function of bounded variation is regulated.
(c) Show that Lipschitzian functions with values in X and monotone functions with values in $\mathbb{R}$ are of bounded variation.
(d) Check that the function f defined by $f(0) := 0$, $f(x) := x^2 \sin(1/x^2)$ for $x \in \mathbb{R} \setminus \{0\}$ is not of bounded variation on $T := [0,1]$ although it has a derivative at each point of T.
(e) Given $a < b < c$ in $\mathbb{R}$ and $v \in BV([a,c],X)$, show that $V_a^c(v) = V_a^b(v) + V_b^c(v)$ and that $s \mapsto V_a^s(v)$ is a nondecreasing function.
(d) Prove that for all $v \in BV(T) := BV(T,\mathbb{R})$ there exist nondecreasing functions v_1, v_2 such that $v = v_1 - v_2$. [Hint: Take $v_1 := (1/2)(w+v)$, $v_2 := (1/2)(w-v)$ with $w(t) := V_a^t(v)$ for $t \in T$.]

7. (Stieltjes integral) Given a function $v \in BV(T) := BV(T,\mathbb{R})$ for $T := [a,b]$ and a (right-) normalized step function f from T to a Banach space X, let $I_v(f) := \Sigma_{1 \le i \le k} V_a^{t_i}(v) c_i$ if $f := \Sigma_{1 \le i \le k} c_i e_{t_i}$, where $c_i \in X$ and e_{t_i} is defined as in Exercise 5.

(a) Show that $I_v(f)$ does not depend on the decomposition of f. Check that $\|I_v(f)\| \le V_a^b(v) \|f\|_\infty$.
(b) Deduce from the inequality above that the map $f \mapsto I_v(f)$ can be extended to a linear map from the space $R_n(T,X)$ of normalized regulated functions with values in X into X satisfying the same inequality. This map is called the *Stieltjes integral* of f relative to v.
(c) Conversely, given a continuous linear form f^* on the space $R_n(T) := R_n(T,\mathbb{R})$, let $v(t) := f^*(e_t)$, where e_t is defined in Exercise 5. Show that v is of bounded variation on T and that $V_a^b(v) \le \|f^*\|$.
(d) Deduce from what precedes a correspondence between the (topological) dual of the space $R_n(T)$ and the space $BV(T)$. [See [692].]

2.3 Directional Differential Calculus

Now let us consider maps from an open subset W of a normed space X into another normed space Y. A natural means of reducing the study of differentiability to the one-variable case consists in taking restrictions to line segments or regular curves in W.

Definition 2.23. Let X, Y be normed spaces, let W be an open subset of X, let $\bar{x} \in W$, and let $f : W \to Y$. We say that f has a *radial derivative* at $\bar{x}$ in the direction $u \in X$ if $(1/t)(f(\bar{x}+tu) - f(\bar{x}))$ has a limit as $t \to 0_+$. We denote by $f_r'(\bar{x}, u)$ or $d_r f(\bar{x}, u)$ this limit. If f has a radial derivative at $\bar{x}$ in every direction u, we say that f is *radially differentiable* at $\bar{x}$. If, moreover, the map $D_r f(\bar{x}) : u \mapsto d_r f(\bar{x}, u)$ is linear and continuous, we say that f is *Gâteaux differentiable* at $\bar{x}$ and call $D_r f(\bar{x})$ the Gâteaux derivative of f at $\bar{x}$.

One often says that f is directionally differentiable at $\bar{x}$, but we prefer to keep this terminology for a slightly more demanding notion that we consider now. In fact, although the notion of radial differentiability is simple and useful, it has several drawbacks; the main one is that this notion does not enjoy a chain rule. This variant does enjoy such a rule and reflects a smoother behavior of f when the direction u is submitted to small changes.

Definition 2.24. Let X, Y be normed spaces, let W be an open subset of X, let $\bar{x} \in W$, and let $f : W \to Y$. We say that f has a *directional derivative* at $\bar{x}$ in the direction $u \in X$, or that f is differentiable at $\bar{x}$ in the direction u, if $(1/t)(f(\bar{x}+tv) - f(\bar{x}))$ has a limit as $(t, v) \to (0_+, u)$. We denote by $f'(\bar{x}, u)$ or $df(\bar{x}, u)$ this limit. If f has a directional derivative at $\bar{x}$ in every direction u, we say that f is *directionally differentiable* at $\bar{x}$. If, moreover, the map $f'(x) := Df(x) : u \mapsto f'(\bar{x}, u)$ is linear and continuous, we say that f is *Hadamard differentiable* at $\bar{x}$.

The concepts of directional derivative and radial derivative are different, as the next example shows. Thus, it is convenient to dispose of two notations.

Example–Exercise. Let $f : \mathbb{R}^2 \to \mathbb{R}$ be given by $f(r, s) = (r^4 + s^2)^{-1} r^3 s$ for $(r, s) \in \mathbb{R}^2 \setminus \{(0,0)\}$, $f(0,0) = 0$. It is Gâteaux differentiable at $(0,0)$ but not directionally differentiable at $(0,0)$. □

The (frequent) use of the same notation for the radial and directional derivatives is justified by the following observation showing the compatibility of the two notions.

Proposition 2.25. *If X and Y are normed spaces, if W is an open subset of X and if $f : W \to Y$ has a directional derivative at $\bar{x}$ in the direction u, then it has a radial derivative at $\bar{x}$ in the direction u and both derivatives coincide. In particular, if f is Hadamard differentiable at $\bar{x}$, then it is Gâteaux differentiable at $\bar{x}$.*

Conversely, if f is Lipschitzian on a neighborhood V of $\bar{x}$, then f is directionally differentiable at $\bar{x}$ in every direction u in which f is radially differentiable.

Proof. The first assertions stem from an application of the definition of a limit.

Let us prove the converse assertion. Let k be the Lipschitz rate of f on V and let $u \in X$ be such that f is radially differentiable at $\bar{x}$ in the direction u. Setting $r(t, v) := f(\bar{x}+tv) - f(\bar{x}) - tf_r'(\bar{x}, u)$, we have $t^{-1} r(t, u) \to 0$ as $t \to 0$, and since $\|t^{-1}(r(t, v) - r(t, u))\| = \|t^{-1}(f(\bar{x}+tv) - f(\bar{x}+tu))\| \le k\|v - u\| \to 0$ as $(t, v) \to (0_+, u)$, we get $t^{-1} r(t, v) \to 0$ as $(t, v) \to (0_+, u)$. □

While radial differentiability of f at $\bar{x}$ in the direction u is equivalent to differentiability of the function $f_{\bar{x},u} : t \mapsto f(\bar{x}+tu)$ at 0, directional differentiability of f at $\bar{x}$ amounts to differentiability of the composition of f with curves issued from $\bar{x}$ with the initial direction u, as the next proposition shows.

Proposition 2.26. *The map $f : W \to Y$ is differentiable at $\bar{x}$ in the direction $u \in X \setminus \{0\}$ if and only if f is radially differentiable at $\bar{x}$ in the direction u and for every $\tau > 0$ and every (continuous) $c : [0, \tau] \to W$ that is right differentiable at 0 with $c'_+(0) = u$, $c(0) = \bar{x}$, the map $f \circ c$ is right differentiable at 0 and $(f \circ c)'_+(0) = d_r f(\bar{x}, u)$.*

Proof. Suppose f is differentiable at $\bar{x}$ in the direction $u \in X$. Given $\tau > 0$ and $c : [0, \tau] \to W$ that is right differentiable at 0 with $c'_+(0) = u$ and $c(0) = \bar{x}$, let us set $v_t := (1/t)(c(t) - c(0))$, so that $v_t \to u$ as $t \to 0_+$. Then

$$\frac{f(c(t)) - f(c(0))}{t} = \frac{f(\bar{x}+tv_t) - f(\bar{x})}{t} \to df(\bar{x}, u) \text{ as } t \to 0_+.$$

Now let us prove the sufficient condition. Suppose f has a radial derivative at $\bar{x}$ in the direction u but is not differentiable at $\bar{x}$ in the direction $u \neq 0$. There exist $\varepsilon > 0$ and some sequence $(t_n, u_n) \to (0_+, u)$ such that $\bar{x} + t_n u_n \in W$ for all $n \in \mathbb{N}$ and

$$\left\| \frac{f(\bar{x}+t_n u_n) - f(\bar{x})}{t_n} - d_r f(\bar{x}, u) \right\| \geq \varepsilon. \tag{2.5}$$

We may assume that $t_{n+1} \leq (1/2)t_n$. Then let us define $c : [0, t_0] \to X$ by $c(0) := \bar{x}$,

$$c(t) := \bar{x} + (t_n - t_{n+1})^{-1} [(t_n - t)t_{n+1}u_{n+1} + (t - t_{n+1})t_n u_n]$$

for $t \in [t_{n+1}, t_n)$. Then one sees that $(1/t)(c(t) - c(0)) \to u$ as t goes to 0, but since $c(t_n) = \bar{x} + t_n u_n$, in view of (2.5), $f \circ c$ is not differentiable at 0 with derivative $d_r f(\bar{x}, u)$.   □

Corollary 2.27. *Let X, Y be normed spaces, let T, W be open subsets of $\mathbb{R}$ and X respectively, let $c : T \to X$ be differentiable at $\bar{t} \in T$ and let $f : W \to Y$ be Hadamard differentiable at $\bar{x} \in W$ and such that $c(T) \subset W$, $\bar{x} = c(\bar{t})$. Then $f \circ c$ is differentiable at $\bar{t}$ and*

$$(f \circ c)'(\bar{t}) = Df(\bar{x})(c'(\bar{t})).$$

Thus, $Df(\bar{x})$ appears as the continuous linear map transforming velocities.

It is easy to show that every linear combination of maps having radial (resp. directional) derivatives at $\bar{x}$ in some direction u has a radial (resp. directional) derivative at $\bar{x}$ in the direction u. In particular, every linear combination of two Gâteaux (resp. Hadamard) differentiable maps is Gâteaux (resp. Hadamard) differentiable. One also deduces from Proposition 2.3 that if f has a directional (resp. radial) derivative at $\bar{x}$ in the direction u and if $A : Y \to Z$ is a continuous linear map, then $A \circ f$ has a directional (resp. radial) derivative at $\bar{x}$ in the direction u and $(A \circ f)'(\bar{x}, u) = A(f'(\bar{x}, u))$.

The preceding example–exercise shows that the composition of two radially differentiable maps is not necessarily radially differentiable. However, one does have a chain rule for directionally differentiable maps. These facts show that Hadamard differentiability is a more interesting property than Gâteaux differentiability.

Theorem 2.28. *Let X, Y, Z be normed spaces, let U and V be open subsets of X and Y respectively, and let $f : U \to Y$, $g : V \to Z$ be directionally differentiable maps at $\overline{x} \in W := f^{-1}(V)$ and $\overline{y} := f(\overline{x}) \in V$ respectively. Then $h := g \circ f$ is directionally differentiable at $\overline{x}$ and*

$$d\,(g \circ f)\,(\overline{x}, u) = dg(f(\overline{x}), df(\overline{x}, u)).$$

In particular, if f is Hadamard differentiable at $\overline{x}$ and g is Hadamard differentiable at $\overline{y} := f(\overline{x})$, then $h := g \circ f$ is Hadamard differentiable at $\overline{x}$ and

$$D(g \circ f)(\overline{x}) = Dg(\overline{y}) \circ Df(\overline{x}).$$

Proof. More generally, let us show that if f has a directional derivative at $\overline{x}$ in the direction $u \in X$ and if g has a directional derivative at $f(\overline{x})$ in the direction $v := df(\overline{x}, u)$, then $h := g \circ f$ has a directional derivative at $\overline{x}$ in the direction u. For (t, u') close enough to $(0, u)$ one has $\overline{x} + tu' \in W$. Let $q(t, u') := (1/t)(f(\overline{x} + tu') - f(\overline{x}))$. Then $q(t, u') \to v := df(\overline{x}, u)$ as $(t, u') \to (0_+, u)$. Therefore

$$\frac{h(\overline{x} + tu') - h(\overline{x})}{t} = \frac{g(\overline{y} + tq(t, u')) - g(\overline{y})}{t} \to dg(\overline{y}, v) \ \text{ as } (t, u') \to (0_+, u).$$

The statement can also be proved using Proposition 2.26.   □

The notion of radial differentiability is sufficient to get a mean value theorem. Recall that the *segment* $[a, b]$ (respectively (a, b)) with endpoints a, b in a normed space is the set $\{(1 - t)a + tb : t \in [0, 1]\}$ (respectively $\{(1 - t)a + tb : t \in (0, 1)\}$).

Proposition 2.29. *If $f : W \to Y$ is radially differentiable at each point of a segment $[w, x]$ contained in W, then*

$$\|f(x) - f(w)\| \le \sup_{t \in (0,1)} \|d_r f(w + t(x - w), x - w)\|.$$

Proof. Let $h : [0, 1] \to Y$ be given by $h(t) := f((1 - t)w + tx)$; it is right differentiable on $(0, 1)$, with right derivative $h'_+(t) = d_r f((1 - t)w + tx, x - w)$, and continuous on $[0, 1]$. Corollary 2.9 then yields the estimate.   □

A variant can be deduced when f is Gâteaux differentiable at each point of $S := (a, b)$, since then one has $\|d_r f(z, x - w)\| \le \|D_r f(z)\| \cdot \|x - w\|$ for all $z \in S$, $w, x \in X$.

Proposition 2.30. *Let X and Y be normed spaces, let W be an open subset of X containing the segment $[w, x]$, and let $f : W \to Y$ be continuous on $[w, x]$ and Gâteaux* *differentiable at each point of* $S := (w, x)$, *with* $m := \sup_{z \in S} \|D_r f(z)\| < +\infty$. *Then one has*

$$\|f(x) - f(w)\| \leq m \|x - w\|.$$

Corollary 2.31. *Let X and Y be normed spaces, let W be a convex open subset of X, and let $f : W \to Y$ be Gâteaux differentiable at each point of W and such that for some $c \in \mathbb{R}$ one has $\|D_r f(w)\| \leq c$ for every $w \in W$. Then f is Lipschitzian with rate c: for all $x, x' \in W$ one has*

$$\left\| f(x) - f(x') \right\| \leq c \left\| x - x' \right\|.$$

In particular, if $D_r f(w) = 0$ for every $w \in W$, then f is constant on W. Such a result is also valid if W is connected instead of convex. An extension of the estimate of Proposition 2.30 is also valid in the case that W is connected, provided one replaces the usual distance with the geodesic distance d_W in W defined as in Exercise 5.

In the usual case in which $X_0 = X$, the following corollary gives an approximation of f in the case that one has an approximate value of the derivative of f around $\bar{x}$.

Corollary 2.32. *Let X and Y be normed spaces, let X_0 be a linear subspace of X, let W be a convex open subset of X, and let $f : W \to Y$ be Gâteaux differentiable at each point of W and such that for some $c \in \mathbb{R}$ and some $\ell \in L(X_0, Y)$ one has $\|D_r f(x)(u) - \ell(u)\| \leq c \|u\|$ for every $x \in W$, $u \in X_0$. Then for every $x, x' \in W$ such that $x - x' \in X_0$, one has*

$$\left\| f(x) - f(x') - \ell(x - x') \right\| \leq c \left\| x - x' \right\|.$$

This result (obtained by changing f into $f - \ell$ in the preceding corollary) will serve to get Fréchet differentiability from Gâteaux differentiability. For the moment, let us point out another passage from Gâteaux differentiability to Hadamard differentiability.

Proposition 2.33. *Let W be an open subset of X. If $f : W \to Y$ is radially differentiable on a neighborhood V of $\bar{x}$ in W and if for some $u \in X \setminus \{0\}$, its radial derivative $d_r f : V \times X \to Y$ is continuous at $(\bar{x}, u)$, then f is directionally differentiable at $\bar{x}$ in the direction u.*

In particular, if f is Gâteaux differentiable on V and if $d_r f : V \times X \to Y$ is continuous at each point of $\{\bar{x}\} \times X$, then f is Hadamard differentiable at $\bar{x}$.

Proof. Without loss of generality, we may suppose u has norm 1. Given $\varepsilon > 0$, let $\delta \in (0, 1)$ be such that $\|f_r'(x, v) - f_r'(\bar{x}, u)\| \leq \varepsilon$ for all $(x, v) \in B(\bar{x}, 2\delta) \times B(u, \delta)$, with $B(\bar{x}, 2\delta) \subset V$. Setting $r(t, v) := f(\bar{x} + tv) - f(\bar{x}) - tf_r'(\bar{x}, u)$, we observe that for every $v \in B(u, \delta)$ the map $r_v := r(\cdot, v)$ is differentiable on $[0, \delta]$ and $\|r_v'(t)\| = \|f_r'(\bar{x} + tv, v) - f_r'(\bar{x}, u)\| \leq \varepsilon$. Since $r_v(0) = 0$, Corollary 2.9 yields $\|r(t, v)\| \leq \varepsilon t$. That shows that f has $f_r'(\bar{x}, u)$ as a directional derivative at $\bar{x}$ in the direction u. The last assertion is an immediate consequence. $\square$

The importance of this continuity condition leads us to introduce a definition.

Definition 2.34. Given normed spaces X, Y and an open subset W of X, a map $f : W \to Y$ is said to be of *class* D^1 at $\overline{w}$ (resp. on W) if it is Hadamard differentiable around $\overline{w}$ (resp. on W) and if $df : W \times X \to Y$ is continuous at $(\overline{w}, v)$ for all $v \in X$ (resp. on $W \times X$). We say that f is of class D^k with $k \in \mathbb{N}$, $k > 1$, if f is of class D^1 and if df is of class D^{k-1}.

We denote by $D^1(W, Y)$ the space of maps of class D^1 from W to Y and by $BD^1(W, Y)$ the space of maps $f \in D^1(W, Y)$ that are bounded and such that f' is bounded from W to $L(X, Y)$. Let us note the following two properties.

Proposition 2.35. *For every $f \in D^1(W, Y)$ the map $f' : w \mapsto Df(w) := df(w, \cdot)$ is locally bounded.*

Proof. Suppose, to the contrary, that there exist $w \in W$ and a sequence $(w_n) \to w$ such that $(r_n) := (\|Df(w_n)\|) \to +\infty$. For each $n \in \mathbb{N}$ one can pick some unit vector $u_n \in X$ such that $\|df(w_n, u_n)\| > r_n - 1$. Setting (for $n \in \mathbb{N}$ large) $x_n := r_n^{-1} u_n$, we see that $((w_n, x_n)) \to (w, 0)$ but $(\|df(w_n, x_n)\|) \to 1$, a contradiction. $\square$

Corollary 2.36. *Let $f : W \to Y$ be a Hadamard (or Gâteaux) differentiable function. Then f is of class D^1 if and only if f' is locally bounded and for all $u \in X$ the map $x \mapsto f'(x)u$ is continuous. In particular, if $Y = \mathbb{R}$ and if $f \in D^1(W, \mathbb{R})$, the derivative is continuous when X^* is provided with the topology of uniform convergence on compact sets (the bw^* topology).*

Proof. The necessary condition stems from the preceding proposition. The sufficient condition follows from the inequalities

$$\|f'(w)v - f'(x)u\| \le \|f'(w)(v - u)\| + \|f'(w)u - f'(x)u\| \le m\varepsilon/(2m) + \varepsilon/2 = \varepsilon,$$

when for some $m > 0$ and a given $\varepsilon > 0$ one can find a neighborhood V of x in W such that $\|f'(w)\| \le m$ for $w \in V$ and $\|f'(w)u - f'(x)u\| \le \varepsilon/2$ for $w \in V$, $w \in B(u, \varepsilon/2m)$. $\square$

Proposition 2.37. *If X, Y, Z are normed spaces, if U and V are open subsets of X and Y respectively, and if $f \in D^1(U, Y)$, $g \in D^1(V, Z)$, then $h := g \circ f \in D^1(W, Z)$ for $W := f^{-1}(V)$.*

Proof. This conclusion is an immediate consequence of the formula $dh(w, x) = dg(f(w), df(w, x))$ for all $(w, x) \in W \times X$. $\square$

Under a differentiability assumption, convex functions, integral functionals, and Nemitskii operators are important examples of maps of class D^1.

Example (Nemitskii operators). Let $(S, \mathscr{F}, \mu)$ be a measure space, let X, Y be Banach spaces, let $f : S \times X \to Y$ be a measurable map of class D^1 in its second variable and such that $g : (s, x, v) \mapsto df_s(x, v)$ is measurable, f_s being the map $x \mapsto f(s, x)$. Then, if for $p, q \in [1, +\infty)$, the Nemitskii operator $F : L_p(S, X) \to$

$L_q(S,Y)$ given by $F(u) := f(\cdot, u(\cdot))$ for $u \in L_p(S,X)$ is well defined and Gâteaux differentiable, with derivative given by $D_r F(u)(v) = df(\cdot, u(\cdot), v(\cdot))$, then F is of class D^1. This follows from the following result applied to $g := df(\cdot, u(\cdot), v(\cdot))$ (see [37]).

Lemma 2.38 (Krasnoselskii's theorem). *Let $(S, \mathscr{F}, \mu)$ be a measure space, let W, Z be Banach spaces, and let $g : S \times W \to Z$ be a measurable map such that for all $s \in S \setminus N$, where N has null measure, the map $g(s, \cdot)$ is continuous. If for some $p, q \in [1, +\infty)$ and all $u \in L_p(S, W)$ the map $g(\cdot, u(\cdot))$ belongs to $L_q(S, Z)$, then the Nemitskii operator $G : L_p(S, W) \to L_q(S, Z)$ given by $G(u) := g(\cdot, u(\cdot))$ for $u \in L_p(S, X)$ is continuous.*

Exercises

1. Let X, Y be normed spaces and let W be an open subset of X. Prove that $f : W \to Y$ is Hadamard differentiable at $\bar{x}$ if and only if there exists a continuous linear map $\ell : X \to Y$ such that the map q_t given by $q_t(v) := (1/t)(f(\bar{x} + tv) - f(\bar{x}))$ converges to ℓ as $t \to 0_+$, uniformly on compact subsets of X. Deduce another proof of Proposition 2.50 below from this characterization.

2. Prove that if $f : W \to Y$ is radially differentiable at $\bar{x}$ in the direction u and if f is *directionally steady* at $\bar{x}$ in the direction u in the sense that $(1/t)(f(\bar{x} + tv) - f(\bar{x} + tu)) \to 0$ as $(t, v) \to (0_+, u)$, then f is directionally differentiable at $\bar{x}$ in the direction u. Give an example showing that this criterion is more general than the Lipschitz condition of Proposition 2.25.

3. Let $f : \mathbb{R}^2 \to \mathbb{R}$ be given by $f(r,s) := r^2 s(r^2 + s^2)^{-1}$ for $(r,s) \in \mathbb{R}^2 \setminus \{(0,0)\}$, $f(0,0) = 0$. Show that f has a radial derivative (which is in fact a bilateral derivative) but is not Gâteaux differentiable at $(0,0)$.

4. Let E be a Hilbert space and let $X := D^1(T, E)$, where $T := [0,1]$. Endow X with the norm $\|x\| := \sup_{t \in T} \|x(t)\| + \sup_{t \in T} \|x'(t)\|$. Define the length of a curve $x : [0,1] \to E$ by

$$\ell(x) := \int_0^1 \|x'(t)\| \, dt.$$

(a) Show that ℓ is a continuous sublinear functional on X with Lipschitz rate 1.
(b) Let W be the set of $x \in X$ such that $x'(t) \neq 0$ for all $t \in [0,1]$. Show that W is an open subset of X and that ℓ is Gâteaux differentiable on W.
(c) Show that ℓ is of class D^1 on W [Hint: Use convergence results for integrals.] In order to prove that ℓ is of class C^1 one may use the following questions.
(d) Let $E_0 := E \setminus \{0\}$ and let $D : E_0 \to E$ be given by $D(v) := \|v\|^{-1} v$. Given $u, v \in E_0$ show that $\|D(u) - D(v)\| \leq 2\|u\|^{-1}\|u - v\|$.
(e) Deduce from the preceding inequality that ℓ' is continuous.

5. Prove the assertion following Corollary 2.31, defining the geodesic distance $d_W(x,x')$ between two points x,x' of W as the infimum of the lengths of curves joining x to x'.

6. Prove that if $f : W \to Y$ has a directional derivative at some point $\bar{x}$ of the open subset W of X, then its derivative $Df(\bar{x}) : u \mapsto df(\bar{x}, u)$ is continuous if it is linear.

7. Prove Proposition 2.29 by deducing it from the classical mean value theorem (Lemma 2.6) for real-valued functions, using the Hahn–Banach theorem. [Hint: Take y^* with norm one such that $\langle y^*, y \rangle = \|y\|$ for $y := f(x) - f(w)$, set $g(t) := \langle y^*, f(x+t(w-x)) \rangle$, and pick $\theta \in (0,1)$ such that $g(1) - g(0) = g'_+(\theta)$.]

8. Show that the norm $x \mapsto \|x\| := \sup_{t \in T} |x(t)|$ on the Banach space $X := C(T)$ of continuous functions on $T := [0,1]$ is Hadamard differentiable at $\bar{x} \in X$ if and only if the function $t \mapsto |x(t)|$ attains its maximum on T at a single point.

9. (a) Let a, b be two points of a normed space X. Show that the function g given by $g(t) := \|a + tb\|$ has a right derivative and a left derivative at all points of $\mathbb{R}$.

   (b) Let $f : T \to X$, where T is an interval of $\mathbb{R}$. Show that if f has a right derivative $f'_+(t)$ at some $t \in T$, then $g \circ f$ has a right derivative at t and $(g \circ f)'_+(t) \leq \|f'_+(t)\|$.

10. Use the preceding exercise to deduce a mean value theorem from Lemma 2.6.

2.4   Fréchet Differential Calculus

Nonlinear maps are difficult to study. The main purpose of differential calculus consists in getting some information using an affine approximation to a given nonlinear map around a given point. Of course, the meaning of the word "approximation" has to be made precise. For that purpose, we define remainders. Fréchet differentiability consists in an approximation by a continuous affine map.

Definition 2.39. Given normed spaces X and Y, we denote by $o(X,Y)$ the set of maps $r : X \to Y$ such that $r(x)/\|x\| \to 0$ as $x \to 0$ in $X \setminus \{0\}$. The elements of $o(X,Y)$ will be called *remainders*.

Thus, $r : X \to Y$ is a remainder if and only if there exists some map $\alpha : X \to Y$ satisfying $\alpha(x) \to 0$ as $x \to 0$ and $r(x) = \|x\| \alpha(x)$. Moreover, $r \in o(X,Y)$ if and only if there exists a modulus $\mu : \mathbb{R}_+ \to \mathbb{R}_+ \cup \{+\infty\}$ such that $\|r(x)\| \leq \mu(\|x\|)\|x\|$ (recall that $\mu : \mathbb{R}_+ \to \mathbb{R}_+ \cup \{+\infty\}$ is a *modulus* when μ is nondecreasing, $\mu(0) = 0$, and μ is continuous at 0). Such a case occurs when there exist $c > 0$ and $p > 1$ such that $\|r(x)\| \leq c \|x\|^p$. Following Landau, remainders are often denoted by $o(\cdot)$, and different remainders are often denoted by the same letters, since they are considered as inessential for the assigned purposes.

If $r, s : X \to Y$ are two maps that coincide on some neighborhood V of 0 in X, then s belongs to $o(X,Y)$ if and only if r belongs to $o(X,Y)$. Thus if $p : V \to Y$ is defined on some neighborhood V of 0 in X, we consider that p is a remainder if some extension r of p to all of X is a remainder. The preceding observation shows that this property does not depend on the choice of the extension.

The following result is a direct consequence of the rules for limits.

Lemma 2.40. *For every pair of normed spaces X, Y, the set $o(X,Y)$ of remainders is a linear space.*

The class of remainders is stable under composition by continuous linear maps.

Lemma 2.41. *For all normed spaces W, X, Y, Z, for every $r \in o(X,Y)$ and all continuous linear maps $A : W \to X$, $B : Y \to Z$ one has $r \circ A \in o(W,Y)$ and $B \circ r \in o(X,Z)$ (hence $B \circ r \circ A \in o(W,Z)$).*

Proof. Let $\alpha : X \to Y$ be such that $\alpha(x) \to 0$ as $x \to 0$ and $r(x) = \|x\| \alpha(x)$. Then if $A : W \to X$ is *stable* at 0, i.e., is such that there exists some $c > 0$ for which $\|A(w)\| \leq c \|w\|$ for w in a neighborhood of 0 in W, in particular if A is linear and continuous, then one has $\|r(A(w))\| = \|A(w)\| \|\alpha(A(w))\| \leq c \|w\| \|\alpha(A(w))\|$ and $\alpha(A(w)) \to 0$ as $w \to 0$, so that $r \circ A \in o(W,Y)$. Similarly, if $B : Y \to Z$ is stable at 0, then $B \circ r \in o(X,Z)$. The assertion about $B \circ r \circ A$ is a combination of the two other cases. $\square$

The proof of the next lemma is an easy consequence of the rules for limits.

Lemma 2.42. *Given normed spaces $X, Y_1, \ldots, Y_k$, $Y := Y_1 \times \cdots \times Y_k$, a map $r : X \to Y$ is a remainder if and only if its components $r_1, \ldots, r_k$ are remainders.*

We are ready to define differentiability in the Fréchet sense; this notion is so usual that one often writes "differentiable" instead of "Fréchet differentiable."

Definition 2.43. Given normed spaces X, Y and an open subset W of X, a map $f : W \to Y$ is said to be *(Fréchet) differentiable* (or firmly differentiable, or just differentiable) at $\bar{x} \in W$ if there exist a continuous linear map $\ell : X \to Y$ and a remainder $r \in o(X,Y)$ such that for $x \in W$ one has

$$f(x) = f(\bar{x}) + \ell(x - \bar{x}) + r(x - \bar{x}). \tag{2.6}$$

It is often convenient to write the preceding relation in the form

$$f(\bar{x} + u) - f(\bar{x}) = \ell(u) + r(u)$$

for u close to 0. Here the continuous affine map $x \mapsto f(\bar{x}) + \ell(x - \bar{x})$ can be viewed as an approximation of f that essentially determines the behavior of f around $\bar{x}$. The continuous linear map ℓ is called the *derivative* of f at $\bar{x}$ and is denoted by $Df(\bar{x})$ or $f'(\bar{x})$. It is unique: given two approximations ℓ_1, ℓ_2 of $f(\bar{x} + \cdot)$ around 0 and two remainders r_1, r_2 such that $f(\bar{x} + u) - f(\bar{x}) = \ell_1(u) + r_1(u) = \ell_2(u) + r_2(u)$, one has

$\ell_1 = \ell_2$, since $\ell := \ell_1 - \ell_2$ is the remainder $r := r_2 - r_1$; in fact, for every $u \in X$ and every $t > 0$ small enough, one has

$$\ell(u) = \frac{1}{t} r(tu) = \frac{1}{t} \alpha(tu) \|tu\| = \alpha(tu) \|u\| \to 0 \text{ as } t \to 0,$$

so that $\ell(u) = 0$. Thus $L(X,Y) \cap o(X,Y) = \{0\}$. Uniqueness is also a consequence of Corollary 2.50 below and of the fact that the directional derivative is unique, since it is obtained as a limit.

When $Y := \mathbb{R}$, the derivative $Df(\bar{x})$ of f at $\bar{x}$ belongs to the dual X^* of X. When X is a Hilbert space with scalar product $(\cdot \mid \cdot)$ it may be convenient to use the *Riesz isometry* $R : X \to X^*$ given by $\langle R(x), y \rangle = (x \mid y)$ to get an element $\nabla f(\bar{x})$ of X, called the *gradient* of f at x, by setting $\nabla f(\bar{x}) := R^{-1}(Df(\bar{x}))$. It allows one to visualize the derivative, but in some respects, it is preferable to work with the derivative.

Proposition 2.44. *If $f : W \to Y$ is differentiable at $\bar{x} \in W$, then it is continuous at $\bar{x}$.*

Proof. This follows from the fact that every remainder is continuous at 0. □

Proposition 2.45. *If $f, g : W \to Y$ are differentiable at $\bar{x} \in W$, then for every $\lambda, \mu \in \mathbb{R}$ the map $h := \lambda f + \mu g$ is differentiable at $\bar{x}$ and $Dh(\bar{x}) = \lambda Df(\bar{x}) + \mu Dg(\bar{x})$.*

Proof. If $r(x) := f(\bar{x} + x) - f(\bar{x}) - f'(\bar{x})(x)$, $s(x) := g(\bar{x} + x) - g(\bar{x}) - g'(\bar{x})x$, one has $h(\bar{x} + x) = h(\bar{x}) + \lambda f'(\bar{x})(x) + \mu g'(\bar{x})(x) + t(x)$, where $t := \lambda r + \mu s \in o(X,Y)$. Thus h is differentiable at $\bar{x}$ and $h'(\bar{x}) = \lambda f'(\bar{x}) + \mu g'(\bar{x})$. □

Examples. (a) A constant map is everywhere differentiable and its derivative is 0.
(b) A continuous linear map $\ell \in L(X,Y)$ is differentiable at every point $\bar{x}$ and its derivative at $\bar{x}$ is ℓ since $\ell(\bar{x} + x) = \ell(\bar{x}) + \ell(x)$.
(c) A continuous affine map $f := \ell + c$, where $\ell \in L(X,Y)$ and $c \in Y$, is differentiable at every $\bar{x} \in X$ and $Df(\bar{x}) = \ell$.
(d) If $f : X := X_1 \times X_2 \to Y$ is a continuous bilinear map, then f is differentiable at every point $\bar{x} := (\bar{x}_1, \bar{x}_2) \in X$, and for $x = (x_1, x_2)$, one has $Df(\bar{x})(x) = f(x_1, \bar{x}_2) + f(\bar{x}_1, x_2)$, since $f(\bar{x} + x) - f(\bar{x}) = f(x_1, \bar{x}_2) + f(\bar{x}_1, x_2) + f(x_1, x_2)$. Here f is a remainder since $\|f(x)\| \leq \|f\| \|x_1\| \|x_2\| \leq \|f\| \|x\|^2$ whenever $\|x\| \geq \|x\|_\infty := \max(\|x_1\|, \|x_2\|)$.
(e) If $f : X \to Y$ is a continuous *quadratic map* in the sense that there exists a continuous bilinear map $b : X \times X \to Y$ such that $f(x) = b(x,x)$, then f is differentiable at every point $\bar{x} \in X$ and $Df(\bar{x})(x) = b(\bar{x},x) + b(x,\bar{x})$ for $x \in X$. This follows from the chain rule below and the preceding example. Alternatively, one may observe that $r := f$ is a remainder, since for every $x \in X$ one has $\|f(x)\| \leq \|b\| \|x\|^2$ and $f(\bar{x} + x) = f(\bar{x}) + b(\bar{x},x) + b(x,\bar{x}) + f(x)$.
(f) If $f : T \to Y$ is defined on an open interval T of $\mathbb{R}$, then f is differentiable at $\bar{x} \in T$ if and only if f has a derivative at $\bar{x}$ and $Df(\bar{x})$ is the linear map $r \mapsto rf'(\bar{x})$, whence $f'(\bar{x}) = Df(\bar{x})(1)$. The key point in this example is illuminated in the following exercise. □

Exercise. Show that for every normed space Y the space $L(\mathbb{R}, Y)$ is isomorphic (and even isometric) to Y via the evaluation map $\ell \mapsto \ell(1)$, whose inverse is the map $v \mapsto \ell_v$, where $\ell_v \in L(\mathbb{R}, Y)$ is defined by $\ell_v(r) := rv$ for $r \in \mathbb{R}$. $\qquad\square$

The following characterization will be helpful.

Lemma 2.46. *Given an open subset W of X, a map $f : W \to Y$ is differentiable at $\bar{x}$ if and only if there exists a map $F : W \to L(X, Y)$ that is continuous at $\bar{x}$ and such that $f(x) - f(\bar{x}) = F(x)(x - \bar{x})$ for all $x \in W$.*

Proof. Suppose there is a map $F : W \to L(X, Y)$ continuous at $\bar{x}$ such that $f(x) = f(\bar{x}) + F(x)(x - \bar{x})$ for all $x \in W$. Then $f(x) - f(\bar{x}) = F(\bar{x})(x - \bar{x}) + r(x)$, where r is the remainder defined by $r(x) := (F(\bar{x} + x) - F(\bar{x}))(x)$, so that f is differentiable at $\bar{x}$ and $Df(\bar{x}) = F(\bar{x})$. To prove the converse, using the Hahn–Banach theorem, for $x \in W$ we pick $\ell_x \in X^*$ such that $\|\ell_x\| = 1$ and $\ell_x(x) = \|x\|$. Then, setting $A := Df(\bar{x})$ and writing the remainder r appearing in (2.6) in the form $r(u) = \alpha(u)\|u\| = \alpha(u)\ell_u(u)$ with $\alpha(u) \to 0$ as $u \to 0$, we get

$$f(\bar{x} + u) - f(\bar{x}) = (A + \alpha(u)\ell_u)(u),$$

or $f(x) - f(\bar{x}) = F(x)(x - \bar{x})$ for $F(x) := A + \alpha(x - \bar{x})\ell_{x - \bar{x}} \to A = F(\bar{x})$ as $x \to \bar{x}$. $\qquad\square$

Let us give a chain rule. It is a cornerstone of differential calculus.

Theorem 2.47 (Chain rule). *Let X, Y, Z be normed spaces, let U, V be open subsets of X and Y respectively, and let $f : U \to Y$, $g : V \to Z$ be differentiable at $\bar{x} \in U$ and $\bar{y} = f(\bar{x})$ respectively and be such that $f(U) \subset V$. Then $h := g \circ f$ is differentiable at $\bar{x}$ and*

$$Dh(\bar{x}) = Dg(\bar{y}) \circ Df(\bar{x}). \tag{2.7}$$

Proof. Let $\ell := Df(\bar{x})$, $m := Dg(\bar{y})$ and let $r \in o(X, Y)$, $s \in o(Y, Z)$ be defined by

$$r(x) := f(\bar{x} + x) - f(\bar{x}) - \ell(x), \quad s(y) := g(\bar{y} + y) - g(\bar{y}) - m(y).$$

Then, setting $y := \ell(x) + r(x)$ for $x \in U - \bar{x}$, so that $f(\bar{x} + x) = \bar{y} + y$, we get

$$h(\bar{x} + x) - h(\bar{x}) - m(\ell(x)) = g(\bar{y} + y) - g(\bar{y}) - m(y - r(x)) = s(y) + m(r(x)). \tag{2.8}$$

Lemma 2.41 ensures that $m \circ r \in o(X, Z)$. Now, given $c > \|\ell\|$, there exists some $\rho > 0$ such that for $x \in B(0, \rho)$ one has $\|r(x)\| \leq (c - \|\ell\|)\|x\|$ and hence $\|\ell(x) + r(x)\| \leq c\|x\|$. Then the proof of Lemma 2.41 ensures that $s \circ (\ell + r) \in o(X, Z)$. Thus, the right-hand side $s \circ (\ell + r) + m \circ r$ of (2.8) is a remainder, and we conclude that h is differentiable at $\bar{x}$ with derivative the continuous linear map $m \circ \ell$. $\qquad\square$

The following corollary is a consequence of the fact that the derivative of a continuous linear map ℓ at an arbitrary point is ℓ itself.

Corollary 2.48. *Let X, Y, Z be normed spaces, let U, V be open subsets of X and Y respectively, and let $f : U \to Y$, $g : V \to Z$ be such that $f(U) \subset V$ and let $h := g \circ f$.*

(a) If f is differentiable at $\bar{x}$ and $V := Y$, $g \in L(Y, Z)$, then h is differentiable at $\bar{x}$ and $Dh(\bar{x}) = g \circ Df(\bar{x})$.

(b) If g is differentiable at $\bar{y} := f(\bar{x})$ and $U := X$, $f \in L(X, Y)$, then h is differentiable at $\bar{x}$ and $Dh(\bar{x}) = Dg(\bar{y}) \circ f$.

Corollary 2.49. *The differentiability of $f : W \to Y$ (with W open in X) at $\bar{x}$ does not depend on the choices of the norms on X and Y within their equivalences classes.*

In fact, changing the norm amounts to composing with a continuous linear map.

Corollary 2.50. *Let X, Y be normed spaces, let W be an open subset of X, and let $f : W \to Y$. If f is Fréchet differentiable at $\bar{x} \in W$, then f is Hadamard differentiable at $\bar{x}$. If X is finite-dimensional, the converse holds.*

Thus, the mean value theorems of Sect. 2.1.2 are in force for Fréchet differentiability. Also, the interpretation of the derivative as a rule for the transformation of velocities remains valid for the Fréchet derivative.

Proof. The first assertion follows from the definitions or from Theorem 2.47 and Proposition 2.26.

Assuming that X is finite-dimensional, let us prove that if f is directionally differentiable at $\bar{x}$, and if its directional derivative $f'(\bar{x}, \cdot)$ is continuous, then r given by

$$r(w) := f(\bar{x} + w) - f(\bar{x}) - f'(\bar{x}, w)$$

is a remainder. Adding the assumption that $f'(\bar{x}, \cdot)$ is linear will prove the converse assertion. Suppose, to the contrary, that there exist $\varepsilon > 0$ and a sequence $(w_n) \to 0$ such that for all $n \in \mathbb{N}$, $\|r(w_n)\| > \varepsilon \|w_n\|$. Then $t_n := \|w_n\|$ is positive; setting $u_n := t_n^{-1} w_n$, we may suppose the sequence (u_n) converges to some vector u of the unit sphere of X. Then, given $\varepsilon' \in (0, \varepsilon)$, we can find $k \in \mathbb{N}$ such that for $n \geq k$ we have $\|f'(\bar{x}, u_n) - f'(\bar{x}, u)\| \leq \varepsilon - \varepsilon'$, so that

$$\left\| f(\bar{x} + t_n u_n) - f(\bar{x}) - t_n f'(\bar{x}, u) \right\| > \varepsilon t_n \|u_n\| - t_n \left\| f'(\bar{x}, u_n) - f'(\bar{x}, u) \right\| \geq \varepsilon' t_n,$$

a contradiction to the assumption that f is differentiable at $\bar{x}$ in the direction u. $\square$

Another link between directional differentiability and firm differentiability is pointed out in the next statement. A direct proof using Corollary 2.32 is easy. We present a proof in the case that f' is continuous around $\bar{x}$.

Proposition 2.51. *If f is Gâteaux differentiable on W and if $f' : W \to L(X, Y)$ is continuous at $\bar{x} \in W$, then f is Fréchet differentiable at $\bar{x}$.*

Proof. Without loss of generality, replacing Y by its completion, we may suppose Y is complete; replacing W by a ball centered at $\bar{x}$, we may also suppose W is convex. Then for $x \in W$ one has $f(x) - f(\bar{x}) = F(x)(x - \bar{x})$ with

$$F(x) := \int_0^1 Df(\bar{x}+t(x-\bar{x}))dt,$$

and F is continuous at $\bar{x}$, so that the criteria of Lemma 2.46 apply. □

This result shows that it may be a sensible strategy to start with radial differentiability in order to prove that a map is of *class* C^1, i.e., that it is differentiable with a continuous derivative. For instance, if one deals with an integral functional

$$f(x) := \int_S F(s,x(s))ds,$$

where S is some measure space and x belongs to some space of measurable maps, it is advisable to use Lebesgue's theorem to differentiate inside the integral (under appropriate assumptions) by taking the limit in the quotient

$$\frac{1}{t}[f(\bar{x}+tu)-f(\bar{x})] = \int_S \frac{1}{t}[F(s,\bar{x}(s)+tu(s)) - F(s,\bar{x}(s))]ds.$$

Continuity arguments may be invoked later, for instance using Krasnoselskii's criterion.

Let us note other consequences of Theorem 2.47.

Proposition 2.52. *Let X, $Y_1,\ldots,Y_n$ be normed spaces, let W be an open subset of X, and let $f := (f_1,\ldots,f_n) : W \to Y := Y_1 \times \cdots \times Y_n$. Then f is differentiable at $\bar{x} \in W$ if and only if its components $f_i : W \to Y_i$ ($i = 1,\ldots,n$) are differentiable at $\bar{x}$ and for $v \in X$,*
$$Df(\bar{x})(v) = (Df_1(\bar{x})(v),\ldots,Df_n(\bar{x})(v)).$$

Proof. Let $p_i : Y \to Y_i$ denote the ith canonical projection. If f is differentiable at $\bar{x}$, then Corollary 2.48 ensures that $f_i := p_i \circ f$ is differentiable at $\bar{x}$ and $Df_i(\bar{x}) = p_i \circ Df(\bar{x})$. Conversely, suppose that $f_1,\ldots,f_n$ are differentiable at $\bar{x}$. Let $r_i \in o(X,Y_i)$ be given by $r_i(x) = f_i(\bar{x}+x) - f_i(\bar{x}) - Df_i(\bar{x})(x)$. Then by Lemma 2.42, we have that $r := (r_1,\ldots,r_n) \in o(X,Y)$ and $r(x) = f(\bar{x}+x) - f(\bar{x}) - \ell(x)$ for $\ell \in L(X,Y)$ given by $\ell(x) := (Df_1(\bar{x})(x),\ldots,Df_n(\bar{x})(x))$. Thus f is differentiable at $\bar{x}$, with derivative ℓ. □

Now, let us consider the case in which the source space X is a product $X_1 \times \cdots \times X_n$ and W is an open subset of X. One says that $f : W \to Y$ has a *partial derivative at $\bar{x} \in W$ relative to X_i* for some $i \in \mathbb{N}_n$ if the map $f_{i,\bar{x}} : x_i \mapsto f(\bar{x}_1,\ldots,\bar{x}_{i-1},x_i,\bar{x}_{i+1},\ldots,\bar{x}_n)$ is differentiable at $\bar{x}_i$. Then one denotes by $D_i f(\bar{x})$ or $\frac{\partial f}{\partial x_i}(\bar{x})$ the derivative of the map $f_{i,\bar{x}}$ at $\bar{x}_i$. Let $j_i \in L(X_i,X)$ be the insertion given by $j_i(x_i) := (0,\ldots,0,x_i,0,\ldots,0)$. Since the map $f_{i,\bar{x}}$ is just the composition of the affine map $x_i \mapsto j_i(x_i - \bar{x}_i) + \bar{x} = (\bar{x}_1,\ldots,\bar{x}_{i-1},x_i,\bar{x}_{i+1},\ldots,\bar{x}_n)$ with f, from Corollary 2.48 (b) and the fact that $v = j_1(v_1) + \cdots + j_n(v_n)$, while $D_i f(\bar{x}) = Df_{i,\bar{x}}(\bar{x}_i) = Df(\bar{x}) \circ j_i$, one gets the following proposition.

Proposition 2.53. *If $f : W \to Y$ is defined on an open subset W of a product space $X := X_1 \times \cdots \times X_k$ and if f is differentiable at $\overline{x}$, then for $i = 1, \ldots, k$, the map f has a partial derivative at $\overline{x}$ relative to X_i and*

$$\forall v := (v_1, \ldots, v_k), \qquad Df(\overline{x})(v) = D_1 f(\overline{x}) v_1 + \cdots + D_k f(\overline{x}) v_k.$$

When $X := \mathbb{R}^m$, $Y := \mathbb{R}^n$, the matrix $(D_i f_j(\overline{x}))$ of $Df(\overline{x})$ formed with the partial derivatives of the components $(f_j)_{1 \le j \le n}$ of f is called the *Jacobian matrix* of f at $\overline{x}$. It determines $Df(\overline{x})$.

Note that it may happen that f has partial derivatives at $\overline{x}$ with respect to all its variables but is not differentiable at $\overline{x}$.

Example. Let $f : \mathbb{R}^2 \to \mathbb{R}$ be given by $f(r,s) := rs(r^2 + s^2)^{-1}$ for $(r,s) \neq (0,0)$ and $f(0,0) = 0$. Since $f(r,0) = 0 = f(0,s)$, f has partial derivatives with respect to its two variables at $(0,0)$. However, f is not continuous at $(0,0)$, hence is not differentiable at $(0,0)$. □

Now let us introduce a reinforced notion of differentiability that allows us to formulate several results with assumptions weaker than continuous differentiability.

Definition 2.54. Let X and Y be normed spaces, let W be an open subset of X, and let $\overline{x} \in W$. A map $f : W \to Y$ is said to be *circa-differentiable* (or peri-differentiable, or strictly differentiable) at $\overline{x}$ if there exists some continuous linear map $\ell \in L(X,Y)$ such that for every $x, x' \in W$ one has

$$\frac{\|f(x) - f(x') - \ell(x - x')\|}{\|x - x'\|} \to 0 \text{ as } x, x' \to \overline{x} \text{ with } x' \neq x. \tag{2.9}$$

If X_0 is a linear subspace of X, we say that f is *circa-differentiable* (or strictly differentiable) *at $\overline{x}$ with respect to X_0* if there exists some continuous linear map $\ell \in L(X_0, Y)$ such that (2.9) holds whenever $x, x' \in W$ satisfy $x - x' \in X_0$.

Let us relate the preceding notion to continuous differentiability. Taking $x' = \overline{x}$ in relation (2.9), one sees that if f is circa-differentiable at $\overline{x}$, then f is differentiable at $\overline{x}$ and $Df(\overline{x}) = \ell$.

Definition 2.55. The map $f : W \to Y$ will be said to be continuously differentiable at $\overline{x} \in W$, or of *class C^1* at $\overline{x}$, and we write $f \in C^1_{\overline{x}}(W, Y)$, if f is differentiable on some neighborhood V of $\overline{x}$ and if the derivative $f' : V \to L(X,Y)$ of f given by $f'(x) := Df(x)$ for $x \in V$ is continuous at $\overline{x}$. If f is of class C^1 at each point x of W, then f is said to be of class C^1 on W and one writes $f \in C^1(W, Y)$.

One says that f is of class C^k with $k \in \mathbb{N}$, $k > 1$, if f is of class C^1 and if f' is of class C^{k-1}. Then one writes $f \in C^k(W, Y)$.

Proposition 2.56. *Let X and Y be normed spaces, let W be an open subset of X and let $\overline{x} \in W$. A map $f : W \to Y$ that is differentiable on a neighborhood $U \subset W$ of $\overline{x}$ is circa-differentiable at $\overline{x} \in W$ if and only if $f \in C^1_{\overline{x}}(W, Y)$.*

Proof. Suppose $f \in C^1_{\bar{x}}(W, Y)$ and let $\ell := Df(\bar{x})$. Given $\varepsilon > 0$ one can find $\delta > 0$ such that $B(\bar{x}, \delta) \subset W$ and for $x \in B(\bar{x}, \delta)$ one has $\|Df(x) - \ell\| \leq \varepsilon$. Then using Corollary 2.32, for $x, x' \in B(\bar{x}, \delta)$, one has

$$\|f(x') - f(x) - \ell(x' - x)\| \leq \varepsilon \|x' - x\|,$$

so that f is circa-differentiable at $\bar{x}$.

Conversely, suppose f is circa-differentiable at $\bar{x}$ and is differentiable on a neighborhood V of $\bar{x}$ contained in W. Given $u \in X$ and $\varepsilon > 0$, assuming that the preceding inequality holds whenever $x, x' \in B(\bar{x}, \delta) \subset V$, one gets for all $x \in B(\bar{x}, \delta)$, $u \in X$

$$\|Df(x)(u) - \ell(u)\| = \lim_{t \to 0_+} t^{-1} \|f(x + tu) - f(x) - \ell(tu)\| \leq \varepsilon \|u\|,$$

so that $\|Df(x) - \ell\| \leq \varepsilon$ and $f' : x \mapsto Df(x)$ is continuous at $\bar{x}$. $\square$

We are now in a position to give a converse of Proposition 2.53.

Proposition 2.57. *If $f : W \to Y$ is defined on an open subset W of a product space $X := X_1 \times \cdots \times X_k$, if for $i = 1, \ldots, k$, f has a partial derivative at $\bar{x} \in W$ relative to X_i, and if f is circa-differentiable at $\bar{x}$ with respect to $X_1, \ldots, X_{i-1}, X_{i+1}, \ldots, X_k$, then f is differentiable at $\bar{x}$. In particular, if f has partial derivatives on some neighborhood of $\bar{x}$ all of which but one are continuous at $\bar{x}$, then f is differentiable at $\bar{x}$.*

Proof. It suffices to give the proof for $k = 2$; an induction yields the general case.

Thus, let f be circa-differentiable at $\bar{x}$ with respect to X_1 and have a partial derivative at $\bar{x}$ relative to X_2. The first assumption means that there exists some $\ell_1 \in L(X_1, Y)$ such that for every $\varepsilon > 0$ one can find some $\delta > 0$ such that $B(\bar{x}, 2\delta) \subset W$ and for $x := (x_1, x_2) \in B(\bar{x}, \delta)$, $u_1 \in X_1$, $\|u_1\| \leq \delta$ one has

$$\|f(x_1 + u_1, x_2) - f(x_1, x_2) - \ell_1(u_1)\| \leq \varepsilon \|u_1\|. \tag{2.10}$$

Setting $\ell_2 := D_2 f(\bar{x})$ and taking a smaller $\delta > 0$ if necessary, we may suppose that

$$\|f(\bar{x}_1, \bar{x}_2 + u_2) - f(\bar{x}_1, \bar{x}_2) - \ell_2(u_2)\| \leq \varepsilon \|u_2\|$$

for every $u_2 \in X_2$ satisfying $\|u_2\| \leq \delta$. Then, taking $(x_1, x_2) := (\bar{x}_1, \bar{x}_2 + u_2)$ in (2.10) with $u := (u_1, u_2) \in B(0, \delta)$, we get

$$\|f(\bar{x} + u) - f(\bar{x}) - \ell_1(u_1) - \ell_2(u_2)\|$$
$$\leq \|f(\bar{x} + u) - f(\bar{x}_1, \bar{x}_2 + u_2) - \ell_1(u_1)\| + \|f(\bar{x}_1, \bar{x}_2 + u_2) - f(\bar{x}_1, \bar{x}_2) - \ell_2(u_2)\|$$
$$\leq \varepsilon \|u_1\| + \varepsilon \|u_2\| = \varepsilon \|(u_1, u_2)\|$$

if one takes the norm on X given by $\|(u_1, u_2)\| := \|u_1\| + \|u_2\|$. $\square$

Corollary 2.58. *A map $f : W \to Y$ defined on an open subset W of a product space $X := X_1 \times \cdots \times X_k$ is of class C^1 on W if and only if f has partial derivatives on W that are jointly continuous.*

Now let us give a result dealing with the interchange of limits and differentiation.

Theorem 2.59. *Let (f_n) be a sequence of Fréchet (resp. Hadamard) differentiable functions from a bounded, convex, open subset W of a normed space X to a Banach space Y. Suppose*

(a) There exists some $\overline{x} \in W$ such that $(f_n(\overline{x}))$ converges in Y
(b) The sequence (f_n') uniformly converges on W to some map $g : W \to L(X, Y)$

Then (f_n) uniformly converges on W to some map f that is Fréchet (resp. Hadamard) differentiable on W. Moreover, $f' = g$.

Proof. Let us prove the first assertion. Let $r > 0$ be such that W is contained in the ball $B(\overline{x}, r)$. Given n, p in $\mathbb{N}$, Corollary 2.31 yields, for every $x \in W$,

$$\left\| f_p(x) - f_p(\overline{x}) - (f_n(x) - f_n(\overline{x})) \right\| \leq \sup_{w \in W} \left\| f_p'(w) - f_n'(w) \right\| \cdot \|x - \overline{x}\| \leq r \left\| f_p' - f_n' \right\|_\infty, \tag{2.11}$$

$$\left\| f_p(x) - f_n(x) \right\| \leq \left\| f_p(\overline{x}) - f_n(\overline{x}) \right\| + r \left\| f_p' - f_n' \right\|_\infty. \tag{2.12}$$

Since $\left\| f_p' - f_n' \right\|_\infty \to 0$ as $n, p \to \infty$ and since $(f_p(\overline{x}) - f_n(\overline{x})) \to 0$ as $n, p \to \infty$, we see that $(f_n(x))$ is a Cauchy sequence, hence has a limit in the complete space Y; we denote it by $f(x)$. Passing to the limit on p in (2.12) we see that the limit is uniform on W.

Now, given $x \in W$, let us prove that f is differentiable at x with derivative $g(x)$. Given $\varepsilon > 0$, we can find $k \in \mathbb{N}$ such that for $p > n \geq k$ one has $\left\| f_p' - f_n' \right\|_\infty \leq \varepsilon/3$, hence $\|g' - f_n'\|_\infty \leq \varepsilon/3$. Using again Corollary 2.31 with $x' := x + u \in W$, we get

$$\left\| (f_p(x+u) - f_p(x)) - (f_n(x+u) - f_n(x)) \right\| \leq (\varepsilon/3) \|u\|,$$

and passing to the limit on p, we obtain

$$\|f(x+u) - f(x) - (f_n(x+u) - f_n(x))\| \leq (\varepsilon/3) \|u\|. \tag{2.13}$$

In the Fréchet differentiable case, we can find $\delta > 0$ such that $B(x, \delta) \subset W$ and for all $u \in \delta B_X$,

$$\|f_k(x+u) - f_k(x) - g(x)(u)\| \leq \left\| f_k(x+u) - f_k(x) - f_k'(x)(u) \right\|$$
$$+ \left\| f_k'(x)(u) - g(x)(u) \right\| \leq (\varepsilon/3) \|u\| + (\varepsilon/3) \|u\|.$$

Combining this estimate with relation (2.13), in which we take $n = k$, we get

$$\forall u \in \delta B_X, \qquad \|f(x+u) - f(x) - g(x)(u)\| \leq \varepsilon \|u\|,$$

so that f is Fréchet differentiable at x with derivative $g(x)$.

In the Hadamard differentiable case, given $\varepsilon > 0$ and a unit vector u, we take $\delta \in (0,1)$ such that $B(x, 2\delta) \subset W$ and for $t \in (0, \delta)$, $v \in B(u, \delta)$,

$$\|f_k(x+tv) - f_k(x) - g(x)(tu)\|$$
$$\leq \|f_k(x+tv) - f_k(x) - f_k'(x)(tu)\| + \|f_k'(x)(tu) - g(x)(tu)\| \leq (\varepsilon/3)t + (\varepsilon/3)t.$$

Gathering this estimate with relation (2.13), in which we take $n = k$, $u = tv$, we get

$$\forall (t,v) \in (0, \delta) \times B(u, \delta), \qquad \|f(x+tv) - f(x) - g(x)(tu)\| \leq \varepsilon t,$$

so that f is Hadamard differentiable at x and $f'(x) = g(x)$.   □

Corollary 2.60. *Let X, Y be normed spaces, Y being complete, and let W be an open subset of X. The space $B^1(W, Y)$ (resp. $BC^1(W, Y)$) of bounded, Lipschitzian, differentiable (resp. of class C^1) maps from W to Y is complete for the norm $\|\cdot\|_{1,\infty}$ given by*

$$\|f\|_{1,\infty} := \sup_{x \in W} \|f(x)\| + \sup_{x \in W} \|f'(x)\|.$$

Here we use the fact that if f is Lipschitzian and differentiable, its derivative is bounded.

Proof. Let (f_n) be a Cauchy sequence of $(B^1(W, Y), \|\cdot\|_{1,\infty})$. Then (f_n') is a Cauchy sequence of the space $B(W, L(X, Y))$ of bounded maps from W into $L(X, Y)$ for the uniform norm; thus it converges and its limit is continuous if $f_n \in BC^1(W, Y)$. Similarly, (f_n) converges in $B(W, Y)$. The theorem ensures that the limit f of (f_n) is Fréchet differentiable and its derivative is the limit of (f_n'), hence is bounded. Thus f belongs to $B^1(W, Y)$ and $(f_n) \to f$ for $\|\cdot\|_{1,\infty}$. If (f_n) is contained in $BC^1(W, Y)$, then f' is continuous, whence $f \in BC^1(W, Y)$.   □

A directional version follows similarly from Theorem 2.59.

Corollary 2.61. *Let X, Y be normed spaces, Y being complete, and let W be an open subset of X. The space $BH^1(W, Y)$ of bounded, Lipschitzian, Hadamard differentiable maps from W to Y is complete for the norm $\|\cdot\|_{1,\infty}$. The same is true for its subspace $BD^1(W, Y)$ formed by bounded, Lipschitzian maps of class D^1.*

Now let us derive the important Borwein–Preiss smooth variational principle from the Deville–Godefroy–Zizler theorem (Theorem 1.152). When $Y := \mathbb{R}$, we simplify the notation $B^1(X, Y)$ into $B^1(X)$, and we adopt similar simplifications for the other spaces.

Theorem 2.62 (Borwein–Preiss variational principle). *Let X be a Banach space and let $F := B^1(X)$ (resp. $BH^1(X)$, $BC^1(X)$, $BD^1(X)$) with the norm $\|\cdot\|_{1,\infty}$ defined above. Suppose there exists some nonnull function $b \in F$ with bounded support.*

Then, given a lower semicontinuous function $f : X \to \mathbb{R}_\infty$ that is bounded below, the set G of $g \in F$ such that $f + g$ is well-posed is generic in F.

Moreover, there exists some $\kappa > 0$ depending only on X such that for every $\varepsilon > 0$ and every $u \in X$ satisfying $f(u) < \inf f(X) + \kappa \varepsilon^2$ one can find some $g \in F$ satisfying $\|g\|_{1,\infty} \leq \varepsilon$ and some minimizer v of $f + g$ belonging to $B(u, \varepsilon)$.

Note that one has $f(v) - \varepsilon \leq f(v) + g(v) \leq f(u) + g(u) \leq f(u) + \varepsilon$, hence $f(v) \leq f(u) + 2\varepsilon$.

Proof. Conditions (b) and (c) of Theorem 1.152 are obviously satisfied, whereas (a) is part of our assumptions (here we have changed W into F in order to avoid confusion with what precedes). Moreover, $(F, \|\cdot\|)$ is complete by the preceding corollary. The last assertion follows from the corresponding localization property in Theorem 1.152 and the relation $\|g(t\cdot)\| \leq t \|g\|$ for $t \geq 1$, $g \in F$.   □

Exercises

1. **(a)** Show that $r : X \to Y$ is a remainder if and only if there exists a remainder ρ on $\mathbb{R}$ such that $\|r(x)\| \leq \rho(\|x\|)$ for all x close to 0.
   (b) Prove the other two characterizations of remainders that follow the definition.

2. Define a notion of directional remainder that could be used for the study of Hadamard differentiability.

3. Show that when $f : W \to Y$ is Fréchet differentiable at $\bar{x}$, then it is *stable* at $\bar{x}$ in the sense that there exists $c > 0$ such that $\|f(\bar{x} + x) - f(\bar{x})\| \leq c \|x\|$ for $\|x\|$ small enough.

4. Give a direct proof that Fréchet differentiability implies Hadamard differentiability.

5. Show that if $f : X_1 \times X_2 \to Y$ is circa-differentiable at $\bar{x} := (\bar{x}_1, \bar{x}_2)$ with respect to X_1 and X_2, then it is circa-differentiable at $\bar{x}$.

6. In Theorem 2.59, when W is not bounded, assuming that (f_n') converges to g uniformly on bounded subsets of W, get a similar interchange result in which the convergence of (f_n) to f is uniform on bounded subsets of the open convex set W.

7. In Theorem 2.59, assuming that W is a connected open subset of X and that the convergence of (f_n') is locally uniform (in the sense that for every $x \in W$ there exists some ball with center x contained in W on which the convergence of (f_n') is uniform), prove that (f_n) is locally uniformly convergent and that its limit f is differentiable with derivative g.

8. Give a direct proof of Proposition 2.51.

9. With the hypothesis of Proposition 2.51, show that the map f is circa-differentiable at $\bar{x}$. Is it of class C^1 at $\bar{x}$?

10. Express the chain rule for differentiable maps between $\mathbb{R}^m$, $\mathbb{R}^n$, $\mathbb{R}^p$ in terms of a matrix product for the Jacobians of f and g.

11. Using the Hahn–Banach theorem, show that $f : W \to Y$ is circa-differentiable at $a \in W$ if and only if there exists a map $F : W \times W \to L(X,Y)$ continuous at (a,a) such that $f(u) - f(v) = F(u,v)(u-v)$. Then $f'(a) = F(a,a)$.

12. Show that if X is finite-dimensional, then $f : U \to Y$, with U open in X, is of class D^1 if and only if f is of class C^1. [Hint: For every element e of a basis of X the map $x \mapsto Df(x)(e)$ is continuous when f is of class D^1.]

13. Given normed spaces X,Y and a topology $\mathscr{T}$ (or a convergence) on the space of maps from B_X to Y, one can define a notion of $\mathscr{T}$-semiderivative at $\bar{x}$ of a map $f : B(\bar{x},r) \to Y$: it consists in requiring that the family of maps $(f_t)_{0<t<r}$ from B_X to Y given by $f_t(v) := t^{-1}(f(\bar{x}+tv) - f(\bar{x}))$ have a limit as $t \to 0_+$. If the limit is the restriction to B_X of a continuous linear map, one speaks of a $\mathscr{T}$-derivative. Interpret Gâteaux, Hadamard, and Fréchet derivatives with the help of the topologies of uniform convergence on the families of finite subsets, compact subsets, and bounded subsets. Observe that such a process also applies to some other families of sets, such as the family of weakly compact subsets of B_X.

14. Show that the norm $x \mapsto \|x\| := \sup_{t \in T} |x(t)|$ on the Banach space $X := C(T)$ of continuous functions on $T := [0,1]$ is not Fréchet differentiable at any point. Compare with Exercise 8 of the preceding section.

15. Let X and Y be normed spaces, let $\bar{x} \in X$, c, $r > 0$, $W := B(\bar{x},r)$, $f : W \to Y$ be of class C^1 and such that $\|f'(x) - f'(\bar{x})\| \leq c\|x - \bar{x}\|$ for all $x \in W$.

(a) Show that $\|f(x) - f(\bar{x}) - f'(\bar{x})(x-\bar{x})\| \leq (c/2)\|x-\bar{x}\|^2$ for all $x \in W$.
(b) Suppose that f' is Lipschitzian with rate c on W. Show that for all $w,x \in W$ one has $\|f(x) - f(w) - f'(w)(x-w)\| \leq (c/2)\|x-w\|^2$.

2.5 Inversion of Differentiable Maps

In the present section, we show that simple methods linked with differentiability concepts lead to efficient ways of solving nonlinear systems or vectorial equations

$$f(x) = 0. \qquad (2.14)$$

Here X and Y are Banach spaces, W is an open subset of X, and $f : W \to Y$ is a map. We start with a classical constructive algorithm.

2.5.1 Newton's Method

Newton's method is an iterative process that relies on a notion of approximation by a linear map. We formulate it as follows.

Definition 2.63. The map $f : W \to Y$ has a *Newton approximation* at $\bar{x} \in W$ if there exist $r > 0$, $\alpha > 0$ and a map $A : B(\bar{x}, r) \to L(X, Y)$ such that $B(\bar{x}, r) \subset W$ and

$$\forall x \in B(\bar{x}, r), \qquad \|f(x) - f(\bar{x}) - A(x)(x - \bar{x})\| \le \alpha \|x - \bar{x}\|. \qquad (2.15)$$

A map $A : V \to L(X, Y)$ is a *slant derivative* of f at $\bar{x}$ if V is a neighborhood of $\bar{x}$ contained in W and if for every $\alpha > 0$ there exists some $r > 0$ such that $B(\bar{x}, r) \subset V$ and relation (2.15) holds.

Thus f is differentiable at $\bar{x}$ if and only if f has a slant derivative at $\bar{x}$ that is constant on some neighborhood of $\bar{x}$. But condition (2.15) is much less demanding, as the next lemma shows.

Lemma 2.64. *The following assertions are equivalent:*

(a) f has a Newton approximation A that is bounded near $\bar{x}$
(b) f has a slant derivative A at $\bar{x}$ that is bounded on some neighborhood of $\bar{x}$
(c) f is stable at $\bar{x}$, i.e., there exist $c > 0$, $r > 0$ such that

$$\forall x \in B(\bar{x}, r), \qquad \|f(x) - f(\bar{x})\| \le c \|x - \bar{x}\|. \qquad (2.16)$$

Proof. (a)$\Rightarrow$(c) If for some α, $\beta > 0$ and some $r > 0$ a map $A : B(\bar{x}, r) \to L(X, Y)$ is such that (2.15) holds with $\|A(x)\| \le \beta$ for all $x \in B(\bar{x}, r)$, then by the triangle inequality, relation (2.16) holds with $c := \alpha + \beta$.

(c)$\Rightarrow$(b) We use a corollary of the Hahn–Banach theorem asserting the existence of some map $s : X \to X^*$ such that $s(x)(x) = \|x\|$ and $\|s(x)\| = 1$ for all $x \in X$. Suppose (2.16) holds. Then setting $A(\bar{x}) = 0$ and for $w \in W \setminus \{\bar{x}\}$, $x \in X$,

$$A(w)(x) = \langle s(w - \bar{x}), x \rangle \frac{f(w) - f(\bar{x})}{\|w - \bar{x}\|},$$

we easily check that $\|A(w)\| \le c$ for all $w \in W$ and that $A(x)(x - \bar{x}) = f(x) - f(\bar{x})$ for all $x \in W$, so that (2.15) holds with $\alpha = 0$ and A is a slant derivative of f at $\bar{x}$.

(b)$\Rightarrow$(a) is obvious, a slant derivative of f at $\bar{x}$ being a Newton approximation of f at $\bar{x}$. $\qquad \square$

In the elementary Newton method that follows, we first assume that (2.14) has a solution $\bar{x}$.

Proposition 2.65. *Let $\bar{x}$ be a solution to (2.14), let α, $\beta, r > 0$ satisfy $\gamma := \alpha\beta < 1$, and let $A : B(\bar{x}, r) \to L(X, Y)$ be such that (2.15) holds, $A(x)$ being invertible with $\|A(x)^{-1}\| \le \beta$ for all $x \in B(\bar{x}, r)$. Then the sequence (x_n) given by*

$$x_{n+1} := x_n - A(x_n)^{-1}(f(x_n)) \qquad (2.17)$$

is well defined for every initial point $x_0 \in B(\bar{x}, r)$ and converges linearly to $\bar{x}$ with rate γ.

The last assertion means that $\|x_{n+1} - \bar{x}\| \le \gamma \|x_n - \bar{x}\|$, hence $\|x_n - \bar{x}\| \le c\gamma^n$ for some $c > 0$ (in fact $c := \|x_0 - \bar{x}\|$). Thus, if A is a slant derivative of f at $\bar{x}$, then (x_n) converges superlinearly to $\bar{x}$: for all $\varepsilon > 0$ there is some $k \in \mathbb{N}$ such that $\|x_{n+1} - \bar{x}\| \le \varepsilon \|x_n - \bar{x}\|$ for all $n \ge k$.

Proof. Using the fact that $f(\bar{x}) = 0$, so that

$$x_{n+1} - \bar{x} = A(x_n)^{-1}\left(f(\bar{x}) - f(x_n) + A(x_n)(x_n - \bar{x})\right),$$

we inductively obtain that

$$\|x_{n+1} - \bar{x}\| \le \beta \|f(x_n) - f(\bar{x}) - A(x_n)(x_n - \bar{x})\| \le \alpha\beta \|x_n - \bar{x}\|,$$

so that $x_{n+1} \in B(\bar{x}, r)$: the whole sequence (x_n) is well defined and converges to $\bar{x}$. $\qquad \square$

Under reinforced assumptions, one can show the existence of a solution.

Theorem 2.66 (Kantorovich). *Let $x_0 \in W$, α, $\beta > 0$, $r > 0$ with $\gamma := \alpha\beta < 1$, $B(x_0, r) \subset W$ and let $A : B(x_0, r) \to L(X, Y)$ be such that for all $x \in B(x_0, r)$ the map $A(x) : X \to Y$ has a right inverse $B(x) : Y \to X$ satisfying $\|B(x)(\cdot)\| \le \beta \|\cdot\|$ and*

$$\forall w, x \in B(x_0, r), \qquad \|f(w) - f(x) - A(x)(w - x)\| \le \alpha \|w - x\|. \quad (2.18)$$

If $\|f(x_0)\| < \beta^{-1}(1 - \gamma)r$ and if f is continuous, the sequence given by the Newton iteration

$$x_{n+1} := x_n - B(x_n)(f(x_n)) \qquad (2.19)$$

is well defined and converges to a solution $\bar{x}$ of (2.14). Moreover, one has $\|x_n - \bar{x}\| \le r\gamma^n$ for all $n \in \mathbb{N}$ and $\|\bar{x} - x_0\| \le \beta(1 - \gamma)^{-1}\|f(x_0)\| < r$.

Here $B(x)$ is a *right inverse* of $A(x)$ if $A(x) \circ B(x) = I_Y$; $B(x)$ is not assumed to be linear.

Proof. Let us prove by induction that $x_n \in B(x_0, r)$, $\|x_{n+1} - x_n\| \le \beta\gamma^n \|f(x_0)\|$, and $\|f(x_n)\| \le \gamma^n \|f(x_0)\|$. For $n = 0$ these relations are obvious. Assuming that they are valid for $n < k$, we get

$$\|x_k - x_0\| \le \sum_{n=0}^{k-1} \|x_{n+1} - x_n\| \le \beta \|f(x_0)\| \sum_{n=0}^{\infty} \gamma^n = \beta \|f(x_0)\| (1 - \gamma)^{-1} < r,$$

or $x_k \in B(x_0, r)$, and since $f(x_{k-1}) + A(x_{k-1})(x_k - x_{k-1}) = 0$, from (2.18), (2.19), we have

$$\|f(x_k)\| \leq \|f(x_k) - f(x_{k-1}) - A(x_{k-1})(x_k - x_{k-1})\| \leq \alpha \|x_k - x_{k-1}\| \leq \gamma^k \|f(x_0)\|$$

and

$$\|x_{k+1} - x_k\| \leq \beta \|f(x_k)\| \leq \beta \gamma^k \|f(x_0)\|.$$

Since $\gamma < 1$, the sequence (x_n) is a Cauchy sequence, hence converges to some $\bar{x} \in X$ satisfying $\|\bar{x} - x_0\| \leq \beta \|f(x_0)\| (1 - \gamma)^{-1} < r$. Moreover, by the continuity of f, we get $f(\bar{x}) = \lim_n f(x_n) = 0$. Finally,

$$\|x_n - \bar{x}\| \leq \lim_{p \to +\infty} \|x_n - x_p\| \leq \lim_{p \to +\infty} \sum_{k=n}^{p-1} \|x_{k+1} - x_k\| \leq r\gamma^n. \qquad \square$$

We deduce from Kantorovich's theorem a result that is the root of important estimates in nonlinear analysis.

Theorem 2.67 (Lyusternik–Graves theorem). *Let X and Y be Banach spaces, let W be an open subset of X, and let $g : W \to Y$ be circa-differentiable at some $\bar{x} \in W$ with a surjective derivative $Dg(\bar{x})$. Then g is open at $\bar{x}$. More precisely, there exist some $\rho, \sigma, \kappa > 0$ such that g has a right inverse $h : B(g(\bar{x}), \sigma) \to W$ satisfying $\|h(y) - \bar{x}\| \leq \kappa \|g(\bar{x}) - y\|$ for all $y \in B(g(\bar{x}), \sigma)$ and*

$$\forall (w, y) \in B(\bar{x}, \rho) \times B(g(\bar{x}), \sigma) \quad \exists x \in W : \ g(x) = y, \ \|x - w\| \leq \kappa \|g(w) - y\|. \tag{2.20}$$

Proof. Let $A : W \to L(X, Y)$ be the constant map with value $A := Dg(\bar{x})$ (we use a familiar abuse of notation). The open mapping theorem yields some $\beta > 0$ and some right inverse $B : Y \to X$ of A such that $\|B(\cdot)\| \leq \beta \|\cdot\|$. Let $\alpha, r > 0$ be such that $\gamma := \alpha\beta < 1$, $B(\bar{x}, 2r) \subset W$ and

$$\forall w, x \in B(\bar{x}, 2r), \quad \|g(w) - g(x) - Dg(\bar{x})(w - x)\| \leq \alpha \|w - x\|. \tag{2.21}$$

Let $\sigma, \tau > 0$ be such that $\sigma + \tau < \beta^{-1}(1 - \gamma)r$, and let $\rho \in (0, r]$ be such that $g(w) \in B(g(\bar{x}), \tau)$ for all $w \in B(\bar{x}, \rho)$. Given $w \in B(\bar{x}, \rho)$, $y \in B(g(\bar{x}), \sigma)$, let us set $f(x) := g(x) - y$ for $x \in B(\bar{x}, \rho)$, so that $\|f(w)\| \leq \|g(w) - g(\bar{x})\| + \|g(\bar{x}) - y\| < \beta^{-1}(1 - \gamma)r$, and by (2.21), we have that (2.18) holds in the ball $B(x_0, r)$, with $x_0 := w$. Using the estimate $\|x - x_0\| \leq \beta \|f(x_0)\| (1 - \gamma)^{-1} < r$ obtained in the proof of Kantorovich's theorem for a solution x of the equation $f(x) = 0$, we get some $x \in W$ such that $g(x) = y$, $\|x - w\| \leq \kappa \|g(w) - y\|$ with $\kappa := \beta(1 - \gamma)^{-1}$. The right inverse h is obtained by taking $w := \bar{x}$ in (2.20). $\qquad \square$

Exercises

1. Let X and Y be Banach spaces, let $\bar{x} \in X$, b, c, $r > 0$, $W := B(\bar{x}, r)$, $f : W \to Y$ be of class C^1 and such that f' is Lipschitzian with rate c on W and $\|f'(w)^{\mathsf{T}}(y^*)\| \geq$

$b\|y^*\|$ for all $w \in W$, $y^* \in Y^*$. Let $b > cr$. Using Kantorovich's theorem, prove that for all $y \in B(f(\bar{x}), (b - cr)r)$ there exists $x \in W$ satisfying $f(x) = y$ and $\|x - \bar{x}\| \le b^{-1}\|y - f(\bar{x})\|$. [Hint: Use the Banach–Schauder theorem to find a right inverse $B(w)$ of $A(w) := f'(w)$ for all $w \in W$ satisfying $\|B(w)(\cdot)\| \le b^{-1}\|\cdot\|$ and use Exercise 15 of Sect. 2.4 to check condition (2.18)]

2. Using Exercise 15 of Sect. 2.4 to establish a refined version of Kantorovich's theorem and prove that the conclusion of the preceding result can be extended to every $y \in B(f(\bar{x}), br)$.

3. (**Convexity of images of small balls** [842]). Let X be a Hilbert space, let Y be a normed space, let $a \in X$, c, ρ, $\sigma > 0$, $W := B(a, \rho)$, and let $f : W \to Y$ be differentiable and such that f' is Lipschitzian with rate c on W and $\|f'(a)^{\mathsf{T}}(y^*)\| \ge \sigma\|y^*\|$ for all $y^* \in Y^*$. Prove that for $r > 0$, $r < \min(\rho, \sigma/2c)$, the image $f(B)$ of $B := B(a, r)$ by the nonlinear map f is convex. [Hint: Given $x_0, x_1 \in B$, $y_0 := f(x_0)$, $y_1 := f(x_1)$, $y := (1/2)(y_0 + y_1)$, $\bar{x} := (1/2)(x_0 + x_1)$, show that $\|f'(w)^{\mathsf{T}}(y^*)\| \ge b\|y^*\|$ for all $w \in W$, $y^* \in Y^*$ for $b := \sigma - cr$ and apply the preceding exercise.]

4. Extend the (surprising!) result of the preceding exercise to the case that X is a Banach space with a uniformly convex norm.

2.5.2 The Inverse Mapping Theorem

The inverse mapping theorem is a milestone of differential calculus. It shows the interest and the power of derivatives. It has numerous applications in differential geometry, differential topology, and the study of dynamical systems.

When $f : T \to \mathbb{R}$ is a continuous function on some open interval T of $\mathbb{R}$, one can use the order of $\mathbb{R}$ and the intermediate value theorem to get results about invertibility of f. If, moreover, f is differentiable at some $r \in T$ and if $f'(r)$ is nonnull, one can conclude that $f(T)$ contains some neighborhood of $f(r)$. When f is a map of several variables, one would like to know whether such a conclusion is valid, and even more, whether f induces a bijection from some neighborhood of a given point $\bar{x}$ onto some neighborhood of $f(\bar{x})$. Of course, one cannot expect a global result without further assumptions, since the derivative is a local notion.

Following René Descartes's advice, we will reach our main results, concerning the possibility of inverting nonlinear maps, through several small steps; some of them have an independent interest.

First, given a bijection f between two metric spaces X, Y, we would like to know whether a map close enough to f is still a bijection. When X and Y are finite-dimensional normed spaces and f is a linear isomorphism, we know that every linear map g that is close enough to f for some norm on the space $L(X, Y)$ of linear continuous maps from X into Y is still an isomorphism: taking bases in X and Y, we see that if g is close enough to f, its determinant will remain different from 0. A similar result holds in infinite-dimensional spaces: the set of linear continuous maps that are isomorphisms onto their images is open in the space $L(X,Y)$. When X and Y are complete, a more precise result can be given.

Proposition 2.68. *Let f be a linear isomorphism between two Banach spaces X and Y. Then every $g \in L(X,Y)$ such that $\|f - g\| < \|f^{-1}\|^{-1}$ is an isomorphism.*

Proof. Let us first consider the case $X = Y$, $f = I_X$. Let $u := I_X - g$, so that $u \in L(X,X)$ satisfies $\|u\| < 1$. Since the map $(v,w) \mapsto w \circ v$ is continuous, since

$$I_X - u^{n+1} = (I_X - u) \circ \left(\sum_{k=0}^{n} u^k \right) = \left(\sum_{k=0}^{n} u^k \right) \circ (I_X - u),$$

and since the series $\sum_{k=0}^{\infty} u^k$ is absolutely convergent (since $\|u^k\| \le \|u\|^k$), we get that its sum is a right and left inverse of $I_X - u$. Thus $I_X - u$ is invertible.

The general case can be deduced from this special case. Given $g \in L(X,Y)$ such that $\|f - g\| < r := \|f^{-1}\|^{-1}$, setting $u := I_X - f^{-1} \circ g$, we observe that $\|u\| \le \|f^{-1} \circ (f - g)\| \le \|f^{-1}\| \cdot \|f - g\| < 1$. Therefore, by what precedes, $f^{-1} \circ g = I_X - u$ is invertible. It follows that g is invertible, with inverse $(I_X - u)^{-1} \circ f^{-1}$.   □

Now let us turn to a nonlinear situation. Let us first observe that if $f : U \to V$ is a bijection between two open subsets of normed spaces X and Y respectively, it may occur that f is differentiable at some $a \in U$ whereas its inverse g is not differentiable at $b = g(a)$: take $U = V = \mathbb{R}$, f given by $f(x) = x^3$, whose inverse $y \mapsto y^{1/3}$ is not differentiable at 0. However, if f is differentiable at some $a \in X$ and if its inverse g is differentiable at $b := g(a)$, then the derivative of g at b is the inverse $f'(a)^{-1}$ of the derivative $f'(a)$ of f at a. This fact simply follows from the chain rule: from $g \circ f = I_U$ and $f \circ g = I_V$ one deduces that $g'(b) \circ f'(a) = I_X$ and $f'(a) \circ g'(b) = I_Y$.

Our first step is not as obvious as the preceding observation, since one of its assumptions is now a conclusion.

Lemma 2.69. *Let U and V be two open subsets of normed spaces X and Y respectively. Assume that $f : U \to V$ is a homeomorphism that is differentiable at $a \in U$ and such that $f'(a)$ is an isomorphism. Then the inverse g of f is differentiable at $b = f(a)$ and $g'(b) = f'(a)^{-1}$.*

Proof. Using translations if necessary, we may suppose $a = 0$, $f(a) = 0$ without loss of generality. Changing f into $h^{-1} \circ f$, where $h := f'(a)$, we may also suppose $Y = X$ and $f'(a) = I_X$. Then setting $s(y) := g(y) - y$, we have to show that $s(y)/\|y\| \to 0$ as $y \to 0$, $y \ne 0$. Let us set $r(x) := f(x) - x$. Given $\varepsilon \in (0,1)$, we can find $\rho > 0$ such that $\|r(x)\| \le (\varepsilon/2) \|x\|$ for $x \in \rho B_X$. Since g is continuous, we can find $\sigma > 0$ such that $\|g(y)\| \le \rho$ for $y \in \sigma B_Y$. Then for $y \in \sigma B_Y$ and $x := g(y)$, we have $y = f(x) = x + r(x)$, and hence

$$\|y\| \ge \|x\| - \|r(x)\| \ge (1/2) \|x\|,$$
$$\|s(y)\| = \|g(y) - y\| = \|r(x)\| \le (\varepsilon/2) \|x\| \le \varepsilon \|y\|. \qquad \square$$

In order to get a stronger result in which the invertibility of f is part of the conclusion instead of being an assumption, we will use the reinforced differentiability property of Definition 2.54. Recall that a map $f : W \to Y$ from an open subset W of a normed space X into another normed space Y is *circa-differentiable* (or strictly differentiable) at $a \in W$ if there exists a continuous linear map $\ell : X \to Y$ such that the map $r = f - \ell$ is Lipschitzian with arbitrarily small Lipschitz rate on sufficiently small neighborhoods of a: for every $\varepsilon > 0$ there exists $\rho > 0$ such that $B(a, \rho) \subset W$ and

$$\forall w, w' \in B(a, \rho), \quad \left\| f(w) - f(w') - \ell(w - w') \right\| \le \varepsilon \left\| w - w' \right\|.$$

The criterion for circa-differentiability given in Proposition 2.56 uses continuous differentiability or slightly less. Thus, the reader who is not interested in refinements may suppose throughout that f is of class C^1.

Our next step is a perturbation result. We formulate it in a general framework.

Lemma 2.70. *Let (U, d) be a metric space, let Y be a normed space, let $j, h : U \to Y$ be such that*

(a) j is injective and its inverse $j^{-1} : j(U) \to U$ is Lipschitzian with rate γ;
(b) h is Lipschitzian with rate λ.

Then if $\gamma\lambda < 1$, the map $f := j + h$ is still injective and its inverse $f^{-1} : f(U) \to U$ is Lipschitzian with rate $\gamma(1 - \gamma\lambda)^{-1}$.

Note that the Lipschitz rate of the inverse of the perturbed map f is close to the Lipschitz rate of j^{-1} when λ is small. It may be convenient to reformulate this lemma by saying that a map $e : X \to Y$ between two metric spaces is *expansive with rate $c > 0$* if for all $x, x' \in X$ one has

$$d(e(x), e(x')) \ge c d(x, x').$$

This property amounts to

$$d(e^{-1}(y), e^{-1}(y')) \le c^{-1} d(y, y')$$

for every $y, y' \in e(X)$, i.e., e is injective and its inverse is Lipschitzian on the image $e(X)$ of e. Thus the lemma can be rephrased as follows:

Lemma. *Let X be a metric space and let Y be a normed space. Let $e : X \to Y$ be expansive with rate $c > 0$ and let $h : X \to Y$ be Lipschitzian with rate $\ell < c$. Then $g := e + h$ is expansive with rate $c - \ell$.*

Proof. The lemma results from the following relations, valid for every $x, x' \in X$:

$$\left\| g(x) - g(x') \right\| \ge \left\| e(x) - e(x') \right\| - \left\| h(x) - h(x') \right\| \ge c d(x, x') - \ell d(x, x').$$

Note that for $c = \gamma^{-1}$, $\ell = \lambda$ one has $(c - \ell)^{-1} = \gamma(1 - \gamma\lambda)^{-1}$. $\square$

Since we have defined differentiability only on open subsets, it will be important to ensure that $f(U)$ is open in order to apply Lemma 2.69. We reach this conclusion in two steps. The first one relies on the Banach–Picard contraction theorem.

Lemma 2.71. *Let W be an open subset of a Banach space Y and let $k : W \to Y$ be a Lipschitzian map with rate $c < 1$. Then the image of W by $f := I_W + k$ is open.*

Proof. We will prove that for every $a \in W$ and for every closed ball $B[a,r]$ contained in W, the closed ball $B[f(a),(1-c)r]$ is contained in the set $f(W)$, and in fact in the set $f(B[a,r])$. Without loss of generality, we may suppose $a = 0$, $k(a) = 0$, using translations if necessary. Given $y \in (1-c)rB_Y$ we want to find $x \in rB_Y$ such that $y = f(x)$. This equation can be written $y - k(x) = x$. We note that $x \mapsto y - k(x)$ is Lipschitzian with rate $c < 1$ and that it maps rB_Y into itself, since

$$\|y - k(x)\| \le \|y\| + \|k(x)\| \le (1-c)r + cr = r.$$

Since rB_Y is a complete metric space, the contraction theorem yields some fixed point x of this map. Thus $y = f(x) \in f(W)$. $\qquad\square$

Lemma 2.72. *Let (U,d) be a metric space, let Y be a Banach space, let $\gamma > 0, \lambda > 0$ with $\gamma\lambda < 1$, and let $j, h : U \to Y$ be such that $W := j(U)$ is open and*

(a) j is injective and its inverse $j^{-1} : W \to U$ is Lipschitzian with rate γ;
(b) h is Lipschitzian with rate λ.

Then the map $f := j + h$ is injective, its inverse is Lipschitzian, and $f(U)$ is open.

Proof. Let $k := h \circ j^{-1}$, so that $f \circ j^{-1} = I_W + k$ and k is Lipschitzian with rate $\gamma\lambda < 1$. Then Lemma 2.71 shows that $f(U) = f(j^{-1}(W)) = (I+k)(W)$ is open. $\quad\square$

We are ready to state the inverse mapping theorem.

Theorem 2.73 (Inverse mapping theorem). *Let X and Y be Banach spaces, let W be an open subset of X, and let $f : W \to Y$ be circa-differentiable at $a \in W$ and such that $f'(a)$ is an isomorphism from X onto Y. Then there exist neighborhoods U of a and V of $b := f(a)$ such that $U \subset W$ and such that f induces a homeomorphism from U onto V whose inverse is differentiable at b.*

Proof. In the preceding lemma, let us take $j := f'(a)$, $h = f - j$. Since j is an isomorphism, its inverse is Lipschitzian with rate $\|j^{-1}\|$. Let U be a neighborhood of a such that h is Lipschitzian with rate $\lambda < 1/\|j^{-1}\|$. Then by the preceding lemma, $V := f(U)$ is open and $f \mid U$ is a homeomorphism from U onto V, and by Lemma 2.69, its inverse is differentiable at b. $\qquad\square$

Exercise. Show that the inverse of f is in fact circa-differentiable at b.

Exercise (Square root of an operator). Let E be a Banach space and let $X := L(E,E)$. Considering the map $f : X \to X$ given by $f(u) := u^2 := u \circ u$, show that there exist a neighborhood V of I_E in X and a differentiable map $g : V \to X$ such that $g(v)^2 := g(v) \circ g(v) = v$ for all $v \in V$.

The following classical terminology is helpful.

Definition 2.74. A C^k-*diffeomorphism* between two open subsets of normed spaces is a homeomorphism that is of class C^k, as is its inverse ($k \geq 1$).

The following example plays an important role in the sequel, so that we make it a lemma.

Lemma 2.75. *Let X and Y be Banach spaces. Then the set $\mathrm{Iso}(X,Y)$ of isomorphisms from X onto Y is open in $L(X,Y)$ and the map $i : \mathrm{Iso}(X,Y) \to \mathrm{Iso}(Y,X)$ given by $i(u) = u^{-1}$ is a C^∞-diffeomorphism, i.e., a C^k-diffeomorphism for all $k \geq 1$.*

Proof. The first assertion has been proved in Proposition 2.68. Let us prove the second assertion by first considering the case $X = Y$ and by showing that i is differentiable at the identity map I_X, with derivative $Di(I_X)$ given by $Di(I_X)(v) = -v$. Taking $\rho \in (0,1)$, this follows from the expansion

$$\forall v \in L(X,X), \ \|v\| \leq \rho, \qquad (I_X + v)^{-1} = I_X - v + s(v),$$

with $s(v) := v^2 \circ \sum_{k=0}^{\infty} (-1)^k v^k$: s defines a remainder, since $\|(-1)^k v^k\| \leq \rho^k$ and $\|s(v)\| \leq (1-\rho)^{-1}\|v\|^2$. Thus i is differentiable at I_X.

Now in the general case, for $u \in \mathrm{Iso}(X,Y)$, $w \in L(X,Y)$ satisfying $\|w\| < 1/\|u^{-1}\|$, $v := u^{-1} \circ w$, one has $u + w = u \circ (I_X + v) \in \mathrm{Iso}(X,Y)$,

$$i(u+w) = \left[u \circ (I_X + u^{-1} \circ w) \right]^{-1} = (I_X + u^{-1} \circ w)^{-1} \circ u^{-1}$$
$$= \left(I_X - u^{-1} \circ w + s(v) \right) \circ u^{-1},$$

and one sees that i is differentiable at u, with

$$Di(u)(w) = -u^{-1} \circ w \circ u^{-1}. \tag{2.22}$$

Thus the derivative $i' : \mathrm{Iso}(X,Y) \to L(L(X,Y), L(Y,X))$ is obtained by composing i with the map $k : L(Y,X) \to L(L(X,Y), L(Y,X))$ given by $k(z)(w) := -z \circ w \circ z$ for $z \in L(Y,X)$, $w \in L(X,Y)$, which is continuous and quadratic, hence is of class C^1. It follows that i' is continuous and i is of class C^1. Then i' is of class C^1. By induction, we obtain that i is of class C^k for all $k \geq 1$. Since i is a bijection with inverse $i^{-1} : \mathrm{Iso}(Y,X) \to \mathrm{Iso}(X,Y)$ given by $i^{-1}(z) = z^{-1}$, we get that i is a C^∞-diffeomorphism.   □

Note that formula (2.22) generalizes the usual case $i(t) = t^{-1}$ on $\mathbb{R} \setminus \{0\}$ for which $i'(u) = -u^{-2}$ and $Di(u)(w) = -u^{-2}w$.

Corollary 2.76. *Let X and Y be Banach spaces, let W be an open subset of X, and let $f : W \to Y$ be of class C^k ($k \geq 1$) and such that $f'(a)$ is an isomorphism from X onto Y for some $a \in W$. Then there exist neighborhoods U of a and V of $b := f(a)$ such that $U \subset W$ and such that $f \mid U$ is a C^k-diffeomorphism between U and V.*

Proof. Let us first consider the case $k = 1$. The inverse mapping theorem ensures that f induces a homeomorphism from a neighborhood U of a onto a neighborhood V of b. Since f' is continuous at a and since the set $\mathrm{Iso}(X, Y)$ of isomorphisms from X onto Y is open in $L(X, Y)$, taking a smaller U if necessary, we may assume that $f'(x)$ is an isomorphism for all $x \in U$. Then Lemma 2.69 guarantees that $g := f^{-1}$ is differentiable at $f(x)$. Moreover, one has

$$g'(y) = (f'(g(y)))^{-1}.$$

Since the map $i : u \mapsto u^{-1}$ is of class C^1 on $\mathrm{Iso}(X, Y)$, $g' = i \circ f' \circ g$ is continuous. Thus g is of class C^1.

Now suppose by induction that g is of class C^k if f is of class C^k, and let us prove that when f is of class C^{k+1}, then g is of class C^{k+1}. That follows from the expression $g' = i \circ f' \circ g$, which shows that g' is of class C^k as a composite of maps of class C^k.   □

Let us give a global version of the inverse mapping theorem.

Corollary 2.77. *Let X and Y be Banach spaces, let W be an open subset of X, and let $f : W \to Y$ be an injection of class C^k such that for every $x \in W$, the linear map $f'(x)$ is an isomorphism from X onto Y. Then $f(W)$ is open and f is a C^k-diffeomorphism between W and $f(W)$.*

Proof. The inverse mapping theorem ensures that $f(W)$ is open in Y. Thus f is a continuous bijection from W onto $f(W)$ and its inverse is locally of class C^k, hence is of class C^k.   □

Exercise. Let $f : T \to \mathbb{R}$ be a continuous function on some open interval T of $\mathbb{R}$. Show that if f is differentiable at some $r \in T$ with $f'(r)$ nonnull, then $f(T)$ contains some neighborhood of $f(r)$. Show by an example that it may happen that there is no neighborhood of r on which f is injective.

Example–Exercise (Polar coordinates). Let $W := (0, +\infty) \times (-\pi, \pi) \subset \mathbb{R}^2$ and let $f : W \to \mathbb{R}^2$ be given by $f(r, \theta) = (r\cos\theta, r\sin\theta)$. Then f is a bijection from W onto $\mathbb{R}^2 \setminus D$, with $D := (-\infty, 0] \times \{0\}$ and the Jacobian matrix of f at (r, θ) is

$$\begin{pmatrix} \cos\theta & -r\sin\theta \\ \sin\theta & r\cos\theta \end{pmatrix}.$$

Its determinant (called the *Jacobian* of f) is $r(\cos^2\theta + \sin^2\theta) = r > 0$; hence f is a diffeomorphism of class C^∞ from W onto $f(W)$. Using the relation $\tan(\theta/2) = 2\sin(\theta/2)\cos(\theta/2)/2\cos^2(\theta/2) = \sin\theta/(1 + \cos\theta)$, show that its inverse is given by

$$(x, y) \mapsto \left(\sqrt{x^2 + y^2}, 2\mathrm{Arc}\tan \frac{y}{x + \sqrt{x^2 + y^2}} \right).$$

Example–Exercise (Spherical coordinates). Let $W := (0,+\infty) \times (-\pi,\pi) \times (\frac{-\pi}{2}, \frac{\pi}{2})$ and let $f : W \to \mathbb{R}^3$ be given by $f(r,\theta,\omega) = (r\cos\theta \sin\omega, r\sin\theta \sin\omega, r\cos\omega)$. Show that f is a diffeomorphism from W onto its image. The angles θ,ω are known as *Euler angles*. On the globe, they can serve to measure latitude and longitude.

Example–Exercise. Is $f : \mathbb{R}^2 \to \mathbb{R}^2$ given by $f(x,y) := (x^2 - y^2, 2xy)$ a diffeomorphism? Give an interpretation by considering $z \mapsto z^2$, with $z := x + iy$, identifying $\mathbb{C}$ with $\mathbb{R}^2$.

2.5.3 The Implicit Function Theorem

Functions are sometimes defined in an implicit, indirect way. For example, in economics, the famous Phillips curve is defined through the equation

$$1.39u(w + 0.9) = 9.64,$$

where u is the rate of unemployment and w is the annual rate of variation of nominal wages; in such a case one can express u in terms of w and vice versa. However, given Banach spaces X,Y,Z, an open subset W of $X \times Y$, and a map $f : W \to Z$, it is often impossible to determine an explicit map $h : X_0 \to Y$ from an open subset X_0 of X such that $(x,h(x)) \in W$ and $f(x,h(x)) = 0$ for all $x \in X_0$. When the existence of such a map is known (but not necessarily in an explicit form), one says that it is an *implicit function* determined by f. The following result guarantees the existence and regularity of such a map.

Theorem 2.78. *Let X,Y,Z be Banach spaces, let W be an open subset of $X \times Y$, and let $f : W \to Z$ be a map of class C^1 at $(a,b) \in W$ such that $f(a,b) = 0$ and the second partial derivative $D_Y f(a,b)$ is an isomorphism from Y onto Z. Then there exist open neighborhoods U of (a,b) and V of a in W and X respectively and a map $h : V \to Y$ of class C^1 at a such that $h(a) = b$ and*

$$((x,y) \in U,\ f(x,y) = 0) \iff (x \in V,\ y = h(x)). \tag{2.23}$$

If f is of class C^k with $k \geq 1$ on W, then h is of class C^k on V. Moreover,

$$Dh(a) = -D_Y f(a,b)^{-1} \circ D_X f(a,b). \tag{2.24}$$

Proof. Let $F : W \to X \times Z$ be the map given by $F(x,y) := (x, f(x,y))$. Then F is of class C^1 at (a,b), as are its components, and

$$DF(a,b)(x,y) = (x, D_X f(a,b)x + D_Y f(a,b)y).$$

It is easy to check that $DF(a,b)$ is invertible and that its inverse is given by

$$(DF(a,b))^{-1}(x,z) = (x, -(D_Y f(a,b))^{-1} \circ D_X f(a,b)x + (D_Y f(a,b))^{-1} z).$$

Therefore, the inverse mapping theorem yields open neighborhoods U of (a,b) in W and U' of $(a,0)$ in $X \times Z$ such that F induces a homeomorphism from U onto U' of class C^1 at (a,b). Its inverse G is of class C^1 at $(a,0)$, satisfies $G(a,0) = (a,b)$, and has the form $(x,z) \mapsto (x,g(x,z))$. Let $V := \{x \in X : (x,0) \in U'\}$ and let $h : V \to Y$ be given by $h(x) = g(x,0)$. Then the equivalence

$$((x,y) \in U, \ (x,z) = (x, f(x,y))) \Leftrightarrow ((x,z) \in U', \ (x,y) = (x,g(x,z)))$$

entails, by definition of V and h,

$$((x,y) \in U, \ f(x,y) = 0) \Leftrightarrow (x \in V, \ y = h(x)).$$

When f is of class C^k on W, with $k \geq 1$, F is of class C^k; hence G and h are of class C^k on U' and V respectively. Moreover, the computation of the inverse $DF(a,b)^{-1}$ we have done shows that

$$Dh(a) = D_X g(a,0) = -D_Y f(a,b)^{-1} \circ D_X f(a,b).$$

$\square$

Example. Let X be a Hilbert space, and for $Y := \mathbb{R}$, let $f : X \times Y \to \mathbb{R}$ be given by $f(x,y) = \|x\|^2 + y^2 - 1$. Then f is of class C^∞ and for $(a,b) := (0,1)$ one has

$$Df(a,b)(u,v) = 2(a \mid u) + 2bv = 2v,$$

whence $D_Y f(a,b) = 2I_Y$ is invertible and $D_Y f(a,b)^{-1} = (1/2)I_Z$. Here we can take $U := B(a,1) \times (0,+\infty)$, $V := B(a,1)$, and the implicit function is given by $h(x) = (1 - \|x\|^2)^{1/2}$. As mentioned above, it is not always the case that U and h can be described explicitly as in this classical parameterization of the upper hemisphere.

When Z is finite-dimensional, the regularity assumption on f can be relaxed in two ways.

Theorem 2.79. *Let X,Y,Z be Banach spaces, Y and Z being finite-dimensional, let W be an open subset of $X \times Y$, and let $f : W \to Z$ be Fréchet differentiable at $(a,b) \in W$ such that $f(a,b) = 0$ and the partial derivative $D_Y f(a,b)$ is an isomorphism from Y onto Z. Then there exist open neighborhoods U of (a,b) and V of a in W and X respectively and a map $h : V \to Y$ Fréchet differentiable at a such that $h(a) = b$ and*

$$\forall x \in V, \qquad\qquad f(x,h(x)) = 0.$$

Differentiating this relation, we recover the value of $Dh(a)$:

$$Dh(a) = -D_Y f(a,b)^{-1} \circ D_X f(a,b).$$

The proof below is slightly simpler when $A := D_X f(a,b) = 0$; one can reduce it to that case by a linear change of variables.

Proof. Using translations and composing f with $D_Y f(a,b)^{-1}$, we may suppose $(a,b) = (0,0)$, $Z = Y$, and $D_Y f(a,b) = I_Y$. Let $r : W \to Y$ be a remainder such that

$$f(x,y) := Ax + y + r(x,y).$$

For $\varepsilon \in (0, 1/2]$ let $\delta := \delta(\varepsilon) > 0$ be such that $\delta B_{X \times Y} \subset W$, $\|r(x,y)\| \leq \varepsilon(\|x\| + \|y\|)$ for all $(x,y) \in \delta B_{X \times Y}$. Let $\beta := \delta/2$, $\alpha := (2\|A\| + 1)^{-1} \beta$, and for $x \in \alpha B_X$ let $k_x : \beta B_Y \to Y$ be given by

$$k_x(y) := -Ax - r(x,y).$$

Then k_x maps βB_Y into itself, since for $y \in \beta B_Y$ we have $\|k_x(y)\| \leq \|A\| \alpha + (1/2)(\alpha + \beta) \leq \beta$. The Brouwer fixed-point theorem ensures that k_x has a fixed point $y_x \in \beta B_Y$: $-Ax - r(x, y_x) = y_x$. Then setting $h(x) := y_x$, we have $f(x, h(x)) = Ax + h(x) + r(x, h(x)) = 0$. It remains to show that h is differentiable at 0. Since

$$\|h(x)\| = \|k_x(h(x))\| \leq \|A\| \|x\| + \varepsilon \|x\| + \varepsilon \|h(x)\|,$$

so that $\|h(x)\| \leq (1 - \varepsilon)^{-1}(\|A\| + \varepsilon) \|x\|$, we get

$$\|h(x) + Ax\| = \|r(x, h(x))\| \leq \varepsilon \|x\| + \varepsilon \|h(x)\| \leq \varepsilon(1 - \varepsilon)^{-1}(\|A\| + 1) \|x\|.$$

This shows that h is differentiable at 0 with derivative $-A$. $\square$

A similar (and simpler) proof yields the first assertion of the next statement.

Theorem 2.80 [785]. *Let X and Y be normed spaces, Y being finite-dimensional, and let $f : X \to Y$ be continuous on a neighborhood of $a \in X$ and differentiable at a, with $f'(a)(X) = Y$. Then there exist a neighborhood V of $b := f(a)$ in Y and a right inverse $g : V \to X$ that is differentiable at a and such that $g(b) = a$.*

If C is a convex subset of X, if $a \in C$, and if $f'(a)(cl(\mathbb{R}_+(C - a))) = Y$, one can even get that $g(V) \subset C$ if one does not require that the directional derivative of g at b be linear.

The second weakening of the assumptions concerns the kind of differentiability.

Theorem 2.81. *Let X, Y, Z be Banach spaces, Y and Z being finite-dimensional, let W be an open subset of $X \times Y$, and let $f : W \to Z$ be a map of class D^1 at $(a,b) \in W$ such that $f(a,b) = 0$ and the partial derivative $D_Y f(a,b)$ is an isomorphism from Y onto Z. Then there exist open neighborhoods U of (a,b) and V of a in W and X respectively and a map $h : V \to Y$ of class D^1 such that $h(a) = b$ and*

$$((x,y) \in U, \ f(x,y) = 0) \iff (x \in V, \ y = h(x)).$$

Proof. We may suppose W is a ball $B((a,b),\rho_0)$, $Y = Z$, $D_Y f(a,b) = I_Y$. With the notation of the preceding proof, using the compactness of the unit ball of Y, we may suppose the remainder r satisfies, for $\rho \in (0,\rho_0)$ and every $x \in \rho B_X$, $y, y' \in \rho B_Y$,

$$\|r(x,y) - r(x,y')\| = \left\| \int_0^1 (D_Y f(x, (1-t)y + ty') - I_Y)(y - y')dt \right\| \le c(\rho) \|y - y'\|,$$

where $c(\rho) \to 0$ as $\rho \to 0_+$. Taking ρ_0 small enough, we see that the map k_x is a contraction with rate $c(\rho_0) \le 1/2$. Picking $\alpha \in (0, \rho_0)$ so that for $x \in \alpha B_X$, $\|k_x(0)\| = \|-f(x,0)\| \le \rho/2$, the Banach–Picard contraction theorem ensures that k_x has a unique fixed point y_x in the ball ρB_Y. Then setting $h(x) := y_x$, we have $f(x, h(x)) = 0$, and y_x is the unique solution of the equation $f(x,y) = 0$ in the ball ρB_Y. Moreover, h is continuous as a uniform limit of continuous maps given by iterations. Restricting f to $X_1 \times Y$, where X_1 is an arbitrary finite-dimensional subspace of X, we get that h is Gâteaux differentiable. Since $\mathrm{Iso}(Y)$ is an open subset of $L(Y,Y)$ and since $(x,y) \mapsto D_Y f(x,y)$ is continuous for the norm of $L(Y,Y)$ by the above argument, we obtain from the relation

$$Dh(x)v = -D_Y f(x, h(x))^{-1}(D_X f(x, h(x))v)$$

that $(x,v) \mapsto Dh(x)v$ is continuous. $\square$

Exercises

1. Show that the inverse mapping theorem can be deduced from the implicit mapping theorem by considering the map $(x,y) \mapsto y - f(x)$.

2. Let $f : \mathbb{R}^4 \to \mathbb{R}^3$ be given by

$$f(w,x,y,z) = (w + x + y + z, w^2 + x^2 + y^2 + z - 2, w^3 + x^3 + y^3 + z).$$

Show that there exist a neighborhood V of $a := 0$ in $\mathbb{R}$ and a map $h : V \to \mathbb{R}^3$ of class C^∞ such that $h(0) = (0, -1, 1)$ and $f(h(z), z) = 0$ for every $z \in V$. Compute the derivative of h at 0.

3. Let X be the space of square $n \times n$ matrices and let $f : X \times \mathbb{R} \to \mathbb{R}$ be given by $f(A, r) = \det(A - rI)$. Let $r \in \mathbb{R}$ be such that $f(A, r) = 0$ and $D_2 f(A, r) \ne 0$. Show that there exist an open neighborhood U of A in X and a function $\lambda : U \to \mathbb{R}$ of class C^∞ such that for each B in U, $\lambda(B)$ is a simple eigenvalue of B.

4. Given Banach spaces $W, X, Z, Y := Z^*$, maps $f : W \times X \to \mathbb{R}$, $g : W \times X \to Z$ of class C^2, consider the parameterized mathematical programming problem

$$(\mathscr{P}_w) \quad \text{minimize } f(w,x) \text{ subject to } g(w,x) = 0$$

and let $p(w)$ be its value. Suppose that for some $\overline{w} \in W$ and a solution $\overline{x} \in X$ of $(\mathscr{P}_{\overline{w}})$ the derivative $B := D_X g(\overline{w}, \overline{x})$ is surjective and its kernel N has a topological supplement M. Let ℓ be the *Lagrangian* of $(\mathscr{P}_{\overline{w}})$:

$$\ell(w, x, y) := f(w, x) + \langle y, g(w, x) \rangle,$$

and let $\overline{y}$ be a multiplier at $\overline{x}$, i.e., an element of Y such that $D_X \ell(\overline{w}, \overline{x}, \overline{y}) = 0$. Suppose $D_X^2 \ell(\overline{w}, \overline{x}, \overline{y}) \mid N$ induces an isomorphism from N onto $N^* \simeq M^\perp$. Let $A := D_X^2 \ell(\overline{w}, \overline{x}, \overline{y})$.

(a) Show that for every $(x^*, z) \in X^* \times Z$ the system

$$Au + B^\mathsf{T} v = x^*,$$

$$Bu = z,$$

has a unique solution $(u, v) \in X \times Y$ continuously depending on (x^*, z).

(b) Show that the Karush–Kuhn–Tucker system

$$D_X f(w, x) + y \circ D_X g(w, x) = 0,$$

$$g(w, x) = 0,$$

determines $(x(w), y(w))$ as an implicit function of w in a neighborhood of $\overline{w}$ with $x(\overline{w}) = \overline{x}$, $y(\overline{w}) = \overline{y}$, the multiplier at $\overline{x}$.

(c) Suppose $x(w)$ is a solution to $(\mathscr{P}_w)$ for w close to $\overline{w}$. Show that p is of class C^1 near $\overline{w}$. Using the relations $p(w) = \ell(w, x(w), y(w))$, $D_X \ell(w, x(w), y(w)) = 0$, $D_Y \ell(w, x(w), y(w)) = 0$, show that $Dp(w) = D_w \ell(w, x(w), y(w))$.

(d) Deduce from what precedes that p is of class C^2 around $\overline{w}$ and give the expression of $D^2 p(\overline{w}) := (p'(\cdot))'(\overline{w})$.

2.5.4   The Legendre Transform

As an application of inversion results, let us give an account (and even a refinement) of the classical notion of Legendre function of class C^k. We will see that the Legendre transform enables one to pass from the Euler–Lagrange equations of the calculus of variations to the Hamilton equations, which are explicit (rather than implicit) differential equations of first order (instead of second order). Recall that a map $g : U \to V$ between two metric spaces is *stable* or is *Stepanovian* if for every $\overline{u} \in U$ there exist some $r > 0$, $c \in \mathbb{R}_+$ such that for every $u \in B(\overline{u}, r)$ one has

$$d(g(u), g(\overline{u})) \leq c\, d(u, \overline{u}).$$

Such an assumption is clearly a weakening of the Lipschitz condition.

Definition 2.82. A function $f : U \to \mathbb{R}$ on an open subset U of a normed space X is a (classical) *Legendre function* if it is differentiable, and its derivative $f' : U \to Y := X^*$ is a Stepanovian bijection onto an open subset V of Y whose inverse h is Stepanovian.

Then one defines the *Legendre transform* of f as the function $f^L : V \to \mathbb{R}$ given by

$$f^L(y) := \langle h(y), y \rangle - f(h(y)), \qquad y \in V.$$

Since h is just a Stepanovian function, it is surprising that f^L is in fact of class C^1.

Lemma 2.83. *If f is a Legendre function on U, then its Legendre transform f^L is of class C^1 on $V := f'(U)$ and of class C^k ($k \geq 1$) if f is of class C^k. Moreover, f^L is a Legendre function, $\left(f^L \right)^L = f$ and for all $(u, v) \in U \times V$ one has*

$$v = Df(u) \Leftrightarrow u = Df^L(v).$$

Proof. Given $v := Df(u) \in V$, let $y \in V - v$, let $x := h(v+y) - h(v) \in U - u$, and let $r(x) = f(u+x) - f(u) - Df(u)x$. Then since $h(v) = u$, $h(v+y) = u+x$, one has

$$f^L(v+y) - f^L(v) - \langle u, y \rangle = \langle u+x, v+y \rangle - f(u+x) - \langle u, v \rangle + f(u) - \langle u, y \rangle$$

$$= \langle x, v+y \rangle - Df(u)(x) - r(x) = \langle x, y \rangle - r(x).$$

Since there exists $c \in \mathbb{R}_+$ such that $\|x\| \leq c \|y\|$ for $\|y\|$ small enough, the last right-hand side is a remainder as a function of y. Thus f^L is differentiable at v and $Df^L(v) = u = h(v)$. Therefore $(f^L)' = h$ is a bijection with inverse f' and f^L is a Legendre function. Now

$$\left(f^L \right)^L (u) = \langle Df^L(v), v \rangle - f^L(v) = \langle u, v \rangle - (\langle u, v \rangle - f(u)) = f(u).$$

When f is of class C^k, $(f^L)' = h$ is of class C^{k-1}, as an induction shows, thanks to the Stepanov property of f' and h. □

Exercise. Let X be a normed space, let $A : X \to X^*$ be a linear isomorphism, let $b \in X^*$, and let f be given by $f(x) := (1/2)\langle Ax, x \rangle + \langle b, x \rangle$ for $x \in X$. Show that f is a Legendre function and compute f^L.

2.5.5 Geometric Applications

When looking at familiar objects such as forks, knives, funnels, roofs, spires, one sees that some points are smooth, while some other points of the objects present ridges or peaks or cracks. Mathematicians have found concepts that enable them to deal with such cases.

The notions of (regular) curve, surface, hypersurface, and so on can be embodied in a general framework in which some differential calculus can be done. The underlying idea is the possibility of straightening a piece of the set; for this purpose, some forms of the inverse mapping theorem will be appropriate.

We first define a notion of smoothness for a subset S of a normed space X around some point a.

Definition 2.84. A subset S of a normed space X is said to be C^k-*smooth* around a point $a \in S$ if there exist normed spaces Y, Z, an open neighborhood U of a in X, an open neighborhood V of 0 in $Y \times Z$, and a C^k-diffeomorphism $\varphi : U \to V$ such that $\varphi(a) = 0$ and

$$\varphi(U \cap S) = (Y \times \{0\}) \cap V. \tag{2.25}$$

A subset S of a normed space X is said to be a *submanifold of class* C^k if it is C^k-smooth around each of its points.

Thus, φ straightens $U \cap S$ onto the piece $(Y \times \{0\}) \cap V$ of the linear space $Y \times \{0\}$, which can be identified with a neighborhood of 0 in Y. The map φ is called a *chart*, and a collection $\{\varphi_i\}$ of charts whose domains form a covering of S is called an *atlas*. When Y is of dimension d, one says that S is of dimension d around a. When Z is of dimension c, one says that S is of codimension c around a.

The following example can be seen as a general model.

Example. Let $X := Y \times Z$, where Y, Z are normed spaces, let W be an open subset of Y, and let $f : W \to Z$ be a map of class C^k. Then its graph $S := \{(w, f(w)) : w \in W\}$ is a C^k-submanifold of X: taking $U := V := W \times Z$, and setting $\varphi(w, z) := (w, z - f(w))$, we define a C^k-diffeomorphism from U onto V with inverse given by $\varphi^{-1}(w, z) = (w, z + f(w))$ for which (2.25) is satisfied.   □

When in the preceding example we take $Z := \mathbb{R}$ and the epigraph $E := \{(w, y) \in W \times \mathbb{R} : y \geq f(w)\}$ of f, we get a model for the notion of submanifold with boundary. We just give a formal definition in which a subset Z_+ of a normed space Z is said to be a *half-space* of Z if there exists some $h \in Z^* \setminus \{0\}$ such that $Z_+ := h^{-1}(\mathbb{R}_+)$.

Definition 2.85. A subset S of a normed space X is said to be a C^k-*submanifold with boundary* if for every point a of S, either S is C^k-smooth around a or there exist normed spaces Y, Z, a half-space Z_+ of Z, an open neighborhood U of a in X, an open neighborhood V of 0 in $Y \times Z$, and a C^k-diffeomorphism $\varphi : U \to V$ such that $\varphi(a) = 0$ and

$$\varphi(U \cap S) = (Y \times Z_+) \cap V.$$

Such a notion is useful for giving a precise meaning to the expression "S is a regular open subset of $\mathbb{R}^d$" (an improper expression, since usually one considers the closure of such a set).

There are two usual ways of obtaining submanifolds: through equations and through parameterizations. For instance, the graph S of the preceding example can be defined either as the image under $(I_W, f) : w \mapsto (w, f(w))$ of the parameter space

W or as the set of points $(y, z) \in Y \times Z$ satisfying $y \in W$ and the equation $z - f(y) = 0$. As a more concrete example, we observe that for given $a, b \in \mathbb{P}$, the ellipse

$$E := \left\{ (x, y) \in \mathbb{R}^2 : \frac{x^2}{a^2} + \frac{y^2}{b^2} = 1 \right\}$$

can be seen as the image of the parameterization $f : \mathbb{R} \to \mathbb{R}^2$ given by $f(t) := (a \cos t, b \sin t)$.

Exercise. Give parameterizations for the ellipsoid $\left\{ (x, y, z) \in \mathbb{R}^3 : \frac{x^2}{a^2} + \frac{y^2}{b^2} + \frac{z^2}{c^2} = 1 \right\}$ and do the same for the other surfaces of $\mathbb{R}^3$ defined by quadratic forms.

Even if S is not smooth around $a \in S$, one can get an idea of its shape around a using an approximation. The concept of tangent cone offers such an approximation; it can be seen as a geometric counterpart to the directional derivative.

Definition 2.86. The *tangent cone* (or *contingent cone*) to a subset S of a normed space X at some point a in the closure of S is the set $T(S, a)$ of vectors $v \in X$ such that there exist sequences $(v_n) \to v$, $(t_n) \to 0_+$ for which $a + t_n v_n \in S$ for all $n \in \mathbb{N}$.

Equivalently, one has $v \in T(S, a)$ if and only if there exist sequences (a_n) in S, $(t_n) \to 0_+$ such that $(v_n) := (t_n^{-1}(a_n - a)) \to v$: v is the limit of a sequence of secants to S issued from a.

Some rules for dealing with tangent cones are given in the next lemma, whose elementary proof is left as an exercise.

Lemma 2.87. *Let X be a normed space, let S, S' be subsets of X such that $S \subset S'$. Then for every $a \in S$ one has $T(S, a) \subset T(S', a)$.*

If U is an open subset of X, then for every $a \in S \cap U$ one has $T(S, a) = T(S \cap U, a)$.

If X' is another normed space, if $g : U \to X'$ is Hadamard differentiable at a, and if $S' \subset X'$ contains $g(S \cap U)$, then one has $Dg(a)(T(S, a)) \subset T(S', g(a))$.

If $\varphi : U \to V$ is a C^k-diffeomorphism between two open subsets of normed spaces X, X' and if S is a subset of X containing a, then for $S' := \varphi(S \cap U)$ and $a' := \varphi(a)$, one has $T(S', a') = D\varphi(a)(T(S, a))$.

Exercise. Deduce from the second assertion of the lemma that for $g : U \to X'$ Hadamard differentiable at a, $b := g(a)$, $S := g^{-1}(b)$ one has $T(S, a) \subset \ker Dg(a)$. Moreover, if for some $c > 0$, $\rho > 0$ one has $d(x, g^{-1}(b)) \leq cd(g(x), b)$ for all $x \in B(a, \rho)$, then one has $T(S, a) = \ker Dg(a)$.

Exercise. Let $S := \{(x, y) \in \mathbb{R}^2 : x^3 = y^2\}$. Check that $T(S, (0, 0)) = \mathbb{R}_+ \times \{0\}$.

When S is smooth around $a \in S$ in the sense of Definition 2.84, one can give an alternative characterization of $T(S, a)$ in terms of velocities.

Proposition 2.88. *If S is C^1-smooth around $a \in S$, then the tangent cone $T(S, a)$ to S at a coincides with the set $T^l(S, a)$ of $v \in X$ such that there exist $\tau > 0$ and $c : [0, \tau] \to X$ right differentiable at 0 with $c'_+(0) = v$ and satisfying $c(0) = a$, $c(t) \in S$ for all $t \in [0, \tau]$. Moreover, if $\varphi : U \to V$ is a C^1-diffeomorphism such that $\varphi(a) = 0$*

and $\varphi(S \cap U) = (Y \times \{0\}) \cap V$, then one has $T(S,a) = (D\varphi(a))^{-1}(Y \times \{0\})$, and $T(S,a)$ is a closed linear subspace of X.

Proof. The result follows from Lemma 2.87 and the observation that if S is an open subset of some closed linear subspace L of X then $T(S,a) = L = T^l(S,a)$. $\qquad\square$

Now let us turn to sets defined by equations. We need the following result.

Theorem 2.89 (Submersion theorem). *Let X and Z be Banach spaces, let W be an open subset of X, and let $g : W \to Z$ be a map of class C^k with $k \geq 1$ such that for some $a \in W$ the map $Dg(a)$ is surjective and its kernel N has a topological supplement M in X. Then there exist an open neighborhood U of a in W and a diffeomorphism φ of class C^k from U onto a neighborhood V of $(0, g(a))$ in $N \times Z$ such that $\varphi(a) = (0, g(a))$,*

$$g \mid U = p \circ \varphi,$$

where p is the canonical projection from $N \times Z$ onto Z. In particular, g is open around a in the sense that for every open subset U' of U, the image $g(U')$ is open.

This result shows that the nonlinear map g has been straightened into a simple continuous linear map, a projection, using the diffeomorphism φ.

Proof. Let $F : W \to N \times Z$ be given by $F(x) = (p_N(x) - p_N(a), g(x))$, where $p_N : X \to N$ is the projection on N associated with the isomorphism between X and $M \times N$. Then F is of class C^k and $DF(a)(x) = (p_N(x), Dg(a)(x))$. Clearly $DF(a)$ is injective: when $p_N(x) = 0$, $Dg(a)(x) = 0$, one has $x \in M \cap N$, hence $x = 0$. Let us show that $DF(a)$ is surjective: given $(y, z) \in N \times Z$, there exists $v \in X$ such that $Dg(a)(v) = z$, and since $y - p_N(v) \in N$, for $x := v + y - p_N(v)$, we have that $Dg(a)(x) = Dg(a)(v) = z$ and $p_N(x) = p_N(y) = y$. Thus, by the Banach isomorphism theorem, we have that $DF(a)$ is an isomorphism of X onto $N \times Z$. The inverse mapping theorem ensures that the restriction φ of F to some open neighborhood U of a is a C^k-diffeomorphism onto some neighborhood V of $(0, g(a))$. $\qquad\square$

Note that for $Z := \mathbb{R}$, the condition on g reduces to the following: g is of class C^k and $g'(a) \neq 0$. Note also that when $N := \{0\}$, we recover the inverse function theorem.

The application we have in view follows readily.

Corollary 2.90. *Let X and Z be Banach spaces, let W be an open subset of X, and let $g : W \to Z$ be a map of class C^k with $k \geq 1$. Let*

$$S := \{x \in W : g(x) = 0\}.$$

Suppose that for some $a \in S$ the map $g'(a) := Dg(a)$ is surjective and its kernel N has a topological supplement in X. Then S is C^k-smooth around a. Moreover, $T(S,a) = \ker g'(a)$.

Proof. Using the notation of the submersion theorem, setting $Y := N$, we see that Definition 2.84 is satisfied, noting that for $x \in U$ we have $x \in S \cap U$ iff $p(\varphi(x)) =$

$g(x) = 0$, iff $\varphi(x) \in (Y \times \{0\}) \cap V$. Now, the preceding proposition asserts that $T(S,a) = (\varphi'(a))^{-1}(Y \times \{0\})$. But since $g \mid U = p \circ \varphi$, we have $g'(a) = p \circ \varphi'(a)$, $\ker g'(a) = (\varphi'(a))^{-1}(\ker p) = (\varphi'(a))^{-1}(Y \times \{0\})$. Hence $T(S,a) = \ker g'(a)$. □

The regularity condition on g can be relaxed thanks to the Lyusternik–Graves theorem.

Proposition 2.91 (Lyusternik). *Let X and Y be Banach spaces, let W be an open subset of X, and let $g : W \to Y$ be circa-differentiable at $a \in S := \{x \in W : g(x) = 0\}$, with $g'(a)(X) = Y$. Then $T(S,a) = \ker g'(a)$.*

Proof. The inclusion $T(S,a) \subset \ker g'(a)$ follows from Lemma 2.87. Conversely, let $v \in \ker g'(a)$. Theorem 2.67 yields some $\kappa, \rho > 0$ such that for all $w \in B(a,\rho)$ there exists some $x \in W$ such that $g(x) = y := 0$, $\|x - w\| \le \kappa \|g(w)\|$. Taking $w := a + tv$ with $t > 0$ so small that $w \in B(a,\rho)$, we get some $x_t \in S$ satisfying $\|x_t - (a + tv)\| \le o(t) := \kappa \|g(\bar{x} + tv)\|$. Thus $v \in T(S,a)$ and even $v \in T^l(S,a)$. □

In the following example, we use the fact that when $Y = \mathbb{R}$, the surjectivity condition on $g'(a)$ reduces to $g'(a) \ne 0$ (or $\nabla g(a) \ne 0$ if X is a Hilbert space).

Example–Exercise. Let X be a Hilbert space and let $g : X \to \mathbb{R}$ be given by $g(x) := \frac{1}{2}(A(x) \mid x) - \frac{1}{2}$, where A is a linear isomorphism from X onto X that is symmetric, i.e., such that $(Ax \mid y) = (Ay \mid x)$ for every $x, y \in X$. Let $S := g^{-1}(\{0\})$. For all $a \in S$ one has $\nabla g(a) = A(a) \ne 0$, since $(A(a) \mid a) = 1$. Thus S is a C^∞-submanifold of X. Taking $X = \mathbb{R}^2$ and appropriate isomorphisms A, find the classical conic curves; then take $X = \mathbb{R}^3$ and find the classical conic surfaces, including the sphere, the ellipsoid, the paraboloid, and the hyperboloid.

A variant of the submersion theorem can be given with differentiability instead of circa-differentiability when the spaces are finite-dimensional. Its proof (we skip) relies on the Brouwer fixed-point theorem rather than on the contraction theorem.

Proposition 2.92. *Let X and Z be Banach spaces, Z being finite-dimensional, let W be an open subset of X, and let $g : W \to Z$ be Hadamard differentiable at $a \in W$, with $Dg(a)(X) = Z$. Then there exist open neighborhoods U of a in W, V of $g(a)$ in Z and a map $h : V \to U$ that is differentiable at $g(a)$ and such that $h(g(a)) = a$, $g \circ h = I_V$. In particular, g is open at a.*

Now let us turn to representations via parameterizations. We need the following result.

Theorem 2.93 (Immersion theorem). *Let P and X be Banach spaces, let O be an open subset of P, and let $f : O \to X$ be a map of class C^k with $k \ge 1$ such that for some $\overline{p} \in O$ the map $Df(\overline{p})$ is injective and its image Y has a topological supplement Z in X. Then there exist open neighborhoods U of $a := f(\overline{p})$ in X, Q of $\overline{p}$ in O, W of 0 in Z and a C^k-diffeomorphism $\psi : V := Q \times W \to U$ such that $\psi(q,0) = f(q)$ for all $q \in Q$.*

Again the conclusion can be written in the form of a commutative diagram, since $f|Q = \psi \circ j$, where $j : Q \to Q \times W$ is the canonical injection $y \mapsto (y,0)$. Again the nonlinear map f has been straightened by ψ into a linear map $j = \psi^{-1} \circ (f|Q)$.

Proof. Let $F : O \times Z \to X$ be given by $F(p,z) = f(p) + z$. Then F is of class C^k and $F'(\overline{p},0)(p,z) = f'(\overline{p})(p) + z$ for $(p,z) \in P \times Z$, so that $F'(\overline{p},0)$ is an isomorphism from $P \times Z$ onto $Y + Z = X$. The inverse mapping theorem asserts that F induces a C^k-diffeomorphism ψ from some open neighborhood of $(\overline{p},0)$ onto some open neighborhood U of $f(\overline{p})$. Taking a smaller neighborhood of $(\overline{p},0)$ if necessary, we may suppose it has the form of a product $Q \times W$. Clearly, $\psi(q,0) = f(q)$ for $q \in Q$. $\qquad \square$

Example–Exercise. Let $P := \mathbb{R}^2$, $O := (-\pi,\pi) \times (-\pi/2,\pi/2)$, $X := \mathbb{R}^3$, and let f be given by $f(\varphi,\theta) := (\cos\theta\cos\varphi, \cos\theta\sin\varphi, \sin\varphi)$. Identify the image of f.

Exercise. Let us note that the image $f(O)$ of f is not necessarily a C^k-submanifold of X. Find a counterexample with $P := \mathbb{R}$, $X := \mathbb{R}^2$.

A topological assumption ensures that the image $f(O)$ is a C^k-submanifold of X.

Corollary 2.94 (Embedding theorem). *Let P and X be Banach spaces, let O be an open subset of P, and let $f : O \to X$ be a map of class C^k with $k \geq 1$ such that for every $p \in O$ the map $f'(p)$ is injective and its image has a topological supplement in X. Then if f is a homeomorphism from O onto $f(O)$, its image $S := f(O)$ is a C^k-submanifold of X.*

Moreover, for every $p \in O$ one has $T(S, f(p)) = f'(p)(P)$.

One says that f is an embedding of O into X and that S is parameterized by O.

Proof. Given $a := f(p)$ in S, with $p \in O$, we take $Q_a \subset O$, $U_a \subset X$, $W_a \subset Z$ and a C^k-diffeomorphism $\psi_a : V_a := Q_a \times W_a \to U_a$ such that $\psi_a(q,0) = f(q)$ for all $q \in Q_a$ as in the preceding theorem. Performing a translation in P, we may suppose $p = 0$. Using the assumption that f is a homeomorphism from O onto $S = f(O)$, we can find an open subset U_a' of X such that $f(Q_a) = S \cap U_a'$. Let $U := U_a \cap U_a'$, $V := \psi_a^{-1}(U)$, $\varphi := \psi_a^{-1}|U$, $Y := P$, so that $\varphi(a) = (0,0)$. Let us check relation (2.25), i.e., $\varphi(S \cap U) = (Y \times \{0\}) \cap V$. For all $(y,0) \in (Y \times \{0\}) \cap V$, we have $x := \varphi^{-1}(y,0) = \psi_a(y,0) = f(y) \in S$, hence $x \in S \cap U$; conversely, when $x \in S \cap U = f(Q_a)$ there is a unique $q \in Q_a$ such that $x = f(q)$, so that $x = \psi_a(q,0) = \varphi^{-1}(q,0)$ and $\varphi(x) = (q,0) \in (Y \times \{0\}) \cap V$.

Then $T(S,a) = T(S \cap U, a) = (\varphi'(a))^{-1}(T((Y \times \{0\}) \cap V, 0))$, and, since $T((Y \times \{0\}) \cap V, 0) = \psi_a'(0)(P \times \{0\}) = Y \times \{0\}$, we get $T(S,a) = Y = f'(p)(P)$. $\qquad \square$

Exercises

1. (Conic section) Let $S \subset \mathbb{R}^3$ be defined by the equations $x^2 + y^2 - 1 = 0$, $x - z = 0$. Show that S is a submanifold of $\mathbb{R}^3$ of class C^∞ (it has been known since Apollonius that S is an ellipse). Find an explicit diffeomorphism (in fact linear isomorphism) sending S onto an ellipse of the plane $\mathbb{R}^2 \times \{0\}$.

2. (Viviani's window) Let S be the subset of $\mathbb{R}^3$ defined by the system $x^2 + y^2 = x$, $x^2 + y^2 + z^2 - 1 = 0$. Show that S is a submanifold of $\mathbb{R}^3$ of class C^∞.

3. (The torus) Let $r > s > 0$, let $O := (0, 2\pi) \times (0, 2\pi)$, and let $f : O \to \mathbb{R}^3$ be given by $f(\alpha, \beta) = ((r + s\cos\beta)\cos\alpha, (r + s\cos\beta)\sin\alpha, s\sin\beta)$. Show that f is an embedding onto the torus $\mathbb{T}$ deprived from its greatest circle and from the set $\mathbb{T} \cap (\mathbb{R}_+ \times \{0\} \times \mathbb{R})$, where

$$\mathbb{T} := \left\{ (x, y, z) \in \mathbb{R}^3 : \left(\sqrt{x^2 + y^2} - r \right)^2 + z^2 = s^2 \right\}.$$

4. Using the submersion theorem, show that $\mathbb{T}$ is a C^∞-submanifold of $\mathbb{R}^3$.

5. (a) (Beltrami's tractricoid) Let $f : \mathbb{R} \to \mathbb{R}^2$ be given by $f(t) := (1/\cosh t, t - \tanh t)$. Determine the points of $T := f(\mathbb{R})$ that are smooth.
(b) (Beltrami's pseudosphere) Let $g(s, t) := (\cos s / \cosh t, \sin s / \cosh t, t - \tanh t)$. Determine the points of $S := g(\mathbb{R}^2)$ that are smooth. They form a surface of (negative) constant Gaussian curvature. It can serve as a model for hyperbolic geometry.

6. Study the **Roman surface** of equation $x^2 y^2 + y^2 z^2 + z^2 x^2 - xyz = 0$. Consider its parameterization $(\theta, \varphi) \mapsto (\cos\theta \cos\varphi \sin\varphi, \sin\theta \cos\varphi \sin\varphi, \cos\theta \sin\theta \cos^2\varphi)$.

7. Study the **cross-cap surface** $\{(1 + \cos v)\cos u, (1 + \cos v)\sin u, \tanh(u - \pi)\sin v) : (u, v) \in [0, 2\pi] \times [0, 1]\}$ and compare it with the *self-intersecting disk*, the image of $[0, 2\pi] \times [0, 1]$ by the parameterization $(u, v) \mapsto (v\cos 2u, v\sin 2u, v\cos u)$.

8. Study **Whitney's umbrella** $\{(uv, u, v^2) : (u, v) \in \mathbb{R}^2\}$. Check that it is determined by the equation $x^2 - y^2 z = 0$. Such a surface is of interest in the theory of singularities. For this surface or the preceding one, make some drawings if you can or find some on the Internet.

9. Let $O := (0, 1) \cup (1, \infty) \subset \mathbb{R}$, $f : O \to \mathbb{R}^2$ being given by $f(t) = (t + t^{-1}, 2t + t^{-2})$. Show that f is an embedding, but that its continuous extension to $(0, +\infty)$ given by $f(1) = (2, 3)$ is of class C^k but is not an immersion.

10. Let X be a normed space and let $f : X \to \mathbb{R}$ be Lipschitzian around $x \in X$. Show that f is Hadamard differentiable at $x \in X$ iff the tangent cone to the graph G of f at $(x, f(x))$ is a hyperplane.

11. Show that the fact that the tangent cone at $(x, f(x))$ to the epigraph E of f is a half-space does not imply that f is Hadamard differentiable at x.

2.5.6   The Method of Characteristics

Let us consider the partial differential equation

$$F(w, Du(w), u(w)) = 0, \qquad w \in W_0, \tag{2.26}$$

where W is a reflexive Banach space, W_0 is an open subset of W whose boundary ∂W_0 is a submanifold of class C^2, and $F : (w, p, z) \mapsto F(w, p, z)$ is a function of class C^2 on $W_0 \times W^* \times \mathbb{R}$. We look for a solution u of class C^2 satisfying the boundary condition

$$u \mid \partial W_0 = g, \tag{2.27}$$

where $g : \partial W_0 \to \mathbb{R}$ is a given function of class C^2. We leave aside the question of compatibility conditions for the data (F, g). The method of characteristics consists in associating to (2.26) a system of ordinary differential equations (in which W^{**} is identified with W) called the *system of characteristics*:

$$w'(s) = D_p F(w(s), p(s), z(s)), \tag{2.28}$$

$$p'(s) = -D_w F(w(s), p(s), z(s)) - D_z F(w(s), p(s), z(s)) p(z), \tag{2.29}$$

$$z'(s) = \langle D_p F(w(s), p(s), z(s)), p(s) \rangle. \tag{2.30}$$

Suppose a smooth solution u of (2.26) is known. Let us relate it to a solution $s \mapsto (w(s), p(s), z(s))$ of the system (2.28)–(2.30). Let

$$q(s) := Du(y(s)), \qquad r(s) := u(y(s)),$$

where $y(\cdot)$ is the solution of the differential equation

$$y'(s) := D_p F(y(s), Du(y(s)), u(y(s))), \qquad y(0) = w_0.$$

Then

$$r'(s) = Du(y(s)) \cdot y'(s) = \langle q(s), D_p F(y(s), p(s), z(s)) \rangle.$$

For all $e \in W$, identifying W^{**} and W, we have

$$q'(s) \cdot e = D^2 u(y(s)) \cdot y'(s) \cdot e = \langle D_p F(y(s), p(s), z(s)), D^2 u(y(s)) \cdot e \rangle.$$

Now, taking the derivative of the function $F(\cdot, Du(\cdot), u(\cdot))$ and writing u, Du instead of $u(w), Du(w)$, we have

$$D_w F(w, Du, u) e + D_p F(w, Du, u) D^2 u(w) \cdot e + D_z F(w, Du, u) Du(w) e = 0.$$

Thus, replacing (w, Du, u) by $(y(s), q(s), r(s))$ and noting that e is arbitrary in W, we get

$$q'(s) = -D_w F(w(s), q(s), r(s)) - D_z F(w(s), q(s), r(s))q(s).$$

It follows that $s \mapsto (y(s), q(s), r(s))$ is a solution of the characteristic system. Taking the same initial data $(w_0, p_0, g(w_0))$, by uniqueness of the solution of the characteristic system, we get $y(s) = w(s)$, $p(s) = q(s)$, and $z(s) = r(s) := u(w(s))$. This means that knowing the solution of the characteristic system, we get the value of u at $w(s)$. If around some point $\overline{w} \in W_0$ we can represent every point w of a neighborhood of $\overline{w}$ as the value $w(s)$ for the solution of (2.28)–(2.30) issued from some initial data, then we get u around $\overline{w}$. In the following classical example, the search for the initial data is particularly simple.

Example. Let $W := \mathbb{R}^n$, $W_0 := \mathbb{R}^{n-1} \times \mathbb{P}$, F being given by $F(w, p, z) := p \cdot b(w, z) - c(w, z)$, where $b : W_0 \times \mathbb{R} \to W$, $c : W_0 \times \mathbb{R} \to \mathbb{R}$. Then, taking into account the relation $D_p F(w(s), p(s), z(s)) \cdot p(s) = p(s) \cdot b(w(s), z(s)) = c(w(s), z(s))$, (2.28), (2.30) of the characteristic system read as a system in (w, z):

$$w'(s) = b(w(s), z(s)),$$

$$z'(s) = c(w(s), z(s)).$$

In the case that $b := (b_1, \ldots, b_n)$ is constant with $b_n \neq 0$ and $c(w, z) := z^{k+1}/k$, with $k > 0$, the solution of this system with initial data $((v, 0), g(v)) \in \mathbb{R}^n \times \mathbb{P}$ is given by

$$w_i(s) = b_i s + v_i \ (i = 1, \ldots, n-1), \quad w_n(s) = b_n s, \quad z(s) = \frac{g(v)}{(1 - g(v)^k s)^{1/k}}.$$

It is defined for s in the interval $S := [0, g(v)^{-k})$. Given $x := (x_1, \ldots, x_n) \in W_0$ near $\overline{x} \in W_0$, the initial data v is found by solving the equations $b_i s + v_i = x_i \ (i \in \mathbb{N}_{n-1})$, $x_n = b_n s$: $v_i = x_i - a_i x_n$ with $a_i := b_i/b_n$. What precedes shows that u is given by

$$u(x) = \frac{g(x_1 - a_1 x_n, \ldots, x_{n-1} - a_{n-1} x_n)}{(1 - g(x_1 - a_1 x_n, \ldots, x_{n-1} - a_{n-1} x_n)^k x_n/b_n)^{1/k}}$$

and is defined in the set $\{(x_1, \ldots, x_n) : x_n g(x_1 - a_1 x_n, \ldots, x_{n-1} - a_{n-1} x_n)^k < b_n\}$. □

A special case of (2.26) is of great importance. It corresponds to the case $w := (x, t) \in W_0 := U \times (0, \tau)$ for some $\tau \in (0, +\infty]$ and some open subset U of a hyperplane X of W and $F((x, t), (y, v), z) := v + H(x, t, y, z)$, so that (2.26) and the boundary condition (2.27) take the form

$$D_t u(x, t) + H(x, t, D_x u(x, t), u(x, t)) = 0, \quad (x, t) \in W_0 \times (0, \tau), \tag{2.31}$$

$$u(x, 0) = g(x), \quad x \in W_0. \tag{2.32}$$

Such a system is called a *Hamilton–Jacobi equation*.

Let us note that as in the example of quasilinear equations, the general case can be reduced to this form under a mild condition. First, since W_0 is the interior of a smooth manifold with boundary, taking a chart, we may assume for a local study that $W_0 = U \times (0, \tau)$ for some $\tau > 0$ and some open subset U of a hyperplane X of W. Now, using the implicit function theorem around $\overline{w} \in \partial W_0$, F can be reduced to the form $F((x,t),(y,v),z) := v + H(x,t,y,z)$, provided $D_v F(\overline{w}, \overline{p}, \overline{z}) \neq 0$. Such a condition can be expressed intrinsically (i.e., without using the chart) by finding a vector $\overline{v}$ transverse to ∂W_0 at $\overline{w}$ such that $D_w F(\overline{w}, \overline{y}, \overline{z}) \cdot \overline{v} \neq 0$.

The characteristic system associated with (2.31) can be reduced to

$$x'(s) = D_y H(x(s), s, y(s), z(s)), \tag{2.33}$$

$$y'(s) = -D_x H(x(s), s, y(s), z(s)) - D_z H(x(s), s, y(s), z(s)) y(s), \tag{2.34}$$

$$z'(s) = D_y H(x(s), s, y(s), z(s)) \cdot y(s) - H(x(s), s, y(s), z(s)), \tag{2.35}$$

by dropping the equation $t'(s) = 1$ and observing that we do not need an equation for $D_t u(x(s), t(s))$, since this derivative is known to be $-H(x(s), s, y(s), z(s))$. In order to take into account the dependence on the initial condition $(v, Dg(v), g(v))$, the one-jet of g at $v \in U \subset X$, let us denote by $s \mapsto (\widehat{x}(s, v), \widehat{y}(s, v), \widehat{z}(s, v))$ the solution to the system (2.33)–(2.35). Since the right-hand side of this system is of class C^1, the theory of differential equations ensures that the solution is a mapping of class C^1 in (s, v). In view of the initial data, we have

$$\forall v \in U, \ v' \in X, \qquad D_v \widehat{x}(0, v) v' = v'.$$

It follows that for all $\overline{v} \in U$ there exist a neighborhood V of $\overline{v}$ in U and some $\sigma \in (0, \tau)$ such that for $s \in (0, \sigma)$, the map $\widehat{x}_s : v \mapsto \widehat{x}(s, v)$ is a diffeomorphism from V onto $V_s := \widehat{x}(s, V)$. From the analysis that precedes, we get that for $x \in V_s$ one has $u(x, s) = \widehat{z}(s, v)$ with $v := (\widehat{x}_s)^{-1}(x)$. Thus we get a local solution to the system (2.31)–(2.32). In general, one cannot get a global solution with such a method: it may happen that for two values v_1, v_2 of v the characteristic curves issued from v_1 and v_2 take the same value for some $t > 0$.

Exercises

1. Write down the characteristic system for the **conservation law**

$$D_t u(x, t) + D_x u(x, t) \cdot b(u(x, t)) = 0, \qquad u(v, 0) = g(v),$$

where $b : \mathbb{R} \to X$, $g : X \to \mathbb{R}$ are of class C^1. Check that its solution satisfies $\widehat{x}(s, v) = v + sb(g(v))$, $\widehat{z}(s, v) = g(v)$. Compute $D_v \widehat{x}(s, v)$ and check that for all $\overline{v} \in X$, this element of $L(X, X)$ is invertible for (s, v) close enough to $(0, \overline{v})$. Deduce a local solution of the equation of conservation law from this property.

2. (Haar's uniqueness theorem) Suppose $X = \mathbb{R}$ and $H : X \times \mathbb{R} \times X^* \times \mathbb{R} \to \mathbb{R}$
satisfies the Lipschitz condition with constants k, ℓ:

$$\forall (t,x,y,y',z,z') \in T \times \mathbb{R}^4, \qquad \left| H(x,t,y,z) - H(x,t,y',z') \right| \leq k \left| y - y' \right| + \ell \left| z - z' \right|,$$

where T is the triangle $T := \{(x,t) \in X \times [0,a] : x \in [b + \ell t, c - \ell t]\}$, for some
constants a, b, c. Show that if u_1, u_2 are two solutions of class C^1 in T of the
system (2.31)–(2.32), then $u_1 = u_2$. [For a generalization to $X := \mathbb{R}^n$ see [925,
Theorem 1.6], [960].]

3. Suppose $X = \mathbb{R}$, $g = I_X$, and $H : X \times \mathbb{R} \times X^* \times \mathbb{R} \to \mathbb{R}_\infty$ is given by $H(x,t,y,z) :=$
$|t - 1|^{-1/2} y$ for $t \in [0,1)$, $+\infty$ otherwise. Using the method of characteristics, show
that a solution to the system (2.31)–(2.32) is given by $u(x,t) = x - 2 + 2\sqrt{1 - t}$ for
$(x,t) \in X \times (0,1)$.

4. Suppose $X = \mathbb{R}$, and that g and H are given by $H(x,t,y,z) := -y^2/2$, $g(x) :=$
$x^2/2$. Using the method of characteristics, show that a solution to the system (2.31)–
(2.32) is given by $u(x,t) = x^2/2(1 - t)$ for $(x,t) \in X \times (0,1)$.

5. Suppose $X = \mathbb{R}$, and that g and H are given by $H(x,t,y,z) := e^{-3t} yz(a'(t)e^{2t} +$
$b'(t)z^2) - z$, $g(x) := x$, where a and b are nonnegative functions of class C^1 satisfying
$a(0) = 1$, $b(0) = 0$, $a + b > 0$. Show that the characteristics associated with the
system (2.31)–(2.32) satisfy $\widehat{x}(t,v) = a(t)v + b(t)v^3$, $\widehat{z}(t) = e^t v$, so that $v \mapsto \widehat{x}(t,v)$ is
a bijection. Assuming that there exists some $\tau > 0$ such that $a(t) = 0$ for $t \geq \tau$, show
that $u(x,t) = e^t b(t)^{-1/3} x^{1/3}$ for $(x,t) \in X \times [\tau, \infty)$, so that u is not differentiable at
$(0,t)$.

6. Suppose $X = \mathbb{R}$, and that g and H are given by $g(x) := x^2/2$, $H(x,t,y,z) :=$
$a'(t)e^{-t}y^2/2 + b'(t)e^{-3t}y^4 - z$, where a and b are as in the preceding exercise. Show
that the characteristics associated with the system (2.31)–(2.32) satisfy $\widehat{x}(t,v) =$
$a(t)v + 4b(t)v^3$, $\widehat{z}(t) = e^t (a(t)v^2/2 + 3b(t)v^4)$, so that for $t \geq \tau$, $v \mapsto \widehat{x}(t,v)$ is a
bijection on a neighborhood of 0, in spite of the fact that $D_v\widehat{x}(t,0) = 0$ and $u(x,t) =$
$3.4^{-4/3}b(t)x^{4/3}$, so that u is of class C^1 but not C^2 around $(0,t)$.

2.6 Applications to Optimization

We will formulate necessary optimality conditions for the problem with constraint

$$(\mathscr{P}) \qquad \text{minimize } f(x) \text{ under the constraint } x \in F,$$

where F is a nonempty subset of the normed space X called the *feasible set* or the
admissible set. These conditions will involve the concept of normal cone.

2.6.1   Normal Cones, Tangent Cones, and Constraints

In fact, we will use some variants of the concept of normal cone that fit different differentiability assumptions on the function f. When the feasible set is a convex set these variants coincide (Exercise 6) and the concept is very simple.

Definition 2.95. The *normal cone* $N(C,\bar{x})$ to a convex subset C of X at $\bar{x} \in C$ is the set of $\bar{x}^* \in X^*$ that attain their maximum on C at $\bar{x}$:

$$N(C,\bar{x}) := \{\bar{x}^* \in X^* : \forall x \in C, \ \langle \bar{x}^*, x - \bar{x} \rangle \leq 0\}.$$

Thus, when C is a linear subspace, $N(C,\bar{x}) = C^\perp$, where $C^\perp$ is the orthogonal of C (or annihilator of C) in X^*:

$$C^\perp := \{\bar{x}^* \in X^* : \forall x \in C, \ \langle \bar{x}^*, x \rangle = 0\}.$$

When C is a cone, one has $N(C,0) = C^0$, where C^0 is the polar cone of C.

In the nonconvex case the preceding definition has to be modified by introducing a remainder in the inequality in order to allow a certain curvature or inaccuracy.

Definition 2.96. The *firm or Fréchet normal cone* $N_F(F,\bar{x})$ to a subset F of X at $\bar{x} \in F$ is the set of $\bar{x}^* \in X^*$ for which there exists a remainder $r(\cdot)$ such that $\bar{x}^*(\cdot) + r(\cdot - \bar{x})$ attains its maximum on F at $\bar{x}$:

$$\bar{x}^* \in N_F(F,\bar{x}) \iff \exists r \in o(X,\mathbb{R}) \quad \forall x \in F, \ \langle \bar{x}^*, x - \bar{x} \rangle + r(x - \bar{x}) \leq 0.$$

In other words, $\bar{x}^* \in X^*$ is a firm normal to F at $\bar{x}$ iff for every $\varepsilon > 0$ there exists $\delta > 0$ such that for all $x \in F \cap B(\bar{x},\delta)$ one has $\langle \bar{x}^*, x - \bar{x} \rangle \leq \varepsilon \|x - \bar{x}\|$.

Equivalently,

$$\bar{x}^* \in N_F(F,\bar{x}) \quad \iff \quad \limsup_{x \to \bar{x}, \ x \neq \bar{x}} \frac{1}{\|x - \bar{x}\|} \langle \bar{x}^*, x - \bar{x} \rangle \leq 0.$$

We will give some properties and calculus rules in the next subsection. For the moment it is important to convince oneself that this notion corresponds to the intuitive idea of an "exterior normal" to a set, for instance by making drawings in simple cases. We shall present a necessary condition using this concept without delay. In it we say that f attains a *local maximum* (resp. *local minimum*) on F at $\bar{x}$ if $f(x) \leq f(\bar{x})$ (resp. $f(x) \geq f(\bar{x})$) for all x in some neighborhood of $\bar{x}$ in F. It is convenient to say that $\bar{x}$ is a *local maximizer* (resp. *local minimizer*) of f on F.

Theorem 2.97 (Fermat's rule). *Suppose f attains a local maximum on F at $\bar{x}$ and is Fréchet differentiable at $\bar{x}$. Then*

$$f'(\bar{x}) \in N_F(F,\bar{x}).$$

If f attains a local minimum on F at $\bar{x}$ and is Fréchet differentiable at $\bar{x}$ then

$$0 \in f'(\bar{x}) + N_F(F,\bar{x}).$$

Proof. Suppose f attains a local maximum on F at $\bar{x}$ and is differentiable at $\bar{x}$. Set

$$f(x) = f(\bar{x}) + \langle \bar{x}^*, x - \bar{x} \rangle + r(x - \bar{x})$$

with r a remainder, $\bar{x}^* := f'(\bar{x})$, so that for $x \in F$ close enough to $\bar{x}$ one has

$$\langle \bar{x}^*, x - \bar{x} \rangle + r(x - \bar{x}) = f(x) - f(\bar{x}) \le 0.$$

Hence $\bar{x}^* \in N_F(F,\bar{x})$. Changing f into $-f$, one obtains the second assertion. □

The second formula shows how the familiar rule $f'(\bar{x}) = 0$ of unconstrained minimization has to be changed by introducing an additional term involving the normal cone. Without such an additional term the condition would be utterly invalid.

Example. The identity map $f = I_\mathbb{R}$ on $\mathbb{R}$ attains its minimum on $F := [0,1]$ at 0 but $f'(0) = 1$.

Example. Suppose F is the unit sphere of the Euclidean space $\mathbb{R}^3$ representing the surface of the earth and suppose f is a smooth function representing the temperature. If f attains a local minimum on F at $\bar{x}$, in general $\nabla f(\bar{x})$ is not 0; however, $\nabla f(\bar{x})$ is on the downward vertical at $\bar{x}$, and if one can increase one's altitude at that point, one usually experiences a decrease of the temperature. □

When the objective function f is not Fréchet differentiable but just Hadamard differentiable, an analogue of Fermat's rule can still be given by introducing a variant of the notion of firm normal cone. It goes as follows; although this variant appears to be more technical than the concept of Fréchet normal cone, it is a general and important notion. It can be formulated with the help of the notion of directional remainder: $r : X \to Y$ is a *directional remainder* if for all $u \in X \setminus \{0\}$ one has $r(tv)/t \to 0$ as $t \to 0_+$, $v \to u$; we write $r \in o_D(X,Y)$.

Definition 2.98. The *normal cone* (or *directional normal cone*) to the subset F at $\bar{x} \in \mathrm{cl}(F)$ is the set $N(F,\bar{x}) := N_D(F,\bar{x})$ of $\bar{x}^* \in X^*$ for which there exists a directional remainder $r(\cdot)$ such that $\bar{x}^*(\cdot) + r(\cdot - \bar{x})$ attains its maximum on F at $\bar{x}$:

$$\bar{x}^* \in N(F,\bar{x}) := N_D(F,\bar{x}) \iff \exists r \in o_D(X,\mathbb{R}) \quad \forall x \in F, \quad \langle \bar{x}^*, x - \bar{x} \rangle + r(x - \bar{x}) \le 0.$$

In other words, $\bar{x}^* \in X^*$ is a normal to F at $\bar{x}$ iff for all $u \in X \setminus \{0\}$, $\varepsilon > 0$ there exists $\delta > 0$ such that $\langle \bar{x}^*, v \rangle \le \varepsilon$ for every $(t,v) \in (0,\delta] \times B(u,\delta)$ satisfying $\bar{x} + tv \in F$:

$$\bar{x}^* \in N(F,\bar{x}) := N_D(F,\bar{x}) \iff \forall u \in X, \quad \limsup_{(t,v) \to (0_+,u),\, \bar{x}+tv \in F} \frac{1}{t} \langle \bar{x}^*, (\bar{x}+tv) - \bar{x} \rangle \le 0.$$

Let us note that the case $u = 0$ can be discarded in the preceding reformulation because the condition is automatically satisfied in this case with $\delta = \varepsilon \min(1, \|\bar{x}^*\|^{-1})$. This cone often coincides with the Fréchet normal cone and it always contains it, as the preceding reformulations show.

Lemma 2.99. *For every subset F and every $\bar{x} \in \mathrm{cl}(F)$ one has $N_F(F,\bar{x}) \subset N(F,\bar{x})$.*

The duality property we prove now compensates the complexity of the definition of the (directional) normal cone compared to the definition of the firm normal cone.

Proposition 2.100. *The normal cone to F at $\bar{x}$ is the polar cone to the tangent cone to F at $\bar{x}$:*

$$(\bar{x}^* \in N(F,\bar{x})) \Leftrightarrow (\forall u \in T(F,\bar{x}), \quad \langle \bar{x}^*, u \rangle \le 0).$$

Proof. Given $\bar{x}^* \in N(F,\bar{x})$ and $u \in T(F,\bar{x}) \setminus \{0\}$, for every $\varepsilon > 0$, taking $\delta \in (0,\varepsilon)$ such that $\langle \bar{x}^*, v \rangle \le \varepsilon$ for every $(t,v) \in (0,\delta] \times B(u,\delta)$ satisfying $\bar{x} + tv \in F$ and observing that such a pair (t,v) exists since $u \in T(F,\bar{x})$, we get $\langle \bar{x}^*, u \rangle \le \langle \bar{x}^*, v \rangle + \|\bar{x}^*\| \|u - v\| \le \varepsilon + \varepsilon \|\bar{x}^*\|$. Since ε is arbitrarily small, we get $\langle \bar{x}^*, u \rangle \le 0$.

Conversely, given $\bar{x}^*$ in the polar cone of $T(F,\bar{x})$, given $u \in T(F,\bar{x})$, and given $\varepsilon > 0$, taking $\delta > 0$ such that $\delta \|\bar{x}^*\| \le \varepsilon$, the inequality $\langle \bar{x}^*, v \rangle \le \varepsilon$ holds whenever $t \in (0,\delta)$, $v \in t^{-1}(F - \bar{x}) \cap B(u,\delta)$, since

$$\langle \bar{x}^*, v \rangle \le \langle \bar{x}^*, u \rangle + \langle \bar{x}^*, v - u \rangle \le \|\bar{x}^*\| \|u - v\| \le \delta \|\bar{x}^*\| \le \varepsilon.$$

If $u \in X \setminus T(F,\bar{x})$ we can find $\delta > 0$ such that no such pair (t,v) exists. Thus, we have $\langle \bar{x}^*, v \rangle \le \varepsilon$ for every $(t,v) \in (0,\delta] \times B(u,\delta)$ satisfying $\bar{x} + tv \in F$: $\bar{x}^* \in N(F,\bar{x})$. $\square$

Theorem 2.101 (Fermat's rule). *Suppose f attains a local maximum on F at $\bar{x} \in F$ and is Hadamard differentiable at $\bar{x}$. Then for all $v \in T(F,\bar{x})$ one has $f'(\bar{x})v \le 0$:*

$$f'(\bar{x}) \in N(F,\bar{x}).$$

If f attains a local minimum on F at $\bar{x}$, then for all $v \in T(F,\bar{x})$ one has $f'(\bar{x})v \ge 0$:

$$0 \in f'(\bar{x}) + N(F,\bar{x}).$$

Proof. Let V be an open neighborhood of $\bar{x}$ in X such that $f(x) \le f(\bar{x})$ for all $x \in F \cap V$. Given $v \in T(F,\bar{x})$, let $(v_n) \to v$, $(t_n) \to 0_+$ be sequences such that $\bar{x} + t_n v_n \in F$ for all $n \in \mathbb{N}$. For n large enough, we have $\bar{x} + t_n v_n \in F \cap V$, hence $f(\bar{x} + t_n v_n) - f(\bar{x}) \le 0$. Dividing by t_n and passing to the limit, the (Hadamard) differentiability of f at $\bar{x}$ yields $f'(\bar{x})(v) \le 0$. $\square$

It is possible to give a third version of Fermat's rule that does not suppose that f is differentiable; it is set in the space X instead of its dual X^*. In it, we use the *directional (lower) derivative* (or *contingent derivative*) of f given by

$$f^D(\bar{x}, u) := \liminf_{(t,v)\to(0_+,u)} \frac{1}{t}(f(\bar{x}+tv) - f(\bar{x}))$$

and the tangent cone to F at $\bar{x}$ as introduced in Definition 2.86.

In view of their fundamental character, we will return to these notions of tangent and normal cones. For the moment, the definition itself suffices to give the primal version of Fermat's rule we announced. Note that this version entails the preceding theorem, since $f^D(\bar{x}, \cdot) = f'(\bar{x})$ when f is Hadamard differentiable at $\bar{x}$.

Theorem 2.102. *Suppose f attains a local maximum on F at $\bar{x}$. Then*

$$f^D(\bar{x}, u) \le 0 \text{ for all } u \in T(F, \bar{x}).$$

Proof. Let $u \in T(F, \bar{x})$. There exist $(t_n) \to 0_+$, $(u_n) \to u$ such that $\bar{x}+t_n u_n \in F$ for all $n \in \mathbb{N}$. For n large enough we have $f(\bar{x}+t_n u_n) \le f(\bar{x})$, so that

$$f^D(\bar{x}, u) \le \liminf_n \frac{1}{t_n}(f(\bar{x}+t_n u_n) - f(\bar{x})) \le 0. \qquad \square$$

For minimization problems, a variant of the tangent cone is required, since the rule $f^D(\bar{x}, u) \ge 0$ for $u \in T(F, \bar{x})$ is not valid in general.

Example. Let $F := \{0\} \cup \{2^{-2n} : n \in \mathbb{N}\} \subset \mathbb{R}$ and let $f : \mathbb{R} \to \mathbb{R}$ be even and given by $f(x) = 0$ for every $x \in F$, $f(2^{-2k+1}) = -2^{-2k+1}$, f being affine on each interval $[2^{-j}, 2^{-j+1}]$. Show that $f^D(\bar{x}, 1) = -1$ for $\bar{x} := 0$, although $f(\bar{x}) = \min f(F)$.

Definition 2.103. The *incident cone* (or *adjacent cone*) to F at $\bar{x} \in \mathrm{cl}(F)$ is the set

$$T^I(F, \bar{x}) := \{u \in X : \forall(t_n) \to 0_+, \exists(u_n) \to u, \quad \bar{x}+t_n u_n \in F \quad \forall n\}$$
$$= \{u \in X : \forall(t_n) \to 0_+, \exists(x_n) \to \bar{x}, (t_n^{-1}(x_n - \bar{x})) \to u, \quad x_n \in F \quad \forall n\}.$$

It is easy to show that

$$u \in T^I(F, \bar{x}) \Leftrightarrow \lim_{t\to 0_+} \frac{1}{t} d(\bar{x}+tu, F) = 0.$$

Let us also introduce the *incident derivative* of a function f at $\bar{x}$ by

$$f^I(\bar{x}, u) := \inf\{r \in \mathbb{R} : (u, r) \in T^I(E_f, \bar{x}_f)\},$$

where E_f is the epigraph of f and $\bar{x}_f := (\bar{x}, f(\bar{x}))$.

Proposition 2.104. *Suppose f is directionally stable at $\bar{x}$ in the sense that for all $u \in X \setminus \{0\}$ one has $(1/t)(f(\bar{x}+tv) - f(\bar{x}+tu)) \to 0$ as $(t, v) \to (0, u)$. If f attains a local minimum on F at $\bar{x}$, then*

$$f^I(\bar{x}, u) \geq 0 \, \text{for all } u \in T(F, \bar{x}),$$

$$f^D(\bar{x}, u) \geq 0 \, \text{for all } u \in T^I(F, \bar{x}).$$

Proof. Suppose, to the contrary, that there exists some $u \in T(F, \bar{x})$ such that $f^I(\bar{x}, u) < 0$. Then there exists some $r < 0$ such that $(u, r) \in T^I(E_f, \bar{x}_f)$; thus, if $(t_n) \to 0_+$ and $(u_n) \to u$ are such that $\bar{x} + t_n u_n \in F$ for all $n \in \mathbb{N}$, one can find a sequence $((v_n, r_n)) \to (u, r)$ such that $\bar{x}_f + t_n(v_n, r_n) \in E_f$ for all $n \in \mathbb{N}$. Then $f(\bar{x}) + t_n r_n \geq f(\bar{x} + t_n v_n)$ for all $n \in \mathbb{N}$ and

$$0 > r \geq \limsup_n (1/t_n)(f(\bar{x} + t_n v_n) - f(\bar{x})) = \limsup_n (1/t_n)(f(\bar{x} + t_n u_n) - f(\bar{x})) \geq 0,$$

a contradiction. The proof of the second assertion is similar.   □

Exercises

1. Given an element $\bar{x}$ of the closure of a subset F of a normed space X, show that the tangent cone and the incident cone can be expressed in terms of limits of sets:

$$T(F, \bar{x}) = \limsup_{t \to 0_+} (1/t)(F - \bar{x}), \qquad T^I(F, \bar{x}) = \liminf_{t \to 0_+} (1/t)(F - \bar{x}).$$

2. Deduce from Exercise 1 that $v \in T(F, \bar{x})$ iff $\liminf_{t \to 0_+} (1/t)d(\bar{x} + tv, F) = 0$ and that $v \in T^I(F, \bar{x})$ if and only if $\lim_{t \to 0_+} (1/t)d(\bar{x} + tv, F) = 0$.

3. Find a subset F of $\mathbb{R}$ such that $1 \in T(F, 0)$ but $T^I(F, 0) = \{0\}$.

4. Show that if X is a finite-dimensional normed space, then for every subset F of X and every $\bar{x} \in \text{cl}(F)$, one has $N(F, \bar{x}) = N_F(F, \bar{x})$.

5. Show that for every subset F of a normed space and every $\bar{x} \in \text{cl}(F)$, the cones $N(F, \bar{x})$ and $N_F(F, \bar{x})$ are convex and closed.

6. Show that for every convex subset C of a normed space X and every $\bar{x} \in \text{cl}(C)$ the cones $N(C, \bar{x})$ and $N_F(C, \bar{x})$ coincide with the normal cone in the sense of convex analysis described in Definition 2.95.

7. Let $f : \mathbb{R} \to \mathbb{R}$ be differentiable at $a \in \mathbb{R}$ and such that a is a minimizer of f on some interval $[a, b]$ with $b > a$. Check that $f'(a) \geq 0$.

8. Show that the incident cone $T^I(F, \bar{x})$ can be called the velocity cone of F at $\bar{x}$ since $v \in T^I(F, \bar{x})$ iff there exists some $c : [0, 1] \to F$ such that $c(0) = \bar{x}$, c is right differentiable at 0, and $c'_+(0) = v$.

2.6.2 Calculus of Tangent and Normal Cones

We devote this subsection to some calculus rules for normal cones. These rules will enable us to compute the normal cones to sets defined by equalities and inequalities, an important topic for the application to concrete optimization problems.

In order to show that the two notions of normal cone we introduced correspond to the classical notion in the smooth case, let us make some easy but useful observations.

Proposition 2.105. *The notions of normal cone and of Fréchet normal cone are local notions: if F and G are two subsets such that $F \cap V = G \cap V$ for some neighborhood V of $\bar{x}$, then $N(F,\bar{x}) = N(G,\bar{x})$ and $N_F(F,\bar{x}) = N_F(G,\bar{x})$.*

Proposition 2.106. *Given normed spaces X, Y and $\bar{x} \in F \subset X$, $\bar{y} \in G \subset Y$, one has*

$$N(F \times G, (\bar{x},\bar{y})) = N(F,\bar{x}) \times N(G,\bar{y}),$$
$$N_F(F \times G, (\bar{x},\bar{y})) = N_F(F,\bar{x}) \times N_F(G,\bar{y}).$$

Proposition 2.107. *The normal cone and the firm normal cone are antitone: for $F \subset G$ and every $\bar{x} \in \mathrm{cl}F$ one has $N(G,\bar{x}) \subset N(F,\bar{x})$ and $N_F(G,\bar{x}) \subset N_F(F,\bar{x})$. Moreover, if F is a finite union, $F = \bigcup_{i \in I} F_i$, then*

$$N(F,\bar{x}) = \bigcap_{i \in I} N(F_i,\bar{x}), \quad N_F(F,\bar{x}) = \bigcap_{i \in I} N_F(F_i,\bar{x}).$$

This fact helps in the computation of normal cones, as the next example shows.

Example. Let $F := \{(r,s) \in \mathbb{R}^2 : rs = 0\}$, so that $F = F_1 \cup F_2$ with $F_1 := \mathbb{R} \times \{0\}$, $F_2 := \{0\} \times \mathbb{R}$. Then since F_i is a linear subspace, one has $N(F_i,0) = F_i^{\perp}$; hence $N(F,0) = F_1^{\perp} \cap F_2^{\perp} = \{0\}$.

However, the computations of normal cones to intersections are not obvious. One may just have the inclusions

$$N(F \cap G,\bar{x}) \supset N(F,\bar{x}) \cup N(G,\bar{x}), \qquad N_F(F \cap G,\bar{x}) \supset N_F(F,\bar{x}) \cup N_F(G,\bar{x}).$$

Example. Let $X := \mathbb{R}^2$ with its usual Euclidean norm and let $F := B_X + e$, $G := B_X - e$, where $e = (0,1)$. Then $N(F \cap G, 0) = \mathbb{R}^2$, whereas $N(F,0) \cup N(G,0) = \{0\} \times \mathbb{R}$.

Now let us show that the notions of normals and firm normals are invariant under differentiable transformations (diffeomorphisms).

Proposition 2.108. *Let $g : U \to V$ be a map between two open subsets of the normed spaces X and Y respectively and let $B \subset U$, $C \subset V$ be such that $g(B) \subset C$. Then if g is F-differentiable, respectively H-differentiable, at $\bar{x} \in B$, then for $\bar{y} := g(\bar{x})$, one has respectively*

$$N_F(C,\bar{y}) \subset (g'(\bar{x})^\mathsf{T})^{-1}(N_F(B,\bar{x})), \tag{2.36}$$

$$N(C,\bar{y}) \subset (g'(\bar{x})^\mathsf{T})^{-1}(N(B,\bar{x})).$$

Relation (2.36) is an equality when $C = g(B)$ and there exist $\rho > 0$, $c > 0$ such that

$$\forall y \in C \cap B(\bar{y},\rho), \qquad d(\bar{x}, g^{-1}(y) \cap B) \le cd(y,\bar{y}). \tag{2.37}$$

Proof. Let $\bar{y}^*$ be an element of $N_F(C,\bar{y})$: for some remainder $r(\cdot)$ and for all $y \in C$ we have $\langle \bar{y}^*, y - \bar{y} \rangle \le r(\|y - \bar{y}\|)$. The differentiability of g at $\bar{x}$ can be written for some remainder s:

$$g(x) - g(\bar{x}) = A(x - \bar{x}) + s(\|x - \bar{x}\|), \tag{2.38}$$

where $A := g'(\bar{x})$. Taking $x \in B$, since $y := g(x) \in C$, we get

$$\langle A^\mathsf{T}(\bar{y}^*), x - \bar{x} \rangle = \langle \bar{y}^*, g(x) - g(\bar{x}) - s(\|x - \bar{x}\|) \rangle$$
$$\le r(\|g(x) - g(\bar{x})\|) - \langle \bar{y}^*, s(\|x - \bar{x}\|) \rangle := t(\|x - \bar{x}\|),$$

where t is a remainder, since $\|g(x) - g(\bar{x})\| \le (\|A\| + 1)\|x - \bar{x}\|$ for x close enough to $\bar{x}$. The proof for the normal cone is similar. It can also be deduced from the inclusion $g'(\bar{x})(T(B,\bar{x})) \subset T(C,\bar{y})$.

Now suppose $C = g(B)$ and relation (2.37) holds for some $\rho > 0$, $c > 0$. Then for all $y \in C \cap B(\bar{y},\rho)$, there exists some $x_y \in g^{-1}(y) \cap B$ satisfying $\|x_y - \bar{x}\| \le 2c\|y - \bar{y}\|$. Let $\bar{y}^* \in Y^*$ be such that $\bar{x}^* := g'(\bar{x})^\mathsf{T}(\bar{y}^*) \in N_F(B,\bar{x})$. Then there exists a remainder $r(\cdot)$ such that

$$\forall x \in B, \qquad \langle \bar{y}^*, g'(\bar{x})(x - \bar{x}) \rangle = \langle \bar{x}^*, x - \bar{x} \rangle \le r(x - \bar{x}).$$

Taking into account (2.38), we get for all $y \in C \cap B(\bar{y},\rho)$,

$$\langle \bar{y}^*, y - \bar{y} \rangle = \langle \bar{y}^*, g(x_y) - g(\bar{x}) \rangle \le r(\|x_y - \bar{x}\|) + \|\bar{y}^*\| s(\|x_y - \bar{x}\|),$$

and since $\|x_y - \bar{x}\| \le 2c\|y - \bar{y}\|$, we conclude that $\bar{y}^* \in N_F(C,\bar{y})$.   □

Corollary 2.109. *Let $g : U \to V$ be a bijection between two open subsets of the normed spaces X and Y respectively such that g and $h := g^{-1}$ are H-differentiable, respectively F-differentiable, at $\bar{x}$ and $\bar{y} := g(\bar{x})$ respectively, and let $B \subset U$, $C = g(B)$. Then we have respectively*

$$N(B,\bar{x}) = g'(\bar{x})^\mathsf{T}(N(C,\bar{y}))$$

and

$$N_F(B,\bar{x}) = g'(\bar{x})^\mathsf{T}(N_F(C,\bar{y})).$$

Proof. Since $h'(\bar{y})^{\mathsf{T}}$ is the inverse of $g'(\bar{x})^{\mathsf{T}}$, one has the inclusions of Proposition 2.108 and their analogues in which $h, \bar{y}, C$ take the roles of $g, \bar{x}, B$, respectively.   □

For an inverse image, it is possible to ensure equality in the inclusions of Proposition 2.108. However, a technical assumption called a qualification condition should be added, for otherwise, the result may be invalid, as the following example shows.

Example. Let $X = Y = \mathbb{R}$, $g(x) = x^2$, $C = \{0\}$, $B = g^{-1}(C)$. Then $N(B,0) = \mathbb{R} \neq g'(\bar{x})^{\mathsf{T}}(N(C,0)) = \{0\}$.

The factorization of Lemma 1.108 will be helpful for handling inverse images.

Proposition 2.110 (Lyusternik). *Let X, Y be Banach spaces, let U be an open subset of X, and let $g : U \to Y$ be circa-differentiable at $\bar{x} \in U$ with $g'(\bar{x})(X) = Y$. Then for $S := g^{-1}(\bar{y})$ with $\bar{y} := g(\bar{x})$ one has $N(S,\bar{x}) = N_F(S,\bar{x}) = g'(\bar{x})^{\mathsf{T}}(Y^*)$.*

Proof. Proposition 2.108 ensures that $g'(\bar{x})^{\mathsf{T}}(Y^*) \subset N_F(S,\bar{x}) \subset N(S,\bar{x})$. Now, given $x^* \in N(S,\bar{x})$, for all $v \in T(S,\bar{x}) = \ker g'(\bar{x}) = -T(S,\bar{x})$ we have $g'(\bar{x})v = 0$, so that Lemma 1.108 yields some $y^* \in Y^*$ such that $x^* = y^* \circ g'(\bar{x}) = g'(\bar{x})^{\mathsf{T}}(y^*)$.   □

A more general case is treated in the next theorem.

Theorem 2.111. *Let X, Y be Banach spaces, let U be an open subset of X, and let $g : U \to Y$ be a map that is circa-differentiable at $\bar{x} \in U$ with $A := g'(\bar{x})$ surjective. Then if C is a subset of Y and if $\bar{x} \in B := g^{-1}(C)$, $\bar{y} := g(\bar{x}) \in C$, one has*

$$N(B,\bar{x}) = g'(\bar{x})^{\mathsf{T}}(N(C,\bar{y})),$$

$$N_F(B,\bar{x}) = g'(\bar{x})^{\mathsf{T}}(N_F(C,\bar{y})).$$

Proof. We prove the Fréchet case only, leaving the directional case to the reader. The Lyusternik–Graves theorem (Theorem 2.67) asserts the existence of $\sigma > 0$, $c > 0$ such that for all $y \in B(\bar{y}, \sigma)$ there exists $x_y \in g^{-1}(y)$ satisfying $\|x_y - \bar{x}\| \le c \|y - \bar{y}\|$. When $y \in C \cap B(\bar{y}, \sigma)$ we have $x_y \in g^{-1}(C) = B$; hence $d(\bar{x}, g^{-1}(y) \cap B) \le d(\bar{x}, x_y) \le cd(\bar{y}, y)$. Moreover, setting $V := B(\bar{y}, \sigma)$, $U := g^{-1}(V)$, $B' := B \cap U$, $C' := C \cap V$, we have $g(B') = C'$ and $N_F(B,\bar{x}) = N_F(B',\bar{x})$ and $N_F(C,\bar{y}) = N_F(C',\bar{y})$. Thus, we can replace B with B' and C with C'. Then Proposition 2.108 ensures that $N_F(B,\bar{x}) = g'(\bar{x})^{\mathsf{T}}(N_F(C,\bar{y}))$.   □

2.6.3 Lagrange Multiplier Rule

As observed above, the usual necessary condition $f'(a) = 0$ in order that a function $f : X \to \mathbb{R}$ attain at a its minimum when it is differentiable there has to be modified when some restrictions are imposed. In the present section we consider the frequent case of constraints defined by equalities and we present a practical rule. The case of inequalities will be dealt with later on. The famous Lagrange multiplier rule is a direct consequence of Fermat's rule and Proposition 2.110.

Theorem 2.112 (Lagrange multiplier rule). *Let X, Y be Banach spaces, let W be an open subset of X, let $f : W \to \mathbb{R}$ be differentiable at a, and let $g : W \to Y$ be circa-differentiable at a with $g'(a)(X) = Y$. Let $b := g(a)$. Suppose that f attains on $S := g^{-1}(b)$ a local minimum at a. Then there exists some $y^* \in Y^*$ (called the Lagrange multiplier) such that*

$$f'(a) = y^* \circ g'(a).$$

Example. Let us find the shape of a box having a given volume $v > 0$ and minimal area. Denoting by x, y, z the lengths of the sides of the box, we are led to minimize

$$f(x,y,z) := 2(xy + yz + zx) \quad \text{subject to } g(x,y,z) := xyz - v = 0, \ x,y,z > 0.$$

First, we secure the existence of a solution by showing that f is coercive on $S := g^{-1}(0)$. In fact, if $w_n := (x_n, y_n, z_n) \in S$ and $(\|w_n\|) \to +\infty$, one of the components of w_n, say x_n, converges to $+\infty$; then, since $y_n + z_n \geq 2\sqrt{y_n z_n} = 2\sqrt{v/x_n}$, we get

$$f(w_n) \geq 2x_n(y_n + z_n) \geq 4\sqrt{vx_n} \to +\infty.$$

Now let (x,y,z) be a minimizer of f on S. Since the derivative of g is nonzero at (x,y,z), the Lagrange multiplier rule yields some $\lambda \in \mathbb{R}$ such that

$$2(y+z) = \lambda yz,$$
$$2(z+x) = \lambda zx,$$
$$2(x+y) = \lambda xy.$$

Then multiplying each side of the first equation by x, and doing similar operations with the other two equations, we get

$$\lambda v = \lambda xyz = 2x(y+z) = 2y(z+x) = 2z(x+y),$$

whence by summation, $3\lambda v = 4(xy + yz + zx) > 0$. Subtracting the above equations one from another, we get

$$2(y-x) = \lambda z(y-x), \quad 2(z-y) = \lambda x(z-y), \quad 2(x-z) = \lambda y(x-z).$$

Since λ, x, y, z are positive, considering the various cases, we get $x = y = z$. Since the unique solution of the necessary condition is $w := (v^{1/3}, v^{1/3}, v^{1/3})$, we conclude that w is the solution of the problem and the optimal box is a cube. We also note that the least area is $a(v) := f(w) = 6v^{2/3}$ and that $\lambda = 4v^{-1/3}$ is exactly the derivative of the function $v \mapsto a(v)$, a general fact we will explain later on that shows that the artificial multiplier λ has in fact an important interpretation as the measure for the change of the optimal value when the parameter v varies.

Example–Exercise. Let X be some Euclidean space and let $A \in L(X,X)$ that is symmetric. Let f and g be given by $f(x) = (Ax \mid x)$, $g(x) = \|x\|^2 - 1$. Take $v \in S_X$ such that f attains its minimum on the unit sphere S_X at v. Then show that there exists some $\lambda \in \mathbb{R}$ such that $Av = \lambda v$. Deduce from this result that every symmetric square matrix is diagonalizable.

Exercises

1. (Simplified **Karush–Kuhn–Tucker theorem**) Let X,Y be Banach spaces, let $g : X \to Y$ be circa-differentiable at $\overline{x}$ with $g'(\overline{x})(X) = Y$, and let $C \subset Z$ be a closed convex cone of Y. Suppose $\overline{x} \in F := g^{-1}(C)$ is a minimizer on F of a function $f : X \to \mathbb{R}$ that is differentiable at $\overline{x}$. Use Theorem 2.111 and Fermat's rule in order to get the existence of some $\overline{y}^* \in C^0$ such that $\langle \overline{y}^*, g(\overline{x}) \rangle = 0$, $f'(\overline{x}) + \overline{y}^* \circ g'(\overline{x}) = 0$.

2. **(a)** Compute the tangent cone at $(0,0)$ to the set

$$F := \left\{ (r,s) \in \mathbb{R}^2 : s \geq |r| (1+r^2)^{-1} \right\}.$$

(b) Use Fermat's rule to give a necessary condition in order that $(0,0)$ be a local minimizer of a function f on F, assuming that f is differentiable at $(0,0)$.

(c) Rewrite F as $F = \left\{ (r,s) \in \mathbb{R}^2 : g_1(r,s) \leq 0, g_2(r,s) \leq 0 \right\}$ with g_1, g_2 given by $g_1(r,s) = r(1+r^2)^{-1} - s$, $g_2(r,s) = -r(1+r^2)^{-1} - s$ and apply the Karush–Kuhn–Tucker theorem to get the condition obtained in **(b)**.

3. **(a)** Compute the tangent cone to the set $F = F' \cup F''$, where

$$F' := \left\{ (r,s) \in \mathbb{R}^2 : r^4 + s^4 - 2rs = 0 \right\},$$

$$F'' := \left\{ (r,s) \in \mathbb{R}^2 : r^4 + s^4 + 2rs = 0 \right\},$$

first for some point $a \neq (0,0)$, then for $a = (0,0)$. [Hint: First study the symmetry properties of F and set $s = tr$.]

(b) Write a necessary condition in order that a differentiable function $f : \mathbb{R}^2 \to \mathbb{R}$ attains on F a local minimizer at $(0,0)$. Assuming that f is twice differentiable at $(0,0)$, write a second-order necessary condition.

4. Give the dimensions of a cylindrical can that has a given volume v and minimal area $a(v)$. Give an interpretation of the multiplier in terms of the derivative of $a(\cdot)$.

5. Give the dimensions of a cylindrical can that has a given area a and maximal volume $v(a)$. Give an interpretation of the multiplier in terms of the derivative of $v(\cdot)$.

6. Give the dimensions of a box without lid that has a given volume v and minimal area $a(v)$. Give an interpretation of the multiplier in terms of the derivative of $a(\cdot)$.

7. Give the dimensions of a box without lid that has a given area a and maximal volume $v(a)$. Give an interpretation of the multiplier in terms of the derivative of $v(\cdot)$.

2.7 Introduction to the Calculus of Variations

The importance of the calculus of variations stems from its role in the history of the development of analysis and from its efficacy in presenting general principles that govern a number of physical phenomena. Among these are Fermat's principle governing the path of a ray of light and the Euler–Maupertuis principle of least action governing mechanics. Historically, the calculus of variations appeared at the end of the seventeenth century with the *brachistochrone problem*, solved in 1696 by Johann Bernoulli. This problem consists in determining a curve joining two given points along which a frictionless bead slides under the action of gravity in minimal time. The novelty of such a problem lies in the fact that the unknown is a geometrical object, a curve or a function, not a real number or a finite sequence of real numbers. Thus, such a topic brings to the fore the use of functional spaces, even if one limits one's attention to one-dimensional problems.

In fact, the choice of an appropriate space of functions is part of the problem. Several choices are possible. The most general one involves absolutely continuous maps and Lebesgue null sets and is a bit technical; for many problems piecewise C^1 curves would suffice. We adopt an intermediate choice.

Let E be a Banach space and let T be a compact interval of $\mathbb{R}$ (we will not consider higher-dimensional problems, in spite of the fact that problems such as the problem of minimal surfaces are important and although many partial differential equations are derived from problems in the calculus of variations). Without loss of generality, we may suppose $T := [0,1]$. We will use the space $X := R^1(T,E)$ of functions $x : T \to E$ that are primitives of (normalized) regulated functions from T to E; this means that there exists a function $x' : T \to E$ that is right continuous on $[0,1)$ and has a left limit $x'(t_-)$ for all $t \in (0,1]$ with $x'_-(1) = x'(1_-)$ such that

$$x(t) = x(0) + \int_0^t x'(s)\mathrm{d}s, \qquad t \in T.$$

Then x' is determined by x, since for each $t \in [0,1)$, $x'(t)$ is the right derivative of x at t and $x'(1)$ is the left derivative of x at 1. We endow X with the norm

$$\|x\| = \sup_{t \in T} \|x(t)\| + \sup_{t \in T} \|x'(t)\|.$$

It is equivalent to the norm $x \mapsto \|x(0)\| + \sup_{t \in T} \|x'(t)\|$, as is easily seen. Then X is a Banach space (use Theorem 2.59).

Given $(e_0, e_1) \in E \times E$, an open subset U of $E \times E \times T$, and a continuous function $L : U \to \mathbb{R}$, the problem consists in minimizing the function j given by

$$j(x) = \int_0^1 L(x(t), x'(t), t) dt$$

over the set $W(e_0, e_1)$ of elements x of X such that $x(0) = e_0$, $x(1) = e_1$, and $(x(t), x'(t), t) \in U$ for each $t \in T$. We note that since L is continuous, the function $t \mapsto L(x(t), x'(t), t)$ is regulated, so that the integral is well defined. We have more.

Lemma 2.113. *Given U, L, and j as above, the set $W := \{x \in X : \mathrm{cl}(J^1 x(T)) \subset U\}$, where $J^1 x(T) := \{(x(t), x'(t), t) : t \in T\}$, is open in X and j is continuous on W.*

Proof. By Proposition 2.16, for all $x \in W$, the set $\mathrm{cl}(J^1 x(T))$ is a compact subset of $E \times E \times T$. Thus, there exists some $r > 0$ such that $B(J^1 x(T), r) \subset U$. Then for all $w \in X$ satisfying $\|w - x\| < r$ one has $w \in W$. Thus W is open in X.

Moreover, L being continuous is uniformly continuous around $\mathrm{cl}(J^1 x(T))$ in the sense that for every $\varepsilon > 0$ one can find $\delta > 0$ such that for all $(e, v, t) \in \mathrm{cl}(J^1 x(T))$ and all $(e', v', t') \in B((e, v, t), \delta)$ one has $|L(e', v', t') - L(e, v, t)| \leq \varepsilon$. Therefore, for all $w \in X$ satisfying $\|w - x\| \leq \delta$, one has $|L(w(t), w'(t), t) - L(x(t), x'(t), t)| \leq \varepsilon$, hence $|j(w) - j(x)| \leq \varepsilon$. $\square$

Proposition 2.114. *Suppose L is continuous on U and has partial derivatives with respect to its first and second variables that are continuous on U. Then j is Hadamard differentiable on W and for $\bar{x} \in W$, $x \in X$ one has*

$$j'(\bar{x})x = \int_0^1 [D_1 L(\bar{x}(t), \bar{x}'(t), t)x(t) + D_2 L(\bar{x}(t), \bar{x}'(t), t)x'(t)] dt.$$

Proof. Let us set $L_t(e, v) = L(e, v, t)$ for $(e, v, t) \in U$ and

$$Y := \{(e_1, e_2, v_1, v_2, t) : \forall s \in [0, 1], ((1 - s)e_1 + se_2, (1 - s)v_1 + sv_2, t) \in U\},$$

$$Z := \{(w_1, w_2) \in W^2 : \forall t \in T, (w_1(t), w_2(t), w_1'(t), w_2'(t), t) \in Y\},$$

and

$$K(e_1, e_2, v_1, v_2, t) := DL_t((1 - s)e_1 + se_2, (1 - s)v_1 + sv_2).$$

The compactness of $[0, 1]$ easily yields that Y is open in $E^2 \times E^2 \times T$. Then a proof similar to that of Lemma 2.113 shows that Z is open in $X \times X$ and that setting

$$J(w_1, w_2, x) := \int_0^1 K(w_1(t), w_2(t), w_1'(t), w_2'(t), t) \cdot (x(t), x'(t)) dt$$

for $(w_1, w_2, x) \in Z \times X$, the map J is continuous from $Z \times X$ into $\mathbb{R}$. Now, since

$$L(e_1, v_1, t) - L(e_2, v_2, t) = K(e_1, e_2, v_1, v_2, t),$$

substituting w_1 and w_2 and integrating over T, we get

$$j(w_1) - j(w_2) = J(w_1, w_2, w_1 - w_2).$$

Since J is continuous, the function j is of class D^1 on W. In particular, it is Hadamard differentiable on W and for $\bar{x} \in W$, $x \in X$ one has $j'(\bar{x})(x) = J(\bar{x}, \bar{x}, x)$. $\qquad \square$

Exercise. Prove that j is Gâteaux differentiable using the definition and an interchange theorem between integration and derivation.

Exercise. Prove that in fact j is Fréchet differentiable.

Proposition 2.115. *Suppose L satisfies the assumptions of the preceding proposition and $\bar{x}$ is a local minimizer of j on $W(e_0, e_1)$. Then $\bar{x}$ is a critical point of j on $W(e_0, e_1)$ in the following sense:*

$$j'(\bar{x})v = 0 \qquad \forall v \in X_0 := W(0,0) := \{x \in X : x(0) = 0 = x(1)\}.$$

Proof. Let N be a neighborhood of $\bar{x}$ in X such that $j(w) \geq j(\bar{x})$ for every $w \in N \cap W(e_0, e_1)$. Given $v \in X_0$, for $r \in \mathbb{R}$ with $|r|$ small enough, we have $w := \bar{x} + rv \in W$ by Lemma 2.113 and $w(0) = e_0$, $w(1) = e_1$. Thus $w \in N \cap W(e_0, e_1)$, hence $j(\bar{x} + rv) \geq j(\bar{x})$ for $|r|$ small enough. It follows that $j'(\bar{x})v = 0$. $\qquad \square$

Theorem 2.116 (Euler–Lagrange condition). *Suppose L satisfies the assumptions of Proposition 2.114 and $\bar{x} \in W$ is a critical point of j on $W(e_0, e_1)$. Then the function $D_1 L(\bar{x}(\cdot), \bar{x}'(\cdot), \cdot)$ is a primitive of $D_2 L(\bar{x}(\cdot), \bar{x}'(\cdot), \cdot)$: for every $t \in [0,1)$ the right derivative of $D_2 L(\bar{x}(\cdot), \bar{x}'(\cdot), \cdot)$ exists and is such that*

$$\frac{d}{dt}\left(D_2 L(\bar{x}(t), \bar{x}'(t)), t\right) = D_1 L(\bar{x}(t), \bar{x}'(t), t). \tag{2.39}$$

The solutions of this equation are called *extremals*.

We break the proof into three steps. Taking $A(t) := D_2 L(t, \bar{x}(t), \bar{x}'(t))$, $B(t) := D_1 L(t, \bar{x}(t), \bar{x}'(t))$ in the last one, we shall get the result. The first step is as follows.

Lemma 2.117. *Let f be a nonnegative element of the space $R_n(T, \mathbb{R})$ of normalized regulated functions on T such that $\int_0^1 f(t)dt = 0$. Then $f = 0$.*

Proof. Suppose, to the contrary, that there exists some $r \in T$ such that $f(r) > 0$. When $r < 1$, using the right continuity of f at r we can find some $\alpha, \delta > 0$ such that $r + \delta < 1$ and $f(s) \geq \alpha$ for $s \in [r, r+\delta]$. Then we get $\int_0^1 f(t)dt \geq \int_r^{r+\delta} f(t)dt \geq \alpha\delta > 0$, a contradiction. If $r = 1$, a similar argument using the left continuity of f at 1 also leads to a contradiction. $\qquad \square$

Lemma 2.118. *Let $F \in R_n(T, E^*)$ be such that for all $x \in X_0 := \{x \in X : x(0) = 0 = x(1)\}$ one has $\int_0^1 F(t) \cdot x'(t)dt = 0$. Then $F(\cdot)$ is constant.*

More precisely, for $e^* := \int_0^1 F(t)\mathrm{d}t$ one has $F(t) = e^*$ for all $t \in T$.

Proof. Since $\int_0^1 e^* \cdot x'(t)\mathrm{d}t = 0$ for all $x \in X_0$, subtracting from F its means e^*, we are reduced to showing that $F(\cdot) = 0$ when $\int_0^1 F(t) \cdot x'(t)\mathrm{d}t = 0$ for every $x \in X_0$. Given $e \in E$, let us introduce $f, g : T \to \mathbb{R}$, $v, x : T \to E$ given by $g(t) = F(t)(e) := \langle F(t), e \rangle$, $f(t) = (g(t))^2$, $v(s) := F(s)(e)e := \langle F(s), e \rangle e$, $x(t) = \int_0^t v(s)\mathrm{d}s$. We see that $x(0) = 0$, $x'_+(t) = v(t)$ for $t \in [0,1)$, $x(1) = \int_0^1 v(t)\mathrm{d}t = \langle \int_0^1 F(t)\mathrm{d}t, e \rangle e = 0$, since the means of F is 0, so that $x \in X_0$. Our assumption yields

$$\int_0^1 f(t)\mathrm{d}t = \int_0^1 \langle F(t), e \rangle F(t)(e)\,\mathrm{d}t = \int_0^1 F(t)(\langle F(t), e \rangle e)\,\mathrm{d}t = \int_0^1 F(t) \cdot x'(t)\mathrm{d}t = 0.$$

The preceding lemma ensures that $f(t) = 0$ for every $t \in T$. Since e is arbitrary in E, we get $F(t) = 0$ for every $t \in T$.   □

Lemma 2.119 (Dubois–Reymond lemma). *Let $A, B \in R_n(T, E^*)$ be such that*

$$\forall x \in X_0, \qquad \int_0^1 \left[A(t)x(t) + B(t)x'(t) \right] \mathrm{d}t = 0.$$

Then B is a primitive of A: for every $t \in T$ one has $B(t) = B(0) + \int_0^t A(s)\mathrm{d}s$.

Proof. Let us set $C(t) := B(0) + \int_0^t A(s)\mathrm{d}s$. Then for each $x \in X_0$ the function $t \mapsto C(t)x(t)$ has a right derivative $t \mapsto A(t)x(t) + C(t)x'(t)$, and by assumption,

$$0 = \int_0^1 \left[A(t)x(t) + B(t)x'(t) \right] \mathrm{d}t = \int_0^1 \left[\frac{\mathrm{d}}{\mathrm{d}t}(C(t)x(t)) + (B(t) - C(t))x'(t) \right] \mathrm{d}t$$

$$= C(1)x(1) - C(0)x(0) + \int_0^1 (B(t) - C(t))x'(t)\mathrm{d}t = \int_0^1 (B(t) - C(t))x'(t)\mathrm{d}t.$$

Lemma 2.118 ensures that $B - C$ is constant. Since $B(0) - C(0) = 0$, $B = C$.   □

Corollary 2.120. *Suppose the Lagrangian L is independent of e: $L(e, v, t) = \widehat{L}_t(v)$. Then for every extremal $\bar{x}(\cdot)$, the function $t \mapsto D\widehat{L}_t(\bar{x}'(t))$ is a constant.*

Proof. Since $D_1 L = 0$, (2.39) is reduced to $\frac{\mathrm{d}}{\mathrm{d}t}D_2 L(t, \bar{x}(t), \bar{x}'(t)) = 0$, and hence $\widehat{L}(\cdot, \bar{x}'(\cdot))$ is constant.   □

When L is of class C^2, the Euler–Lagrange equation (2.39) is an implicit ordinary differential equation of order two. Let us show how it can be reduced to an explicit first-order differential system under the assumption that for $(e, t) \in T \times E$ the function $L_{e,t} : v \mapsto L(e, v, t)$ is a Legendre function on $U_{e,t} := \{v \in E : (e, v, t) \in U\}$. We set $V_{e,t} := DL_{e,t}(U_{e,t})$ and we denote by V the union of the sets $\{e\} \times V_{e,t} \times \{t\}$ and by $H : V \to \mathbb{R}$ the *Hamiltonian* given by

$$H(e, p, t) = \langle p, v \rangle - L(e, v, t) \qquad \text{for } p := D_2 L(e, v, t),$$

so that $H_{e,t} := H(e,\cdot,t)$ is the Legendre transform of $L_{e,t}$. We have seen that

$$D_2H(e,p,t) = v \Longleftrightarrow p = D_2L(e,v,t).$$

Assuming that D_2L is of class C^1, with $D_2^2L(e,v,t)$ invertible, we get that the function $v(e,p,t)$ determined by the implicit equation

$$p - D_2L(e,v(e,p,t),t) = 0$$

is differentiable with respect to e. Then in view of the expression of H and of the preceding relation, abbreviating $v(e,p,t)$ into v, for all $e' \in E$, one has

$$D_1H(e,p,t)e' = \langle p, D_1v(e,p,t).e' \rangle - D_1L(e,v,t)e' - D_2L(e,v,t)(D_1v(e,p,t)e')$$
$$= -D_1L(e,v(e,p,t),t)e',$$

or

$$D_1H(e,p,t) = -D_1L(e,v(e,p,t),t). \tag{2.40}$$

Theorem 2.121 (Hamilton). *Suppose that for all $(e,t) \in T \times E$, the map $D_2L(e,\cdot,t)$ is a diffeomorphism from $U_{e,t}$ onto its image $V_{e,t}$. Let $\bar{x}$ be an extremal and let $\bar{y}(t) := D_2L(\bar{x}(t),\bar{x}'(t),t)$. Then the pair $(\bar{x},\bar{y})$ satisfies the Hamilton differential system*

$$\bar{x}'(t) = D_2H(\bar{x}(t),\bar{y}(t),t),$$
$$\bar{y}'(t) = -D_1H(\bar{x}(t),\bar{y}(t),t).$$

Proof. Plugging $e = \bar{x}(t)$, $v = \bar{x}'(t)$, $p := \bar{y}(t)$ into the relation $v = D_2H(e,p,t)$, we get the first equation. By the Euler–Lagrange equation (2.39) and relation (2.40), we have

$$\bar{y}'(t) := \frac{\mathrm{d}}{\mathrm{d}t}\left(D_2L(\bar{x}(t),\bar{x}'(t),t)\right) = D_1L(\bar{x}(t),\bar{x}'(t),t) = -D_1H(\bar{x}(t),\bar{y}(t),t).$$

Exercises

1. (Geodesics in a Hilbert space) Let E be a Hilbert space, $U := T \times E \times (E \setminus \{0\})$, with $T := [0,1]$, and let L be the Lagrangian given by $L(e,v,t) := \|v\|$. Given e_0, $e_1 \in E$, show that if $\bar{x}: T \to E$ is an extremal over the set $W(e_0,e_1) := \{x \in R^1(T,E) : x'(T) \subset E \setminus \{0\}, x(0) = e_0, x(1) = e_1\}$, then $t \mapsto \bar{x}'(t)/\|\bar{x}'(t)\|$ is a constant vector u. Setting $s(t) := \int_0^t \|x'(r)\| \, \mathrm{d}r$, show that $\bar{x}(t) = e_0 + s(t)u$ with $u = (e_1 - e_0)/s(1)$, so that $\bar{x}$ runs along the segment $[e_0,e_1]$.

2. (**Classical mechanics**) Let us consider a solid with mass m whose position is determined by parameters $(q_1,\ldots,q_n) \in \mathbb{R}^n$ subject to a force $F(q_1,\ldots,q_n)$ deriving from a potential $U(q_1,\ldots,q_n)$ in the sense that $F(q_1,\ldots,q_n) = \nabla U(q_1,\ldots,q_n)$. Its kinetic energy is given by $T(v_1,\ldots,v_n) = (1/2)m(v_1^2 + \cdots + v_n^2)$. Setting

$$L(q_1,\ldots,q_n,v_1,\ldots,v_n) = T(v_1,\ldots,v_n) + U(q_1,\ldots,q_n),$$

show that the Euler–Lagrange equations turn out to be the **Newton equation**

$$mq''(t) = F(q(t)),$$

in which $q''(t) := (q_1''(t),\ldots,q_n''(t))$ is the acceleration.

3. Suppose that the Lagrangian L is independent of t. Show that if $x(\cdot)$ is an extremal, then the function $t \mapsto L(x(t),x'(t)) - D_1L(x(t),x'(t)) \cdot x'(t)$ is constant on T.

4. Let $(e,v) \mapsto L(e,v)$ be a nonnegative Lagrangian on some open subset of $E \times E$. Using Exercise **3**, show that every extremal of L is also an extremal of $\sqrt{L}$.

  Conversely, show that if $x(\cdot)$ is an extremal of $\sqrt{L}$ such that for some reparameterization $s \mapsto \theta(s)$, the Lagrangian $L(y(s),y'(s))$ is constant, where $y(s) := x(\theta(s))$, then y is an extremal of L.

5. Let E be a Hilbert space and let L be the Lagrangian given by $L(e,v) := \|v\|^2$. Show that if $\bar{x} : T := [0,1] \to E$ minimizes $j : x \mapsto \int_0^1 \|x'(t)\|^2\,dt$ over the set $W(e_0,e_1) := \{x \in R^1(T,E) : x(0) = e_0,\ x(1) = e_1,\ x'(T) \subset E \setminus \{0\}\}$, then $t \mapsto \bar{x}'(t)$ is constant on T and $\bar{x}$ is also an extremal of the length functional $\ell : x \mapsto \int_0^1 \|x'(t)\|\,dt$ over the set $W(e_0,e_1)$. Use the preceding exercise to show that conversely, if $\bar{x}$ is an extremal of the length functional ℓ and if for some reparameterization θ the function $s \mapsto \|\bar{x}'(\theta(s))\|$ is constant, then $\bar{x} \circ \theta$ is an extremal of j.

6. **Fermat's principle** states that the trajectory of light is an extremal of the travel time functional T, associated with the Lagrangian L given by $L(e,v,t) := 1/\|v\|$. Derive the **Descartes–Snell law** of light refraction on the boundary of two media of constant indices c_i ($i = 1,2$) separated by a hyperplane.

7. (**Lobachevskian geometry**) Find the extremals of the length function

$$\ell(x) := \int_0^1 \frac{\sqrt{x_1'(t)^2 + x_2'(t)^2}}{x_2(t)}\,dt,$$

i.e., the *geodesics*, on the Poincaré half-plane $P := \mathbb{R} \times (0,+\infty)$ endowed with the Riemannian metric $L(e,v) = \|v\|/e_2$, where e_2 is the second component of $e \in P$. [Hint: Show that the half-circles with centers in $\mathbb{R} \times \{0\}$ are geodesics, as well as the half-lines issued from $(0,0)$.]

8. (**Brachistochrone problem**) Show that for all $a,b > 0$, the *cycloids* given by $x(t) := (a(t - \sin t) + b, a(1 - \cos t))$ are extremals of the integral functional whose Lagrangian $L : P \times \mathbb{R}^2 \to \mathbb{R}$ is given by $L(e,v) := (e_2)^{-1/2}\|v\|,\ P := \mathbb{R} \times \mathbb{P}$.

9. (Minimal surfaces of revolution) Show that the *catenaries* $x(t) = c \cosh(t/c)$ $(c > 0)$ are extremals of the integral functional whose Lagrangian $L : \mathbb{R} \times \mathbb{R} \to \mathbb{R}$ is given by $L(e,v) := e\sqrt{v^2 + 1}$. They can be seen as sections of minimal surfaces of revolution used in power stations.

2.8 Notes and Remarks

Differential calculus is part of every course in analysis, so that numerous textbooks are devoted to it. Here we have been inspired by the books of Cartan [197], Dieudonné [294], Lang [611, 612], which were among the first to give modern presentations of the theory. A detailed study of the theory in topological linear spaces are the papers by Averbukh and Smolyanov [49, 50]; see also [946, 947], which contain interesting historical views. These works show that the notion of differentiability has many variants. Mappings of class D^1 were introduced in [779, 805]. Richard Hamilton showed the importance of such a class for implicit function theorems in Fréchet spaces [466]. Theorems 2.79–2.81 are in the line of results in [461,462] and [785, Theorem 4.1] but have new features. The terminology "circa-differentiable" is not traditional but it reflects the nature of the concept and it fits the notion of circa subdifferential (or subdifferential in the sense of Clarke). The initial terminology was "strongly differentiable" [755] and was turned into "strictly differentiable," despite the fact that there is no strict inequality in the definition.

The paper of Dolecki and Greco [307] shows the difficulties in giving due credit with the example of the contribution of Peano [778], that remained in shadow for a long time. Another example is the credit given to Hadamard here that should be confirmed [459].

The version of the Borwein–Preiss variational principle we present slightly differs from the original one in [128]; it covers other cases, but the perturbation is not given a precise form as in [128].

The name of Kantorovich is associated with Newton's method in view of the improvements made by this author (see [584]). The last exercise of Sect. 2.5.3 is inspired by [404], which contains several applications of the result. A proof of the submersion theorem in the case that the image space is finite-dimensional can be found in [462, 785].

A breakthrough in differential geometry was the book [611], by Serge Lang, that introduced in a neat manner differentiable manifolds modeled on Banach spaces. Lyusternik is considered a pioneer in the computation of the tangent cone to an inverse image using metric estimates (see [693, 694]). The subject was greatly extended with the works of Ioffe [511, 531, 538, and others].

As mentioned above, the calculus of variations was a strong incentive for the development of differential calculus and analysis. Books on the topic abound. In particular, [192, 197, 418, 549, 988] can be recommended as introductions. A historical account is given in [450].

Chapter 3
Elements of Convex Analysis

How many goodly creatures are there here!
How beauteous mankind is! O brave new world,
That has such people in't!

—William Shakespeare, *The Tempest*, V, 1

The class of convex functions is an important class that enjoys striking and useful properties. A homogenization procedure makes it possible to reduce this class to the subclass of sublinear functions. This subclass is next to the family of linear functions in terms of simplicity: the epigraph of a sublinear function is a convex cone, a notion almost as simple and useful as the notion of linear subspace. These two facts explain the rigidity of the class, and its importance.

Besides striking continuity and differentiability properties, the class of convex functions exhibits a substitute for the derivative that serves as a prototype for nonsmooth analysis. The main differences with classical analysis are the one-sided character of the subdifferential and the fact that a bunch of linear forms is substituted for the derivative. Still, nice calculus rules can be devised. Some of them, for instance for the subdifferential of the maximum of two functions, go beyond usual calculus rules. Besides classical rules of convex analysis, we illuminate some fuzzy rules for the calculus of subdifferentials. In doing so, we pave the way to similar rules in the nonconvex case. Thus it appears that convex analysis is not an isolated subject, but is part of a more general field. In fact, with differential calculus, it constitutes one of the two roots of nonsmooth analysis.

In the case of convex analysis, there is no restriction on the spaces for what concerns subdifferential rules, even if the case of reflexive spaces is somewhat simpler than the case of general Banach spaces. Subdifferential calculus makes it possible to formulate optimality conditions that have the precious particularity of being both necessary and sufficient. Moreover, subdifferentials are closely linked with duality, so that we provide a short account of this important topic. We also gather some elements of the geometry of normed spaces that will be useful later on.

J.-P. Penot, *Calculus Without Derivatives*, Graduate Texts in Mathematics 266,
DOI 10.1007/978-1-4614-4538-8_3, © Springer Science+Business Media New York 2013

Even if we do not insist on the point, it appears that duality plays some role in the interplay between convexity and differentiability of norms or powers of norms.

The class of convex functions also illustrates a typical feature of nonsmooth analysis that shows a spectacular difference with classical analysis: the study of functions of this class is intimately tied to the study of (convex) sets. The many passages from functions to sets and vice versa represent a fruitful and attractive approach that exemplifies the unity and the flexibility of mathematics and shows how lively the field is: most of the developments of what is known as convex analysis occurred during the second half of the twentieth century, and the topic is still under development. As pointed out in the books [99, 126, 137, 871, 872], convexity is a simple notion with much power and complexity.

3.1   Continuity Properties of Convex Functions

Convex functions enjoy nice properties for what concerns optimization. A simple example is as follows.

Proposition 3.1. *Every local minimizer of a convex function* $f : X \to \mathbb{R}_\infty :=$ $\mathbb{R} \cup \{+\infty\}$ *on a normed space (or topological vector space) X is a global minimizer.*

Proof. Let $\bar{x} \in X$ and let V be a neighborhood of $\bar{x}$ such that $f(\bar{x}) \le f(v)$ for all $v \in V$. Given $x \in X$, one can find $t \in (0,1)$ such that $v := \bar{x} + t(x - \bar{x}) \in V$. Then by convexity, we have $t f(x) + (1-t)f(\bar{x}) \ge f(v) \ge f(\bar{x})$, hence $f(x) \ge f(\bar{x})$.   □

For convex functions, one has remarkably simple continuity criteria.

Proposition 3.2. *Let $f : X \to \mathbb{R}_\infty$ be a convex function on a normed space (or topological vector space) X. If f is finite at some $\bar{x} \in X$, the following assertions are equivalent:*

(a) *f is bounded above on some neighborhood V of $\bar{x}$;*
(b) *f is upper semicontinuous at $\bar{x}$;*
(c) *f is continuous at $\bar{x}$.*

Proof. The implications $(c) \Rightarrow (b) \Rightarrow (a)$ are obvious. Let us prove $(a) \Rightarrow (b)$ and $(b) \Rightarrow (c)$. We may suppose that $\bar{x} = 0$, $f(\bar{x}) = 0$ by performing a translation and adding a constant. Given $\varepsilon > 0$, let $m \ge \sup f(V)$, $m \ge \varepsilon$. Let $U := \varepsilon m^{-1}V$. Then for $u \in U$, setting $v := \varepsilon^{-1} m u \in V$, we have

$$f(u) \le \varepsilon m^{-1} f(v) + (1 - \varepsilon m^{-1})f(0) \le \varepsilon,$$

and since U is a neighborhood of 0, f is upper semicontinuous at 0. In order to prove from (b) that f is continuous at 0, we observe that for $w \in W := U \cap (-U)$ we have $0 = f(0) \le \frac{1}{2}f(w) + \frac{1}{2}f(-w) \le \frac{1}{2}f(w) + \frac{1}{2}\varepsilon$, whence $f(w) \ge -\varepsilon$.   □

Remark. If V is the ball $B[\bar{x}, r]$ and $\sup f(V) \leq m$, then for $c := r^{-1}(m - f(\bar{x}))$ one has

$$\forall x \in B[0, r], \qquad f(\bar{x} + x) - f(\bar{x}) \leq c \|x\|,$$

since for $x \in B[0, r]$, setting $t := r^{-1}\|x\|$, taking u such that $\|u\| = r$, $x = tu$, one gets

$$f(\bar{x} + x) - f(\bar{x}) = f((1 - t)\bar{x} + t(\bar{x} + u)) - f(\bar{x})$$

$$\leq t\left(f(\bar{x} + u) - f(\bar{x})\right) \leq r^{-1}(m - f(\bar{x})) \|x\|,$$

a property called *quietness* at $\bar{x}$. In fact, for all $x \in B[0, r]$, since $f(\bar{x} + x) - f(\bar{x}) \geq -\left(f(\bar{x} - x) - f(\bar{x})\right) \geq -c\|x\|$, we have $|f(\bar{x} + x) - f(\bar{x})| \leq c\|x\|$, a stability property we will strengthen later on into a local Lipschitz property.   □

The following results illustrate the uses of the preceding criteria.

Proposition 3.3. *Suppose $f : X \to \mathbb{R}_\infty$ is a convex function on a finite-dimensional space X. Then f is continuous on the interior of its domain $D_f := \mathrm{dom} f := f^{-1}(\mathbb{R})$.*

Proof. Given $\bar{x} \in \mathrm{int}D_f$, let $x_1, \ldots, x_n \in D_f$ be such that $\bar{x}$ belongs to the interior of the convex hull C of $\{x_1, \ldots, x_n\}$ (for instance, one can take for C a ball with center $\bar{x}$ for some polyhedral norm, X being identified with some $\mathbb{R}^d$). Then f is bounded above on C by $m := \max(f(x_1), \ldots, f(x_n))$, hence is continuous at $\bar{x}$.   □

Proposition 3.4. *Let $f : X \to \mathbb{R}_\infty$ be a lower semicontinuous convex function on a Banach space X. Then f is continuous on the core of its domain D_f (which coincides with the interior of D_f).*

Proof. Given $\bar{x} \in \mathrm{core}D_f$, let $m > f(\bar{x})$ and let $C := \{x \in X : f(x) \leq m\}$. Again we may suppose $\bar{x} = 0$. Then C is a closed convex subset of X that is absorbing: for all $x \in X$ we can find $r > 0$ such that $rx \in D_f$ and for $s > 0$ small enough we have $f(rsx) \leq (1 - s)f(0) + sf(rx) < m$, so that $rsx \in C$. Thus C is a neighborhood of 0 by Lemma 1.59, and f is continuous at 0 by Proposition 3.2   □

Convex functions enjoy an almost "miraculous" propagation property.

Proposition 3.5. *Let $f : X \to \mathbb{R}_\infty$ be a convex function on a normed space X. If f is continuous at some $\bar{x} \in D_f := \mathrm{dom} f$ then f is continuous on the interior of D_f.*

Proof. Given $x_0 \in \mathrm{int}D_f$, let us prove that f is continuous at x_0. Using a translation, we may suppose $x_0 = 0$. Then since D_f is a neighborhood of 0, there exists some $r > 0$ such that $\bar{y} := -r\bar{x} \in D_f$. Let V be a neighborhood of 0 such that f is bounded above by some m on $\bar{x} + V$. Then by convexity, f is bounded above on

$$r(1 + r)^{-1}(\bar{x} + V) + (1 + r)^{-1}\bar{y} = r(1 + r)^{-1}V \in \mathcal{N}(0)$$

by $r(1 + r)^{-1}m + (1 + r)^{-1}f(\bar{y})$. Then by Proposition 3.2, f is continuous at x_0.   □

A crucial semicontinuity property of convex functions is the following.

Theorem 3.6. *If f is a convex function that is lower semicontinuous on a normed space X, then f is lower semicontinuous on X endowed with the weak topology.*

Proof. This is an immediate consequence of Mazur's theorem: for every real number r the sublevel set $[f \leq r] := \{x \in X : f(x) \leq r\}$ of f is closed and convex, hence weakly closed.   □

The preceding proof shows that the same property holds for quasiconvex functions, i.e., functions whose sublevel sets are convex.

Corollary 3.7. *A coercive lower semicontinuous convex function f on a reflexive Banach space X attains its infimum.*

Proof. The result is obvious if f takes only the value $+\infty$ (i.e., $f = +\infty^X$). For $f \neq +\infty^X$ we pick $x_0 \in \text{dom} f$ and use coercivity to get some $r > 0$ such that $f(x) > f(x_0)$ for $x \in X \setminus rB_X$. Since X is reflexive, rB_X is weakly compact. Since f is weakly lower semicontinuous, there exists some $\bar{x} \in rB_X$ such that $f(\bar{x}) \leq f(x)$ for all $x \in rB_X$, in particular $f(\bar{x}) \leq f(x_0)$, since $x_0 \in rB_X$. Then $f(\bar{x}) \leq f(x)$ for all $x \in X$.   □

Convexity and local boundedness entail a regularity property stronger than continuity: a local Lipschitz property. In fact, the result is not just a local one: the following statement and its corollary give a precise content to this assertion: the corollary shows that a Lipschitz property is available on balls that may be big, provided the function is bounded above on a larger ball. One even gets a quantitative estimate of the Lipschitz rate.

Proposition 3.8. *Let f be a convex function on a convex subset C of a normed space X and let $\alpha, \beta \in \mathbb{R}$, $\rho > 0$. Suppose f is bounded below by β on a subset B of C and is bounded above by α on a subset A of C such that $B + \rho U_X \subset A$, where U_X is the open unit ball of X. Then f is Lipschitzian on B with rate $\rho^{-1}(\alpha - \beta)$.*

Proof. Given $x, y \in B$ and $\delta > \|x - y\|$, let $z := y + \rho \delta^{-1}(y - x) \in A$, since $B + \rho U_X \subset A$. Then $y = x + t(z - x)$, where $t := \delta(\delta + \rho)^{-1} \in [0, 1]$; hence

$$f(y) - f(x) \leq t(f(z) - f(x)) \leq t(\alpha - \beta) \leq \delta \rho^{-1}(\alpha - \beta).$$

Interchanging the roles of x and y and taking the infimum on δ in $(\|x - y\|, +\infty)$, we get $| f(y) - f(x) | \leq \rho^{-1}(\alpha - \beta)\|x - y\|$.   □

The preceding statement is versatile enough to apply in a variety of geometric cases. The simplest one is the case of balls.

Corollary 3.9. *Suppose the convex function f on the normed space X is bounded above by α on some ball $B(\bar{x}, r)$. Then for every $s \in (0, r)$ the function f is Lipschitzian on the ball $B(\bar{x}, s)$ with rate $2(r - s)^{-1}(\alpha - f(\bar{x}))$.*

Proof. Taking $A := B(\bar{x}, r), B := B(\bar{x}, s), \rho := r - s, \beta := 2f(\bar{x}) - \alpha$, it suffices to observe that for all $x \in B$ one has $f(x) \geq \beta$ by convexity.   □

Corollary 3.10. *Every convex function that is continuous on an open convex subset U of a normed space is locally Lipschitzian on U.*

Now let us turn to some links between convex functions and continuous affine functions. Hereinafter we say that a convex function is *closed* if it is lower semicontinuous and either it is identically equal to $-\infty$ (in which case we denote it by $-\infty^X$) or it takes its values in $\mathbb{R}_\infty := \mathbb{R} \cup \{+\infty\}$. Recall that $f \in \overline{\mathbb{R}}^X$ is *proper* if f does not take the value $-\infty$ and if it is not the constant function $+\infty^X$. Then its epigraph is a proper subset of $X \times \mathbb{R}$ (i.e., is nonempty and different from the whole space).

We observed that a lower semicontinuous convex function assuming the value $-\infty$ cannot take a finite value (Exercise 1 of Sect. 1.4.1). Thus a lower semicontinuous convex function $f \in \overline{\mathbb{R}}^X$ taking a finite value is either proper or $+\infty^X$. Note that given a closed convex subset C of X, the function given by $f(x) = -\infty$ for $x \in C$, $f(x) = +\infty$ for $x \in X \setminus C$ is an example of a lower semicontinuous convex function that is not closed and not proper.

If f is the supremum of a nonempty family of continuous affine functions, then f is either $+\infty^X$ or a closed proper convex function. In both cases, and in the case of $f = -\infty^X$ (which corresponds to the empty family), it is a closed convex function. A remarkable converse holds.

Theorem 3.11. *Every closed convex function is the supremum of a family of continuous affine functions (the ones it majorizes). If f is proper, this family is nonempty.*

Clearly, if $f = +\infty^X$, one can take the family of all continuous affine functions on X, while if $f = -\infty^X$ one takes the empty family. The following lemma is the first step of the proof of this result for the case $f \neq -\infty^X$.

Lemma 3.12. *For every lower semicontinuous convex function $f : X \to \mathbb{R}_\infty$ there exists a continuous affine function g such that $g \leq f$. Moreover, if $w \in \operatorname{dom} f$ and $r < f(w)$, we may require that $g(w) > r$.*

Proof. The case $f = +\infty^X$ is obvious. Let us suppose $f \neq +\infty^X$, so that the epigraph E_f of f is nonempty. Let $w \in \operatorname{dom} f$ and $r < f(w)$. The Hahn–Banach theorem allows us to separate the compact set $\{(w,r)\}$ from the closed convex set E_f: there exist $(h,c) \in X^* \times \mathbb{R} = (X \times \mathbb{R})^*$ and $b \in \mathbb{R}$ such that

$$\forall (x,s) \in E_f, \qquad \langle h, x \rangle + cs > b > \langle h, w \rangle + cr. \qquad (3.1)$$

Taking $x = w$, $s > f(w) > r$, we see that $c > 0$. Dividing both sides of the first inequality by c, we get

$$s > -c^{-1}h(x) + c^{-1}b \qquad \forall x \in \operatorname{dom} f, \ \forall s \geq f(x).$$

It follows that $f \geq g$ for g given by $g(x) := -c^{-1}h(x) + c^{-1}b$. Moreover, the second inequality in relation (3.1) can be written $g(w) > r$. $\qquad \square$

Now let us prove Theorem 3.11. Again, the cases $f = +\infty^X$, $f = -\infty^X$ being obvious, we may suppose $\operatorname{dom} f \neq \varnothing$. Let $w \in X$ and $r < f(w)$. If $w \in \operatorname{dom} f$, the preceding lemma provides us with a continuous affine function $g \leq f$ with $g(w) > r$.

Now let us consider the case $w \in X \setminus \mathrm{dom}\, f$. Separating $\{(w,r)\}$ from E_f, we get some $(h,c) \in (X \times \mathbb{R})^*$ and $b \in \mathbb{R}$ such that relation (3.1) holds. Taking $x \in \mathrm{dom}\, f$ and s large, we see that $c \geq 0$. If $c > 0$, we can conclude as in the preceding proof. If $c = 0$, observing that $b - h(w) > 0$, taking a continuous affine function k such that $k \leq f$ (such a function exists, by the lemma) and setting

$$g := k + n(b - h),$$

with $n > (b - h(w))^{-1}(r - k(w))$, we see that $g(w) > r$ and $g \leq f$, since $k \leq f$ and $b - h(x) \leq 0$ for $x \in \mathrm{dom}\, f$ by relation (3.1) with $c = 0$. □

Since lower semicontinuity is stable by the operation of taking suprema, one can deduce Theorem 3.6 from Theorem 3.11.

3.1.1 Supplement: Another Proof of the Robinson–Ursescu Theorem

We are in a position to prove the Robinson–Ursescu theorem in the reflexive case without using the notion of ideally convex set.

Theorem 3.13. *Let W, X be Banach spaces, and let $F : W \rightrightarrows X$ be a multimap with closed convex graph. If W is reflexive, then for every $(\overline{w}, \overline{x})$ in (the graph of) F such that $X = \mathbb{R}_+(F(W) - \overline{x})$, i.e., $\overline{x} \in \mathrm{core}\, F(W)$, the multimap F is open at $(\overline{w}, \overline{x})$. In fact, F is open at $(\overline{w}, \overline{x})$ with a linear rate in the sense that there exist some $c > 0$, $\overline{r} > 0$ such that*

$$\forall r \in (0, \overline{r}), \qquad B(\overline{x}, r) \subset F(B(\overline{w}, cr)).$$

Proof. Let us define a function $f : X \to \mathbb{R}_\infty$ by

$$f(x) := d(\overline{w}, F^{-1}(x)) := \inf\{\|w - \overline{w}\| : w \in W, \ x \in F(w)\}, \qquad x \in X,$$

with the convention that $\inf \varnothing = +\infty$. Since $f(x) = \inf\{\|w - \overline{w}\| + \iota_F(w,x) : w \in W\}$ and since $(w,x) \mapsto \|w - \overline{w}\| + \iota_F(w,x)$ is convex, f is convex. Let us prove that f is lower semicontinuous on X by showing that for every $r \in \mathbb{R}$, its sublevel set $S_f(r) := f^{-1}((-\infty, r])$ is closed. Let (x_n) be a sequence of $S_f(r)$ converging to some $x \in X$. Since X is reflexive and for all $n \in \mathbb{N}$ the set $F^{-1}(x_n)$ is closed, convex, hence weakly closed, there exists some $w_n \in F^{-1}(x_n)$ such that $\|w_n - \overline{w}\| = f(x_n)$. The sequence (w_n), being contained in the sequentially weakly compact ball $B[\overline{w}, r]$, has a subsequence that weakly converges to some $w \in B[\overline{w}, r]$. Since F is weakly closed in $W \times X$, we have $(w,x) \in F$, hence $f(x) \leq \|w - \overline{w}\| \leq \liminf_n \|w_n - \overline{w}\| \leq r$. Thus $S_f(r)$ is closed and f is lower semicontinuous, hence is continuous on the core of its domain $F(X)$ by Proposition 3.4. In fact, f is locally Lipschitzian around $\overline{x}$, so that there exist $c > 0$, $\overline{r} > 0$ such that $f(x) = |f(x) - f(\overline{x})| \leq cd(x, \overline{x})$ for all $x \in B(\overline{x}, \overline{r})$. Thus for $r \in (0, \overline{r})$ and $x \in B(\overline{x}, r)$, one can find $w \in F^{-1}(x)$ with $\|w - \overline{w}\| < cr$: the last assertion is proved. □

The openness property of Theorem 3.13 can be strengthened to openness at a linear rate around $(\overline{w}, \overline{x})$.

Corollary 3.14. *Let $F : W \rightrightarrows X$ be a multimap with convex graph between two normed spaces. Suppose that for some $\rho, r > 0$ and some $(\overline{w}, \overline{x}) \in F$ one has $B(\overline{x}, r) \subset F(B(\overline{w}, \rho))$. Then for every $s \in (0, r/3]$ there exists some $c > 0$ such that*

$$\forall (w, x) \in B(\overline{w}, \rho) \times B(\overline{x}, s), \qquad d(w, F^{-1}(x)) \leq cd(x, F(w)).$$

Proof. Given $s \in (0, r/3]$, let $c := 4(r - s)^{-1}\rho$ and let $(w, z) \in B(\overline{w}, \rho) \times B(\overline{x}, s)$ with $z \in F(w)$. Then for every $x \in B(\overline{x}, s)$, $y \in F(w) \setminus B(\overline{x}, r)$ one has

$$d(x, y) \geq d(y, \overline{x}) - d(\overline{x}, x) > r - s \geq 2s > d(x, z),$$

hence $d(x, F(w)) \leq d(x, z) < 2s \leq d(x, F(w) \setminus B(\overline{x}, r))$ and $d(x, F(w) \cap B(\overline{x}, r)) = d(x, F(w))$. Let $f : X \to \mathbb{R}$ be the function defined by

$$f(x) := d(w, F^{-1}(x)), \qquad x \in X,$$

with the usual convention $\inf \varnothing = +\infty$. Since $f(x) = \inf\{\|w - w'\| + \iota_F(w', x) : w' \in W\}$, f is convex. Since $B(\overline{x}, r) \subset F(B(\overline{w}, \rho))$, f is bounded above by $\alpha := 2\rho$ on $B(\overline{x}, r)$ and $f(\overline{x}) \geq 0$. Thus Corollary 3.9 gives $f(x) \leq f(z) + cd(x, z)$. Since $f(z) = 0$, taking the infimum over $z \in F(w) \cap B(\overline{x}, r)$, we get the announced inequality. $\qquad \square$

Exercises

1. Let X be a separable Hilbert space with Hilbertian basis $\{e_n : n \in \mathbb{N}\}$ and let the function $f : X \to \mathbb{R}$ be given by

$$f(x) := \sum_{n=0}^{\infty} |x_n|^{n+2} \quad \text{for } x = \sum_{n=0}^{\infty} x_n e_n.$$

(a) Show that f is well defined on X, bounded above by 1 on the unit ball, and everywhere bounded below by 0.

(b) Show that the Lipschitz rate of f around e_k is at least $k + 2$.

(c) Deduce from what precedes that f is not Lipschitzian on the ball rB_X with $r > 1$. Observe that f is not bounded above on such a ball. [Inspired by [852].]

2. Using the data and the notation of Corollary 3.9 and noting that f is bounded above on $B(\overline{x}, s)$ by $(1 - r^{-1}s)f(\overline{x}) + r^{-1}s\alpha$, hence is bounded below by $\beta := (1 + r^{-1}s)f(\overline{x}) - r^{-1}s\alpha$ on this ball, show that the Lipschitz rate of f on $B(\overline{x}, s)$ is at most $(1 + r^{-1}s)(r - s)^{-1}(\alpha - f(\overline{x}))$.

3. Prove a similar estimate of the Lipschitz rate of f when one supposes that f is bounded above by some α on the sphere with center $\bar{x}$ and radius r.

4. (**a**) Let $f : X \to \mathbb{R}$ be a uniformly continuous function on a normed space. Show that for every $\delta > 0$ there exists $k > 0$ such that $d(f(x), f(y)) \leq kd(x,y)$ for all $x, y \in X$ satisfying $d(x,y) \geq \delta$ [Hint: Use a subdivision of the segment $[x,y]$ by points u_i such that $d(u_i, u_{i+1}) \leq \alpha$, where $\alpha > 0$ is such that $d(f(u), f(v)) \leq 1$ whenever $u, v \in X$ satisfy $d(u,v) \leq \alpha$.]
(**b**) Prove that every uniformly continuous convex function f on X is Lipschitzian. [Hint: Use (a) and Proposition 3.8].

5. (**The log barrier**) Prove that $f : \mathbb{R}^{n^2} \to \mathbb{R}_\infty$ given by $f(u) = -\log(\det u)$ if u is a symmetric positive definite matrix, $+\infty$ otherwise, is a convex function.

6. Deduce from Proposition 3.4 that for every closed convex subset of a Banach space one has $\operatorname{int} C = \operatorname{core} C$. [Hint: Use the indicator function ι_C of C.]

7. Prove that on the dual X^* of a nonreflexive Banach space one can find a convex function f that is continuous for the topology associated with the dual norm, but that is not lower semicontinuous for the weak* topology. [Hint: Take $f \in X^{**} \setminus X$.]

3.2 Differentiability Properties of Convex Functions

Convex functions have particular differentiability properties. The case of one-variable functions, which is our starting point, will be our first piece of evidence. However, it is a substitute for the derivative that will be the main point of this section. Later on, we will see that this new object, called the subdifferential, enjoys useful calculus rules. The idea of replacing a linear functional by a set of linear forms will be our leading thread in all that follows.

3.2.1 Derivatives and Subdifferentials of Convex Functions

We first observe that if $f : I \to \mathbb{R}$ is a finite convex function on some interval I of $\mathbb{R}$, then for $r < s < t$ in I the following inequalities hold:

$$\frac{f(s) - f(r)}{s - r} \leq \frac{f(t) - f(r)}{t - r} \leq \frac{f(t) - f(s)}{t - s}. \tag{3.2}$$

They express that the slope of the secant to the graph of f is a nondecreasing function of the abscissas of its extremities and stem from the convexity inequality

$$f(s) = f\left(\frac{t-s}{t-r}r + \frac{s-r}{t-r}t\right) \leq \frac{t-s}{t-r}f(r) + \frac{s-r}{t-r}f(t)$$

(since the coefficients of $f(r)$ and $f(t)$ are in $[0,1]$ and have sum 1), yielding

$$f(s) - f(r) \leq \frac{s-r}{t-r}(f(t) - f(r)), \qquad f(t) - f(s) \geq \frac{t-s}{t-r}(f(t) - f(r)).$$

Lemma 3.15. *If $f : I \to \mathbb{R}$ is a finite convex function on some interval I of $\mathbb{R}$, then for every $s \in I \setminus \{\sup I\}$ the right derivative $D_r f(s) := f'_+(s)$ of f at s exists in $\mathbb{R} \cup \{-\infty\}$ and is given by*

$$D_r f(s) := \lim_{t \to s_+} \frac{f(t) - f(s)}{t - s} = \inf_{t > s} \frac{f(t) - f(s)}{t - s}.$$

If, moreover, s is in the interior of I, then $D_r f(s)$ is finite, the left derivative $D_\ell f(s)$ exists and is finite, and $D_\ell f(s) \leq D_r f(s)$. Furthermore, the functions $s \mapsto D_r f(s)$ and $s \mapsto D_\ell f(s)$ are nondecreasing.

Proof. The first assertion is a direct consequence of the existence of a limit for the nondecreasing function $t \mapsto (t - s)^{-1}(f(t) - f(s))$ on $(s, \sup I)$. The second assertion stems from the fact that when $s \in \text{int} I$, the limit is finite, since by (3.2), for $r < s$ this quotient is bounded below by $(s - r)^{-1}(f(s) - f(r))$. Thus $(s - r)^{-1}(f(s) - f(r)) \leq D_r f(s) \leq (t - s)^{-1}(f(t) - f(s))$. Similarly, changing (s,t) into (r,s), we have $D_r f(r) \leq (s - r)^{-1}(f(s) - f(r))$, hence $D_r f(r) \leq D_r f(s)$. Changing f into g given by $g(t) := f(-t)$, we get the assertions about the left derivative. The inequality $D_\ell f(s) \leq D_r f(s)$ is obtained by a passage to the limit as $t \to s_+$ and $r \to s_-$ in relation (3.2). $\qquad\square$

It may happen that the left derivative $D_\ell f$ of a convex function f does not coincide with the right derivative (consider $r \mapsto |r|$). Relation (3.2) shows that for $r < t$ one has $D_r f(r) \leq D_\ell f(t)$. Thus if $D_\ell f(t) < D_r f(t)$, one gets $\lim_{r \to t_-} D_r f(r) \leq D_\ell f(t) < D_r f(t)$, and $D_r f(\cdot)$ has a jump at t. Since $D_r f(\cdot)$ is nondecreasing, such points of discontinuity of $D_r f(\cdot)$ are at most countable. Since f is nondifferentiable at t if and only if $D_\ell f(t) < D_r f(t)$, we get the next result.

Proposition 3.16. *Let $f : I \to \mathbb{R}$ be a convex function on an open interval of $\mathbb{R}$. Then the set of points at which f is not differentiable is at most countable.*

The following characterizations of convexity are classical and useful.

Proposition 3.17. *Let $f : I \to \mathbb{R}$ be a differentiable function on an open interval of $\mathbb{R}$. Then f is convex if and only if its derivative is nondecreasing. If f is twice differentiable, then f is convex if and only if for all $r \in I$ one has $f''(r) \geq 0$.*

Proof. The necessary condition is a consequence of Lemma 3.15. Let us prove the sufficient condition. Let f be differentiable with nondecreasing derivative. Given $r, t \in I$ and $s \in (r, t)$, we have $s = ar + bt$ with $a = (t - r)^{-1}(t - s) \geq 0$, $b = (t - r)^{-1}(s - r) \geq 0$, $a + b = 1$. The mean value theorem yields some $p \in (r, s)$

and some $q \in (s,t)$ such that $(s-r)^{-1}(f(s)-f(r)) = f'(p)$, $(t-s)^{-1}(f(t)-f(s)) = f'(q)$. Since $f'(p) \le f'(q)$, rearranging terms, we get

$$(t-r)f(s) \le (t-s)f(r) + (s-r)f(t).$$

This is equivalent to $f(s) \le af(r) + bf(t)$. Thus f is convex. The last assertion is given by elementary calculus.   □

Now suppose $f : X \to \mathbb{R}_\infty$ is defined on a vector space X.

Proposition 3.18. *If $f : X \to \mathbb{R}_\infty := \mathbb{R} \cup \{+\infty\}$ is a convex function on a vector space X, then for all $x \in \operatorname{dom} f$ and for all $v \in X$ the radial derivative*

$$d_r f(x,v) := \lim_{t \to 0_+} \frac{f(x+tv) - f(x)}{t}$$

exists and is equal to $\inf_{t>0} t^{-1}(f(x+tv) - f(x))$. It is finite if $x \in \operatorname{core}(\operatorname{dom} f)$.

Proof. Let g be given by $g(t) = f(x+tv)$. Then g is convex and its right derivative at 0 is $d_r f(x,v)$. It exists in $[-\infty, +\infty)$ if $(x + (0, \infty)v) \cap \operatorname{dom} f$ is nonempty, and it is $+\infty$ otherwise. Even in the latter case, this right derivative is $\inf_{t>0} t^{-1}(g(t) - g(0)) = \inf_{t>0} t^{-1}(f(x+tv) - f(x))$. When x belongs to $\operatorname{core}(\operatorname{dom} f)$, for every $v \in X$, 0 is in the interior of $\operatorname{dom} g$, and we can conclude with Lemma 3.15.   □

Proposition 3.19. *If $f : X \to \mathbb{R}_\infty$ is a convex function on a vector space X, then for all $x \in \operatorname{dom} f$, the radial derivative $d_r f(x, \cdot)$ is a sublinear function.*

Proof. Clearly $d_r f(x, \cdot)$ is positively homogeneous. Let us prove that it is subadditive: for every $v, w \in X$ we have $f(x + \frac{1}{2}t(v+w)) \le \frac{1}{2}f(x+tv) + \frac{1}{2}f(x+tw)$; hence

$$d_r f(x, v+w) = \lim_{t \to 0_+} \frac{2}{t}\left[f\left(x + \frac{t}{2}(v+w)\right) - f(x)\right]$$

$$\le \lim_{t \to 0_+} \frac{1}{t}(f(x+tv) - f(x)) + \lim_{t \to 0_+} \frac{1}{t}(f(x+tw) - f(x))$$

$$= d_r f(x,v) + d_r f(x,w).$$ □

The preceding statement can also be justified by checking that

$$d_r f(x,v) = \inf\{s : (v,s) \in T^r(E_f, x_f)\},$$

where E_f is the epigraph of f, $x_f := (x, f(x))$, and $T^r(E_f, x_f)$ is the radial tangent cone to E_f at x_f, where the *radial tangent cone* to a convex set C at $z \in C$ is the set

$$T^r(C,z) := \mathbb{R}_+(C - z).$$

When X is a normed space, $T^r(C,z)$ is not closed in general, as simple examples show. Therefore, it is advisable to replace it with the *tangent cone* $T(C,z)$ to C at z. In the case that C is convex, $T(C,z)$ is just the closure of $T^r(C,z)$. In the case that C

is the epigraph of a convex function f finite at x and $z := x_f := (x, f(x))$, the tangent cone $T(C, z)$ is the epigraph of the *directional derivative of* of f at x defined by

$$f'(x, v) := df(x, v) := \liminf_{(t, u) \to (0_+, v)} \frac{f(x + tu) - f(x)}{t}.$$

Since $f'(x, \cdot) = df(x, \cdot)$ is lower semicontinuous, it has better duality properties than $d_r f(x, \cdot)$, and it is as closely connected to the following fundamental notion as $d_r f(x, \cdot)$ is.

Definition 3.20. If $f : X \to \mathbb{R}_\infty$ is a function on a normed space X and $x \in X$, then the *Moreau–Rockafellar subdifferential* of f at x is the empty set if $x \in X \setminus \operatorname{dom} f$, and if $x \in \operatorname{dom} f$, it is the set $\partial f(x) := \partial_{MR} f(x)$ of $x^* \in X^*$ such that

$$\forall w \in X, \qquad f(w) \geq f(x) + \langle x^*, w - x \rangle. \tag{3.3}$$

This is a global notion that is very restrictive for an arbitrary function. For a convex function it turns into a crucial tool that is a useful substitute for the derivative, as we will shortly see. A strong advantage of the Moreau–Rockafellar subdifferential is that it yields a characterization of minimizers.

Proposition 3.21. *A function f on a normed space X attains its minimum at $x \in \operatorname{dom} f$ if and only if $0 \in \partial f(x)$.*

The result is an immediate consequence of the definition. Calculus rules will make it efficient. In particular, they enable us to give optimality conditions for problems with constraints.

A first consequence of the following result is that the subdifferential of a convex function f is not just a global notion, but also a local notion.

Theorem 3.22. *If f is a convex function on a normed space X and $x \in \operatorname{dom} f$, then*

$$x^* \in \partial f(x) \iff \forall v \in X \ \langle x^*, v \rangle \leq df(x, v)$$
$$\iff \forall v \in X \ \langle x^*, v \rangle \leq d_r f(x, v).$$

If $x \in \operatorname{core}(\operatorname{dom} f)$ and f is Gâteaux differentiable at x, then $\partial f(x) = \{Df(x)\}$.

Recall that f is said to be *Gâteaux differentiable* at x with derivative $Df(x) := \ell \in X^*$ if f is finite at x and for all $v \in X$,

$$\frac{f(x + tv) - f(x)}{t} \to \ell(v) \text{ as } t \to 0, \quad t \neq 0.$$

Proof. Given $x^* \in \partial f(x)$, for every $t > 0$, $u \in X$ we have

$$\langle x^*, tu \rangle \leq f(x + tu) - f(x).$$

Dividing by t and taking the lim inf as $(t,u) \to (0_+,v)$, we get $\langle x^*,v \rangle \le df(x,v) \le d_r f(x,v)$. Now if f is convex and if x^* satisfies the inequality $\langle x^*,v \rangle \le d_r f(x,v)$ for all $v \in X$, then for $v \in X$, $t \in (0,1)$, by the monotonicity observed in relation (3.2), we have

$$\langle x^*,v \rangle \le d_r f(x,v) \le \frac{1}{t}(f(x+tv) - f(x)) \le f(x+v) - f(x).$$

Setting $v = w - x$, we obtain relation (3.3).   □

A simple interpretation of the subdifferential of a function can be given in terms of the normal cone to its epigraph. The *normal cone* to a convex subset C of a normed space X at some $z \in C$ is defined as the set $N(C,z)$ of $z^* \in X^*$ such that $\langle z^*, w - z \rangle \le 0$ for every $w \in C$; thus it is the polar cone to the radial tangent cone $T^r(C,z)$, and also, by density, it is the polar cone to $T(C,z)$.

Proposition 3.23. *For a convex function f on a normed space X and $x \in \mathrm{dom}\, f$, one has the following equivalence in which E_f is the epigraph of f and $x_f := (x,f(x))$:*

$$x^* \in \partial f(x) \iff (x^*,-1) \in N(E_f,x_f).$$

The proof is immediate from the definition of $\partial f(x)$:

$$(x^*,-1) \in N(E_f,x_f) \iff \forall (w,r) \in E_f \ \langle x^*,w-x \rangle - (r - f(x)) \le 0 \iff x^* \in \partial f(x).$$

Let us describe the notion of normal to a convex set in terms of subdifferentials.

Proposition 3.24. *For a convex subset C of a normed space X, the normal cone to C at $x \in C$ is the subdifferential of the indicator function ι_C to C at x. It is also the cone $\mathbb{R}_+ \partial d_C(x)$ generated by the subdifferential of the distance function to C at x.*

Proof. By definition, $x^* \in N(C,x)$ iff $\langle x^*, w - x \rangle \le 0$ for all $x \in C$. Since $\iota_C(w) = 0$ for $w \in C$ and $\iota_C(w) = \infty$ for $w \in X \setminus C$, this property is equivalent to $x^* \in \partial \iota_C(x)$. The inclusion $\mathbb{R}_+ \partial d_C(x) \subset N(C,x)$ is obvious: when $r \in \mathbb{R}_+$, $z^* \in \partial d_C(z)$, one has

$$\forall w \in C, \qquad \langle rz^*, w - x \rangle \le rd_C(w) - rd_C(x) = 0.$$

Conversely, when $x^* \in N(C,x)$, the function $-x^*$ attains its infimum on C at x, and is Lipschitzian with rate $c = \|x^*\|$, so that by the penalization lemma, $-x^* + cd_C$ attains its infimum on X at x; then $0 \in \partial(-x^* + cd_C)(x)$, which is equivalent to $x^* \in c\partial d_C(x)$.   □

The last argument shows the interest in having calculus rules at one's disposal. Such rules will be considered in the next section.

A simple consequence of the *subdifferentiability* of a convex function f at a point x (i.e., of the nonemptiness of $\partial f(x)$) is the lower semicontinuity of f at x. More interesting are the following criteria for subdifferentiability.

Theorem 3.25 (Moreau). *If a convex function f on a normed space X is finite and continuous at x, then $\partial f(x)$ is nonempty and weak* compact. Moreover, for all $u \in X$,*

$$f'(x,u) = \max\{\langle x^*,u\rangle : x^* \in \partial f(x)\}.$$

Proof. For every $r > f(x)$ there exists a neighborhood V of x such that $V \times (r,\infty)$ is contained in the epigraph E_f of f. Thus the interior of E_f is convex and nonempty. It does not contain $x_f := (x, f(x))$, since for $s < f(x)$ close to $f(x)$ one has $(x,s) \notin E_f$. The geometric Hahn–Banach theorem yields some $(u^*,c) \in (X \times \mathbb{R})^*$ such that

$$\langle u^*,w\rangle + cr > \langle u^*,x\rangle + cf(x) \qquad \forall (w,r) \in \mathrm{int} E_f.$$

This implies (by taking $w = x$, $r = f(x)+1$) that $c > 0$ and, by Lemma 1.56, that

$$\langle u^*,w-x\rangle + c(r - f(x)) \geq 0 \qquad \forall (w,r) \in E_f.$$

In turn, this relation, which can be written

$$f(w) - f(x) \geq \langle -c^{-1}u^*, w-x\rangle \qquad \forall w \in X,$$

shows that $x^* := -c^{-1}u^* \in \partial f(x)$. Thus $\partial f(x)$ is nonempty.

Since $\partial f(x)$ is the intersection of the weak* closed half-spaces

$$D_w := \{x^* \in X^* : \langle x^*, w-x\rangle \leq f(w) - f(x)\}, \qquad\qquad w \in \mathrm{dom} f,$$

it is always weak* closed. When f is continuous at x, taking $\rho > 0$ such that $f(w) \leq f(x)+1$ for $w \in B(x,\rho)$, for all $x^* \in \partial f(x)$, we have $\|x^*\| = \rho^{-1}\sup\{\langle x^*, w-x\rangle : w \in B(x,\rho)\} \leq \rho^{-1}$. The second assertion will be proved with the alternative proof that follows. $\qquad\square$

Alternative proof. By the remark following Proposition 3.2 we can find $c \in \mathbb{R}_+$ and $r > 0$ such that $|f(x+v) - f(x)| \leq c\|v\|$ for $v \in B(0,r)$. It follows that $|f'(x,w)| \leq c\|w\|$ for $w \in X$. Then given $u \in X$, the Hahn–Banach theorem yields a linear functional x^* such that $x^*(u) = f'(x,u)$ and $x^* \leq f'(x,\cdot) \leq c\|\cdot\|$. Thus x^* is continuous and $x^* \in \partial f(x)$. $\qquad\square$

Remarks. (a) Without the continuity assumption, $\partial f(x)$ may be unbounded. It is the case for the indicator function of $\mathbb{R}_+$ on $X = \mathbb{R}$, for which $\partial f(0) = -\mathbb{R}_+$.

(b) It may happen that $\partial f(x)$ is empty at some point x of the domain of f: this fact occurs for $X = \mathbb{R}$, $x = 1$, $f(u) = -\sqrt{1-u^2}$ for $u \in [-1,1]$, $f(u) = +\infty$ for $u \in \mathbb{R} \setminus [-1,1]$, although f is continuous on $(-1,1)$. $\qquad\square$

Examples. (a) For $f := \|\cdot\|$ one has $\partial \|\cdot\|(x) = \{x^* \in X^* : \|x^*\| = 1, \langle x^*,x\rangle = \|x\|\}$.

(b) Let X be a normed space and let $j(\cdot) := \frac{1}{2}\|\cdot\|^2$. Then $\partial j(x) = J(x)$, the *duality (multi)map* defined by $J(x) := \{x^* \in X^* : \|x^*\| = \|x\|, \langle x^*,x\rangle = \|x\|^2\}$, and $J(x)$ is nonempty, as shown by applying Corollary 1.72 or Theorem 3.25. $\qquad\square$

Corollary 3.26. *Let* $f : X \to \mathbb{R}_\infty$ *be a convex function finite and continuous at* $x \in X$. *If* $\partial f(x)$ *is a singleton* $\{x^*\}$, *then* f *is Gâteaux and Hadamard differentiable at* x *and* $Df(x) = x^*$.

Proof. The preceding theorem ensures that $f'(x, \cdot) = x^*$. Thus f is Gâteaux differentiable. Since f is Lipschitzian around x, it is Hadamard differentiable. □

Corollary 3.27. *Let* f *be a convex function on a normed space* X. *Suppose the restriction of* f *to the affine subspace* A *generated by* $\operatorname{dom} f$ *is continuous at* $x \in \operatorname{dom} f$. *Then* $\partial f(x)$ *is nonempty.*

Proof. Without loss of generality, we may suppose $x = 0$, so that A is the vector subspace generated by $\operatorname{dom} f$. The preceding theorem ensures that the restriction $f \mid A$ of f to A is subdifferentiable at 0. Then every continuous linear extension of every element of $\partial(f \mid A)(0)$ belongs to $\partial f(0)$, and such extensions exist by the Hahn–Banach theorem. □

Corollary 3.28. *Let* f *be a convex function on a finite-dimensional normed space* X *and let* $x \in \operatorname{ri} \operatorname{dom} f$ *(i.e., be such that* $\mathbb{R}_+(\operatorname{dom} f - x)$ *is a linear subspace). Then* $\partial f(x)$ *is nonempty.*

Proof. Recall that for a subset D of X, $\operatorname{ri} D$ is the set of points that belong to the interior of D in the affine subspace Y generated by D. Taking $D = \operatorname{dom} f$, we have that the restriction g of f to Y is continuous at x. The preceding corollary applies. □

It will be proved later that for every closed proper convex function f on a Banach space X, the set of points $x \in X$ such that $\partial f(x)$ is nonempty is dense in $\operatorname{dom} f$.

Let us give a subdifferentiability criterion using the concept of calmness. A function $f : X \to \mathbb{R}_\infty$ finite at $\overline{x} \in X$ is said to be *calm at* $\overline{x}$ if $-f$ is quiet at $\overline{x}$, i.e., if there exist $c \in \mathbb{R}_+$ and a neighborhood V of $\overline{x}$ such that $f(x) - f(\overline{x}) \geq -c \|x - \overline{x}\|$ for all $x \in V$. If one can take $V = X$, one says that f is *globally calm at* $\overline{x}$. The *calmness rate* of f at $\overline{x}$ is the infimum $\gamma_f(\overline{x})$ of the constants $c > 0$ for which the preceding inequality is satisfied on some neighborhood of $\overline{x}$. The *remoteness* of a nonempty subset S of X or X^* is the number $\rho(S) := \inf\{\|s\| : s \in S\}$.

Proposition 3.29. *A convex function* $f : X \to \mathbb{R}_\infty$ *finite at some* $\overline{x} \in X$ *is subdifferentiable at* $\overline{x}$ *iff it is globally calm at* $\overline{x}$, *if and only if it is calm at* $\overline{x}$. *Moreover, the calmness rate of* f *at* $\overline{x}$ *is equal to the remoteness* $\rho(\partial f(\overline{x}))$ *of* $\partial f(\overline{x})$.

Proof. If $\partial f(\overline{x})$ is nonempty, for every element $\overline{x}^* \in \partial f(\overline{x})$ one can take $c = \|\overline{x}^*\|$ to get global calmness, so that $\gamma_f(\overline{x}) \leq \rho(\partial f(\overline{x}))$. Conversely, if one can find $c \in \mathbb{R}_+$ such that $f(\overline{x} + x) - f(\overline{x}) \geq -c \|x\|$ for all $x \in X$, then the sandwich theorem yields some $x^* \in X^*$ such that $f(\overline{x} + x) - f(\overline{x}) \geq \langle x^*, x \rangle \geq -c \|x\|$ for all $x \in X$. Then $x^* \in \partial f(\overline{x})$ and $x^* \in cB_{X^*}$, so that $\rho(\partial f(\overline{x})) \leq \gamma_f(\overline{x})$ and equality holds. □

Exercise. Establish the inequality $xy \leq p^{-1}x^p + q^{-1}y^q$ for every $x, y \in \mathbb{R}_+$, $p, q > 1$ satisfying $p^{-1} + q^{-1} = 1$ by minimizing the function $x \mapsto p^{-1}x^p - xy$ for a fixed $y > 0$. Deduce from that inequality *Hölder's inequality*:

$$\forall a := (a_i),\ b := (b_i) \in \mathbb{R}^n, \qquad \sum_{i=1}^{n} |a_i b_i| \le \left(\sum_{i=1}^{n} |a_i|^p \right)^{1/p} \left(\sum_{i=1}^{n} |b_i|^q \right)^{1/q}.$$

[Hint: Set $s_i := a_i/\|a\|_p$, $t_i := b_i/\|b\|_q$, with $\|a\|_p := (\Sigma_{1 \le i \le n} |a_i|^p)^{1/p}$, $\|b\|_q := (\Sigma_{1 \le i \le n} |b_i|^q)^{1/q}$ and note that $\Sigma_{1 \le i \le n} |s_i|^p = 1$, $\Sigma_{1 \le i \le n} |t_i|^q = 1$, $\Sigma_{1 \le i \le n} |s_i t_i| \le \Sigma_{1 \le i \le n} (p^{-1} |s_i|^p + q^{-1} |t_i|^p)$.]

Exercise. (a) Let A be a positive definite symmetric matrix, let λ_1 (resp. λ_n) be its smallest (resp. largest) eigenvalue, and let $\lambda := \sqrt{\lambda_1 . \lambda_n}$. Check that the function $f : t \mapsto t/\lambda + \lambda/t$ is convex on $[\lambda_1, \lambda_n]$ and satisfies $f(\lambda_1) = \sqrt{\lambda_1/\lambda_n} + \sqrt{\lambda_n/\lambda_1} = f(\lambda_n)$, whence $f(t) \le \sqrt{\lambda_1/\lambda_n} + \sqrt{\lambda_n/\lambda_1}$ for all $t \in [\lambda_1, \lambda_n]$.
(b) Show that μ is an eigenvalue of $\lambda^{-1} A + \lambda A^{-1}$ if and only if μ is an eigenvalue of A. [Hint: Reduce A to a diagonal form.]
(c) Using the inequality $2\sqrt{ab} \le a + b$ for $a, b > 0$, show that $2\sqrt{\langle Ax, x \rangle . \langle A^{-1}x, x \rangle} \le \lambda^{-1}\langle Ax, x \rangle + \lambda \langle A^{-1}x, x \rangle$ for all $x \in \mathbb{R}^n$.
(d) Deduce from this **Kantorovich's inequality**:

$$\forall x \in \mathbb{R}^n, \quad \langle Ax, x \rangle \cdot \langle A^{-1}x, x \rangle \le (1/4)(\sqrt{\lambda_1/\lambda_n} + \sqrt{\lambda_n/\lambda_1})\|x\|^2.$$

3.2.2 Differentiability of Convex Functions

For differentiability questions, too, convex functions enjoy special properties. A first instance is the following result, which displays an easy test for differentiability using the functions

$$r_x(w) := f(x+w) + f(x-w) - 2f(x), \tag{3.4}$$

$$\sigma_x(t,u) := (1/t)(f(x+tu) + f(x-tu) - 2f(x)). \tag{3.5}$$

Proposition 3.30. *A convex function $f : X \to \mathbb{R}_\infty$ finite and continuous at some point $x \in X$ is Fréchet (resp. Hadamard) differentiable at x if and only if r_x is a remainder (resp. if for all $u \in S_X$ one has $\sigma_x(t,u) \to 0$ as $t \to 0$).*

Proof. Necessity is obtained by addition directly from the definitions. Let us prove sufficiency in the Fréchet case. Since f is finite and continuous at x, $\partial f(x)$ is nonempty. Let $x^* \in \partial f(x)$. Then the definition of $\partial f(x)$ and (3.4) yield

$$0 \le f(x+w) - f(x) - \langle x^*, w \rangle = f(x) - f(x-w) + \langle x^*, -w \rangle + r_x(w) \le r_x(w).$$

That shows that f is Fréchet differentiable at x with derivative x^*. The Gâteaux case follows by a reduction to one-dimensional subspaces; since f is continuous at x, it is Lipschitzian around x, so that Gâteaux differentiability coincides with Hadamard differentiability. $\square$

Two other instances arise with automatic continuity properties of derivatives.

Proposition 3.31. *If $f : W \to \mathbb{R}$ is continuous and convex on an open convex subset W of a normed space X and if f is Gâteaux differentiable at $x \in W$, then f is Hadamard differentiable at x and df is continuous at (x,v) for all $v \in X$. If, moreover, f is Gâteaux differentiable around x, then f is of class D^1 around x.*

Proof. For every $r > df(x,v)$ one can find $s > 0$ such that $r > s^{-1}[f(x+sv) - f(x)]$. Thus for (x',v') close enough to (x,v) one has $r > s^{-1}[f(x'+sv') - f(x')] \geq df(x',v')$, so that

$$df(x,v) \geq \limsup_{(x',v') \to (x,v)} df(x',v').$$

Since $df(x',v') \geq -df(x',-v')$, the linearity of $df(x,\cdot)$ implies that

$$\liminf_{(x',v') \to (x,v)} df(x',v') \geq - \limsup_{(x',v') \to (x,v)} df(x',-v') \geq -df(x,-v) = df(x,v).$$

These inequalities prove our continuity assertion. Hadamard differentiability ensues (and can be deduced from the local Lipschitz property of f). $\qquad\square$

In the next statement the continuity of the derivative of f is strengthened, and for a subset A of X^* and $r \in \mathbb{P}$, we use the notation $B(A,r) := \{x^* : d(x^*,A) < r\}$.

Proposition 3.32. *Let $f : W \to \mathbb{R}$ be a convex function on some open convex subset W of a normed space X. If f is Fréchet differentiable at some $x \in W$ and Gâteaux differentiable on W, then its derivative is continuous at x.*
More generally, if f is Fréchet differentiable at some $x \in W$, then its subdifferential ∂f is continuous at x: for all $\varepsilon > 0$, there exists $\eta > 0$ such that $\partial f(w) \cap B(Df(x),\varepsilon) \neq \varnothing$ and $\partial f(w) \subset B(Df(x),\varepsilon)$ for all $w \in B(x,\eta)$.

Proof. It suffices to prove the second assertion. The differentiability of f at $\bar{x}$ entails continuity of f on W, hence that $\partial f(w) \neq \varnothing$ for all $w \in W$. Let $x^* := Df(x)$. Given $\varepsilon \in (0, d(x, X \setminus W))$, $\alpha \in (0,\varepsilon)$, let $\delta > 0$ be such that

$$\forall u \in B(0,\delta), \qquad f(x+u) - f(x) - \langle x^*, u \rangle \leq \alpha \|u\|. \qquad (3.6)$$

Let $c := \alpha \varepsilon^{-1} \in (0,1)$. For all $w \in B(x,(1-c)\delta)$, $w^* \in \partial f(w)$, $v \in X$ one has

$$f(w) - f(w+v) + \langle w^*, v \rangle \leq 0.$$

Setting $u := w - x + v$ in (3.6) with $v \in B(0,c\delta)$, one has $u \in B(0,\delta)$, $x+u = w+v$, and adding the respective sides of the preceding inequalities, one gets

$$f(w) - f(x) - \langle x^*, u \rangle + \langle w^*, v \rangle \leq \alpha \|u\|.$$

Using the relation $\langle x^*, u - v \rangle = \langle x^*, w - x \rangle \le f(w) - f(x)$, this inequality yields

$$\langle w^* - x^*, v \rangle \le \alpha \|u\| \le \alpha \delta.$$

Taking the supremum over $v \in B(0, c\delta)$, one gets $\|w^* - x^*\| \le c^{-1}\alpha = \varepsilon$. □

Corollary 3.33. *A Fréchet differentiable convex function on an open convex subset of a normed space is of class C^1.*

Let us give density and structure properties of the set of points of differentiability of a convex function.

Theorem 3.34. *(a) Let $f : W \to \mathbb{R}$ be a continuous convex function on some open convex subset W of a normed space X. Then the set F of points in W of Fréchet differentiability of f is a (possibly empty) $\mathscr{G}_\delta$ subset of W.*

*(b) (**Asplund, Lindenstrauss**) If X is a Banach space whose dual is separable, then F is dense in W.*

*(c) (**Mazur**) If X is separable, the set H of Hadamard differentiability of f is also a $\mathscr{G}_\delta$ subset. If X is separable and complete, then H is dense in W.*

Proof. (a) For $u \in S_X$, let $\sigma_x(\cdot, u)$ be the function of relation (3.5), and let

$$G_n := \left\{ x \in W : \exists t > 0 : \sup_{u \in S_X} \sigma_x(t, u) < \frac{1}{n} \right\}.$$

Since for all $u \in S_X$ the function $t \mapsto \sigma_x(t, u)$ is nondecreasing, the same is true for $\tau_x : t \mapsto \sup\{\sigma_x(t, u) : u \in S_X\}$. Thus by Proposition 3.30, $F = \cap_n G_n$, and it suffices to prove that G_n is open for every $n \in \mathbb{N} \setminus \{0\}$. Now, if $x \in G_n$ and if $r > 0$ is the radius of a ball with center x in W on which f is Lipschitzian with rate κ, we can pick $t \in (0, r)$ such that $\tau_x(t) < 1/n$. Since $|\tau_x(t) - \tau_w(t)| \le 4t^{-1}\kappa\|x - w\|$, we have $\tau_w(t) < 1/n$ for $\|x - w\|$ small enough: G_n contains an open ball with center x.

(b) For the proof we refer to [20, 668], and [98, Theorem 4.17], [376, Theorem 8.21].

(c) Let us pick a countable dense subset $\{u_m : m \in \mathbb{N}\}$ of S_X and observe that $u \mapsto \sigma_w(t, u)$ is Lipschitzian on S_X, uniformly for w in a small ball with center x if $t > 0$ is small enough. Thus the set H of Gâteaux (or Hadamard) differentiability of f is $\cap_{m,n} H_{m,n}$, where

$$H_{m,n} := \{x \in W : \exists t > 0 : |f(x + tu_m) + f(x - tu_m) - 2f(x)| < t/n\},$$

and this set is again open. It is dense in W because for all given $\bar{x} \in W$ and $\varepsilon > 0$ the one-variable function $f_m : t \mapsto f(\bar{x} + tu_m)$ is convex continuous on some open interval containing 0, hence is differentiable at some point $s \in (-\varepsilon, \varepsilon)$ by Proposition 3.16, so that $\bar{x} + su_m \in H_{m,n} \cap B(\bar{x}, \varepsilon)$ by Proposition 3.30. When X is complete, W is a Baire space (Lemma 1.26), so that $H = \cap_{m,n} H_{m,n}$ is dense. □

Exercises

1. (a) Let $f : \mathbb{R} \to \mathbb{R}$ be given by $f(x) = |x|$. Show that $\partial f(0) = [-1,1]$.
(b) Check that the subdifferential at 0 of a sublinear function f on a normed space X is given by $\partial f(0) = \{x^* \in X^* : x^* \le f\}$. Prove that $\partial f(x) = \{x^* \in X^* : x^* \le f, \langle x^*, x \rangle = f(x)\}$ for $x \in X$.

2. For a convex function f on $\mathbb{R}$ continuous at x show that $\partial f(x) = [D_\ell f(x), D_r f(x)]$.

3. Prove that the closure of the radial tangent cone at $x \in C$ to a convex subset of a normed space coincides with the tangent cone to C as defined in Chap. 2.

4. Prove that the normal cone $N(C,x)$ to a convex subset C of a normed space coincides with the normal cone to C as defined in Chap. 2.

5. (Ubiquitous convex sets) Exhibit a proper convex subset C of a Banach space X such that $T(C,\bar{x}) = X$ for some $\bar{x} \in$ Bdry C. Show that X must be infinite-dimensional. [Hint: Take for X a separable Hilbert space with Hilbertian basis (e_n) and set $C := \{x = \Sigma_n x^n e_n : |x^n| \le 2^{-n} \, \forall n\}, \bar{x} = 0$.]

6. Let $f : \mathbb{R}^2 \to \mathbb{R}_\infty$ be given by $f(x_1,x_2) := \max(|x_1|, 1 - \sqrt{x_2})$ for $(x_1,x_2) \in \mathbb{R} \times \mathbb{R}_+$, $+\infty$ otherwise. Prove that f is convex but that $\mathrm{dom}\partial f$ is not convex.

7. Let X be a Hilbert space, let C be a nonempty closed convex subset of X, and let $f : X \to \mathbb{R}$ be given by $f(x) := (1/2)[\|x\|^2 - \|x - P(x)\|^2]$, where P is the *metric projection* of X onto C: $P(x) := \{u\}$, where $u \in C$, $\|x - u\| = d(x,C)$. Show that f is convex and that f is everywhere Fréchet differentiable, with gradient given by $\nabla f(x) = P(x)$ for all $x \in X$. [Hint: Note that $f(x) = \sup\{\langle x, y \rangle - (1/2)\|y\|^2 : y \in C\}$, i.e., f is the conjugate of $(1/2)\|\cdot\|^2 + \iota_C(\cdot)$; use the estimates $\|x + u - P(x + u)\|^2 \le \|x + u - P(x)\|^2$ and $\|x - P(x)\|^2 \le \|x - P(x + u)\|^2$ to prove that f is differentiable at x.]

3.3 Calculus Rules for Subdifferentials

Convex functions enjoy several subdifferential calculus rules that are akin to the usual rules of differential calculus. Nonetheless, there are some differences: in many cases a technical assumption is needed to get the interesting inclusion. Moreover, one does not have $\partial(-f)(x) = -\partial f(x)$ in general. On the other hand, some rules of convex analysis have no analogues in the differentiable case. An example of these new rules is the following obvious observation.

Lemma 3.35. *Suppose $f \le g$ and $f(\bar{x}) = g(\bar{x})$ for some $\bar{x} \in X$. Then $\partial f(\bar{x}) \subset \partial g(\bar{x})$.*

This observation easily yields the following (rather inessential) rule for infima.

Lemma 3.36. *Let $(f_i)_{i \in I}$ be a finite family of functions and let $\bar{x} \in \bigcap_{i \in I} \operatorname{dom} f_i$. If $f := \inf_{i \in I} f_i$ and if $f_i(\bar{x}) = f(\bar{x})$ for all $i \in I$, then $\partial f(\bar{x}) = \bigcap_{i \in I} \partial f_i(\bar{x})$.*

Proof. The inclusion $\partial f(\bar{x}) \subset \bigcap_{i \in I} \partial f_i(\bar{x})$ stems from the preceding lemma. For the opposite inclusion, we note that for all $\bar{x}^* \in \bigcap_{i \in I} \partial f_i(\bar{x})$, for all $i \in I$, and for every $x \in X$ one has $f_i(\bar{x}) + \langle \bar{x}^*, x - \bar{x} \rangle \le f_i(x)$, hence $f(\bar{x}) + \langle \bar{x}^*, x - \bar{x} \rangle \le f(x)$ by our assumption.   □

When the space of parameters is a normed space, a different formulation can be given. It is useful for duality theory.

Proposition 3.37. *Let $f : W \times X \to \mathbb{R}_\infty$, where W and X are normed spaces. Let p be the performance function given by $p(w) := \inf\{f(w,x) : x \in X\}$ and let $S : W \rightrightarrows X$ be the solution multimap given by $S(w) := \{x \in X : f(w,x) = p(w)\}$. Suppose that for some $\overline{w} \in X$ one has $S(\overline{w}) \ne \varnothing$. Then one has the equivalence*

$$\overline{w}^* \in \partial p(\overline{w}) \iff \forall \bar{x} \in S(\overline{w}) \quad (\overline{w}^*, 0) \in \partial f(\overline{w}, \bar{x})$$

$$\iff \exists \bar{x} \in S(\overline{w}) \quad (\overline{w}^*, 0) \in \partial f(\overline{w}, \bar{x}).$$

Proof. For all $\bar{x} \in S(\overline{w})$, $(w,x) \in W \times X$, one has $f(\overline{w}, \bar{x}) = p(\overline{w})$, $f(w,x) \ge p(w)$, whence

$$\overline{w}^* \in \partial p(\overline{w}) \iff \forall w \in W, \quad p(w) \ge p(\overline{w}) + \langle \overline{w}^*, w - \overline{w} \rangle$$

$$\implies \forall (w,x) \in W \times X, \quad f(w,x) \ge f(\overline{w}, \bar{x}) + \langle (\overline{w}^*, 0), (w - \overline{w}, x - \bar{x}) \rangle,$$

or $(\overline{w}^*, 0) \in \partial f(\overline{w}, \bar{x})$. Conversely, if this last relation holds for some $\bar{x} \in S(\overline{w})$ and some $\overline{w}^* \in W^*$, then taking the infimum over $x \in X$ in the last inequality, one gets

$$\forall w \in W, \qquad p(w) \ge p(\overline{w}) + \langle \overline{w}^*, w - \overline{w} \rangle,$$

i.e., $\overline{w}^* \in \partial p(\overline{w})$.   □

The case of the supremum of a finite family of convex functions is more likely to occur than the case of the infimum. In the next supplement, a generalization to an arbitrary family is studied.

Proposition 3.38. *Let $(f_i)_{i \in I}$ be a finite family of convex functions on a normed space X and let $f := \sup_{i \in I} f_i$. Let $\bar{x} \in \bigcap_{i \in I} \operatorname{dom} f_i$ and let $I(\bar{x}) := \{i \in I : f_i(\bar{x}) = f(\bar{x})\}$. Suppose that for all $i \in I$ the function f_i is continuous at $\bar{x}$. Then one has*

$$f'(\bar{x}, \cdot) = \max_{i \in I(\bar{x})} f_i'(\bar{x}, \cdot), \tag{3.7}$$

$$\partial f(\bar{x}) = \operatorname{co}\left(\bigcup_{i \in I(\bar{x})} \partial f_i(\bar{x}) \right). \tag{3.8}$$

Proof. Let $u \in X$. Since f is continuous at $\bar{x}$, we have $f'(\bar{x}, u) = d_r f(\bar{x}, u)$, and a similar equality for f_i. For $i \in I(\bar{x})$, since $f_i \le f$ and $f_i(\bar{x}) = f(\bar{x})$, we have $f_i'(\bar{x}, u) \le$

$f'(\bar{x}, u)$. Thus $s := \max_{i \in I(\bar{x})} f_i'(\bar{x}, u) \leq f'(\bar{x}, u)$ and equality holds when $s = +\infty$. Let us suppose that $s < +\infty$ and let us show that for every $r > s$ we have $r \geq f'(\bar{x}, u)$; that will prove that $s = f'(\bar{x}, u)$. For $i \in I(\bar{x})$, let $t_i > 0$ be such that

$$(1/t)\,(f_i(\bar{x} + tu) - f_i(\bar{x})) < r \qquad \text{for } t \in (0, t_i).$$

Since for $j \in J(\bar{x}) := I \setminus I(\bar{x})$, the function f_j is continuous at $\bar{x}$, given $\varepsilon > 0$ such that $f_j(\bar{x}) + \varepsilon < f(\bar{x})$ for all $j \in J(\bar{x})$, we can find $t_j > 0$ such that

$$f_j(\bar{x} + tu) < f(\bar{x}) - \varepsilon \qquad \text{for } t \in (0, t_j).$$

Then for $t \in (0, t_0)$, with $t_0 := \min(|r|^{-1}\varepsilon, \min_{j \in J(\bar{x})} t_j)$, we have $-\varepsilon \leq tr$; hence

$$f(\bar{x} + tu) = \max_{i \in I} f_i(\bar{x} + tu) \leq \max(\max_{i \in I(\bar{x})}(f_i(\bar{x}) + tr), f(\bar{x}) - \varepsilon) = f(\bar{x}) + tr.$$

Thus $f'(\bar{x}, u) \leq r$ and $f'(\bar{x}, u) = \max_{i \in I(\bar{x})} f_i'(\bar{x}, u)$.

For $i \in I(\bar{x})$, the inclusion $\partial f_i(\bar{x}) \subset \partial f(\bar{x})$ follows from Lemma 3.35 or from the inequality $f_i'(\bar{x}, \cdot) \leq f'(\bar{x}, \cdot)$. Denoting by C the right-hand side of (3.8), and observing that $\partial f(\bar{x})$ is convex, the inclusion $C \subset \partial f(\bar{x})$ ensues. Let us show that assuming that there exists some $\bar{w}^* \in \partial f(\bar{x}) \setminus C$ leads to a contradiction. Since C is weak* closed (in fact weak* compact), the Hahn–Banach theorem yields some $c \in \mathbb{R}$ and $u \in X$ (the dual of X^* endowed with the weak* topology in view of Proposition 1.4) such that

$$\langle \bar{w}^*, u \rangle > c \geq \langle x^*, u \rangle \qquad \forall x^* \in C.$$

Since $f'(\bar{x}, u) \geq \langle \bar{w}^*, u \rangle$, we get

$$f'(\bar{x}, u) > c \geq \sup_{x^* \in C} \langle x^*, u \rangle = \sup_{i \in I(\bar{x})} \sup_{x^* \in \partial f_i(\bar{x})} \langle x^*, u \rangle = \sup_{i \in I(\bar{x})} f_i'(\bar{x}, u),$$

a contradiction to the equality we established.   □

Now let us give a classical and convenient sum rule.

Theorem 3.39. *Let f and g be convex functions on a normed space X. If f and g are finite at $\bar{x}$ and if f is continuous at some point of $\operatorname{dom} f \cap \operatorname{dom} g$, then*

$$\partial(f + g)(\bar{x}) = \partial f(\bar{x}) + \partial g(\bar{x}).$$

Proof. The inclusion $\partial f(\bar{x}) + \partial g(\bar{x}) \subset \partial(f + g)(\bar{x})$ is an immediate consequence of the definition of the subdifferential. Let us prove the reverse inclusion under the assumptions of the theorem. Let $\bar{x}^* \in \partial(f + g)(\bar{x})$. Replacing f and g by the functions f_0 and g_0 given respectively by

$$f_0(x) = f(\overline{x}+x) - f(\overline{x}) - \langle \overline{x}^*, x \rangle,$$
$$g_0(x) = g(\overline{x}+x) - g(\overline{x}),$$

we may suppose $\overline{x} = 0$, $\overline{x}^* = 0$, $f(\overline{x}) = g(\overline{x}) = 0$. Then we have $f(x) + g(x) \geq 0$ for every $x \in X$ and $f(0) = 0 = g(0)$. The interior C of the epigraph E of f is nonempty and contained in the strict epigraph of f, hence is disjoint from the hypograph

$$H := \{(x,s) \in X \times \mathbb{R} : s \leq -g(x)\}$$

of $-g$. Let $(u^*, c) \in (X \times \mathbb{R})^* \setminus \{(0,0)\}$, which separates C and H:

$$\langle u^*, w \rangle + cr > 0 \geq \langle u^*, x \rangle + cs \quad \forall (w,r) \in C, \ \forall (x,s) \in H$$

(we use the fact that $0 \in \mathrm{cl}(C) \cap H$). Let u be a point of $\mathrm{dom}\,g$ at which f is finite and continuous. Taking $w = x = u$ and $r \in (f(u), +\infty)$ large enough, we see that $c \geq 0$. If we had $c = 0$, taking $(x,s) = (u, -g(u))$, we would have $\langle u^*, w \rangle > \langle u^*, u \rangle$ for all w in a neighborhood of u, an impossibility. Thus $c > 0$. Since $E \subset \mathrm{cl}(C)$ we get

$$r \geq \langle -c^{-1}u^*, w \rangle \ \forall (w,r) \in E, \quad g(x) \geq \langle c^{-1}u^*, x \rangle \ \forall x \in \mathrm{dom}\,g,$$

and since $f(0) = 0$, $g(0) = 0$, we get $x^* := -c^{-1}u^* \in \partial f(0)$, $-x^* \in \partial g(0)$. □

We deduce a chain rule from the sum rule, although a direct proof can be given.

Theorem 3.40 (Chain rule). *Let X and Y be normed spaces, let $A : X \to Y$ be a linear continuous map, and let $g : Y \to \mathbb{R}_\infty$ be finite at $\overline{y} := A(\overline{x})$ and continuous at some point of $A(X)$. Then for $f := g \circ A$ one has*

$$\partial f(\overline{x}) = A^\mathsf{T}(\partial g(\overline{y})) := \partial g(\overline{y}) \circ A.$$

Proof. The inclusion $\partial g(\overline{y}) \circ A \subset \partial f(\overline{x})$ is immediate, without any assumption on g. Let us first observe that the reverse inclusion is valid without any assumption in the case $X := W \times Y$, $\overline{x} := (\overline{w}, \overline{y})$ and A is the projection $p_Y : (w,y) \mapsto y$: then $f(w,y) = g(y)$, and for every $\overline{x}^* := (\overline{w}^*, \overline{y}^*) \in \partial f(\overline{x})$ one must have $\overline{w}^* = 0$, $\overline{y}^* \in \partial g(\overline{y})$, as is easily checked by observing that

$$\langle \overline{w}^*, w - \overline{w} \rangle \leq f(w, \overline{y}) - f(\overline{w}, \overline{y}) = 0 \quad \forall w \in W,$$
$$\langle \overline{y}^*, y - \overline{y} \rangle \leq f(\overline{w}, y) - f(\overline{w}, \overline{y}) = g(y) - g(\overline{y}) \quad \forall y \in Y.$$

This special case will be used later, and we return now to the general case.
Let ι_G be the indicator function of the graph G of A and let $h := g \circ p_Y$. Then

$$f(x) = \inf\{g(y) + \iota_G(x,y) : y \in Y\} = \inf\{h(x,y) + \iota_G(x,y) : y \in Y\}$$

and $f(\bar{x}) = k(\bar{x}, \bar{y})$, where k is the function $(x, y) \mapsto h(x, y) + \iota_G(x, y)$. Given $\bar{x}^* \in \partial f(\bar{x})$, Proposition 3.37 ensures that $(\bar{x}^*, 0)$ is in the subdifferential of k at $(\bar{x}, \bar{y})$. Since h is finite and continuous at some point of the domain of ι_G, Theorem 3.39 and the preceding special case yield some $\bar{y}^* \in \partial g(\bar{y})$, $(\bar{u}^*, \bar{v}^*) \in \partial \iota_G(\bar{x}, \bar{y})$ such that

$$(\bar{x}^*, 0) = (0, \bar{y}^*) + (\bar{u}^*, \bar{v}^*).$$

Since $(\bar{u}^*, \bar{v}^*) \in N(G, (\bar{x}, \bar{y}))$ or $\bar{u}^* = -\bar{v}^* \circ A$, as is easily checked, we have $\bar{x}^* = \bar{u}^* = -\bar{v}^* \circ A = \bar{y}^* \circ A$.   □

In Banach spaces, one can get rid of the continuity assumptions in the preceding two rules, replacing them by some qualification condition. These results can be obtained through duality and use of the Robinson–Ursescu theorem, as will be shown later.

3.3.1 Supplement: Subdifferentials of Marginal Convex Functions

A generalization of the rule for the subdifferential of the supremum of a finite family of convex functions can be given. Let X be a normed space, let $(f_s)_{s \in S}$ be a family of convex functions $f_s : X \to \overline{\mathbb{R}}$ parameterized by a set S, and let $f := \sup_{s \in S} f_s$. Suppose $f(x) > -\infty$ for all $x \in X$. Given $\bar{x}, x \in f^{-1}(\mathbb{R})$, $\varepsilon \in \mathbb{R}_+$, we set

$$S(x, \varepsilon) := \{s \in S : f_s(x) \geq f(x) - \varepsilon\}, \qquad S(\varepsilon) := S(\bar{x}, \varepsilon).$$

In general, the set $S(0)$ may be empty, so that one has to use the nonempty sets $S(\varepsilon)$ for $\varepsilon \in \mathbb{P} := (0, +\infty)$. The family $\mathcal{M} := \{S(\varepsilon) : \varepsilon \in \mathbb{P}\}$ is a *filter base* (called the *maximizing filter base* of $s \mapsto f_s(\bar{x})$). This means that for all $M, M' \in \mathcal{M}$ one can find $M'' \in \mathcal{M}$ such that $M'' \subset M \cap M'$; in fact, for $\varepsilon, \varepsilon' \in \mathbb{P}$ one has $S(\varepsilon) \cap S(\varepsilon') = S(\varepsilon'')$ for $\varepsilon'' := \min(\varepsilon, \varepsilon')$. Since the family $\mathcal{N}(\bar{x})$ of neighborhoods of $\bar{x}$ is a filter base too, the family $\mathcal{M} \times \mathcal{N}(\bar{x}) := \{S(\varepsilon) \times V : S(\varepsilon) \in \mathcal{M}, V \in \mathcal{N}(\bar{x})\}$ also is a filter base. Given a function $g : S \times X \to \overline{\mathbb{R}}$, we set

$$\limsup_{\mathcal{M} \times \mathcal{N}(\bar{x})} g(s, x) := \inf_{M \times V \in \mathcal{M} \times \mathcal{N}(\bar{x})} \sup_{(s, x) \in M \times V} g(s, x) = \inf_{\varepsilon > 0} \inf_{\rho > 0} \sup_{x \in B(\bar{x}, \rho)} \sup_{s \in S(\varepsilon)} g(s, x)$$

$$= \inf_{\varepsilon > 0} \sup_{(s, x) \in S(\varepsilon) \times B(\bar{x}, \varepsilon)} g(s, x),$$

since we can replace ε and ρ with $\min(\varepsilon, \rho)$. In the following proposition, given $u \in X$, we take $g(s, x) := f_s'(x, u) := \lim_{t \to 0_+} (1/t)(f_s(x + tu) - f_s(x))$, changing the notation for the radial derivative for the sake of simplicity.

Recall that $\overline{\mathrm{co}}^*(A)$ denotes the weak* closed convex hull of a subset A of X^*.

Proposition 3.41 (Valadier). *Let $f := \sup_{s \in S} f_s$, as above. For $\bar{x} \in \operatorname{dom} f$ and $u \in X$ such that $f'(\bar{x}, u) < +\infty$ one has*

$$f'(\bar{x}, u) \leq \limsup_{\mathcal{M} \times \mathcal{N}(\bar{x})} f'_s(x, u) = \inf_{\varepsilon > 0} \sup_{(s,x) \in S(\varepsilon) \times B(\bar{x}, \varepsilon)} f'_s(x, u). \tag{3.9}$$

If $\bar{x} \in \operatorname{core}(\operatorname{dom} f)$ and if for some $\bar{\varepsilon} > 0$ and all $s \in S(\bar{\varepsilon})$, the functions f_s are continuous at $\bar{x}$, one has

$$\partial f(\bar{x}) \subset \bigcap_{\varepsilon > 0} C_\varepsilon, \qquad \text{where } C_\varepsilon := \overline{\operatorname{co}}^* \left(\bigcup_{(s,x) \in S(\varepsilon) \times B(\bar{x}, \varepsilon)} \partial f_s(x) \right). \tag{3.10}$$

If, moreover, f is continuous at $\bar{x}$, (3.9) and (3.10) are equalities.

Proof. Since for all $V \in \mathcal{N}(\bar{x})$ we have $\bar{x} + tu \in V$ for $t > 0$ small enough, to prove (3.9) it suffices to show that

$$f'(\bar{x}, u) \leq \inf_{t > 0} \inf_{\varepsilon > 0} \sup_{s \in S(\varepsilon)} f'_s(\bar{x} + tu, u). \tag{3.11}$$

This inequality being obvious when $f'(\bar{x}, u) = -\infty$, we may suppose $f'(\bar{x}, u) \in \mathbb{R}$. We have to prove that for every $\alpha > 0$ and every $t, \varepsilon > 0$ there exists $s \in S(\varepsilon)$ such that $f'_s(\bar{x} + tu, u) \geq f'(\bar{x}, u) - \alpha$. Since $S(\cdot)$ and $t \mapsto f'_s(\bar{x} + tu, u)$ are nondecreasing, we may suppose $\varepsilon \leq \alpha, t < 1/4$ and

$$\frac{1}{2t} (f(\bar{x} + 2tu) - f(\bar{x})) \leq f'(\bar{x}, u) + \varepsilon. \tag{3.12}$$

Then we pick $s \in S(\bar{x} + tu, \varepsilon t)$, i.e., $s \in S$ such that

$$f_s(\bar{x} + tu) \geq f(\bar{x} + tu) - \varepsilon t \geq f(\bar{x} + tu) - \alpha t. \tag{3.13}$$

Since $f(\bar{x} + tu) \geq f(\bar{x}) + t f'(\bar{x}, u)$, we have $f_s(\bar{x} + tu) \geq f(\bar{x}) + t f'(\bar{x}, u) - \varepsilon t$. Moreover, relation (3.12) ensures that

$$f_s(\bar{x} + 2tu) \leq f(\bar{x} + 2tu) \leq f(\bar{x}) + 2t f'(\bar{x}, u) + 2t\varepsilon,$$

whence by the convexity relation $f_s(\bar{x}) \geq 2f_s(\bar{x} + tu) - f_s(\bar{x} + 2tu)$ and $t < 1/4$,

$$f_s(\bar{x}) \geq 2(f(\bar{x}) + t f'(\bar{x}, u) - \varepsilon t) - (f(\bar{x}) + 2t f'(\bar{x}, u) + 2\varepsilon t) \geq f(\bar{x}) - 4\varepsilon t \geq f(\bar{x}) - \varepsilon,$$

i.e., $s \in S(\varepsilon)$. By the inequalities used for Lemma 3.15 and relation (3.13) we get

$$f'_s(\bar{x} + tu, u) \geq \frac{f_s(\bar{x} + tu) - f_s(\bar{x})}{t} \geq \frac{f(\bar{x} + tu) - \alpha t - f(\bar{x})}{t} \geq f'(\bar{x}, u) - \alpha.$$

That proves (3.11) and (3.9).

In order to prove inclusion (3.10) under the additional assumption, let us show that if $\bar{x}^* \notin C_{\varepsilon}$ for some $\varepsilon > 0$, then $\bar{x}^* \notin \partial f(\bar{x})$. Since $C_{\alpha} \subset C_{\beta}$ for $\alpha < \beta$, we may suppose $\varepsilon \leq \bar{\varepsilon}$, so that f_s is continuous at $\bar{x}$ for all $s \in S(\varepsilon)$. The Hahn–Banach theorem yields some $u \in X \setminus \{0\}$ such that

$$\langle \bar{x}^*, u \rangle > \sup_{(s,x) \in S(\varepsilon) \times B(\bar{x},\varepsilon)} \{ \langle x^*, u \rangle : x^* \in \partial f_s(x) \} = \sup_{(s,x) \in S(\varepsilon) \times B(\bar{x},\varepsilon)} f_s'(x,u).$$

Then since $f'(\bar{x}, u) < +\infty$, because $\bar{x} \in \mathrm{core}(\mathrm{dom}\, f)$, inequality (3.9) yields $\langle \bar{x}^*, u \rangle > f'(\bar{x}, u)$ and $\bar{x}^* \notin \partial f(\bar{x})$.

Now let us show that when f is continuous at $\bar{x}$, given $u \in X$, we have

$$f'(\bar{x}, u) \geq \limsup_{\mathcal{M} \times \mathcal{N}(\bar{x})} f_s'(x,u) := \inf_{\varepsilon > 0} \inf_{\rho > 0} \sup_{x \in B(\bar{x},\rho)} \sup_{s \in S(\varepsilon)} f_s'(x,u).$$

Given $\alpha > 0$, let us find some $\varepsilon, \rho > 0$ such that

$$f'(\bar{x}, u) + \alpha \geq \sup_{x \in B(\bar{x},\rho)} \sup_{s \in S(\varepsilon)} f_s'(x,u). \tag{3.14}$$

Let $m > 0$ and let $V \in \mathcal{N}(\bar{x})$ be open and such that $f(x) \leq m$ for all $x \in V$. Let $t > 0$ be such that $x_t := \bar{x} + tu \in V$ and

$$\frac{1}{t}(f(\bar{x}+tu) - f(\bar{x})) \leq f'(\bar{x}, u) + \frac{\alpha}{2}.$$

Let $\beta := t\alpha/4$. Since f is continuous at x_t, there exists a neighborhood V_t of x_t contained in V such that f is bounded above by $f(x_t) + \beta$ on V_t. Let us pick $\varepsilon \in (0, \beta)$ and $r > 0$ such that $r(m - f(\bar{x}) + \varepsilon) < \beta - \varepsilon$. Let us show that for all $x \in \bar{x} - r(V - \bar{x}) \in \mathcal{N}(\bar{x})$, $s \in S(\varepsilon)$, we have

$$f_s(x) \geq f(\bar{x}) - \beta. \tag{3.15}$$

Let $y \in V$ be such that $x = \bar{x} - r(y - \bar{x})$, so that $\bar{x} = (1+r)^{-1}x + r(1+r)^{-1}y$. Thus we have $f_s(y) \leq m$ and $f_s(\bar{x}) \leq (1+r)^{-1}f_s(x) + r(1+r)^{-1}f_s(y)$, and hence

$$f_s(x) \geq (1+r)f_s(\bar{x}) - rf_s(y) \geq (1+r)(f(\bar{x}) - \varepsilon) - rm \geq f(\bar{x}) - \beta.$$

Now let us pick $\rho > 0$ such that $B(x_t, \rho) \subset V_t$ and $B(0, \rho) \subset r(V - \bar{x})$. For $s \in S(\varepsilon)$ and $x \in B(\bar{x}, \rho)$ we have $x \in \bar{x} - r(V - \bar{x})$, and hence $f_s(x) \geq f(\bar{x}) - \beta$ by (3.15) and $x + tu = \bar{x} + tu + (x - \bar{x}) \in B(x_t, \rho) \subset V_t$, so that $f_s(x + tu) \leq f(x_t) + \beta$. Therefore

$$f_s'(x,u) \leq \frac{1}{t}(f_s(x+tu) - f_s(x)) \leq \frac{1}{t}(f(x_t) - f(\bar{x}) + 2\beta)$$

$$\leq f'(\bar{x}, u) + \frac{\alpha}{2} + \frac{2}{t}\beta = f'(\bar{x}, u) + \alpha.$$

Thus (3.14) is established, and since α is arbitrarily small, we get equality in relation (3.9). The bijection between closed convex sets and support functions being a lattice isomorphism, equality in (3.10) ensues.   □

The preceding result can be simplified if one uses compactness assumptions.

Proposition 3.42 (Rockafellar). *Let S be a compact topological space, let $(f_s)_{s \in S}$ be a family of convex functions on some convex open subset U of a normed space X, let $f := \sup_{s \in S} f_s$, and let $\bar{x} \in \mathrm{dom}\, f$. Suppose that for some neighborhood V of $\bar{x}$ in U the following assumptions are satisfied:*

(a) For all $x \in V$ the function $s \mapsto f_s(x)$ is upper semicontinuous and finite
(b) For all $s \in S$, f_s is upper semicontinuous at $\bar{x}$

Then $S(\bar{x}) := \{s \in S : f_s(\bar{x}) = f(\bar{x})\}$ is nonempty, and for all $u \in X$ one has

$$f'(\bar{x}, u) = \max_{s \in S(\bar{x})} f_s'(\bar{x}, u), \tag{3.16}$$

$$\partial f(\bar{x}) = \overline{\mathrm{co}}^* \Big(\bigcup_{s \in S(\bar{x})} \partial f_s(\bar{x}) \Big). \tag{3.17}$$

Proof. Assumption (a) ensures that $S(\bar{x}) = \cap_{\varepsilon > 0} S(\varepsilon)$ is nonempty and compact. Moreover, given $u \in X$ and $s \in S(\bar{x})$, since $f_s'(\bar{x}, u) = \inf_{t > 0}(1/t)(f_s(\bar{x} + tu) - f(\bar{x}))$, the function $s \mapsto f_s'(\bar{x}, u)$ is upper semicontinuous, and we can write max instead of sup in relation (3.16). Let us prove relation (3.16). For all $s \in S(\bar{x})$, since $f_s \le f$ and $f_s(\bar{x}) = f(\bar{x})$, we have $f_s'(\bar{x}, u) \le f'(\bar{x}, u)$, hence

$$\max\{f_s'(\bar{x}, u) : s \in S(\bar{x})\} \le f'(\bar{x}, u).$$

Let us prove the reverse inequality. Let $r < f'(\bar{x}, u)$. Let $t_u > 0$ be such that $\bar{x} + tu \in V$ for all $t \in [0, t_u]$. For every $t \in (0, t_u]$, the set

$$S_u(t) := \{s \in S : f_s(\bar{x} + tu) \ge f(\bar{x}) + rt\}$$

is nonempty, since $(1/t)(f(\bar{x} + tu) - f(\bar{x})) \ge f'(\bar{x}, u) > r$. By (a), $S_u(t)$ is closed. The convexity of f_s ensures that for $t \in (0, t_u]$, $\theta \in (0, 1)$, $t' = \theta t$, one has $S_u(t') \subset S_u(t)$, since for $s \in S_u(t')$ one has $f(\bar{x}) \ge f_s(\bar{x})$, hence

$$\theta f_s(\bar{x} + tu) + (1 - \theta)f(\bar{x}) \ge f_s(\bar{x} + \theta tu) \ge f(\bar{x}) + \theta rt$$

and $f_s(\bar{x} + tu) \ge f(\bar{x}) + rt$ after simplification. Thus $\cap_t S_u(t)$ is nonempty. Let $\bar{s} \in \cap_t S_u(t)$. Since $f_{\bar{s}}$ is upper semicontinuous at $\bar{x}$, a passage to the limit in the definition of $S_u(t)$ shows that $f_{\bar{s}}(\bar{x}) \ge f(\bar{x})$, i.e., $\bar{s} \in S(\bar{x})$. Then since $f_{\bar{s}}(\bar{x}) = f(\bar{x})$ and $\bar{s} \in S_u(t)$ for all $t \in (0, t_u]$, one gets $(1/t)(f_{\bar{s}}(\bar{x} + tu) - f_{\bar{s}}(\bar{x})) \ge r$, hence $f_{\bar{s}}'(\bar{x}, u) \ge r$. Thus relation (3.16) holds.

Let C be the right-hand side of relation (3.17). For all $s \in S(\bar{x})$ the inequalities $f_s'(\bar{x}, \cdot) \le f'(\bar{x}, \cdot)$ entail the inclusion $\partial f_s(\bar{x}) \subset \partial f(\bar{x})$, hence $C \subset \partial f(\bar{x})$. Now for every $\bar{x}^* \in X^* \setminus C$ one can find some $u \in X \setminus \{0\}$ such that

$$\langle \bar{x}^*, u \rangle > \sup_{x^* \in C} \langle x^*, u \rangle \geq \sup_{s \in S(\bar{x})} \sup_{x^* \in \partial f_s(\bar{x})} \langle x^*, u \rangle = \sup_{s \in S(\bar{x})} f_s'(\bar{x}, u) = f'(\bar{x}, u),$$

f_s being continuous at $\bar{x}$. Thus $\bar{x}^* \notin \partial f(\bar{x})$ and (3.17) holds.

Exercises

1. Let f, g be two convex functions on a normed space X that are finite at some $\bar{x} \in X$. Suppose g is Fréchet differentiable at $\bar{x}$. Then show that $\partial(f+g)(\bar{x}) = \partial f(\bar{x}) + g'(\bar{x})$.

2. Let f be a convex function on a normed space X that is finite at some $\bar{x} \in X$. Suppose there exists some $\ell \in X^*$ such that r defined by $r(x) := \max(f(\bar{x}+x) - f(\bar{x}) - \ell(x), 0)$ is a remainder. Show that f is Fréchet differentiable at $\bar{x}$.

3. Recall Proposition 3.29: if a convex function f on a normed space X is finite at $x \in X$ then f is subdifferentiable at x iff it is calm at x iff there exists $c > 0$ such that $f(w) \geq f(x) - c\|w - x\|$ for all $w \in X$. Show that in such a case one has $\partial f(x) \cap cB_{X^*} \neq \varnothing$ but that one may have $\partial f(x) \not\subset cB_{X^*}$.

4. Without compactness of some $S(\bar{x}, \varepsilon)$, relation (3.17) may not hold, even when $S(\bar{x})$ is nonempty or even when $S(\bar{x}) = S$. Check that it fails for $S := (1, 2]$, $X := \mathbb{R}$, $f_s(x) := |x|^s$, so that $f(x) = |x|$ for $x \in [-1, 1]$.

5. Prove that a differentiable function $f : W \to \mathbb{R}$ defined on an open convex subset of a normed space X is convex if and only if $f' : W \to X^*$ is *monotone*, i.e., satisfies $\langle f'(w) - f'(x), w - x \rangle \geq 0$ for all $w, x \in W$.

6. Show that for a convex function $f : X \to \mathbb{R}_\infty$ on a normed space X, the multimap $\partial f : X \rightrightarrows X^*$ is *monotone*, i.e., satisfies $\langle w^* - x^*, w - x \rangle \geq 0$ for all $w, x \in X$, $w^* \in \partial f(w)$, $x^* \in \partial f(x)$.

3.4   The Legendre–Fenchel Transform and Its Uses

There are several instances in mathematics in which a duality can be used to transform a given problem into an associated one called the dual problem. The dual problem may appear to be more tractable and may yield useful information about the original problem and even help to solve it entirely. For optimization problems, the Legendre–Fenchel conjugacy is certainly the most useful duality. It is intimately linked with the calculus of subdifferentials; for this reason, the study of this transform is fully justified here.

3.4.1 The Legendre–Fenchel Transform

Given a normed space X in duality with its topological dual X^* through the usual pairing $\langle \cdot, \cdot \rangle$ and a function $f : X \to \overline{\mathbb{R}}$, the knowledge of the performance function

$$f_*(x^*) := \inf_{x \in X}(f(x) - \langle x^*, x \rangle) \tag{3.18}$$

associated with the natural perturbation of f by continuous linear forms is likely to give precious information about f, at least when f is closed proper convex. Since f_* is concave (it is called the *concave conjugate* of f) and upper semicontinuous, one usually prefers to deal with the convex conjugate or *Legendre–Fenchel conjugate* (or simply Fenchel conjugate) f^* of f given by $f^* = -f_*$:

$$f^*(x^*) := \sup_{x \in X}(\langle x^*, x \rangle - f(x)). \tag{3.19}$$

We note that whenever the domain of f is nonempty, f^* takes its values in $\mathbb{R}_\infty :=$ $\mathbb{R} \cup \{+\infty\}$. We also observe that f^* is convex and lower semicontinuous for the weak* topology on X^* as a supremum of continuous affine functions. Notice that we could replace X^* with another space Y in duality with X.

The computation of conjugates is eased by the calculus rules we give below. The following examples illustrate the interest of this transformation.

Examples. (a) Let f be the indicator function ι_C of some subset C of X. Then f^* is the *support function* h_C or σ_C of C given by $h_C(x^*) := \sigma_C(x^*) := \sup_{x \in C}\langle x^*, x \rangle$.
(b) Let $h_S : X^* \to \mathbb{R}_\infty$ be the support function of $S \subset X$. Then $h_S^* = \iota_C$, where $C :=$ $\mathrm{clco}(S)$ is the closed convex hull of S.
(c) If f is linear and continuous, then f^* is the indicator function of $\{f\}$.
(d) For $X = \mathbb{R}$, $f = \frac{1}{p}|\cdot|^p$ with $p \in (1, \infty)$ one has $f^* = \frac{1}{q}|\cdot|^q$ with $q := (1 - \frac{1}{p})^{-1}$.
(e) If $f = \|\cdot\|$, then $f^* = \iota_{B^*}$, the indicator function of the closed unit ball B^* of X^*.
(f) More generally, if f is positively homogeneous and $f(0) = 0$, then f^* is the indicator function of $\partial f(0)$.

Other examples are proposed in exercises. Examples (e), (f) point out a connection between subdifferentials and conjugates; we will consider this question with more generality later on. Examples (a) and (b) illustrate the close relationships between functions and sets; these links are of great importance for this book. Let us point out the potential generality of Example (a), which shows that the computation of conjugate functions can be reduced to the calculus of support functions: for every function $f : X \to \overline{\mathbb{R}}$ with epigraph E_f, the value at x^* of f^* satisfies the relation

$$f^*(x^*) = \sigma_{E_f}(x^*, -1), \tag{3.20}$$

as an immediate interpretation of the definition shows (see also Exercise 1 below).

The Fenchel transform enjoys nice properties. We leave their easy proofs as exercises.

Proposition 3.43. *The Fenchel transform satisfies the following properties:*

It is antitone: for every pair of functions f, g with $f \leq g$ one has $f^* \geq g^*$.
For every function f and $c \in \mathbb{R}$, $(f + c)^* = f^* - c$;
For every function f and $c > 0$, $(cf)^*(x^*) = cf^*(c^{-1}x^*)$ for all $x^* \in X^*$;
For every function f and $c > 0$, if $g := f(c\cdot)$, then $g^* = f^*(c^{-1}\cdot)$;
For every function f and $\ell \in X^*$, $(f + \ell)^* = f^*(\cdot - \ell)$;
For every function f and $\overline{x} \in X$, $(f(\cdot + \overline{x}))^* = f^* - \langle \cdot, \overline{x} \rangle$.

For every pair of functions f, g one has $(f \square g)^* = f^* + g^*$, where the infimal convolution $f \square g$ is defined by $(f \square g)(x) := \inf_w f(x - w) + g(w)$.

Let us examine whether f^* enables one to recover f. For this purpose, we introduce the *biconjugate* of f as the function $f^{**} := (f^*)^*$. Here we use the same symbol for the conjugate g^* of a function g on X^*:

$$g^*(x) := \sup_{x^* \in X^*} (\langle x^*, x \rangle - g(x^*)).$$

In doing so we commit some abuse of notation, since in fact we consider the restriction of g^* to $X \subset X^{**}$. However, the notation is compatible with the choice of the pairing between X and X^*. In fact, our study could be cast in the framework of topological vector spaces X, Y in separated duality; taking for Y the dual of X endowed with the weak* topology, one would get X as the dual of Y.

Theorem 3.44. *For every function $f : X \to \overline{\mathbb{R}}$ one has $f^{**} \leq f$. If f is closed proper convex (or if $f = +\infty^X$ or $f = -\infty^X$, the constant functions with values $+\infty$ and $-\infty$ respectively) then $f^{**} = f$.*

Proof. Given $x \in X$, for every function $f : X \to \overline{\mathbb{R}}$ and every $x^* \in X^*$ we have $-f^*(x^*) \leq f(x) - \langle x^*, x \rangle$ hence $f^{**}(x) = \sup\{\langle x^*, x \rangle - f^*(x^*) : x^* \in X^*\} \leq f(x)$.

Let us suppose f is closed proper convex. For every $w \in X$ and $r < f(w)$ we can find $x^* \in X^*$ and $c \in \mathbb{R}$ such that $r < \langle x^*, w \rangle - c$, $\langle x^*, x \rangle - c \leq f(x)$ for all $x \in X$. Then we have $f^*(x^*) \leq c$, hence $f^{**}(w) \geq \langle x^*, w \rangle - c > r$. Therefore $f^{**} \geq f$, hence $f^{**} = f$. The cases of the constant functions $-\infty^X$, $+\infty^X$ with values $-\infty$ and $+\infty$ respectively are immediate. □

Corollary 3.45. *For every function $f : X \to \overline{\mathbb{R}}$ bounded below by a continuous affine function and with nonempty domain, the greatest closed proper convex function on X bounded above by f is $f^{**} \mid X$. If f is not bounded below by a continuous affine function, then $f^{**} = -\infty^X$.*

Proof. The last assertion is obvious, since $f^* = +\infty^{X^*}$ when f is not bounded below by a continuous affine function (since $f^*(w^*) < c$ for some $w^* \in X^*$, $c \in \mathbb{R}$ implies that $f(x) \geq \langle w^*, x \rangle - c$ for all $x \in X$). If g is a closed proper convex function satisfying $g \leq f$, we have $g^* \geq f^*$, since the Fenchel transform is antitone; then $g = g^{**} \leq f^{**}$. Thus when $f \neq +\infty^X$ and f is bounded below by a continuous affine function, f^{**} is proper and clearly lower semicontinuous and convex, hence closed proper convex, and f^{**} is the greatest such function bounded above by f. □

Corollary 3.46. *For every function $f : X \to \overline{\mathbb{R}}$ one has $f^{***} = f^*$.*

Proof. The result is obvious if $f^* = +\infty^{X^*}$ or if $f^* = -\infty^{X^*}$; otherwise, f^* is closed proper convex. □

A crucial relationship between the Fenchel conjugate and the Moreau–Rockafellar subdifferential is given by the Young–Fenchel relation that follows.

Theorem 3.47 (Young–Fenchel). *For every function $f : X \to \overline{\mathbb{R}}$ and for every $x \in X$, $x^* \in X^*$ one has $f(x) + f^*(x^*) \geq \langle x^*, x \rangle$.*

When $f(x) \in \mathbb{R}$ equality holds if and only if $x^* \in \partial f(x)$. Moreover, $x^* \in \partial f(x)$ implies $x \in \partial f^*(x^*)$.

Proof. The first assertion is a direct consequence of the definition. When $f(x) \in \mathbb{R}$, the equality $f(x) + f^*(x^*) = \langle x^*, x \rangle$ is equivalent to each of the following assertions:

$$f(x) + f^*(x^*) \leq \langle x^*, x \rangle,$$
$$f(x) - f(w) + \langle x^*, w \rangle \leq \langle x^*, x \rangle \qquad \forall w \in X,$$
$$x^* \in \partial f(x).$$

Moreover, they imply the inequality $f^{**}(x) + f^*(x^*) \leq \langle x^*, x \rangle$, equivalent to $x \in \partial f^*(x^*)$. □

Theorem 3.48. *For every function $f : X \to \overline{\mathbb{R}}$ one has $f^{**}(x) = f(x)$ whenever $\partial f(x) \neq \varnothing$.*

Moreover, when $f^{**}(x) = f(x) \in \mathbb{R}$, one has $\partial f(x) = \partial f^{**}(x)$ and $x^* \in \partial f(x)$ if and only if $x \in \partial f^*(x^*)$.

Proof. Given $x^* \in \partial f(x)$, let $g : w \mapsto \langle x^*, w - x \rangle + f(x)$. Then g is a continuous affine function satisfying $g \leq f$, so that $g \leq f^{**}$ and $g(x) = f(x) \geq f^{**}(x)$, so that $f^{**}(x) = f(x)$ and $x^* \in \partial f^{**}(x)$. Moreover, when $f^{**}(x) = f(x) \in \mathbb{R}$, the reverse inclusion $\partial f^{**}(x) \subset \partial f(x)$ follows from the relations $f^{**} \leq f$, $f^{**}(x) = f(x)$. □

Corollary 3.49. *When $f = f^{**}$ the multimap ∂f^* is the inverse of the multimap ∂f:*

$$x^* \in \partial f(x) \Leftrightarrow x \in \partial f^*(x^*).$$

The following special case is of great importance for dual problems.

Corollary 3.50. *When $f^{**}(0) = f(0) \in \mathbb{R}$ the set of minimizers of f^* is $\partial f(0)$. For every function g with finite infimum, the set $\partial g^*(0)$ is the set of minimizers of g^{**}.*
When $f^{**} = f$ and $f^*(0)$ is finite, the set $\partial f^*(0)$ is the set of minimizers of f.

Proof. The first assertion follows from the equivalences $x^* \in \partial f(0) \Leftrightarrow 0 \in \partial f^*(x^*)$ $\Leftrightarrow x^*$ is a minimizer of f^*. The second one ensues because $g^*(0) = -\inf g(X)$ and $g^{***} = g^*$. Taking $g := f$, one gets the last assertion. □

Exercises

1. Check that for every function $f : X \to \overline{\mathbb{R}}$ with nonempty domain, the support function of the epigraph E_f of f satisfies $h_{E_f}(x^*, -1) = f^*(x^*)$ and

$$h_{E_f}(x^*, r) = -rf^*(-r^{-1}x^*) \quad \text{for } r < 0,$$

$$h_{E_f}(x^*, 0) = h_{\text{dom} f}(x^*),$$

$$h_{E_f}(x^*, r) = +\infty \qquad \text{for } r > 0.$$

2. Show that for every function $f : X \to \overline{\mathbb{R}}$, the greatest lower semicontinuous convex function bounded above by f is either f^{**} or the *valley function* υ_C associated with the closed convex hull C of $\text{dom} f$, given by $\upsilon_C(x) = -\infty$ if $x \in C$, $\upsilon_C(x) = +\infty$ if $x \notin C$.

3. If X is a normed space and $f = g \circ \|\cdot\|$, where $g : \mathbb{R}_+ \to \mathbb{R}_\infty$ is extended by $+\infty$ on $\mathbb{R}_-$, show that $f^* = g^* \circ \|\cdot\|_*$, where $\|\cdot\|_*$ is the dual norm of $\|\cdot\|$.

4. For $X = \mathbb{R}$ and $f(x) = \exp x$, check that $f^*(y) = y \log y - y$ for $y > 0$, $f^*(0) = 0$, $f(y) = +\infty$ for $y < 0$.

5. Let $f : \mathbb{R} \to \mathbb{R}_\infty$ be given by $f(x) := -\ln x$ for $x \in \mathbb{P}$, $f(x) := +\infty$ for $x \in \mathbb{R}_-$. Check that $f^*(x^*) = -\ln|x^*| - 1$ for $x^* \in -\mathbb{P}$, $f(x) := +\infty$ for $x \in \mathbb{R}_+$.

6. Let $f : X \to \mathbb{R}_\infty$ and let g be the convex hull of f. Show that $g^* = f^*$.

7. Let $f : X \to \mathbb{R}_\infty$ and let h be the lower semicontinuous hull of f. Show that $h^* = f^*$.

3.4.2 The Interplay Between a Function and Its Conjugate

Let us give examples of the information one can draw from the study of the conjugate function.

We first study the transfer to f^* of growth properties of an arbitrary function f. In order to obtain symmetry in the properties below, we assume that we have two normed spaces X, Y in *metric duality*, i.e., that there exists a continuous bilinear coupling $c := \langle \cdot, \cdot \rangle : X \times Y \to \mathbb{R}$ such that $\|y\| = \sup\{\langle x, y \rangle : x \in B_X\}$ for all $y \in Y$ and $\|x\| = \sup\{\langle x, y \rangle : y \in B_Y\}$ for all $x \in X$. Such is the case when Y is the dual of X or when X is the dual of Y.

Lemma 3.51. *Let $f : X \to \mathbb{R}_\infty$ be proper and let $r, c \in \mathbb{R}_+$, $a, b \in \mathbb{R}$.*

(a) If f is such that $f \geq a$ on rB_X and $f(\cdot) \geq c\|\cdot\| - b$ on $X \setminus rB_X$, then for $y \in cB_Y$ one has $f^*(y) \leq r\|y\| - \min(a, cr - b)$.

(b) If f is supercoercive *in the sense that* $\alpha_f := \liminf_{\|x\| \to +\infty} f(x)/\|x\| > 0$ *and if f is bounded below on bounded sets, then for all* $c \in (0, \alpha_f)$, f^* *is bounded above on* cB_Y.

(c) *If* f^* *is bounded above by* b *on* cB_Y, *then* $f(\cdot) \geq c\,\|\cdot\| - b$ *on* X.

(d) f *is bounded below on bounded sets and* f *is* hypercoercive *in the sense that* $f(x)/\|x\| \to +\infty$ *as* $\|x\| \to +\infty$ *if and only if* f^* *is bounded above on bounded sets.*

Proof. (a) For $y \in cB_Y$, setting $s := \|x\|$, one has

$$f^*(y) \leq \max(\ \sup_{x \in rB_X}\ (\langle y, x \rangle - a),\ \sup_{x \in X \setminus rB_X}\ (\langle y, x \rangle - c\,\|x\| + b))$$

$$\leq \max(r\,\|y\| - a, \sup_{s \geq r} s(\|y\| - c) + b) = r\,\|y\| + \max(-a, b - rc).$$

(b) For all $c \in (0, \alpha_f)$ one can find $r > 0$ such that $f(x)/\|x\| \geq c$ for all $x \in X \setminus rB_X$. Setting $b = 0$ and $a := \inf f(rB_X)$ in (a), one gets $f^*(\cdot) \leq r\,\|\cdot\| - \min(a, cr) \leq \max(cr - a, 0)$ on cB_Y.

(c) If $f^* \leq b$ on cB_Y, then for all $x \in X$ one has $f(x) \geq f^{**}(x) \geq \sup_{y \in cB_Y} (\langle y, x \rangle - b) = c\,\|x\| - b$.

(d) When f is hypercoercive, i.e., when $\alpha_f = +\infty$, and f is bounded below on bounded sets, assertion (b) ensures that for all $c \in \mathbb{R}_+$, f^* is bounded above on cB_Y. The converse follows from (c). □

For a closed proper convex function, the relationships between growth properties of f and boundedness properties of f^* are more striking.

Proposition 3.52. *Let X be a normed space, let $f : X \to \mathbb{R}_\infty$ be closed convex proper, and let $c \in \mathbb{R}_+$, $a, b \in \mathbb{R}$. Then the following assertions are equivalent:*

(a) f *is supercoercive:* $\alpha_f := \liminf_{\|x\| \to +\infty} f(x)/\|x\| > 0$
(b) *There exist* $b \in \mathbb{R}$, $c \in \mathbb{P}$ *such that* $f \geq c\,\|\cdot\| - b$
(c) f *is coercive in the sense that* $f(x) \to +\infty$ *as* $\|x\| \to +\infty$
(d) *The sublevel sets of f are bounded*
(e) f^* *is bounded above on a neighborhood of* 0
(f) $0 \in \mathrm{int}(\mathrm{dom}(f^*))$

Proof. (a)$\Rightarrow$(b) Since f is bounded below by a continuous affine function, it is bounded below on balls. Given $c \in (0, \alpha_f)$, we can find $r > 0$ such that $f(\cdot) \geq c\,\|\cdot\|$ on $X \setminus rB_X$ and $a \in \mathbb{R}$ such that $f(\cdot) \geq a$ on rB_X. Taking $b := (cr - a)^+ := \max(cr - a, 0)$, we get $f(\cdot) \geq c\,\|\cdot\| - b$ on rB_X and $X \setminus rB_X$ hence on X.

(b)$\Rightarrow$(c) is obvious and (c)$\Leftrightarrow$(d) is easy.

(d)$\Rightarrow$(a) Suppose $\alpha_f \leq 0$. Given a sequence $(\varepsilon_n) \to 0_+$ in $(0, 1)$, one can find $x_n \in X$ such that $\|x_n\| \geq n/\varepsilon_n$ and $f(x_n) \leq \varepsilon_n\,\|x_n\|$. Let $t_n := 1/(\varepsilon_n\,\|x_n\|) \leq 1/n$. Then, given $w \in \mathrm{dom} f$, for $u_n := (1 - t_n)w + t_n x_n$, one has

$$f(u_n) \leq (1 - t_n)f(w) + t_n f(x_n) \leq |f(w)| + 1,$$

but (u_n) is unbounded, since $\|u_n\| \geq t_n \|x_n\| - (1 - t_n)\|w\| \geq 1/\varepsilon_n - \|w\|$, a contradiction to (d).

(b)$\Leftrightarrow$(e) has been proved in the preceding lemma and (e)$\Rightarrow$(f) is obvious.

(f)$\Rightarrow$(e) is a consequence of Proposition 3.4, since f^* is convex and lower semicontinuous and $Y := X^*$ is complete.   □

Now let us point out relationships between rotundity properties of f and smoothness of f^*. A function $f : X \to \mathbb{R}_\infty$ is said to be *strictly convex*, respectively *uniformly convex* with constant $c > 0$, if for every distinct $x_0, x_1 \in X, t \in (0, 1)$ one has respectively

$$f((1-t)x_0 + tx_1) < (1-t)f(x_0) + tf(x_1), \tag{3.21}$$

$$f((1-t)x_0 + tx_1) < (1-t)f(x_0) + tf(x_1) - ct(1-t)\|x_0 - x_1\|^2. \tag{3.22}$$

Theorem 3.53. *Let f be a closed proper convex function finite and continuous at $\bar{x} \in X$. If f^* is strictly convex (resp. uniformly convex), then f is Hadamard (resp. Fréchet) differentiable at $\bar{x}$.*

Proof. For $\bar{x}^* \in \partial f(\bar{x})$ one has $\bar{x} \in \partial f^*(\bar{x}^*)$ by Theorem 3.48, whence $0 \in \partial (f^* - \bar{x})(\bar{x}^*)$ and $\bar{x}^*$ is a minimizer of $f^* - \bar{x}$. When f^* is strictly convex, $f^* - \bar{x}$ is strictly convex, too, and it has at most one minimizer. Thus $\partial f(\bar{x})$ is a singleton and f is Hadamard differentiable at $\bar{x}$ in view of Corollary 3.26. We leave the Fréchet case as an exercise (see [61, 984], where quantitative information is provided).   □

Exercises

1. Let X be a Hilbert space with scalar product $(\cdot \mid \cdot)$ and let $A : X \to X$ be a symmetric, linear, continuous map such that the quadratic form q associated with A is positive on $X \setminus \{0\}$. Let $b \in X$ and let f be given by $f(x) = q(x) - (b \mid x)$.
(a) Check that A and the square root $A^{1/2}$ of A are injective and that their images satisfy $R(A) \subset R(A^{1/2})$.
(b) Using Theorem 3.40 and the relation $q = g \circ A^{1/2}$ for $g := \frac{1}{2}\|\cdot\|^2$, show that $q^*(x^*) = \frac{1}{2}\|(A^{1/2})^{-1}(x^*)\|^2$ for $x^* \in R(A^{1/2})$, $q^*(x^*) = +\infty$ otherwise.
(c) Check that if $b \in R(A)$, then f attains its minimum at $A^{-1}(b)$.
(d) Check that if $b \in R(A^{1/2}) \setminus R(A)$, then f is bounded below but does not attain its infimum.
(e) Check that if $b \notin R(A^{1/2})$, then $\inf_{x \in X} f(x) = -\infty$.
(f) Deduce from the preceding questions that when $R(A)$ is closed, then $R(A^{1/2}) = R(A)$. [Hint: When $R(A) \neq X$ take $b \in X \setminus R(A)$ and pick some $u \in R(A)^\perp$ such that $(b \mid u) > 0$; then check that $\inf_{r>0} f(ru) = -\infty$.]

2. (a) Show that if f is such that $f \geq b$ and $\mathrm{dom} f \subset rB_X$, then one has $f^*(\cdot) \leq r\|\cdot\| - b$.

(b)] Show that if f is such that $f \geq b$ and $f(\cdot) \geq c \|\cdot\|$ on $X \setminus rB_X$, then one has $cB_{X^*} \subset \operatorname{dom} f^*$.

3. Give an example of a coercive function that is not supercoercive.

4. Give an example of a supercoercive function that is not hypercoercive.

5. If $f(x) = \frac{1}{p} \|x\|^p$ with $p \in (1, \infty)$, show that $f^*(x^*) = \frac{1}{q} \|x^*\|_*^q$ with $q := (1 - \frac{1}{p})^{-1}$, where $\|\cdot\|_*$ is the dual norm. Observe that for $p = 2$ one has $q = 2$.

6. Let X be a Hilbert space identified with its dual. Show that $f^* = f$ iff $f := \frac{1}{2} \|\cdot\|^2$.

7. (Legendre transform) Let $f : X \to \mathbb{R}_\infty$ be a lower semicontinuous proper convex function that is differentiable on its open domain W and such that its derivative $f' : W \to X^*$ realizes a bijection between W and $W^* := f'(W)$, with inverse h. Let $f^L : W^* \to \mathbb{R}$ be the Legendre transform of f: $f^L(w^*) := \langle w^*, h(w^*) \rangle - f(h(w^*))$. Show that f^L coincides with the restriction to W^* of the conjugate f^* of f.

3.4.3   A Short Account of Convex Duality Theory

Let us give a short account of the usefulness of duality for solving optimization problems. A general approach for dealing with the optimization problem

$$(\mathscr{P}) \qquad \text{minimize} \qquad f(x), \qquad x \in X,$$

where X is a set and $f : X \to \mathbb{R}_\infty$, consists in embedding it in a family of problems

$$(\mathscr{P}_w) \qquad \text{minimize} \qquad P_w(x), \qquad x \in X,$$

where w is an element of a normed space W and $P_w := P(w, \cdot) : X \to \mathbb{R}_\infty$ is a family of objective functions deduced from a *perturbation function* (or parameterization function) $P : W \times X \to \mathbb{R}_\infty$ in such a way that $f = P_0$. We associate to P the *performance function* (or value function) p given by

$$p(w) := \inf_{x \in X} P(w, x), \qquad w \in W.$$

The inequality $p^{**}(0) \leq p(0)$ gives an estimate of the value $p(0)$ of $(\mathscr{P})$. Since $p^{**}(0) = \sup\{-p^*(w^*) : w^* \in W^*\}$, this estimate involves the *dual problem*, which is the maximization problem

$$(\mathscr{D}) \qquad \text{maximize} \qquad -p^*(w^*), \qquad w^* \in Y := W^*.$$

When X is a normed space, it can be expressed in terms of the perturbation function P itself, since for all $w^* \in W^*$,

$$p^*(w^*) = \sup_{w \in W} (\langle w^*, w \rangle - \inf_{x \in X} P(w,x))$$

$$= \sup_{(w,x) \in W \times X} (\langle (w^*,0),(w,x) \rangle - P(w,x)) = P^*(w^*,0).$$

One can put $(\mathscr{D})$ in the form of the minimization of p^*. This convex problem can be called the *adjoint problem*, and when X is a normed space, it is expressed as

$$(\mathscr{P}^*) \qquad \text{minimize} \qquad P^*(w^*,0), \qquad w^* \in Y := W^*.$$

When the value $p^{**}(0)$ of $(\mathscr{D})$ coincides with the value $p(0)$ of $(\mathscr{P}^*)$, one says that *weak duality* holds or that there is no duality gap. We know that p is convex whenever P is convex. Its subdifferentiability at 0 yields the *strong duality relation*

$$\inf (\mathscr{P}) = -\min (\mathscr{P}^*),$$

where min is taken in the usual sense that if $\inf (\mathscr{P}^*)$ is finite, then $(\mathscr{P}^*)$ has a solution.

Proposition 3.54. *Suppose the performance function p is convex and finite at 0. Then there is no duality gap if and only if p is lower semicontinuous at 0.*

Proof. Since $p^{**} \leq p$ and p^{**} is lower semicontinuous, the equality $p^{**}(0) = p(0)$ entails that p is lower semicontinuous at 0. Conversely, when p is convex, finite, and lower semicontinuous at 0, its lower semicontinuous hull $\overline{p}$ satisfies $\overline{p}(0) = p(0)$. Then since $\overline{p}$ is lower semicontinuous, convex, proper (since $\overline{p}(0) \in \mathbb{R}$), one has $p^{**} = \overline{p}$. In particular, one has $p^{**}(0) = \overline{p}(0) = p(0)$. $\qquad\square$

In the following proposition we do not require any convexity assumption, but we use the *Moreau–Rockafellar subdifferential* of p given by

$$\partial_{MR}p(\overline{w}) := \partial p(\overline{w}) := \{w^* \in W^* : \forall w \in W \; p(w) \geq p(\overline{w}) + \langle w^*, w - \overline{w} \rangle\},$$

a stringent notion when p is nonconvex.

Proposition 3.55. *If the Moreau–Rockafellar subdifferential $\partial p(0)$ of p at 0 is nonempty, then strong duality holds: one has $\inf(\mathscr{P}) = \max(\mathscr{D})$, and $(\mathscr{D})$ has optimal solutions. More precisely, the set S^* of solutions of $(\mathscr{D})$ is $\partial p(0)$.*

Proof. Let $\overline{w}^* \in \partial p(0)$: for all $w \in W$ one has $p(w) \geq p(0) + \langle \overline{w}^*, w \rangle$. Thus $-p(0) \geq p^*(\overline{w}^*)$; hence $p(0) \leq -p^*(\overline{w}^*) \leq \sup_{w^* \in W^*} -p^*(w^*) = p^{**}(0)$, and $\overline{w}^*$ is a solution to $(\mathscr{D})$, $p(0) = p^{**}(0)$. Conversely, if $-p^*(\overline{w}^*) = \sup_{w^* \in W^*} -p^*(w^*) = p(0)$, for all $w \in W$ one has $p(w) - \langle \overline{w}^*, w \rangle \geq p(0)$. That means that $\overline{w}^* \in \partial p(0)$. $\qquad\square$

Corollary 3.56. *Suppose p is convex and $\inf(\mathscr{P})$ is finite. Suppose there exists some $\overline{x} \in X$ such that $P(\cdot,\overline{x})$ is finite and continuous at 0. More generally, denoting by V the vector space generated by $\mathrm{dom}\,p$, suppose there exist some $r > 0$, $m \in \mathbb{R}$ and some map $w \mapsto x(w)$ from $B(0,r) \cap V$ to X such that $P(w,x(w)) \leq m$ for all $w \in B(0,r) \cap V$. Then $p \,|\, V$ is continuous, p is subdifferentiable at 0, and strong duality holds.*

Proof. Under the general assumption, p is majorized on $B(0,r) \cap V$, since for $w \in B(0,r) \cap V$ one has $p(w) \le P(w,x(w)) \le m$. Thus $p \mid V$ is continuous, and by Corollary 3.27, p is subdifferentiable at 0. □

The preceding corollary makes it possible to get the subdifferentiability rules under continuity assumptions we have seen previously (exercise). We rather prove new subdifferentiability rules under semicontinuity assumptions and algebraic assumptions that are quite convenient. They derive from the following corollary.

Corollary 3.57. *Let W, X be Banach spaces and let p be the performance function associated to a perturbation $P : W \times X \to \mathbb{R}_\infty$ that is convex, lower semicontinuous, and such that*

$$Z := \bigcup_{x \in X} \mathbb{R}_+ \operatorname{dom} P(\cdot, x) = -Z = \operatorname{cl}(Z). \tag{3.23}$$

Then if $p(0) \in \mathbb{R}$, p is subdifferentiable at 0, and strong duality holds.

Note that assumption (3.23) means that Z is a closed vector subspace of W.

Proof. By Corollary 3.27, we may suppose $Z = W$. The set

$$F := \{(x, r, w) \in X \times \mathbb{R} \times W : P(w, x) \le r\},$$

being the image of the epigraph of P under an isomorphism (an interchange of components), is closed and convex. Relation (3.23) means that the projection $: C := p_W(F)$ of F is absorbing, i.e., $0 \in \operatorname{core} C$. The Robinson–Ursescu theorem ensures that F, considered as a multimap from $X \times \mathbb{R}$ to W, is open at every $(\bar{x}, \bar{r}) \in X \times \mathbb{R}$ such that $(\bar{x}, \bar{r}, 0) \in F$; more precisely, there exists some $c > 0$ such that

$$\forall t \in (0, 1], \qquad B(0, tc) \subset F(B((\bar{x}, \bar{r}), t)).$$

Setting $t = 1$, we obtain that for all $w \in B(0, c)$ there exists some $(x, r) \in B((\bar{x}, \bar{r}), 1)$ such that $(x, r, w) \in F$, i.e., $P(w, x) \le r \le s := |\bar{r}| + 1$. Thus p is bounded above by s on $B(0, c)$, hence is continuous at 0 and subdifferentiable at 0. Strong duality ensues. □

A case of special interest is the minimization problem

$$(\mathscr{P}) \qquad \text{minimize} \qquad f(x) + h(g(x)), \quad x \in D,$$

where $f : X \to \mathbb{R}_\infty$, $g : D \to W$, $h : W \to \mathbb{R}_\infty$, with X, W Banach spaces and D a subset of X. When h is the indicator function ι_C of a convex subset C of W, $(\mathscr{P})$ amounts to the minimization of f over $D \cap g^{-1}(C)$. When $W := \mathbb{R}^{k+m}$, $C := \mathbb{R}^k_- \times \{0\}$, one gets the classical mathematical programming problem.

It is usual to associate to $(\mathscr{P})$ the perturbation $P : W \times X \to \mathbb{R}_\infty$ given by

$$P(w, x) := f(x) + h(g(x) + w) + \iota_D(x).$$

When $h = \iota_C$, with $C := \mathbb{R}_-^k \times \{0\}$ as above, one has

$$p(w) = \inf\{f(x) : x \in D, \ g_i(x) + w_i \leq 0, \ g_j(x) + w_j = 0, \ i \in \mathbb{N}_k, \ j \in \mathbb{N}_m\}.$$

The objective function $-p^*$ of $(\mathcal{D})$ can easily be expressed in terms of the data:

$$-p^*(y) = \inf_{w \in W}(p(w) - \langle y, w \rangle) = \inf_{w \in W} \inf_{x \in D}(f(x) + h(g(x) + w) - \langle y, w \rangle)$$

$$= \inf_{x \in D} \inf_{w \in W}(f(x) + h(g(x) + w) - \langle y, g(x) + w \rangle + \langle y, g(x) \rangle)$$

$$= \inf_{x \in D}[f(x) + \langle y, g(x) \rangle + \inf_{z \in W}(h(z) - \langle y, z \rangle)]$$

$$= \inf_{x \in D}[f(x) + \langle y, g(x) \rangle] - h^*(y).$$

When $h = \iota_C$, with C a convex cone in W, h^* is the indicator function of the polar cone C^0, and the function ℓ given by

$$\ell(x, y) := f(x) + \langle y, g(x) \rangle + \iota_D(x) - \iota_{C^0}(y)$$

is called the *Lagrangian*.

To obtain duality results, one may require that P be convex. Such an assumption is akin to convexity requirements on f, g, and h. In fact, duality results can be obtained under the much weaker assumption that the performance function p associated to P is convex. A criterion for this can be given by considering the set

$$E_{f,g} := \{(w, r) \in W \times \mathbb{R} : \exists x \in g^{-1}(w) \cap D, \ f(x) < r\}.$$

The pair (f, g) is said to be *convexlike* if $E_{f,g}$ is convex. We observe that $E_{f,g}$ is the strict epigraph of the performance function q given by

$$q(w) := \inf\{f(x) : x \in D, g(x) = w\}.$$

Therefore (f, g) is convexlike if and only if q is convex.

Lemma 3.58 (Bourass–Giner [163]). *If (f, g) is convexlike and h is convex, then p is convex. In fact, p is the infimal convolution $h \square \tilde{q}$ of h and $\tilde{q}$, where $\tilde{q}(w) := q(-w)$ for $w \in W$. In particular, when h is the indicator function of a convex subset C of W, one has $p(w) = \inf_{v \in C} q(v - w)$.*

Proof. For all $w \in W$ we have

$$(h \square \tilde{q})(w) = \inf_{v \in W}(h(w + v) + q(v)) = \inf_{w \in W} \inf\{h(w + g(x)) + f(x) : x \in D, \ g(x) = v\},$$

and the right-hand side is just $p(w)$. $\square$

Exercises

1. Let C be a closed convex subset of a normed space X, and let σ_C be the support function of C given by $\sigma_C(x^*) := \sup\{\langle x^*, x \rangle : x \in C\}$ for $x^* \in X^*$. Prove that $d_C = (\sigma_C + \iota_{B_{X^*}})^*$.

2. If C is a subset of a normed space X, the *signed distance* to C is the function $d_C^{\pm}$ given by $d_C^{\pm}(x) := d_C(x)$ if $x \in X \setminus C$, $d_C^{\pm}(x) := -d_{X \setminus C}(x)$ for $x \in C$.
(a) Show that $d_C^{\pm}$ is convex when C is convex.
(b) Let C be a closed convex subset of X, let ι_S be the indicator function of the unit sphere in X^*, and let σ_C be the support function of C. Prove that $d_C^{\pm} = (\sigma_C + \iota_S)^*$.
(c) Suppose C is a nonempty open convex subset of X and let $w \in C$. Let $s : x \mapsto 2w - x$. Check that $C \cap s(C) = \varnothing$ and use a separation theorem. Prove the relation

$$\inf\{\|w - x\| : x \in X \setminus C\} = -\sup\{\langle x^*, w \rangle - \sigma_C(x^*) : x^* \in X^* \setminus B(0,1)\}.$$

(d) Show that if the infimum is attained at some $\bar{x} \in X \setminus C$, then there exists some $\bar{x}^* \in X^*$ such that $\bar{x}^* \in S(\bar{x} - w) := \{x^* \in S_{X^*} : \langle x^*, \bar{x} - w \rangle = \|\bar{x} - w\|\}$. (See [173].)

3. Show that the weak duality inequality $\inf(\mathscr{P}) + \inf(\mathscr{P}^*) \geq 0$ stems from the Fenchel inequality $P(0,x) + P^*(w^*,0) \geq \langle 0,x \rangle + \langle w^*,0 \rangle = 0$.

4. Show that the dual problem of the **linear programming** problem

$$(\mathscr{P}) \text{ minimize} \qquad \langle c,x \rangle \text{ under the constraints } \quad x \in \mathbb{R}^n_+, Ax \leq b$$

can be written

$$(\mathscr{D}) \text{ maximize} \qquad \langle b,y \rangle \text{ under the constraints } \quad y \in \mathbb{R}^m_+, A^\mathsf{T} y \leq -c.$$

5. Show that the dual problem of the **quadratic programming** problem

$$(\mathscr{P}) \text{ minimize} \qquad \frac{1}{2}\langle Qx,x \rangle + \langle c,x \rangle \text{ under the constraints } \quad x \in \mathbb{R}^n_+, Ax \leq b$$

when Q is positive definite can be written

$$(\mathscr{D}) \text{ maximize} \quad -\frac{1}{2}\langle Q^{-1}(A^\mathsf{T} y + c), A^\mathsf{T} y + c \rangle - \langle y,b \rangle \text{ under the constraint } y \in \mathbb{R}^m_+.$$

6. (**General Fenchel equality**) Given a family $f_1, \ldots, f_k$ of convex lower semicontinuous functions that are finite and continuous at some point of X, prove that

$$\inf_{x \in X}(f_1(x) + \cdots + f_k(x)) = \inf\{f_1^*(x_1^*) + \cdots + f_k^*(x_k^*) : x_1^* + \cdots + x_k^* = 0\}.$$

7. (**Geometric programming**) Let $G(X)$ be the class of functions on X that are finite sums of functions of the form $x \mapsto c \log(\exp\langle a_1^*, x \rangle + \cdots + \exp\langle a_m^*, x \rangle)$ for some

$a_i^* \in X^*$ ($i \in \mathbb{N}_m$), $c > 0$. Given $g_0, g_1, \ldots, g_k$ in $G(X)$, write down a Lagrangian dual problem for the problem of minimizing $g_0(x)$ under the constraints $g_i(x) \leq 0$ ($i \in \mathbb{N}_k$) and give a duality result.

3.4.4 Duality and Subdifferentiability Results

Consequences for subdifferential calculus will be the final aims of this section. The following theorem, containing both a sum rule and a composition rule, is a step in such a direction. It generalizes Theorems 3.39 and 3.40 (Exercise 1). Again, for a function $h : Z \to \mathbb{R}$ and $r \in \mathbb{R}$, we set $\{h \leq s\} := h^{-1}((-\infty, s])$.

Theorem 3.59 (Fenchel–Rockafellar). *Let X, Y be normed spaces, let $A : X \to Y$ be a continuous linear map, and let $f : X \to \mathbb{R}_\infty$, $g : Y \to \mathbb{R}_\infty$ be convex functions such that there exist $r > 0$, $s \in \mathbb{R}_+$ for which*

$$rB_Y \subset A(\{f \leq s\} \cap sB_X) - \{g \leq s\}. \tag{3.24}$$

Then for all $x^* \in X^*$ one has

$$(f + g \circ A)^*(x^*) = \min_{y^* \in Y^*} (f^*(x^* - A^\mathsf{T} y^*) + g^*(y^*)). \tag{3.25}$$

Moreover, for every $x \in \operatorname{dom} f \cap A^{-1}(\operatorname{dom} g)$ one has

$$\partial(f + g \circ A)(x) = \partial f(x) + A^\mathsf{T}(\partial g(Ax)). \tag{3.26}$$

Proof. Let $W := Y$, let $x^* \in X^*$, and let $P : W \times X \to \mathbb{R}_\infty$ be given by

$$P(w, x) := f(x) - \langle x^*, x \rangle + g(Ax + w).$$

For all $w \in rB_W$, (3.24) yields $x_w \in \{f \leq s\} \cap sB_X$ and $y_w \in \{g \leq s\}$ such that $w = y_w - Ax_w$. Then the performance function p given by $p(w) := \inf_{x \in X} P(w, x)$ satisfies

$$p(w) \leq P(w, x_w) = f(x_w) - \langle x^*, x_w \rangle + g(y_w) \leq 2s + s\|x^*\|,$$

and strong duality holds. Now, for $y^* \in Y^*$, setting $y := Ax + w$, one has

$$P^*(y^*, 0) = \sup_{(w,x) \in W \times X} (\langle y^*, w \rangle - \langle x^*, x \rangle - f(x) - g(Ax + w))$$

$$= \sup_{x \in X} (\langle x^*, x \rangle - \langle y^*, Ax \rangle - f(x)) + \sup_{y \in W} (\langle y^*, y \rangle - g(y))$$

$$= f^*(x^* - A^\mathsf{T} y^*) + g^*(y^*),$$

so that (3.25) follows from the relation

$$(f + g \circ A)^*(x^*) = - \inf_{x \in X} P(0,x) = - \inf(\mathscr{P})$$

$$= \min(\mathscr{P}^*) = \min_{y^* \in Y^*} P^*(y^*, 0) = \min_{y^* \in Y^*} (f^*(x^* - A^\mathsf{T} y^*) + g^*(y^*)).$$

Now if $x^* \in \partial h(x)$, with $h := f + g \circ A$, one has $h(x) + h^*(x^*) - \langle x^*, x \rangle = 0$, and there exists some $y^* \in Y^*$ such that $h^*(x^*) = f^*(x^* - A^\mathsf{T} y^*) + g^*(y^*)$, whence

$$0 = (f(x) + f^*(x^* - A^\mathsf{T} y^*) - \langle x^* - A^\mathsf{T} y^*, x \rangle) + (g(Ax) + g^*(y^*) - \langle A^\mathsf{T} y^*, x \rangle).$$

Since both terms in parentheses are nonnegative, they are null. Thus $x^* - A^\mathsf{T} y^* \in \partial f(x)$, $A^\mathsf{T} y^* \in \partial g(Ax)$, and the nontrivial inclusion of equality (3.26) holds. □

Theorem 3.60 (Attouch–Brézis). *Let X, Y be Banach spaces, let $A : X \to Y$ be a continuous linear map, and let $f : X \to \mathbb{R}_\infty$, $g : Y \to \mathbb{R}_\infty$ be closed proper convex functions such that the following cone is closed and symmetric (i.e., $Z = -Z = \mathrm{cl}Z$):*

$$Z := \mathbb{R}_+ (A(\mathrm{dom}\, f) - \mathrm{dom}\, g).$$

Then the conclusions of the Fenchel–Rockafellar theorem hold.

Note that the assumption on Z means that Z is a closed vector subspace. It is obviously satisfied when the simple algebraic condition that follows is fulfilled:

$$Y = \mathbb{R}_+ (A(\mathrm{dom}\, f) - \mathrm{dom}\, g).$$

Proof. Taking $W := Y$, we define the perturbation function P as in the preceding proof. Then for $x \in X$, we have $w \in \mathrm{dom}\, P(\cdot, x)$ if and only if $x \in \mathrm{dom}\, f$ and $w \in \mathrm{dom}\, g - Ax$, so that the cone generated by the union over x of $\mathrm{dom}\, P(\cdot, x)$ is $\mathbb{R}_+ (\mathrm{dom}\, g - A(\mathrm{dom}\, f))$, the closed linear subspace Z. Then Corollary 3.57 ensures that strong duality holds, and the proof can be finished like the preceding one. □

Exercises

1. Show that Theorem 3.59 generalizes Theorems 3.39 and 3.40.

2. Let P be a closed convex cone of a Banach space X, let Q be its polar cone, and let B be the closed unit ball. Prove that the distance function to Q and the support function to $P \cap B$ are equal.

3. Show that if C is a nonempty closed convex subset of X containing the origin, the conjugate μ_C^* of the gauge function μ_C of C is given by $\mu_C^* = \iota_{C^0}$, where $C^0 := \{x^* \in X^* : \langle x^*, x \rangle \leq 1\}$ is the polar set of C and μ_C is given by $\mu_C(x) := \inf\{r \in \mathbb{P} : x \in rC\}$.

4. Let $A : X \to W$ be a continuous linear operator. Suppose W is ordered by a closed convex cone W_+ and $Y := W^*$ is ordered by the cone $Y_+ = -W_+^0$. Let $b \in W$ and let $f : X \to \mathbb{R}_\infty$.

(a) Find the dual problem of the mathematical programming problem

$$(\mathscr{P}) \qquad \text{minimize} f(x), \qquad x \in X, \ Ax \le b,$$

using the perturbation function P given by $P(w,x) := f(x) + \iota_{F(w)}(x)$, where $F(w) := \{x \in X : Ax \le b - w\}$.

(b) Show that the function $L : X \times Y \to \overline{\mathbb{R}}$ given by $L(x,y) := -(P(\cdot,x))^*(y)$ is a Lagrangian of $(\mathscr{P})$ in the sense that $\sup\{L(x,y) : y \in Y\} = f(x) + \iota_{F(0)}(x)$.

(c) (**Linear programming**) Give an explicit form of the dual problem when $X = \mathbb{R}^n$, $W := \mathbb{R}^m$, $W_+ = \mathbb{R}^m_+$, and f is a linear form on X.

(d) (**Quadratic programming**) Give an explicit form of the dual problem when $X = \mathbb{R}^n$, $W := \mathbb{R}^m$, $W_+ = \mathbb{R}^m_+$, and f is a quadratic form: $f(x) = (1/2)\langle Qx,x\rangle + \langle q,x\rangle$, with Q positive definite. Generalize to the case in which Q is positive semidefinite. (See [692].)

5. Given a function $f : X \to \mathbb{R}_\infty$, check that

$$\text{epi} f^* \times \{-1\} = (S(Q))^0 \cap (X^* \times \mathbb{R} \times \{-1\}),$$

where $Q := \mathbb{R}_+(\text{epi} f \times \{-1\})$ and S is the map $(x,r,s) \mapsto (x,s,r)$, a linear isometry.

6. Let X and Y be Banach spaces, let $f,g : X \times Y \to \mathbb{R}_\infty$ be proper, convex, lower semicontinuous functions, and let $p,q : X \to \mathbb{R}_\infty$ be given by $p(x) := \inf\{f(x,y) : y \in Y\}$, $q(x) := \inf\{g(x,y) : y \in Y\}$. Suppose that $L := \mathbb{R}_+(\text{dom}\, p - \text{dom}\, q)$ is a closed vector subspace. Prove that for $h : X \times Y \to \mathbb{R}_\infty$ given as follows, its conjugate h^* has a similar form:

$$h(x,y) := (f(x,\cdot)\square g(x,\cdot))(y) := \inf\{f(x,u) + g(x,v) : u + v = y\},$$

$$h^*(x^*,y^*) := (f^*(\cdot,y^*)\square g^*(\cdot,y^*))(x^*) := \inf\{f^*(u^*,y^*) + g(v^*,y^*) : u^* + v^* = x^*\}.$$

(See [894, Theorem 4.2].)

7. Given a Banach space X and a convex function f defined on it, show that the Fenchel conjugate f^* of f is Gâteaux differentiable at some x^* iff any sequence (x_n) such that $(f(x_n) - x^*(x_n)) \to \inf(f - x^*)$ converges.

3.5 General Convex Calculus Rules

While the subdifferential calculus rules we have seen suffice for most uses, it is of interest to look for calculus rules that do not require additional assumptions. Such rules exist, but involve some fuzziness. This approximative character is typical of the calculus rules of nonsmooth analysis. Thus this section will prepare for similar developments in the nonconvex case that will be dealt with later on.

In the sequel, given a function f on a normed space X and a net $(x_i)_{i \in I}$ of X with limit $\bar{x}$, it will be convenient to write $(x_i) \to_f \bar{x}$ instead of $(x_i) \to \bar{x}$ and $(f(x_i)) \to f(\bar{x})$. This means that for every $\varepsilon > 0$ one can find some $k \in I$ such that for all $i \ge k$ one has $x_i \in B_f(\bar{x},\varepsilon) := B(\bar{x},\varepsilon,f) := \{x \in B(\bar{x},\varepsilon) : |f(x) - f(\bar{x})| < \varepsilon\}$.

For a net $(x_i^*)_{i \in I}$ weak* converging to some $\bar{x}^*$ in X^* we write $(x_i^*)_{i \in I} \overset{*}{\to} \bar{x}^*$. We first note a stability property of the subdifferential to a convex function.

Proposition 3.61. *Let f be a convex function on a normed space X, let $x \in \mathrm{dom}\, f$, $(x_i)_{i \in I} \to_f x$, and let $x_i^* \in \partial f(x_i)$ be such that $(x_i^*)_{i \in I} \overset{*}{\to} x^*$ and $(\langle x_i^*, x_i - x \rangle)_{i \in I} \to 0$. Then $x^* \in \partial f(x)$.*

Note that the assumption $(\langle x_i^*, x_i - x \rangle)_{i \in I} \to 0$ is satisfied when $(x_i^*)_{i \in I}$ is bounded.

Proof. It suffices to observe that for all $w \in X$ one has

$$\langle x^*, w - x \rangle = \lim_i \langle x_i^*, w - x \rangle = \lim_i \langle x_i^*, w - x_i \rangle \le \lim_i (f(w) - f(x_i)) = f(w) - f(x).$$

$\square$

Taking for f the indicator function of a convex set C, we get the following consequence, which can be given an easy direct proof.

Corollary 3.62. *Let C be a convex subset of a normed space X, let $(x_i)_{i \in I}$ be a net in C with limit $x \in C$, and let $x_i^* \in N(C, x_i)$ be such that $(x_i^*)_{i \in I}$ weak* converges to some x^* and $(\langle x_i^*, x_i - x \rangle)_{i \in I} \to 0$. Then $x^* \in N(C, x)$.*

3.5.1 Fuzzy Calculus Rules in Convex Analysis

Now let us turn to calculus rules. Before giving a fuzzy rule for a composite function, let us start with a characterization of the normal cone to the inverse image of a convex set. For the sake of simplicity, we first restrict our attention to the case of reflexive Banach spaces. In the general case one has to replace sequences by nets, and strong convergence in dual spaces has to be replaced by weak* convergence.

Theorem 3.63. *Let X and Y be reflexive Banach spaces, let $A : X \to Y$ be linear and continuous, and let $C := A^{-1}(D)$, where D is a closed convex subset of Y. Let $\bar{x} \in C$, $\bar{y} := A(\bar{x})$. Then $\bar{x}^* \in N(C, \bar{x})$ if and only if there exist sequences $(x_n) \to \bar{x}$, $(y_n) \to \bar{y} := A\bar{x}$ in D, (y_n^*) in Y^* such that $y_n^* \in N(D, y_n)$ for all n, and*

$$(\|A^{\mathsf{T}} y_n^* - \bar{x}^*\|)_n \to 0, \tag{3.27}$$

$$(\|y_n^*\| \cdot \|y_n - Ax_n\|)_n \to 0. \tag{3.28}$$

Relation (3.28) can be considered as an additional information supplementing (3.27) that is a fuzzy version of the equality $\bar{x}^* = A^{\mathsf{T}} \bar{y}^*$ for some $\bar{y}^* \in N(D, \bar{y})$.

Proof. Sufficiency: given (x_n), (y_n), (y_n^*) as in the statement, for all $x \in C$ we have

$$\langle \bar{x}^*, x - \bar{x} \rangle - \langle y_n^*, Ax - Ax_n \rangle = \langle \bar{x}^* - A^{\mathsf{T}} y_n^*, x \rangle + \langle \bar{x}^*, x_n - \bar{x} \rangle + \langle A^{\mathsf{T}} y_n^* - \bar{x}^*, x_n \rangle \to 0;$$

since $(\langle y_n^*, y_n - Ax_n \rangle) \to 0$, we get $\langle \bar{x}^*, x - \bar{x} \rangle = \lim \langle y_n^*, Ax - y_n \rangle \le 0$: $\bar{x}^* \in N(C, \bar{x})$.

Now let us prove the necessary condition. Let $\bar{x}^* \in N(C,\bar{x})$. Without loss of generality we may suppose $\bar{x} = 0$. Let us introduce the penalized decoupling function $p_n : X \times Y \to \mathbb{R}_\infty$ given by

$$p_n(x,y) := \iota_D(y) - \langle \bar{x}^*, x \rangle + n \|Ax - y\|^2 + \|x\|^2.$$

Noting that p_n is weakly lower semicontinuous, let (x_n, y_n) be a minimizer of p_n on $B_{X \times Y}$ that is weakly compact. The relations $p_n(x_n, y_n) \le p_n(0,0) = 0$ yield $y_n \in D$,

$$- \langle \bar{x}^*, x_n \rangle + n \|Ax_n - y_n\|^2 + \|x_n\|^2 \le 0. \tag{3.29}$$

Let (x_∞, y_∞) be the limit of a weakly convergent subsequence of $((x_n, y_n))_n$. Since $y_n \in D$ for all n and D is weakly closed, we have $y_\infty \in D$. Now $(n \|Ax_n - y_n\|^2)$ is bounded, so that $(Ax_n - y_n) \to 0$; hence $Ax_\infty = y_\infty$ and $x_\infty \in C$. Since $\bar{x}^* \in N(C,\bar{x})$ with $\bar{x} = 0$, (3.29) yields

$$- \langle \bar{x}^*, x_\infty \rangle + \|x_\infty\|^2 \le \liminf_n - \langle \bar{x}^*, x_n \rangle + n \|Ax_n - y_n\|^2 + \|x_n\|^2 \le 0 \le - \langle \bar{x}^*, x_\infty \rangle.$$

Thus $x_\infty = 0$ and the whole sequence (x_n) weakly converges to 0. Using again (3.29), we get that $(\|x_n\|) \to 0$. Thus (y_n) converges to $\lim_n Ax_n = 0$. Then for n large enough, (x_n, y_n) is in the interior of $B_{X \times Y}$, and using the rules of convex analysis, the optimality condition $(0,0) \in \partial p_n(x_n, y_n)$ can be written, for some $y_n^* \in N(D, y_n)$, $x_n^* \in \partial \|\cdot\|^2 (x_n)$, and $z_n^* \in \partial \|\cdot\|^2 (Ax_n - y_n)$, as

$$(0,0) = (x_n^* - \bar{x}^* + nz_n^* \circ A, y_n^* - nz_n^*).$$

This relation yields $\bar{x}^* = x_n^* + nz_n^* \circ A$, $y_n^* = nz_n^*$. Thus $A^\mathsf{T} y_n^* = nz_n^* \circ A = \bar{x}^* - x_n^*$. Moreover, the properties of the duality mapping $J := \frac{1}{2}\partial \|\cdot\|^2$ yield $\|x_n^*\| = 2\|x_n\| \to 0$, so that $(\|A^\mathsf{T} y_n^* - \bar{x}^*\|) = (\|x_n^*\|) \to 0$. Similarly, the properties of the duality mapping of Y yield $\|y_n^*\| = \|nz_n^*\| = 2n \|Ax_n - y_n\|$, so that $(\|y_n^*\| \|Ax_n - y_n\|) = (2n \|Ax_n - y_n\|^2) \to 0$, as (3.29) shows, since $(x_n) \to 0$. $\qquad\square$

The following result and the next one applied to indicator functions show that one can drop reflexivity in Theorem 3.63. Conversely, using epigraphs, one can deduce them from the nonreflexive version of Theorem 3.63.

Theorem 3.64. *Let X and Y be Banach spaces, X being reflexive, let $A \in L(X,Y)$, and let $f := g \circ A$, where $g : Y \to \mathbb{R}_\infty$ is lower semicontinuous and convex. Let $\bar{x} \in \mathrm{dom}\, f$, $\bar{x}^* \in X^*$. Then $\bar{x}^* \in \partial f(\bar{x})$ if and only if there exist sequences $(x_n) \to \bar{x}$ in X, $(y_n) \to_g \bar{y} := A\bar{x}$ in Y, (y_n^*) in Y^* such that $y_n^* \in \partial g(y_n)$ for all n and*

$$(\|A^\mathsf{T} y_n^* - \bar{x}^*\|)_n \to 0, \tag{3.30}$$

$$(\|y_n^*\| \cdot \|y_n - Ax_n\|)_n \to 0. \tag{3.31}$$

Assertion (3.31) can be viewed an additional information that somewhat compensates the fuzziness of (3.30), which replaces the missing relation $A^* \bar{y}^* = \bar{x}^*$ for some $\bar{y}^* \in \partial g(\bar{y})$. When (y_n^*) is bounded, this additional information is superfluous; this happens when g is continuous at $\bar{y}$. It can be noted that together with (3.30), condition (3.31) implies the condition

$$(\langle y_n^*, y_n - \bar{y}\rangle)_n \to 0. \tag{3.32}$$

In fact, $\langle y_n^*, y_n - \bar{y}\rangle = \langle y_n^*, y_n - Ax_n\rangle + \langle A^\mathsf{T}y_n^*, x_n - \bar{x}\rangle$, and each term converges to 0.

Proof. Let us first observe that when $\bar{x}^*$ satisfies the above conditions, it belongs to $\partial f(\bar{x})$, since for all $x \in X$, by (3.32) we have

$$f(x) - f(\bar{x}) = g(A(x)) - g(\bar{y}) = \lim_n (g(A(x)) - g(y_n))$$

$$\geq \liminf_n \langle y_n^*, A(x) - y_n\rangle = \liminf_n \langle y_n^*, A(x) - \bar{y}\rangle$$

$$= \liminf_n \langle A^\mathsf{T}y_n^*, x - \bar{x}\rangle = \langle \bar{x}^*, x - \bar{x}\rangle.$$

Now let us prove the converse when X and Y are both reflexive. Let C, D be the epigraphs of f and g respectively and let $B := A \times I_\mathbb{R} : X \times \mathbb{R} \to Y \times \mathbb{R}$, so that $C = B^{-1}(D)$. Given $\bar{x}^* \in \partial f(\bar{x})$, we have $(\bar{x}^*, -1) \in N(C, \bar{x}_f)$, where $\bar{x}_f := (\bar{x}, f(\bar{x}))$. Theorem 3.63 yields sequences $((x_n, r_n)) \to \bar{x}_f$, $((y_n, s_n)) \to (\bar{y}, g(\bar{y}))$ in D, $((z_n^*, s_n^*))$ such that $((z_n^*, -s_n^*)) \in N(D, (y_n, s_n))$ for all n and $\lim_n \|((A^\mathsf{T}z_n^*, -s_n^*) - (\bar{x}^*, -1)\| = 0$, $\lim_n \|(z_n^*, -s_n^*)\| \cdot \|(y_n, s_n) - (Ax_n, r_n)\| = 0$. These last relations give $(s_n^*) \to 1$, $(\|A^\mathsf{T}y_n^* - \bar{x}^*\|)_n \to 0$ for $y_n^* := z_n^*/s_n^*$, $(\|y_n^*\| \cdot \|y_n - Ax_n\|)_n \to 0$. We easily see that $(y_n^*, -1) \in N(D, (y_n, g(y_n)))$, i.e., $y_n^* \in \partial g(y_n)$. Now, since g is lower semicontinuous and $g(y_n) \leq s_n$ with $(s_n) \to g(\bar{y})$, we get $(g(y_n)) \to g(\bar{y})$.

Proof of the converse in the case Y is arbitrary. Without loss of generality, we suppose $\bar{x} = 0$. Let $\bar{x}^* \in \partial f(\bar{x})$ and let μ be a modulus of lower semicontinuity of $(x, y) \mapsto g(y) - \langle \bar{x}^*, x\rangle$ at $(\bar{x}, A\bar{x}) = (0, 0)$, i.e., a modulus μ such that $g(y) - \langle \bar{x}^*, x\rangle \geq g(0) - \mu(\|(x, y)\|)$ for $(x, y) \in X \times Y$, and let $\rho > 0$ be such that $\mu(r) \leq 1$ for $r \in [0, \rho]$. For $n \geq 1$, let us introduce the penalized function $p_n : \rho B_{X \times Y} \to \mathbb{R}_\infty$ given by

$$p_n(x, y) := g(y) - \langle \bar{x}^*, x\rangle + n\|Ax - y\|^2 + \|x\|^2.$$

The function p_n being bounded below and jointly lower semicontinuous when $\rho B_X \times \rho B_Y$ is endowed with the product topology of the weak topology on ρB_X with the strong topology on ρB_Y, and ρB_X being compact, given a sequence (t_n) in $(0, 1)$ with limit 0, Corollary 1.90 yields a pair $(x_n, y_n) \in \rho B_X \times \rho B_Y$ such that

$$\forall (x, y) \in \rho B_X \times \rho B_Y, \quad p_n(x_n, y_n) \leq p_n(x, y) + t_n\|y - y_n\|. \tag{3.33}$$

Then the relations

$$p_n(x_n, y_n) \leq p_n(0, 0) + t_n\|y_n\| = g(0) + t_n\|y_n\|$$

$$\leq g(y_n) - \langle \bar{x}^*, x_n\rangle + \mu(\|(x_n, y_n)\|) + t_n\|y_n\|$$

yield

$$n \|y_n - Ax_n\|^2 + \|x_n\|^2 \le \mu(\|(x_n, y_n)\|) + t_n \rho \le 1 + \rho, \tag{3.34}$$

so that $(y_n - Ax_n) \to 0$ and $\|x_n\|^2 \le 1 + \rho$. Let us show that $(x_n) \to 0$; this will also imply $(y_n) \to 0$. Let us assume, to the contrary, that for some $\alpha > 0$ and some infinite subset N of $\mathbb{N}$ we have $\|x_n\| \ge \alpha$ for all $n \in N$. Using the reflexivity of X we get an infinite subset P of N such that $(x_n)_{n \in P} \to x_\infty$ weakly for some $x_\infty \in (1+\rho)B_X$. Since A is weakly continuous and $(\|y_n - Ax_n\|) \to 0$, it follows that $((x_n, y_n))_{n \in P}$ weakly converges to (x_∞, y_∞), with $y_\infty = A(x_\infty)$. Taking limits in the relations $p_n(x_n, y_n) \le p_n(0,0) + t_n \|y_n\| = g(0) + t_n \|y_n\| = f(0) + t_n \|y_n\|$ and using the fact that $\overline{x}^* \in \partial f(\overline{x})$ and that g is weakly lower semicontinuous, we get

$$g(y_\infty) - \langle \overline{x}^*, x_\infty \rangle + \alpha^2 \le \liminf_{n \in P} \left(g(y_n) - \langle \overline{x}^*, x_n \rangle + n \|Ax_n - y_n\|^2 + \|x_n\|^2 \right) \le f(0)$$

$$\le f(x_\infty) - \langle \overline{x}^*, x_\infty \rangle = g(y_\infty) - \langle \overline{x}^*, x_\infty \rangle,$$

a contradiction. Hence $(x_n) \to 0$, and since $(y_n - Ax_n) \to 0$, we get $(y_n) \to 0$. Then $\limsup_n g(y_n) \le \limsup_n p(x_n, y_n) \le g(0)$, and since g is lower semicontinuous, $(g(y_n)) \to g(0)$. Moreover, by (3.33), for n large enough, $(x_n, y_n) \in \rho B_X \times \rho B_Y$ is a local minimizer of the function

$$q_n : (x,y) \mapsto p_n(x,y) + t_n \|y - y_n\| = g(y) - \langle \overline{x}^*, x \rangle + n \|Ax - y\|^2 + \|x\|^2 + t_n \|y - y_n\|.$$

It follows from Theorem 3.39 that we can find $y_n^* \in \partial g(y_n)$, $x_n^* \in \partial \|\cdot\|^2 (x_n)$, $w_n^* \in B_{Y^*}$, and $z_n^* \in \partial \|\cdot\|^2 (Ax_n - y_n)$ such that

$$(0,0) = (x_n^* - \overline{x}^* + nz_n^* \circ A, y_n^* - nz_n^* + t_n w_n^*),$$

or $\overline{x}^* = x_n^* + nz_n^* \circ A$, $y_n^* = nz_n^* - t_n w_n^*$. Thus $A^{\mathsf{T}} y_n^* = y_n^* \circ A = \overline{x}^* - x_n^* - t_n A^{\mathsf{T}} w_n^*$. Again, the properties of the duality mapping yield $\|x_n^*\| = 2\|x_n\| \to 0$, so that $(\|A^{\mathsf{T}} y_n^* - \overline{x}^*\|) = (\|x_n^* + t_n A^{\mathsf{T}} w_n^*\|) \to 0$. Similarly, the properties of the duality mapping of Y yield $\|y_n^*\| = \|nz_n^* - t_n w_n^*\| \le 2n\|Ax_n - y_n\| + t_n$, and setting $r_n = \|(x_n, y_n)\|$, $s_n = \mu(r_n) + t_n \rho$, it follows from inequality (3.34) that

$$\|y_n^*\| \|Ax_n - y_n\| \le 2n \|Ax_n - y_n\|^2 + t_n \|Ax_n - y_n\| \le 2s_n + t_n \|Ax_n - y_n\| \to 0.$$

$\square$

In arbitrary Banach spaces one can get rules that are similar to those we proved in reflexive Banach spaces; however, since closed balls are no longer weakly compact, in order to use compactness arguments, one has to take restrictions to finite-dimensional subspaces. Such a process brings nets into the picture (the family of finite-dimensional subspaces being directed, but not countable in general).

Theorem 3.65. *Let X, Y be Banach spaces, $A : X \to Y$ a continuous linear map, and let $f := g \circ A$, where $g : Y \to \mathbb{R}_\infty$ is lower semicontinuous and convex. Let $\overline{x} \in$*

dom f, $\bar{x}^* \in X^*$. *Then $\bar{x}^* \in \partial f(\bar{x})$ if and only if there exist nets* $(x_i)_{i \in I} \to \bar{x}$ *in X,* $(y_i)_{i \in I} \to_g \bar{y} := A\bar{x}$ *in Y,* $(y_i^*)_{i \in I}$ *in Y^* such that $y_i^* \in \partial g(y_i)$ for all $i \in I$ and*

$$(A^\mathsf{T} y_i^*)_{i \in I} \overset{*}{\to} \bar{x}^*, \tag{3.35}$$

$$(\|y_i^*\| \cdot \|y_i - Ax_i\|)_{i \in I} \to 0, \tag{3.36}$$

$$(\langle y_i^*, y_i - \bar{y} \rangle)_{i \in I} \to 0. \tag{3.37}$$

More precisely, if $\bar{x}^* \in \partial f(\bar{x})$, then for every finite-dimensional subspace W of X containing $\bar{x}$ there exist sequences $(x_n) \to \bar{x}$ in W, $(y_n) \to_g \bar{y} := A\bar{x}$ in Y, (y_n^*) in Y^* such that $y_n^* \in \partial g(y_n)$ for each n and

$$(\|A^\mathsf{T} y_n^* |_W - \bar{x}^* |_W \|)_n \to 0, \tag{3.38}$$

$$(\|y_n^*\| \cdot \|y_n - Ax_n\|)_n \to 0. \tag{3.39}$$

If $(y_i^*)_{i \in I}$ has a bounded subnet $(y_j^*)_{j \in J}$, for every weak*-cluster point $\bar{y}^*$ of $(y_j^*)_{j \in J}$ one has $\bar{x}^* = A^\mathsf{T}(\bar{y}^*)$ and $\bar{y}^* \in \partial g(\bar{x})$, but in general the net $(y_i^*)_{i \in I}$ is unbounded.

Proof. The proof of the sufficient condition is similar to the one given above. The second assertion is a simple application of Theorem 3.64, denoting by $B :$ $W \to X$ the canonical inclusion and observing that $f \circ B = g \circ (A \circ B)$, that $\bar{x}^* |_W \in \partial (f \circ B)(\bar{x})$, and that for every $x^* \in X^*$, $y^* \in Y^*$ one has $x^* |_W = B^\mathsf{T} x^*$, $A^\mathsf{T} y^* |_W = (A \circ B)^\mathsf{T}(y^*)$.

Now let us make clear why the first assertion stems from the second one. We denote by $\mathbb{P}$ the set of positive numbers, by $\mathscr{W}$ the set of finite-dimensional linear subspaces of X ordered by inclusion, and we provide the product $I := \mathscr{W} \times \mathbb{P}$ with the order $(W, r) \le (W', r')$ if $W \subset W'$, $r' \le r$. Thus I is directed. The second assertion (with the axiom of choice) makes it possible to pick for every $i := (W, r) \in I$ some $(x_i, y_i, y_i^*) \in X \times Y \times Y^*$ such that $\|x_i - \bar{x}\| < r$, $\|y_i - \bar{y}\| < r$, $|g(y_i) - g(\bar{y})| < r$, $y_i^* \in \partial g(y_i)$, $\|A^\mathsf{T} y_i^* |_W - \bar{x}^* |_W \| < r$, $\|y_i^*\| \cdot \|y_i - Ax_i\| < r$, $|\langle y_i^*, y_i - \bar{y} \rangle| < r$. These choices provide the required nets. In fact, given $\varepsilon > 0$ and a finite set $F := \{a_1, \dots, a_k\}$ of unit vectors in X, denoting by W_F the linear space generated by F and setting $i_\varepsilon := (W_F, \varepsilon)$, for $i \ge i_\varepsilon$, we have $|\langle A^\mathsf{T} y_i^* - \bar{x}^*, a_j \rangle| < \varepsilon$ for $i \ge i_\varepsilon$, $j = 1, \dots, k$, so that $(A^\mathsf{T} y_i^*)_{i \in I} \overset{*}{\to} \bar{x}^*$; we also have $\|y_i^*\| \cdot \|y_i - Ax_i\| < \varepsilon$ for $i \ge i_\varepsilon$. $\square$

A sum rule can be deduced from the chain rule. It holds for every finite family $(f_j)_{1 \le j \le k}$ of lower semicontinuous convex functions on X, with a similar proof, replacing X^2 by X^k and (f_1, f_2) by (f_j).

Theorem 3.66. *Let f_1, f_2 be lower semicontinuous proper convex functions on a Banach space X and let $f := f_1 + f_2$ be finite at $\bar{x} \in X$. Let $\bar{x}^* \in X^*$. Then $\bar{x}^* \in \partial f(\bar{x})$ if and only if for every finite-dimensional linear subspace W of X and for $j = 1, 2$, there exist sequences $(x_{j,n})_n \to \bar{x}$ in W, $(x_{j,n}^*)_n$ in X^* such that $x_{j,n}^* \in \partial f_j(x_{j,n})$ for all j, n, $(f_j(x_{j,n}))_n \to f_j(\bar{x})$ for $j = 1, 2$ and*

$$\left(\left\|\left\|x_{1,n}^*\left|w\right.+x_{2,n}^*\left|w\right.-\bar{x}^*\left|w\right.\right\|\right\|\right)_n \to 0, \tag{3.40}$$

$$\left(\left\|x_{1,n}-x_{2,n}\right\|\cdot\left(\left\|x_{1,n}^*\right\|+\left\|x_{2,n}^*\right\|\right)\right)_n \to 0. \tag{3.41}$$

Thus $\bar{x}^* \in \partial f(\bar{x})$ if and only if for $j = 1,2$, there exist nets $(x_{j,i})_{i\in I} \to_{f_j} \bar{x}$, $(x_{j,i}^*)_{i\in I}$ such that $x_{j,i}^* \in \partial f_j(x_{j,i})$ for all $i \in I$, $j \in \{1,2\}$ and

$$(x_{1,i}^* + x_{2,i}^*)_{i\in I} \xrightarrow{*} \bar{x}^*, \tag{3.42}$$

$$\left(\left\|x_{1,i}-x_{2,i}\right\|\cdot\left(\left\|x_{1,i}^*\right\|+\left\|x_{2,i}^*\right\|\right)\right)_{i\in I} \to 0. \tag{3.43}$$

Proof. The sufficient condition is a simple verification. We first note that

$$\langle x_{1,i}^*,x_{1,i}-\bar{x}\rangle + \langle x_{2,i}^*,x_{2,i}-\bar{x}\rangle = \langle x_{1,i}^*+x_{2,i}^*,x_{1,i}-\bar{x}\rangle + \langle x_{2,i}^*,x_{2,i}-x_{1,i}\rangle,$$

so that (3.42) and (3.43) imply

$$\left(\langle x_{1,i}^*,x_{1,i}-\bar{x}\rangle + \langle x_{2,i}^*,x_{2,i}-\bar{x}\rangle\right)_i \to 0. \tag{3.44}$$

Given $\bar{x}^* \in X^*$ satisfying the above conditions, we have $(f_1(x_{1,i}) + f_2(x_{2,i}))_i \to f(\bar{x})$; hence for all $x \in X$, the inequalities $f_j(x) - f_j(x_{j,i}) \ge \langle x_{j,i}^*, x - x_{j,i}\rangle$ for $i \in I$, $j = 1,2$, (3.42), and (3.43) imply $\bar{x}^* \in \partial f(\bar{x})$, since

$$f(x) - f(\bar{x}) \ge \liminf_i \left(\langle x_{1,i}^*, x - x_{1,i}\rangle + \langle x_{2,i}^*, x - x_{2,i}\rangle\right)$$

$$\ge \liminf_i \left(\langle x_{1,i}^* + x_{2,i}^*, x - \bar{x}\rangle\right) = \langle \bar{x}^*, x - \bar{x}\rangle.$$

Now let us prove the necessary condition. Let $Y = X^2$ endowed with the supremum norm and let $g : Y \to \mathbb{R}_\infty$ be given by $g(x_1,x_2) := f_1(x_1) + f_2(x_2)$. For $A : X \to Y$ given by $Ax := (x,x)$, we have $f = g \circ A$. Applying Theorem 3.65, we get nets (or sequences when X is reflexive) $(x_i)_i \to \bar{x}$ in X, $(y_i)_i \xrightarrow{g} \bar{y} := A\bar{x}$ in Y, $(y_i^*)_i$ in Y^* with $(A^\mathsf{T}y_i^*) \xrightarrow{*} \bar{x}^*$, $(\|y_i^*\|\cdot\|y_i - Ax_i\|) \to 0$ such that $y_i^* \in \partial g(y_i)$ for all i. Setting $y_i := (x_{1,i},x_{2,i})$, $y_i^* := (x_{1,i}^*,x_{2,i}^*)$, we easily see that $y_i^* \in \partial g(y_i)$ means that $x_{j,i}^* \in \partial f_j(x_{j,i})$ for $j = 1,2$ and all i. Since $\|x_{1,i}-x_{2,i}\| \le \|x_{1,i}-x_i\| + \|x_i-x_{2,i}\| \le 2\|y_i - Ax_i\|$ and $A^\mathsf{T}y_i^* = x_{1,i}^* + x_{2,i}^*$, we get $(x_{1,i}^* + x_{2,i}^*)_{i\in I} \xrightarrow{*} \bar{x}^*$ and

$$\left(\|x_{1,i}-x_{2,i}\|\right)\cdot\left(\|x_{1,i}^*\| + \|x_{2,i}^*\|\right) \le 2\|y_i - Ax_i\|\cdot\|y_i^*\| \to 0.$$

Moreover, since for each $j = 1,2$, f_j is lower semicontinuous at $\bar{x}$, we have $\liminf_i f_j(x_{j,i}) \ge f_j(\bar{x})$ and

$$\limsup_i f_1(x_{1,i}) = \limsup_i \left(g(y_i) - f_2(x_{2,i})\right) \le g(\bar{y}) - \liminf_i f_2(x_{2,i})$$

$$\le g(A\bar{x}) - f_2(\bar{x}) = f_1(\bar{x}),$$

so that $(f_1(x_{1,i}))_i \to f_1(\bar{x})$, and similarly, $(f_2(x_{2,i}))_i \to f_2(\bar{x})$, as announced. □

Remark. Since $(f_j(x_{j,i}))_i \to f_j(\bar{x})$ and $x^*_{j,i} \in \partial f_j(x_{j,i})$ for each $j \in \{1,2\}$ and all i, we deduce from (3.44) that $(\langle x^*_{j,i}, x_{j,i} - \bar{x}\rangle) \to 0$ for each $j \in \{1,2\}$. □

Remark. The fuzzy sum rule is in fact equivalent to the fuzzy composition rule. Let us show that we can deduce the latter from the former. Given $f := g \circ A$, let $F : X \times Y \to \mathbb{R}_\infty$ be given by $F(x,y) := g(y) + \iota_{G(A)}(x,y)$, where $G(A)$ is the graph of A and $\iota_{G(A)}$ is the indicator function of $G(A)$, so that for all $x \in X$ we have

$$f(x) = \inf_{y \in Y} F(x,y).$$

Given $\bar{x}^* \in \partial f(\bar{x})$, since $F(\bar{x}, A\bar{x}) = f(\bar{x})$, Proposition 3.37 ensures that $(\bar{x}^*, 0) \in \partial F(\bar{x}, A\bar{x})$. Now, by Theorem 3.66, there exist nets $(w_i, y_i) \to (\bar{x}, \bar{y})$, $(x_i, z_i)_i \to (\bar{x}, A\bar{x})$, (w^*_i, y^*_i), (x^*_i, z^*_i) such that $(x_i, z_i) \in G(A)$, i.e., $z_i = Ax_i$, $w^*_i = 0$, $y^*_i \in \partial g(y_i)$, $(x^*_i, z^*_i) \in \partial \iota_{G(A)}(x_i, z_i) = N(G(A), (x_i, z_i))$ for all i and

$$(0, y^*_i) + (x^*_i, z^*_i) \xrightarrow{*} (\bar{x}^*, 0), \qquad (3.45)$$

$$\max(\|x_i - w_i\|, \|z_i - y_i\|)(\|x^*_i\| + \|z^*_i\| + \|y^*_i\|) \to 0. \qquad (3.46)$$

Since $(x^*_i, z^*_i) \in N(G(A), (x_i, z_i))$, we have $x^*_i = -A^\mathsf{T} z^*_i$. Thus (3.45), (3.46) yields (3.35), (3.36). □

A variant of the fuzzy sum rule will be useful, in particular in the case that the second summand is the indicator function of a weakly compact convex subset.

Proposition 3.67. *Let $h := f + g$, where f, g are convex, lower semicontinuous, finite at $\bar{x} \in X$. Suppose there exists some $\gamma > 0$ such that $K := \{x \in \bar{x} + \gamma B_X : g(x) \le g(\bar{x}) + \gamma\}$ is weakly compact. If $\bar{x}^* \in \partial h(\bar{x})$, then there exist sequences $(w_n) \to_f \bar{x}$, $(z_n) \to_g \bar{x}$, (w^*_n), (z^*_n) such that $w^*_n \in \partial f(w_n)$, $z^*_n \in \partial g(z_n)$ for all $n \in \mathbb{N}$ and*

$$(\|w^*_n + z^*_n - \bar{x}^*\|)_n \to 0,$$

$$((\|w^*_n\| + \|z^*_n\|) \cdot \|w_n - z_n\|)_n \to 0.$$

Proof. Changing f into $f - \bar{x}^*$ and performing a translation, we may suppose $\bar{x}^* = 0$ and $\bar{x} = 0$. Let μ be a modulus of lower semicontinuity of $(w, z) \mapsto f(w) + g(z)$ at $(0, 0)$, i.e., a modulus μ such that $f(w) + g(z) \ge f(0) + g(0) - \mu(\|(w,z)\|)$ for $(w, z) \in X^2$ and let $\rho \in (0, \gamma)$ be such that $\mu(r) \le \gamma/2$ for $r \in [0, \rho]$. For $n \ge 1$, let us introduce the penalized decoupling function $p_n : \rho B_{X^2} \to \mathbb{R}_\infty$ given by

$$p_n(w, z) := f(w) + g(z) + n\|w - z\|^2 + \|w\|^2.$$

For all $(w, z) \in \rho B_{X^2}$ we have $f(w) \ge f(0) - \gamma/2$, so that if (w, z) satisfies $p_n(w, z) \le p_n(0, 0) + \gamma/2$, hence $g(z) \le f(0) + g(0) - f(w) + \gamma/2 \le g(0) + \gamma$, we get $z \in K$. Since p_n is bounded below and jointly lower semicontinuous when $\rho B_X \times \rho B_X$ is endowed with the product topology of the strong topology with the weak topology in which K is compact, given a sequence (t_n) in $(0,1)$ with limit 0, the partial Ekeland theorem yields a pair $(w_n, z_n) \in \rho B_X \times \rho B_X$ such that

$$\forall (w,z) \in \rho B_X \times \rho B_Y, \qquad p_n(w_n, z_n) \le p_n(w,z) + t_n \|w - w_n\|.$$

In particular, $p_n(w_n, z_n) \le p_n(0,0) + t_n \rho < p_n(0,0) + \gamma/2$ for n large and $z_n \in K$. Then, since $p_n(0,0) = h(0)$, the relations

$$p_n(w_n, z_n) \le p_n(0,0) + t_n \|w_n\| \le f(w_n) + g(z_n) + \mu(\|(w_n, z_n)\|) + t_n \rho$$

yield

$$n \|z_n - w_n\|^2 + t_n \|w_n\|^2 \le \mu(\|(w_n, z_n)\|) + t_n \rho \le \gamma + \rho, \qquad (3.47)$$

so that $(z_n - w_n) \to 0$. Let us show that $(w_n) \to 0$. Suppose, to the contrary, that there are $\alpha > 0$ and an infinite subset N of $\mathbb{N}$ such that $\|w_n\| \ge \alpha$ for all $n \in N$. Using the weak compactness of K, we get a weak limit point z of $(z_n)_{n \in N}$, and taking limits in the relations $p_n(w_n, z_n) \le p_n(0,0) + t_n \|w_n\| \le h(0) + t_n \rho$, since $\bar{x}^* = 0$, we get

$$f(z) + g(z) + \alpha^2 \le \liminf_{n \in N} p_n(w_n, z_n) \le h(0) \le f(z) + g(z),$$

a contradiction. Thus $(w_n) \to 0$ and $(z_n) \to 0$. Then for n large enough, (w_n, z_n) is a local minimizer of the convex function $q_n : (w,z) \mapsto p_n(w,z) + t_n \|w - w_n\|$. It follows that we can find $w_n^* \in \partial f(w_n)$, $z_n^* \in \partial g(z_n)$, $u_n^* \in \partial \|\cdot\|^2 (w_n)$, $v_n^* \in B_{X^*}$, and $x_n^* \in \partial \|\cdot\|^2 (w_n - z_n)$ such that

$$(0,0) = (w_n^* + n x_n^* + u_n^* + t_n v_n^*, z_n^* - n x_n^*).$$

This relation means that $z_n^* = n x_n^* = -w_n^* - u_n^* - t_n v_n^*$. The properties of the duality mappings yield $\|u_n^*\| = 2\|w_n\| \to 0$, $\|x_n^*\| = 2\|w_n - z_n\| \to 0$, so that $(\|z_n^* + w_n^*\|) = (\|u_n^* + t_n v_n^*\|) \to 0$. Thus $\|z_n^*\| = \|n x_n^*\| \le 2n \|w_n - z_n\|$, and setting $r_n = \|(w_n, z_n)\|$, $s_n = \mu(r_n)$, it follows from inequality (3.47) that

$$\|z_n^*\| \cdot \|w_n - z_n\| \le 2n \|w_n - z_n\|^2 \le 2s_n + 2t_n \rho \to 0,$$

$$\|w_n^*\| \cdot \|w_n - z_n\| \le \|w_n^* + z_n^*\| \cdot \|w_n - z_n\| + \|z_n^*\| \cdot \|w_n - z_n\| \to 0.$$

Finally, since g is lower semicontinuous,

$$\limsup_n f(w_n) + g(0) \le \limsup_n f(w_n) + \liminf_n g(z_n)$$

$$\le \limsup_n (f(w_n) + g(z_n)) \le \limsup_n p(w_n, z_n) \le f(0) + g(0),$$

so that $(f(z_n)) \to f(0)$. Similarly, $(g(z_n)) \to g(0)$. $\qquad\qquad\qquad \square$

Corollary 3.68 (Brøndsted–Rockafellar). *For a closed proper convex function f on a Banach space X, the set of points $x \in X$ such that $\partial f(x)$ is nonempty is dense in $\mathrm{dom}\, f$. More precisely, for every $\bar{x} \in \mathrm{dom}\, f$ there exists a sequence $(x_n) \to_f \bar{x}$ such that $\partial f(x_n) \neq \varnothing$ for all $n \in \mathbb{N}$. One can even find $(x_n) \to_f \bar{x}$ and (x_n^*) such that $(\|x_n^*\| . \|x_n - \bar{x}\|) \to 0$ and $x_n^* \in \partial f(x_n)$ for all $n \in \mathbb{N}$.*

Proof. Given $\bar{x} \in \mathrm{dom}\, f$, let $S := \{\bar{x}\}$ and $g := \iota_S$; then g satisfies the compactness assumption of Proposition 3.67. Since $f_S := f + \iota_S$ attains its minimum at $\bar{x}$, one has $0 \in \partial f_S(\bar{x})$. Then the fuzzy sum rule of that proposition yields the result. □

3.5.2 Exact Rules in Convex Analysis

Technical assumptions, called qualification conditions, can be given in order to ensure the expected equalities $\partial(f + g)(x) = \partial f(x) + \partial g(x)$ and $\partial f(x) = A^\mathsf{T}(\partial g(Ax))$ in the composition rule for $f = g \circ A$. They involve the asymptotic subdifferentials of the functions, where for $x \in \mathrm{dom}\, f$, the *asymptotic subdifferential* of f at x or *singular subdifferential* of f at x is defined as follows:

$$\partial_\infty f(x) := \{x^* \in X^* : (x^*, 0) \in N(E_f, x_f)\}$$

with $x_f := (x, f(x))$, E_f being the epigraph of f. The terminology and the notation are justified by the following observations. Here, the *asymptotic cone* $T_\infty(C)$ of a nonempty subset of a normed space Y is the set $\limsup_{t \to \infty} C/t$ of limits of sequences (x_n/t_n), where $(t_n) \to +\infty$, $x_n \in C$ for all n. Thus if C is bounded, one has $T_\infty(C) = \{0\}$. When C is a closed convex subset, $T_\infty(C)$ is the set of $y \in Y$ such that for all $y_0 \in C$ and $t \in \mathbb{R}_+$ one has $y_0 + ty \in C$.

Proposition 3.69. *Let $f : X \to \mathbb{R}_\infty$ be a convex function on a normed space X and let $x \in \mathrm{dom}\, f$.*

(a) *$u^* \in \partial_\infty f(x)$ whenever there exist nets $(t_i)_{i \in I} \to 0_+$, $(x_i)_{i \in I} \to_f x$, $(x_i^*)_{i \in I}$ in X^* with $(t_i x_i^*) \xrightarrow{*} u^*$, $(\langle t_i x_i^*, x_i - x \rangle)_{i \in I} \to 0$, and $x_i^* \in \partial f(x_i)$ for all $i \in I$.*
(b) *If $\partial f(x)$ is nonempty, then $\partial_\infty f(x)$ is the asymptotic cone of $\partial f(x)$.*
(c) *If f is continuous at x, then $\partial_\infty f(x) = \{0\}$*

Proof. (a) Given $(t_i)_{i \in I}$, $(x_i)_{i \in I}$, $(x_i^*)_{i \in I}$ as in the statement, setting $z_i := (x_i, f(x_i))$, $z := x_f := (x, f(x))$, $z_i^* := (t_i x_i^*, -t_i) \in N(E_f, z_i)$, one has $(\langle z_i^*, z_i - z \rangle)_{i \in I} \to 0$, and hence $(u^*, 0) \in N(E_f, x_f)$, in view of Proposition 3.62.

(b) Let $x_0^* \in \partial f(x)$. Then for all $u^* \in \partial_\infty f(x)$ and all $t \in \mathbb{R}_+$, since $N(E_f, z)$ is a convex cone, one has $(x_0^* + tu^*, -1) = (x_0^*, -1) + t(u^*, 0) \in N(E_f, z)$, and hence u^* belongs to the asymptotic cone $T_\infty(\partial f(x))$ of $\partial f(x)$. Conversely, if $u^* \in T_\infty(\partial f(x))$, for all $t > 0$ one has $(x_0^* + tu^*, -1) \in N(E_f, z)$, hence $(u^*, 0) = \lim_{t \to \infty} t^{-1}(x_0^* + tu^*, -1) \in N(E_f, z)$, i.e., $u^* \in \partial_\infty f(x)$.

(c) If f is continuous at x, then $\partial f(x)$ is bounded and nonempty; hence $\partial_\infty f(x) = \{0\}$ by (b). □

We need the following compactness notions.

Definition 3.70. A convex subset C of X is *normally compact* at $\bar{x} \in C$ if a net $(u_i^*)_{i \in I}$ has a nonnull weak* cluster point whenever there exists a net $(x_i)_{i \in I} \to_C \bar{x}$ such that $u_i^* \in N(C, x_i) \cap S_{X^*}$ for all $i \in I$.

A convex function $f : X \to \mathbb{R}_\infty$ is *normally compact* at $\bar{x} \in \mathrm{dom} f$ if its epigraph is normally compact at $\bar{x}_f := (\bar{x}, f(\bar{x}))$. A convex function $f : X \to \mathbb{R}_\infty$ is said to be *subdifferentially compact* at $\bar{x} \in \mathrm{dom} f$ if for every net $(x_i, x_i^*)_{i \in I}$ such that $(x_i)_{i \in I} \to_f \bar{x}$, $(\langle x_i^*, x_i - \bar{x} \rangle)_{i \in I} \to 0$, $(\|x_i^*\|) \to +\infty$, and $x_i^* \in \partial f(x_i)$ for all $i \in I$, the net $(\|x_i^*\|^{-1} x_i^*)_{i \in I}$ has a nonnull weak* cluster point.

Clearly, in a finite-dimensional normed space, every convex set is normally compact, and every function is normally compact. Moreover, one has the following criteria.

Lemma 3.71. *(a) If the interior of a convex set C is nonempty, then C is normally compact at all $x \in C$.*

(b) If a convex function f is normally compact at $\bar{x}$, then f is subdifferentially compact at $\bar{x}$.

(c) If a convex function f is continuous at some point of its domain, then f is normally compact at each point of its domain.

(d) A convex set C is normally compact at $\bar{x}$ if and only if its indicator function ι_C is subdifferentially compact at $\bar{x}$.

Proof. (a) Let $a \in C$ and $r > 0$ be such that the ball $B(a, r)$ is contained in C. Then for every net $(x_i)_{i \in I}$ of C with limit $x \in C$, every net $(u_i^*)_{i \in I}$ such that $u_i^* \in N(C, x_i)$, $\|u_i^*\| = 1$ for all $i \in I$, for all $u \in B_X$ one has $\langle u_i^*, a + ru - x_i \rangle \leq 0$, hence

$$\langle u_i^*, a - x_i \rangle \leq -r \|u_i^*\| = -r,$$

so that every weak* cluster point u^* of $(u_i^*)_{i \in I}$ satisfies $\langle u^*, a - x \rangle \leq -r$, hence is nonnull.

(b) Let $((x_i, x_i^*))_{i \in I}$ be a net in ∂f with $(x_i) \to_f \bar{x}$, $(\|x_i^*\|) \to \infty$. Let $z_i := (x_i, f(x_i))$, and let $t_i := \|(x_i^*, -1)\|$. For $i \in I$, $(u_i^*, r_i) := (t_i^{-1} x_i^*, -t_i^{-1}) \in N(E_f, z_i)$ is a unit vector and $(z_i)_{i \in I} \to_{E_f} \bar{x}_f$. Then if (u^*, r) is a nonnull weak* cluster point of $((u_i^*, r_i))$, one has $r = 0$, hence $u^* \neq 0$, and since $(t_i^{-1} \|x_i^*\|) \to 1$, the net $(\|x_i^*\|^{-1} x_i^*)$ has a nonnull weak* cluster point u^*.

Assertion (c) follows from assertion (a).

(d) Suppose ι_C is subdifferentially compact at $\bar{x}$ and let $(x_i)_{i \in I} \to \bar{x}$ in C, $x_i^* \in N(C, x_i) \cap S_{X^*}$ for all $i \in I$. Let $r_i := \langle x_i^*, x_i - \bar{x} \rangle$ so that $(r_i) \to 0$. Suppose first that $J := \{ j \in I : r_j = 0 \}$ is bounded above by some $\bar{i}$. Then for $i \in I' := I \backslash J$, let $s_i := r_i^{-1/2}$ and $y_i^* := s_i x_i^* \in \partial \iota_C(y_i)$ for $y_i := x_i$, so that $(\|y_i^*\|)_{i \in I'} \to \infty$ and $(\langle y_i^*, y_i - \bar{x} \rangle)_{i \in I'} \to 0$, and we conclude that $(x_i^*) = (y_i^* / \|y_i^*\|)$ has a nonnull weak* cluster point. Now suppose J is not majorized, i.e., J is cofinal. Then for $k := (j, n) \in K := J \times \mathbb{N}$, setting $x_k^* := n x_j^* \in \partial \iota_C(x_k)$ for $x_k := x_j$, one has $\langle x_k^*, x_k - \bar{x} \rangle = 0$, and we conclude again that $(x_j^*)_{j \in J} = (x_k^* / \|x_k^*\|)_{k \in K}$ has a nonnull weak* cluster point.

The reverse implication is immediate: given a net $(x_i, x_i^*)_{i \in I}$ in the graph of $\partial \iota_C$ such that $(x_i)_{i \in I} \to \bar{x}$, $(\|x_i^*\|) \to +\infty$, the net $(\|x_i^*\|^{-1} x_i^*)_{i \in I}$ has a nonnull weak* cluster point, since $u_i^* := \|x_i^*\|^{-1} x_i^* \in N(C, x_i) \cap S_{X^*}$ for all $i \in I$. $\qquad \square$

A characterization of normal compactness can be given.

Proposition 3.72. *A convex set C is normally compact at $\bar{x} \in C$ if and only if for all nets $(x_i)_{i \in I} \to_C \bar{x}$ and $(x_i^*)_{i \in I}$ such that $x_i^* \in N(C, x_i)$ for all $i \in I$ one has*

$$(x_i^*)_{i \in I} \xrightarrow{*} 0 \implies (\|x_i^*\|)_{i \in I} \to 0. \tag{3.48}$$

Proof. Suppose C is not normally compact at $\bar{x}$. Let $(x_i)_{i \in I} \to_C \bar{x}$ and let $(x_i^*)_{i \in I}$ in S_{X^*} satisfying $x_i^* \in N_F(C, x_i)$ for all $i \in I$ and such that 0 is the only weak* cluster value of (x_i^*). Since B_{X^*} is weak* compact, the net (x_i^*) weak* converges to 0. Then if relation (3.48) holds, $(\|x_i^*\|) \to 0$, a contradiction to $\|x_i^*\| = 1$ for all i.

Conversely, suppose the property of the statement is not satisfied. Then there exist nets $(x_i) \to_C \bar{x}$ and $(x_i^*) \xrightarrow{*} 0$ satisfying $x_i^* \in N_F(C, x_i)$ for all $i \in I$ such that $(\|x_i^*\|)$ does not converge to 0. Taking a subnet if necessary, we can assume that there exists some $r > 0$ such that $r_i := \|x_i^*\| \geq r$ for all i. Then $(x_i^*/r_i) \xrightarrow{*} 0$, so that it cannot have a nonnull weak* cluster value and C is not normally compact at $\bar{x}$. $\qquad \square$

Using the notion of normal compactness, one can give an exact version of Theorem 3.63 and exact subdifferential rules.

Theorem 3.73. *Let f, g be closed proper convex functions finite at $\bar{x}$. If f (or g) is subdifferentially compact at $\bar{x}$, one has $\partial(f + g)(\bar{x}) = \partial f(\bar{x}) + \partial g(\bar{x})$, provided*

$$\partial_\infty f(\bar{x}) \cap (-\partial_\infty g(\bar{x})) = \{0\}. \tag{3.49}$$

Proof. By Theorem 3.39, given $\bar{x}^* \in \partial(f + g)(\bar{x})$, there exist nets $(w_i) \to_f \bar{x}$, $(z_i) \to_g \bar{x}$, (w_i^*), (z_i^*) such that $w_i^* \in \partial f(w_i)$, $z_i^* \in \partial g(z_i)$ for all i and

$$(\|w_i^* + z_i^* - \bar{x}^*\|)_i \to 0, \tag{3.50}$$

$$\langle w_i^*, w_i - \bar{x} \rangle_i \to 0, \qquad \langle z_i^*, z_i - \bar{x} \rangle_i \to 0. \tag{3.51}$$

Let $r_i := \|w_i^*\|$. Suppose $(r_i)_{i \in I}$ has a subnet $(r_j)_{j \in J}$ with limit $+\infty$. Taking a subnet if necessary and using the subdifferential compactness of f at $\bar{x}$, we may suppose $(r_j^{-1} w_j^*)_{j \in J}$ has a nonnull weak* limit $w^* \in \partial_\infty f(\bar{x})$. Then $(r_j^{-1} \bar{x}^*)_{j \in J} \to 0$, whence by (3.50), $(r_j^{-1} z_j^*)_{j \in J} \xrightarrow{*} -w^*$, whence using again (3.51), we have $-w^* \in \partial_\infty g(\bar{x})$, a contradiction to assumption (3.49). Thus $(\|w_i^*\|)_{i \in I}$ is eventually bounded, and a subnet $(w_j^*)_{j \in J}$ of $(w_i^*)_{i \in I}$ has a weak* limit $w^* \in \partial f(\bar{x})$ by Corollary 3.61. Then a subnet $(z_j^*)_{j \in J}$ of $(z_i^*)_{i \in I}$ has a weak* limit $z^* \in \partial g(\bar{x})$ and $w^* + z^* = \bar{x}^*$. $\qquad \square$

Theorem 3.74. *Let X, Y be Banach spaces, let $A \in L(X, Y)$, and let $f := g \circ A$, where $g : Y \to \mathbb{R}_\infty$ is lower semicontinuous and convex. Let $\bar{x} \in \mathrm{dom}\, f$, $\bar{x}^* \in X^*$,*

$\bar{y} := A(\bar{x})$. *Suppose g is subdifferentially compact at $\bar{y}$. Then $\partial f(\bar{x}) = A^{\mathsf{T}}(\partial g(\bar{y}))$,*
provided

$$\partial_\infty g(\bar{y}) \cap \ker A^{\mathsf{T}} = \{0\}. \tag{3.52}$$

Proof. It suffices to prove that $\partial f(\bar{x}) \subset A^{\mathsf{T}}(\partial g(\bar{y}))$. Given $\bar{x}^* \in \partial f(\bar{x})$, let $(x_i)_{i \in I}$,
$(y_i)_{i \in I}$, $(y_i^*)_{i \in I}$ be as in Theorem 3.65. Let $r_i := \|y_i^*\|$. Suppose $(r_i)_{i \in I}$ has a subnet
$(r_j)_{j \in J}$ with limit $+\infty$. Taking a subnet if necessary and using the subdifferential
compactness of g at $\bar{y}$, we may suppose $(r_j^{-1} y_j^*)_{j \in J}$ has a nonnull weak* limit z^*.
Then, since $x^* = w^* - \lim_j A^{\mathsf{T}}(y_j^*)$, one has $A^{\mathsf{T}}(z^*) = w^* - \lim_j r_j^{-1} A^{\mathsf{T}}(y_j^*) = 0$, a
contradiction to relation (3.52) and the fact that $z^* \in \partial_\infty g(\bar{y})$ by Proposition 3.69.
Thus $(\|y_i^*\|)_{i \in I}$ is eventually bounded, and a subnet $(y_j^*)_{j \in J}$ of $(y_i^*)_{i \in I}$ has a weak*
limit y^*. Then one gets $x^* = A^{\mathsf{T}}(y^*)$, and Corollary 3.61 entails $y^* \in \partial g(\bar{y})$.   □

Exercises

1. **Composition rule with openness.** Let X, Y be Banach spaces, let $A \in L(X, Y)$
be surjective, and let $f := g \circ A$, where $g : Y \to \mathbb{R}_\infty$ is lower semicontinuous, convex,
and finite at $\bar{y} := A\bar{x}$. Given $\bar{x}^* \in \partial f(\bar{x})$, deduce from the definition of $\partial f(\bar{x})$ that
for every $u \in \ker A$ one has $\langle \bar{x}^*, u \rangle = 0$, so that there exists some $\bar{y}^* \in Y^*$ satisfying
$\bar{x}^* = \bar{y}^* \circ A$. Conclude from the surjectivity of A that $\bar{y}^* \in \partial g(\bar{y})$.

2. **Subdifferential determination of convex functions.** Given two lower semicontinuous proper convex functions f, g on a Banach space X such that $\partial f \subset \partial g$, prove
that there exists some $c \in \mathbb{R}$ such that $f(\cdot) = g(\cdot) + c$. [Hint: Reduce the question to
the case $X = \mathbb{R}$ by taking composition with an affine map from $\mathbb{R}$ to X; see [918].]

3. **Mixed calculus rule.** Given reflexive Banach spaces X and Y, $A \in L(X, Y)$,
lower semicontinuous functions $f : X \to \mathbb{R}_\infty$, $g : Y \to \mathbb{R}_\infty$ finite at $\bar{x} \in X$ and
$\bar{y} := A\bar{x}$ respectively, show that $\bar{x}^* \in \partial(f + g \circ A)(\bar{x})$ if and only if there exist
sequences $((x_n, x_n^*))$, $((y_n, y_n^*))$ in the graphs of ∂f and ∂g respectively such that
$(x_n^* + A^{\mathsf{T}}(y_n^*)) \to \bar{x}^*$, $(x_n) \to_f \bar{x}$, $(y_n) \to_g \bar{y}$, and

$$((\|x_n^*\| + \|y_n^*\|) \cdot \|y_n - Ax_n\|) \to 0. \tag{3.53}$$

$$(\langle x_n^*, x_n - \bar{x} \rangle) \to 0, \quad (\langle y_n^*, y_n - \bar{y} \rangle) \to 0. \tag{3.54}$$

4. **Equivalence of complementary conditions.** With the data of the preceding
exercise, show that when $x_n^* \in \partial f(x_n)$, $y_n^* \in \partial g(y_n)$, $(x_n^* + A^{\mathsf{T}}(y_n^*)) \to \bar{x}^*$, $(x_n) \to_f \bar{x}$,
$(y_n) \to_g \bar{y}$, condition (3.53) is equivalent to condition (3.54). [See [918].]

5. Examine the relationships between the Attouch–Brézis qualification condition
and the qualification condition of Theorem 3.74.

6. **Subdifferential of a distance function.** Let C be a closed convex subset of a
Banach space X and let $w \in X \setminus C$, $w^* \in \partial d_C(w)$. Show that for every $\varepsilon > 0$ there

exist $x \in C$, $x^* \in N(C,x)$ such that $\|x - w\| \leq d_C(w) + \varepsilon^2$, $\|x^* - w^*\| \leq \varepsilon$. [Hint: Pick $v \in C$ such that $\|v - w\| \leq d_C(w) + \varepsilon^2$ and use the Ekeland variational principle for the function f defined on $X \times X$ by $f(u,x) := \|u - x\| - \langle w^*, u \rangle + \iota_C(x)$.]

7. Deduce from Exercise 6 that when C is a closed convex subset of a Banach space X and $w \in X \setminus C$ has a best approximation x in C, then one has $\partial d_C(w) \subset N(C,x)$.

3.5.3   Mean Value Theorems

Mean value theorems can be deduced from the fuzzy calculus rules we established. They may serve as introductions to similar results in the nonconvex case.

Theorem 3.75 (Fuzzy mean value theorem). *Let X be a Banach space and let $f : X \to \mathbb{R}_\infty$ be lower semicontinuous, convex, and finite at $\bar{x} \in X$. Then, for every $\bar{y} \in X \setminus \{\bar{x}\}$ and for every $r \in \mathbb{R}$ such that $f(\bar{y}) \geq r$, there exist $u \in [\bar{x}, \bar{y})$ and sequences $(u_n) \to_f u$, (u_n^*) such that $(\langle u_n^*, u_n - u \rangle) \to 0$, $u_n^* \in \partial f(u_n)$ for all n and*

$$\liminf_n \langle u_n^*, \bar{y} - \bar{x} \rangle \geq r - f(\bar{x}), \tag{3.55}$$

$$(\|u_n^*\| \, d(u_n, [\bar{x}, \bar{y}])) \to 0, \tag{3.56}$$

$$\liminf_n \langle u_n^*, x - u_n \rangle \geq \frac{\|x - u\|}{\|\bar{y} - \bar{x}\|}(r - f(\bar{x})) \quad \forall x \in (\bar{x} + \mathbb{R}_+(\bar{y} - \bar{x})) \setminus [\bar{x}, u). \tag{3.57}$$

Proof. Let $e^* \in X^*$ be such that $\langle e^*, \bar{y} - \bar{x} \rangle = f(\bar{x}) - r$ and let $g : X \to \mathbb{R}_\infty$ be defined by $g(x) := f(x) + \langle e^*, x \rangle + \iota_S(x)$, where S is the segment $[\bar{x}, \bar{y}]$. Since $g(\bar{y}) \geq g(\bar{x})$, the lower semicontinuous function g attains its minimum on S and X at a point $u \in S$, $u \neq \bar{y}$. Writing $g := h + \iota_S$ with $h := f + e^*$, Proposition 3.67 yields sequences $(u_n) \to_h u$, $(v_n) \to_S u$, (u_n^*), (v_n^*) such that $(u_n^* + e^* + v_n^*) \to 0$, $(\|v_n^*\| \cdot \|v_n - u_n\|) \to 0$, $u_n^* \in \partial f(u_n)$, $v_n^* \in \partial \iota_S(v_n)$ for all n. Then $v_n \in S \setminus \{\bar{y}\}$ for n large enough, $\bar{y} - v_n = t_n(\bar{y} - \bar{x})$ for some $t_n \in (0,1)$, and since $\langle v_n^*, \bar{y} - v_n \rangle \leq 0$, we get $\langle v_n^*, \bar{y} - \bar{x} \rangle \leq 0$. Thus

$$\liminf_n \langle u_n^*, \bar{y} - \bar{x} \rangle \geq \liminf_n \langle u_n^* + v_n^*, \bar{y} - \bar{x} \rangle = -\langle e^*, \bar{y} - \bar{x} \rangle = r - f(\bar{x}).$$

Since $v_n \in S$, we deduce from $(\|u_n^*\| \cdot \|v_n - u_n\|) \to 0$ that $\|u_n^*\| \, d(u_n, [\bar{x}, \bar{y}]) \to 0$. Finally, given $x \in (\bar{x} + \mathbb{R}_+(\bar{y} - \bar{x})) \setminus [\bar{x}, u)$, setting $x := u + t(\bar{y} - \bar{x})$ with $t \in \mathbb{R}_+$, and observing that $\lim_n (\langle u_n^* + e^*, u_n - u \rangle + \langle v_n^*, v_n - u \rangle) = 0$ and $\langle v_n^*, u - v_n \rangle \leq 0$ for all n, we get $\liminf_n \langle u_n^*, u - u_n \rangle \geq 0$ and

$$\liminf_n \langle u_n^*, x - u_n \rangle \geq \liminf_n \langle u_n^*, u - u_n \rangle + \liminf_n t \langle u_n^*, \bar{y} - \bar{x} \rangle \geq t(r - f(\bar{x})),$$

hence relation (3.57). $\square$

Corollary 3.76 (Usual mean value theorem). *Let W be an open convex subset of a Banach space and let $f : W \to \mathbb{R}$ be convex and continuous. Then for every $x, y \in W$, there exist $u \in [\bar{x}, \bar{y})$ and $u^* \in \partial f(u)$, such that*

$$\langle u^*, \bar{y} - \bar{x} \rangle \geq f(\bar{y}) - f(\bar{x}). \tag{3.58}$$

Proof. We extend f by $+\infty$ on $X \setminus V$, where V is a closed convex neighborhood of $[\bar{x}, \bar{y}]$ contained in W, and we pass to the limit in (3.55), using the fact that the multimap ∂f is locally bounded and closed by Proposition 3.61.   □

More powerful forms of the preceding result can be given. Here, instead of considering a function f on a segment $[\bar{x}, \bar{y}]$ of X, we suppose f is defined on a neighborhood of a "drop" D:

$$D = [\bar{x}, C] := \bigcup_{y \in C} [\bar{x}, y] := \{(1 - t)\bar{x} + ty : y \in C, \, t \in [0, 1]\},$$

where C is a closed convex subset of X and $\bar{x} \in X$. We observe that D is the convex hull of $C \cup \{\bar{x}\}$. In the present section, we limit our study to the case of a compact convex set C. We start with a generalization of Rolle's theorem.

Theorem 3.77 (Multidirectional Rolle's theorem, compact case). *Let C be a weakly compact convex subset of a Banach space X and let $\bar{x} \in X \setminus C$, $D := [\bar{x}, C]$. Suppose $f : X \to \mathbb{R}_\infty$ is convex, lower semicontinuous, finite at $\bar{x}$, and $\inf f(C) \geq f(\bar{x})$. Then there exists $u \in D \setminus C$ such that $f(u) = \inf f(D)$, and for every $\varepsilon > 0$ one can find $w \in B(u, \varepsilon)$ and $w^* \in \partial f(w)$ such that $|f(w) - f(u)| < \varepsilon$, $|\langle w^*, w - u \rangle| < \varepsilon$,*

$$\langle w^*, y - \bar{x} \rangle > -\varepsilon \|y - \bar{x}\| \qquad \forall y \in C, \tag{3.59}$$

$$\langle w^*, x - w \rangle \geq -\varepsilon \|x - w\| - \varepsilon \qquad \forall x \in \bar{x} + \mathbb{R}_+(C - \bar{x}), \tag{3.60}$$

$$\|w^*\| d(w, D) < \varepsilon. \tag{3.61}$$

Proof. In fact, the result is valid for every $u \in D \setminus C$ such that $f(u) = \inf f(D)$, as we shall see. Without loss of generality, using the translation by $-\bar{x}$, we may assume $\bar{x} = 0$. Since D is weakly compact and f is lower semicontinuous, we can find a minimizer u of f on D. Since $\inf f(C) \geq f(\bar{x})$, we may suppose $u \in D \setminus C$ and reduce ε if necessary to have $\varepsilon < d(u, C)$. Then $0 \in \partial(f + g)(u)$, where $g := \iota_D$, the indicator function of D, satisfies the compactness assumption of Proposition 3.67. Its conclusion yields $w, z \in B(u, \varepsilon)$, with $|f(w) - f(u)| < \varepsilon$, $z \in D$, and $w^* \in \partial f(w)$, $z^* \in N(D, z)$ such that

$$\|w^* + z^*\| < \varepsilon, \qquad \|w^*\| . \|w - z\| < \varepsilon, \qquad \|z^*\| . \|w - z\| < \varepsilon.$$

Relation (3.61) stems from the inequalities $\|w^*\| . \|w - z\| < \varepsilon$, $d(w, D) \leq \|w - z\|$. Since $\varepsilon < d(u, C)$, and since $\|z - u\| < \varepsilon$, we have $z \in D \setminus C$. Then since $[1, +\infty)C$ is closed, as is easily seen, and since $D \setminus C = \mathbb{R}_+ C \cap (X \setminus [1, +\infty)C)$, we have $N(D, z) \subset$

$N(D \setminus C, z) = N(\mathbb{R}_+ C, z)$, whence $\langle -z^*, rc - z \rangle \geq 0$ for all $r \in \mathbb{R}_+$, $c \in C$. Then relation (3.60) holds: for $x := rc \in \mathbb{R}_+ C$ we have

$$\langle w^*, x - w \rangle \geq \langle -z^*, x - w \rangle - \varepsilon \|x - w\|$$
$$\geq \langle -z^*, x - z \rangle + \langle -z^*, z - w \rangle - \varepsilon \|x - w\| \geq -\varepsilon - \varepsilon \|x - w\|.$$

Given $y \in C$, let $c := ty + (1 - t)\bar{y} \in C$, where $t \in (0, 1]$ and $\bar{y} \in C$ are such that $z = t\bar{x} + (1 - t)\bar{y}$. Using the inequality $0 \leq \langle -z^*, c - z \rangle = \langle -z^*, t(y - \bar{x}) \rangle$ and the relation $\|w^* + z^*\| < \varepsilon$, after dividing by t we obtain (3.59):

$$\langle w^*, y - \bar{x} \rangle \geq \langle -z^*, y - \bar{x} \rangle - \varepsilon \|y - \bar{x}\| \geq -\varepsilon \|y - \bar{x}\|.$$

$\square$

Theorem 3.78 (Multidirectional mean value theorem, compact case). *Let C be a weakly compact convex subset of a Banach space X and let $\bar{x} \in X \setminus C$, $D := [\bar{x}, C]$. Suppose $f : X \to \mathbb{R}_\infty$ is convex, lower semicontinuous, finite at $\bar{x}$, and let $r \in \mathbb{R}$ be such that $r \leq \inf f(C)$. Then there exist $\bar{y} \in C$, $s \in [0, 1)$, $u := (1 - s)\bar{x} + s\bar{y} \in D \setminus C$, and sequences $(u_n) \to u$, (u_n^*) in X^* such that $(f(u_n)) \to f(u)$, $(\langle u_n^*, u - u_n \rangle) \to 0$, $u_n^* \in \partial f(u_n)$ for all $n \in \mathbb{N}$, and*

$$\liminf_n \langle u_n^*, y - \bar{x} \rangle \geq r - f(\bar{x}) \qquad \forall y \in C, \tag{3.62}$$

$$\liminf_n \langle u_n^*, x - \bar{x} \rangle \geq (t - s)(r - f(\bar{x})) \qquad \forall t \in \mathbb{R}_+, \forall x \in \bar{x} + t(C - \bar{x}), \tag{3.63}$$

$$\|u_n^*\| d(u_n, D) \to 0. \tag{3.64}$$

Proof. Let $C' := C \times \{1\} \subset X' := X \times \mathbb{R}$, $\bar{x}' := (\bar{x}, 0)$, and let $f' : X' \to \mathbb{R}_\infty$ be given by $f'(x, t) := f(x) + t(f(\bar{x}) - r)$, so that $\inf f'(C') \geq f'(\bar{x}')$. Given a sequence $(\varepsilon_n) \to 0_+$, let us apply Rolle's theorem above to f' with C, $\bar{x}$, ε replaced by C', $\bar{x}'$, ε_n. We get $w := (u, s) \in (D \setminus C) \times [0, 1)$, sequences $(w_n) := ((u_n, s_n)) \to (u, s)$, $(w_n^*) := (u_n^*, s_n^*)$ satisfying $w_n^* \in \partial f'(w_n)$ for all $n \in \mathbb{N}$, i.e., $u_n^* \in \partial f(u_n)$, $s_n^* = f(\bar{x}) - r$, $(f(u_n) + s_n(f(\bar{x}) - r)) \to f(u) + s(f(\bar{x}) - r)$, whence $(f(u_n)) \to f(u)$, $(\langle u_n^*, u - u_n \rangle + s_n^*(s_n - s)) \to 0$, whence $(\langle u_n^*, u - u_n \rangle) \to 0$, and for all $y \in C$, $t \in \mathbb{R}_+$, setting $y' := (y, 1)$, $x' := \bar{x}' + t(y' - \bar{x}')$,

$$\langle u_n^*, y - \bar{x} \rangle + f(\bar{x}) - r > -\varepsilon_n \|y - \bar{x}\| - \varepsilon_n \qquad \forall y \in C,$$

$$\langle u_n^*, x - u \rangle + (t - s_n)(f(\bar{x}) - r) \geq -\varepsilon_n \|x - w\| - \varepsilon_n \qquad \forall t \in \mathbb{R}_+, \forall x \in \bar{x} + t(C - \bar{x}),$$

$$\|u_n^*\| d(u_n, D) < \varepsilon_n.$$

Passing to the limit, we get the announced relations. $\square$

One can get rid of the compactness condition on C. We shall show this later, dropping the convexity assumptions, too, at the expense of changing the Moreau–Rockafellar subdifferential into an adapted subdifferential.

Exercises

1. Let $f : X \to \mathbb{R}_\infty$ be convex, lower semicontinuous, finite at $\bar{x}, \bar{y} \in X$ with $f(\bar{x}) = f(\bar{y})$. Show that there exist $u \in (\bar{x}, \bar{y})$ and sequences $(u_n) \to_f u$, (u_n^*) such that $u_n^* \in \partial f(u_n)$ for all n, $(f(u_n)) \to f(u)$, $(\langle u_n^*, \bar{y} - \bar{x}\rangle) \to 0$, $(\|u_n^*\| d(u_n, [\bar{x}, \bar{y}])) \to 0$, and $(\langle u_n^*, u - u_n\rangle) \to 0$.

2. Show that the mean value theorem is a special case of the multidirectional mean value theorem in the compact case.

3. Use the mean value theorem to show the equivalence of the following assertions about an arbitrary lower semicontinuous convex function f on a normed space X:

(a) There exist constants a, b such that for all $x \in X$ one has $f(x) \le a \|x\|^2 + b$.
(b) There exist constants c, d such that for all $x \in X$, $x^* \in \partial f(x)$ one has $\|x^*\| \le c \|x\| + d$. [Hint: See [449, Proposition 4.3].]

3.5.4   Application to Optimality Conditions

Let us apply the above calculus rules to the constrained optimization problem

$$(\mathscr{C}) \quad \text{minimize} f(x) \quad \text{subject to } x \in C,$$

where $f : X \to \mathbb{R}_\infty$ is convex and C is a convex subset of X. We assume that f takes at least one finite value on C, so that $\inf(\mathscr{C})$ is not $+\infty$. Then $(\mathscr{C})$ is equivalent to the minimization of $f_C := f + \iota_C$ on X.

Optimality conditions for problem $(\mathscr{C})$ involve the notion of normal cone to C at some $\bar{x} \in C$; in the convex case we are dealing with here, its simple definition was given before Proposition 3.23. We recall it for the reader's convenience: the *normal cone* to C at $\bar{x} \in C$ is the set $N(C, \bar{x})$ of continuous linear forms on X that attain their maximum on C at $\bar{x}$:

$$N(C, \bar{x}) := \partial \iota_C(\bar{x}) := \{x^* \in X^* : \forall x \in C \ \langle \bar{x}^*, x - \bar{x}\rangle \le 0\}.$$

Example. If $\bar{x}$ is in the interior of C, one has $N(C, \bar{x}) = \{0\}$, since a continuous linear form that has a local maximum is null. $\square$

Example. Let $g \in X^* \setminus \{0\}$, $c \in \mathbb{R}$, and let $D := \{x \in X : g(x) \le c\}$. Then if $\bar{x}$ is such that $g(\bar{x}) < c$, one has $\bar{x} \in \text{int} D$, hence $N(D, \bar{x}) = \{0\}$, while for all $\bar{x}$ such that $g(\bar{x}) = c$, one has $N(D, \bar{x}) = \mathbb{R}_+ g$. In fact, for every $r \in \mathbb{R}_+$ and all $x \in D$ one has $rg(x - \bar{x}) \le 0$, hence $rg \in N(D, \bar{x})$. Conversely, let $h \in N(D, \bar{x})$. Then for all $u \in \text{Ker} g$, one has $\bar{x} + u \in D$, hence $h(u) \le 0$. Changing u into $-u$, we see that $\text{Ker} g \subset \text{Ker} h$, so that there exists $r \in \mathbb{R}$ such that $h = rg$: picking $u \in X$ satisfying $g(u) = 1$ (this is possible since $g \ne 0$), we have $r = h(u)$, and since $\bar{x} - u \in D$, we get that $-r = h(-u) = h((\bar{x} - u) - \bar{x}) \le 0$, hence $r \in \mathbb{R}_+$. $\square$

Theorem 3.79. *A sufficient condition for $\bar{x} \in C$ to be a solution to $(\mathscr{C})$ is*

$$0 \in \partial f(\bar{x}) + N(C, \bar{x}).$$

Under one of the following assumptions, this condition is necessary:

(a) *f is finite and continuous at some point of C;*
(b) *f is finite at some point of the interior of C;*
(c) *f is lower semicontinuous, $C = \mathrm{cl}(C)$, $\mathbb{R}_+(\mathrm{dom}\, f - C) = -\mathrm{cl}(\mathbb{R}_+(\mathrm{dom}\, f - C))$, and X is complete.*

Proof. Suppose $\bar{x} \in C$ is such that $0 \in \partial f(\bar{x}) + N(C, \bar{x})$. Let $\bar{x}^* \in \partial f(\bar{x})$ be such that $-\bar{x}^* \in N(C, \bar{x})$. Then $f(\bar{x})$ is finite and for all $x \in C$, one has $f(x) - f(\bar{x}) \geq \langle \bar{x}^*, x - \bar{x} \rangle \geq 0$: $\bar{x}$ is a solution to $(\mathscr{C})$.

The necessary condition stems from the relations $0 \in \partial(f + \iota_C)(\bar{x}) = \partial f(\bar{x}) + \iota_C(\bar{x})$, valid under each of the assumptions (a)–(c). $\qquad \square$

One can also give a necessary and sufficient optimality condition that does not require additional assumptions. For simplicity, we restrict our attention to the case that X is reflexive, although the general case is similar, using nets and weak* convergence instead of sequences and strong convergence.

Theorem 3.80. *A necessary and sufficient condition for $\bar{x} \in C$ to be a solution to $(\mathscr{C})$ is that there exist sequences $(x_n), (w_n) \to \bar{x}$, $(w_n^*), (x_n^*)$ such that $(f(x_n)) \to f(\bar{x})$, $(w_n^* + x_n^*) \to 0$, $(\|w_n - x_n\| \cdot (\|w_n^*\| + \|x_n^*\|)) \to 0$, $w_n \in C$, $w_n^* \in N(C, w_n)$, $x_n^* \in \partial f(x_n)$ for all $n \in \mathbb{N}$.*

Proof. The condition stems from the fuzzy sum rule used in transcribing the inclusion $0 \in \partial(f + \iota_C)(\bar{x})$ that characterizes $\bar{x}$ as a minimizer of $g + \iota_C$. Let us give a direct proof of sufficiency. Given sequences as in the statement, noting that

$$|\langle x_n^*, x - x_n \rangle - \langle w_n^*, w_n - x \rangle| \leq \|w_n^* + x_n^*\| (\|x\| + \|w_n\|) + \|x_n^*\| \cdot \|x_n - w_n\| \to 0,$$

for all $x \in C$, we get, since $w_n^* \in N(C, w_n)$ for all $n \in \mathbb{N}$,

$$f(x) \geq \liminf_n (f(x_n) + \langle x_n^*, x - x_n \rangle) \geq f(\bar{x}) + \liminf_n (\langle -w_n^*, x - w_n \rangle) \geq f(\bar{x}),$$

so that $\bar{x}$ is a solution to $(\mathscr{C})$. $\qquad \square$

In order to apply the conditions of Theorem 3.79 to the important case in which C is defined by inequalities, let us give a means to compute the normal cone to C in such a case. We start with the case of a single inequality, generalizing the second example of this subsection.

Lemma 3.81. *Let $g : X \to \mathbb{R}_\infty$ be a convex function and let $C := \{x \in X : g(x) \leq 0\}$, $\bar{x} \in g^{-1}(0)$. Suppose $C' := \{x \in X : g(x) < 0\}$ is nonempty and g is continuous at each point of C'. Then for $x' \in C'$ one has $N(C, x') = \{0\}$ and $N(C, \bar{x}) = \mathbb{R}_+ \partial g(\bar{x})$.*

Proof. For all $x' \in C'$ the set C is a neighborhood of x', so that $N(C,x') = \{0\}$. The inclusion $N(C,\bar{x}) \supset \mathbb{R}_+ \partial g(\bar{x})$ is obvious: given $r \in \mathbb{R}_+$ and $\bar{x}^* \in \partial g(\bar{x})$, for all $x \in C$ one has $\langle r\bar{x}^*, x - \bar{x} \rangle \leq r(g(x) - g(\bar{x})) \leq 0$, hence $r\bar{x}^* \in N(C,\bar{x})$.

Conversely, let $\bar{x}^* \in N(C,\bar{x}) \setminus \{0\}$. The interior of C is nonempty, since it contains C'. Since $\langle \bar{x}^*, x \rangle \leq \langle \bar{x}^*, \bar{x} \rangle$ for all $x \in C$, we have $\langle \bar{x}^*, x \rangle < \langle \bar{x}^*, \bar{x} \rangle$ for all $x \in \text{int}(C)$ (otherwise, $\bar{x}^*$ would have a local maximum, hence would be 0). In particular, $g(x) < 0$ implies $\langle \bar{x}^*, x \rangle < \langle \bar{x}^*, \bar{x} \rangle$. Thus $g(x) \geq 0$ for all x such that $\langle \bar{x}^*, x \rangle \geq \langle \bar{x}^*, \bar{x} \rangle$. Therefore $\bar{x}$ is a minimizer of g on $D := \{x \in X : \langle \bar{x}^*, x \rangle \geq \langle \bar{x}^*, \bar{x} \rangle\}$. Since g is continuous at $\bar{x} \in D$, we have $0 \in \partial g(\bar{x}) + N(D,\bar{x})$ by assertion (a) of the preceding theorem. But the second example of the present section, with $g := -\bar{x}^*$, $c := \langle -\bar{x}^*, \bar{x} \rangle$, ensures that $N(D,\bar{x}) = -\mathbb{R}_+ \bar{x}^*$. Since $0 \notin \partial g(\bar{x})$ because $\bar{x}$ is not a minimizer of g, we get some $s > 0$ such that $s\bar{x}^* \in \partial g(\bar{x})$, hence $\bar{x}^* \in s^{-1} \partial g(\bar{x})$. □

The case of a finite number of inequalities is a consequence of Lemma 3.81 and of the following rule for the calculus of normal cones.

Lemma 3.82. *Let $C_1, \ldots, C_k$ be convex subsets of X and let $\bar{x} \in C := C_1 \cap \cdots \cap C_k$. Then*

$$N(C,\bar{x}) = N(C_1,\bar{x}) + \cdots + N(C_k,\bar{x})$$

whenever one of the following assumptions is satisfied:

(a) There exist $j \in \{1, \ldots, k\}$ and some $z \in C_j$ that belongs to $\text{int}\,C_i$ for all $i \neq j$;

(b) X is a Banach space, $C_1, \ldots, C_k$ are closed, and for $D := \{(x, \ldots, x) : x \in X\}$, $P := C_1 \times \cdots \times C_k$, the cone $\mathbb{R}_+(P - D)$ is a closed linear subspace of X^k.

Proof. Assumption (a) ensures that $\partial(\iota_{C_1} + \cdots + \iota_{C_k})(\bar{x}) = \partial \iota_{C_1}(\bar{x}) + \cdots + \partial \iota_{C_k}(\bar{x})$, since for $i \neq j$ the function ι_{C_i} is finite and continuous at $z \in \text{dom}\,\iota_{C_j}$. The Attouch–Brézis theorem gives the conclusion under assumption (b), since if A is the diagonal map $x \mapsto (x, \ldots, x)$ from X into X^k, one has $C = A^{-1}(P)$, hence $\iota_C = \iota_P \circ A$ and

$$\partial \iota_C(\bar{x}) = A^{\mathsf{T}}(\partial \iota_P(\bar{x})) = A^{\mathsf{T}}(\partial \iota_{C_1}(\bar{x}) \times \cdots \times \partial \iota_{C_k}(\bar{x})) = \partial \iota_{C_1}(\bar{x}) + \cdots + \partial \iota_{C_k}(\bar{x}),$$

as is easily checked. □

The next example shows the necessity of requiring some additional assumptions, traditionally called "qualification conditions."

Example. For $i = 1, 2$, let $C_i := B[c_i, 1]$ with $c_i := (0, (-1)^i)$ in $X := \mathbb{R}^2$ endowed with the Euclidean norm. Then $C := C_1 \cap C_2 = \{\bar{x}\}$, with $\bar{x} := (0,0)$, hence $N(C,\bar{x}) = X^*$, but $N(C_i, \bar{x}) = \{0\} \times (-1)^{i+1} \mathbb{R}_+$ and $N(C_1, \bar{x}) + N(C_2, \bar{x}) = \{0\} \times \mathbb{R}$. □

Lemma 3.83. *Let $g_i : X \to \mathbb{R}_\infty$ be convex, let $C_i := \{x \in X : g_i(x) \leq 0\}$ for $i \in I := \{1, \ldots, k\}$, let $\bar{x} \in C := C_1 \cap \cdots \cap C_k$, and let $I(\bar{x}) := \{i \in I : g_i(\bar{x}) = 0\}$. Suppose that for all $i \in I \setminus I(\bar{x})$, g_i is continuous at $\bar{x}$ and that for all $i \in I(\bar{x})$, g_i is continuous on $C_i' := \{x \in X : g_i(x) < 0\}$. Suppose there exists some $x_0 \in C_i'$ for all $i \in I(\bar{x})$. Then for $\bar{x}^* \in X^*$, one has $\bar{x}^* \in N(C,\bar{x})$ if and only if there exist $y_1, \ldots, y_k$ in $\mathbb{R}_+$ such that*

$$\bar{x}^* \in y_1 \partial g_1(\bar{x}) + \cdots + y_k \partial g_k(\bar{x}), \quad y_1 g_1(\bar{x}) = 0, \ldots, y_k g_k(\bar{x}) = 0. \qquad (3.65)$$

Proof. The sufficient condition is immediate: if $\bar{x}^* = y_1\bar{x}_1^* + \cdots + y_k\bar{x}_k^*$ with $\bar{x}_i^* \in \partial g_i(\bar{x})$ and $y_i \in \mathbb{R}_+$ with $y_i g_i(\bar{x}) = 0$, for all $x \in C$ we get $\langle \bar{x}^*, x - \bar{x} \rangle \leq 0$, since the sum of the terms $y_i \langle \bar{x}_i^*, x - \bar{x} \rangle$ is less than or equal to 0, since $x \in C_i$ and $\bar{x}_i^* \in \partial g_i(\bar{x})$.

Let us suppose now that $\bar{x}^* \in N(C, \bar{x})$. For $i \in I \setminus I(\bar{x})$, since g_i is continuous at $\bar{x}$ and $g_i(\bar{x}) < 0$, one has $\bar{x} \in \text{int}(C_i)$, hence $N(C, \bar{x}) = N(C', \bar{x})$, where C' is the intersection of the family (C_i) for $i \in I(\bar{x})$. Given $x_0 \in C_i'$ for all $i \in I(\bar{x})$, for $t \in (0, 1)$ let $x_t := (1 - t)x_0 + t\bar{x}$. $i \in I(\bar{x})$ we have $g_i(x_t) \leq (1 - t)g_i(x_0) + tg_i(\bar{x}) = (1 - t)g_i(x_0) < 0$, and since g_i is continuous on C_i', we have $x_t \in \text{int}(C_i)$. Thus Lemma 3.82 yields some $\overline{w}_i^* \in N(C_i, \bar{x})$ such that $\bar{x}^* = \overline{w}_1^* + \cdots + \overline{w}_k^*$ (with $\overline{w}_i^* = 0$ for $i \in I \setminus I(\bar{x})$). For $i \in I(\bar{x})$, Lemma 3.81 provides some $y_i \in \mathbb{R}_+$ and some $\bar{x}_i^* \in \partial g_i(\bar{x})$ satisfying $\overline{w}_i^* = y_i \bar{x}_i^*$. Since for $i \in I \setminus I(\bar{x})$, g_i is continuous at $\bar{x}$, we can write $\overline{w}_i^* = y_i \bar{x}_i^*$ with $y_i = 0$, $\bar{x}_i^*$ arbitrary in $\partial g_i(\bar{x})$, which is nonempty by Theorem 3.25. Thus relation (3.65) holds. □

This characterization and Theorem 3.79 give immediately a necessary and sufficient optimality condition for the mathematical programming problem

$$(\mathcal{M}) \quad \text{minimize} f(x) \quad \text{subject to } x \in C := \{x \in X : g_1(x) \leq 0, \ldots, g_k(x) \leq 0\},$$

where f and $g_1, \ldots, g_k$ are as above.

Theorem 3.84 (Karush–Kuhn–Tucker theorem). *Let $f : X \to \mathbb{R}_\infty$, $g_1, \ldots, g_k$ be as in the preceding lemma and let $\bar{x} \in C$. Suppose f is continuous at $\bar{x}$ and the* Slater condition *holds: there exists some $x_0 \in \text{dom} f$ such that $g_i(x_0) < 0$ for $i \in I(\bar{x})$. Then $\bar{x}$ is a solution to $(\mathcal{M})$ if and only if there exist $\bar{y}_1, \ldots, \bar{y}_k$ in $\mathbb{R}_+$, such that*

$$0 \in \partial f(\bar{x}) + \bar{y}_1 \partial g_1(\bar{x}) + \cdots + \bar{y}_k \partial g_k(\bar{x}), \quad \bar{y}_1 g_1(\bar{x}) = 0, \ldots, \bar{y}_k g_k(\bar{x}) = 0.$$

Introducing the *Lagrangian function* ℓ by

$$\ell(x, y) := \ell_y(x) := f(x) + y_1 g_1(x) + \cdots + y_k g_k(x), \quad x \in X, \ y \in \mathbb{R}^k,$$

and the set $K(\bar{x})$ of *Karush–Kuhn–Tucker multipliers at $\bar{x}$,*

$$K(\bar{x}) := \{y := (y_1, \ldots, y_k) \in \mathbb{R}_+^k, \ 0 \in \partial \ell_y(\bar{x}), \ y.g(\bar{x}) = 0\},$$

the above condition can be written $\bar{y} \in K(\bar{x})$. Here we use the fact that $\bar{y}_i g_i(\bar{x}) \leq 0$ for all i, so that $\bar{y}_1 g_1(\bar{x}) + \cdots + \bar{y}_k g_k(\bar{x}) = 0$ is equivalent to $\bar{y}_i g_i(\bar{x}) = 0$ for all i; we also use the continuity assumption on the g_i's at $\bar{x}$. Thus in order to take the constraints into account, the condition $0 \in \partial f(\bar{x})$ of the unconstrained problem has been replaced by a similar condition with $\ell_{\bar{y}}$ instead of f. Despite this justification, the multipliers $\bar{y}_i$ seem to be artificial ingredients (in [819], by analogy with the theater, they were given the name *deus ex machinae*). However, they cannot be neglected, as shown by Exercise 1 below, even if in solving practical problems one is led to get rid of them as soon as possible. In fact, the "marginal" interpretation we provide below shows that their knowledge is not without interest, since they provide useful information about the behavior of the value of perturbed problems. In order to shed some light on such an interpretation, let us introduce for $w := (w_1, \ldots, w_k) \in \mathbb{R}^k$ the perturbed problem

$(\mathcal{M}_w)$ minimize $f(x)$ subject to $x \in C_w := \{x \in X : g_i(x) + w_i \leq 0,\ i = 1, \ldots, k\}$

and set $G := \{(x, w) \in X \times \mathbb{R}^k : g_1(x) + w_1 \leq 0, \ldots, g_k(x) + w_k \leq 0\}$,

$$p(w) := \inf\{f(x) : x \in C_w\}.$$

Since $p(w) = \inf_{x \in X} P(w, x)$, with $P(w, x) := f(x) + \iota_G(x, w)$, p is convex, G and P being convex.

Let us also introduce the set M of *Lagrange multipliers*:

$$M := \left\{ y \in \mathbb{R}_+^k : \inf_{x \in C} f(x) = \inf_{x \in X} \ell_y(x) \right\}.$$

Theorem 3.85. *Suppose $p(0)$ is finite. Then the set M of Lagrange multipliers coincides with $\partial p(0)$. Moreover, for all $\bar{x}$ in the set S of solutions to $(\mathcal{M})$ one has $K(\bar{x}) = M$.*

It follows that the set $K(\bar{x})$ is independent of the choice of $\bar{x}$ in S.

Proof. Let $y \in M$. Given $w \in \mathbb{R}^k$, then for all $x \in C_w$ and $i = 1, \ldots, k$, we have $y_i g_i(x) \leq -y_i w_i$, since $y_i \in \mathbb{R}_+$ and $g_i(x) \leq -w_i$. Thus by definition of M,

$$p(0) = \inf_{x \in X} \ell_y(x) \leq \inf_{x \in X} f(x) - \langle y, w \rangle \leq \inf_{x \in C_w} f(x) - \langle y, w \rangle = p(w) - \langle y, w \rangle,$$

so that $y \in \partial p(0)$.

Conversely, assume that the functions g_i are finite and let $y \in \partial p(0)$. We first observe that $y \in \mathbb{R}_+^k$, since for $w \in \mathbb{R}_+^k$ we have $C \subset C_{-w}$, hence $p(-w) \leq p(0)$, so that taking for w the elements of the canonical basis of $\mathbb{R}^k$, the inequalities $\langle y, -w \rangle \leq p(-w) - p(0) \leq 0$ imply that the components of y are nonnegative. Now, given $x \in X$, either $y_i g_i(x) = +\infty$ for some i and $\ell_y(x) \geq p(0)$, else taking $w_i := -g_i(x)$ for $i = 1, \ldots, k$, one has $x \in C_w$, hence $f(x) \geq p(w)$ and

$$f(x) + \langle y, g(x) \rangle \geq p(w) + \langle y, g(x) \rangle \geq p(0) + \langle y, w \rangle + \langle y, g(x) \rangle = p(0),$$

so that $\inf_{x \in X} \ell_y(x) \geq p(0)$. Since for $x \in C$ one has $\langle y, g(x) \rangle \leq 0$, hence $\inf_{x \in X} \ell_y(x) \leq \inf_{x \in C} \ell_y(x) \leq \inf_{x \in C} f(x) = p(0)$, we get $\inf_{x \in X} \ell_y(x) = p(0)$, hence $y \in M$.

Finally, let $\bar{x} \in S$ and let $\bar{y} \in K(\bar{x})$. Then since $0 \in \partial \ell_{\bar{y}}(\bar{x})$ or $\ell_{\bar{y}}(\bar{x}) = \inf_{x \in X} \ell_{\bar{y}}(x)$ and $\bar{y} \cdot g(\bar{x}) = 0$, we have $\ell_{\bar{y}}(\bar{x}) = f(\bar{x}) = \inf_{x \in C} f(x)$ and we get $\bar{y} \in M$. Conversely, let $\bar{y} \in M$. Then $f(\bar{x}) = p(0) = \inf_{x \in X} \ell_{\bar{y}}(x) \leq \ell_{\bar{y}}(\bar{x})$, so that $\bar{y} \cdot g(\bar{x}) \geq 0$. Since for all $i = 1, \ldots, k$ we have $\bar{y}_i \geq 0$ and $g_i(\bar{x}) \leq 0$, the reverse inequality holds; hence $\bar{y} \cdot g(\bar{x}) = 0$. Moreover, the relations $\inf_{x \in X} \ell_{\bar{y}}(x) = p(0) = f(\bar{x}) = \ell_{\bar{y}}(\bar{x})$ imply that $0 \in \partial \ell_{\bar{y}}(\bar{x})$. Therefore $\bar{y} \in K(\bar{x})$.   □

The bearing of multipliers on the changes of p can be made more explicit [692].

Corollary 3.86. *Let w, $w' \in \mathbb{R}^k$ and let y, y' be multipliers for the problems of minimizing $f(x)$ under the constraints $g(x) + w \leq 0$ and $g(x) + w' \leq 0$ respectively. Then the values $p(w)$ and $p(w')$ of these problems satisfy the relations*

$$\langle y, w' - w \rangle \leq p(w') - p(w) \leq \langle y', w' - w \rangle.$$

Proof. Given $w \in \mathbb{R}^k$, let $h : x \mapsto g(x) + w$ and let $q : z \mapsto \inf\{f(x) : h(x) + z \leq 0\}$. If y is a multiplier for this problem, one has $y \in \partial q(0)$, so that for $w' \in \mathbb{R}^k$, $z := w' - w$,

$$\langle y, w' - w \rangle = \langle y, z \rangle \leq q(z) - q(0) = p(w') - p(w).$$

Interchanging the roles of w and w', we get the second inequality.   □

Exercises

1. **(a)** Compute the normal cone to $\mathbb{R}_+$.
(b) Given a convex function $f : \mathbb{R} \to \mathbb{R}$, give a necessary and sufficient condition for it to attain its minimum on $C := \{x \in \mathbb{R} : -x \leq 0\}$ at 0. Taking $f(x) = x$, check that the condition $f'(0) = 0$ is not satisfied.
(c) Compute the normal cone to $\mathbb{R}_+^n$ at some $\bar{x} \in \mathbb{R}_+^n$.

2. Show that the sufficient condition of the Karush–Kuhn–Tucker theorem holds without the Slater condition and continuity assumptions.

3. State and prove a necessary and sufficient optimality condition for a program including equality constraints given by continuous affine functions.

4. **(a)** Use the Lagrangian $\ell : X \times \mathbb{R}^k \to \mathbb{R} \cup \{-\infty, +\infty\}$ given by

$$\ell(x, y) = \begin{cases} f(x) + y_1 g_1(x) + \cdots + y_k g_k(x) & \text{if } (x, y) \in X \times \mathbb{R}_+^k, \\ -\infty & \text{if } (x, y) \in X \times (\mathbb{R}^k \setminus \mathbb{R}_+^k), \end{cases}$$

to formulate optimality conditions for the problem ($\mathcal{M}$).
(b) Introduce a Lagrangian $\ell : X \times Y \to \mathbb{R}$ adapted to the problem of minimizing a convex function f under the constraint $x \in C := \{x \in X : g(x) \in -Z_+\}$, where Z_+ is a closed convex cone in a Banach space Z and $g : X \to Z$ is a map whose epigraph $E := \{(x, z) \in X \times Z : z \in Z_+ + g(x)\}$ is closed and convex.

5. Let $X := \mathbb{R}$, let $f : X \to \overline{\mathbb{R}}$, $g : X \to \mathbb{R}$ be given by $f(x) := -x^\alpha$ for $x \in \mathbb{R}_+$, with $\alpha \in (0, 1)$, $f(x) := +\infty$ for $x < 0$, $g(x) = x$ for $x \in \mathbb{R}$. Show that there is no Karush–Kuhn–Tucker multiplier at the solution of ($\mathcal{M}$) with such data.

6. (Minimum-volume ellipsoid problem) Let $(e_1, \ldots, e_n)$ be the canonical basis of $\mathbb{R}^n$ and let S_{++}^n be the set of positive definite matrices of format (n,n).
(a) Show that the identity matrix I is the unique optimal solution of the problem

$$\text{Minimize} - \log \det u, \quad u \in S_{++}^n, \quad \|u(e_i)\|^2 - 1 \leq 0, \quad i = 1, \ldots, n.$$

[Hint: Use Theorem 3.84 and some compactness argument; see [126, pp. 32, 48].]
(b) Deduce from (a) the following special form of *Hadamard's inequality*: for $u \in S_{++}^n$ and $u_i := u(e_i)$, one has $\det(u_1, \ldots, u_n) \leq \|u_1\| \cdots \|u_n\|$.

7. Characterize the tangent cone to the positive cone $L_p(S)_+$ of $L_p(S)$ for $p \in [1, \infty)$, S being a finite measured space. [See [226].]

3.6 Smoothness of Norms

In order to choose the space on which we state a given problem (when it is possible) and the substitute for the derivatives of functions, we need to know whether a space has sufficiently many smooth nontrivial functions. In particular, it is useful to know whether a power $\|\cdot\|^p$ ($p > 1$) of the norm is smooth.

In order to give some versatility to the following famous differentiability test for a norm, we adopt the framework of normed spaces in metric duality. Let us recall that two normed spaces are said to be in *metric duality* if there exists a continuous bilinear coupling $c := \langle \cdot, \cdot \rangle : X \times Y \to \mathbb{R}$ such that $\|y\| = \sup\{\langle x, y \rangle : x \in B_X\}$ for all $y \in Y$ and $\|x\| = \sup\{\langle x, y \rangle : y \in B_Y\}$ for all $x \in X$. Such a presentation enables us to treat simultaneously the case in which Y is the dual of X and the case in which X is the dual of Y. We say that a sequence (y_n) of Y c-*weakly converges* (or simply weakly converges) to $y \in Y$ if for every $x \in X$ we have $(\langle x, y_n \rangle) \to \langle x, y \rangle$. This notion coincides with weak* convergence when $Y := X^*$ and with weak convergence when $X := Y^*$.

Proposition 3.87 (Šmulian test). *Let X and Y be normed spaces in metric duality and let $\bar{x} \in S_X$. The following assertions (a) and (b) are equivalent, and if Y is the dual of X, then (a), (b), and (c) are equivalent:*

(a) *The norm of X is Fréchet (resp. Hadamard) differentiable at $\bar{x}$;*
(b) *For every pair of sequences (y_n), (z_n) of S_Y such that $(\langle \bar{x}, y_n \rangle) \to 1$, $(\langle \bar{x}, z_n \rangle) \to 1$, one has $(\|y_n - z_n\|) \to 0$ (resp. $(y_n - z_n)$ c-weakly converges to 0);*
(c) *A sequence (y_n) of S_Y is convergent (resp. c-weakly convergent) whenever $(\langle \bar{x}, y_n \rangle) \to 1$.*

Proof. (a)$\Rightarrow$(b) Suppose the norm $\|\cdot\|$ of X is Hadamard differentiable at $\bar{x} \in S_X$. By Lemma 3.30, for every given $\varepsilon > 0$ and any $u \in S_X$ there exists some $\delta > 0$ such that $\|\bar{x} + tu\| + \|\bar{x} - tu\| \leq 2 + \varepsilon t$ when $t \in [-\delta, \delta]$. Let (y_n) and (z_n) be sequences of S_Y such that $(\langle \bar{x}, y_n \rangle) \to 1$ and $(\langle \bar{x}, z_n \rangle) \to 1$. Then for $t := \delta$, one can find $k \in \mathbb{N}$ such that for all $n \geq k$ one has

$$t\langle u, y_n - z_n \rangle = \langle \bar{x} + tu, y_n \rangle + \langle \bar{x} - tu, z_n \rangle - \langle \bar{x}, y_n \rangle - \langle \bar{x}, z_n \rangle$$

$$\leq \|\bar{x} + tu\| + \|\bar{x} - tu\| - 2 + 2\delta\varepsilon \leq 3\delta\varepsilon.$$

Thus $\langle u, y_n - z_n \rangle \leq 3\varepsilon$ for $n \geq k$. Changing u into $-u$, we see that $(\langle u, y_n - z_n \rangle) \to 0$. The Fréchet case is similar, using uniformity in $u \in S_X$.

(b)$\Rightarrow$(a) Suppose the norm $\|\cdot\|$ of X is not Hadamard differentiable at $\bar{x} \in S_X$. By Lemma 3.30 there exist some $u \in S_X$, some $\varepsilon > 0$, and some sequence $(t_n) \to 0_+$ such that $\|\bar{x} + t_n u\| + \|\bar{x} - t_n u\| - 2 \geq 3t_n\varepsilon$ for all n. Let us pick y_n, z_n in S_Y such that

$$\langle \bar{x} + t_n u, y_n \rangle \geq \|\bar{x} + t_n u\| - t_n\varepsilon, \qquad \langle \bar{x} - t_n u, z_n \rangle \geq \|\bar{x} - t_n u\| - t_n\varepsilon. \tag{3.66}$$

Then $\langle \bar{x}, y_n \rangle \geq \|\bar{x} + t_n u\| - t_n\varepsilon - t_n \|u\| \cdot \|y_n\|$ and $\langle \bar{x}, y_n \rangle \leq 1$, so that $(\langle \bar{x}, y_n \rangle) \to 1$, and similarly, $(\langle \bar{x}, z_n \rangle) \to 1$. Since $\|\bar{x}\| = 1$, $\|y_n\| = 1$, $\|z_n\| = 1$, we get

$$t_n\langle u, y_n - z_n \rangle = \langle \bar{x} + t_n u, y_n \rangle + \langle \bar{x} - t_n u, z_n \rangle - \langle \bar{x}, y_n \rangle - \langle \bar{x}, z_n \rangle$$

$$\geq \|\bar{x} + t_n u\| + \|\bar{x} - t_n u\| - 2t_n\varepsilon - \|\bar{x}\| \cdot \|y_n\| - \|\bar{x}\| \cdot \|z_n\| \geq t_n\varepsilon,$$

and hence $\langle u, y_n - z_n \rangle \geq \varepsilon$, a contradiction to the assumption that $(y_n - z_n)$ c-weakly converges to 0.

When the norm $\|\cdot\|$ of X is not Fréchet differentiable at $\bar{x} \in S_X$, one can find $\varepsilon > 0$ and sequences $(t_n) \to 0_+$, (u_n) in S_X such that $\|\bar{x} + t_n u_n\| + \|\bar{x} - t_n u_n\| - 2 \geq 3t_n\varepsilon$ for all $n \in \mathbb{N}$. Then taking (y_n), $(z_n) \in S_Y$ as in relation (3.66), the preceding computation reads $\langle u_n, y_n - z_n \rangle \geq \varepsilon$, hence $\|y_n - z_n\| \geq \varepsilon$, a contradiction to the assumption that $(y_n - z_n) \to 0$.

(b)$\Rightarrow$(c) when $Y = X^*$. Let $\bar{y} := \|\cdot\|'(\bar{x})$. One has $\|\bar{y}\| \leq 1$, since the norm is Lipschitzian with rate 1 and, by homogeneity, $\langle \bar{x}, \bar{y} \rangle = \lim_{t\to 0}(1/t)(\|\bar{x} + t\bar{x}\| - \|\bar{x}\|) = 1$, so that $\bar{y} \in S_Y$ and we can take $z_n := \bar{y}$ in assertion (b). Thus (c) holds.

(c)$\Rightarrow$(b) when $Y = X^*$. Let (y_n), (z_n) be sequences of S_Y such that $(\langle x, y_n \rangle) \to 1$, $(\langle x, z_n \rangle) \to 1$. Let $w_n = y_p$ when $n := 2p$, $w_n = z_p$ when $n := 2p + 1$. Then $(\langle x, w_n \rangle) \to 1$, so that by (c), (w_n) converges (resp. c-weakly converges). Thus $(y_n - z_n) \to 0$ (resp. c-weakly converges to 0). □

The following notions will appear as dual to differentiability of the norm.

Definition 3.88. A norm $\|\cdot\|$ on a vector space X is said to be *rotund* (or *strictly convex*) if every point u of its unit sphere S_X is an extremal point of the unit ball B_X in the sense that it cannot be the midpoint of a segment of B_X not reduced to a singleton.

It is said to be *locally uniformly rotund* (LUR), or *locally uniformly convex*, if for all x, $x_n \in X$ satisfying $(\|x_n\|) \to \|x\|$, $(\|x + x_n\|) \to 2\|x\|$ one has $(x_n) \to x$.

Let us exhibit some characterizations of these properties.

Lemma 3.89. *For a normed space* $(X, \|\cdot\|)$ *the following assertions are equivalent:*

(a) $\|\cdot\|$ *is rotund;*
(b) If $x, y \in S_X$ *satisfy* $\|x + y\| = 2$, *then* $x = y$;
(c) If $x, y \in X$ *satisfy* $\|x + y\|^2 = 2\|x\|^2 + 2\|y\|^2$, *then* $x = y$;
(d) If $x, y \in X \setminus \{0\}$ *satisfy* $\|x + y\| = \|x\| + \|y\|$, *then* $x = \lambda y$ *for some* $\lambda \in \mathbb{R}_+$.

Proof. (a)$\Leftrightarrow$(b) is a reformulation, since $\|x + y\| = 2$ means that $\frac{1}{2}(x + y) \in S_X$.
  (c)$\Rightarrow$(b) is immediate. (b)$\Rightarrow$(c) For $x, y \in X$, since

$$2\|x\|^2 + 2\|y\|^2 - \|x + y\|^2 \geq 2\|x\|^2 + 2\|y\|^2 - (\|x\| + \|y\|)^2 = (\|x\| - \|y\|)^2,$$

the relation $2\|x\|^2 + 2\|y\|^2 - \|x + y\|^2 = 0$ implies $\|x\| = \|y\|$. Setting $x := ru$, $y := rv$ with $r := \|x\| = \|y\|$, $u, v \in S_X$, for $r > 0$ we get $\|u + v\| = 2$, so that $u = v$ and $x = y$, whereas for $r = 0$ we have $x = y = 0$.
  (b)$\Rightarrow$(d) Suppose $\|x + y\| = \|x\| + \|y\|$ and $r := \|x\| \leq s := \|y\|$. Then

$$2 \geq \|r^{-1}x + s^{-1}y\| \geq r^{-1}\|x + y\| - \|r^{-1}y - s^{-1}y\|$$

$$= r^{-1}(\|x\| + \|y\|) - (r^{-1} - s^{-1})\|y\| = r^{-1}\|x\| + s^{-1}\|y\| = 2.$$

Thus $\|r^{-1}x + s^{-1}y\| = 2$ and $r^{-1}x = s^{-1}y$. (d)$\Rightarrow$(b) is immediate. $\square$

Lemma 3.90. *For a normed space* $(X, \|\cdot\|)$ *the following assertions are equivalent:*

(a) $\|\cdot\|$ *is locally uniformly rotund;*
(b) If $x, x_n \in S_X$ *for* $n \in \mathbb{N}$ *satisfy* $(\|x + x_n\|) \to 2$, *then* $(x_n) \to x$;
(c) If $x, x_n \in X$ *satisfy* $(2\|x\|^2 + 2\|x_n\|^2 - \|x + x_n\|^2) \to 0$, *then* $(x_n) \to x$.

Proof. (a)$\Rightarrow$(b) is obvious. The converse is obtained by considering (in the nontrivial case $x \neq 0$) $u := x/\|x\|$, $u_n := x_n/\|x_n\|$ (for n large enough).
  (c)$\Rightarrow$(b) is immediate. (b)$\Rightarrow$(c) For $x, x_n \in X$, since

$$2\|x\|^2 + 2\|x_n\|^2 - \|x + x_n\|^2 \geq 2\|x\|^2 + 2\|x_n\|^2 - (\|x\| + \|x_n\|)^2 = (\|x\| - \|x_n\|)^2,$$

the relation $\lim_n (2\|x\|^2 + 2\|x_n\|^2 - \|x + x_n\|^2) = 0$ implies $(\|x_n\|) \to \|x\|$. (c) follows by considering $(x_n/\|x_n\|)$. $\square$

The LUR property has interesting consequences, as the next proposition shows.

Proposition 3.91. *If* $\|\cdot\|$ *is a LUR norm, then X has the (sequential) Kadec–Klee property: a sequence* (x_n) *of X converges to $x \in X$ whenever it weakly converges to x and* $(\|x_n\|) \to \|x\|$.

Proof. Let $x \in X$ and let $(x_n)_{n \in I}$ be a weakly convergent sequence whose limit x is such that $(\|x_n\|) \to \|x\|$. Then $\limsup_n \|x + x_n\| \leq \limsup_n (\|x\| + \|x_n\|) = 2\|x\|$. On the other hand, since the norm is weakly lower semicontinuous, we have $\liminf_n \|x + x_n\| \geq \|2x\|$. Thus $(\|x + x_n\|) \to 2\|x\|$, and since the norm is LUR, we get $(x_n) \to x$. $\square$

Let us turn to duality results.

Proposition 3.92. *Let $\|\cdot\|$ be a norm on X and let $\|\cdot\|_*$ be its dual norm.*

(a) If $\|\cdot\|_*$ is a rotund norm, then $\|\cdot\|$ is Hadamard differentiable on $X \setminus \{0\}$.
(b) If $\|\cdot\|_*$ is Hadamard differentiable on $X^* \setminus \{0\}$, then $\|\cdot\|$ is a rotund norm.

In particular, a compatible norm on a reflexive Banach space X is Hadamard differentiable on $X \setminus \{0\}$ if and only if its dual norm is strictly convex.

Proof. (a) By Corollary 3.26, it suffices to show that for every $x \in X \setminus \{0\}$,

$$S(x) := \partial \|\cdot\| (x) = \{x^* \in X^* : \|x^*\|_* = 1, \langle x^*, x \rangle = \|x\|\}$$

is a singleton. Let $x^*, y^* \in S(x)$. Then $2\|x\| = \langle x^*, x \rangle + \langle y^*, x \rangle \le \|x^* + y^*\|_* \cdot \|x\| \le 2\|x\|$, and by assertion (b) of Lemma 3.89, we have $x^* = y^*$.

(b) If $\|\cdot\|$ is not rotund, one can find $x, y \in S_X$ such that $x \ne y$ and $(1-t)x + ty \in S_X$ for all $t \in [0,1]$. Taking $t := 1/2$ and $f \in S_{X^*}$ such that $f((1-t)x + ty) = 1$, we see that $1 = f((1-t)x + ty) = (1-t)f(x) + tf(y) \le 1$, so that this inequality is an equality and $f(x) = f(y) = 1$. Viewing x and y as elements of X^{**}, we have $x, y \in \partial \|\cdot\|_* (f)$, so that $\|\cdot\|_*$ is not differentiable at f. $\qquad\square$

Proposition 3.93. *Let $(X, \|\cdot\|)$ be a normed space. If the dual norm $\|\cdot\|_*$ is LUR, then $\|\cdot\|$ is Fréchet differentiable on $X \setminus \{0\}$.*

Proof. We use Šmulian test (c). Let $x \in S_X$. Using a corollary of the Hahn–Banach theorem, we pick $f \in S_{X^*}$ such that $f(x) = 1$. Let (f_n) be a sequence of S_{X^*} such that $(f_n(x)) \to 1$. Since

$$2 \ge \|f + f_n\|_* \ge (f + f_n)(x) \to 2,$$

we have $\lim_n (2\|f\|_*^2 + 2\|f_n\|_*^2 - \|f + f_n\|_*^2) = 0$; hence by the LUR property, $(f_n) \to f$. Then by Proposition 3.87, $\|\cdot\|$ is Fréchet differentiable at x, hence on $(0, +\infty)x$. $\qquad\square$

So, it will be useful to detect when a norm on the dual of X is a dual norm.

Lemma 3.94. *An equivalent norm $\|\cdot\|$ on the dual X^* of a Banach space X is the dual norm of an equivalent norm $\|\cdot\|_X$ on X if and only if it is weak* lower semicontinuous.*

Proof. If $\|\cdot\|$ is the dual norm of an equivalent norm $\|\cdot\|_X$, then $\|\cdot\| = \sup\{\langle x, \cdot \rangle : x \in X, \|x\|_X = 1\}$ is weak* lower semicontinuous as a supremum of weak* continuous linear forms.

Conversely, if $\|\cdot\|$ is weak* lower semicontinuous, its unit ball B^* is convex, weak* closed, hence coincides with its bipolar. Then one can see that $\|\cdot\|$ is the dual norm of the Minkowski gauge of the polar set of B^*, a compatible norm on X. $\qquad\square$

The following renorming theorem is of interest. We refer to [98, p. 89], [289], [292, p. 113], [376, Theorem 8.20] for the proof of its second assertion.

Theorem 3.95. (a) *Every separable Banach space X has an equivalent norm that is Hadamard differentiable on* $X \setminus \{0\}$.

   (b) *Every Banach space X whose dual is separable has an equivalent norm that is Fréchet differentiable on* $X \setminus \{0\}$.

Proof. (a) Let $(e_n)_{n \in \mathbb{N}}$ be a countable dense subset of B_X. Define a norm by

$$\|f\| = \left[\|f\|_0^2 + \sum_{n=0}^{\infty} 2^{-n} f^2(e_n) \right]^{1/2}, \qquad f \in X^*,$$

where $\|\cdot\|_0$ is the original norm of X^*. The norm $\|\cdot\|$ is easily seen to be weak* lower semicontinuous, so that it is the dual norm of some norm $\|\cdot\|_*$ on X. In view of Lemma 3.92, it suffices to show that $\|\cdot\|$ is rotund. Let $f, g \in X^*$ be such that $\|f + g\|^2 = 2\|f\|^2 + 2\|g\|^2$. Since $2\|f\|_0^2 + 2\|g\|_0^2 \geq \|f + g\|_0^2$ and $2f^2(e_n) + 2g^2(e_n) \geq (f + g)^2(e_n)$ for all n, we get that these last inequalities are equalities, so that $(f - g)^2(e_n) = 2f^2(e_n) + 2g^2(e_n) - (f + g)^2(e_n) = 0$ for all n. Thus $f(e_n) = g(e_n)$ for all n, and by density, $f = g$.   □

Exercises

1. Using the Šmulian test, show that if the norm of a normed space is Fréchet differentiable on $X \setminus \{0\}$, then it is of class C^1 there.

2. Show that a normed space X is strictly convex if and only if each point x of its unit sphere S_X is an *exposed point* of the unit ball B_X, i.e., for each $x \in S_X$ there exists $f \in X^*$ such that $f(x) > f(u)$ for all $u \in B_X \setminus \{x\}$.

3. Let S be a locally compact topological space and let $X := C_0(S)$ be the space of bounded continuous functions on S converging to 0 at infinity: $x \in C_0(S)$ iff $x(\cdot)$ is bounded, continuous on S, and if for every $\varepsilon > 0$, one can find a compact subset K of S such that $\sup |x(S \setminus K)| \leq \varepsilon$.
(a) Show that the supremum norm $\|\cdot\|_\infty$ is Hadamard differentiable at $x \in X$ if and only if $M_x := \{s \in S : |x(s)| = \|x\|_\infty\}$ is a singleton.
(b) Show that $\|\cdot\|_\infty$ is Fréchet differentiable at $x \in X$ if and only if M_x is a singleton $\{\bar{s}\}$ such that $\bar{s}$ is an isolated point of S. (See [289, p. 5].)

4. Let $X := \ell_1(I)$ be the space of absolutely summable families $x := (x_i)_{i \in I}$ endowed with the norm $\|x\|_1 := \sum_{i \in I} |x_i|$.
(a) Show that $\|\cdot\|_1$ is nowhere Hadamard differentiable if I is uncountable.
(b) If $I := \mathbb{N}$, show that $\|\cdot\|_1$ is Hadamard differentiable at x if and only if $x_i \neq 0$ for all $i \in I$.
(c) If $I := \mathbb{N}$, show that $\|\cdot\|_1$ is nowhere Fréchet differentiable. (See [289, p. 6].)

5. A normed space $(X, \|\cdot\|)$ is said to be *uniformly smooth* if the function $\sigma_X : \mathbb{R}_+ \to \mathbb{R}$ given by

$$\sigma_X(t) := \sup\{(1/2)(\|x+tu\|+\|x-tu\|)-\|x\| : x,u \in S_X\}$$

is a remainder (i.e., $\sigma_X(t)/t \to 0$ as $t \to 0_+$).
(a) Show that $(X,\|\cdot\|)$ is uniformly smooth iff $\|\cdot\|$ is uniformly differentiable on S_X. (See [376].)
(b) Show that if $(X,\|\cdot\|)$ is uniformly smooth then the derivative S of the norm satisfies

$$\|S(x)-S(y)\| \leq \sigma_X(2\|x/\|x\|-y/\|y\|\|)\|x/\|x\|-y/\|y\|\|$$

for $x,y \in X \setminus \{0\}$.

6. A normed space $(X,\|\cdot\|)$ is said to be *uniformly rotund* if for every pair of sequences (x_n), (y_n) of B_X such that $(\|x_n+y_n\|) \to 2$ one has $(\|x_n-y_n\|) \to 0$.
(a) Show that $(X,\|\cdot\|)$ is *uniformly rotund* iff $\gamma_X : \mathbb{R}_+ \to \mathbb{R}_\infty$ given by

$$\gamma_X(s) := \inf\{1-\|(1/2)(x+y)\| : x,y \in S_X, \|(1/2)(x-y)\| \geq s\}, \qquad s \in [0,1],$$

$\gamma_X(s) := +\infty$ for $s \in \mathbb{R} \setminus [0,1]$ is a *gage*, i.e., $(s_n) \to 0$ whenever $(\gamma_X(s_n)) \to 0$.
(b) Show that $s \mapsto \gamma_X(s)/s$ is nondecreasing. (See [394].)
(c) Show that $(X,\|\cdot\|)$ is uniformly rotund iff X^* is uniformly smooth and

$$\sigma_{X^*}(t) = \sup_{t>0}(st-\gamma_X(s)), \qquad t > 0.$$

(d)] Show that $(X,\|\cdot\|)$ is reflexive if it is either uniformly rotund or uniformly smooth.

7. Show that the space $X := L_p(S,\mu)$ $(p > 1)$ is uniformly rotund and uniformly smooth with

$$\gamma_X(s) = (p-1)s^2/2 + o(s^2) \quad \text{for } p \in (1,2), \gamma_X(s) = s^p/p + o(s^p) \text{ for } p \geq 2,$$

$$\sigma_X(t) = t^p/p + o(t^p) \quad \text{for } p \in (1,2], \sigma_X(t) = (p-1)t^2/2 + o(t^2) \text{ for } p \geq 2.$$

(See [79, 467].)

3.7 Favorable Classes of Spaces

We need to single out classes of Banach spaces on which continuous convex functions have sufficiently many points of differentiability. That justifies the following definition.

Definition 3.96. A Banach space X is called an *Asplund space* (resp. a *Mazur space*) if every continuous convex function f defined on an open convex subset W of X is Fréchet (resp. Hadamard) differentiable on a dense $\mathcal{G}_\delta$ subset D of W.

These terminologies are not the original ones: initially, Asplund spaces were called differentiability spaces, and usually Mazur spaces are called weak Asplund spaces.

By Theorem 3.34, separable Banach spaces are Mazur spaces. A stronger separability assumption ensures that the space is Asplund.

Theorem 3.97. *A Banach space X whose dual is separable is an Asplund space.*

Proof. Let f be a continuous convex function on an open convex subset W of X. For all $x \in W$ let $g_x \in \partial f(x)$ and $\delta(x) := d(x, X \setminus W)$. The set $A := W \setminus F$ of points of W at which f is nondifferentiable is the union over $m \in \mathbb{N} \setminus \{0\}$ of the sets

$$A_m := \{x \in W : \forall r \in (0, \delta(x))\ \exists v \in rB_X,\ f(x+v) - f(x) - g_x(v) > (6/m)\|v\|\}.$$

Since X^* is separable, for all $m \in \mathbb{N} \setminus \{0\}$ there is a countable cover $\mathscr{B}_m := \{B_{m,n} : n \in \mathbb{N}\}$ of X^* by balls with radius $1/m$. Let $A_{m,n} := \{x \in A_m : g_x \in B_{m,n}\}$. Since W is a Baire space, and since A is the union of the sets $A_{m,n}$, it suffices to show that the closure of $A_{m,n}$ in W has an empty interior. Given an element w of this closure, let us show that for every $\varepsilon \in (0, \delta(w))$, the ball $B(w, \varepsilon)$ is not contained in the closure of $A_{m,n}$ in W. We will show that there exists some $y \in B(w, \varepsilon)$ that has a neighborhood V disjoint from $A_{m,n}$. Without loss of generality, taking a smaller ε if necessary, we may suppose f is Lipschitzian on $B(w, \varepsilon)$ with rate k for some $k > 1/m$. Since w is in the closure of $A_{m,n}$, we can find some $x \in A_{m,n} \cap B(w, \varepsilon)$. By definition of A_m, taking $r \in (0, \varepsilon - d(w, x))$, there exists $v \in rB_X$ such that

$$f(x+v) - f(x) > g_x(v) + (6/m)\|v\|. \tag{3.67}$$

We will show that for $y := x + v$, $s := \|v\|$, $V := B(y, s/km) \cap B(w, \varepsilon) \in \mathscr{N}(y)$, we have $V \cap A_{m,n} = \varnothing$. Suppose, to the contrary, that one can find some $z \in V \cap A_{m,n}$. Then by definition of $A_{m,n}$, we have $\|g_x - g_z\| < 2/m$, and since $g_z \in \partial f(z)$,

$$f(x) - f(z) \geq g_z(x - z).$$

Combining this inequality with relation (3.67), we obtain

$$f(x+v) - f(z) \geq g_x(v + x - z) - g_{::}(x - z) + g_z(x - z) + (6/m)\|v\|. \tag{3.68}$$

Since $\|v + x - z\| = \|y - z\| < s/km$ and $\|g_x\| \leq k$, we have $|g_x(v + x - z)| < s/m$. Moreover, the inequalities $\|x - z\| = \|y - v - z\| \leq \|y - z\| + \|v\| < s/km + s < 2s$, $|g_x(x - z) - g_z(x - z)| \leq 2s\|g_x - g_z\| < 4s/m$, and (3.68) yield

$$f(x+v) - f(z) > -s/m - 4s/m + (6/m)\|v\| = s/m,$$

a contradiction to the inequality $\|v + x - z\| < s/km$ and the fact that f is Lipschitzian with rate k on the ball $B(w, \varepsilon)$, which contains both $y := x + v$ and z. $\qquad\square$

The importance of Asplund spaces for generalized differentiation is illuminated by the following deep result, which is outside the scope of this book.

Theorem 3.98 (Preiss [850]). *Every locally Lipschitzian function f on an open subset U of an Asplund space is Fréchet differentiable on a dense subset of U.*

The class of Asplund spaces can be characterized in a number of different ways and satisfies interesting stability and duality properties. In particular, it is connected with the Radon–Nikodým property described below. Let us just mention the following facts in this connection, referring to [98, 289, 832] for proofs and additional information.

Proposition 3.99. *(a) A Banach space X is an Asplund space if and only if every separable subspace of X is an Asplund space.*

(b) A Banach space X is an Asplund space if and only if the dual of every separable subspace of X is separable.

In particular, every reflexive Banach space is an Asplund space. On the other hand, $C([0,1])$, $L_1([0,1])$, $\ell_1(\mathbb{N})$, and $\ell_\infty(\mathbb{N})$ are not Asplund spaces.

Let us state some permanence properties.

Proposition 3.100. *(a) The class of Asplund spaces is closed under isomorphisms; that is, if X and Y are isomorphic Banach spaces and if X is Asplund, then Y is Asplund.*

(b) Every closed linear subspace of an Asplund space is an Asplund space.

(c) Every quotient space of an Asplund space is an Asplund space.

(d) The class of Asplund spaces is closed under extensions: if X is a Banach space and Y is an Asplund subspace of X such that the quotient space X/Y is Asplund, then X is Asplund.

Corollary 3.101. *If X is an Asplund space, then for all $n \in \mathbb{N} \setminus \{0\}$, X^n is an Asplund space.*

Proof. Let us prove the result by induction. Assume that X^n is Asplund. The graph Y of the map $s : (x_1, \ldots, x_n) \mapsto x_1 + \cdots + x_n$ is isomorphic to X^n, hence is Asplund by assumption. The quotient of X^{n+1} by Y is isomorphic to X, since s is onto. Thus X^{n+1} is Asplund.   □

The following result clarifies the relationships between Asplund spaces and spaces that can be renormed by a Fréchet differentiable norm. It will be explained and proved in the next chapter (Corollary 4.66).

Theorem 3.102 (Ekeland–Lebourg [352]). *If a Banach space can be renormed by a norm that is Fréchet differentiable off 0 (or more generally, if it admits a Fréchet differentiable bump function), then it is an Asplund space.*

For separable spaces, there is a partial converse, but in general the converse fails: R. Haydon has exhibited a compact space T such that $C(T)$ is Asplund but cannot be renormed by a Fréchet (or even Gâteaux) differentiable norm.

Proposition 3.103. *For every separable Asplund space X there exists a norm inducing the topology of X that is Fréchet differentiable on $X \setminus \{0\}$.*

Proof. This is a consequence of Proposition 3.99 and Theorem 3.95.   □

On the other hand, on any Banach space that is not Asplund, one can find a Lipschitzian convex function that is nowhere differentiable. One can even take for it an equivalent norm, as explained in the next proposition. In order to prove this and give a dual characterization of Asplund spaces, let us define a *weak* slice* of a nonempty set $A \subset X^*$ as a subset of A of the form

$$S(x,A,\alpha) = \{x^* \in A : \langle x^*, x \rangle > \sigma_A(x) - \alpha\},$$

where $x \in X \setminus \{0\}$, $\alpha > 0$, and where σ_A is the support function of A:

$$\sigma_A(x) = \sup\{\langle x^*, x \rangle : x^* \in A\}.$$

The subset A is said to be weak* *dentable* if it admits weak* slices of arbitrarily small diameter. The space X^* is said to have the dual *Radon–Nikodým property* if every nonempty bounded subset A of X^* is weak* dentable. This property is important in functional analysis, in particular for vector measures (see [98, 832, 941]).

Theorem 3.104 ([832, Theorem 5.7]). *A Banach space X is an Asplund space if and only if its dual space X^* has the dual Radon–Nikodým property.*

The following proposition shows the implication that the dual of an Asplund space has the dual Radon–Nikodým property. We omit the reverse implication.

Proposition 3.105. *Let $(X, \|\cdot\|)$ be a Banach space whose dual space does not have the dual Radon–Nikodým Property. Then there exist $c > 0$ and an equivalent norm $\|\cdot\|'$ on X such that for all $x \in X$,*

$$\limsup_{w \to 0,\, w \neq 0} \frac{1}{\|w\|'} \left(\|x + w\|' + \|x - w\|' - 2\|x\|' \right) > c. \tag{3.69}$$

In particular $\|\cdot\|'$ is nowhere Fréchet differentiable.

Proof. Since X^* does not have the dual Radon–Nikodým property, there exist $c > 0$ and a nonempty bounded subset A of X^* whose weak* slices have diameter greater than $3c$. In particular, for all $x \in X \setminus \{0\}$, the weak* slice $S(x,A,c)$ of A has diameter greater than $3c$. Let us pick $y^*, z^* \in S(x,A,c)$ such that $\|y^* - z^*\| > 3c$. We can find $u \in S_X$ such that $\langle y^* - z^*, u \rangle > 3c$. Because the definition of $S(x,A,c) \subset A$ ensures that $\langle y^*, x \rangle > \sigma_A(x) - c$, $\langle z^*, x \rangle > \sigma_A(x) - c$, we get

$$\sigma_A(x+u) + \sigma_A(x-u) \geq \langle y^*, x+u \rangle + \langle z^*, x-u \rangle$$
$$\geq (\sigma_A(x) - c + \langle y^*, u \rangle) + (\sigma_A(x) - c - \langle z^*, u \rangle)$$
$$\geq 2\sigma_A(x) + \langle y^* - z^*, u \rangle - 2c > 2\sigma_A(x) + c.$$

Let $b > \sup\{\|a\| : a \in A\}$. Since for every sublinear function h, in particular $h = \|\cdot\|$ and $h = \sigma_{-A}$, one has $h(x+u) + h(x-u) - 2h(x) \geq 0$, we get that $\|\cdot\|' := b\|\cdot\| +$

$\sigma_A + \sigma_{-A}$ satisfies the announced relation and is a norm equivalent to $\|\cdot\|$. For $x = 0$ and $c < 2$ this relation is obvious. $\square$

The class we introduce now has much interest in relation to Hadamard derivatives.

Definition 3.106. A Banach space X is a *weakly compactly generated space*, a WCG space for short, if there is a weakly compact set $Q \subset X$ such that X coincides with the closure of span Q (the smallest linear subspace of X containing Q).

Thus any separable space is WCG (take $Q = \{0\} \cup \{n^{-1}x_n, \, n = 1, 2, \dots\}$, where $\{x_n : n \in \mathbb{N}\}$ is a dense countable subset of the unit sphere) and any reflexive space X is WCG (take $Q = B_X$). The space $\ell_1(I)$ described below is a WCG space iff I is countable. The space $L_1(T, \mu)$ is a WCG space iff μ is σ-finite. As noted above, some important separable (hence WCG) spaces are not Asplund. Though the definition of a WCG space is purely topological (in contrast to the definition of Asplund spaces, which is analytic), the two classes have a substantial intersection. We shall see that the class of spaces that are both WCG and Asplund is a convenient framework for developing nonconvex subdifferential calculus.

Characterizations of WCG spaces are mentioned in the next theorem. Here, given a set I, we denote by $c_0(I)$ the set of families $y := (y_i)_{i \in I}$ such that for all $r > 0$ the set $\{i \in I : |y_i| > r\}$ is finite, and we endow the space $c_0(I)$ with the norm $\|y\|_\infty :=$ $\sup_{i \in I} |y_i|$ for $y := (y_i)_{i \in I}$. This space is complete, and the subspace $c_{00}(I) := \mathbb{R}^{(I)}$ of families $y := (y_i)_{i \in I}$ whose support if finite is dense in $c_0(I)$, the support of $(y_i)_{i \in I}$ being the set of $i \in I$ such that $y_i \neq 0$. Thus the dual of $c_0(I)$ can be identified with $\ell_1(I)$, the space of absolutely summable families $y := (y_i)_{i \in I}$ with the norm $y \mapsto \|y\|_1 := \sum_i |y_i|$, the supremum of finite sums $\sum_{j \in J} |y_j|$ over the family of finite subsets J of I. Identifying $y := (y_i)_{i \in I}$ with a function $i \mapsto y(i) := y_i$, we see that the weak topology on $c_0(I)$ coincides with the topology of pointwise convergence on I. Taking $Q := \{e_j : j \in I\} \cup \{0\}$, where $e_j(i) = 1$ if $i = j$, 0 otherwise, we see that $c_0(I)$ is a WCG space. It is even an archetype of a WCG space:

Theorem 3.107 (Amir–Lindenstrauss [9, 258, 364]). *For a Banach space X the following properties are equivalent:*

(a) *X is a WCG space;*
(b) *There are a reflexive Banach space W and an injective, linear, continuous map $j : W \to X$ with dense image;*
(c) *There exist a set I and an injective, linear, continuous map $h : X^* \to c_0(I)$ that is continuous from the weak* topology on X^* to the weak topology on $c_0(I)$.*

The two facts we need about WCG spaces are the next result and Corollary 3.110.

Theorem 3.108 ([376, Theorems 11.16, 11.20], [447, 937]). *If X is a WCG space, then there is an equivalent norm on X that is both locally uniformly rotund and Hadamard differentiable off the origin.*

An important consequence of Theorem 3.107 for our aims is the next result.

Theorem 3.109 (Borwein–Fitzpatrick [116]). *Let X be a WCG space, let T be a subset of a topological space P, and let 0 be a point in the closure of T such that 0 has a countable basis of neighborhoods. If $F : T \rightrightarrows X^*$ has a bounded image, then one has*

$$\mathrm{w}^* - \mathrm{seq} - \limsup_{t(\in T) \to 0} F(t) = \mathrm{w}^* - \limsup_{t(\in T) \to 0} F(t).$$

Proof. Let $(U_n)_{n \geq 1}$ be a countable basis of neighborhoods of 0. Setting $F_n := F(T \cap U_n)$ and denoting by F_∞ the set of all weak* limits of sequences (x_n^*) such that $x_n^* \in F_n$ for all $n \geq 1$, we see that the result amounts to the relation

$$F_\infty = \bigcap_{n \geq 1} \mathrm{cl}^*(F_n)$$

for every nonincreasing sequence (F_n) of bounded subsets of X^*. Let $h : X^* \to Y := c_0(I)$ be a linear continuous injection that is weak* to weak continuous as given by Theorem 3.107. For $n \geq 1$, let H_n be the weak closure of $h(F_n)$, so that $h(\mathrm{cl}^*(F_n)) \subset H_n$. Let $r > 0$ be such that $F_n \subset rB_{X^*}$ for all n. Since h induces a homeomorphism from rB_{X^*} endowed with the weak* topology onto its image endowed with the weak topology, the set $H_\infty := h(F_\infty)$ is the set of weak limits of sequences (y_n) such that $y_n \in h(F_n)$ for all $n \geq 1$ and it suffices to show that H_∞ contains the intersection H of the family (H_n), the opposite inclusion being obvious.

Given $y_0 := (y_0(i))_{i \in I} \in H$, let us construct inductively a sequence (y_n) weakly converging to y_0 such that $y_n \in h(F_n)$ for all $n \geq 1$. Let us start with an arbitrary element y_1 of $h(F_1)$. Let us assume that $y_1, \ldots, y_{n-1}$ have been chosen, and for $k \in \{0, \ldots, n-1\}$, let $I_k := \{i_{j,k} : j \in \mathbb{N}\}$ be the support of y_k, which is countable. Taking into account the fact that y_0 is in the weak closure of $h(F_n)$, we pick $y_n := (y_n(i))_{i \in I} \in h(F_n)$ such that $|y_n(i) - y_0(i)| \leq 1/n$ for all $i \in J_n := I_{n,0} \cup \cdots \cup I_{n,n-1}$, where $I_{n,k} := \{i_{0,k}, \ldots, i_{n,k}\}$ for $k = 0, \ldots, n-1$. Let us show that (y_n) weakly converges to y_0. For i in the union J of the sets J_n (for $n \geq 1$), it is clear that $(y_n(i))_n \to y_0(i)$. Since J is also the union of the sets I_n (for $n \geq 0$), for $i \in I \setminus J$ one has $y_0(i) = 0$, $y_n(i) = 0$. Thus $(y_n) \overset{*}{\to} y_0$. $\qquad\square$

Taking for F a constant multimap, we get the following consequence.

Corollary 3.110. *If B is a bounded subset of the dual of a WCG space X, then for the weak* topology the sequential closure of B coincides with the closure of B. In particular, every sequentially weak* closed bounded subset of X^* is weak* closed.*

Corollary 3.111. *The closed unit ball of the dual X^* of a WCG space is sequentially compact for the weak* topology in the sense that every sequence of B_{X^*} has a weak* convergent subsequence.*

Proof. Given a bounded sequence (x_n^*) of X^*, let $F(n) := \{x_p : p \geq n\}$ for $n \in \mathbb{N}$ and let x^* be a weak* cluster point of (x_n^*), i.e., a point in $\mathrm{cl}^*(F(n))$ for all $n \in \mathbb{N}$. Theorem 3.109 yields a sequence $(y_n^*) \overset{*}{\to} x^*$ such that $y_n^* \in F(n)$ for all n. It is the required subsequence of (x_n^*). $\qquad\square$

In fact, this sequential compactness property is valid in a class of spaces larger than the class of WCG spaces. We have a general criterion ensuring such a property.

Lemma 3.112. *Let T be a compact topological space such that every nonempty closed subset S of T has a $\mathscr{G}_\delta$-point s, i.e., a point $s \in S$ such that $\{s\} = \cap_n S_n$, where S_n is an open subset of S. Then T is sequentially compact.*

Proof. Let (t_n) be a sequence of T. For $m \in \mathbb{N}$, let $T_m := \mathrm{cl}(\{t_n : n \geq m\})$. Then $S := \cap_m T_m$ is the set of cluster points of (t_n), hence is closed and nonempty. By assumption, there exist $s \in S$ and a sequence (G_n) of open subsets of T such that $\{s\} = \cap_n S_n$, where $S_n := G_n \cap S$. We may suppose the sequence (G_n) is decreasing, and since T is regular, we may even suppose that $\mathrm{cl}(G_{n+1}) \subset G_n$ for all $n \in \mathbb{N}$. Since $s \in G_n \cap T_m$ for all $m, n \in \mathbb{N}$, we have $G_n \cap \{t_n : n \geq m\} \neq \varnothing$. Therefore, we can define inductively an increasing sequence $(k(n))_n$ of $\mathbb{N}$ such that $t_{k(n)} \in G_n$ for all n. Let t be a cluster point of the sequence $(t_{k(n)})_n$. Then t is a cluster point of (t_n); hence $t \in S$ and $t \in \cap_n \mathrm{cl}(G_n) = \cap_n G_n$. Thus $t = s$ and s is the only cluster point of $(t_{k(n)})$. It follows that the subsequence $(t_{k(n)})$ of (t_n) converges to s. $\qquad\square$

Theorem 3.113 (Hagler, Johnson). *Let X be a Banach space such that every continuous sublinear function on X has a point of Gâteaux differentiability. Then the closed unit ball B_{X^*} of X^* is sequentially compact for the weak* topology.*

In particular, for every Mazur (or Asplund) space X the dual unit ball B_{X^*} of X^* is sequentially compact for the weak topology.*

Proof. In view of the lemma, it suffices to show that every closed nonempty subset S of the closed unit ball T of X^* endowed with the weak* topology has a $\mathscr{G}_\delta$-point. Let $h : X \to \mathbb{R}$ be the support function of S: $h(x) := \sup\{\langle x, y \rangle : y \in S\}$. Since h is a continuous sublinear function, it has by assumption a Gâteaux differentiability point $\overline{x}$. Let $\overline{y} := h'(\overline{x})$ and for $n \in \mathbb{N}$, let

$$G_n := \{y \in T : \langle \overline{x}, y \rangle > h(\overline{x}) - 2^{-n}\}.$$

Since $\langle \overline{x}, \overline{y} \rangle = h'(\overline{x}) \cdot \overline{x} = h(\overline{x})$, we have $\overline{y} \in S_n := G_n \cap S$ for all $n \in \mathbb{N}$ and S_n is open in S. Let us show that $\cap_n S_n = \{\overline{y}\}$. Take $z \in \cap_n S_n$. Then $z \in S$, $\langle \overline{x}, z \rangle = h(\overline{x})$, and for all $x \in X$ we have

$$\langle x, z - \overline{y} \rangle = \lim_{t \to 0_+} \frac{\langle \overline{x} + tx, z \rangle - \langle \overline{x}, z \rangle}{t} - \langle x, \overline{y} \rangle \leq \lim_{t \to 0_+} \frac{h(\overline{x} + tx) - h(\overline{x})}{t} - \langle x, \overline{y} \rangle = 0.$$

Thus $z = \overline{y}$, x being arbitrary in X, and $\overline{y}$ is a $\mathscr{G}_\delta$-point of S. $\qquad\square$

Combining fuzzy sum rules with Theorem 3.113, one gets other consequences.

Corollary 3.114 (Hagler–Sullivan [460], Stegall [903]). *If X has a Gâteaux differentiable norm compatible with its topology, or if X has a Lipschitzian Hadamard differentiable bump function, in particular if X is a subspace of a WCG space, then the unit ball in X^* is sequentially compact for the weak-star topology.*

Exercises

1. Show that the weak topology of a Banach space need not be sequential. [Hint: In a separable Hilbert space X with orthonormal base (e_n) show that 0 is in the weak closure of the set $S := \{e_m + m e_n : m, n \in \mathbb{N}, m < n\}$ but no sequence in S weakly converges to 0.]

2. (**Šmulian's theorem**). Prove that every sequence of a weakly compact subset of a Banach space has a weakly convergent subsequence.

3. Let I be an infinite set and let $X := \ell_\infty(I)$ be the space of bounded functions on I with the supremum norm. Show that the unit ball B_{X^*} of X^* contains a weak* compact subset that has no weak* convergent sequence besides those that are eventually constant.

4. Let I be an uncountable set and let $X := \ell_\infty(I)$ be as in Exercise 3. Show that the unit ball B_{X^*} of X^* is weak* compact but not weak* sequentially compact.

5. Show that on the space $X := \ell_\infty := \ell_\infty(\mathbb{N})$ of bounded sequences there are continuous sublinear functions that are nowhere Gâteaux differentiable. [Hint: By Theorem 3.113, it suffices to show that there is a sequence (f_n) of B_{X^*} that has no convergent subsequence. Define f_n by $f_n(x) := x_n$, where $x := (x_n) \in \ell_\infty$. Given an increasing sequence $(k(n))$ of $\mathbb{N}$, let $x := (x_n) \in \ell_\infty$ be defined by $x_{k(n)} := (-1)^n$ and $x_p = 0$ for $p \notin k(\mathbb{N})$. Then $(f_{k(n)})_n$ cannot weak* converge, since $\langle x, f_{k(n)} \rangle = (-1)^n$.]

6. Show that the class $\mathscr{W}$ of Banach spaces having weak* sequentially compact dual balls is stable under the following operations: (a) taking dense continuous linear images; (b) taking quotients; (c) taking subspaces.

[see [293, p. 227]]

7. (**Davis–Figiel–Johnson–Pelczynski theorem**) Let Q be a weakly compact symmetric convex subset of a Banach space X. Show that there exists a weakly compact symmetric convex subset P of X containing Q such that the linear span Y of P endowed with the gauge of P is a reflexive space.

3.8   Notes and Remarks

A number of important topics of convex analysis have been left aside in the present chapter: algorithms [75, 497, 711], approximation theory [506, 619], geometric aspects [99], mechanics [156], optimal control, the study of special classes of convex sets and functions, in particular polyhedral convex sets, among others. We refer to the monographs [52, 126, 353, 497, 498, 506, 507, 549, 619, 871, 872] and their bibliographies for a wider view.

The Fenchel conjugacy appeared in [696]; but it was the lecture notes [389] by Fenchel and the famous book [871] that made it popular. The lecture notes [735] were the main starting point of convex analysis in the general setting of infinite-dimensional spaces; they were followed by [37, 39, 52, 198, 353, 619, 692] and others.

The first results linking coercivity of a function with boundedness of its conjugate appeared in [23, 407, 735]. Many researchers have related rotundity of a function to smoothness of the conjugate; see [61, 98, 984].

The concise introduction to duality we adopted is a short diversion that does not reflect the importance of the topic. We refer to [353, 619, 692, 711, 872] for less schematic expositions. In [776, 822] no linear structure is required on the decision space X and extensions to nonconvex dualities are presented; see also their references.

The fuzzy (or so-called asymptotic) calculus rules for subdifferentials were discovered by Thibault in [916] using some previous results of Hiriart-Urruty and Phelps [500]; more light was shed on this topic in [917, 918] and [804], where the connection with nonsmooth analysis was pointed out, as well as some normal compactness conditions. In [138] convex calculus is rather seen as an output of nonsmooth analysis.

The qualification condition (b) of Lemma 3.82 bears some analogy with the transversality condition of differential topology for differentiable manifolds, as observed in [785]. Recent contributions have enlarged our views on qualification conditions (see [150, 151, 180, 563, 564, 983] among others articles).

The study of marginal functions presented in the supplement to Sect. 3.5 is inspired by the study in Valadier's thesis [945].

The section about differentiability of norms is just a short account in order to prepare the introduction of special classes of Banach spaces. The Šmulian test could be deduced from differentiability results of $(1/2) \|\cdot\|^2$ and general duality results between differentiability of a function and rotundity of its conjugate as in [61, 984].

Chapter 4
Elementary and Viscosity Subdifferentials

> *"Excellent!" I cried. "Elementary," said he.*
>
> —*Sir Arthur Conan Doyle, "The Crooked Man"*

We devote the present chapter to some fundamental notions of nonsmooth analysis upon which some other constructions can be built. Their main features are easy consequences of the definitions. Normal cones have already been considered in connection with optimality conditions. Here we present their links with subdifferentials for nonconvex, nonsmooth functions. When possible, we mention the corresponding notions of tangent cones and directional derivatives; then one gets a full picture of four related objects that can be considered the four pillars of nonsmooth analysis, or even the six pillars if one considers graphical derivatives and coderivatives of multimaps. In the present framework, in contrast to the convex objects defined in Chap. 5, the passages from directional derivatives and tangent cones to subdifferentials and normal cones respectively are one-way routes, because the first notions are nonconvex, while a dual object exhibits convexity properties. On the other hand, the passages from analytical notions to geometrical notions and the reverse passages are multiple and useful. These connections are part of the attractiveness of nonsmooth analysis.

We observe that the calculus rules that are available in such an elementary framework are rather poor, although some of them (such as calculus rules for suprema, infima, and value functions) go beyond the possibilities of ordinary differential calculus. However, the most useful calculus rules such as sum rules and composition rules exist under an approximate form in adapted spaces. This crucial fact is an incentive to introduce limiting subdifferentials and normal cones for which the preceding fuzzy rules will become exact rules, while the concepts will turn out to be more robust, at the expense of a loss of precision. That will be done in Chaps. 6 and 7. In the present chapter, most of our study is limited to the framework of "smooth" Banach spaces, i.e., to Banach spaces in which smooth bump functions exist. This is not the most general class of spaces on which fuzzy calculus can be

J.-P. Penot, *Calculus Without Derivatives*, Graduate Texts in Mathematics 266,   263
DOI 10.1007/978-1-4614-4538-8_4, © Springer Science+Business Media New York 2013

obtained. But on such spaces, smooth variational principles are available. In the case of Fréchet subdifferentials, all the rules are valid in the class of Asplund spaces. But to reach this conclusion, one has to use more sophisticated results such as separable reduction. We present this passage in Sect. 4.6, but it can be skipped on a first reading. We note that it is not known whether there are Asplund spaces that are not Fréchet smooth.

Whereas the use of directional concepts is convenient (because one has notions of directional derivative and tangent cone at one's disposal), it cannot be as precise and powerful as the use of firm (or Fréchet) notions. Thus, as in differential calculus, directional notions can be considered first steps. However, even if calculus with directional notions is poorer and not as precise as with firm notions, it can be developed to a similar extent, provided one keeps track of minimization properties or uses a variant of the directional subdifferential called the viscosity directional subdifferential. We endeavor to present common properties of the two subdifferentials we study in a unified manner. They can be embedded in a full family using bornological subdifferentials. We point out this concept, but for the sake of simplicity, we refrain from taking steps outside the two main concepts. Keeping track of their analogies and differences is already a challenge for the reader.

When some continuity property (or "sleekness" or "regularity" property) is available, one gets exact rules and strong properties that can be compared to what occurs with mappings of class C^1 or of class D^1. Throughout, X, Y are (possibly infinite-dimensional) Banach spaces, but the concepts can be introduced for general normed spaces.

4.1 Elementary Subderivatives and Subdifferentials

It is the purpose of this section to present some concepts of subdifferentials that encompass both the notion of derivative and the notion of subdifferential in the convex case. The main advantages of these concepts are their close relationships with corresponding notions of derivatives and the fact that they provide rather accurate approximations.

4.1.1 Definitions and Characterizations

The following definition introducing the Fréchet subdifferential is obtained as a simple one-sided modification of the concept of Fréchet derivative.

Definition 4.1. Given a normed space X and a function $f : X \to \overline{\mathbb{R}}$ finite at $\overline{x} \in X$, the *firm (or Fréchet) subdifferential* of f at $\overline{x}$ is the set $\partial_F f(\overline{x})$ of $\overline{x}^* \in X^*$ satisfying the following property: for every $\varepsilon > 0$ there exists some $\delta > 0$ such that

$$\forall x \in B(\bar{x}, \delta), \qquad f(x) - f(\bar{x}) - \langle \bar{x}^*, x - \bar{x} \rangle \geq -\varepsilon \|x - \bar{x}\|. \qquad (4.1)$$

In other words, $\bar{x}^* \in \partial_F f(\bar{x})$ if and only if

$$\lim_{\|v\| \to 0_+} \inf \frac{1}{\|v\|} [f(\bar{x} + v) - f(\bar{x}) - \langle \bar{x}^*, v \rangle] \geq 0. \qquad (4.2)$$

Setting for $r \in \mathbb{P} := (0, +\infty)$,

$$\mu_f(r) := (1/r) \sup\{f(\bar{x}) - f(\bar{x} + v) + \langle \bar{x}^*, v \rangle : v \in rB_X\},$$

we see that $\bar{x}^* \in \partial_F f(\bar{x})$ if and only if $\mu := \mu_f$ is a *modulus*, i.e., a nondecreasing function $\mu : \mathbb{R}_+ \to \overline{\mathbb{R}}_+$ such that $\mu(0) = 0$ and $\mu(t) \to 0$ as $t \to 0_+$. Thus, $\bar{x}^* \in \partial_F f(\bar{x})$ if and only if there exists an element μ of the set $\mathcal{M}$ of moduli such that

$$\forall v \in X, \qquad f(\bar{x} + v) - f(\bar{x}) - \langle \bar{x}^*, v \rangle \geq -\mu(\|v\|) \|v\|. \qquad (4.3)$$

In fact, μ_f defined as above is the smallest modulus μ satisfying (4.3); it can be called the modulus of firm subdifferentiability of f at $\bar{x}$ for $\bar{x}^*$.

Equivalently, $\bar{x}^* \in \partial_F f(\bar{x})$ if and only if there exists a *remainder r*, i.e., a function $r : \mathbb{R}_+ \to \overline{\mathbb{R}}_+$ satisfying $\mu(t) := t^{-1} r(t) \to 0$ as $t \to 0_+$ and

$$\forall v \in X, \qquad f(\bar{x} + v) - f(\bar{x}) - \langle \bar{x}^*, v \rangle \geq -r(\|v\|). \qquad (4.4)$$

This definition belongs to the realm of unilateral analysis (or one-sided analysis): the equality, or double inequality, in the definition of the Fréchet derivative in terms of limits has been replaced by a single inequality; moreover, in the formulation (4.2), the limit is replaced with a limit inferior.

Example. The Moreau–Rockafellar subdifferential of f at $\bar{x}$, i.e., the set $\partial_{MR} f(\bar{x})$ of $\bar{x}^*$ such that $f(x) - f(\bar{x}) - \langle \bar{x}^*, x - \bar{x} \rangle \geq 0$ for every $x \in X$, is obviously contained in $\partial_F f(\bar{x})$. If f is convex, the two subdifferentials coincide (Exercise 1). Clearly, if f is not convex, the Moreau–Rockafellar subdifferential is a very restrictive notion that cannot be very useful, since it is global instead of local.

Example. Given a remainder π, the *π-proximal subdifferential* of f at $\bar{x}$ is the set $\partial_\pi f(\bar{x})$ of $\bar{x}^* \in X^*$ such that there exist $c > 0$ and $\rho > 0$ for which

$$\forall x \in B(\bar{x}, \rho), \qquad f(x) - f(\bar{x}) - \langle \bar{x}^*, x - \bar{x} \rangle \geq -c\pi(\|x - \bar{x}\|).$$

For $\pi(\cdot) = (\cdot)^2$, one denotes by $\partial_P f(\bar{x})$ the set $\partial_\pi f(\bar{x})$ and simply calls it the *proximal subdifferential*. Clearly, this set is contained in $\partial_F f(\bar{x})$. One of its advantages is that it can be used without any knowledge of limits or limits inferior. It is well suited for geometrical questions such as best approximations, at least in Hilbert spaces. However, this subdifferential cannot be seen as a first-order notion, as the following example shows.

Example. Let $f : \mathbb{R} \to \mathbb{R}$ be given by $f(x) = -|x|^{3/2}$, $\bar{x} = 0$. Then f is of class C^1 but $\partial_P f(\bar{x})$ is empty, as is easily seen.

Thus one may have $f = g + h$ with g of class C^1, $\partial_P h(\bar{x}) \neq \varnothing$, and $\partial_P f(\bar{x}) = \varnothing$.

□

Remark. Obviously, if f is Fréchet differentiable at $\bar{x}$, then $f'(\bar{x}) \in \partial_F f(\bar{x})$ and $f'(\bar{x}) \in \tilde{\partial}_F f(\bar{x}) := -\partial_F(-f)(\bar{x})$, the Fréchet *superdifferential* at $\bar{x}$. Conversely, if $\bar{x}^* \in \partial_F f(\bar{x})$ and if $\tilde{x}^* \in \tilde{\partial}_F f(\bar{x})$, then $\tilde{x}^* = \bar{x}^*$ and f is Fréchet differentiable at $\bar{x}$, with derivative this linear form. In fact, one can find remainders r_+, r_- such that

$$\forall v \in X, \qquad \langle \tilde{x}^*, v \rangle + r_+(\|v\|) \geq f(\bar{x} + v) - f(\bar{x}) \geq \langle \bar{x}^*, v \rangle - r_-(\|v\|);$$

hence by homogeneity,

$$\forall v \in X, \qquad \langle \tilde{x}^*, v \rangle - \langle \bar{x}^*, v \rangle \geq -\lim_{t \to 0_+} \frac{1}{t}(r_+(\|tv\|) + r_-(\|tv\|)) = 0,$$

and $\tilde{x}^* = \bar{x}^*$. Then, using the remainder $\max(r_+, r_-)$, the assertion about the differentiability of f is immediate.

Let us observe that it may happen that $\partial_F f(\bar{x})$ is reduced to a singleton although f is not differentiable at $\bar{x}$, as the following example shows.

Example. Let $f : \mathbb{R} \to \mathbb{R}$ be given by $f(0) = 0$, $f(x) = |x| \sin^2(1/x)$ for $x \neq 0$. Then $\partial_F f(0) = \{0\}$, but f is not differentiable at 0.

A closely related notion is the notion of directional or contingent subdifferential.

Definition 4.2. The *directional subdifferential* (or Dini–Hadamard or Bouligand or contingent subdifferential) of $f : X \to \overline{\mathbb{R}}$ at $\bar{x} \in f^{-1}(\mathbb{R})$ is the set $\partial_D f(\bar{x})$ of $\bar{x}^* \in X^*$ satisfying the following property: for every $u \in X$ and $\varepsilon > 0$ there exists some $\delta > 0$ such that

$$\forall (t, v) \in (0, \delta) \times B(u, \delta), \qquad f(\bar{x} + tv) - f(\bar{x}) - \langle \bar{x}^*, tv \rangle \geq -\varepsilon t. \tag{4.5}$$

In other words, $\bar{x}^* \in \partial_D f(\bar{x})$ iff for all $u \in X$,

$$\liminf_{(t,v) \to (0_+, u)} \frac{1}{t}[f(\bar{x} + tv) - f(\bar{x}) - \langle \bar{x}^*, tv \rangle] \geq 0, \tag{4.6}$$

iff for all $u \in X$, there exists a modulus α such that

$$f(\bar{x} + tv) - f(\bar{x}) - \langle \bar{x}^*, tv \rangle \geq -t\alpha(t + \|v - u\|). \tag{4.7}$$

Thus, one has

$$\partial_F f(\bar{x}) \subset \partial_D f(\bar{x}). \tag{4.8}$$

Although the definition of the firm subdifferential is simpler than the definition of the directional subdifferential, it is often easier to check whether a linear form belongs to the second set. The reason lies in the following connection with a directional derivative, which yields a convenient characterization.

Proposition 4.3. *A continuous linear form* $\bar{x}^*$ *belongs to* $\partial_D f(\bar{x})$ *iff it is bounded above by the* lower directional *(or contingent or Hadamard)* (sub)*derivate* $f^D(\bar{x}, \cdot)$ *defined by*

$$f^D(\bar{x}, u) := \liminf_{(t,v) \to (0_+, u)} \frac{1}{t}\left(f(\bar{x} + tv) - f(\bar{x})\right).$$

Proof. This follows from (4.6), since $\lim_{(t,v) \to (0_+, u)} t^{-1}(\langle \bar{x}^*, tv \rangle) = \langle \bar{x}^*, u \rangle$. □

A necessary condition for directional *subdifferentiability* of f at $\bar{x}$, i.e., nonemptiness of $\partial_D f(\bar{x})$, is a one-sided Lipschitz-like property called calmness at $\bar{x}$. A function f on a normed space X is said to be *calm at* $\bar{x}$ if $f(\bar{x})$ is finite and if there exist $c > 0$, $\rho > 0$ such that $f(\bar{x} + u) \geq f(\bar{x}) - c\|u\|$ for all u in the ball ρB_X. Calmness is closely related to properties of the lower directional derivate $f^D(\bar{x}, \cdot)$.

Proposition 4.4. *For* $f : X \to \overline{\mathbb{R}}$ *finite at* $x \in X$, *the following assertions are equivalent and are satisfied when* $\partial_D f(x)$ *is nonempty:*

(a) f is calm at x;
(b) There exists some $c \in \mathbb{R}_+$ such that $f^D(x, u) \geq -c\|u\|$ for all $u \in X$;
(c) The function $f^D(x, \cdot)$ is proper;
(d) $f^D(x, 0) = 0$.

If in addition f is tangentially convex *at x in the sense that $f^D(x, \cdot)$ is sublinear, these assertions are equivalent to the nonemptiness of $\partial_D f(x)$.*

Proof. The implications (a)⇒(b)⇒(c)⇒(d) are immediate, taking into account the fact that either $f^D(x, 0) = 0$ or $-\infty$. Let us prove that (d)⇒(a). If (a) does not hold, there exists a sequence $(u_n) \to 0$ such that $f(x + u_n) - f(x) < -n^2\|u_n\|$. Taking a subsequence if necessary, we may suppose that $(t_n) := (n\|u_n\|) \to 0$. Then

$$t_n^{-1}\left(f(x + t_n(t_n^{-1}u_n)) - f(x)\right) < -n,$$

and since $(t_n^{-1}u_n) \to 0$, passing to the liminf, we get $f^D(x, 0) = -\infty$.

When $f^D(x, \cdot)$ is sublinear and (c) holds, a consequence of the Hahn–Banach theorem asserts that $f^D(x, \cdot)$ is the supremum of the family $\partial_D f(x)$ of its minorants; thus this set is nonempty. On the other hand, if $x^* \in \partial_D f(x)$, then (b) holds with $c := \|x^*\|$. □

In the following corollary, f is said to be *quiet at* $\bar{x}$ if $-f$ is calm at $\bar{x}$, i.e., if there exist some $c, \rho > 0$ such that $f(x) - f(\bar{x}) \leq c\|x - \bar{x}\|$ for all $x \in B(\bar{x}, r)$; then c is called a *rate of quietness* at $\bar{x}$. This is the case if f is Lipschitzian with rate c around $\bar{x}$, or more generally, if f is *stable*, or *Stepanovian*, with rate c at $\bar{x}$, i.e., if there exists some $\rho > 0$ such that $|f(x) - f(\bar{x})| \leq c\|x - \bar{x}\|$ for all $x \in B(\bar{x}, \rho)$.

Corollary 4.5. *The function f is Hadamard differentiable at $\bar{x}$ iff both $\partial_D f(\bar{x})$ and $\widetilde{\partial}_D f(\bar{x}) := -\partial_D(-f)(\bar{x})$, the directional superdifferential of f at $\bar{x}$, are nonempty. Then $\partial_D f(\bar{x}) \cap \widetilde{\partial}_D f(\bar{x}) = \{f'(\bar{x})\}$.*

If f is quiet with rate c at $\bar{x}$, then $\partial_D f(\bar{x}) \subset cB_{X^*}$.

Proof. If $\bar{x}^* \in \partial_D f(\bar{x})$ and $\tilde{x}^* \in -\partial_D(-f)(\bar{x})$, then for all $u \in X$, one has

$$\langle \bar{x}^*, u \rangle \leq \liminf_{(t,v)\to(0_+,u)} \frac{1}{t}(f(\bar{x}+tv) - f(\bar{x})) \leq \limsup_{(t,v)\to(0_+,u)} \frac{1}{t}(f(\bar{x}+tv) - f(\bar{x})) \leq \langle \tilde{x}^*, u \rangle,$$

so that $\tilde{x}^* = \bar{x}^*$ and the differential quotient has a limit, i.e., f is Hadamard differentiable at $\bar{x}$. The converse is obvious.

When f is quiet with rate c at $\bar{x}$ one has $f^D(\bar{x}, \cdot) \leq c\|\cdot\|$, so that $\|x^*\| \leq c$ whenever $x^* \in X^*$ is majorized by $f^D(\bar{x}, \cdot)$. $\qquad\square$

The inclusion (4.8) may be strict, as the following example shows.

Example. Let f be a function that is Hadamard differentiable at some point $\bar{x} \in X$, but not Fréchet differentiable. Then $\partial_D f(\bar{x}) = \{f'(\bar{x})\}$, $\partial_D(-f)(\bar{x}) = \{-f'(\bar{x})\}$ but $\partial_F f(\bar{x})$ or $\partial_F(-f)(\bar{x})$ is empty in view of the preceding remark. $\qquad\square$

Proposition 4.6. *If X is finite-dimensional, then $\partial_F f(\bar{x}) = \partial_D f(\bar{x})$.*

Proof. Suppose, to the contrary, that there exists $\bar{x}^* \in \partial_D f(\bar{x}) \setminus \partial_F f(\bar{x})$. By definition of $\partial_F f(\bar{x})$, there exists $\varepsilon > 0$ such that for all $n \in \mathbb{N} \setminus \{0\}$ one can find $x_n \in B(\bar{x}, n^{-1})$ satisfying

$$f(x_n) - f(\bar{x}) - \langle \bar{x}^*, x_n - \bar{x} \rangle < -\varepsilon\|x_n - \bar{x}\|.$$

Then $t_n := \|x_n - \bar{x}\| > 0$. Let $u_n := t_n^{-1}(x_n - \bar{x})$. Since (u_n) is a sequence of the unit sphere that is compact, taking a subsequence if necessary, one may assume that it converges to some unit vector u. The preceding inequality yields

$$f^D(\bar{x}, u) \leq \liminf_n t_n^{-1}(f(\bar{x}+t_n u_n) - f(\bar{x})) \leq \lim_n \langle \bar{x}^*, u_n \rangle - \varepsilon = \langle \bar{x}^*, u \rangle - \varepsilon,$$

and we get a contradiction to the characterization of Proposition 4.3. $\qquad\square$

The following characterizations may be useful. They build a bridge toward the notion of viscosity subdifferential in which the test function φ is smooth throughout the space (or in some neighborhood of the point). This last notion is widely used in the theory of Hamilton–Jacobi equations. It will be considered later.

Proposition 4.7. *For every normed space X and every function f finite at $\bar{x}$, the firm (resp. directional) subdifferential of f at $\bar{x}$ coincides with the set of derivatives at $\bar{x}$ of functions φ that are Fréchet (resp. Hadamard) differentiable at $\bar{x}$ and such that $\varphi \leq f$, $\varphi(\bar{x}) = f(\bar{x})$. Equivalently, $\partial_F f(\bar{x})$ (resp. $\partial_D f(\bar{x})$) is the set of derivatives at $\bar{x}$ of functions ψ that are Fréchet (resp. Hadamard) differentiable at $\bar{x}$ and such that $f - \psi$ attains its minimum at $\bar{x}$.*

Proof. Clearly, if φ is Fréchet (resp. Hadamard) differentiable and such that $\varphi \leq f$, $\varphi(\bar{x}) = f(\bar{x})$, one has $\varphi'(\bar{x}) \in \partial_F f(\bar{x})$ (resp. $\varphi'(\bar{x}) \in \partial_D f(\bar{x})$). Conversely, given $\bar{x}^* \in \partial_F f(\bar{x})$ and a modulus μ satisfying (4.3), setting

$$\varphi(x) := \min(f(x), g(x)), \text{ with } g(x) := f(\bar{x}) + \langle \bar{x}^*, x - \bar{x} \rangle,$$

one sees that $\varphi \leq f$, $\varphi(\bar{x}) = f(\bar{x})$, and

$$0 \geq \varphi(x) - f(\bar{x}) - \langle \bar{x}^*, x - \bar{x} \rangle \geq -\mu(\|x - \bar{x}\|) \|x - \bar{x}\|,$$

so that φ is Fréchet differentiable at $\bar{x}$ with derivative $\bar{x}^*$. The case of the Hadamard subdifferential is similar, using (4.7) instead of (4.3).

The passage from the first characterization to the second one is obvious, taking $\psi = \varphi$. The reverse passage amounts to replacing ψ by $\varphi := \psi - \psi(\bar{x}) + f(\bar{x})$. $\square$

We will see some calculus rules after pointing out simple properties and delineating links with geometrical objects.

Exercises

1. (a) Prove that $\bar{x}^* \in \partial_F f(\bar{x})$ iff $\lim_{\|v\| \to 0_+} \frac{1}{\|v\|} [f(\bar{x} + v) - f(\bar{x}) - \langle \bar{x}^*, v \rangle]^- = 0$, where the negative part of a real number r is denoted by $r^- := \max(-r, 0)$.
(b) Prove a similar statement for $\partial_D f(\bar{x})$.

2. For $f : \mathbb{R} \to \overline{\mathbb{R}}$ finite at $\bar{x}$, give a condition in order that $\partial_D f(\bar{x})$ be nonempty in terms of the Dini lower derivatives $df(\bar{x}, 1)$ and $df(\bar{x}, -1)$.

3. Show that for $\bar{x} \in f^{-1}(\mathbb{R})$ the set $\partial_F f(\bar{x})$ is convex, but not always weak* closed.

4. Given a subset E of a normed space X, let $0_E = 1 - \chi_E$, where χ_E is the *characteristic function* of E, i.e., 0_E is given by $0_E(x) = 0$ for $x \in E$, $0_E(x) = 1$ for $x \in X \setminus E$ and let ι_E be the *indicator function* of E given by $\iota_E(x) = 0$ for $x \in E$, $\iota_E(x) = +\infty$ for $x \in X \setminus E$. Show that for all $x \in E$ one has $0_E^D(x, \cdot) = \iota_E^D(x, \cdot)$ and that $\partial_D 0_E(x) = N(E, x)$.

5 (Characterization of $\partial_D f(\bar{x})$ with the notion of sponge [930]). A subset S of a normed space X is called a *sponge* if for all $u \in X \setminus \{0\}$ there exists $\delta > 0$ such that the drop $[0, \delta] B(u, \delta)$ is contained in S.

(a) Show that every neighborhood of 0 is a sponge, and that the converse holds if X is finite-dimensional.
(b) Show that in an infinite-dimensional space there are sponges that are not neighborhoods of 0.
(c) Prove that $\bar{x}^* \in \partial_D f(\bar{x})$ iff f is calm at $\bar{x}$ and for every $\varepsilon > 0$ there exists a sponge S such that

$$\forall x \in S + \bar{x}, \quad f(x) - f(\bar{x}) - \langle \bar{x}^*, x - \bar{x} \rangle \geq -\varepsilon \|x - \bar{x}\|.$$

6. Show that in every infinite-dimensional normed space X there exists a function $f : X \to \mathbb{R}$ finite at $\overline{x} = 0$ such that $f^D(\overline{x}, u) = 0$ for all $u \in X \setminus \{0\}$ but $\partial_F f(\overline{x}) = \varnothing$. Describe a sponge S such that f attains its minimum on S at 0. [Hint: Using the fact that the unit sphere S_X is not precompact, take a sequence (e_n) of S_X and $\delta > 0$ such that $\|e_n - e_p\| \geq \delta$ for all $n \neq p$ and define f by $f(x) = -2^{-n}$ if $x = 4^{-n}e_n$ for some $n \in \mathbb{N}$, $f(x) = 0$ otherwise, and observe that f is not calm at 0.]

7. On the space $X := C(T)$ of continuous functions on $T := [0, 1]$, endowed with the supremum norm, consider the function $f : X \to \mathbb{R}$ given by $f(x) = \min x(T)$.

(a) Prove that the Fréchet subdifferential of f is empty at every point.
(b) Prove that $\partial_D f(x)$ is nonempty if and only if x attains its minimum on T at a unique t, if and only if f is Hadamard differentiable at x.

4.1.2   Some Elementary Properties

We have seen some relationships with differentiability. Let us consider some other elementary properties of the subdifferentials we introduced. The first one is obvious.

Proposition 4.8. *The subdifferentials ∂_F and ∂_D are local in the sense that if f and g coincide on some neighborhood of $\overline{x}$, then $\partial_F f(\overline{x}) = \partial_F g(\overline{x})$ and $\partial_D f(\overline{x}) = \partial_D g(\overline{x})$.*

Proposition 4.9. *If f is convex, then $\partial_F f(\overline{x})$ and $\partial_D f(\overline{x})$ coincide with the Moreau–Rockafellar subdifferential $\partial_{MR} f(\overline{x})$:*

$$\partial_F f(\overline{x}) = \partial_D f(\overline{x}) = \partial_{MR} f(\overline{x}) := \{\overline{x}^* \in X^* : \forall x \in X, \quad f(x) \geq f(\overline{x}) + \langle \overline{x}^*, x - \overline{x} \rangle\}.$$

Proof. It is clear that $\partial_F f(\overline{x})$ and $\partial_D f(\overline{x})$ contain $\partial_{MR} f(\overline{x})$. Let f be convex and let $\overline{x}^* \in \partial_D f(\overline{x})$. Then $\overline{x}^*$ is bounded above by $f^D(\overline{x}, \cdot)$, hence belongs to $\partial_{MR} f(\overline{x})$ by Theorem 3.22.   □

Proposition 4.10. *For every function f finite at $\overline{x}$, $\partial_F f(\overline{x})$ (resp. $\partial_D f(\overline{x})$) is a closed (resp. weak* closed) convex subset of X^*.*

Proof. The weak* closedness and convexity of $\partial_D f(\overline{x})$ stem from Proposition 4.3. Let $\overline{x}^* \in X^*$ be in the closure of $\partial_F f(\overline{x})$. For every $\varepsilon > 0$ there exists $x^* \in \partial_F f(\overline{x})$ such that $\|\overline{x}^* - x^*\| < \varepsilon/2$; let $\delta > 0$ be such that for every $x \in B(\overline{x}, \delta)$ one has

$$f(x) - f(\overline{x}) - \langle x^*, x - \overline{x} \rangle \geq -(\varepsilon/2)\|x - \overline{x}\|.$$

Then, for every $x \in B(\overline{x}, \delta)$, one has

$$f(x) - f(\overline{x}) - \langle \overline{x}^*, x - \overline{x} \rangle \geq -\varepsilon\|x - \overline{x}\|.$$

Thus $\overline{x}^* \in \partial_F f(\overline{x})$: this set is closed. The convexity of $\partial_F f(\overline{x})$ is obvious.   □

The sets $\partial_F f(\overline{x})$ and $\partial_D f(\overline{x})$ may be empty, even for a Lipschitzian function.

Example. For $X = \mathbb{R}$, $f(x) := -|x|$, $\bar{x} := 0$ one has $\partial_F f(\bar{x}) = \partial_D f(\bar{x}) = \varnothing$. □

Such a fact may appear to be a drawback. On the other hand, it makes the optimality condition of the next theorem a nontrivial test.

In finite dimensions, Rademacher's theorem ensures that the subdifferential $\partial_F f(x)$ of a locally Lipschitzian function f is nonempty for x in the complement of a set of null measure. An extension has been given to some infinite-dimensional cases by D. Preiss. We will give a simpler density result later (Theorem 4.65).

The preceding subdifferentials fail to enjoy the most useful calculus rules. In particular, the inclusion $\partial(f + g)(\bar{x}) \subset \partial f(\bar{x}) + \partial g(\bar{x})$ is not valid in general, as the following example shows.

Example. For $X = \mathbb{R}$, $f(x) := -|x|$, $g = -f$, $\bar{x} := 0$ one has $\partial_F f(\bar{x}) = \partial_D f(\bar{x}) = \varnothing$ but $\partial_F(f + g)(0) = \partial_D(f + g)(0) = \{0\}$. □

In the sequel we strive to get some form of this desirable inclusion. The subdifferentials introduced in the next chapters will be more suitable for such an objective. However, they will lose the accuracy of the directional and firm subdifferentials. Also, the following obvious, but precious, property will no longer be valid.

Proposition 4.11. *The firm and the directional subdifferentials are homotone in the sense that for $f \geq g$ with $f(\bar{x}) = g(\bar{x})$ finite one has*

$$\partial_F g(\bar{x}) \subset \partial_F f(\bar{x}), \qquad \partial_D g(\bar{x}) \subset \partial_D f(\bar{x}).$$

An immediate but useful consequence is the following necessary condition.

Theorem 4.12. *If f attains at $\bar{x}$ a local minimum, then one has $0 \in \partial_F f(\bar{x})$ and $0 \in \partial_D f(\bar{x})$.*

Proof. Let g be the constant function with value $f(\bar{x})$. Then $f \geq g$ with $f(\bar{x}) = g(\bar{x})$ and the preceding proposition applies. □

Exercises

1. Prove Proposition 4.11 with the help of Proposition 4.7.

2. Show that the proximal subdifferential at 0 of $f : \mathbb{R} \to \mathbb{R}$ given by $f(x) := |x| - |x|^{3/2}$ is the open interval $(-1, 1)$, hence is not closed.

3. Given a subset S of X, $x \in S$ and $\lambda \in [0, 1]$, show that $\lambda \partial d_S(x) \subset \partial d_S(x)$ for $\partial := \partial_F$ and $\partial = \partial_D$. [Hint: Use Proposition 4.11 or the fact that d_S attains its infimum at x and $\partial d_S(x)$ is convex.]

4. Give a detailed justification of Proposition 4.8.

5. Let X,Y be normed spaces and let $f = h \circ g$, where $h : Y \to \mathbb{R}$ is finite at $\bar{y} :=$ $g(\bar{x})$ and $g : X \to Y$ is a bijection that is Fréchet differentiable (resp. Hadamard differentiable) at $\bar{x}$ whose inverse has the same property at $\bar{y}$. Prove that $\partial_F f(\bar{x}) = g'(\bar{x})^\mathsf{T}(\partial_F h(\bar{y})) := \partial_F h(\bar{y}) \circ g'(\bar{x})$ (resp. $\partial_D f(\bar{x}) = g'(\bar{x})^\mathsf{T}(\partial_D h(\bar{y}))$). Give a localized version of such a result, assuming that g is a bijection of a neighborhood U of $\bar{x}$ onto a neighborhood V of $g(\bar{x})$.

6. Let X be a finite-dimensional Euclidean space and let $f : X \to \mathbb{R}$ be a lower semicontinuous function. Show that the domain of $\partial_F f$ is dense in X. [Hint: Given $\bar{x} \in X$ and $\varepsilon > 0$, show that for some $\rho \in (0, \varepsilon)$ and some $t > 0$ large enough the function $x \mapsto f(x) + t \|x - \bar{x}\|^2$ attains its infimum on $B[\bar{x}, \rho]$ at a point in $B(\bar{x}, \rho)$.]

7. Devise calculus rules for calm, tangentially convex functions. (See [853].)

4.1.3 Relationships with Geometrical Notions

Since the concepts of subdifferential introduced in the preceding subsections can be used without any regularity condition on the function, we may apply them to the case of an indicator function. Recall that the *indicator function* ι_E of a subset E of X is the function that takes the value 0 on E and the value $+\infty$ on $X \setminus E$. Such a function is useful in dealing with feasible sets in optimization. We have seen in Chap. 2 that the notions of normal cone are also useful in formulating optimality conditions. It is important and easy to relate these notions to the notions of subdifferential we introduced.

Proposition 4.13. *For every subset E of a normed space X and for every $\bar{x} \in \mathrm{cl}E$, the firm normal cone and the directional normal cone to E at $\bar{x}$ coincide with the corresponding subdifferentials of the indicator function of E: we have respectively*

$$N_F(E, \bar{x}) = \partial_F \iota_E(\bar{x}) \quad and \quad N_D(E, \bar{x}) = \partial_D \iota_E(\bar{x}).$$

Moreover, one has

$$\mathbb{R}_+ \partial_F d_E(\bar{x}) \subset N_F(E, \bar{x}), \qquad \mathbb{R}_+ \partial_D d_E(\bar{x}) \subset N_D(E, \bar{x}). \qquad (4.9)$$

The first of the last inclusions is in fact an equality, as we show below.

Proof. The equalities stem from the definitions of the normal cones $N_F(E, \bar{x})$ and $N_D(E, \bar{x})$ adopted in Chap. 2. They can be made more explicit in the following way (in the second equivalence, u is an arbitrary nonnull vector):

$$\bar{x}^* \in N_F(E, \bar{x}) \Leftrightarrow \exists \delta(\cdot) : x \in E \cap B(\bar{x}, \delta(\varepsilon)) \Rightarrow \langle \bar{x}^*, x - \bar{x} \rangle \le \varepsilon \|x - \bar{x}\|,$$

$$\bar{x}^* \in N_D(E, \bar{x}) \Leftrightarrow \exists \delta(\cdot, \cdot) : t \in (0, \delta(\varepsilon, u)), \ v \in \frac{E - \bar{x}}{t} \cap B(u, \delta(\varepsilon, u)) \Rightarrow \langle \bar{x}^*, v \rangle \le \varepsilon.$$

The inclusions are consequences of Proposition 4.11, taking $f = \iota_E$ and $g = cd_E$ for $c \in \mathbb{R}_+$ arbitrary. Since d_E is Lipschitzian with rate 1, we get $\partial_F d_E(\bar{x}) \subset N_F(E,\bar{x}) \cap B_{X^*}$ and a similar relation with the directional notions. □

Combined with the chain rules we will establish, the preceding characterizations enable one to recover the calculus rules for normal cones. As an example, we observe that when $E := g^{-1}(C)$ with g Fréchet differentiable at $\bar{x} \in E$, using the fact that $\iota_E = \iota_C \circ g$, one gets

$$N_F(E,\bar{x}) = \partial_F(\iota_C \circ g)(\bar{x}) \supset \partial_F \iota_C(g(\bar{x})) \circ g'(\bar{x}) = (g'(\bar{x}))^\top (N_F(C, g(\bar{x}))).$$

Although the characterization of the directional normal cone seems to be rather involved, the directional normal cone deserves to be called, for short, the normal cone in view of the fact that it is the polar cone to the tangent cone. For the firm normal cone, an analogous relationship is seldom used, since it is more subtle (Exercise 7).

In turn, the lower directional derivative f^D of a function f can be interpreted geometrically in a simple way using the notion of tangent cone.

Proposition 4.14. *The tangent cone at $\bar{x}_f := (\bar{x}, f(\bar{x}))$ to the epigraph E_f of f is the epigraph of the lower (or contingent) subderivate $f^D(\bar{x}, \cdot)$:*

$$T^D(E_f, \bar{x}_f) = \left\{ (u,r) \in X \times \mathbb{R} : r \ge f^D(\bar{x}, u) \right\},$$

$$f^D(\bar{x}, u) = \min\left\{ r : (u,r) \in T^D(E_f, \bar{x}_f) \right\}.$$

As usual, min means that if the infimum is finite, then it is attained.

Proof. We have $(u,r) \in T^D(E_f, \bar{x}_f)$ iff there exist sequences $(t_n) \to 0_+$, $(u_n) \to u$, $(r_n) \to r$ such that $(\bar{x}, f(\bar{x})) + t_n(u_n, r_n) \in E_f$, or equivalently, $f(\bar{x}) + t_n r_n \ge f(\bar{x} + t_n u_n)$ iff there exist sequences $(t_n) \to 0_+$, $(u_n) \to u$ such that $r \ge \liminf_n t_n^{-1}(f(\bar{x} + t_n u_n) - f(\bar{x}))$ iff $r \ge \liminf_{(t,v) \to (0_+, u)} t^{-1}(f(\bar{x} + tv) - f(\bar{x})) := f^D(\bar{x}, u)$. The second relation is a consequence of the fact that $T^D(E_f, \bar{x}_f)$ is closed and stable under addition of vectors of the form $(0, p)$ with $p \in \mathbb{R}_+$. □

Corollary 4.15. *The directional subdifferential $\partial_D f(\bar{x})$ of f at $\bar{x} \in \mathrm{dom} f$ and the normal cone $N_D(E_f, \bar{x}_f)$ to the epigraph E_f of f at $\bar{x}_f := (\bar{x}, f(\bar{x}))$ are related by the following equivalence:*

$$\bar{x}^* \in \partial_D f(\bar{x}) \Leftrightarrow (\bar{x}^*, -1) \in N_D(E_f, \bar{x}_f). \tag{4.10}$$

Moreover, $\bar{x}^* \in \partial_D f(\bar{x})$ whenever there exists $c > 0$ such that $(\bar{x}^*, -1) \in c \partial_D d_{E_f}(\bar{x}_f)$.
If f is Lipschitzian around $\bar{x}$ with rate c and if $X \times \mathbb{R}$ is endowed with the norm given by $\|(x,r)\|_c := c\|x\| + |r|$, one has

$$\bar{x}^* \in \partial_D f(\bar{x}) \Leftrightarrow (\bar{x}^*, -1) \in c \partial_D d_{E_f}(\bar{x}_f).$$

Proof. Relation (4.10) follows from the previous characterizations:

$$\bar{x}^* \in \partial_D f(\bar{x}) \Leftrightarrow \forall u \in X, \quad \langle \bar{x}^*, u \rangle \leq f^D(\bar{x}, u)$$

$$\Leftrightarrow \forall (u,r) \in T^D(E_f, \bar{x}_f), \quad \langle \bar{x}^*, u \rangle \leq r$$

$$\Leftrightarrow \forall (u,r) \in T^D(E_f, \bar{x}_f), \quad \langle (\bar{x}^*, -1), (u,r) \rangle \leq 0$$

$$\Leftrightarrow (\bar{x}^*, -1) \in N_D(E_f, \bar{x}_f).$$

The second assertion is a consequence of the relation $c\partial_D dE_f(\bar{x}_f) \subset N_D(E_f, \bar{x}_f)$. A proof of the last equivalence is presented in the supplement on bornological subdifferentials. □

A similar relationship holds for the Fréchet subdifferential.

Proposition 4.16. *The Fréchet subdifferential $\partial_F f(\bar{x})$ of f at $\bar{x} \in \mathrm{dom} f$ and the Fréchet normal cone $N_F(E_f, \bar{x}_f)$ to the epigraph E_f of f at $\bar{x}_f := (\bar{x}, f(\bar{x}))$ are related by the following equivalence:*

$$\bar{x}^* \in \partial_F f(\bar{x}) \Leftrightarrow (\bar{x}^*, -1) \in N_F(E_f, \bar{x}_f). \tag{4.11}$$

If f is Lipschitzian around $\bar{x}$ with rate c and if $X \times \mathbb{R}$ is endowed with the norm $\|\cdot\|_c$, one has

$$\bar{x}^* \in \partial_F f(\bar{x}) \Leftrightarrow (\bar{x}^*, -1) \in c\partial_F dE_f(\bar{x}_f).$$

Proof. If $\bar{x}^* \in \partial_F f(\bar{x})$, then one has $(\bar{x}^*, -1) \in N_F(E_f, \bar{x}_f)$: given $\varepsilon > 0$ one can find $\delta > 0$ such that for all $(x,r) \in B(\bar{x}_f, \delta) \cap E_f$ one has

$$\langle (\bar{x}^*, -1), (x - \bar{x}, r - f(\bar{x})) \rangle \leq \langle \bar{x}^*, x - \bar{x} \rangle - f(x) + f(\bar{x})$$

$$\leq \varepsilon \|x - \bar{x}\| \leq \varepsilon \|(x,r) - (\bar{x}, f(\bar{x}))\|.$$

Now let $(\bar{x}^*, -1) \in N_F(E_f, \bar{x}_f)$. Let us first show that for $c > b := \|\bar{x}^*\|$ and some $\rho > 0$ we have the calmness property

$$\forall x \in B(\bar{x}, \rho), \quad f(x) - f(\bar{x}) \geq -c \|x - \bar{x}\|. \tag{4.12}$$

By definition of the Fréchet normal cone, given $\varepsilon \in (0,1)$ with $\varepsilon < c - b$ we can find $\eta > 0$ such that for all $(x,r) \in E_f \cap B(\bar{x}_f, \eta)$ we have

$$\langle (\bar{x}^*, -1), (x - \bar{x}, r - f(\bar{x})) \leq \varepsilon(c+1)^{-1} \max(\|x - \bar{x}\|, |r - f(\bar{x})|). \tag{4.13}$$

Let us show that for $\rho := \eta/(c+1)$ relation (4.12) is satisfied. If the opposite inequality holds for some $x \in B(\bar{x}, \rho)$, setting $s := \|x - \bar{x}\| > 0$ and taking $r := f(\bar{x}) - cs$ in inequality (4.13), so that $(x,r) \in E_f \cap B(\bar{x}_f, \eta)$, we get the following contradictory inequalities:

$$\varepsilon s < -bs + cs \leq \langle \bar{x}^*, x - \bar{x} \rangle - r + f(\bar{x}) \leq \varepsilon(c+1)^{-1}(s+cs).$$

Now let us show that for all $x \in B(\bar{x}, \rho)$ we have

$$f(x) - f(\bar{x}) \geq \langle \bar{x}^*, x - \bar{x} \rangle - \varepsilon \|x - \bar{x}\|.$$

Suppose, to the contrary, that for some $x \in B(\bar{x}, \rho)$ the opposite inequality is satisfied:

$$f(x) - f(\bar{x}) < \langle \bar{x}^*, x - \bar{x} \rangle - \varepsilon \|x - \bar{x}\|. \tag{4.14}$$

Then $f(x) - f(\bar{x}) < \|\bar{x}^*\| \|x - \bar{x}\|$, and by (4.12), we get

$$|f(x) - f(\bar{x})| \leq c \|x - \bar{x}\| \leq c\rho.$$

Thus we can take $r = f(x)$ in inequality (4.13), and we obtain

$$\langle \bar{x}^*, x - \bar{x} \rangle - f(x) + f(\bar{x}) \leq \varepsilon \|x - \bar{x}\|,$$

a contradiction to (4.14). The proof of the last assertion is left as an exercise; it can be adapted from the exercise in the supplement on bornological subdifferentials. $\qquad \square$

Corollary 4.17. *Let f be finite at $\bar{x}$ and let ∂ (resp. N) stand either for ∂_D or ∂_F (resp. N_D or N_F). If $\partial f(\bar{x})$ is nonempty, then with the preceding notation,*

$$N(E_f, \bar{x}_f) = \mathrm{cl}(\mathbb{R}_+(\partial f(\bar{x}) \times \{-1\})).$$

Proof. Since $N(E_f, \bar{x}_f)$ is a closed convex cone, the inclusion $\mathrm{cl}(\mathbb{R}_+(\partial f(\bar{x}) \times \{-1\})) \subset N(E_f, \bar{x}_f)$ stems from the previous corollary and proposition. Let us prove the reverse inclusion when one can pick some $\bar{x}^* \in \partial f(\bar{x})$. Let $(x^*, -r) \in N(E_f, \bar{x}_f)$; since $\{0\} \times \mathbb{R}_+ \subset T(E_f, \bar{x}_f)$, we have $r \in \mathbb{R}_+$. When r is positive, we have $w^* := r^{-1} x^* \in \partial f(\bar{x})$ by the last two statements and $(x^*, -r) = r(w^*, -1)$. When $r = 0$, setting $x_t^* := \bar{x}^* + t^{-1}(1 - t)x^*$, we have $(x^*, 0) = \lim_{t \to 0_+} t(x_t^*, -1) \in N(E_f, \bar{x}_f)$, since $(t x_t^*, -t) = t(\bar{x}^*, -1) + (1 - t)(x^*, 0) \in N(E_f, \bar{x}_f)$ by convexity, and hence $x_t^* \in \partial f(\bar{x})$. $\qquad \square$

Proposition 4.18. *Let $N = N_D$ (resp. $N = N_F$) and $\partial = \partial_D$ (resp. $\partial = \partial_F$). If $f : X \to \overline{\mathbb{R}}$ is finite at $\bar{x}$ one has $N(E_f, \bar{x}_f) = \mathbb{R}_+(\partial f(\bar{x}) \times \{-1\}) \cup (\partial^\infty f(\bar{x}) \times \{0\})$, where*

$$\partial^\infty f(\bar{x}) := \{x^* \in X^* : (x^*, 0) \in N(E_f, \bar{x}_f)\}.$$

The cone $\partial_D^\infty f(\bar{x})$ (resp. $\partial_F^\infty f(\bar{x})$) is called the directional (resp. firm or Fréchet) *asymptotic subdifferential* or *singular subdifferential* of f at $\bar{x}$. This terminology is justified when $\partial f(\bar{x})$ is nonempty: for every $x^* \in \partial^\infty f(\bar{x})$, $\bar{x}^* \in \partial f(\bar{x})$, and every $t > 0$ one has $(\bar{x}^*, -1) + (t x^*, 0) \in N(E_f, \bar{x}_f)$ since $N(E_f, \bar{x}_f)$ is a convex cone, whence $\bar{x}^* + t x^* \in \partial f(\bar{x})$; conversely, if for a given $\bar{x}^* \in \partial f(\bar{x})$ this inclusion holds for every $t > 0$, then one has $(x^*, 0) = \lim_{t \to \infty} t^{-1}(\bar{x}^* + t x^*, -1) \in N(E_f, \bar{x}_f)$, which is closed, and hence $x^* \in \partial^\infty f(\bar{x})$.

Proof. Since $\bar{x}_f + \mathbb{R}_+(0,1)$ is contained in E_f, one has $\{0\} \times \mathbb{R}_+ \subset T^D(E_f, \bar{x}_f)$, hence $N_F(E_f, \bar{x}_f) \subset N_D(E_f, \bar{x}_f) \subset X^* \times \mathbb{R}_-$. The result ensues: given $(x^*, -r) \in N(E_f, \bar{x}_f)$, either $r = 0$ and $x^* \in \partial^\infty f(\bar{x})$ or $r > 0$ and $(r^{-1}x^*, -1) \in N(E_f, \bar{x}_f)$, so that $x^* = r(r^{-1}x^*) \in r\partial f(\bar{x})$. The opposite inclusion is obvious. $\qquad \square$

Example. Let $f : \mathbb{R} \to \mathbb{R}$ be given by $f(x) = x$ for $x \in \mathbb{R}_-$, $f(x) := \sqrt{x}$ for $x \in \mathbb{R}_+$. Then $\partial f(0) = [1, +\infty)$, $\partial^\infty f(0) = \mathbb{R}_+$ for $\partial = \partial_D$ and ∂_F.

Let us note some other facts concerning normal cones to epigraphs.

Lemma 4.19. *Let $f : X \to \overline{\mathbb{R}}$ be finite at $\bar{x}$, let $\bar{x}_f := (\bar{x}, f(\bar{x}))$, $w := (\bar{x}, r)$, $z := (\bar{x}, s)$ be in the epigraph E_f of f with $r < s$. Then for $N = N_D$ or $N = N_F$ one has*

$$N(E_f, z) \subset N(E_f, w) \subset N(E_f, \bar{x}_f) \subset X^* \times \mathbb{R}_-, \qquad N(E_f, z) \subset X^* \times \{0\}.$$

Proof. The first inclusion entails the second one. It follows from the relations $E_f + z - w \subset E_f$ and $N(E_f + z - w, z) = N(E_f, w)$. Since $\bar{x}_f + \mathbb{R}_+(0,1) \subset E_f$, $z + \{0\} \times (r - s, s - r) \subset E_f$, one has $\mathbb{R}_+(0,1) \subset T(E_f, \bar{x}_f)$, $\{0\} \times \mathbb{R} \subset T(E_f, z)$, hence $N_F(E_f, \bar{x}_f) \subset N_D(E_f, \bar{x}_f) \subset X^* \times \mathbb{R}_-$, $N(E_f, z) \subset X^* \times \{0\}$. $\qquad \square$

As above, we say that f is *quiet* at $\bar{x}$ if $-f$ is calm at $\bar{x}$ (hence $f(\bar{x}) \in \mathbb{R}$).

Lemma 4.20. *If $f : X \to \overline{\mathbb{R}}$ is quiet at $\bar{x}$, then for every $(x^*, r^*) \in N_D(E_f, \bar{x}_f) \setminus \{(0,0)\}$ one has $r^* < 0$ and $(-r^*)^{-1}x^* \in \partial_D f(\bar{x})$. If, moreover, $(x^*, r^*) \in N_F(E_f, \bar{x}_f)$, then one has $(-r^*)^{-1}x^* \in \partial_F f(\bar{x})$.*

Proof. If f is quiet at $\bar{x}$ with rate $c > 0$ in the sense that $f(x) - f(\bar{x}) \le c\|x - \bar{x}\|$ for all x near $\bar{x}$, then for all $u \in X$ one has $f^D(\bar{x}, u) \le c\|u\|$, whence for all $(x^*, r^*) \in N_D(E_f, \bar{x}_f)$, $\langle x^*, u \rangle + r^*c\|u\| \le 0$. Thus $\|x^*\| \le -r^*c$ and $r^* < 0$ when $(x^*, r^*) \ne (0,0)$, since $r^* = 0$ would imply $x^* = 0$. Then $((-r^*)^{-1}x^*, -1) \in N_D(E_f, \bar{x}_f)$, since $N_D(E_f, \bar{x}_f)$ is a cone. Then Corollary 4.15 ensures that $\bar{x}^* := (-r^*)^{-1}x^* \in \partial_D f(\bar{x})$. Proposition 4.16 asserts that $\bar{x}^* \in \partial_F f(\bar{x})$ when $(x^*, r^*) \in N_F(E_f, \bar{x}_f)$. $\qquad \square$

The Fréchet subdifferential being closely related to the norm of the space, it enjoys a pleasant property about distance functions.

Lemma 4.21. *For a subset E of X, its distance function d_E, and $w \in \mathrm{cl}E$ one has*

$$\partial_F d_E(w) = N_F(E, w) \cap B_{X^*}, \tag{4.15}$$

$$N_F(E, w) = \mathbb{R}_+ \partial_F d_E(w). \tag{4.16}$$

Proof. Since d_E is Lipschitzian with rate 1, one has $\partial_F d_E(w) \subset B_{X^*}$. Moreover, as already observed in Proposition 4.13, one has $\partial_F d_E(w) \subset N_F(E, w)$. Conversely, given $w^* \in N_F(E, w) \cap B_{X^*}$ and $\varepsilon > 0$, one can find $\delta > 0$ such that $\langle w^*, x - w \rangle \le \varepsilon\|x - w\|$ for all $x \in E \cap B[w, \delta]$. Then w is a minimizer on $E_\delta := E \cap B[w, \delta]$ of the function $f : x \mapsto \varepsilon\|x - w\| - \langle w^*, x \rangle$, which is Lipschitzian with rate at most $1 + \varepsilon$. The penalization lemma (Lemma 1.121) ensures that w is a minimizer of the function $f + (1 + \varepsilon)d(\cdot, E_\delta)$ on $B[w, \delta]$. Since $d(\cdot, E_\delta)$ coincides with $d(\cdot, E) := d_E$

on $B(w, \delta/2)$, as is easily checked, this implies that for all $u \in B(w, \delta/2)$

$$d_E(u) - d_E(w) - \langle w^*, u - w \rangle \geq -\varepsilon d_E(u) - \varepsilon \|u - w\| \geq -2\varepsilon \|u - w\|.$$

Since $\varepsilon > 0$ is arbitrary, one gets $w^* \in \partial_F d_E(w)$.

The second relation in the statement is an easy consequence of the first one. $\quad\square$

The case $w \in X \setminus E$ is not as simple; we consider it under additional assumptions. Later on, we shall give an approximate version. Again, the result is specific to the Fréchet subdifferential. Recall that the norm of X is said to have the *Kadec–Klee property* if the topology induced on the unit sphere S_X of X coincides with the topology induced by the norm. This property holds when X is a Hilbert space. Given $w \in X$, we set

$$S(w) := \{w^* \in X^* : \langle w^*, w \rangle = \|w\|, \|w^*\| = 1\} = \partial \|\cdot\| (w).$$

Proposition 4.22 (Borwein and Giles). *Let E be a nonempty closed subset of a normed space X and let $w \in X \setminus E$. Then for all $w^* \in \partial_F d_E(w)$ one has $\|w^*\| = 1$. If $x \in E$ is such that $\|x - w\| = d_E(w)$, then $w^* \in S(w - x)$ and $w^* \in \partial_F d_E(x) \subset N_F(E, x)$. If the norm of X is Hadamard (resp. Fréchet) differentiable at $w - x$, then d_E is Hadamard (resp. Fréchet) differentiable at w.*

If X is reflexive, there exists some $z \in X$ such that $w^* \in S(w - z)$ and $\|w - z\| = d_E(w)$. If, moreover, the norm of X has the Kadec–Klee property, then z is a best approximation of w in E, so that $w^* \in \partial_F d_E(z) \cap S(w - z)$.

Proof. Let $w^* \in \partial_F d_E(w)$ and let $\varepsilon > 0$ be given. For a given sequence $(t_n) \to 0_+$ in $(0, 1)$, let $x_n \in E$ be such that $\|x_n - w\| \leq d_E(w) + t_n^2$. Since $\{x_n : n \in \mathbb{N}\}$ is a bounded set, one can find $\delta \in (0, \varepsilon/2)$ such that when $t_n < \delta$ one has

$$\langle w^*, x_n - w \rangle \leq t_n^{-1} [d_E(w + t_n(x_n - w)) - d_E(w)] + \varepsilon/2$$
$$\leq t_n^{-1} [\|w + t_n(x_n - w) - x_n\| - \|w - x_n\| + t_n^2] + \varepsilon/2$$
$$\leq -\|w - x_n\| + \varepsilon.$$

Thus $\liminf_n \langle w^*, w - x_n \rangle \geq \lim_n \|w - x_n\| = d_E(w)$ and $\|w^*\| \geq 1$, whence $\|w^*\| = 1$, d_E being Lipschitzian with rate 1. Moreover, since $\langle w^*, w - x_n \rangle \leq \|w^*\| \cdot \|w - x_n\| \leq d_E(w) + t_n^2$, one has $(\langle w^*, w - x_n \rangle)_n \to d_E(w)$.

If w has a best approximation x in E, one can take $x_n := x$ in what precedes, so that $\langle w^*, w - x \rangle = \|w - x\|$ and $w^* \in S(w - x)$. On the other hand, from the obvious relations $d_E(x + v) \geq d_E(w + v) - \|w - x\| = d_E(w + v) - d_E(w)$, $d_E(x) = 0$, we get $\partial_F d_E(w) \subset \partial_F d_E(x)$. Suppose the norm of X is Hadamard (resp. Fréchet) differentiable at $w - x$. Then for all $u \in X$ one has

$$\langle w^*, u \rangle \le \liminf_{(t,v) \to (0_+,u)} \frac{1}{t} [d_E(w+tv) - d_E(w)]$$

$$\le \limsup_{(t,v) \to (0_+,u)} \frac{1}{t} [\|w+tv-x\| - \|w-x\|] = \langle S(w-x), u \rangle,$$

hence $w^* = S(w-x)$ and $(t^{-1}(d_E(w+tu) - d_E(w))) \to S(w-x)(u)$, so that d_E is Hadamard differentiable at w. The proof of the Fréchet case is similar.

If X is reflexive, the sequence (x_n) has a subsequence that weakly converges to some $z \in w + rB_X$, with $r := d_E(w)$. Then $\langle w^*, w-z \rangle = \lim_n \langle w^*, w-x_n \rangle = r$, so that $\|w-z\| = r$ and $w^* \in S(w-z)$. If, moreover, the norm of X has the Kadec–Klee property, one gets $(w-x_n) \to w-z$ in norm, so that $z \in E$ and z is a best approximation of w in E.   □

The rules we have seen for the calculus of normal cones entail some rules for subdifferentials. A more systematic study will be undertaken in the next sections.

Exercises

1. Given a subset E of a normed space X and $x \in E$, show that $N_F(E,x)$ (resp. $N_D(E,x)$) is the set of derivatives $f'(x)$ of functions f that are differentiable (resp. directionally differentiable) at x and attain their maximum on E at x.

2. Let f be a Lipschitzian function on X with rate $c \ge 1$ and let $X \times \mathbb{R}$ be endowed with the norm given by $\|(x,r)\|_c := c\|x\| + |r|$. Let E_f be the epigraph of f and let $\bar{x}_f := (\bar{x}, f(\bar{x}))$. Show that one has

$$\bar{x}^* \in \partial_F f(\bar{x}) \Leftrightarrow (\bar{x}^*, -1) \in \partial_F d_{E_f}(\bar{x}_f).$$

[Hint: Given $\bar{x}^* \in \partial_F f(\bar{x})$, one has $(\bar{x}^*, -1) \in N_F(E_f, \bar{x}_f) \cap B_{X^* \times \mathbb{R}}$ by Proposition 4.16 and the fact that f is Lipschitzian with rate $c \ge 1$. Thus $(\bar{x}^*, -1) \in \partial_F d_{E_f}(\bar{x}_f)$. The reverse implication is immediate from Proposition 4.16.]

3. Let X be a normed space and let E be a closed subset of X. Let $w \in X \setminus E$. Show that in the relation $\partial_F d_E(w) \subset S_{X^*}$ one cannot replace ∂_F with ∂_D. [Hint: Take in ℓ_2 the complement to the set $\{x = (x_1, x_2, \ldots) : -(1+(2/n)) < x_n < 1+(1/n)\}$ and $w = 0$.]

4. Let f be a lower semicontinuous function on X, let $(x,r) \in X \times \mathbb{R}$, and let $(x^*, r^*) \in \partial_D d_{\text{epi} f}(x,r)$.

(a) Prove that $r^* \le 0$ and $r^* = 0$ if $r > f(x)$.
(b) Show that r^* may still be equal to zero even if $r < f(x)$.
(c) Assume that $\partial_D d_{\text{epi} f}(x,r) \ne \varnothing$ for all $r < f(x)$. Show that $r^* < 0$ if r is sufficiently close to $f(x)$.

5. Let $S \subset X$, $\bar{x} \in S$. Show that the cone generated by $\partial_D d_S(\bar{x})$ may be a proper subset of $N_D(S,\bar{x})$. [Hint: Take in ℓ_2 the set $S = \cup S_k$, where $S_1 = \{x = (x_1, x_2, \dots) : x_1 \geq 0\}$ and $S_k = \{x = (x_1, x_2, \dots) : x_1 \geq -(1/k), x_k \geq 1/k^2\}$ and take $\bar{x} = 0$.]

6. Let $S \subset X$ be a convex set and let $\bar{x} \in S$. Using the distance function d_S, prove that

$$N_F(S,\bar{x}) = N_D(S,\bar{x}) = \{\bar{x}^* : \langle \bar{x}^*, u - \bar{x} \rangle \leq 0, \ \forall u \in S\}.$$

[Hint: $\bar{x}^*$ belongs to the normal cone to S at $\bar{x}$ in the sense of convex analysis if and only if either $\bar{x}^* = 0$ or $\|\bar{x}^*\|^{-1}\bar{x}^* \in \partial d_S(\bar{x})$ (in the sense of convex analysis).]

7. Show that the Fréchet normal cone $N_F(E,\bar{x})$ to a subset E of X at $\bar{x} \in \mathrm{cl}(E)$ is the polar cone (in the duality between X^* and $X^{**} := (X^*)^*$) of the weak** tangent cone

$$T^{**}(E,\bar{x}) := \mathrm{w}^{**} - \limsup_{t \to 0_+} E_t \qquad \text{for } E_t := \frac{1}{t}(E - \bar{x})$$

when E_t is considered as a subset of X^{**}. Here $\mathrm{w}^{**} - \limsup_{t \to 0_+} E_t$ denotes the set of weak** cluster points of bounded nets $(v_t)_{t>0}$ (or sequences (v_{t_n}) with $(t_n) \to 0$) with $v_t \in E_t$ for all $t > 0$. In particular, when X is reflexive, $N_F(E,\bar{x})$ is w*-closed.

4.1.4 Coderivatives

The notions of tangent cone and normal cone enable one to introduce concepts of generalized derivatives for multimaps. We first define a primal notion and then a dual one. Again, we identify a multimap with its graph, using the *transpose* $H^{\mathsf{T}} : Y^* \rightrightarrows X^*$ of a positively homogeneous multimap $H : X \rightrightarrows Y$ between two normed spaces defined by

$$H^{\mathsf{T}}(y^*) := \{x^* \in X^* : \forall (x,y) \in H, \quad \langle x^*, x \rangle - \langle y^*, y \rangle \leq 0\}.$$

Definition 4.23. The *directional* or *contingent derivative* at $z := (x,y)$ of a multimap $F : X \rightrightarrows Y$ between two normed spaces is the multimap $DF(x,y) : X \rightrightarrows Y$ whose graph is the tangent cone $T(F,z) := T^D(F,z)$ to the graph of F at z:

$$DF(x,y)(u) := D_D F(x,y)(u) := \{v \in Y : (u,v) \in T(F,z)\}.$$

Definition 4.24. The *directional* (or *contingent*) *coderivative* of $F : X \rightrightarrows Y$ at $z := (x,y)$ is the multimap $D^*F(x,y) := D_D^*F(x,y) : Y^* \rightrightarrows X^*$ that is the transpose of $DF(x,y)$:

$$D^*F(x,y)(y^*) = \{x^* \in X^* : \langle x^*, u \rangle - \langle y^*, v \rangle \leq 0 \ \forall u \in X, \ \forall v \in DF(z)(u)\}$$
$$= \{x^* \in X^* : (x^*, -y^*) \in N_D(F,z)\}.$$

The *firm (or Fréchet) coderivative* of F at (x,y) is the multimap $D_F^* F(x,y) : Y^* \rightrightarrows X^*$ given by

$$D_F^* F(x,y)(y^*) := \{x^* \in X^* : (x^*, -y^*) \in N_F(F,z)\}.$$

Since $N_F(F,z) \subset N_D(F,z)$, one has $D_F^* F(x,y)(y^*) \subset D^* F(x,y)(y^*)$ for all $y^* \in Y^*$. When $F(x)$ is a singleton $\{y\}$, one writes $DF(x)$ instead of $DF(x,y)$ and $D^* F(x)$ (resp. $D_F^* F(x)$) instead of $D^* F(x,y)$ (resp. $D_F^* F(x,y)$). When F is a mapping that is Hadamard differentiable at x, one has $DF(x) = F'(x)$ and $D^* F(x,y) = F'(x)^\mathsf{T}$, the transpose of the derivative $F'(x)$ of F at x, as is easily checked. Similarly, when F is Fréchet differentiable at x, one has $D_F^* F(x) = F'(x)^\mathsf{T}$. When $Y = \mathbb{R}$ and $F(\cdot) := [f(\cdot), +\infty)$ for some function $f : X \to \mathbb{R}_\infty$, one has

$$\partial_D f(x) = D^* F(x, f(x))(1)$$

in view of Proposition 4.15, which asserts that $x^* \in \partial_D f(x)$ if and only if $(x^*, -1) \in N_D(\mathrm{epi}\, f, x_f) = N_D(F, x_f)$ for $x_f := (x, f(x))$; similarly, $\partial_F f(x) = D_F^* F(x_f)(1)$.

The calculus rules we have given for normal cones entail calculus rules for coderivatives. We also have the following scalarization result. Here we say that a map $g : X \to Y$ between two normed spaces is *tangentially compact at* $\bar{x} \in X$ if for every $u \in X \setminus \{0\}$, $(u_n) \to u$, $(t_n) \to 0_+$ the sequence $(t_n^{-1}(g(\bar{x}+t_n u_n) - g(\bar{x})))$ has a convergent subsequence. This condition is satisfied if g is directionally differentiable at $\bar{x}$ or if Y is finite-dimensional and if g is *directionally stable* at $\bar{x}$ in the sense that for every $u \in X \setminus \{0\}$ there exist $\varepsilon > 0$ and $c \in \mathbb{R}_+$ such that $\|g(\bar{x}+tv) - g(\bar{x})\| \leq ct$ for all $t \in (0, \varepsilon)$, $v \in B(u, \varepsilon)$. The latter condition is satisfied when g is *stable* (or Stepanovian) at $\bar{x}$ in the sense that there exist $r > 0$ and $k \in \mathbb{R}_+$ such that $\|g(x) - g(\bar{x})\| \leq k\|x - \bar{x}\|$ for all $x \in B(\bar{x}, r)$; for $Y = \mathbb{R}$ this definition coincides with the one given above for functions.

Proposition 4.25 (Scalarization). *For every map $g : X \to Y$ between two normed spaces and for every $\bar{x} \in X$, $y^* \in Y^*$ one has the following inclusions. The first one is an equality if g is tangentially compact at $\bar{x}$; the second one is an equality if g is stable at $\bar{x}$:*

$$\partial_D(y^* \circ g)(\bar{x}) \subset D^* g(\bar{x})(y^*), \qquad \partial_F(y^* \circ g)(\bar{x}) \subset D_F^* g(\bar{x})(y^*).$$

Proof. Let $h := y^* \circ g$, let $x^* \in \partial_D h(\bar{x})$, and let G be the graph of g. Then for every $(u,v) \in T^D(G, (\bar{x}, \bar{y}))$, where $\bar{y} := g(\bar{x})$, we can find sequences $(t_n) \to 0_+$, $(u_n) \to u$, $(v_n) \to v$ such that $\bar{y} + t_n v_n = g(\bar{x} + t_n u_n)$ for all n, hence

$$\langle y^*, v \rangle = \left\langle y^*, \lim_n \frac{1}{t_n}(g(\bar{x}+t_n u_n) - g(\bar{x})) \right\rangle$$

$$= \lim_n \frac{1}{t_n}[\langle y^*, g(\bar{x}+t_n u_n)\rangle - \langle y^*, g(\bar{x})\rangle] \geq h^D(\bar{x}, u) \geq \langle x^*, u\rangle,$$

so that $x^* \in D^*g(\bar{x})(y^*)$. If $x^* \in \partial_F h(\bar{x})$, then for every $\varepsilon > 0$ we can find some $\delta > 0$ such that for all $x \in B(\bar{x}, \delta)$ (hence for all x such that $(x, g(x)) \in B((\bar{x}, \bar{y}), \delta))$, we have

$$\langle (x^*, -y^*), (x, g(x)) - (\bar{x}, \bar{y}) \rangle = \langle x^*, x - \bar{x} \rangle - (h(x) - h(\bar{x})) \le \varepsilon \|x - \bar{x}\|.$$

Since $\|x - \bar{x}\| \le \|(x - \bar{x}, g(x) - \bar{y})\|$, we get $(x^*, -y^*) \in N_F(G, (\bar{x}, \bar{y})): x^* \in D_F^* g(\bar{x})(y^*)$.

Let g be tangentially compact at $\bar{x}$ and let $x^* \in D^*g(\bar{x})(y^*)$. Then for every $u \in X$ and every sequence $((u_n, t_n)) \to (u, 0_+)$ such that $(t_n^{-1}(h(\bar{x} + t_n u_n) - h(\bar{x}))) \to h^D(\bar{x}, u)$ we can find $v \in Y$ that is a cluster point of the sequence $(t_n^{-1}(g(\bar{x} + t_n u_n) - g(\bar{x})))$. Then $(u, v) \in T^D(G, (\bar{x}, \bar{y}))$, $h^D(\bar{x}, u) = \langle y^*, v \rangle$ and $(x^*, -y^*) \in N_D(G, (\bar{x}, \bar{y}))$, whence

$$\langle x^*, u \rangle - h^D(\bar{x}, u) = \langle x^*, u \rangle - \langle y^*, v \rangle \le 0,$$

so that $x^* \in \partial_D h(\bar{x})$.

Finally, suppose g is stable at $\bar{x}$ and $x^* \in D_F^* g(\bar{x})(y^*)$. Let $c \in \mathbb{R}_+, \rho > 0$ be such that $\|g(x) - g(\bar{x})\| \le c \|x - \bar{x}\|$ for all $x \in B(\bar{x}, \rho)$. Since $(x^*, -y^*) \in N_F(G, (\bar{x}, \bar{y}))$, given $\varepsilon > 0$ we can find $\delta \in (0, \rho)$ such that for all $(x, y) \in G \cap B((\bar{x}, \bar{y}), \delta)$ one has

$$\langle x^*, x - \bar{x} \rangle - \langle y^*, y - \bar{y} \rangle \le \varepsilon(c + 1)^{-1}(\|x - \bar{x}\| + \|y - \bar{y}\|).$$

Since $y - \bar{y} = g(x) - g(\bar{x})$ and $\|g(x) - g(\bar{x})\| \le c \|x - \bar{x}\|$, this relation can be written

$$\langle x^*, x - \bar{x} \rangle \le \langle y^*, g(x) \rangle - \langle y^*, g(\bar{x}) \rangle + \varepsilon \|x - \bar{x}\|,$$

so that $x^* \in \partial_F (y^* \circ g)(\bar{x})$. $\qquad\square$

The following example shows that one cannot drop the stability assumption.

Example. Let $g : \mathbb{R} \to \mathbb{R}^2$ be given by $g(x) = (x, \sqrt{|x|})$. Then $N(G, (0, 0, 0)) = \mathbb{R} \times \mathbb{R} \times \mathbb{R}_-$, so that for $y^* := (1, 0)$ one has $D^*g(0)(y^*) = \mathbb{R}$ but $\partial(y^* \circ g)(0) = \{1\}$.

The following result shows how a classical property of differential calculus can be extended to multimaps using coderivatives. A converse in adapted spaces will be established later.

Proposition 4.26. *Let V, W be open subsets of normed spaces Y and Z respectively and let $M : Y \rightrightarrows Z$ be a multimap that is pseudo-Lipschitzian on $V \times W$ with rate c in the sense that*

$$\forall v, v' \in V, \ w \in W \cap M(v), \quad d(w, M(v')) \le cd(v, v'). \tag{4.17}$$

Then for all $(v, w) \in V \times W$, $z^* \in Z^*$, $y^* \in D_F^* M(v, w)(z^*)$ one has $\|y^*\| \le c \|z^*\|$.

Defining the *norm* of a *process*, i.e., a positively homogeneous multimap $H : Z^* \rightrightarrows Y^*$ by

$$\|H\| := \sup\{\|y^*\| : y^* \in H(z^*), \ z^* \in S_{Z^*}\},$$

the conclusion can be written $\|D_F^* M(v, w)\| \le c$ for all $(v, w) \in V \times W$.

Proof. Let $(v,w) \in V \times W$, $z^* \in Z^*$, $y^* \in D_F^* M(v,w)(z^*)$ and let r be a remainder such that

$$(y,z) \in M - (v,w) \implies \langle y^*, y \rangle + \langle -z^*, z \rangle \le r(\|y\| + \|z\|).$$

Given $c' > c$, relation (4.17) ensures that for all $y \in Y$ small enough (so that $v' := v + y \in V$) one can find $z \in c' \|y\| B_Z$ such that $w + z \in M(v + y)$. Then we get

$$\langle y^*, y \rangle \le \langle z^*, z \rangle + r(\|y\| + \|z\|) \le c' \|y\| \|z^*\| + r((c'+1) \|y\|),$$

so that $\|y^*\| \le c' \|z^*\|$. Since c' is arbitrarily close to c, we get $\|y^*\| \le c \|z^*\|$.   □

Calculus rules for coderivatives will be given later. Here we just point out how they can be derived from calculus rules for functions. Let us consider the case of the composition $H := G \circ F$ of two multimaps $F : X \rightrightarrows Y$, $G : Y \rightrightarrows Z$. We have $x^* \in D^* H(\bar{x}, \bar{z})(z^*)$ if and only if $(x^*, -z^*) \in \partial \iota_{\mathrm{gph}H}(\bar{x}, \bar{z})$ and

$$\iota_{\mathrm{gph}H}(\bar{x}, \bar{z}) = \inf\{\iota_{\mathrm{gph}F}(\bar{x}, y) + \iota_{\mathrm{gph}G}(y, \bar{z}) : y \in Y\}. \tag{4.18}$$

Thus, sums and performance functions are involved. A geometric approach can also be taken considering the graph of H, that is the projection on $X \times Z$ of the intersection $(\mathrm{gph}F \times Z) \cap (X \times \mathrm{gph}G)$.

4.1.5   Supplement: Incident and Proximal Notions

The preceding relationships between analytical and geometrical notions incite us to present variants of the directional derivative and directional subdifferential. They are not as important as the notions expounded above, but they have some interest as the exercises below show. They arise from a variant of the contingent cone introduced in Definition 2.153 we recall here.

Definition 4.27. Given a subset S of a normed space X and $x \in S$, a vector $v \in X$ is said to be an *incident vector* to S at x if for every sequence $(t_n) \to 0_+$ there exists a sequence $(v_n) \to v$ such that $x + t_n v_n \in S$ for all $n \in \mathbb{N}$. Thus, the set $T^I(S,x)$ of incident vectors to S at x, called the *incident cone* (or *inner tangent cone*) to S at x, is

$$T^I(S,x) := \liminf_{t \to 0_+} \frac{1}{t}(S - x).$$

When $T^I(S,x)$ coincides with the contingent cone $T^D(S,x) = \limsup_{t \to 0_+} \frac{1}{t}(S-x)$, one says that S is *derivable* at x; then the tangent cone to S at x is $\lim_{t \to 0_+} \frac{1}{t}(S-x)$.

We define the *incident normal cone* to S at x as the polar cone $N_I(S,x) := (T^I(S,x))^0$ of the incident cone to S at x.

Exercise. Show that if S is convex, or if S is a differentiable submanifold of X, then S is derivable at each of its points.

Exercise. Show that the set $S := \{0\} \cup \{2^{-n} : n \in \mathbb{N}\}$ is not derivable at 0.

Exercise. Show that the set $T^I(S,x)$ of incident vectors to S at x is also the set of velocities of curves in S issued from x: $v \in T^I(S,x)$ iff there exists $c : [0,1] \to S$ such that $c(0) = x$, $c'(0) := \lim_{t\to 0_+}(1/t)(c(t) - x)$ exists and is equal to v.

Exercise. Show that if A (resp. B) is a subset of a normed space X (resp. Y), then

$$T^I(A,x) \times T^I(B,y) = T^I(A \times B, (x,y)),$$

$$T^I(A,x) \times T^D(B,y) \subset T^D(A \times B, (x,y)) \subset T^D(A,x) \times T^D(B,y).$$

Deduce from these relations that $A \times B$ is derivable at (x,y) iff A and B are derivable at x and y respectively.

Exercise. Show that if A (resp. B) is a subset of a normed space X (resp. Y), then

$$N_D(A \times B, (x,y)) = N_D(A,x) \times N_D(B,y),$$

$$N_I(A \times B, (x,y)) = N_I(A,x) \times N_I(B,y),$$

$$N_F(A \times B, (x,y)) = N_F(A,x) \times N_F(B,y).$$

By analogy with the contingent (or lower directional) derivate of a function f at x, we define the *incident derivate* (or *inner derivate*) of f at x by

$$f^I(x,u) := \inf\{r \in \mathbb{R} : (u,r) \in T^I(E_f, x_f)\},$$

where E_f is the epigraph of f and $x_f := (x, f(x))$.

Similarly, we define the *incident subdifferential* of f at x as

$$\partial_I f(x) := \{x^* \in X^* : \langle x^*, \cdot \rangle \le f^I(x, \cdot)\}.$$

Exercise. With the preceding notation and $E_f := \mathrm{epi} f$, $x_f := (x, f(x))$, show that

$$x^* \in \partial_I f(x) \iff (x^*, -1) \in N_I(E_f, x_f).$$

Exercise. Show that f is *epi-differentiable* at x in the sense that $f^I(x, \cdot) = f^D(x, \cdot)$ if and only if the epigraph E_f of f is derivable at x_f. Show that this occurs when f has a directional derivative at x.

For questions connected with distance functions, the notion of proximal normal is an appropriate tool, at least in Hilbert spaces.

Definition 4.28. Given a subset S of a normed space X and $x \in S$, a vector $v \in X$ is said to be a *primal proximal normal* to S at $x \in S$, and one writes $v \in N^P(S,x)$, if there exists some $r > 0$ such that $d_S(x + rv) = r\|v\|$.

This relation can be translated geometrically as $B(x+rv,r\|v\|)\cap S=\varnothing$, since the latter equality means that $d_S(x+rv)=r\|v\|$. Since for $t\in[0,r]$ one has $B(x+tv,t\|v\|)\subset B(x+rv,r\|v\|)$ by the triangle inequality, one also has

$$v\in N^P(S,x)\Leftrightarrow\exists r>0:\forall t\in[0,r],\quad B(x+tv,t\|v\|)\cap S=\varnothing$$
$$\Leftrightarrow\exists r>0:\forall t\in[0,r],\quad d_S(x+tv)=t\|v\|.$$

The set of *proximal normals* to S at x is $N_P(S,x):=J_X(N^P(S,x))$, where J_X is the *duality multimap* of X given by $J_X(v):=\{v^*\in X^*:\langle v^*,v\rangle=\|v\|^2=\|v^*\|^2\}$. When X is a Hilbert space, J_X is the Riesz isometry, which allows one to identify X^* with X and J_X with I_X, the identity map. Then the following geometric characterization may be useful:

$$v\in N_P(S,x)\Leftrightarrow\exists r>0:\forall t\in[0,r],\quad(x+tv+t\|v\|B_X)\cap S=\{x\}$$

In fact, if v satisfies this last condition, for $t\in[0,r]$ one has $B(x+tv,t\|v\|)\cap S=\varnothing$, hence $v\in N_P(S,x)$. Conversely, suppose $d_S(x+rv)=r\|v\|$ for some $r>0$, i.e., $\|x+rv-s\|^2\geq r^2\|v\|^2$ for all $s\in S$, or equivalently, after expanding $\|x+rv-s\|^2$,

$$\|x-s\|^2\geq 2r(v\mid s-x)\quad\forall s\in S.$$

Then for all $t\in[0,r)$, $s\in S\setminus\{x\}$, taking into account the sign of $(v\mid s-x)$, one has

$$\|x-s\|^2>2t(v\mid s-x),$$

hence $\|x+tv-s\|^2>t^2\|v\|^2$. Thus $(x+tv+t\|v\|B_X)\cap S=\{x\}$.

Proposition 4.29. *For every closed subset of a Hilbert space X and every $x\in S$, the set $N_P(S,x)$ of proximal normals to S at x is convex.*

Proof. Let $v_0,v_1\in N_P(S,x)$. The preceding remark shows that for all $r>0$ small enough, x is the projection of $x+rv_i$ $(i=0,1)$ in S, or equivalently,

$$\|x-s\|^2\geq 2r(v_i\mid s-x)\quad\forall s\in S.$$

Given $t\in[0,1]$, for $v:=(1-t)v_0+tv_1$, we see that the preceding inequality holds with v instead of v_i. Thus x is the projection of $x+rv$ on S and $v\in N_P(S,x)$. $\square$

Proposition 4.30. *For every closed subset S of a Hilbert space X and every $x\in S$, one has $N_P(S,x)\subset N_F(S,x)\subset N_D(S,x)$.*

Proof. Let $v\in N_P(S,x)$, so that for some $r>0$, one has $\|x-s\|^2\geq 2r(v\mid s-x)$ for all $s\in S$. Then, given $\varepsilon>0$, taking $\delta\in(0,2r\varepsilon)$, for all $s\in S\cap B(x,\delta)$ we have $(v\mid s-x)\leq(\delta/2r)\|x-s\|\leq\varepsilon\|x-s\|$. Thus $v\in N_F(S,x)$. $\square$

Exercise. Relate the proximal subdifferential $\partial_P f(\bar x)$ at $\bar x\in\mathrm{dom}f$ of a function $f:X\to\overline{\mathbb{R}}$ to $N_P(\mathrm{epi}f,(\bar x,f(\bar x)))$. Consider the case of the indicator function of some closed subset S of a Hilbert space.

4.1.6 Supplement: Bornological Subdifferentials

The firm subdifferential and the directional subdifferential are two special cases of a general process we briefly describe now. Given a point $\bar{x}$ of a normed space X, to every convergence γ on the set $\overline{\mathbb{R}}^X$ of functions from X to $\overline{\mathbb{R}}$ we can associate a notion of derivative for $f \in \overline{\mathbb{R}}^X$ finite at $\bar{x}$ by taking as derivative of f at $\bar{x}$ the limit, if it exists, of the functions $f_t : u \mapsto t^{-1}(f(\bar{x}+tu) - f(\bar{x}))$ as $t \to 0$, $t \neq 0$. Assuming that the limit is linear and continuous, the Fréchet derivative corresponds to the topology of uniform convergence on bounded subsets, while the directional derivative corresponds to continuous convergence (or uniform convergence on compact subsets if one considers continuous functions). Subdifferentials can be obtained in a similar way using a one-sided convergence: for every convergence γ on $\overline{\mathbb{R}}^X$ one can define $\bar{x}^*$ to be in the subdifferential of f at $\bar{x}$ associated with γ if $(\bar{x}^* - f_t)^+ \to 0$ as $t \to 0_+$ where $r^+ = \max(r, 0)$ is the positive part of the real number r.

A general means for obtaining a convergence on $\overline{\mathbb{R}}^X$ consists in selecting a family $\mathscr{B}$ of subsets of X and in requiring uniform convergence on the members of $\mathscr{B}$. It is usual to require that $\mathscr{B}$ be a *bornology*, i.e., that $\mathscr{B}$ be a covering of X by bounded subsets and that $\mathscr{B}$ be *hereditary* (i.e., that $B \in \mathscr{B}$ whenever $B \subset B'$ for some $B' \in \mathscr{B}$).

Let us rephrase the definition of the subdifferential associated with the convergence defined by $\mathscr{B}$ when $\mathscr{B}$ is a bornology on X (or just a covering of X): if f is a function on X finite at $\bar{x}$, then the subdifferential associated with $\mathscr{B}$ is the set $\partial_{\mathscr{B}} f(\bar{x})$ of $\bar{x}^* \in X^*$ such that for all $B \in \mathscr{B}$ one has

$$\liminf_{t \to 0_+} \inf_{v \in B} \frac{1}{t}(f(\bar{x}+tv) - f(\bar{x}) - \langle \bar{x}^*, tv \rangle) \geq 0,$$

or more explicitly, for every $B \in \mathscr{B}$,

$$\forall \varepsilon > 0, \; \exists \delta > 0 : \forall t \in [0, \delta], \; \forall v \in B, \quad f(\bar{x}+tv) - f(\bar{x}) - \langle \bar{x}^*, tv \rangle \geq -\varepsilon t.$$

When $\mathscr{B}$ is the canonical bornology, i.e., the family of all bounded subsets of X, one gets the Fréchet subdifferential, as is easily seen. But other choices are of interest, for instance the family of finite subsets of X and the family of compact or weakly compact subsets of X. When $\mathscr{B}$ is the family of sets contained in some compact subset, one can show that $\partial_{\mathscr{B}} f(\bar{x}) = \partial_D f(\bar{x})$ (Exercise 1).

In the sequel we suppose that the bornology $\mathscr{B}$ is such that $B \times T \in \mathscr{B}_{X \times \mathbb{R}}$ for all $B \in \mathscr{B}_X$ and all compact intervals T of $\mathbb{R}$; this natural condition is satisfied in the last two examples and in the case of the canonical bornology.

The following observation is a simple consequence of the definitions.

Lemma 4.31. *For every function f finite at $\bar{x}$, the set $\partial_{\mathscr{B}} f(\bar{x})$ is closed in the topology of uniform convergence on the members of $\mathscr{B}$.*

A characterization of $\partial_{\mathscr{B}} f$ when f is Lipschitzian can be given. It uses the following lemma, which is of independent interest.

Lemma 4.32. *Suppose f is Lipschitzian with rate c around some $\bar{x} \in X$ and $X \times \mathbb{R}$ is endowed with the norm given by $\|(x,r)\| = c\|x\| + |r|$. Then for (u,r) near $(\bar{x}, f(\bar{x}))$, the distance of (u,r) to the epigraph E of f satisfies*

$$d_E(u,r) = (f(u) - r)^+ := \max(f(u) - r, 0).$$

Proof. Since $(u, f(u)) \in E$, the inequality $d_E(u,r) \le (f(u) - r)^+$ holds for every function f. When f is Lipschitzian with rate c on a ball $B(\bar{x}, \rho)$ and $\sigma \in (0, \rho/3)$, this inequality cannot be strict when $(u,r) \in B(\bar{x}, \sigma) \times B(f(\bar{x}), c\sigma)$: otherwise, for some $(v,s) \in E$, we would have

$$c\|u - v\| + |r - s| < f(u) - r \le c\|u - \bar{x}\| + |f(\bar{x}) - r| < 2c\sigma,$$

hence $\|u - v\| < 2\sigma$, $\|v - \bar{x}\| \le \|v - u\| + \|u - \bar{x}\| < 3\sigma < \rho$, so that we would get

$$c\|u - v\| + |r - s| < f(u) - r \le f(u) - f(v) + s - r \le |f(u) - f(v)| + |r - s|,$$

a contradiction to the Lipschitz assumption. □

Proposition 4.33. *If ∂ is the subdifferential $\partial_{\mathscr{B}}$ associated with a bornology $\mathscr{B}$, then for every function f on X finite at $\bar{x}$, the implications (a)$\Rightarrow$(b)$\Leftrightarrow$(c)$\Rightarrow$(d) hold among the following assertions, in which E denotes the epigraph of f and $\bar{x}_f := (\bar{x}, f(\bar{x}))$. If f is Lipschitzian with rate c around $\bar{x}$, and if $X \times \mathbb{R}$ is endowed with the norm given by $\|(x,r)\| := c\|x\| + |r|$, then all these assertions are equivalent:*

(a) $(\bar{x}^*, -1) \in \mathbb{R}_+ \partial d_E(\bar{x}_f)$;
(b) $(\bar{x}^*, -1) \in \partial \iota_E(\bar{x}_f)$, *where ι_E is the indicator function of E*;
(c) $\bar{x}^* \in \partial f(\bar{x})$;
(d) $\bar{x}^* \in \partial e_f(\bar{x}_f)$, *where e_f is defined by $e_f(x,r) := \max(f(x) - r, 0)$.*

Note that $\partial_{\mathscr{B}} \iota_E(\bar{x}_f)$ can be considered the $\mathscr{B}$-normal cone to E at $\bar{x}_f$.

Proof. (a)$\Rightarrow$(b) follows from the fact that for all $\lambda \in \mathbb{R}_+$, one has $\lambda d_E \le \iota_E$, $\lambda d_E(\bar{x}_f) = \iota_E(\bar{x}_f)$.

(b)$\Rightarrow$(c) Let us prove that if $\bar{x}^* \notin \partial f(\bar{x})$, then $(\bar{x}^*, -1) \notin \partial \iota_E(\bar{x}_f)$. By assumption, there exist $\alpha \in (0, 1]$, $B \in \mathscr{B}$, sequences (v_n) in B, and $(t_n) \to 0_+$ such that

$$t_n^{-1}\left(f(\bar{x} + t_n v_n) - f(\bar{x}) - \langle \bar{x}^*, t_n v_n \rangle\right) \le -\alpha. \tag{4.19}$$

Let $c := \|\bar{x}^*\| \sup_{v \in B} \|v\| + 1$, let $r_n := t_n^{-1}\left(f(\bar{x} + t_n v_n) - f(\bar{x})\right)$, so that $r_n \le \langle \bar{x}^*, v_n \rangle - \alpha \le c - \alpha$ for all n. Let $r_n' := \max(r_n, -c)$. Then $(v_n, r_n') \in B' := B \times [-c, c] \in \mathscr{B}_{X \times \mathbb{R}}$ and $(\bar{x} + t_n v_n, f(\bar{x}) + t_n r_n') \in E$ for all n. If n is such that $r_n' = r_n$, we deduce from relation (4.19) that

$$t_n^{-1}\left[\iota_E\left(\bar{x} + t_n v_n, f(\bar{x}) + t_n r_n'\right) - \iota_E\left(\bar{x}, f(\bar{x})\right) + t_n r_n' - \langle \bar{x}^*, t_n v_n \rangle\right] \le -\alpha,$$

and the same inequality holds when $r_n' = -c$, since $-c - \langle \bar{x}^*, v_n \rangle \le -1 \le -\alpha$. Thus we cannot have $(\bar{x}^*, -1) \in \partial \iota_E(\bar{x}_f)$.

(c)⇒(b) Let $\overline{x}^* \in \partial f(\overline{x})$. In order to prove that $(\overline{x}^*, -1) \in \partial \iota_E(\overline{x}_f)$, let us show that given $B \in \mathcal{B}$, $c > 0$, $\varepsilon > 0$ we can find $\delta > 0$ such that for all $t \in (0, \delta)$, $(v, r) \in B \times [-c, c]$ with $(\overline{x} + tv, f(\overline{x}) + tr) \in E$ we have

$$\langle (\overline{x}^*, -1), (\overline{x} + tv - \overline{x}, f(\overline{x}) + tr - f(\overline{x})) \rangle \le t\varepsilon.$$

This relation amounts to $tr \ge \langle \overline{x}^*, tv \rangle - t\varepsilon$. Since $tr \ge f(\overline{x} + tv) - f(\overline{x})$, this inequality is satisfied whenever $v \in B$, $t \in (0, \delta)$, where $\delta > 0$ is chosen in such a way that

$$\forall t \in (0, \delta), \ \forall v \in B, \qquad f(\overline{x} + tv) - f(\overline{x}) \ge \langle \overline{x}^*, tv \rangle - \varepsilon t,$$

in accordance with the definition of $\partial f(\overline{x})$.

(c)⇒(d) Since $e_f = \max(h_f, 0)$, where $h_f(x, r) := f(x) - r$, we easily see that for every $\overline{x}^* \in \partial f(\overline{x})$ we have $(\overline{x}^*, -1) \in \partial h_f(\overline{x}_f) \subset \partial e_f(\overline{x}_f)$.

(d)⇒(a) (when f is Lipschitzian) is a consequence of the preceding lemma.   □

Exercises

1. Show that when f is continuous and when $\mathcal{B}$ is the family of sets contained in a compact subset, one has $\partial_{\mathcal{B}} f(\overline{x}) = \partial_D f(\overline{x})$.

2. Given $\varepsilon > 0$ let $\partial_F^\varepsilon f(\overline{x})$ be the set of $\overline{x}^* \in X^*$ such that

$$\liminf_{\|v\| \to 0_+} \frac{1}{\|v\|} [f(\overline{x} + v) - f(\overline{x}) - \langle \overline{x}^*, v \rangle] \ge -\varepsilon.$$

Give elementary calculus rules for these approximate subdifferentials.

3. Given $\varepsilon > 0$ and a subset E of X whose closure contains $\overline{x}$, let $N_F^\varepsilon(E, \overline{x})$ be the set of $\overline{x}^* \in X^*$ such that

$$\limsup_{x \to_E \overline{x}} \frac{1}{\|x - \overline{x}\|} \langle \overline{x}^*, x - \overline{x} \rangle \le \varepsilon.$$

(a) Show that $N_F^\varepsilon(E, \overline{x}) = \partial_F^\varepsilon \iota_E(\overline{x})$.

(b) Let E_f be the epigraph of f and let $\overline{x}_f := (\overline{x}, f(\overline{x}))$. Show that for every $\overline{x}^* \in \partial_F^\varepsilon f(\overline{x})$ one has $(\overline{x}^*, -1) \in N_F^\varepsilon(E_f, \overline{x}_f)$.

(c) Conversely, given $(\overline{x}^*, -1) \in N_F^\varepsilon(E_f, \overline{x}_f)$, find $\alpha := \alpha(\varepsilon)$ such that $\overline{x}^* \in \partial_F^\alpha f(\overline{x})$. (See [525].)

4. Using the concepts of Sect. 4.3, show that the following assertion implies those of Proposition 4.33: $\overline{x}^*$ belongs to the viscosity subdifferential associated with $\mathcal{B}$ in the sense that there exists a function φ on some open neighborhood U of $\overline{x}$ that is such that $\varphi(\overline{x}) = f(\overline{x})$, $\varphi \le f$ on U, φ is $\mathcal{B}$-differentiable on U, with $\varphi'(\overline{x}) = \overline{x}^*$ and such that for every $B \in \mathcal{B}$ and $x \in U$ the function $u \mapsto \sup\{|\langle \varphi'(u) - \varphi'(x), v \rangle| : v \in B\}$ is continuous at x.

Prove that when X has a $\mathscr{B}$-smooth bump function, this assertion is equivalent to the other ones in Proposition 4.33.

5. Show that if $f : X \to Y$ is a Lipschitzian map with rate c between two metric spaces, and if $X \times Y$ is endowed with the metric given by $d((u,v),(x,y)) := cd(u,x) + d(v,y)$, then the distance to the graph G of f satisfies $d((x,y),G) = d(f(x),y)$. [Hint: Mimic the proof of Lemma 4.32.]

4.2 Elementary Calculus Rules

In this section we present some calculus rules that are direct consequences of the definitions. Their interest for optimization problems is limited, since usually one needs inclusions in the reverse direction of that obtained from these rules. However, combined with continuity properties or with other approaches presented in the next chapters, these rules make it possible to get equalities under some regularity conditions. Thus the reader should be aware of them and see them as natural counterparts to the more important fuzzy rules presented in the sequel.

4.2.1 Elementary Sum Rules

The inclusion of the next statement is a direct application of the definitions: the sum of two remainders is a remainder, and a similar stability property holds for multiplication by a nonnegative real number.

In the following results, ∂ stands either for ∂_D or for ∂_F.

Proposition 4.34. *If g and h are finite at $\overline{x}$, and if $r, s \in \mathbb{R}_+$, then one has*

$$r\partial g(\overline{x}) + s\partial h(\overline{x}) \subset \partial(rg + sh)(\overline{x}). \tag{4.20}$$

Although this inclusion is not as useful as the reverse inclusion, it implies a kind of invariance by addition property and a necessary optimality condition for problems with constraints and differentiable objective functions.

Corollary 4.35. *If h is Fréchet, respectively Hadamard, differentiable at $\overline{x} \in \mathrm{dom}g$, then we have respectively*

$$\partial_F(g + h)(\overline{x}) = \partial_F g(\overline{x}) + h'(\overline{x}), \quad \partial_D(g + h)(\overline{x}) = \partial_D g(\overline{x}) + h'(\overline{x}).$$

Another case in which equality occurs in relation (4.20) is the separable case.

Proposition 4.36. *Suppose $X = Y \times Z$, $g(x) = g_1(y)$, $h(x) = h_2(z)$ for $x := (y,z)$ and some functions $g_1 : Y \to \mathbb{R}$, $h_2 : Z \to \mathbb{R}$. Then for $\partial = \partial_D$ and $\partial = \partial_F$ one has*

$$\partial(g+h)(\bar{x}) = \partial g(\bar{x}) + \partial h(\bar{x}). \tag{4.21}$$

Proof. Let $\bar{x} := (\bar{y}, \bar{z}) \in Y \times Z$. For $f := g + h$ the inequality $g_1^D(\bar{y}, u) + h_2^D(\bar{z}, v) \le f^D(\bar{x}, (u,v))$ for all $(u,v) \in Y \times Z$ entails the inclusion $\partial_D g_1(\bar{y}) \times \partial_D h_2(\bar{z}) \subset \partial_D f(\bar{x})$. Conversely, for all $(\bar{y}^*, \bar{z}^*) \in \partial_D f(\bar{x})$, one has $\langle \bar{y}^*, u \rangle \le f^D(\bar{x}, (u,0)) \le g_1^D(\bar{y}, u)$, hence $\bar{y}^* \in \partial_D g_1(\bar{y})$. Similarly $\bar{z}^* \in \partial_D h_2(\bar{z})$. The case of ∂_F is left as an exercise. $\qquad\square$

The special rule of Corollary 4.35 can be applied jointly with the optimality criterion of Theorem 4.12. Thus, we recover Fermat's rule of Chap. 2 and we get an assertion that will be used repeatedly in the sequel.

Proposition 4.37. *If $f + g$ attains a local minimum at $\bar{x}$ and if f is F-differentiable (resp. H-differentiable) at $\bar{x}$, then $-f'(\bar{x}) \in \partial_F g(\bar{x})$ (resp. $-f'(\bar{x}) \in \partial_D g(\bar{x})$).*

Corollary 4.38 (Fermat's rule). *If f attains on a subset F of X a local minimum at $\bar{x} \in F$ and if f is F-differentiable, respectively H-differentiable, at $\bar{x}$ then we have respectively*

$$-f'(\bar{x}) \in N_F(F, \bar{x}), \quad -f'(\bar{x}) \in N_D(F, \bar{x}).$$

Proof. Setting $f_F := f + \iota_F$, where ι_F is the indicator function of F, applying the preceding two propositions and the definitions of normal cones, we get the result.

$\qquad\square$

Exercise. Let $g \in \mathscr{F}(Y)$, $h \in \mathscr{F}(Z)$, $f \in \mathscr{F}(Y \times Z)$ be given by $f(y,z) := g(y) + h(z)$, $\bar{x} := (\bar{y}, \bar{z})$, $u := (v,w) \in Y \times Z$. Give an example showing that one may have $f^D(\bar{x}, u) > g^D(\bar{y}, v) + h^D(\bar{z}, w)$. [Hint: Take $Y = Z = \mathbb{R}$, $\bar{x} = (0,0)$, $t_n := 2^{-4n-2}$, $g(y) := y$ for $|y| \in (t_n/2, 2t_n)$, $g(y) = 0$ otherwise, $h(z) := g(2z)$.]

4.2.2 Elementary Composition Rules

Now let us turn to chain rules. Again, in the general case, the inclusion available is not the most useful one.

Proposition 4.39. *Suppose $f = h \circ g$, where $g : X \to \overline{\mathbb{R}}$ and $h : \overline{\mathbb{R}} \to \overline{\mathbb{R}}$ is a nondecreasing function. If $g(\bar{x})$ and $h(g(\bar{x}))$ are finite, then*

$$\partial_D h(g(\bar{x})) \partial_D g(\bar{x}) \subset \partial_D f(\bar{x}), \quad \partial_F h(g(\bar{x})) \partial_F g(\bar{x}) \subset \partial_F f(\bar{x}).$$

If g is continuous at $\bar{x}$, if the restriction of h to some open interval T containing $\bar{r} := g(\bar{x})$ is (strictly) increasing, and if $(h \mid T)^{-1}$ is differentiable at $h(\bar{r})$, then equality holds in the preceding inclusions.

Proof. Let $\bar{r}^* \in \partial_D h(\bar{r})$, $\bar{y}^* \in \partial_D g(\bar{x})$. There exist maps $\varphi : X \to \overline{\mathbb{R}}$, $\psi : \overline{\mathbb{R}} \to \overline{\mathbb{R}}$ such that $\varphi \le g$, $\varphi(\bar{x}) = g(\bar{x})$, $\psi \le h$, $\psi(\bar{x}) = h(\bar{x})$, which are Hadamard differentiable at $\bar{x}$ and $\bar{r} := g(\bar{x})$ respectively with $\varphi'(\bar{x}) = \bar{x}^*$, $\psi'(\bar{r}) = \bar{r}^*$. Since h is nondecreasing, we have $f = h \circ g \geq h \circ \varphi \geq \psi \circ \varphi$ and $f(\bar{x}) = \psi(\varphi(\bar{x}))$. Now $\psi \circ \varphi$ is Hadamard differentiable at $\bar{x}$ and $\bar{r}^*\bar{y}^* = (\psi \circ \varphi)'(\bar{x}) \in \partial_D f(\bar{x})$. A similar proof is valid for ∂_F. The last assertion is obtained by noting that g coincides with $(h \mid_T)^{-1} \circ f$ near $\bar{x}$. □

Similar inclusions can be obtained when g is vector-valued and differentiable.

Proposition 4.40. *Let X, Y be normed spaces, and let $f = h \circ g$, where $g : X \to Y$ is Hadamard, respectively Fréchet, differentiable at $\bar{x}$ and $h : Y \to \overline{\mathbb{R}}$ is finite at $\bar{y} := g(\bar{x})$. Then we have respectively*

$$g'(\bar{x})^\top(\partial_D h(\bar{y})) := \partial_D h(\bar{y}) \circ g'(\bar{x}) \subset \partial_D(h \circ g)(\bar{x}), \tag{4.22}$$

$$g'(\bar{x})^\top(\partial_F h(\bar{y})) := \partial_F h(\bar{y}) \circ g'(\bar{x}) \subset \partial_F(h \circ g)(\bar{x}). \tag{4.23}$$

Proof. Let us first consider the Fréchet case. Given $\bar{y}^* \in \partial_F h(\bar{y})$, there exists some function $\psi : Y \to \mathbb{R}$ that is Fréchet differentiable at $\bar{y}$ and such that $\psi \leq h$, $\psi(\bar{x}) = h(\bar{x})$, $\psi'(\bar{x}) = \bar{y}^*$. Then $f \geq \psi \circ g$, $(\psi \circ g)(\bar{x}) = h(\bar{y}) = f(\bar{x})$, and since $\psi \circ g$ is Fréchet differentiable at $\bar{x}$, we get $\bar{y}^* \circ g'(\bar{x}) \in \partial_F f(\bar{x})$.

The proof for ∂_D is similar. One may also observe that for $\bar{y}^* \in \partial_D h(\bar{y})$, $u \in X \setminus \{0\}$, one has $w_{t,v} := \frac{1}{t}(g(\bar{x}+tv) - g(\bar{x})) \to w := g'(\bar{x})(u)$ as $(t,v) \to (0_+, u)$ and

$$\langle \bar{y}^*, g'(\bar{x})u \rangle \leq \liminf_{(t,w') \to (0_+, w)} \frac{1}{t}\left(h(\bar{y}+tw') - h(\bar{y})\right)$$

$$\leq \liminf_{(t,v) \to (0_+, u)} \frac{1}{t}\left(h(\bar{y}+tw_{t,v}) - h(\bar{y})\right) = f'(\bar{x}, u),$$

since $h(\bar{y}+tw_{t,v}) = h(g(\bar{x}+tv))$. Thus $\bar{y}^* \circ g'(\bar{x}) \in \partial_D f(\bar{x})$. □

An extension to multimaps can be devised using coderivatives.

Corollary 4.41. *Let $F := H \circ G$ where $G : X \rightrightarrows Y$ and $H := \{h\}$ is the multimap associated with a single-valued map $h : Y \to Z$ that is Hadamard differentiable at $\bar{y} \in G(\bar{x})$, respectively Fréchet differentiable at $\bar{y}$, and let $\bar{z} := h(\bar{y})$. Then we have respectively*

$$D_D^*(H \circ G)(\bar{x}, \bar{z}) \subset D_D^*G(\bar{x}, \bar{y}) \circ \left(h'(\bar{y})\right)^\top, \tag{4.24}$$

$$D_F^*(H \circ G)(\bar{x}, \bar{z}) \subset D_F^*G(\bar{x}, \bar{y}) \circ \left(h'(\bar{y})\right)^\top. \tag{4.25}$$

Proof. We note that $F = \{(x, h(y)) : y \in G(x)\} = (I_X \times h)(G)$. Using Proposition 2.108, we see that for all $(x^*, -z^*) \in N_D(F, (\bar{x}, \bar{z}))$ we have $(x^*, -(h'(\bar{y}))^\top(z^*)) \in N_D(G, (\bar{x}, \bar{y}))$. Writing this relation in terms of coderivatives, we get (4.24). The Fréchet case is similar. □

Conditions ensuring equalities in inclusions (4.22), (4.23) can be given.

Proposition 4.42. *Let X, Y be Banach spaces and let $f = h \circ g$ be as in the preceding proposition, with $g'(\bar{x})(X) = Y$. Then if g is (strictly or) circa-differentiable at $\bar{x}$, respectively if g is Hadamard differentiable and Y is finite-dimensional, one has respectively*

$$\partial_F (h \circ g)(\bar{x}) = g'(\bar{x})^\mathsf{T}(\partial_F h(\bar{y})) := \partial_F h(\bar{y}) \circ g'(\bar{x}), \tag{4.26}$$

$$\partial_D (h \circ g)(\bar{x}) = g'(\bar{x})^\mathsf{T}(\partial_D h(\bar{y})) := \partial_D h(\bar{y}) \circ g'(\bar{x}). \tag{4.27}$$

Proof. Let $\widehat{g}$ be given by $\widehat{g}(x, r) := (g(x), r)$. Since $E_f = \widehat{g}^{-1}(E_h)$, the result for the Fréchet (resp. Hadamard) case follows from the calculus rule for the normal cone to an inverse image and from the characterization given in Proposition 4.16 (resp. Corollary 4.15). □

Taking indicator functions, we recover the geometric result we used.

Corollary 4.43. *Let X, Y be Banach spaces and let $g : X \to Y$ be (strictly or) circa-differentiable at $\bar{x} \in X$ (resp. Hadamard differentiable at $\bar{x}$ and Y finite-dimensional) with $g'(\bar{x})(X) = Y$. Then for every subset D of Y containing $\bar{y} := g(\bar{x})$, for $C := g^{-1}(D)$ one has $N_F(C, \bar{x}) = g'(\bar{x})^\mathsf{T}(N_F(D, \bar{y}))$ (resp. $N_D(C, \bar{x}) = g'(\bar{x})^\mathsf{T}(N_D(D, \bar{y}))$).*

A different inclusion for a composition is as follows.

Proposition 4.44. *Let $f = h \circ g$, where $g : X \to Y$ and $h : Y \to \overline{\mathbb{R}}$. If g is stable at $\bar{x} \in X$ and if h is finite at $\bar{y} := g(\bar{x})$, then for all $y^* \in \widetilde{\partial}_F h(\bar{y}) := -\partial_F(-h)(\bar{y})$, one has*

$$\partial_F f(\bar{x}) \subset \partial_F (y^* \circ g)(\bar{x}).$$

If g is tangentially compact at $\bar{x}$ and $h(\bar{y}) \in \mathbb{R}$, then for $y^* \in -\partial_D(-h)(\bar{y})$, one has

$$\partial_D f(\bar{x}) \subset \partial_D (y^* \circ g)(\bar{x}).$$

Proof. Let $x^* \in \partial_F f(\bar{x})$, $y^* \in \widetilde{\partial}_F h(\bar{y})$ and let $c \in \mathbb{R}_+$, $\rho > 0$ be such that $\|g(x) - g(\bar{x})\| \le c \|x - \bar{x}\|$ for every $x \in B(\bar{x}, \rho)$. Then for every $\alpha > 0$, $\beta > 0$, one can find $\gamma, \delta \in (0, \rho)$ such that for $x \in B(\bar{x}, \gamma)$, $y \in B(\bar{y}, \delta)$, one has

$$\langle x^*, x - \bar{x} \rangle - \alpha \|x - \bar{x}\| \le f(x) - f(\bar{x}),$$

$$h(y) - h(\bar{y}) - \beta \|y - \bar{y}\| \le \langle y^*, y - \bar{y} \rangle.$$

We may suppose $c\gamma \le \delta$. Then for $x \in B(\bar{x}, \gamma)$ and $y := g(x)$, we have $y \in B(\bar{y}, \delta)$,

$$\langle x^*, x - \bar{x} \rangle - (\alpha + c\beta) \|x - \bar{x}\| \le h(g(x)) - h(g(\bar{x})) - \beta \|g(x) - g(\bar{x})\|$$

$$\le \langle y^*, g(x) \rangle - \langle y^*, g(\bar{x}) \rangle.$$

Since α and β can be arbitrarily small, we have $x^* \in \partial_F (y^* \circ g)(\bar{x})$. We leave the proof of the second assertion as an exercise. □

Now let us give a product rule that generalizes Leibniz rule. We use the directional subdifferential, but a similar result holds with the firm subdifferential.

Proposition 4.45. *Given* $f, g : X \to \mathbb{R}_\infty$ *lower semicontinuous and finite at* $\bar{x} \in X$, *let* $p = f \cdot g$ *and let* $\bar{x}^* \in \partial_D f(\bar{x})$, $\bar{y}^* \in \partial_D g(\bar{x})$. *If* $f(\bar{x}) > 0$, $g(\bar{x}) > 0$, *then one has*

$$g(\bar{x})\bar{x}^* + f(\bar{x})\bar{y}^* \in \partial_D p(\bar{x}).$$

Proof. Let U be a neighborhood of $\bar{x}$ on which f and g are positive, and let $h := \log \circ f \mid U$, $k := \log \circ g \mid U$, setting $\log(\infty) := \infty$. Then Proposition 4.39 yields $(1/f(\bar{x}))\bar{x}^* \in \partial_D h(\bar{x})$, $(1/g(\bar{x}))\bar{y}^* \in \partial_D k(\bar{x})$. Since $p \mid U = \exp \circ (h + k)$, applying again Proposition 4.39 and the sum rule, we get

$$g(\bar{x})\bar{x}^* + f(\bar{x})\bar{y}^* = \exp(h(\bar{x}) + k(\bar{x})) \left(\frac{1}{f(\bar{x})}\bar{x}^* + \frac{1}{g(\bar{x})}\bar{y}^* \right) \in \partial_D p(\bar{x}).$$

4.2.3 Rules Involving Order

The following results have no analogues in differential calculus. Their proofs are easy applications of the definitions and Proposition 4.11.

Proposition 4.46. *Let* $(f_i)_{i \in I}$ *be a finite family of functions on* X *that are finite at* $\bar{x} \in X$. *Let* $g := \inf_{i \in I} f_i$, $I(\bar{x}) := \{i \in I : f_i(\bar{x}) = g(\bar{x})\}$.

(a) If for all $j \in I \setminus I(\bar{x})$, f_j *is lower semicontinuous at* $\bar{x}$, *then*

$$\partial_F g(\bar{x}) = \bigcap_{i \in I(\bar{x})} \partial_F f_i(\bar{x}), \qquad \partial_D g(\bar{x}) = \bigcap_{i \in I(\bar{x})} \partial_D f_i(\bar{x}).$$

(b) Let $h := \sup_{i \in I} f_i$ *and let* $S(\bar{x}) := \{i \in I : f_i(\bar{x}) = h(\bar{x})\}$. *Then*

$$\overline{co} \left(\bigcup_{i \in S(\bar{x})} \partial_F f_i(\bar{x}) \right) \subset \partial_F h(\bar{x}), \qquad \overline{co}^* \left(\bigcup_{i \in S(\bar{x})} \partial_D f_i(\bar{x}) \right) \subset \partial_D h(\bar{x}).$$

These last relations are not equalities in general, as shown by the next example.

Example. Let $f_1 : \mathbb{R} \to \mathbb{R}$ be given by $f_1(x) := 2x$ for $x \in \mathbb{R}_-$, $f_1(x) = x$ for $x \in \mathbb{R}_+$, and let $f_2(x) := f_1(-x)$ for $x \in \mathbb{R}$. Then $h(x) := (f_1 \vee f_2)(x) = |x|$, so that $\partial f(0) = [-1, 1]$, while $\partial f_i(0) = \varnothing$ for $i = 1, 2$. $\square$

The next result is extremely useful. Its proof again is an immediate consequence of Proposition 4.11 (set $g(x, y) := p(x)$ and note that $g \leq f$, $g(\bar{x}, \bar{y}) = f(\bar{x}, \bar{y})$).

Theorem 4.47. *Let X, Y be normed spaces, let $f : X \times Y \to \overline{\mathbb{R}}$ be finite at $(\overline{x}, \overline{y})$ and such that $f(\overline{x}, \overline{y}) = p(\overline{x})$, where $p(x) := \inf_{y \in Y} f(x, y)$ for $x \in X$. Then for $\partial = \partial_F$ or $\partial = \partial_D$,*

$$\overline{x}^* \in \partial p(\overline{x}) \implies (\overline{x}^*, 0) \in \partial f(\overline{x}, \overline{y}).$$

Proposition 4.48. *Let $f := h \circ g$, where $g := (g_1, \ldots, g_m) : X \to \mathbb{R}^m$, $h : \mathbb{R}^m \to \mathbb{R}$ is of class C^1 around $\overline{y} := g(\overline{x})$ and nondecreasing in each of its m arguments near $\overline{y}$, with $h'(\overline{y}) \neq 0$. Then*

$$\partial f(\overline{x}) = h'(\overline{y}) \circ (\partial g_1(\overline{x}), \ldots, \partial g_m(\overline{x})). \tag{4.28}$$

Proof. We give the proof for the firm subdifferential, the proof for the directional subdifferential being similar. We use the fact that for some map $v := (v_1, \ldots, v_m) : \mathbb{R}^m \times \mathbb{R}^m \to \mathbb{R}_+^m$ continuous around $(\overline{y}, \overline{y})$ with $h'(\overline{y}) = (v_1(\overline{y}, \overline{y}), \ldots, v_m(\overline{y}, \overline{y}))$ we have

$$h(y) - h(z) = v(y, z)(y - z).$$

Plugging $y := g(x)$, $z := g(\overline{x})$ into this relation and using the inequalities $g_i(x) - g_i(\overline{x}) \geq \langle x_i^*, x - \overline{x} \rangle - \varepsilon(x) \|x - \overline{x}\|$ for $i \in \mathbb{N}_m$, $x_i^* \in \partial_F g_i(\overline{x})$, where $\varepsilon(x) \to 0$ as $x \to \overline{x}$, we get $h'(\overline{y}) \circ (x_1^*, \ldots, x_m^*) \in \partial_F f(\overline{x})$.

Now let $x^* \in \partial_F f(\overline{x})$, so that by Proposition 4.7 there exists a function $\varphi : X \to \overline{\mathbb{R}}$ differentiable at $\overline{x}$ and satisfying $\varphi'(\overline{x}) = x^*$, $\varphi \leq f$, $\varphi(\overline{x}) = f(\overline{x})$. By assumption, there is some $j \in \mathbb{N}_m$ such that $D_j h(\overline{x}) > 0$; without loss of generality we may suppose $j = m$ in order to simplify the writing. The implicit function theorem ensures that the relation $z = h(y)$ is locally equivalent to a relation $y_m = k(y_1, \ldots, y_{m-1}, z)$, where k is of class C^1. Setting $\psi(x) = k(g_1(x), \ldots, g_{m-1}(x), \varphi(x))$, we get $\varphi(x) = h(g_1(x), \ldots, g_{m-1}(x), \psi(x))$ and we have $\psi(\overline{x}) = g_m(\overline{x})$, $\psi \leq g_m$ around $\overline{x}$, hence $x_m^* := \psi'(\overline{x}) \in \partial_F g_m(\overline{x})$. The first part of the proof shows that for all $x_i^* \in \partial_F g_i(\overline{x})$, for $i \in \mathbb{N}_{m-1}$ we have $h'(\overline{y}) \circ (x_1^*, \ldots, x_m^*) \in \partial_F \varphi(\overline{x}) = \overline{x}^*$. Thus (4.28) holds. □

The next proposition is more special than the preceding general rules.

Proposition 4.49. *Let X and Y be normed spaces, let $f : X \to \overline{\mathbb{R}}$ be a lower semicontinuous function, let $g : Y \to \mathbb{R}$ be Gâteaux differentiable at some $\overline{y} \in Y$ with $g'(\overline{y}) \neq 0$. Let $h : X \times Y \to \overline{\mathbb{R}}$ be given by $h(x, y) := \max(f(x), g(y))$. Suppose that for some $\overline{x} \in X$ one has $f(\overline{x}) = g(\overline{y})$. Then, for $\partial = \partial_D$ or $\partial = \partial_F$ and for $(\overline{x}^*, \overline{y}^*) \in \partial h(\overline{x}, \overline{y})$ with $\overline{y}^* \neq g'(\overline{y})$ there exists $\lambda \in (0, 1]$ such that*

$$(\overline{x}^*, \overline{y}^*) \in (1 - \lambda)\partial f(\overline{x}) \times \lambda \partial g(\overline{y}).$$

Proof. Without loss of generality we assume that $\overline{x} = 0$, $\overline{y} = 0$, $f(\overline{x}) = g(\overline{y}) = 0$. Let $(\overline{x}^*, \overline{y}^*) \in \partial h(\overline{x}, \overline{y})$ with $\overline{y}^* \neq g'(\overline{y}) \neq 0$. Let $v \in Y$ be such that $g'(\overline{y})v = 1$. For $t > 0$ small enough we have $h(0, tv) = g(tv)$, hence $1 = g'(\overline{y})v = \lim_{t \to 0_+} (1/t)h(0, tv) \geq \langle (\overline{x}^*, \overline{y}^*), (0, v) \rangle = \langle \overline{y}^*, v \rangle$. Similarly, for all $w \in Y$ such that $g'(\overline{y})w > 0$ we have $g'(\overline{y})w \geq \langle \overline{y}^*, w \rangle$. The same is true if $g'(\overline{y})w \geq 0$ as follows by taking a sequence $(w_n) \to w$ such that $g'(\overline{y})w_n > 0$ for all n. Thus there exists $\lambda \geq 0$ such that

$v^* - \bar{y}^* = \lambda v^*$ for $v^* := g'(\bar{y})$ and $\bar{y}^* = (1 - \lambda)v^*$. The assumption $\bar{y}^* \neq g'(\bar{y})$ yields $\lambda > 0$. Observing that $h(0, -tv) = 0$ for t small enough, we get $\langle \bar{y}^*, -v \rangle \leq 0$, hence $1 - \lambda = (1 - \lambda)\langle v^*, v \rangle = \langle \bar{y}^*, v \rangle \geq 0$. Now, given $u \in X$, $s \geq f^D(\bar{x}, u)$, let us show that $\lambda s \geq \langle \bar{x}^*, u \rangle$; that will ensure that $(\bar{x}^*/\lambda, -1) \in N_D(\text{epi} f, (\bar{x}, f(\bar{x})))$ or $\bar{x}^*/\lambda \in \partial_D f(\bar{x})$. Taking a sequence $((s_n, t_n, u_n)) \to (s, 0_+, u)$ such that $t_n s_n \geq f(t_n u_n)$, setting $s'_n := t_n^{-1} g(t_n s v)$, we note that $(s'_n) \to s$ and $(s''_n) \to s$ for $s''_n := \max(s_n, s'_n)$. Since $h(t_n u_n, t_n s v) \leq t_n s''_n$ for all n, we get $\langle \bar{x}^*, u \rangle + \langle \bar{y}^*, sv \rangle \leq s$ or $\langle \bar{x}^*, u \rangle \leq \lambda s$. The case $\partial = \partial_D$ is proved.

Now let us consider the case of the Fréchet subdifferential ∂_F. As above, we have $\bar{y}^* = (1 - \lambda)v^*$ for $v^* := g'(\bar{y})$, $\lambda \in]0, 1]$. Suppose $\bar{x}^*/\lambda \notin \partial_F f(\bar{x})$: there exist $\alpha > 0$ and a sequence $(u_n) \to 0$ such that $f(u_n) < s_n := \langle \bar{x}^*/\lambda, u_n \rangle - \alpha \|u_n\|$. Since $(g(s_n v)/s_n) \to 1$, there exists a sequence $(\sigma_n) \to 0$ in $\mathbb{R}_+$ such that $h(u_n, s_n v) \leq (1 + \sigma_n)s_n$. Then, for some sequence $(\varepsilon_n) \to 0_+$ one gets

$$(1 + \sigma_n)s_n \geq \langle \bar{x}^*, u_n \rangle + \langle \bar{y}^*, s_n v \rangle - \varepsilon_n(\|u_n\| + s_n \|v\|)$$
$$\geq \lambda s_n + \alpha\lambda \|u_n\| + (1 - \lambda)s_n - \varepsilon_n(\|u_n\| + s_n \|v\|).$$

Then one has

$$(|\sigma_n| + \varepsilon_n \|v\|)|s_n| \geq (\sigma_n + \varepsilon_n \|v\|)s_n \geq (\alpha\lambda - \varepsilon_n)\|u_n\|$$

and $|s_n| \leq (\|\bar{x}^*\|/\lambda + \alpha)\|u_n\|$, a contradiction since $(|\sigma_n| + \varepsilon_n \|v\|) \to 0$. $\qquad\square$

Remark. The conclusion cannot hold in general when $\bar{y}^* = g'(\bar{y}) = 0$, as shown by the example $X = Y := \mathbb{R}$, $f(x) = \min(x, 0)$, $g(y) = 0$ for all $x \in X$, $y \in Y$, $(\bar{x}, \bar{y}) = (0, 0)$. $\qquad\square$

Exercise. In the case $\partial = \partial_D$ and $0 \notin \partial_D g(\bar{y})$, the differentiability assumption on g can be relaxed to *epi-differentiability* at $\bar{y}$ in the sense that $g^D(\bar{y}, \cdot) = g^I(\bar{y}, \cdot)$, since then $h^D((\bar{x}, \bar{y}), (u, v)) = \max(f^D(\bar{x}, u), g^D(\bar{y}, v))$ for all $(u, v) \in X \times Y$.

4.2.4 Elementary Rules for Marginal and Performance Functions

Nonsmoothness appears when one takes suprema or infima of families of smooth functions. Still, the tools we have presented can be used. In the present subsection, given a normed space X, a topological space S (for instance a finite set with the discrete topology), and $f : X \times S \to \overline{\mathbb{R}}$, we limit our study to elementary rules concerning the *marginal function m* and the *performance function p* given by

$$m(x) := \sup_{s \in S} f(x, s), \qquad p(x) := \inf_{s \in S} f(x, s), \qquad x \in X.$$

We observe that we cannot pass from a result about m to a result about p, because in general, for $\partial := \partial_D$ or $\partial := \partial_F$ one has $\partial p(x) \neq -\partial(-p)(x)$: performance functions are quite different from marginal functions.

Setting $f_s := f(\cdot, s)$ for $s \in S$ and

$$M(x) := \{s \in S : f(x,s) = m(x)\}, \qquad P(x) := \{s \in S : f(x,s) = p(x)\},$$

the following observation follows from Proposition 4.11, noting that $p \leq f_s$, $p(x) = f_s(x)$ when $s \in P(x)$ and $f_s \leq m$, $f_s(x) = m(x)$ when $s \in M(x)$.

Proposition 4.50. *For every $x \in X$, $s \in P(x)$, $x^* \in \partial p(x)$ one has $x^* \in \partial f_s(x)$.*
For every $x \in X$, $s \in M(x)$, $x^* \in \partial f_s(x)$ one has $x^* \in \partial m(x)$.

More precise results can be given under various assumptions. The following result is one of the simplest cases. We assume that $f : X \times S \to \mathbb{R}$ is a lower semicontinuous function such that the following assumptions hold:

(P1) f is differentiable at $\bar{x}$ with respect to its first variable, uniformly with respect to the second variable (or equivalently, the family $(f_s)_{s \in S}$ is equi-differentiable at $\bar{x}$): there exists a modulus μ such that for every $(v,s) \in X \times S$ one has

$$|f_s(\bar{x} + v) - f_s(\bar{x}) - Df_s(\bar{x})v| \leq \mu(\|v\|) \|v\|.$$

(P2) The mapping $s \mapsto Df_s(\bar{x})$ is continuous from S into X^*.

Proposition 4.51. *If S is compact, under assumptions (P1), (P2), one has*

$$\partial_F m(\bar{x}) = \overline{\mathrm{co}}\{Df_s(\bar{x}) : s \in M(\bar{x})\},$$

$$\partial_F p(\bar{x}) = \bigcap_{s \in P(\bar{x})} \{Df_s(\bar{x})\}.$$

The proof is left as an exercise that the reader can tackle while reading Sect. 4.7.1

Exercises

1. Show that if X is a normed space and $f := g \circ \ell$, where $\ell : X \to Y$ is a continuous and open linear map with values in another normed space and $g : Y \to \mathbb{R}$ is locally Lipschitzian, then $\partial_D f(x) = \partial_D g(\ell(x)) \circ \ell$. Can one replace ℓ by a differentiable map? by a differentiable map that is open at x?

2. Show by an example that the inclusion $\partial(g+h)(\bar{x}) \subset \partial g(\bar{x}) + \partial h(\bar{x})$ is not valid in general. Explain why this inclusion would be more desirable than the reverse one.

3. Prove Proposition 4.34.

4. Prove Corollary 4.35. [Hint: If $f := g + h$, apply relation (4.20) to $g = f + (-h)$.]

5. Give a rule for the subdifferential of a quotient.

6. (a) With the assumptions and notation of Proposition 4.46, show that for all $v \in X$ one has $g^D(\overline{x}, v) = \min_{i \in I(\overline{x})} f_i^D(\overline{x}, v)$, $h^D(\overline{x}, v) = \max_{i \in S(\overline{x})} f_i^D(\overline{x}, v)$.
(b) Show that in general, the inclusions of Proposition 4.46 are strict.

7. With the assumptions and notation of Proposition 4.42, show that when Y is finite-dimensional and g is differentiable, relation (4.26) can be deduced from (4.27).

4.3 Viscosity Subdifferentials

In the sequel, given two normed spaces X, Y and an open subset W of X, we say that a map $h : W \to Y$ is *F-smooth at* $\overline{x} \in X$ (resp. H-smooth at $\overline{x}$) if h is of class C^1 (resp. D^1) at $\overline{x}$, i.e., if h is Fréchet (resp. Hadamard) differentiable on an open neighborhood of $\overline{x}$ and if h' is continuous at $\overline{x}$ (resp. $dh : (x, v) \mapsto h'(x)v$ is continuous at $(\overline{x}, u)$ for all $u \in X$). We gather both cases by saying that h is *smooth at* $\overline{x}$. We say that h is *smooth* if it is smooth at each point of W. We say that a Banach space X is F-smooth (resp. H-smooth) if there is some F-smooth (resp. H-smooth) function $j := j_X : X \to \mathbb{R}_+$ such that $j(0) = 0$ and

$$((j(x_n))_n \to 0) \implies ((\|x_n\|)_n \to 0). \tag{4.29}$$

We gather these two cases by saying that X is *smooth* and we call j a *forcing function*. Note that in replacing j by j^2, we get the implication

$$((\|x_n\|)_n \to 0) \implies ((j'(x_n))_n \to 0).$$

Condition (4.29) is more general than the requirement that an equivalent norm on X be Fréchet (resp. Hadamard) differentiable on $X \setminus \{0\}$. On a first reading, the reader may assume that j is the square of such a norm, although such an assumption is not as general. Note that when j is a forcing function, the function $k : X \times X \to \mathbb{R}$ given by $k(x, x') := j(x - x')$ is a forcing bifunction in the sense of Sect. 1.6.

We first show that the existence of a smooth Lipschitzian bump function ensures condition (4.29). Recall that b is a *bump function* if it is nonnegative, not identically equal to zero, and null outside some bounded set (which can be taken to be B_X).

Proposition 4.52. *If X has a smooth forcing function, then it has a smooth bump function. If there is on X a Lipschitzian (resp. smooth) bump function b_0, then there is a Lipschitzian (resp. smooth) bump function b such that $b(X) \subset [0, 1]$, $b(0) = 1$ and $(x_n) \to 0$ whenever $(b(x_n)) \to 1$, so that $j := 1 - b$ is a forcing function. If b_0 is Lipschitzian and smooth, then b can be chosen to be Lipschitzian and smooth.*

Proof. If j is a smooth forcing function, then there exists some $s > 0$ such that $j(x) \geq s$ for all $x \in X \setminus B_X$. Composing j with a smooth function $h : \mathbb{R} \to \mathbb{R}_+$ such that $h(0) = 1$, $h(r) = 0$ for $r \geq s$, we get a smooth bump function $b := h \circ j$.

Let b_0 be a Lipschitzian (resp. smooth) bump function with support in B_X. We first note that we can assume $b_0(0) > c := (1/2) \sup b_0 > 0$: if it is not the case, we replace b_0 by b_1 given by $b_1(x) := b_0(kx + a)$, where $a \in B_X$ is such that $b_0(a) > (1/2) \sup b_0$ and $k > \|a\| + 1$. We can also assume that b_0 attains its maximum at 0 and that $b_0(0) = 1$. If it is not the case, we replace b_0 by $\theta \circ b_1$, where θ is a smooth function on $\mathbb{R}$ satisfying $\theta(r) = r$ for $r \leq c$, $\theta(r) = 1$ for $r \geq b_0(0)$. Now, given $q \in (0,1)$ we set

$$b(x) := (1 - q^2) \sum_{n=0}^{\infty} q^{2n} b_0(q^{-n}x),$$

with $q^0 := 1$, so that $b(0) = 1 \geq b(x)$ for all x, b is null on $X \setminus B_X$, b is smooth on $X \setminus \{0\}$ as the sum is locally finite on $X \setminus \{0\}$. Moreover, since b_0' is bounded, the series $\sum_{n=0}^{\infty} q^n b_0'(q^{-n}x)$ is uniformly convergent, so that b is of class C^1 (resp. of class D^1). Furthermore, b is Lipschitzian on X when b_0 is Lipschitzian, and for $x \in X \setminus q^k B_X$ one has

$$b(x) \leq (1 - q^2) \sum_{n=0}^{k} q^{2n} < 1 = b_0(0),$$

so that $(x_n) \to 0$ whenever $(b(x_n)) \to 1$. The last assertion is obvious. $\square$

The preceding result can be made more precise (at the expense of simplicity).

Proposition 4.53. *Let X be a normed space. There exists a Lipschitzian smooth bump function on X if and only if the following condition is satisfied:*

H) for all $c > 1$, there exists a function $j : X \to \mathbb{R}$ that is smooth on $X \setminus \{0\}$, with a derivative that is bounded on every bounded subset of $X \setminus \{0\}$ and such that

$$\forall x \in X, \qquad \|x\| \leq j(x) \leq c\|x\|. \tag{4.30}$$

According to the sense given to the word "smooth," we denote this condition by (H_F) or (H_D) whenever it is necessary to be precise. Let us note that replacing j by j^2 to ensure smoothness, condition (H) implies condition (4.29).

The fact that c is arbitrarily close to 1 shows that a result in which one uses the differentiability of the norm on $X \setminus \{0\}$ is likely to be valid under assumption (H).

We have seen that assumption (H_F) (resp. (H_D)) is satisfied when the norm of the dual space of X is locally uniformly rotund (resp. when X is separable), and one can even take for j an equivalent norm.

Proof. Let us first observe that condition (H) ensures the existence of a Lipschitzian smooth bump function: it suffices to take $b := k \circ j^2$, where $k : \mathbb{R} \to \mathbb{R}$ is a Lipschitzian smooth function satisfying $k(0) = 1$ for $r \in (-\infty, \alpha]$ with $\alpha \in (0,1)$ and $k(r) = 0$ for $r \in [1, \infty)$. Considering separately the case in which the dimension of X is 1 and the case in which this dimension is greater than 1, we can prove that j is Lipschitzian on B_X. Since $j^{-1}([0,1]) \subset B_X$, $k \circ j$ is Lipschitzian.

In order to prove the converse, we first define a function j_0 satisfying (4.30) on B_X. Let $q > 1$ with $q^2 < c$. Using a translation and composing a smooth Lipschitzian bump function with a smooth function from $\mathbb{R}$ to $[0,1]$, we may suppose there exists a smooth Lipschitzian bump function $b : X \to [0,1]$ such that $b(x) = 1$ for $x \in (1/q)B_X$, $b(x) = 0$ for $x \in X \setminus B_X$. Let us set $q^0 := 1$, and for $x \in X$,

$$g(x) = \sum_{n=0}^{\infty} q^n b(q^n x), \qquad h_0(x) = \frac{q}{(q-1)g(x)+1}.$$

We have $g(0) = +\infty$, $g(x) \in \mathbb{R}_+$ for all $x \in X \setminus \{0\}$, since for every $\rho > 0$ the sum in the definition of g is finite on $X \setminus \rho B_X$, and g and h_0 are well defined (setting $h_0(0) = 0$), smooth, and Lipschitzian on $X \setminus \rho B_X$. In fact, if $x \in B_X$ and if $m := m(x)$ is the least integer greater than $-(\ln \|x\| / \ln q) - 1$, one has $(1/q) < q^m \|x\| \le 1$ and

$$\frac{q^m - 1}{q - 1} = \sum_{n=0}^{m-1} q^n \le g(x) \le \sum_{n=0}^{m} q^n = \frac{q^{m+1} - 1}{q - 1},$$

so that

$$\|x\| \le q^{-m} \le h_0(x) \le q^{-m+1} \le q^2 \|x\|. \tag{4.31}$$

Since all the terms of the sum defining g except the mth are constant on $q^{-m}B_X \setminus q^{-m-1}B_X$, the derivative of h_0 is

$$h_0'(x) = -q(q-1)((q-1)g(x)+1)^{-2}q^{2m}b'(q^m x).$$

Setting $\beta := \sup_{x \in X} \|b'(x)\|$, we get $\|h_0'(x)\| \le q(q-1)h_0(x)^2 q^{2m-2}\beta \le q(q-1)\beta$. Thus, $j_0 := h_0$ is Lipschitzian on B_X, is smooth on $B_X \setminus \{0\}$, and by (4.31) satisfies relation (4.30) on B_X.

Now let us define j via a " traveling wave" in the following way. We first define a function h_n on $2^n B_X$ by $h_n(x) := 2^n h_0(2^{-n}x)$. Assuming, without loss of generality, that $c^3 \le 2$, we pick a smooth Lipschitzian function $p_n : \mathbb{R}_+ \to [0,1]$ such that $p_n(r) = 0$ for $r \le 2^n c^{-2}$, $p_n(r) = 1$ for $r \ge 2^n c^{-1}$, and starting with $j_0 := h_0$, we inductively define $j_{n+1} : 2^{n+1}B_X \to \mathbb{R}_+$ by $j_{n+1}(x) = h_{n+1}(x)$ for $x \in 2^{n+1}B_X \setminus 2^n B_X$,

$$j_{n+1}(x) := (1 - p_n(h_n(x))) j_n(x) + p_n(h_n(x))h_{n+1}(x) \qquad \text{for } x \in 2^n B_X.$$

Since $h_n(x) \ge 2^n c^{-1}$ when $\|x\| \ge 2^n c^{-1}$, we have $j_{n+1}(x) = h_{n+1}(x)$ for $x \in 2^n B_X \setminus 2^n c^{-1}B_X$, so that j_{n+1} is smooth on $\text{int}(2^{n+1}B_X) \setminus \{0\}$ inasmuch as j_n is smooth on $\text{int}(2^n B_X) \setminus \{0\}$. Moreover, another induction shows that

$$\forall x \in 2^{n+1}B_X, \qquad \|x\| \le j_{n+1}(x) \le c\|x\|.$$

Finally, we observe that j_{n+1} coincides with j_n on $2^n c^{-3} B_X$, hence on $2^{n-1} B_X$, so that $j := \lim_n j_n$ is well defined and smooth on $X \setminus \{0\}$ and satisfies the required estimates. Moreover, the derivatives of h_n and j_n are bounded on $2^n B_X$, so that the derivative of j is bounded on every bounded subset. $\qquad\square$

Now let us introduce the two viscosity subdifferentials we shall consider.

Definition 4.54. Let X be an arbitrary normed space and let $f : X \to \overline{\mathbb{R}}$ be a function finite at $\overline{x} \in X$. The *viscosity Hadamard* (resp. *Fréchet*) *subdifferential* of f at $\overline{x}$ is the set $\partial_H f(\overline{x})$ (resp. $\partial_F^V f(\overline{x})$) of Hadamard (resp. Fréchet) derivatives $\varphi'(\overline{x})$ of functions φ of class D^1 (resp. C^1) on some neighborhood U of $\overline{x}$ minorizing f on U and satisfying $\varphi(\overline{x}) = f(\overline{x})$.

When there exists a bump function of class D^1 (resp. C^1) on X, we may suppose φ is defined on the whole of X in this definition (however, the inequality $\varphi \le f$ is required only near $\overline{x}$). It seems necessary to make a distinction between ∂_D and ∂_H even in smooth spaces. In contrast, since we shall show that $\partial_F^V f = \partial_F f$ for f defined on a Fréchet smooth space, we can keep for a while the heavy notation $\partial_F^V f$. The proof of the coincidence $\partial_F^V f = \partial_F f$ uses the following smoothing result for one-variable functions.

Lemma 4.55. *For $a > 0$, let $r : [0,a] \to \mathbb{R}_+$ be a remainder, i.e., a function with a right derivative at 0 and such that $r(0) = 0$, $r'_+(0) = 0$. Suppose $b := \sup r([0,a]) < +\infty$. Then there exists a nondecreasing remainder $s : [0,a] \to \mathbb{R}_+$ of class C^1 such that $s \ge r$, $s(t) \le \sup r([0,2t])$ for $t \in [0, a/2]$.*

Proof. Let $a_0 = a$, $b_0 := b$, $a_n := 2^{-n}a$, $b_n := \sup r([0, a_{n-1}])$ for $n \ge 1$, so that (b_n) is nonincreasing and $(b_n/a_{n-1}) \to 0$. Let us set $m_n := (1/2)(a_n + a_{n+1})$, $c_n := 2(a_n - a_{n+1})^{-2}(b_n - b_{n+1})$ and construct s by setting $s(0) := 0$,

$$s(t) = b_{n+1} + c_n(t - a_{n+1})^2, \quad t \in [a_{n+1}, m_n],$$

$$s(t) = b_n - c_n(t - a_n)^2, \quad t \in [m_n, a_n],$$

so that $D_\ell s(a_n) = 0$, $D_r s(a_{n+1}) = 0$, s is continuous and derivable at a_n, m_n with

$$s(a_n) = b_n, \quad s'(a_n) = 0, \quad s(m_n) = (b_n + b_{n+1})/2, \quad s'(m_n) = c_n(a_n - a_{n+1}).$$

Thus s is of class C^1 and for $t \in [a_{n+1}, a_n]$, $s(t) \ge b_{n+1} \ge r(t)$, $0 \le s(t)/t \le b_n/a_{n+1} \le 4b_n/a_{n-1}$, so that $s(t)/t \to 0$ as $t \to 0_+$. $\qquad\square$

Theorem 4.56. *Let X be a normed space satisfying condition (H_F). Then for every lower semicontinuous function f on X, $\partial_F^V f(\overline{x})$, the viscosity Fréchet subdifferential of f at $\overline{x}$, coincides with $\partial_F f(\overline{x})$.*

Proof. Without loss of generality we suppose $\overline{x} = 0$. Clearly, $\partial_F^V f(0) \subset \partial_F f(0)$. Given $\overline{x}^* \in \partial_F f(0)$, consider the remainder

$$r(t) := \sup\{f(0) - f(x) + \langle \overline{x}^*, x \rangle : x \in tB_X\}, \qquad t \in \mathbb{R}_+,$$

and we associate with it the remainder s of the preceding lemma, where a is chosen so that $\sup r([0,a]) < +\infty$. Then using j given by (H$_F$), the function φ defined by

$$\varphi(x) := f(0) + \langle \overline{x}^*, x \rangle - s(j(x)), \qquad x \in j^{-1}((-a,a)),$$

is of class C^1 and satisfies $\varphi(0) = f(0)$, $\varphi'(0) = \overline{x}^*$, since $s(t)/t \to 0$ as $t \to 0_+$, $\varphi \le f$, since s is nondecreasing, $s \ge r$, and $\|x\| \le j(x) \le c\|x\|$. Thus $\overline{x}^* \in \partial_F^V f(\overline{x})$. $\qquad\square$

While in spaces satisfying (H$_F$) there is no need to distinguish the Fréchet viscosity subdifferential from the Fréchet subdifferential, the situation is not the same for the Hadamard subdifferential, even when assumption (H$_D$) holds. However, in a finite-dimensional space one has $\partial_H = \partial_D$, since $\partial_F^V = \partial_H \subset \partial_D = \partial_F$.

Let us study some relationships with geometrical notions.

If S is a subset of X and $\overline{x} \in S$, we denote by $N_H(S,\overline{x})$ the *(viscosity) Hadamard normal cone* defined by $N_H(S,\overline{x}) := \partial_H \iota_S(\overline{x})$. In the next statements, we just write $N(S,\overline{x})$ for the viscosity normal cone associated with a subdifferential $\partial \in \{\partial_H, \partial_F^V\}$. If $F : X \rightrightarrows Y$ is a multimap between two normed spaces, the *(viscosity) Hadamard coderivative* $D_H^* F(\overline{x},\overline{y})$ of F at $(\overline{x},\overline{y}) \in \text{gph}(F)$ is defined by

$$\text{gph}(D_H^* F(\overline{x},\overline{y})) := \{(y^*, x^*) : (x^*, -y^*) \in N_H(\text{gph}(F), (\overline{x},\overline{y}))\}.$$

Proposition 4.57. *Let E be a closed subset of a Banach space X and let $\overline{x} \in E$. For both viscosity subdifferentials $\partial = \partial_H$, ∂_F^V, one has $N(E,\overline{x}) = \mathbb{R}_+ \partial d_E(\overline{x})$.*

Proof. Since for every $r \in \mathbb{R}_+$ and every smooth function φ satisfying $\varphi \le rd_E$ around $\overline{x}$, $\varphi(\overline{x}) = rd_E(\overline{x})$ one has $\varphi \le \iota_E$ near $\overline{x}$, we get the inclusion $\mathbb{R}_+ \partial d_E(\overline{x}) \subset N(E,\overline{x})$.

Conversely, let $\overline{x}^* \in N(E,\overline{x})$, so that there exists a smooth function φ minorizing ι_E around $\overline{x}$ and satisfying $\varphi(\overline{x}) = \iota_E(\overline{x})$, $\varphi'(\overline{x}) = \overline{x}^*$. Since φ is locally Lipschitzian, we can find $\rho, r > 0$ such that the Lipschitz rate of $r\varphi$ on $U := B(\overline{x}, 2\rho)$ is 1. Then for $x \in B(\overline{x}, \rho)$ and $u \in E \cap U$, we have $r\varphi(x) \le r\varphi(u) + \|x - u\| \le \|x - u\|$, hence $r\varphi(x) \le d(x, E \cap U) = d(x, E)$ by an easy argument already used. Thus $r\overline{x}^* \in \partial d_E(\overline{x})$. $\qquad\square$

Let us note the following analogue of Corollary 4.15 and Proposition 4.16.

Proposition 4.58. *Let E_f be the epigraph of a lower semicontinuous function f on an arbitrary Banach space X and for $\overline{x} \in \text{dom} f$, let $\overline{x}_f := (\overline{x}, f(\overline{x}))$. Then for both viscosity subdifferentials, one has $\overline{x}^* \in \partial f(\overline{x})$ if and only if $(\overline{x}^*, -1) \in N(E_f, \overline{x}_f)$.*

Proof. Let us first consider the viscosity Fréchet case. Given $\overline{x}^* \in \partial f(\overline{x})$, let φ be a smooth function satisfying $\varphi \le f$ on a neighborhood U of $\overline{x}$ and $\varphi(\overline{x}) = f(\overline{x})$, $\varphi'(\overline{x}) = \overline{x}^*$. Then $\psi : U \times \mathbb{R} \to \mathbb{R}$ given by $\psi(x,r) := \varphi(x) - r$ is smooth, minorizes ι_{E_f}, and satisfies $\psi(\overline{x}_f) = 0 = \iota_{E_f}(\overline{x}_f)$, $(\overline{x}^*, -1) = \psi'(\overline{x}_f) \in N(E_f, \overline{x}_f)$.

Conversely, let $(\bar{x}^*, -1) \in N(E_f, \bar{x}_f)$. Without loss of generality, we suppose $f(\bar{x}) = 0$. Let us pick a smooth function ψ on an open neighborhood W of $\bar{x}_f$ in $E \times \mathbb{R}$ minorizing ι_{E_f} and satisfying $\psi(\bar{x}_f) = 0 = \iota_{E_f}(\bar{x}_f)$, $(\bar{x}^*, -1) = \psi'(\bar{x}_f)$. The implicit function theorem yields $\tau > 0$, an open neighborhood U of $\bar{x}$, and a smooth function $\varphi : U \to (-\tau, \tau)$ such that $\varphi(\bar{x}) = f(\bar{x})$, $(x, r) \in U \times (-\tau, \tau) \subset W$ satisfies $\psi(x, r) = 0$ if and only if $r = \varphi(x)$. By continuity of $\psi'(\cdot)(0, 1)$, we may suppose $\psi(x, \cdot)$ is decreasing on $(-\tau, \tau)$ for all $x \in U$. Since $\psi(\cdot, \varphi(\cdot)) = 0$, for all $r < \varphi(x)$ we get $\psi(x, r) > 0$. Since $\psi \leq \iota_{E_f}$, for all $(x, r) \in E_f \cap (U \times (-\tau, \tau))$ we have $(x, r) \in \text{epi}\,\varphi$. Thus $\varphi \leq f$ on U. Moreover, differentiating the relation $\psi(\cdot, \varphi(\cdot)) = 0$, we get $\bar{x}^* - \varphi'(\bar{x}) = (\bar{x}^*, -1) \circ (I_X, \varphi'(\bar{x})) = \psi'(\bar{x}_f) \circ (I_X, \varphi'(\bar{x})) = 0$, so that $\bar{x}^* \in \partial f(\bar{x})$.

Now let us consider the Hadamard case. The proof that $(\bar{x}^*, -1) \in N(E_f, \bar{x}_f)$ whenever $\bar{x}^* \in \partial f(\bar{x})$ does not need any change. For the reverse implication, we assume again that $f(\bar{x}) = 0$ for the sake of simplicity of notation and we pick a continuous function ψ on a neighborhood $W := U \times [-\tau, \tau]$ of $\bar{x}_f := (\bar{x}, \bar{r})$ in $E \times \mathbb{R}$, smooth on $\text{int}(W)$ and satisfying $\psi \leq \iota_{E_f}$ and $\psi(\bar{x}_f) = 0 = \iota_{E_f}(\bar{x}_f)$, $(\bar{x}^*, -1) = \psi'(\bar{x}_f)$. By continuity of $\psi'(\cdot)(0, 1)$ and ψ, we may suppose $\psi(x, \cdot)$ is decreasing on $[-\tau, \tau]$ for all $x \in U$ and $\psi(x, \tau) > 0$, $\psi(x, -\tau) < 0$ for all $x \in U$. Thus, there exists a unique function $\varphi : U \to (-\tau, \tau)$ such that $\psi(x, \varphi(x)) = 0$ for all $x \in U$. It is easy to see by contradiction that φ is continuous. Theorem 2.81 shows that it is of class D^1; the rest of the proof is similar to the proof in the Fréchet case. $\square$

Several rules for elementary subdifferentials can be extended to the viscosity subdifferentials ∂_H and ∂_F^V. Let us state some of them for later use. Their proofs are obvious.

Proposition 4.59. *The viscosity subdifferentials are homotone in the sense that for $f \geq g$ with $f(\bar{x}) = g(\bar{x})$ finite one has $\partial g(\bar{x}) \subset \partial f(\bar{x})$.*

Proposition 4.60. *Given a lower semicontinuous function f on a Banach space X, $(\bar{x}, \bar{r}) \in \text{epi}\,f$, $(\bar{x}^*, \bar{r}^*) \in N(\text{epi}\,f, (\bar{x}, \bar{r}))$, one has $(\bar{x}^*, \bar{r}^*) \in N(\text{epi}\,f, (\bar{x}, f(\bar{x})))$.*

Proposition 4.61. *(a) Let X and Y be normed spaces, let $g : X \to Y$ be smooth, and let $h : Y \to \overline{\mathbb{R}}$ be finite at $\bar{y}$. Let $f := h \circ g$, $\bar{x} \in X$, $\bar{y} := g(\bar{x})$. Then $g'(\bar{x})^\top(\partial h(\bar{y})) \in \partial f(\bar{x})$.*
(b) If $f : X \times Y \to \overline{\mathbb{R}}$ is given by $f(x, y) := g(y)$ for all $(x, y) \in X \times Y$, then for all $\bar{x} \in X$ one has $(\bar{x}^*, \bar{y}^*) \in \partial f(\bar{x}, \bar{y})$ if and only if $\bar{x}^* = 0$, $\bar{y}^* \in \partial g(\bar{y})$.

Proposition 4.62. *Let X and Y be normed spaces, let $g : X \to Y$ be smooth, let $h : X \to \overline{\mathbb{R}}$ be finite at $\bar{x}$ and $k : Y \to \overline{\mathbb{R}}$ finite at $\bar{y} := g(\bar{x})$ be such that $h - k \circ g$ attains its minimum at $\bar{x}$. Then for all $\bar{y}^* \in \partial k(\bar{y})$ one has $g'(\bar{x})^\top(\bar{y}^*) \in \partial h(\bar{x})$.*

Exercises

1. (**a**) Let f be finite and lower semicontinuous at x. Prove that $x^* \in \partial_D f(x)$ if there is a Lipschitzian Gâteaux differentiable function φ such that $\varphi'(x) = x^*$ and $f - \varphi$ attains a local minimum at x.
(**b**) Show that this implication cannot be reversed: it is possible that $x^* \in \partial_D f(x)$ but that for no Lipschitz Gâteaux differentiable φ the difference $f - \varphi$ attain a local minimum at x. Is this also true if f is Lipschitzian near x?

2. (**a**) Prove that in a space with a Fréchet (resp. Gâteaux) differentiable renorm there is a continuously Fréchet differentiable (resp. Lipschitz and Gâteaux differentiable) bump. [Hint: A convex function Fréchet differentiable at every point of an open set is continuously Fréchet differentiable on the set.]
(**b**) We say that a function φ is a strict bump if $\varphi(0) = 1$, $\varphi(x) = 0$ for x with $\|x\| \geq 1$ and $0 \leq \varphi(x) < 1$ for all $x \neq 0$. Prove that if there is a (continuous, Lipschitz, Fréchet differentiable, Gâteaux differentiable, C^k) bump function on X, then there is also a strict bump with the same properties. [Hint: $\varphi(x + a)$ is a bump for every fixed a, $\psi(\varepsilon, x) = (1 - \varepsilon^2)\varphi(x) + \varepsilon^2 \varphi(x/\varepsilon)$ is a bump (as a function of x) as well as $\int_0^1 \psi(\varepsilon, x)d\varepsilon$.]

4.4 Approximate Calculus Rules

Simple examples show that for $\partial = \partial_D$ or $\partial = \partial_F$, the inclusion $\partial f(x) + \partial g(x) \subset \partial(f + g)(x)$ cannot be reversed in general: take $f = |\cdot|$, $g = -f$ on $X = \mathbb{R}$. That is a pity, because the reverse inclusion would be most useful for calculus. However, we shall show that calculus rules in the most useful direction can be obtained, provided one accepts some fuzziness. We start with minimization rules. We treat the case of composite functions and the case of sums at the same time. In the sequel, a normed space is said to be *smooth* if it has a smooth (i.e., of class C^1 or of class D^1) Lipschitzian bump function. Given a map $g : W \to Z$ between two normed spaces (resp. a function $f : Z \to \overline{\mathbb{R}}$ finite at $\overline{w} \in W$) and $\varepsilon > 0$, we use the respective notations

$$B(\overline{w}, \varepsilon, g) := \{w \in B(\overline{w}, \varepsilon) : \|g(w) - g(\overline{w})\| < \varepsilon\},$$
$$B(\overline{w}, \varepsilon, f) := \{w \in B(\overline{w}, \varepsilon) : |f(w) - f(\overline{w})| < \varepsilon\}.$$

4.4.1 Approximate Minimization Rules

In this subsection ∂ can be either one of the viscosity subdifferentials or one of the elementary subdifferentials ∂_D, ∂_F corresponding to the smoothness of the involved spaces. We shall use the following simple fact: if $f : X \to \mathbb{R}_\infty$, $g : Y \to \mathbb{R}_\infty$, and $k : X \times$

$Y \to \mathbb{R}$ are such that $(x,y) \mapsto f(x) + g(y) - k(x,y)$ attains a finite local minimum at $(\bar{x},\bar{y})$ and if k is differentiable at $(\bar{x},\bar{y})$, then $k'(\bar{x},\bar{y}) \in \partial f(\bar{x}) \times \partial g(\bar{y})$ (since $f - k(\cdot,\bar{y})$ and $g - k(\bar{x},\cdot)$ attain local minima at $\bar{x}$ and $\bar{y}$ respectively).

Theorem 4.63. *Let X and Y be smooth Banach spaces, let $g : X \to Y$ be smooth around $\bar{x} \in X$, and let $f : X \to \mathbb{R}_\infty$, $h : Y \to \mathbb{R}_\infty$ be lower semicontinuous functions finite at $\bar{x}$ and $\bar{y} := g(\bar{x})$ respectively. Suppose $\bar{x}$ is a robust local minimizer of $f + h \circ g$. Then for every $\varepsilon > 0$ there exist $x_\varepsilon \in B(\bar{x},\varepsilon,f)$, $y_\varepsilon \in B(\bar{y},\varepsilon,h)$, $x_\varepsilon^* \in \partial f(x_\varepsilon)$, $y_\varepsilon^* \in \partial h(y_\varepsilon)$ such that $\|y_\varepsilon^*\| \cdot \|y_\varepsilon - g(x_\varepsilon)\| < \varepsilon$ and*

$$x_\varepsilon^* + y_\varepsilon^* \circ g'(x_\varepsilon) \in \varepsilon B_{X^*}.$$

We start with a simple proof that avoids technicalities. Under its additional assumptions it shows that there exists some $c > 0$ such that $\|x_\varepsilon^*\| \le c$, $\|y_\varepsilon^*\| \le c$ for all $\varepsilon > 0$, which is valuable information for passage to the weak* limit. With slight changes it could be adapted to the case that g is a smooth map around $\bar{x}$.

Proof. In the case that X and Y are finite-dimensional and endowed with Euclidean norms, h is Lipschitzian with rate ℓ around $\bar{y}$, and g is linear and continuous. Let us identify X^* with X and Y^* with Y, and let us define a decoupling (or penalized) function p_t, for $t > 0$, by

$$\forall (x,y) \in X \times Y, \qquad p_t(x,y) := f(x) + h(y) + \|x - \bar{x}\|^2 + t^2 \|g(x) - y\|^2.$$

Let $\rho > 0$ be such that $\bar{x}$ is a minimizer of $f + h \circ g$ on $B[\bar{x},\rho]$, f is bounded below by $f(\bar{x}) - 1$ on $B[\bar{x},\rho]$, and h is Lipschitzian with rate ℓ on $g(B[\bar{x},\rho]) \cup B[\bar{y},\rho]$. Let (x_t,y_t) be a minimizer of p_t on $B[\bar{x},\rho] \times B[\bar{y},\rho]$. The inequalities

$$f(\bar{x}) + h(\bar{y}) \le f(x_t) + h(g(x_t)) \le f(x_t) + h(y_t) + \ell\|g(x_t) - y_t\|$$

and $p_t(x_t,y_t) \le p_t(\bar{x},\bar{y}) = f(\bar{x}) + h(\bar{y})$ imply that

$$\|x_t - \bar{x}\|^2 + t^2 \|g(x_t) - y_t\|^2 \le \ell\|g(x_t) - y_t\|.$$

Thus $\|g(x_t) - y_t\| \le \ell t^{-2}$ and $\|x_t - \bar{x}\| \le \ell t^{-1}$, so that for $t > 0$ large enough, $(x_t,y_t) \in \mathrm{int}\,(B[\bar{x},\rho] \times B[\bar{y},\rho])$. Since the last two terms of p_t are smooth, for some $x_t^* \in \partial f(x_t)$, $y_t^* \in \partial h(y_t)$, the optimality condition $(0,0) \in \partial p_t(x_t,y_t)$ can be written

$$x_t^* + 2(x_t - \bar{x}) + 2t^2 g^*(g(x_t) - y_t) = 0, \qquad y_t^* - 2t^2(g(x_t) - y_t) = 0.$$

Plugging $y_t^* = 2t^2(g(x_t) - y_t)$ into the first equation, we get

$$x_t^* + g^*(y_t^*) = -2(x_t - \bar{x}),$$

so that $\|x_t^* + g^*(y_t^*)\| \le 2\ell t^{-1}$ and $\|y_t^*\| \cdot \|g(x_t) - y_t\| \le \ell^2 t^{-2}$, since $\|y_t^*\| \le \ell$, the Lipschitz rate of h being ℓ. The lower semicontinuity of f at $\bar{x}$ (resp. of h at $\bar{y}$) combined with the estimate $f(x_t) \le f(\bar{x}) + h(\bar{y}) - h(y_t)$ shows that $f(x_t) \to f(\bar{x})$ as $t \to +\infty$. $\square$

Proof (general case). We use again a penalization procedure in order to get some decoupling. But in the infinite-dimensional case, we need to invoke the smooth variational principle of Deville–Godefroy–Zizler. The auxiliary function space W we take is the space of bounded smooth functions on $Z := X \times Y$ whose derivatives are bounded; we endow W with the norm

$$k \mapsto \|k\|_W := \sup_{(x,y) \in X \times Y} |k(x,y)| + \sup_{(x,y) \in X \times Y} \|k'(x,y)\|.$$

Since for all $k \in W$, $r \mapsto r^{-1} \|k(r \cdot)\|_W$ is bounded on $[1, \infty)$, there exists some $c_Z > 0$ such that if $z \in Z$ is a c_Z/t^2-approximate minimizer of a function p on Z, there exist a function $k \in W$ with $\|k\|_W \le 1/t$ and a minimizer z_t of $p + k$ in $B(z, 1/t)$.

Let us take some smooth (off 0) forcing functions j_X, j_Y on X and Y respectively satisfying $\|\cdot\| \le j_X \le c\|\cdot\|$, $\|\cdot\| \le j_Y \le c\|\cdot\|$ for some $c > 1$ and let us set

$$p_t(x,y) := f(x) + h(y) + j_X^2(x - \bar{x}) + t^2 j_Y^2(g(x) - y), \qquad t > 0, \ (x,y) \in X \times Y.$$

Given $\varepsilon \in (0,1)$, we may suppose the norm of the derivative of j_X^2 is less than $\varepsilon/3$ on some ball ρB_X. We take $\rho \in (0, \varepsilon)$ such that $\bar{x}$ is a minimizer of $f + h \circ g$ on $B[\bar{x}, \rho]$, f is bounded below by $f(\bar{x}) - \varepsilon/3$ on this ball, h is bounded below by $h(\bar{y}) - \varepsilon/3$ on $B_\rho := g(B[\bar{x}, \rho]) \cup B[\bar{y}, \rho]$, and $\sup\{\|g'(x)\| : x \in B[\bar{x}, \rho]\} \le m$ for some $m \ge 1$. Let $(u_t, v_t) \in B[\bar{x}, \rho] \times B[\bar{y}, \rho]$ be such that

$$p_t(u_t, v_t) \le \inf p_t(B[\bar{x}, \rho] \times B[\bar{y}, \rho]) + c_Z t^{-2}, \qquad p_t(u_t, v_t) \le p_t(\bar{x}, \bar{y}).$$

Theorem 1.133 ensures that $((u_t, v_t))_t$ converges to $(\bar{x}, \bar{y})$ when $t \to +\infty$. In particular, $(u_t, v_t) \in B(\bar{x}, \rho/2) \times B(\bar{y}, \rho/2)$ for $t \ge \tau$ with τ large enough. We take $\tau := \tau(\varepsilon) > \max(2/\rho, 3m/\varepsilon)$ satisfying this requirement. Then for $t \ge \tau$, we have $B[(u_t, v_t), 1/t] \subset B[(\bar{x}, \bar{y}), \rho]$.

Let us apply the Deville–Godefroy–Zizler variational principle with the space W chosen above. It yields some $k_t \in W$ with $\|k_t\|_W \le 1/t$ and a minimizer (x_t, y_t) of $p_t + k_t$ in $B[(u_t, v_t), 1/t]$. Choosing a function $\varepsilon \mapsto t(\varepsilon)$ satisfying $t(\varepsilon) \ge \tau(\varepsilon)$, we simplify the notation $(x_{t(\varepsilon)}, y_{t(\varepsilon)}), t(\varepsilon)$ into $(x_\varepsilon, y_\varepsilon), t$. Since k_t, j_X^2, j_Y^2 are smooth, the optimality condition $(0,0) \in \partial(p_t + k_t)(x_\varepsilon, y_\varepsilon)$ can be written, for some $x_\varepsilon^* \in \partial f(x_\varepsilon)$, $y_\varepsilon^* \in \partial h(y_\varepsilon)$, $u_\varepsilon^* = (j_X^2)'(x_\varepsilon - \bar{x})$, $v_\varepsilon^* = (j_Y^2)'(g(x_\varepsilon) - y_\varepsilon)$, as

$$(x_\varepsilon^* + u_\varepsilon^* + t^2 v_\varepsilon^* \circ g'(x_\varepsilon), y_\varepsilon^* - t^2 v_\varepsilon^*) + k_t'(x_\varepsilon, y_\varepsilon) = (0,0).$$

Then $\left\| y_\varepsilon^* - t^2 v_\varepsilon^* \right\| = \left\| D_2 k_t(x_\varepsilon, y_\varepsilon) \right\| \le 1/t \le \varepsilon/3m$, $\left\| y_\varepsilon^* - t^2 v_\varepsilon^* \right\| \cdot \left\| g'(x_\varepsilon) \right\| \le \varepsilon/3$ and

$$\left\| x_\varepsilon^* + y_\varepsilon^* \circ g'(x_\varepsilon) \right\| \le \left\| u_\varepsilon^* \right\| + \left\| D_1 k_t(x_\varepsilon, y_\varepsilon) \right\| + \left\| y_\varepsilon^* - t^2 v_\varepsilon^* \right\| \cdot \left\| g'(x_\varepsilon) \right\| \le \left\| u_\varepsilon^* \right\| + 2\varepsilon/3.$$

Moreover, we have $p_t(x_\varepsilon, y_\varepsilon) + k_t(x_\varepsilon, y_\varepsilon) \le p_t(\bar{x}, \bar{y}) + k_t(\bar{x}, \bar{y})$, hence

$$\left\| x_\varepsilon - \bar{x} \right\|^2 + t^2 \left\| y_\varepsilon - g(x_\varepsilon) \right\|^2 \le j_X^2(x_\varepsilon - \bar{x}) + t^2 j_Y^2(y_\varepsilon - g(x_\varepsilon)) \le 2/t + 2\varepsilon/3.$$

Denoting by β an upper bound of the norms of j_X' and j_Y' on $B(\bar{x}, \rho)$ and $B(\bar{y}, \rho)$, we get

$$\left\| u_\varepsilon^* \right\| \le 2\beta c \left\| x_\varepsilon - \bar{x} \right\|, \quad \left\| v_\varepsilon^* \right\| \le 2\beta c \left\| g(x_\varepsilon) - y_\varepsilon \right\|, \quad \left\| y_\varepsilon^* \right\| \le 2t^2 \beta c \left\| g(x_\varepsilon) - y_\varepsilon \right\| + 1/t,$$

$$\left\| y_\varepsilon^* \right\| \cdot \left\| g(x_\varepsilon) - y_\varepsilon \right\| \le 2\beta c t^2 \left\| g(x_\varepsilon) - y_\varepsilon \right\|^2 + (1/t) \left\| g(x_\varepsilon) - y_\varepsilon \right\|$$

$$\le 2\beta c(2/t + 2\varepsilon/3) + (1/t^2)(2/t + 2\varepsilon/3)^{1/2}.$$

Changing ε for a smaller $\varepsilon' > 0$, we can make these terms less than ε.
Since $h(y_\varepsilon) \ge h(\bar{y}) - \varepsilon/3$ and $k_t(\bar{x}, \bar{y}) - k_t(x_\varepsilon, y_\varepsilon) \le 2 \left\| k_t \right\|_\infty \le 2 \left\| k_t \right\|_W \le 2\varepsilon/3$,

$$f(x_\varepsilon) \le f(\bar{x}) + h(\bar{y}) - h(y_\varepsilon) + k_t(\bar{x}, \bar{y}) - k_t(x_\varepsilon, y_\varepsilon) \le f(\bar{x}) + \varepsilon,$$

we also ensure that $\left| f(x_\varepsilon) - f(\bar{x}) \right| \le \varepsilon$ and similarly $\left| h(y_\varepsilon) - h(\bar{y}) \right| \le \varepsilon$. $\qquad \square$

Taking for f the null function, we immediately get a rule for composition. Taking $X = Y$, $g := I_X$, the identity map of X, and changing h into g, we get a rule for sums, which we state for our records.

Corollary 4.64. *Let X be a smooth Banach space, let $f, g : X \to \mathbb{R}_\infty$ be lower semicontinuous functions finite at $\bar{x} \in X$. Suppose $\bar{x}$ is a robust local minimizer of $f + g$. Then for every $\varepsilon > 0$ there exist $x_\varepsilon \in B(\bar{x}, \varepsilon, f)$, $y_\varepsilon \in B(\bar{y}, \varepsilon, g)$, $x_\varepsilon^* \in \partial f(x_\varepsilon)$, $y_\varepsilon^* \in \partial g(y_\varepsilon)$ such that*

$$(\left\| x_\varepsilon^* \right\| + \left\| y_\varepsilon^* \right\|) \cdot \left\| x_\varepsilon - y_\varepsilon \right\| \le \varepsilon, \quad x_\varepsilon^* + y_\varepsilon^* \in \varepsilon B_{X^*}.$$

More generally, if $\bar{x}$ is a robust local minimizer of a family $(f_1, \ldots, f_k)$ of lower semicontinuous functions on X that are finite at $\bar{x}$ (in particular if $f_1 + \cdots + f_k$ attains a local minimum at $\bar{x}$ and if either f_1 is inf-compact around $\bar{x}$ or $f_2, \ldots, f_k$ are uniformly continuous near $\bar{x}$), then for every $\varepsilon > 0$, there exist $x_i \in B(\bar{x}, \varepsilon, f_i)$, $x_i^* \in \partial f_i(x_i)$ such that

$$\max_i \left\| x_i^* \right\| \cdot \max_{i,j} \left\| x_i - x_j \right\| \le \varepsilon, \quad x_1^* + \cdots + x_k^* \in \varepsilon B_{X^*}.$$

Proof. To prove the second assertion one considers the diagonal map $x \mapsto (x, \ldots, x)$ from X to X^k and sets $f := 0$, $h(x_1, \ldots, x_k) = f_1(x_1) + \cdots + f_k(x_k)$, noting that $\partial h(x_1, \ldots, x_k) = \partial f_1(x_1) \times \cdots \times \partial f_k(x_k)$. $\qquad \square$

Let us pause to prove a result showing that subdifferentials are often available.

Theorem 4.65 (Ekeland–Lebourg). *Let $f : X \to \mathbb{R}_\infty$ be a lower semicontinuous function on a smooth Banach space X. Then the set $G_\partial := \{(x,r) \in X \times \mathbb{R} : r = f(x),\ \partial f(x) \neq \varnothing\}$ is dense in the graph G of f. In particular, the domain $D_{\partial f} := \{x \in X : \partial f(x) \neq \varnothing\}$ of ∂f is dense in the domain of f.*

Proof. Let $(\bar{x},\bar{r}) \in G$ and let $\varepsilon > 0$ be given. Since f is lower semicontinuous, there exists $\rho \in (0,\varepsilon]$ such that $f(x) > f(\bar{x}) - \varepsilon$ for all $x \in B := B[\bar{x},\rho]$. Ekeland's principle yields some $w \in B[\bar{x},\rho/2]$ that is a minimizer of $f(\cdot) + 2\rho^{-1}\varepsilon\|\cdot - w\|$ on B, hence a local minimizer of this function. Corollary 4.64 yields some $x,y \in B[w,\rho/2]$ and some $x^* \in \partial f(x)$, $y^* \in 2\rho^{-1}\varepsilon B_{X^*}$ such that $\|x^* + y^*\| < \varepsilon$ and $|f(x) - f(w)| < \varepsilon$. Then $\|x - \bar{x}\| \le \rho \le \varepsilon$, and since $f(w) \le f(\bar{x})$, we have $f(x) < f(\bar{x}) + \varepsilon$ and also $f(x) > f(\bar{x}) - \varepsilon$, since $x \in B$. The second assertion follows immediately. □

Corollary 4.66. *Let $g : W \to \mathbb{R}$ be a continuous convex function on an open convex subset of an F-smooth (resp. H-smooth) Banach space X. Then the set D of points of W at which g is Fréchet (resp. Hadamard) differentiable is dense in W. In particular, F-smooth Banach spaces are Asplund spaces.*

Proof. This follows from Corollary 4.65 applied to $f := -g$ (extended by $+\infty$ outside some closed ball $B \subset W$), since a concave function f is Fréchet (resp. Hadamard) differentiable at x whenever $\partial_F f(x)$ (resp. $\partial_D f(x)$) is nonempty. □

The next result is a subdifferential form of the Borwein–Preiss variational principle. It immediately derives from that principle (Theorem 2.62) and from the definitions of the two viscosity subdifferentials $\partial = \partial_F^V, \partial_H$.

Theorem 4.67. *Let X be a smooth Banach space. There exists a constant $\kappa > 0$ such that for every $\varepsilon > 0$, every bounded-below lower semicontinuous function f, and every $u \in X$ such that $f(u) \le \inf f(X) + \kappa\varepsilon^2$, one can find some $z \in B(u,\varepsilon)$ and some $z^* \in \partial f(z)$ satisfying $f(z) \le f(u) + \varepsilon$ and $\|z^*\| \le \varepsilon$.*

We deduce from this result an approximate global minimization rule. For the sake of simplicity, we give it for two functions rather than for k functions.

Theorem 4.68 (Approximate global minimization rule). *Let X be a smooth Banach space and let $f, g \in \mathscr{F}(X)$ be such that $\wedge(f,g)$ is finite. Then for every $\varepsilon > 0$ there exist sequences $((x_n,x_n^*)), ((y_n,y_n^*))$ in the graphs of ∂f and ∂g respectively such that*

$$(\|x_n^* + y_n^*\|) \to 0, \tag{4.32}$$

$$\limsup_n (f(x_n) + g(y_n)) \le \wedge(f,g), \tag{4.33}$$

$$\lim_n \|x_n - y_n\| \cdot (\|x_n^*\| + \|y_n^*\| + 1) = 0. \tag{4.34}$$

Proof. We provide X with a smooth forcing bifunction k_X by setting $k_X(x,y) := j^2(x-y)$, where $j : X \to \mathbb{R}_+$ is smooth on $X \setminus \{0\}$ and such that $\|\cdot\| \le j(\cdot) \le c\|\cdot\|$ for some $c > 1$. It enables us to introduce the decoupling function

$$p_t(x,y) := f(x) + g(y) + t j^2(x-y), \qquad (x,y) \in X^2.$$

Since $\wedge(f,g)$ is finite, there exists $\alpha > 0$ such that $\mu := \inf\{f(x) + g(y) : (x,y) \in \Delta(\alpha)\} > -\infty$, for $\Delta(\alpha) := \{(x,y) \in X^2 : \|x-y\| \le \alpha\}$. Applying Theorem 4.67 to the function $p_n + \iota_{\Delta(\alpha)}$, which is lower semicontinuous and bounded below, we get some $(x_n, y_n) \in \Delta(\alpha)$ and $(u_n^*, v_n^*) \in \partial(p_n + \iota_{\Delta(\alpha)})(x_n, y_n)$ such that $(\|(u_n^*, v_n^*)\|) \to 0$ and $(\gamma_n) \to 0$ for $\gamma_n := p_n(x_n, y_n) - \inf(p_n + \iota_{\Delta(\alpha)})$. Then we have

$$\mu + n j^2(x_n - y_n) \le f(x_n) + g(y_n) + n j^2(x_n - y_n) = \inf(p_n + \iota_{\Delta(\alpha)}) + \gamma_n \le \wedge(f,g) + \gamma_n,$$

so that $(j^2(x_n - y_n)) \to 0$. Thus for n large enough, we have $(x_n, y_n) \in \mathrm{int}(\Delta(\alpha))$, hence $(u_n^*, v_n^*) \in \partial p_n(x_n, y_n)$ or

$$u_n^* = x_n^* + n z_n^*, \quad v_n^* = y_n^* - n z_n^* \quad \text{with } x_n^* \in \partial f(x_n), \, y_n^* \in \partial g(y_n), \, z_n^* := (j^2)'(x_n - y_n).$$

Therefore $(\|x_n^* + y_n^*\|) = (\|u_n^* + v_n^*\|) \to 0$. Relation (4.33) follows from the above string of inequalities. Assuming that j satisfies relation (4.30), so that $\|(j^2)'(x)\| / \|x\|$ is bounded near 0, and observing that $(n j^2(x_n - y_n))$ can be made as small as required, we get (4.34). $\qquad\square$

4.4.2 Approximate Calculus in Smooth Banach Spaces

In this subsection we devise calculus rules for the two viscosity subdifferentials $\partial = \partial_F, \partial_H$, assuming that the spaces are correspondingly smooth. A parallel study with ∂_D is made in a supplement below.

If S is a subset of X, we denote by p_S the seminorm on X^* defined by

$$p_S(x^*) := \sup\{\langle x^*, x \rangle : x \in S \cup (-S)\}.$$

Thus $p_S(x^*) \le \varepsilon$ means that $x^* \in \varepsilon S^0 \cap \varepsilon(-S)^0$. Let us note that the topology associated with the seminorms p_K, where K belongs to the family of compact subsets of X, was given some attention in Chap. 1 for its interest as a substitute for the weak* topology. Let us also recall that a family $(f_1, \ldots, f_k)$ of lower semicontinuous functions on X is quasicoherent around $\bar{x} \in \mathrm{dom} f_i$ ($i \in \mathbb{N}_k := \{1, \ldots, k\}$) whenever one of them is inf-compact around $\bar{x}$ or all but one of them are uniformly continuous around $\bar{x}$.

Theorem 4.69. *Let $(f_1, \ldots, f_k)$ be a family of lower semicontinuous functions on a smooth Banach space X. Suppose $(f_1, \ldots, f_k)$ is quasicoherent around $\bar{x} \in \mathrm{dom} f$ for $f := f_1 + \cdots + f_k$. Then for all $\bar{x}^* \in \partial f(\bar{x})$, there exists some $m > 0$ such that* *the following property holds. Given $\varepsilon > 0$ and a compact subset K of X, there exist $x_i \in B(\bar{x}, \varepsilon, f_i)$ and $x_i^* \in \partial f_i(x_i)$ for $i \in \mathbb{N}_k$ such that $\max_i \|x_i^*\| \cdot \max_{i,j} \|x_i - x_j\| \leq \varepsilon$,*

$$\|x_1^* + \cdots + x_k^*\| \leq m, \qquad p_K(x_1^* + \cdots + x_k^* - \bar{x}^*) \leq \varepsilon.$$

If X is F-smooth and if ∂ is the Fréchet subdifferential, one can require that

$$\|x_1^* + \cdots + x_k^* - \bar{x}^*\| \leq \varepsilon, \qquad \max_i \|x_i^*\| \cdot \max_{i,j} \|x_i - x_j\| \leq \varepsilon.$$

Proof in the Hadamard viscosity case. Let φ be a smooth function such that $\varphi \leq f$ around $\bar{x}$, $\varphi(\bar{x}) = f(\bar{x})$, and $\varphi'(\bar{x}) = \bar{x}^*$. Since φ is of class D^1, there exist some $m > 0$ and $\beta > 0$ such that $\|\varphi'(x)\| + 1 \leq m$ for all $x \in B(\bar{x}, \beta)$ and $f_{k+1} := -\varphi$ is Lipschitzian on $B(\bar{x}, \beta)$. Lemmas 1.124 and 1.125 ensure that $\bar{x}$ is a robust local minimizer of $f_1 + \cdots + f_k + f_{k+1}$. Given $\varepsilon \in (0,1]$ and a compact subset K of X, let $r \geq 1$ be such that $K \subset rB_X$ and let $\alpha \in (0, \beta]$ be such that $p_K(\varphi'(x) - \varphi'(\bar{x})) \leq \varepsilon/2$ for all $x \in B(\bar{x}, \alpha)$. Corollary 4.64 yields $x_i \in B(\bar{x}, \varepsilon, f_i)$, $x_i^* \in \partial f_i(x_i)$ for $i \in \mathbb{N}_{k+1}$ such that $x_1^* + \cdots + x_{k+1}^* \in (\varepsilon/2r)B_{X^*}$, $x_{k+1} \in B(\bar{x}, \alpha)$, $\max_i \|x_i^*\| \cdot \max_{i,j} \|x_i - x_j\| \leq \varepsilon$. Then $\|x_1^* + \cdots + x_k^*\| \leq \|x_{k+1}^*\| + \varepsilon/2r \leq m$, and we have

$$p_K(x_1^* + \cdots + x_k^* - \bar{x}^*) \leq r\|x_1^* + \cdots + x_k^* - \varphi'(x_{k+1})\| + p_K(\varphi'(x_{k+1}) - \varphi'(\bar{x})) \leq \varepsilon.$$

Proof in the Fréchet case. The proof in this case is similar, with K replaced by B_X and φ being of class C^1. One can also take $f_{k+1} := (\varepsilon/2)\|\cdot - \bar{x}\| - \bar{x}^*$ and use the definition of $\partial_F f(\bar{x})$. $\qquad\square$

A corresponding result for composition is as follows.

Theorem 4.70. *Let X and Y be smooth Banach spaces, let $g : X \to Y$ with closed graph G, and let $h : Y \to \mathbb{R}_\infty$ be uniformly continuous around $\bar{y} := g(\bar{x})$ or lower semicontinuous and inf-compact on the image under g of a closed neighborhood U of $\bar{x}$. Then for every $\bar{x}^* \in \partial(h \circ g)(\bar{x})$, there exists some $m > 0$ such that for all compact subsets K of X, L of Y and every $\varepsilon > 0$ there exist some $(x, y) \in B(\bar{x}, \varepsilon, g) \times B(\bar{y}, \varepsilon, h)$, $y^* \in \partial h(y)$, $v^* \in Y^*$, $x^* \in D^*g(x)(v^*)$ such that $\|x^*\| + \|y^* - v^*\| \leq m$ and*

$$p_K(x^* - \bar{x}^*) < \varepsilon, \qquad p_L(y^* - v^*) < \varepsilon. \tag{4.35}$$

If X and Y are F-smooth and if $\partial = \partial_F$, one can require that

$$\|x^* - \bar{x}^*\| < \varepsilon, \quad \|y^* - v^*\| < \varepsilon. \tag{4.36}$$

When g is differentiable at x, from the observation following Definition 4.24, one gets $x^* = v^* \circ Dg(x)$, so that this result is an approximate version of the composition theorem for derivatives. The inf-compactness assumption on h is easily satisfied when X is finite dimensional.

Proof. Let f_1, $f_2 : X \times Y \to \mathbb{R}_\infty$ be given by $f_1(x,y) := \iota_G(x,y)$, $f_2(x,y) := h(y)$, so that $h(g(x)) = \inf\{(f_1 + f_2)(x,y) : y \in Y\}$. Suppose h is uniformly continuous around $\bar{y}$. Then f_2 is uniformly continuous around $(\bar{x}, \bar{y})$, and by Theorem 4.47 and a variant of it in the Hadamard viscosity case, $(\bar{x}^*, 0) \in \partial(f_1 + f_2)(\bar{x}, \bar{y})$. Thus, given $\varepsilon > 0$ and compact subsets K of X, L of Y, the preceding theorem yields some $m > 0$, $(x,v) \in B((\bar{x}, \bar{y}), \varepsilon, f_1)$, $(u,y) \in B((\bar{x}, \bar{y}), \varepsilon, f_2)$, $(x^*, -v^*) \in \partial f_1(x,v)$, $(u^*, y^*) \in \partial f_2(u,y)$ such that $\|(x^*, -v^*) + (u^*, y^*)\| \leq m$ and

$$\sup\{|\langle u^* + x^* - \bar{x}^*, u\rangle| + |\langle y^* - v^*, v\rangle| : (u,v) \in K \times L\} < \varepsilon$$

$$(\text{resp.} \quad \|u^* + x^* - \bar{x}^*\| < \varepsilon, \ \|y^* - v^*\| < \varepsilon).$$

Then $(x,v) \in G$, i.e., $v = g(x)$, $|h(v) - h(\bar{y})| < \varepsilon$, $x^* \in D^* g(x)(v^*)$, $u^* = 0$, $y^* \in \partial h(y)$, $\|x^*\| + \|y^* - v^*\| \leq m$ and (4.35) or (4.36) holds.

When h is inf-compact on $g(U)$ for some closed neighborhood U of $\bar{x}$, we take a smooth function φ such that $\varphi \leq h \circ g$ near $\bar{x}$, $\varphi(\bar{x}) = h(g(\bar{x}))$, $\varphi'(\bar{x}) = \bar{x}^*$, and we set $f_1(x,y) := \iota_G(x,y)$, $f_2(x,y) := h(y)$, $f_3(x,y) := -\varphi(x)$. We easily check that $(\bar{x}, \bar{y}, \bar{y})$ is a robust local minimizer of $f_1 + f_2 + f_3$. Then we conclude as above. □

Now let us consider the important case of performance functions.

Theorem 4.71. *Let V and W be Banach spaces, V being smooth, let $f : V \to \mathbb{R}_\infty$ be a lower semicontinuous function, let $A : V \to W$ be a surjective continuous linear map, and let $p : W \to \overline{\mathbb{R}}$ be the performance function given by*

$$p(w) := \inf\{f(v) : v \in A^{-1}(w)\}.$$

Given $\bar{w} \in p^{-1}(\mathbb{R})$, $\bar{w}^* \in \partial p(\bar{w})$, $\varepsilon > 0$, a compact subset K of W, there exist $v \in V$, $v^* \in \partial f(v)$, $w^* \in W^*$ such that $Av \in B(\bar{w}, \varepsilon)$, $f(v) < p(\bar{w}) + \varepsilon$, $\|v^* - A^\mathsf{T}(w^*)\| < \varepsilon$, $p_K(w^* - \bar{w}^*) < \varepsilon$. If $\bar{v} \in A^{-1}(\bar{w})$ is such that $f(\bar{v}) = p(\bar{w})$, one can take $v \in B(\bar{v}, \varepsilon)$.
If V is F-smooth and if $\partial = \partial_F$, one can require that $\|w^* - \bar{w}^*\| < \varepsilon$.

Exercise. Simplify the statement into the following: given $\bar{w} \in \text{dom} p$, $\bar{w}^* \in \partial p(\bar{w})$, $\varepsilon > 0$, a compact subset M of V, there exist $v \in V$, $v^* \in \partial f(v)$ such that $Av \in B(\bar{w}, \varepsilon)$, $f(v) < p(\bar{w}) + \varepsilon$, $p_M(v^* - A^\mathsf{T}\bar{w}^*) < \varepsilon$. [Hint: To see this, one can omit w^*, pick $\mu > 0$ satisfying $M \subset \mu B_V$, change ε into $\varepsilon' := \varepsilon/(\mu + 1)$, set $K := A(M)$, and note that $p_M(v^* - A^\mathsf{T}w^*) \leq \mu \|v^* - A^\mathsf{T}(w^*)\| < \mu \varepsilon'$, $p_M(A^\mathsf{T}w^* - A^\mathsf{T}\bar{w}^*) = p_K(w^* - \bar{w}^*) < \varepsilon'$, so that the sublinearity of p_M ensures that $p_M(v^* - A^\mathsf{T}\bar{w}^*) \leq \varepsilon'\mu + \varepsilon' = \varepsilon$.]

Proof. Let us first consider the Hadamard viscosity case. Let $c := \max(\|A\|, 1)$ and let ψ be a function of class D^1 such that $\psi(\bar{w}) = p(\bar{w})$, $\psi'(\bar{w}) = \bar{w}^*$, $\psi \leq p$ on $B(\bar{w}, \rho)$ for some $\rho > 0$. Given $\varepsilon > 0$ and a compact subset K of W, let $\lambda > 0$, $\delta \in (0, \rho)$ be such that $\|\psi'(w)\| \leq \lambda$, $p_K(\psi'(w) - \bar{w}^*) \leq \varepsilon$ when $w \in B(\bar{w}, \delta)$. Taking a smaller δ if necessary, we may assume $\lambda\delta \leq \varepsilon$, $\delta \leq \varepsilon$. Let $\bar{v} \in A^{-1}(\bar{w})$ be such that $f(\bar{v}) < p(\bar{w}) + \varepsilon\delta/4c$. Then $\bar{v}$ is an $(\varepsilon\delta/4c)$-minimizer of the function $g : v \mapsto f(v) - \psi(Av)$ on $B(\bar{v}, \delta/c)$, since $f(v) - \psi(Av) \geq f(v) - p(Av) \geq 0$. The Ekeland principle yields some $\bar{u} \in B(\bar{v}, \delta/2c)$ that is a minimizer of $g + (\varepsilon/2)\|\cdot - \bar{u}\|$ on $B[\bar{v}, \delta/c]$

and is such that $g(\overline{u}) \leq g(\overline{v})$. Then Corollary 4.64 yields some $v \in B(\overline{u}, \delta/2c) \subset B(\overline{v}, \delta/c)$, $v^* \in \partial f(v)$, $z^* \in (\varepsilon/2)B_{V^*}$ such that $\|Av - \overline{w}\| < \delta$, $\|v^* - A^\mathsf{T}(w^*) - z^*\| < \varepsilon/2$, $p_K(\overline{w}^* - w^*) < \varepsilon$ for $w^* := \psi'(Av)$, hence $\|v^* - A^\mathsf{T}(w^*)\| < \varepsilon$, and since

$$f(\overline{u}) = g(\overline{u}) + \psi(A\overline{u}) \leq g(\overline{v}) + \psi(A\overline{v}) + \lambda \|A\overline{u} - A\overline{v}\| \leq f(\overline{v}) + \varepsilon/2 < p(\overline{w}) + \varepsilon,$$

we can ensure that $f(v) < p(\overline{w}) + \varepsilon$. When $\overline{v} \in A^{-1}(\overline{w})$ is such that $f(\overline{v}) = p(\overline{w})$, one can choose that $\overline{v}$ and get $v \in B(\overline{v}, \delta/c) \subset B(\overline{v}, \varepsilon)$.

The Fréchet case is similar. Then given $\varepsilon > 0$, we require that $\|\psi'(w) - \overline{w}^*\| < \varepsilon$ for all $w \in B(\overline{w}, \delta)$ so that $\|w^* - \overline{w}^*\| < \varepsilon$ for $w^* := \psi'(Av)$. $\square$

Taking for A a projection, we get the following special cases.

Corollary 4.72. *Let W and X be smooth Banach spaces, let $f : W \times X \to \mathbb{R}_\infty$ be a lower semicontinuous function, and let $p : W \to \mathbb{R}_\infty$ be the performance function given by*

$$p(w) := \inf\{f(w, x) : x \in X\}.$$

Given $\overline{w} \in p^{-1}(\mathbb{R})$, $\overline{w}^* \in \partial p(\overline{w})$, $\varepsilon > 0$, a compact subset K of W, there exist $w \in B(\overline{w}, \varepsilon)$, $x \in X$, $(w^*, x^*) \in \partial f(w, x)$ such that $f(w, x) < p(\overline{w}) + \varepsilon$, $p_K(w^* - \overline{w}^*) < \varepsilon$, $\|x^*\| < \varepsilon$.
If $\partial = \partial_F$ and if W and X are F-smooth, one can take $w^* \in B(\overline{w}^*, \varepsilon)$.

Corollary 4.73. *Let W and X be H-smooth Banach spaces, let $j : W \times X \to \mathbb{R}_\infty$ be a locally Lipschitzian function, let $G : W \rightrightarrows X$ be a multimap with closed graph, and let $p : W \to \mathbb{R}_\infty$ be the performance function given by*

$$p(w) := \inf\{j(w, x) : x \in G(w)\}.$$

Given $\overline{w} \in p^{-1}(\mathbb{R})$, $\overline{w}^* \in \partial_H p(\overline{w})$, $\varepsilon > 0$, a compact subset K of W, there exist $u, w \in B(\overline{w}, \varepsilon)$, $v \in G(u)$, $x \in B(v, \varepsilon)$, $v^* \in X^*$, $u^* \in D_H^* G(u, v)(v^*)$, $(w^*, x^*) \in \partial_H j(w, x)$ such that $|j(w, x) - p(\overline{w})| < \varepsilon$, $p_K(u^* + w^* - \overline{w}^*) < \varepsilon$, $\|x^* - v^*\| \leq \varepsilon$.
If W, X are F-smooth and $\overline{w}^* \in \partial_F p(\overline{w})$, one can require that $u^* \in D_F^* G(u, v)(v^*)$, $(w^*, x^*) \in \partial_F j(w, x)$, and $\|u^* + w^* - \overline{w}^*\| < \varepsilon$.

Proof. In the preceding corollaries, set $f := j + \iota_G$. Applying the fuzzy sum rule would just give $\|x^* - v^*\| \leq m$. Thus one returns to the proof of Theorem 4.71, in which was obtained a minimizer $\overline{u} \in B[\overline{v}, \delta/2c]$ of the function $v \mapsto f(v) - \psi(Av) + (\varepsilon/2)\|v - \overline{u}\|$ on $B[\overline{v}, \delta/c]$. Thus, one can apply Corollary 4.64 to get some (u, v), $(w, x) \in B(\overline{u}, \delta/2c) \subset B(\overline{v}, \delta/c)$, $z \in B(\overline{w}, \delta)$, $(w^*, x^*) \in \partial_H j(w, x)$, $z^* := \psi'(z)$, $(u^*, -v^*) \in N_H(G, (u, v))$, $u_W^* \in (\varepsilon/2)B_{W^*}$, $u_X^* \in (\varepsilon/2)B_{X^*}$, such that

$$\|(w^*, x^*) + (u^*, -v^*) - (z^*, 0) + (u_W^*, u_X^*)\| < \varepsilon/2,$$

and hence $u^* \in D_H^* G(u, v)(v^*)$, $\|x^* - v^*\| < \varepsilon$, $\|w^* + u^* - z^*\| < \varepsilon$, $p_K(z^* - \overline{w}^*) < \varepsilon/2$. Taking $\kappa > 1$ such that $K \subset \kappa B_W$ and changing ε into $\varepsilon/2\kappa$, one gets $p_K(w^* + u^* - \overline{w}^*) < \varepsilon$ using the relation $p_K \leq \kappa \|\cdot\|$ and the sublinearity of p_K. $\square$

The particular case of a distance function deserves special mention.

Theorem 4.74 (Approximate projection theorem). *Let X be an H-smooth Banach space, let E be a closed subset of X, and let $\overline{w} \in X \setminus E$, $\overline{w}^* \in \partial_H d_E(\overline{w})$. Then for every $\varepsilon > 0$ and every compact subset K of X one can find $e \in E$, $e^* \in N_H(E, e) \cap S_{X^*}$ such that $\|e - \overline{w}\| \le d_E(\overline{w}) + \varepsilon$, $p_K(e^* - \overline{w}^*) < \varepsilon$, $\langle e^*, \overline{w} - e \rangle \ge (1 - \varepsilon)\|\overline{w} - e\|$.*

If X is F-smooth and if $\overline{w}^* \in \partial_F d_E(\overline{w})$, then for every $\varepsilon > 0$ one can require that $\|e - \overline{w}\| \le d_E(\overline{w}) + \varepsilon$, $e^* \in \partial_F d_E(e)$, $\|e^* - \overline{w}^*\| < \varepsilon$, $\langle e^*, \overline{w} - e \rangle \ge (1 - \varepsilon)\|\overline{w} - e\|$.

Proof. In the preceding corollary we take $W := X$, G being the multimap with graph $W \times E$, j being given by $j(w, x) := \|w - x\|$, so that $p = d_E$. Given $\overline{w}^* \in \partial_H d_E(\overline{w})$, $\varepsilon \in (0, 1)$, $\varepsilon < d_E(\overline{w})$, and a compact subset K of X, let $\kappa > 0$ be such that $K \subset \kappa B_X$ and let $u, w \in B(\overline{w}, \varepsilon/2)$, $e := v \in E$, $x \in B(v, \varepsilon/2)$, $v^* \in X^*$, $u^* \in D_H^* G(u, v)(v^*)$, $(w^*, x^*) \in \partial_H j(w, x)$ be such that $j(w, x) < p(\overline{w}) + \varepsilon$, $\|x^* - v^*\| < \varepsilon$, $p_K(u^* + w^* - \overline{w}^*) < \varepsilon$, as in the conclusion of Corollary 4.73. Then $u^* = 0$, $e_0^* := -v^* \in N_H(E, e)$, $\|w - x\| < d_E(\overline{w}) + \varepsilon$, $\|e_0^* - w^*\| = \|v^* - x^*\| \le \varepsilon$,

$$\|w - x\| \ge \|\overline{w} - v\| - \|w - \overline{w}\| - \|v - x\| \ge d_E(\overline{w}) - \varepsilon > 0,$$

so that $x^* = -w^* \in S_{X^*}$, $\langle w^*, w - x \rangle = \|w - x\|$. Moreover, $\|e - \overline{w}\| \le \|v - w\| + \|w - \overline{w}\| < d_E(\overline{w}) + 3\varepsilon/2$, $p_K(w^* - \overline{w}^*) < \varepsilon$, $p_K(e_0^* - w^*) \le \kappa \|v^* - x^*\| < \kappa \varepsilon$, whence

$$p_K(e_0^* - \overline{w}^*) \le p_K(e_0^* - w^*) + p_K(w^* - \overline{w}^*) < \varepsilon(\kappa + 1).$$

Furthermore,

$$\langle w^*, \overline{w} - e \rangle \ge \langle w^*, w - x \rangle - \|\overline{w} - w\| - \|x - e\| \ge \|w - x\| - \varepsilon \ge \|\overline{w} - e\| - 3\varepsilon,$$

$$\langle e_0^*, \overline{w} - e \rangle \ge \langle w^*, \overline{w} - e \rangle - \varepsilon \|\overline{w} - e\| \ge \|\overline{w} - e\| - 3\varepsilon - \varepsilon(d_E(\overline{w}) + \varepsilon).$$

Replacing e_0^* with $e^* := e_0^* / \|e_0^*\|$ and ε with some $\varepsilon' < \varepsilon$, one gets the announced inequalities.

When X is F-smooth and $\overline{w}^* \in \partial_F d_E(\overline{w})$, one can substitute B_X for K. □

Exercise. Deduce from Theorem 4.74 that if $\overline{w}^* \in \partial_F d_E(\overline{w})$, one has $\|\overline{w}^*\| = 1$.

4.4.3 Metric Estimates and Calculus Rules

In this subsection we devise subdifferential calculus rules for the two viscosity subdifferentials $\partial = \partial_F, \partial_H$ by relaxing the Lipschitz or uniform continuity assumption of Theorem 4.69. Instead of it, we use some metric estimates. A first rule concerns the normal cone to an intersection. It is an immediate consequence of Propositions 4.57, 4.59 and Theorems 4.69, 4.74.

Theorem 4.75 (Normal cone to an intersection). *Let $(S_1, \ldots, S_k)$ be a family of subsets of a smooth Banach space satisfying the following linear coherence condition at $\overline{x} \in S := S_1 \cap \cdots \cap S_k$: for some $c > 0$, $r > 0$,*

$$\forall x \in B(\overline{x}, r), \qquad d(x, S) \leq cd(x, S_1) + \cdots + cd(x, S_k). \tag{4.37}$$

Let $\overline{x}^* \in N_F(S, \overline{x}) \cap B_{X^*}$. Then for every $\varepsilon > 0$ one can find $x_i \in S_i \cap B(\overline{x}, \varepsilon)$ and $x_i^* \in N_F(S_i, x_i) \cap cB_{X^*}$ such that $\left\| x_1^* + \cdots + x_k^* - \overline{x}^* \right\| \leq \varepsilon$.

When $\overline{x}^* \in N_H(S, \overline{x})$, there exists some $c(\overline{x}) > 0$ such that for every $\varepsilon > 0$ and every compact subset K of X one can find $x_i \in S_i \cap B(\overline{x}, \varepsilon)$ and $x_i^* \in N_H(S_i, x_i) \cap c(\overline{x})B_{X^*}$ such that $p_K(x_1^* + \cdots + x_k^* - \overline{x}^*) \leq \varepsilon$.

Using the fact that the epigraph of the maximum of a finite family of functions is the intersection of the epigraphs of the functions, we get the following rule.

Theorem 4.76. Let $(f_1, \ldots, f_k)$ be a family of lower semicontinuous functions on a smooth Banach space X and let $f := \max(f_1, \ldots, f_k)$ be finite at $\overline{x} \in X$. Let S_i be the epigraph of f_i. Suppose the family (S_i) satisfies the linear coherence condition (4.37) around $(\overline{x}, f(\overline{x}))$ and let $\overline{x}^* \in \partial f(\overline{x})$. Then for all $\varepsilon > 0$ (resp. all $\varepsilon > 0$ and all compact subsets K of X), one can find $x_i \in B(\overline{x}, \varepsilon, f_i)$, $x_i^* \in X^*$ for $i \in \mathbb{N}_k$, a subset I of $\mathbb{N}_k$, $t_i \in \mathbb{P}$ for $i \in I$ such that $x_i^* \in \partial f_i(x_i)$ for $i \in I$, $x_j^* \in \partial^\infty f_j(x_j)$ for $j \in J := \mathbb{N}_k \setminus I$, and respectively

$$\left| \sum_{i \in I} t_i - 1 \right| \leq \varepsilon, \qquad \left\| \sum_{i \in I} t_i x_i^* + \sum_{j \in J} x_j^* - \overline{x}^* \right\| \leq \varepsilon,$$

$$\left| \sum_{i \in I} t_i - 1 \right| \leq \varepsilon, \qquad p_K\left(\sum_{i \in I} t_i x_i^* + \sum_{j \in J} x_j^* - \overline{x}^* \right) \leq \varepsilon.$$

Proof. When deducing this rule from Theorem 4.75 for $S_i := \operatorname{epi} f_i$, we take into account Proposition 4.58 and the lower semicontinuity of f_i to ensure that for some $\rho \in (0, \varepsilon]$ such that $f_i(x) \geq f_i(\overline{x}) - \varepsilon$ for $x \in B(\overline{x}, \rho)$ we can replace a pair $(x_i, r_i) \in \operatorname{epi} f_i \cap B((\overline{x}, f(\overline{x})), \rho)$ with $(x_i, f_i(x_i))$, so that $f_i(\overline{x}) + \varepsilon \geq r_i \geq f_i(x_i) \geq f_i(\overline{x}) - \varepsilon$. Then if $(w_i^*, -t_i) \in N(S_i, (x_i, r_i))$ are such that $\left\| \sum_{i=1}^k (w_i^*, -t_i) - (\overline{x}^*, -1) \right\| \leq \varepsilon$, we take $I := \{i \in \mathbb{N}_k : t_i > 0\}$ and we set $x_i^* := w_i^*/t_i$ for $i \in I$. For $j \in \mathbb{N}_k \setminus I$, we have $x_j^* := w_j^* \in \partial^\infty f_j(x_j)$. The Hadamard case is similar. $\square$

Now let us consider the case of an inverse image $F := g^{-1}(H)$ by a differentiable map $g : X \to Y$, where H is a closed subset of Y. We say that the pair (H, g) is *linearly coherent at* $\overline{x} \in X$ if for some $c > 0$, $r > 0$, one has

$$\forall x \in B(\overline{x}, r), \qquad d(x, F) \leq cd(g(x), H). \tag{4.38}$$

Theorem 4.77. Suppose X, Y are smooth, g is smooth, and the pair (H, g) is linearly coherent at $\overline{x} \in F$ and $\overline{x}^* \in N(F, \overline{x})$. Then, when $N = N_F$, for all $\varepsilon > 0$ there exist $x \in B(\overline{x}, \varepsilon)$, $y \in H \cap B(g(\overline{x}), \varepsilon)$, and $x^* \in B(\overline{x}^*, \varepsilon)$, $y^* \in N(H, y)$, $w^* \in (c + \varepsilon) \|\overline{x}^*\| B_{Y^*}$ such that $x^* = Dg(x)^\mathsf{T}(w^*)$, $\|w^* - y^*\| \leq \varepsilon$.

When $N = N_H$, given $\varepsilon > 0$ and compact subsets K, L of X and Y respectively, one just has $w^* \in Y^*$, $x^* = Dg(x)^\mathsf{T}(w^*)$, $p_K(x^* - \overline{x}^*) < \varepsilon$, $p_L(w^* - y^*) < \varepsilon$.

Proof. We may suppose $r := \|\bar{x}^*\| > 0$ and even that $r = 1$ by homogeneity. Let $\bar{y} := g(\bar{x})$. The linear coherence condition and Proposition 4.59 ensure that $\bar{x}^* \in c\partial(d_H \circ g)(\bar{x})$. Let us first consider the case $\partial = \partial_F$. Then, Theorem 4.70 yields some $(x,z) \in B(\bar{x}, \varepsilon, g) \times B(\bar{y}, \varepsilon/2)$, $x^* \in B(\bar{x}^*, \varepsilon)$, $z^* \in \partial d_H(z)$, $v^* \in B(z^*, \varepsilon/c)$, such that $c^{-1}x^* = Dg(x)^{\mathsf{T}}(v^*)$. Applying the approximate projection theorem, we get some $y \in H$, $y^* \in N(F, y)$ such that $\|y^* - cz^*\| < \varepsilon/2$, $\|y - z\| < d_H(z) + \varepsilon/2 < \|z - \bar{y}\| + \varepsilon/2 \le \varepsilon$. With $w^* := cv^*$, so that $\|w^*\| = c\|v^*\| \le c(\|z^*\| + \varepsilon/c) \le c + \varepsilon$, we get the required elements.

In the case $\partial = \partial_H$, using Proposition 4.57, we take $r > 0$ such that $\bar{x}^*/r \in \partial_H d_F(\bar{x})$ and we make use of the Hadamard versions of the preceding arguments. $\qquad\square$

A composition rule can be readily deduced from this result.

Theorem 4.78. *Suppose $g : X \to Y$ is smooth, $f := h \circ g$ for some lower semicontinuous function h on Y with epigraph H, and the pair $(H, g \times I_{\mathbb{R}})$ is linearly coherent at $\bar{x}_f := (\bar{x}, f(\bar{x}))$. Let $\bar{x}^* \in \partial f(\bar{x})$. Then, when $\partial = \partial_F$, for all $\varepsilon > 0$ there exist some $x \in B(\bar{x}, \varepsilon, f)$, $y \in B(g(\bar{x}), \varepsilon, h)$, $x^* \in B(\bar{x}^*, \varepsilon)$, $y^* \in \partial h(y)$, $w^* \in B(y^*, \varepsilon)$ such that $x^* = Dg(x)^{\mathsf{T}}(w^*)$, $\|y^*\| \cdot \|y - g(x)\| \le \varepsilon$.*

When $N = N_H$, given $\varepsilon > 0$ and compact subsets K, L of X and Y respectively, one just has $p_K(x^* - \bar{x}^*) < \varepsilon$, $p_L(w^* - y^*) < \varepsilon$.

Proof. Since the epigraph F of f satisfies $F = (g \times I_{\mathbb{R}})^{-1}(H)$ and since $(\bar{x}^*, -1) \in N(F, \bar{x}_f)$, taking $\varepsilon' \in (0, 1/2)$, the preceding result provides some $(y, s) \in H \cap B((g(\bar{x}), f(\bar{x})), \varepsilon')$ and some $x^* \in B(\bar{x}^*, \varepsilon')$, $(\hat{y}^*, -s^*) \in N(H, (y, s))$, $w^* \in B(\hat{y}^*, \varepsilon')$ such that $x^* = Dg(x)^{\mathsf{T}}(w^*)$, $|s^* - 1| < \varepsilon'$. We also have $(\hat{y}^*, -s^*) \in N(H, (\hat{y}, h(y)))$, as is easily seen, hence $y^* := \hat{y}^*/s^* \in \partial h(y)$. Changing x^*, w^* into x^*/s^*, w^*/s^* respectively, we get the result. The proof of the Hadamard case is left to the reader. $\qquad\square$

Recall that a family $(f_1, \ldots, f_k)$ of lower semicontinuous functions on X with sum f is said to be *linearly coherent* or to satisfy the *linear metric qualification condition* around some $\bar{x} \in X$ if there exist $c > 0$, $\rho > 0$ such that for all $x \in B(\bar{x}, \rho)$, $(t_1, \ldots, t_k) \in \mathbb{R}^k$ one has

$$d((x, t_1 + \cdots + t_k), \text{epi} f) \le cd((x, t_1), \text{epi} f_1) + \cdots + cd((x, t_k), \text{epi} f_k). \qquad (4.39)$$

Theorem 4.79 (Ioffe). *Let $(f_1, \ldots, f_k)$ be a family of lower semicontinuous functions on X and let $\bar{x}^* \in \partial f(\bar{x})$ for $f := f_1 + \cdots + f_k$. If $(f_1, \ldots, f_k)$ is linearly coherent around $\bar{x} \in X$, then the conclusion of Theorem 4.69 holds.*

Proof. Let $h : X^k \to \mathbb{R}_\infty$ be given by $h(x) = f_1(x_1) + \cdots + f_k(x_k)$ for $x := (x_1, \ldots, x_k)$ and let H be its epigraph. Denoting by $g : X \to X^k$ the diagonal map, we intend to show that (4.39) implies that the pair (H, g) is linearly coherent. Endowing a product with the sum norm, and taking the infimum over the families $(t_1, \ldots, t_k)$ with sum t in (4.39), we see that it suffices to show that for all $(x, t) \in X^k \times \mathbb{R}$ we have

$$\inf\{d((x_1, t_1), \text{epi} f_i) + \cdots + d((x_k, t_k), \text{epi} f_k) : t_1 + \cdots + t_k = t\} \le d((x, t), H).$$

Given $(x,t) \in X^k \times \mathbb{R}$ and $\lambda > d((x,t),H)$, we can find $w := (w_1,\dots,w_k) \in X^k$, $s \geq r := f_1(w_1) + \cdots + f_k(w_k)$ such that $\lambda > \|x_1 - w_1\| + \cdots + \|x_k - w_k\| + (s-t)^+$. Let $r_i := f_i(w_i)$, $t_i := r_i - (r-t)/k$, so that $t_1 + \cdots + t_k = t$, whence $(s-t)^+ \geq (r-t)^+ = (r_1 - t_1)^+ + \cdots + (r_k - t_k)^+$, and $\lambda > d((x_1,t_1),\mathrm{epi}\,f_1) + \cdots + d((x_k,t_k),\mathrm{epi}\,f_k)$. Since λ is arbitrarily close to $d((x,t),H)$, the expected inequality is satisfied.

Then, since h is separable, the conclusion of Theorem 4.78 yields the conclusion of Theorem 4.69. Given $c' > c$, we can even require that $x_i^* \in \partial f_i(x_i)$ with $\|x_i^*\| \leq c'\|\overline{x}^*\|$. $\qquad\square$

Alternative, direct proof. Let c and ρ be as in (4.39). Let $A : X \times \mathbb{R}^k \to X \times \mathbb{R}$ be defined by $A(x,t_1,\dots,t_k) := (x,t_1 + \cdots + t_k)$; its transpose is given by $A^\mathsf{T}(x^*,t^*) = (x^*,t^*,\dots,t^*)$. Let F be the epigraph of f and let

$$h(x,t_1,\dots,t_k) := cd((x,t_1),\mathrm{epi}\,f_1) + \cdots + cd((x,t_k),\mathrm{epi}\,f_k),$$

so that (4.39) can be written $h - d_F \circ A \geq 0$. Let $\overline{t}_i := f_i(\overline{x})$, $\overline{t} := (\overline{t}_1,\dots,\overline{t}_k)$, so that $h - d_F \circ A$ attains it minimum at $(\overline{x},\overline{t})$. Given $\overline{x}^* \in \partial f(\overline{x})$, by Propositions 4.57, 4.58, one has $(\overline{x}^*,-1) \in r\partial d_F(\overline{x},\overline{t})$ for some $r > 0$. Then by Proposition 4.62, in which we take $k := d_F$, $g := A$, one has $A^\mathsf{T}(\overline{x}^*,-1) \in r\partial h(\overline{y})$. Since h is a sum of Lipschitzian functions, given $\varepsilon \in (0,1)$, setting $\alpha := \varepsilon/2(rc+1)$, Theorem 4.69 and Proposition 4.61 yield some $(x_i,t_i) \in B((\overline{x},\overline{t}_i),\varepsilon)$ and $(u_i^*,-s_i^*) \in rc\partial d(\cdot,\mathrm{epi}\,f_i)(x_i,t_i)$ such that

$$\|(u_1^*,-s_1^*,0,\dots,0) + \cdots + (u_k^*,0,\dots,0,-s_k^*) - (\overline{x}^*,-1,\dots,-1)\| \leq \alpha. \qquad (4.40)$$

Thus, $|1 - s_i^*| \leq \alpha$, $s_i^* \geq 1 - \alpha > 1/2$, and for $x_i^* := u_i^*/s_i^*$, one has $x_i^* \in \partial f_i(x_i)$ by Proposition 4.58 and

$$\|x_1^* + \cdots + x_k^* - \overline{x}^*\| \leq \sum_{i=1}^{k} |1/s_i^* - 1|\,\|u_i^*\| + \left\|\sum_{i=1}^{k} u_i^* - \overline{x}^*\right\| \leq \alpha(1-\alpha)^{-1}rc + \alpha \leq \varepsilon.$$

Now let us consider the case $\overline{x}^* \in \partial_H f(\overline{x})$ and X is ∂_H-smooth. Given $\varepsilon > 0$ and a compact subset K of X, setting $\kappa := \sup\{\|x\| : x \in K\}$, $\alpha := \varepsilon/(2rc\kappa + 2)$, Theorem 4.69 and Proposition 4.61 yield some $(x_i,t_i) \in B((\overline{x},\overline{t}_i),\varepsilon)$ and some $(u_i^*,-s_i^*) \in rc\partial_H d(\cdot,\mathrm{epi}\,f_i)$ such that for $L := K \times [-1,1]^k$ one has

$$p_L(u_1^*,-s_1^*,0,\dots,0) + \cdots + (u_k^*,0,\dots,0,-s_k^*) - (\overline{x}^*,-1,\dots,-1)) \leq \alpha,$$

hence $p_K(u_1^* + \cdots + u_k^* - \overline{x}^*) \leq \alpha$ and $|1 - s_1^*| \leq \alpha,\dots,|1 - s_k^*| \leq \alpha$. Then again we have $s_i^* \geq 1/2$, $x_i^* := u_i^*/s_i^* \in \partial_H f_i(x_i)$ by Proposition 4.58 and

$$p_K(x_1^* + \cdots + x_k^* - \overline{x}^*) \leq \sum_{i=1}^{k} \left|\frac{1}{s_i^*} - 1\right| p_K(u_i^*) + p_K\left(\sum_{i=1}^{k} u_i^* - \overline{x}^*\right) \leq \frac{\alpha\kappa rc}{1-\alpha} + \alpha \leq \varepsilon.$$

$\qquad\square$

When applied to a family of indicator functions $f_i := \iota_{S_i}$, the linear coherence condition (4.39) coincides with (4.37), since $d((x,t), \text{epi } \iota_S) = d(x,S) + t^-$ and since $t \mapsto t^- := \max(-t, 0)$ is sublinear.

For a given problem, the verification of the linear coherence condition or other metric estimates may be a difficult task. Thus, it is appropriate to give criteria in terms of subdifferentials. A key to such criteria is given by the following theorem.

Theorem 4.80. *Let X be an F-smooth (resp. H-smooth) Banach space. Then for f in the set $\mathscr{F}(X)$ of lower semicontinuous proper functions on X, the function $\delta_f(x) := \inf\{\|x^*\| : x^* \in \partial f(x)\}$ with $\partial := \partial_F$ (resp. $\partial := \partial_H$) is a decrease index for f.*

Proof. Given $f \in \mathscr{F}(X)$, $x \in X$, $r, c > 0$ such that $f(x) < \inf f(B(x,r)) + cr$, we will find some $u \in B(x,r)$ such that $\delta_f(u) < c$. Let $r' \in (0,r)$, $c' \in (0,c)$ be such that $f(x) < \inf f(B(x,r)) + c'r'$. Ekeland's principle yields some $v \in B(x,r')$ such that $f(v) \le f(w) + c'\|v - w\|$ for all $w \in B[x,r']$. Then v is a local minimizer of the function $w \mapsto f(w) + c'\|w - v\|$, so that Corollary 4.64 asserts that for some $u \in B(v, r - r')$, $u^* \in \partial f(u)$ one has $\|u^*\| < c$. Thus $u \in B(x,r)$ and $\delta_f(u) < c$. $\square$

Exercise. Show that for $f \in \mathscr{F}(U)$, where U is an open subset of an F-smooth Banach space X, one has $\inf_{x \in U} |\nabla|(f)(x) = \inf\{\|x^*\| : x^* \in \partial_F f(U)\}$.

Proposition 4.81 (Fuzzy qualification condition). *Let X be a F-smooth space. (a) Let $(S_1, \ldots, S_k)$ be a family of subsets of X and $\bar{x} \in S := S_1 \cap \cdots \cap S_k$. Condition (4.37) is satisfied whenever the following alliedness condition holds: given $(x_{i,n})_n \to \bar{x}$ in S_i, $(x_{i,n}^*)$ in X^* with $x_{i,n}^* \in N_F(S_i, x_{i,n}) \cap B_{X^*}$ for $i \in \mathbb{N}_k$, $n \in \mathbb{N}$, one has*

$$\left(\left\|x_{1,n}^* + \cdots + x_{k,n}^*\right\|\right)_n \to 0 \implies \left(\left\|x_{i,n}^*\right\|\right)_n \to 0, \; i \in \mathbb{N}_k. \tag{4.41}$$

(b) Let $(f_1, \ldots, f_k)$ be a family of lower semicontinuous functions on X finite at $\bar{x} \in X$. Condition (4.39) is satisfied whenever the following criterion holds: given $(x_{i,n})_n \to \bar{x}$ in X such that $(f_i(x_{i,n}))_n \to f(\bar{x})$, $(x_{i,n}^*, r_{i,n}^*) \in N_F(\text{epi } f_i, (x_{i,n}, f_i(x_{i,n}))) \cap B_{X^* \times \mathbb{R}}$ for $i \in \mathbb{N}_k$, $n \in \mathbb{N}$, one has for $i \in \mathbb{N}_k$,

$$\left(\left\|x_{1,n}^* + \cdots + x_{k,n}^*\right\|\right)_n \to 0, \; (r_{1,n}^* + \cdots + r_{k,n}^*)_n \to 0 \Rightarrow \left(\left\|x_{i,n}^*\right\|\right)_n \to 0, \; (r_{i,n}^*) \to 0.$$

Proof. (a) Let f be given by $f(x) := d(x, S_1) + \cdots + d(x, S_k)$. By the local decrease principle, it suffices to find some $\rho > 0$, $c > 0$ such that for all $w \in B(\bar{x}, 2\rho) \setminus S$ and all $w^* \in \partial_F f(w)$ one has $\|w^*\| \ge c$. We may even replace c and 2ρ by $\varepsilon := \min(c, 2\rho)$. Suppose that this is not possible. Then, given a sequence $(\varepsilon_n) \to 0_+$ in $(0,1)$, for all $n \in \mathbb{N}$, there exist $w_n \in B(\bar{x}, \varepsilon_n) \setminus S$, $w_n^* \in \partial_F f(w_n)$ such that $\|w_n^*\| < \varepsilon_n$. Let $i(n) \in \mathbb{N}_k$ be such that $w_n \notin S_{i(n)}$. Taking a subsequence, one may assume that for some $j \in \mathbb{N}_k$ one has $i(n) = j$ for all $n \in \mathbb{N}$. The fuzzy sum rule yields some $w_{i,n} \in B(w_n, \delta_n)$, with $\delta_n := d(w_n, S_j)$ and $w_{i,n}^* \in \partial_F d(\cdot, S_i)(w_{i,n})$ such that $\|w_{1,n}^* + \cdots + w_{k,n}^*\| < \varepsilon_n$. Applying the approximate projection theorem (Corollary 4.74), for all $n \in \mathbb{N}$, we get some $x_{i,n} \in S_i$ and $x_{i,n}^* \in N(S_i, x_{i,n})$ such that $\|x_{i,n} - w_{i,n}\| < \varepsilon_n$, $\|x_{i,n}^* - w_{i,n}^*\| < \varepsilon/2$, hence $(x_{i,n}) \to \bar{x}$ and $\|x^*_{j,n}\| \geq 1/2$, a contradiction to our assumption, since we can replace $x^*_{i,n}$ with $x^*_{i,n}/r_n$ with $r_n := \max(\|x^*_{1,n}\|, \ldots, \|x^*_{k,n}\|) \geq 1/2$ in order to get elements of $N_F(S_i, x_{i,n}) \cap B_{X^*}$ satisfying $\|x^*_{1,n}/r_n + \cdots + x^*_{k,n}/r_n\| < 2\varepsilon_n$.

Assertion (b) is a consequence of assertion (a). Let us note that in fact assertion (a) is equivalent to assertion (b).   □

Let us turn to a rule for a composition $f := h \circ g$, where $g : X \to Y$ is a map with closed graph between two smooth Banach spaces and h is a lower semicontinuous function on Y, $\bar{x} \in X$, $\bar{y} := g(\bar{x}) \in \mathrm{dom}\, h$, assuming the *coherence condition* that there exist some $c > 0$ and some $\rho > 0$ such that for all $x \in B(\bar{x}, \rho)$, $y \in B(\bar{y}, \rho)$, $r \in \mathbb{R}$ one has

$$d((x,r), \mathrm{epi}\, f) \leq cd((y,r), \mathrm{epi}\, h) + cd((x,y), \mathrm{gph}\, g). \tag{4.42}$$

Theorem 4.82. *Suppose f, h, g satisfy the preceding condition. Then for all $\bar{x}^* \in \partial_F f(\bar{x})$ and all $\varepsilon > 0$ there exist some $(x,y) \in B(\bar{x}, \varepsilon, g) \times B(\bar{y}, \varepsilon, h)$, $y^* \in \partial_F h(y)$, $v^* \in (c + \varepsilon)\|\bar{x}^*\| B_{Y^*}$, $x^* \in B(\bar{x}^*, \varepsilon)$ such that $\|v^* - y^*\| \leq \varepsilon$,*

$$x^* \in D^*_F g(x)(v^*).$$

Proof. Let H be the epigraph of h and let ι_G be the indicator function of the graph G of g. Then for all $(x,y) \in X \times Y$ we have $f(x) \leq j(x,y) := h(y) + \iota_G(x,y)$ with equality for $(x,y) = (\bar{x}, \bar{y})$. Thus, denoting by p_X (resp. p_Y) the canonical projection $(x,y) \mapsto x$ (resp. $(x,y) \mapsto y$), we have $(\bar{x}^*, 0) = (p_X)^{\mathsf{T}}(\bar{x}^*) \in \partial j(\bar{x}, \bar{y})$. In order to apply Theorem 4.79, let us check that $(h \circ p_Y, \iota_G)$ is linearly coherent.

Using the sum norm in $X \times Y \times \mathbb{R}$, we have $d((x,y,t), \mathrm{epi}(h \circ p_Y)) = d_H(y,t)$, $d((x,y,s), \mathrm{epi}\, \iota_G) = d_G(x,y) + s^-$, for $s^- := \max(-s, 0)$, $d_H(y, s+t) \leq d_H(y,t) + s^-$,

$$d((x,y,s+t), \mathrm{epi}\, j) := \inf\{|s+t-r| + \|x-u\| + \|y-v\| : r \geq h(v), (u,v) \in G\}$$
$$\leq d_H(y, s+t) + d_G(x,y) \leq d_H(y,t) + s^- + d_G(x,y),$$

hence

$$d((x,y,s+t), \mathrm{epi}\, j) \leq d((x,y,t), \mathrm{epi}(h \circ p_Y)) + d((x,y,s), \mathrm{epi}\, \iota_G).$$

For each $\varepsilon > 0$, Theorem 4.79 yields some $(x,v) \in B((\bar{x}, \bar{y}), \varepsilon)$, $(u,y) \in B((\bar{x}, \bar{y}), \varepsilon)$, $(0, y^*) \in \partial_F(h \circ p_Y)(u,y)$, $(x^*, -v^*) \in \partial_F \iota_G(x,v)$ such that $|h(y) - h(\bar{y})| < \varepsilon$, and

$$\|(0, y^*) + (x^*, -v^*) - (\bar{x}^*, 0)\| < \varepsilon.$$

That means that $v = g(x)$, $\|x - \bar{x}\| + \|v - \bar{y}\| < \varepsilon$, $\|u - \bar{x}\| + \|y - \bar{y}\| < \varepsilon$, $y^* \in \partial_F h(y)$, $|h(y) - h(\bar{y})| < \varepsilon$, $x^* \in D^* g(x)(v^*)$ and $\max(\|x^* - \bar{x}^*\|, \|y^* - v^*\|) < \varepsilon$.   □

Exercise. Devise a Hadamard version of the preceding result.

4.4.4 Supplement: Weak Fuzzy Rules

Without coherence assumptions, the results one can get are not as precise and one has to replace strong approximations by weak* approximations.

Theorem 4.83. *Let $f_1, \ldots, f_k$ be lower semicontinuous functions on a smooth Banach space X and let $\bar{x}^* \in \partial_D(f_1 + \cdots + f_k)(\bar{x})$ for some $\bar{x} \in X$. Then for every $\varepsilon > 0$ and every weak* neighborhood V of 0 in X^* there exist $x_i \in B(\bar{x}, \varepsilon, f_i)$ and $x_i^* \in \partial f_i(x_i)$ for $i \in \mathbb{N}_k$ (where $\partial = \partial_H$ or $\partial = \partial_F$ according to the smoothness of X) such that*

$$x_1^* + \cdots + x_k^* - \bar{x}^* \in V,$$

$$\mathrm{diam}(x_1, \ldots, x_k) . \max_{1 \leq i \leq k} \|x_i^*\| < \varepsilon.$$

Proof. Without loss of generality, we suppose $\bar{x} = 0$. Given $\varepsilon > 0$ and a weak* neighborhood V of 0 in X^*, there exist $r > 0$ and a finite-dimensional subspace L of X such that $L^\perp + r B_{X^*} \subset V$. Let $\varepsilon', \varepsilon'' > 0$ be such that $\varepsilon' + \varepsilon'' \leq \min(\varepsilon, r)$ and let us denote by f_{k+1} and f_{k+2} the functions given by $f_{k+1}(x) = \varepsilon' \|x\| - \bar{x}^*(x)$, $f_{k+2} := \iota_{L \cap B}$ with $B := \rho B_X$, where $\rho > 0$ is chosen in such a way that $f_1 + \cdots + f_{k+2}$ attains its minimum on B at $\bar{x}$; here we use the fact that the directional subdifferential coincides with the firm subdifferential on the finite-dimensional space L. Since f_{k+2} has compact sublevel sets, $\bar{x}$ is a robust local minimizer of $(f_1, \ldots, f_{k+2})$, so that by Theorem 4.64, there exist $x_i \in B(\bar{x}, \varepsilon'', f_i)$, $x_i^* \in \partial f_i(x_i)$ $(i = 1, \ldots, k+2)$ satisfying

$$x_1^* + \cdots + x_{k+2}^* \in \varepsilon'' B_{X^*}, \qquad \mathrm{diam}(x_1, \ldots, x_{k+2}) . \max_{1 \leq i \leq k+2} \|x_i^*\| < \varepsilon''.$$

Since $f_{k+2}(x_{k+2})$ is finite and since we may take $\varepsilon'' < \rho$, we have $x_{k+2} \in L \cap \mathrm{int} B$ and $x_{k+2}^* \in \partial \iota_L(x_{k+2}) = L^\perp$. Moreover, since f_{k+1} is convex, we have $\|x_{k+1}^* + \bar{x}^*\| \leq \varepsilon'$, hence $x_1^* + \cdots + x_k^* - \bar{x}^* + x_{k+2}^* \in (\varepsilon' + \varepsilon'') B_{X^*} \subset \varepsilon B_{X^*}$. The result follows. □

Taking $k = 1$ one gets an approximation result.

Corollary 4.84. *Let f be a lower semicontinuous function on a smooth space X and let $\bar{x}^* \in \partial_D f(\bar{x})$. Then for every $\varepsilon > 0$ and every weak* neighborhood V of 0 in X^* there exist $x \in B(\bar{x}, \varepsilon, f)$ and $x^* \in \partial f(x)$ such that $x^* - \bar{x}^* \in V$.*

The corresponding result for composition is as follows. It can be deduced from the preceding theorem by a proof similar to that of Theorem 4.70.

Theorem 4.85. *Let X and Y be smooth Banach spaces, let $g : X \to Y$ with closed graph, and let $h : Y \to \mathbb{R}_\infty$ be lower semicontinuous. Then for every $\bar{x}^* \in \partial_D(h \circ g)(\bar{x})$, every $\varepsilon > 0$ and every weak* neighborhoods V, W of 0 in X^* and Y^* respectively, there exist some $(x, y) \in B(\bar{x}, \varepsilon) \times B(\bar{y}, \varepsilon)$, $y^* \in \partial h(y)$, $v^* \in Y^*$, $x^* \in D^* g(x)(v^*)$ such that*

$$x^* - \bar{x}^* \in V, \qquad y^* - v^* \in W, \tag{4.43}$$

$$(\|x^*\| + \|y^*\|) \cdot \|g(x) - y\| < \varepsilon. \tag{4.44}$$

Let us give a variant of Theorem 4.83. Taking $\partial \in \{\partial_D, \partial_H, \partial_F\}$, we say that X is a ∂-*subdifferentiability space* if for every $f \in \mathscr{F}(X)$ the set of points $(x, f(x))$ for which $\partial f(x) \neq \varnothing$ is dense in the graph of f. For $\partial \in \{\partial_H, \partial_F\}$ this requirement is less demanding than ∂-smoothness.

Theorem 4.86 (Ioffe). *Let $\partial \in \{\partial_D, \partial_H, \partial_F\}$ and let f, g be lower semicontinuous functions on a ∂-subdifferentiability space X and let $\bar{x}^* \in \partial_D(f + g)(\bar{x})$. Then for every $\varepsilon > 0$ and every weak* neighborhood V of 0 in X^* there exist $x \in B(\bar{x}, \varepsilon, f)$, $y \in B(\bar{x}, \varepsilon, g)$, $x^* \in \partial f(x)$, $y^* \in \partial g(x)$ such that $x^* + y^* - \bar{x}^* \in V$.*

Proof. (a) We first consider the case that $\bar{x} = 0$ is a local strict minimizer of $f + g$ (i.e., for some $V \in \mathscr{N}(0)$, $f(0) + g(0) < f(x) + g(x)$ for all $x \in V \setminus \{0\}$) and g is locally inf-compact around $\bar{x}$. This means that for some $r > 0$ and all $s \in \mathbb{R}$ the set $\{x \in rB_X : g(x) \leq s\}$ is compact. Let $\rho \in (0, r/2)$, $\rho < \varepsilon/2$ be such that f and g are bounded below on $2\rho B_X$ and $\bar{x} = 0$ is a strict minimizer of $f + g$ on that ball. Let h be given by

$$h(x) := \inf\{f(x + w) + g(w) : w \in \rho B_X\} \quad \text{if } x \in \rho B_X, \qquad h(x) = +\infty \text{ otherwise.}$$

Since $g \mid 2\rho B_X$ is inf-compact, the infimum is attained for all $x \in \rho B_X$, and moreover, h is lower semicontinuous and $h(\bar{x}) = f(\bar{x}) + g(\bar{x})$. Let $(x_n, h(x_n)) \to (\bar{x}, h(\bar{x}))$ be such that $\partial_D h(x_n) \neq \varnothing$ for all n. Let $w_n \in \rho B_X$ be such that $h(x_n) = f(x_n + w_n) + g(w_n)$. Since f and g are bounded below on $2\rho B_X$, $(g(w_n))$ is bounded, and since $g \mid 2\rho B_X$ is lower-compact, we may assume that (w_n) has a limit $\bar{w}$ in ρB_X. By lower semicontinuity of f and g we get $f(\bar{w}) + g(\bar{w}) \leq \lim_n h(x_n) = h(\bar{x}) = f(\bar{x}) + g(\bar{x})$. Since $\bar{x}$ is a strict minimizer in $B(\bar{x}, 2\rho)$ we must have $\bar{w} = 0$. Thus, we can find $x \in B(\bar{x}, \rho, h)$ and $w \in \rho B_X$ such that $\partial_D h(x) \neq \varnothing$ and $f(x + w) + g(w) \leq f(x + w') + g(w')$ for all $w' \in \rho B_X$. By Proposition 4.7, for $x^* \in \partial h(x)$, we can find a function φ Hadamard differentiable at x with $\varphi'(x) = x^*$ such that $\varphi \leq h$, $\varphi(x) = h(x)$. Then for all $u \in X$,

$$\varphi(x + u) - \varphi(x) \leq (f(x + u + w) + g(w)) - (f(x + w) + g(w))$$
$$\leq f(x + u + w) - (f(x + w)),$$
$$\varphi(x + u) - \varphi(x) \leq (f(x + u + w - u) + g(w - u)) - (f(x + w) + g(w))$$
$$\leq g(w - u) - g(w),$$

so that $x^* \in \partial_D f(x + w)$ and $-x^* \in \partial_D g(w)$. When $\partial = \partial_H$ (resp. $\partial = \partial_F$), we can suppose φ is of class D^1 (resp. of class C^1), so that $x^* \in \partial f(x + w)$ and $-x^* \in \partial g(w)$. Since $x + w \in 2\rho B_X$ and $w \in \rho B_X$ and since ρ is arbitrarily small, using the semicontinuity of f and g in the usual way, we get the result in this case. We even have $0 \in \partial f(x + w) + \partial g(w)$.

(b) Now let us turn to the general case. Again, we may assume $\overline{x} = 0$. Changing g into $g - \overline{x}^*$, we may also suppose $\overline{x}^* = 0$. Given a weak* neighborhood V of 0 in X^*, we take a finite-dimensional subspace L of X and $\rho > 0$ such that $\rho B_{X^*} + L^\perp \subset V$. Let $k := g + \iota_L + (\varepsilon/2) \|\cdot\|$, where ε can be assumed to belong to $(0, \rho)$. Clearly, this function is locally inf-compact around $\overline{x}$ and $\overline{x} = 0$ is a strict minimizer of $f + k$. According to step (a), we can find some $v \in B(\overline{x}, \varepsilon/2, f)$, $w \in B(\overline{x}, \varepsilon/2, k)$ and some $v^* \in \partial f(v)$, $w^* \in \partial k(w)$ with $v^* + w^* = 0$. Since w belongs to the domain of k, we have $w \in L$, hence $w + L = L$. Now let us consider the function j given by

$$j(x) := \iota_L(x) - \langle w^*, x - w \rangle + (\varepsilon/2)(\|x - w\| + \|x\|).$$

It is such that $g(x) + j(x) = k(x) - \langle w^*, x - w \rangle + (\varepsilon/2) \|x - w\|$. Since $w^* \in \partial k(w) \subset \partial_D k(w)$ and since the domain of k is contained in L, w is a strict local minimizer of $g + j$. Applying again the first step, we get some $y, z \in B(w, \varepsilon/2) \subset \varepsilon B_X$ and some $y^* \in \partial g(y)$, $z^* := -y^* \in \partial j(z)$. Using the fact that j is convex, we obtain that $z^* \in L^\perp + \varepsilon B_{X^*} - w^*$. Thus $v^* + y^* = -w^* - z^* \in L^\perp + \varepsilon B_{X^*}$ with $v^* \in \partial f(v)$, $y^* \in \partial g(y)$. $\square$

A variant of Theorem 4.74 is as follows.

Theorem 4.87 (Refined approximate projection theorem). *Let X be an H-smooth Banach space, let E be a closed subset of X, and let $\overline{w} \in X \setminus E$, $\overline{w}^* \in \partial_H d_E(\overline{w})$. Then for every $\varepsilon > 0$ and every compact subset K of X one can find $e \in E$, $e^* \in \partial_H d_E(e)$ such that $\|e - \overline{w}\| \leq d_E(\overline{w}) + \varepsilon$, $p_K(e^* - \overline{w}^*) < \varepsilon$.*

Proof. Let ψ be a function of class D^1 around $\overline{w}$ such that $\psi \leq d_E$ and $\psi(\overline{w}) = d_E(\overline{w})$, $\psi'(\overline{w}) = \overline{w}^*$. We may suppose ψ is Lipschitzian with rate $k > 0$. Then for all $(w, x) \in X \times E$ one has $\|w - x\| - \psi(w) \geq 0$. Given $\varepsilon \in (0, 1)$ and a compact subset K of X, let $\rho \in (0, \varepsilon/2)$ be such that $p_K(\psi'(w) - (1 + \rho) \psi'(\overline{w})) < \varepsilon/2$, $p_K(2\rho x^*) < \varepsilon/2$ for all $w \in B(\overline{w}, \rho)$, $x^* \in B_{X^*}$. Let $x_\varepsilon \in E$ be such that $\|\overline{w} - x_\varepsilon\| < d_E(\overline{w}) + \min(\kappa \rho^2, \varepsilon/2)$, where κ is the constant appearing in the Borwein–Preiss variational principle (Theorem 2.62) on X^2 with the space $BD^1(X) \times BD^1(X)$. That result yields some $g, h \in BD^1(X)$ with $\|g\|_{1,\infty} < \rho$, $\|h\|_{1,\infty} < \rho$ and some $(w_\varepsilon, y_\varepsilon) \in B((\overline{w}, x_\varepsilon), \rho)$ such that $(w_\varepsilon, y_\varepsilon)$ is a minimizer of the function $f_E : (w, x) \mapsto \iota_E(x) + f(w, x)$, where $f(w, x) := \|w - x\| - \psi(w) + g(w) + h(x)$: for all $(w, x) \in X \times E$,

$$\|w - x\| - \psi(w) + g(w) + h(x) \geq \|w_\varepsilon - y_\varepsilon\| - \psi(w_\varepsilon) + g(w_\varepsilon) + h(y_\varepsilon).$$

The penalization lemma for the subset $X \times E$ of X^2 endowed with the norm $(w, x) \mapsto (1 + k) \|w\| + \|x\|$ ensures that $(w_\varepsilon, y_\varepsilon)$ is a minimizer of the function $(w, x) \mapsto f(w, x) + (1 + \rho) d_E(x)$ on X^2 (we use the fact that f is Lipschitzian with rate $1 + \rho$ and the relation $d_{X \times E}(w, x) = d_E(x)$ for all $(w, x) \in X^2$). The inequality

$$\forall u \in X, \quad f(w_\varepsilon + u, y_\varepsilon + u) + (1 + \rho) d_E(y_\varepsilon + u) \geq f(w_\varepsilon, y_\varepsilon) + (1 + \rho) d_E(y_\varepsilon)$$

can be expressed as follows: for all $u \in X$,

$$- \psi(w_\varepsilon + u) + g(w_\varepsilon + u) + h(y_\varepsilon + u) + (1 + \rho)d_E(y_\varepsilon + u)$$
$$\geq - \psi(w_\varepsilon) + g(w_\varepsilon) + h(y_\varepsilon) + (1 + \rho)d_E(y_\varepsilon).$$

Since ψ, g, h are Hadamard differentiable, one gets

$$\psi'(w_\varepsilon) - g'(w_\varepsilon) - h'(y_\varepsilon) \in (1 + \rho)\partial_H d_E(y_\varepsilon).$$

Setting $e := y_\varepsilon$, $e^* := (1 + \rho)^{-1}(\psi'(w_\varepsilon) - g'(w_\varepsilon) - h'(y_\varepsilon))$, one gets the result. □

4.4.5 Mean Value Theorems and Superdifferentials

The mean value theorem is known as a cornerstone of differential calculus. It is not less important for subdifferential calculus. In this section ∂ is either ∂_F or ∂_H and X is an F-smooth or H-smooth Banach space. A simpler version valid for soft functions will be given in the next section.

We start with a fuzzy form of the Rolle's theorem.

Theorem 4.88 (Fuzzy Rolle's theorem). *Let $f \in \mathcal{F}(X)$ be finite at $\bar{x} \in X$ and let $\bar{y} \in X \setminus \{\bar{x}\}$ be such that $f(\bar{y}) \geq f(\bar{x})$. Then there exist $u \in [\bar{x}, \bar{y}) := [\bar{x}, \bar{y}] \setminus \{\bar{y}\}$ and sequences $(u_n) \to u$, (u_n^*) such that $(f(u_n)) \to f(u)$, $u_n^* \in \partial f(u_n)$ for all n and*

$$\liminf_n \langle u_n^*, \bar{y} - \bar{x} \rangle \geq 0, \tag{4.45}$$

$$\liminf_n \langle u_n^*, x - u_n \rangle \geq 0 \qquad \forall x \in \bar{x} + \mathbb{R}_+(\bar{y} - \bar{x}), \tag{4.46}$$

$$\lim_n \|u_n^*\| d(u_n, [\bar{x}, \bar{y}]) = 0. \tag{4.47}$$

Proof. Let u be a minimizer of f on the compact set $S := [\bar{x}, \bar{y}]$. Since $f(\bar{y}) \geq f(\bar{x})$, we may suppose $u \neq \bar{y}$. Since $g := \iota_S$ is inf-compact, for every $\varepsilon \in (0, \|u - \bar{y}\|)$ Corollary 4.64 yields $w, z \in B(u, \varepsilon)$, $z \in S$, $w^* \in \partial f(w)$, $z^* \in N(S, v)$ such that

$$\|w^* + z^*\| < \varepsilon, \quad (\|w^*\| + \|z^*\|) \cdot \|w - z\| < \varepsilon, \quad |f(w) - f(u)| < \varepsilon.$$

Since $\|z - u\| < \varepsilon < \|z - \bar{y}\|$, the indicator functions of S and $S' := \bar{x} + \mathbb{R}_+(\bar{y} - \bar{x})$ coincide on a neighborhood of z, and it follows that z^* belongs to the normal cone to S' at z. Therefore, for every $x \in \bar{x} + \mathbb{R}_+(\bar{y} - \bar{x})$ we have $\langle -z^*, x - z \rangle \geq 0$; hence by $\|w^* + z^*\| < \varepsilon$, $\langle w^*, x - z \rangle \geq -\varepsilon \|x - z\|$ and $\langle w^*, x - w \rangle \geq -\varepsilon \|x - z\| - \varepsilon$. Let $t \in [0, 1)$ be such that $z = (1 - t)\bar{x} + t\bar{y}$. Taking $x = \bar{y}$ in the inequality $\langle w^*, x - z \rangle \geq -\varepsilon \|x - z\|$ and dividing by $1 - t$, we get $\langle w^*, \bar{y} - \bar{x} \rangle \geq -\varepsilon \|\bar{y} - \bar{x}\|$. Replacing ε by the general term ε_n of a sequence with limit 0, (w, w^*) by (u_n, u_n^*), we get all the assertions. □

Exercise. Given $f \in \mathcal{F}(X)$, $\bar{x}, \bar{y} \in \mathrm{dom} f$ with $f(\bar{x}) = f(\bar{y})$, show that there exist sequences $(u_n) \to u \in (\bar{x}, \bar{y})$, (u_n^*) such that $(f(u_n)) \to f(u)$, $(\langle u_n^*, \bar{y} - \bar{x} \rangle) \to 0$, $(\|u_n^*\| d(u_n, [\bar{x}, \bar{y}])) \to 0$, $\liminf_n \langle u_n^*, u - u_n \rangle \geq 0$ and $u_n^* \in \partial f(u_n) \cup \partial(-f)(u_n)$ for all n. [Hint: Take for u either a minimizer or a maximizer of f on $[\bar{x}, \bar{y}]$.]

Let us get rid of the restriction $f(\bar{y}) \geq f(\bar{x})$.

Theorem 4.89 (Fuzzy mean value theorem). *Let $f \in \mathcal{F}(X)$ be finite at $\bar{x} \in X$. Then for every $\bar{y} \in X \setminus \{\bar{x}\}$ and for every $r \in \mathbb{R}$ such that $r \leq f(\bar{y})$, there exist $u \in [\bar{x}, \bar{y}]$ and sequences $(u_n) \to u$, (u_n^*) such that $u_n^* \in \partial f(u_n)$ for all n, $(f(u_n)) \to f(u)$,*

$$\liminf_n \langle u_n^*, \bar{y} - \bar{x} \rangle \geq r - f(\bar{x}), \tag{4.48}$$

$$\liminf_n \langle u_n^*, x - u_n \rangle \geq (r - f(\bar{x})) \frac{\|x - u\|}{\|\bar{y} - \bar{x}\|} \quad \forall x \in (\bar{x} + \mathbb{R}_+(\bar{y} - \bar{x})) \setminus [\bar{x}, u), \tag{4.49}$$

$$\liminf_n \left\langle u_n^*, \frac{x - u_n}{\|x - u\|} \right\rangle \geq -\frac{r - f(\bar{x})}{\|\bar{y} - \bar{x}\|} \quad \forall x \in [\bar{x}, u), \tag{4.50}$$

$$\lim_n \|u_n^*\| d(u_n, [\bar{x}, \bar{y}]) = 0. \tag{4.51}$$

Note that for $x = u$, relation (4.49) yields

$$\liminf_n \langle u_n^*, u - u_n \rangle \geq 0. \tag{4.52}$$

Proof. Let $e^* \in X^*$ be such that $\langle e^*, \bar{y} - \bar{x} \rangle = f(\bar{x}) - r$. Setting $h := f + e^*$, we see that $h(\bar{y}) \geq h(\bar{x})$, so that we can apply the Rolle's theorem to h. Observing that $\partial h(u_n) = \partial f(u_n) + e^*$, we obtain (4.48) and (4.49) from (4.45) and (4.46) respectively, since $x - u = q(\bar{y} - \bar{x})$ with $q := \|x - u\| / \|\bar{y} - \bar{x}\|$ when $x \in (\bar{x} + \mathbb{R}_+(\bar{y} - \bar{x})) \setminus [\bar{x}, u)$. Now, given $x \in [\bar{x}, u)$, setting $u = \bar{x} + s(\bar{y} - \bar{x})$ with $s \in [0, 1)$, we have $x = \bar{x} + t(\bar{y} - \bar{x})$ for some $t < s$. Then $u - x = (s - t)(\bar{y} - \bar{x})$, $\|u - x\| = (s - t)\|\bar{y} - \bar{x}\|$, and (4.46) reads

$$\liminf_n \langle u_n^*, x - u_n \rangle \geq \langle e^*, u - x \rangle = \|\bar{y} - \bar{x}\|^{-1} \|x - u\| (f(\bar{x}) - r),$$

so that the proof is complete. $\qquad\qquad\square$

A more powerful version follows, $\bar{y}$ being replaced with a closed convex subset Y of X and the segment $[\bar{x}, \bar{y}]$ being replaced with the "drop"

$$D = [\bar{x}, Y] := \mathrm{co}(\{\bar{x}\} \cup Y) := \bigcup_{y \in Y} [\bar{x}, y] := \{(1 - t)\bar{x} + ty : y \in Y, \, t \in [0, 1]\}.$$

Theorem 4.90 (Multidirectional Rolle's theorem). *Let Y be a closed convex subset of X and let $\bar{x} \in X \setminus Y$, $D := [\bar{x}, Y]$. Suppose $f \in \mathcal{F}(X)$ is lower semicontinuous, finite at $\bar{x}$, and bounded below on $D + \sigma B_X$ for some $\sigma > 0$. Suppose $\wedge_Y f :=$*

$\sup_{r>0} \inf f(Y + rB_X) > f(\bar{x})$, or more generally, $\wedge_Y f > \wedge_D f$. Then for every $\varepsilon > 0$ there exist $u \in D \setminus Y$, $w \in B(u, \varepsilon)$, and $w^* \in \partial f(w)$ such that $|f(w) - \wedge_D f| < \varepsilon$,

$$\langle w^*, y - \bar{x} \rangle \geq -\varepsilon \|y - \bar{x}\| \qquad \forall y \in Y, \tag{4.53}$$

$$\langle w^*, x - u \rangle \geq -\varepsilon \|x - u\| \qquad \forall x \in D, \tag{4.54}$$

$$\|w^*\| \cdot \|w - u\| < \varepsilon. \tag{4.55}$$

Proof. By assumption, $\ell := \wedge_D f := \sup_{r>0} \inf f(D + rB_X)$ is finite (and $\inf f(D + \sigma B_X) \leq \ell \leq f(\bar{x}) < \infty$). Taking $\alpha \in (0, \wedge_Y f - \wedge_D f)$ and $\rho \in (0, \sigma)$ such that $\inf f(Y + \rho B_X) > \wedge_D f + \alpha$, we may assume $\varepsilon < \min(\alpha, \rho)$. The approximate global minimization rule (Theorem 4.68) with $g := \iota_D$ yields some $w \in X$, $w^* \in \partial f(w)$, $u \in D$, $u^* \in N(D, u)$ such that $\|u - w\| < \varepsilon$, $|f(w) - \ell| < \varepsilon$, $\|u^* + w^*\| < \varepsilon$. Since D is convex, $N(D, u)$ is the normal cone in the sense of convex analysis, so that for all $x \in D$, $\langle u^*, x - u \rangle \leq 0$. Combining this inequality with $\|u^* + w^*\| < \varepsilon$, we get (4.54). If we had $u \in Y$, we would have $w \in Y + \rho B_X$, hence $f(w) > \wedge_D f + \alpha$, a contradiction to $f(w) < \ell + \varepsilon$. Thus, $u \in D \setminus Y$. Let $s \in [0, 1)$, $v \in Y$ be such that $u := sv + (1 - s)\bar{x}$. Now, for every $y \in Y$ one has $y' := sv + (1 - s)y \in Y$, $y' - u = (1 - s)(y - \bar{x})$ and hence

$$\langle w^*, y - \bar{x} \rangle = (1 - s)^{-1} \langle w^*, y' - u \rangle > -(1 - s)^{-1} \varepsilon \|y' - \bar{x}\| = -\varepsilon \|y - \bar{x}\|,$$

and (4.53) holds. Relation (4.55) is a consequence of (4.34).   □

Exercise. In the case that Y is compact, use the existence of some $u \in D$ minimizing f on D to get for all $\varepsilon > 0$ some $w \in B(u, \varepsilon)$, $w^* \in \partial f(w)$ satisfying relations (4.53)–(4.55) [Hint: Observe that ι_D is inf-compact and use Corollary 4.64.]

Theorem 4.91 (Multidirectional mean value theorem). *Let Y be a closed convex subset of a smooth Banach space X and let $\bar{x} \in X$, $D := [\bar{x}, Y]$. Suppose $f \in \mathscr{F}(X)$ is finite at $\bar{x}$ and bounded below on $D + \sigma B_X$ for some $\sigma > 0$. Let $r \in \mathbb{R}$, $r \leq \wedge_Y f$. Then there exist sequences (u_n) in D, (y_n) in Y, (w_n) in X with $(u_n - w_n) \to 0$, (w_n^*) in X^* such that $u_n \in [\bar{x}, y_n]$, $w_n^* \in \partial f(w_n)$ for all n and $\limsup_n f(w_n) \leq f(\bar{x})$,*

$$\liminf_n \langle w_n^*, y - \bar{x} \rangle \geq r - f(\bar{x}) \qquad \forall y \in Y, \tag{4.56}$$

$$\|w_n^*\| d(w_n, D) \to 0. \tag{4.57}$$

Proof. We may suppose $\bar{x} = 0$, $f(\bar{x}) = 0$. Let $q_n \in (r - f(\bar{x}) - \varepsilon_n, r - f(\bar{x}))$, where $(\varepsilon_n) \to 0_+$, and let us set $\bar{x}_1 := (\bar{x}, 0) \in X_1 := X \times \mathbb{R}$, $Y_1 := Y \times \{1\}$,

$$f_n(x, t) := f(x) + (1 - t)q_n, \qquad (x, t) \in X_1 := X \times \mathbb{R}. $$

Clearly, $f_n \in \mathscr{F}(X_1)$, is bounded below on $D_1 + \sigma B_{X \times \mathbb{R}}$, where $D_1 := [\bar{x}_1, Y_1]$ and

$$\wedge_{Y_1} f_n := \lim_{\delta \to 0_+} \inf\{f(x) + (1 - t)q_n : d((x, t), Y_1) < \delta\} = \wedge_Y f > q_n = f_n(\bar{x}_1).$$

Given a sequence $(\varepsilon_n) \to 0_+$, applying Theorem 4.90 to f_n with $\varepsilon := \varepsilon_n$, we get some $s_n \in [0,1)$, $y_n \in Y$, $u_n := \overline{x} + s_n(y_n - \overline{x}) \in D$, $(w_n, t_n) \in B((u_n, s_n), \varepsilon_n)$, $(w_n^*, t_n^*) \in \partial f_n(w_n, t_n)$ such that (4.53)–(4.55) hold with u, w, w^*, ε replaced with (u_n, s_n), (w_n, t_n), (w_n^*, t_n^*), ε_n respectively. Then $w_n^* \in \partial f(w_n)$, $t_n^* = -q_n$, and taking $(y, 1)$ with $y \in Y$ in place of y in (4.53), we get

$$\langle w_n^*, y - \overline{x} \rangle - q_n + \varepsilon_n(\|y - \overline{x}\| + 1) \geq 0$$

and (4.56) by passing to the limit. Moreover,

$$d(w_n, D) \leq d((w_n, s_n), D_1) \leq d((w_n, t_n), D_1) + |s_n - t_n|,$$

whence $(\|w_n^*\| d(w_n, D)) \to 0$, since $(\|(w_n^*, q_n)\| \cdot \|(w_n, t_n) - (u_n, s_n)\|) \to 0$, as shown by the proof of the preceding theorem and relation (4.34).

When $\wedge_Y f - f(\overline{x}) > 0$, Rolle's theorem ensures that $|f(w_n) - \wedge_D f| < \varepsilon_n$, hence $f(w_n) < f(\overline{x}) + \varepsilon_n$. When $\wedge_Y f - f(\overline{x}) \leq 0$, similarly we may take (w_n, t_n) such that $f(w_n) + (1 - t_n)q_n < \wedge_{D_1} f_n + \varepsilon_n \leq \wedge_Y f + \varepsilon_n$. Since $\wedge_Y f - f(\overline{x}) - \varepsilon_n < q_n \leq (1 - t_n)q_n$, we get $f(w_n) < f(\overline{x}) + 2\varepsilon_n$ and $\limsup_n f(w_n) \leq f(\overline{x})$. $\qquad\square$

Remark. Setting $x_n := \overline{x} + s_n(y_n - \overline{x})$, we have the additional information that

$$\liminf_n \left\langle w_n^*, \frac{x_n - u_n}{\|x_n - u_n\|} \right\rangle \geq 0.$$

That follows from the choice $(y, t) := (y_n, s_n)$ in (4.54), written here as

$$\langle w_n^*, x - u_n \rangle - q_n(t - s_n) + \varepsilon_n(\|x - u_n\| + |t - s_n|) \geq 0. \tag{4.58}$$

$\qquad\square$

It is sometimes necessary to use the Fréchet (resp. Hadamard) *superdifferential* of a function f, defined as $\widetilde{\partial}_F f(x) := -\partial_F(-f)(x)$ (resp. $\widetilde{\partial}_D f(x) := -\partial_D(-f)(x)$) for $x \in X$. This concept is crucial for the study of Hamilton–Jacobi equations, for instance. It can be related to the subdifferential of f by the following theorem. Again $\overline{co}^*(A)$ denotes the weak* closed convex hull of a subset A of X^*.

Theorem 4.92 (Approximation of superdifferentials). *Let* $f : X \to \mathbb{R}$ *be a lower semicontinuous function. If X is an F-smooth Banach space, then for all $\varepsilon > 0$, $\overline{x} \in X$ one has*

$$\widetilde{\partial}_F f(\overline{x}) \subset \overline{co}^*(\partial_F f(B(\overline{x}, \varepsilon))) + \varepsilon B_{X^*}.$$

Proof. Suppose, to the contrary, that there exist some $\varepsilon > 0$, $\overline{y} \in X$, and $\overline{y}^* \in \widetilde{\partial}_F f(\overline{y})$ such that $\overline{y}^* \notin C + \varepsilon B_{X^*}$, where C denotes the weak* closed convex hull of $\partial_F f(B(\overline{y}, \varepsilon))$. Since εB_{X^*} is weak* compact, the set $C + \varepsilon B_{X^*}$ is weak* closed and convex. Since $\inf\{\langle \varepsilon u^*, v \rangle : u^* \in B_{X^*}\} = -\varepsilon \|v\|$, the Hahn–Banach separation theorem yields some $\alpha > 0$, $v \in X$ with norm 1 such that

$$\langle \overline{y}^*, v \rangle + \alpha < \inf\{\langle y^*, v \rangle : y^* \in C\} - \varepsilon. \tag{4.59}$$

However, by definition of $\partial_F(-f)(\bar{y})$, one can find $\delta \in (0, \varepsilon/2)$ such that

$$\forall y \in B(\bar{y}, 2\delta), \qquad \langle -\bar{y}^*, y - \bar{y} \rangle - \varepsilon \|y - \bar{y}\| \leq f(\bar{y}) - f(y).$$

Applying Theorem 4.89 to $\bar{x} := \bar{y} + \delta v$, $\bar{y}$, and $r := f(\bar{x}) + \langle -\bar{y}^*, \delta v \rangle - \varepsilon \delta \leq f(\bar{y})$, one can find $u \in [\bar{x}, \bar{y}]$ and sequences $(u_n) \to u$, (u_n^*) such that $u_n^* \in \partial f(u_n)$, $\liminf_n \langle u_n^*, \bar{y} - \bar{x} \rangle \geq r - f(\bar{x})$. For n large enough one has $u_n \in B(\bar{y}, 2\delta) \subset B(\bar{y}, \varepsilon)$,

$$\langle u_n^*, -\delta v \rangle \geq r - f(\bar{x}) - \alpha \delta = \langle -\bar{y}^*, \delta v \rangle - \varepsilon \delta - \alpha \delta,$$

or $\langle u_n^*, v \rangle - \varepsilon \leq \langle \bar{y}^*, v \rangle + \alpha$. This is a contradiction to (4.59). $\square$

An estimate of the lower directional derivate can be deduced from Theorem 4.91.

Theorem 4.93 (Subbotin). *Let C be a compact convex subset of an F-smooth Banach space X and let $f : X \to \mathbb{R}_\infty$ be lower semicontinuous, finite at $\bar{x} \in X$. Let $s < \inf\{f^D(\bar{x}, v) : v \in C\}$. Then for every $\varepsilon > 0$ there exist some $x \in B(\bar{x}, \varepsilon, f)$ and $x^* \in \partial_F f(x)$ such that $\langle x^*, v \rangle > s$ for all $v \in C$.*

Proof. We first note that there exists some $\tau > 0$ such that for $t \in (0, \tau]$ we have

$$\inf\{f(\bar{x} + tv + t^2 u) - f(\bar{x}) : u \in B_X, v \in C\} > st + t^2.$$

Otherwise, there would be sequences $(t_n) \to 0_+$, (u_n) in B_X, (v_n) in C such that

$$(1/t_n)(f(\bar{x} + t_n v_n + t_n^2 u_n) - f(\bar{x})) \leq s + t_n.$$

Taking subsequences if necessary, one may assume that (v_n) converges to some $v \in C$, and then one would get $f^D(\bar{x}, v) \leq s$, a contradiction to the choice of s.

Taking a smaller τ if necessary, we may assume that $s\tau + \tau^2 < \varepsilon/2$ and that for all $w \in [0, \tau]C + \tau^2 B_X$ we have $w \in \varepsilon B_X$, $f(\bar{x} + w) > f(\bar{x}) - \varepsilon$, f being lower semicontinuous. Let us apply Theorem 4.91 with $Y := \bar{x} + \tau C$ and $r := f(\bar{x}) + (s\tau + \tau^2)^+ \leq f(\bar{x}) + \varepsilon/2 \leq \inf f(Y + \tau^2 B_X) \leq \wedge_Y f$. Let (u_n), (u_n^*), (s_n), (y_n) be the sequences of that theorem, in X, X^*, $[0, 1]$, and Y respectively, such that $u_n^* \in \partial f(u_n)$ for all n, $(\|(1 - s_n)\bar{x} + s_n y_n - u_n\|) \to 0$, $\limsup_n f(u_n) \leq r$, and $\liminf_n \langle u_n^*, y - \bar{x} \rangle \geq r - f(\bar{x}) \geq s\tau + \tau^2$ for all $y \in Y$. Taking $y = \bar{x} + \tau v$, with $v \in C$, for n large enough we get $\langle u_n^*, \tau v \rangle \geq s\tau$ and for $v_n \in C$ satisfying $\bar{x} + \tau v_n = y_n$,

$$\|\bar{x} + s_n \tau v_n - u_n\| = \|(1 - s_n)\bar{x} + s_n y_n - u_n\| < \tau^2,$$

hence $u_n \in \bar{x} + \varepsilon B_X$, $f(u_n) > f(\bar{x}) - \varepsilon$ and $f(u_n) \leq r + \varepsilon/2 < f(\bar{x}) + \varepsilon$. Thus, we can take $x := u_n$, $x^* := u_n^*$. $\square$

The following consequence shows that Fréchet and Hadamard subdifferentials are intimately related in F-smooth Banach spaces.

Corollary 4.94. *Let X be an F-smooth Banach space and let $f : X \to \mathbb{R}_\infty$ be finite at $\bar{x} \in X$ and lower semicontinuous. Then for every $\varepsilon > 0$ one has*

$$\partial_D f(\bar{x}) \subset \overline{\mathrm{co}}^*(\partial_F f(B(\bar{x}, \varepsilon))).$$

Proof. Let $\varepsilon > 0$ be given. Suppose the announced inclusion does not hold. Let $\bar{x}^* \in \partial_D f(\bar{x}) \setminus \overline{\mathrm{co}}^*(\partial_F f(B(\bar{x}, \varepsilon)))$. Applying the Hahn–Banach theorem, we can find $u \in S_X$ and $s \in \mathbb{R}$ such that

$$\langle \bar{x}^*, u \rangle > s > \sup\{\langle x^*, u \rangle : x^* \in \partial_F f(B(\bar{x}, \varepsilon))\}.$$

Since $f^D(\bar{x}, u) \geq \langle \bar{x}^*, u \rangle > s$, we get a contradiction to the conclusion of the preceding theorem in which we take $C := \{u\}$. $\qquad\square$

Let us give some other consequences of the mean value theorem. The following theorem generalizes a criterion for Lipschitzian behavior.

Theorem 4.95. *Let $f : W \to \mathbb{R}$ be a lower semicontinuous function on an open convex subset W of a smooth Banach space X. Then f is Lipschitzian with rate r if and only if for all $x \in W$ and $x^* \in \partial f(x)$ one has $\|x^*\| \leq r$.*

Proof. Necessity was given in Corollary 4.5. Let us prove sufficiency. Given $\bar{x}, \bar{y} \in W$, Theorem 4.89 yields $u \in [\bar{x}, \bar{y}]$ and sequences $(u_n) \to u$, (u_n^*) such that $f(\bar{y}) - f(\bar{x}) \leq \liminf_n \langle u_n^*, \bar{y} - \bar{x} \rangle \leq r \|\bar{x} - \bar{y}\|$. Exchanging the roles of $\bar{x}$ and $\bar{y}$, one gets $|f(\bar{x}) - f(\bar{y})| \leq r \|\bar{x} - \bar{y}\|$. $\qquad\square$

A connectedness argument yields the following consequence.

Corollary 4.96. *Let $f : W \to \mathbb{R}$ be a lower semicontinuous function on an open connected subset W of a smooth Banach space X. If $\partial f(x) \subset \{0\}$ for all $x \in W$, then f is constant on W.*

Now let us consider some order properties. We use the fact that given a closed convex cone P in a Banach space X we can define a preorder on X by setting $x \leq x'$ if $x' - x \in P$. We say that a map $f : X \to Y$ with values in another preordered Banach space (or in $\overline{\mathbb{R}}$) is *antitone* (resp. *homotone*) if $f(x') \leq f(x)$ (resp. $f(x) \leq f(x')$) whenever $x \leq x'$.

Theorem 4.97. *Let $f : X \to \mathbb{R}_\infty$ be lower semicontinuous on a smooth Banach space X preordered by a closed convex cone P. Suppose that one has $\partial f(x) \subset P^0$ (resp. $\partial f(x) \subset -P^0$) for all $x \in X$. Then f is antitone (resp. homotone).*

Proof. Suppose, to the contrary, that there exist $x, y \in X$ satisfying $x \leq y$ and $f(x) < f(y)$. Let $r \in (f(x), f(y))$. Then Theorem 4.89 yields u near $[x, y]$ and $u^* \in \partial f(u)$ such that $0 < r - f(x) \leq \langle u^*, y - x \rangle$. This is a contradiction to $y - x \in P$, $u^* \in \partial f(u) \subset P^0$. Changing P into $-P$, one obtains a criterion for f to be homotone. $\qquad\square$

The special case obtained by taking $P := \mathbb{R}_-$ (resp. $P := \mathbb{R}_+$) will be useful.

Corollary 4.98. *Let* $f : \mathbb{R} \to \mathbb{R}_\infty$ *be lower semicontinuous and such that* $\partial_F f(x) \subset$ $\mathbb{R}_+$ *(resp.* $\partial_F f(x) \subset \mathbb{R}_-$*) for all* $x \in \mathbb{R}$*. Then* f *is nondecreasing (resp. nonincreasing).*

4.5 Soft Functions

Although one can exhibit wildly nonsmooth functions, in many practical cases the nonsmoothness is tractable. Such a fact leads us to point out a class of functions for which the preceding approximate calculus rules turn out to be exact.

Definition 4.99. A lower semicontinuous function $f : X \to \mathbb{R}_\infty$ is said to be F-*soft* (resp. D-*soft*, resp. H-*soft*) at $\bar{x}$ if $f(\bar{x}) < +\infty$ and if every weak* cluster point of a bounded sequence (x_n^*) of X^* such that there exists a sequence $(x_n) \to_f \bar{x}$ with $x_n^* \in \partial_F f(x_n)$ (resp. $x_n^* \in \partial_D f(x_n)$, resp. $x_n^* \in \partial_H f(x_n)$) for all $n \in \mathbb{N}$ belongs to $\partial_F f(\bar{x})$ (resp. $\partial_D f(\bar{x})$, resp. $\partial_H f(\bar{x})$).

A function is F-soft (resp. D-soft, resp. H-soft) on a subset S of X if it is F-soft (resp. D-soft, resp. H-soft) at each point of S. In the sequel, we often write *soft* instead of F-soft or H-soft, according to the smoothness of X.

With this notion, a link between the three subdifferentials can be pointed out.

Proposition 4.100. *Let* X *be a* ∂_F-*subdifferentiability space and let* $f : X \to \mathbb{R}_\infty$ *be F-soft (resp. H-soft) at* $\bar{x}$ *and Lipschitzian around* $\bar{x}$*. Then* $\partial_D f(\bar{x}) = \mathrm{cl}^*(\partial_F f(\bar{x}))$ *(resp.* $\partial_D f(\bar{x}) = \mathrm{cl}^*(\partial_H f(\bar{x}))$*)*

Proof. Since the set $\partial_D f(\bar{x})$ is weak* closed and contains $\partial_F f(\bar{x})$ and $\partial_H f(\bar{x})$, the inclusions $\mathrm{cl}^*(\partial_F f(\bar{x})) \subset \mathrm{cl}^*(\partial_H f(\bar{x})) \subset \partial_D f(\bar{x})$ are always valid. Let $\bar{x}^* \in \partial_D f(\bar{x})$ and let V be a weak* closed neighborhood of 0 in X^*. Corollary 4.84 yields a sequence $((x_n, x_n^*))$ in the graph of $\partial_F f$ such that $(x_n) \to_f \bar{x}$ and $x_n^* - \bar{x}^* \in V$. The sequence (x_n^*) being bounded has a weak* cluster point x^* that belongs to $\partial_F f(\bar{x})$, since f is F-soft at $\bar{x}$. Then $x^* - \bar{x}^* \in V$. Thus $\bar{x}^* \in \mathrm{cl}^*(\partial_F f(\bar{x}))$. The case of H-softness is similar.   □

It can be shown that important classes of functions are soft, but we just give simple examples.

Example 4.1. If f is of class C^1 at $\bar{x}$, then f is F-soft at $\bar{x}$. If f is of class D^1 at $\bar{x}$, then f is H-soft at $\bar{x}$, since $(f'(x_n)) \to f'(\bar{x})$ for the weak* topology whenever $(x_n) \to \bar{x}$.

Example 4.2. If f is convex, then f is D-soft and F-soft on its domain. Here we use the fact that $\partial_D f$ and $\partial_F f$ coincide with the subdifferential ∂f of convex analysis, so that when $(x_n) \to_f x$ and x^* is a weak* cluster point of a bounded sequence (x_n^*) satisfying $x_n^* \in \partial f(x_n)$ for all n, then for all $w \in X$, one has

$$f(w) \geq \liminf_n (f(x_n) + \langle x_n^*, w - x_n \rangle) \geq f(x) + \langle x^*, w - x \rangle.$$

Example 4.3. The nonconvex function f on $\mathbb{R}$ given by $f(x) = |x| - x^2$ is soft on $\mathbb{R}$.

This last example can be generalized using the following stability properties. They provide many more examples; in particular, sums of convex lower semicontinuous functions with smooth functions are soft.

Theorem 4.101. *Let X and Y be smooth Banach spaces, let $g : X \to Y$ be smooth around $\bar{x} \in X$, and let $f \in \mathscr{F}(X)$ be soft at $\bar{x}$, $h \in \mathscr{F}(Y)$ soft at $\bar{y} := g(\bar{x})$ and Lipschitzian around $\bar{y}$. Then $k := f + h \circ g$ is soft at $\bar{x}$ and*

$$\partial(f + h \circ g)(\bar{x}) = \partial f(\bar{x}) + \partial h(\bar{y}) \circ g'(\bar{x}).$$

Proof. We just treat the case $\partial = \partial_F$. Let $((w_n, w_n^*))$ be a sequence in the graph of ∂k such that $(w_n) \to_k \bar{x}$ and (w_n^*) is bounded and has a weak* limit point $\overline{w}^* \in X^*$. Given a sequence $(\varepsilon_n) \to 0_+$, there exist some $(u_n, x_n, y_n) \in B(w_n, \varepsilon_n) \times B(w_n, \varepsilon_n, f) \times B(g(w_n), \varepsilon_n)$, $x_n^* \in \partial f(x_n)$, $y_n^* \in \partial h(y_n)$, $v_n^* \in B(y_n^*, \varepsilon_n)$ such that $x_n^* + v_n^* \circ g'(u_n) - w_n^* \in \varepsilon_n B_{X^*}$. Setting $c := \sup_n \|g'(u_n)\|$, we may suppose $c < +\infty$ and $z_n^* := x_n^* + y_n^* \circ g'(u_n) - w_n^* \in \alpha_n B_{X^*}$ for $\alpha_n := \varepsilon_n(c + 1)$. Since (y_n^*) is bounded, we can find $\bar{y}^* \in Y^*$ such that $(\overline{w}^*, \bar{y}^*)$ is a weak* cluster point of $((w_n^*, y_n^*))$. Since $(h(y_n)) \to h(\bar{y})$ and h is soft at $\bar{y}$, we have $\bar{y}^* \in \partial h(\bar{y})$. Then $(x_n^*) = (w_n^* - y_n^* \circ g'(u_n) + z_n^*)$ has $\bar{x}^* := \overline{w}^* - \bar{y}^* \circ g'(\bar{x})$ as a weak* cluster point and since f is soft at $\bar{x}$ and $(f(x_n)) \to f(\bar{x})$ (as $(k(w_n)) \to k(\bar{x})$), we get $\bar{x}^* \in \partial f(\bar{x})$. Therefore $\overline{w}^* = \bar{x}^* + \bar{y}^* \circ g'(\bar{x}) \in \partial f(\bar{x}) + \partial h(\bar{y}) \circ g'(\bar{x}) \subset \partial k(\bar{x})$, k is soft at $\bar{x}$, and the inclusion $\partial f(\bar{x}) + \partial h(\bar{y}) \circ g'(\bar{x}) \subset \partial k(\bar{x})$ is an equality. $\square$

Proposition 4.102. *(a) If f is a separable function on $X := X_1 \times \cdots \times X_k$, i.e., if $f(x) := f_1(x_1) + \cdots + f_k(x_k)$ for $x := (x_1, \ldots, x_k)$, where $f_1, \ldots, f_k$ are soft functions, then f is soft.*
(b) Let X be a smooth Banach space and let $f_1, \ldots, f_k$ be lower semicontinuous functions on X that are soft at $\bar{x} \in X$, $f_2, \ldots, f_k$ being Lipschitzian around $\bar{x}$. Then $f := f_1 + \cdots + f_k$ is soft at $\bar{x}$ and

$$\partial(f_1 + \cdots + f_k)(\bar{x}) = \partial f_1(\bar{x}) + \cdots + \partial f_k(\bar{x}).$$

Proof. (a) The assertion stems from the relation $\partial f(x) = \partial f_1(x_1) \times \cdots \times \partial f(x_k)$.
(b) The result stems from Theorem 4.101 using the diagonal map $g : X \to X^{k-1}$ and h defined by $h(y_2, \ldots, y_k) := f_2(y_2) + \cdots + f_k(y_k)$, applying assertion (a) to h. $\square$

Theorem 4.103. *Let $(f_1, \ldots, f_k)$ be a family of continuous functions on a smooth Banach space X. Let $f := \max(f_1, \ldots, f_k)$ and let $\bar{x} \in X$, $I := \{i \in \mathbb{N}_k : f_i(\bar{x}) = f(\bar{x})\}$. If f_i is soft at $\bar{x}$ and Lipschitzian around $\bar{x} \in X$ for all $i \in I$, then f is soft at $\bar{x}$ and*

$$\partial f(\bar{x}) = \mathrm{co}\left(\bigcup_{i \in I} \partial f_i(\bar{x})\right). \tag{4.60}$$

Proof. By continuity, we have $f = \max_{i \in I} f_i$ around $\bar{x}$, so that we may suppose $I = \mathbb{N}_k$. By Proposition 1.129, the family $(f_i)_{i \in I}$ satisfies (4.39) at each point of a neighborhood of $\bar{x}$. By Theorem 4.76, given sequences $(\varepsilon_n) \to 0_+$, $(w_n) \to \bar{x}$ and a weak* cluster point w^* of a bounded sequence (w_n^*) such that $w_n^* \in \partial_F f(w_n)$ for all $n \in \mathbb{N}$, for all $i \in I$ we can find $x_{i,n} \in B(w_n, \varepsilon_n)$, $x_{i,n}^* \in \partial f_i(x_{i,n})$, and $t_{i,n} \in \mathbb{R}_+$ such that

$$\left| t_{1,n} + \cdots + t_{k,n} - 1 \right| \leq \varepsilon_n, \qquad \left\| t_{1,n} x_{1,n}^* + \cdots + t_{k,n} x_{k,n}^* - w_n^* \right\| \leq \varepsilon_n.$$

Since $[0,1]^k \times B_{X^*}^{k+1}$ is compact (for the weak* topology), using subnets, we can find a cluster point (t_i, x_i^*, z^*) of the sequence $((t_{i,n}), (x_{i,n}^*), w_n^*)_n$ with $z^* = w^*$. Then $t_1 + \cdots + t_k = 1$, $w^* = t_1 x_1^* + \cdots + t_k x_k^*$, and since f_i is soft, $x_i^* \in \partial f_i(\bar{x})$. Thus w^* is in the right-hand-side C of (4.60). Since by Proposition 4.59 and convexity of $\partial f(\bar{x})$, C is contained in $\partial f(\bar{x})$, the softness of f and equality (4.60) are proved in the Fréchet case. The Hadamard case is left as an exercise. □

The following version of the mean value theorem is close to the classical statement.

Corollary 4.104. *Suppose f is lower semicontinuous on X and locally Lipschitzian on a segment $[\bar{x}, \bar{y}]$ of X and soft on it. Then there exist $u \in [\bar{x}, \bar{y}]$ and $u^* \in \partial f(u)$ such that $f(u) \leq \max(f(\bar{x}), f(\bar{y}))$ and*

$$\langle u^*, \bar{y} - \bar{x} \rangle \geq f(\bar{y}) - f(\bar{x}).$$

Proof. Let $((u_n, u_n^*))$ be the sequence of ∂f provided by Theorem 4.89. Since (u_n) converges to some point $u \in [\bar{x}, \bar{y}]$, the sequence (u_n^*) is bounded. Since f is soft at u, all of the weak* limit points of (u_n^*) belongs to $\partial f(u)$. Passing to the limit in (4.48), we get the result. □

Exercises

1. A function $f : X \to \mathbb{R}_\infty := \mathbb{R} \cup \{\infty\}$ on a normed space X is said to be *approximately convex at* $\bar{x} \in \mathrm{dom} f$ if for every $\varepsilon > 0$ there exists $\delta > 0$ such that for all $x, y \in B(\bar{x}, \delta), t \in [0,1]$ one has

$$f(tx + (1-t)y) \leq tf(x) + (1-t)f(y) + \varepsilon t(1-t) \|x - y\|.$$

(a) Give examples of classes of functions satisfying such a property.
(b) Show that in a smooth Banach space, f is soft at $\bar{x}$ whenever it is approximately convex at $\bar{x}$.

2. Let us say that a closed subset S of X is *soft* at $\bar{x} \in S$ if its indicator function ι_S is soft at $\bar{x}$. Show that if the distance function d_S to S is F-soft at $\bar{x}$, then S is soft at $\bar{x}$.

3. Prove that if X is F-smooth, and if a closed subset S of X is soft at $\bar{x} \in S$, then the distance function d_S to S of X is soft at $\bar{x}$. [Hint: Use the approximate projection theorem].

4. **(a)** Describe analytically the (filled) corolla C of a flower with n petals (usually with $n = 3, 4, 5, 6, 12$). Observe that the same solid can serve to describe the cupola of the Duomo (Cathedral) of Florence built by Brunelleschi in 1420–1434. [Hint: Let $E = B \cap (r_n B - e_1)$, where B is the unit ball of $\mathbb{R}^2$ for the Euclidean metric, $e_1 := (1, 0)$, $r_n := (2\cos(\pi/n) + 2)^{1/2}$. For $k = 1, \ldots, n$, let R_k be the rotation of angle $2k\pi/n$, let $E_k := R_k(E)$, and let the festooned disk F of $\mathbb{R}^2$ be defined as the intersection of the family $(E_k)_{1 \le k \le n}$. Its boundary is composed of n arcs of circles. Then set $C := \{(w, z) \in F \times \mathbb{R} : 0 \le z \le h - h\|w\|^2\}$.]
(b) Prove that F is a soft subset of the plane and that C is a soft subset of $\mathbb{R}^3$.

4.6 Calculus Rules in Asplund Spaces

The forthcoming separable reduction theorem is an important result of nonsmooth analysis. It enables one to pass from fuzzy sum rules in spaces with smooth norms to fuzzy sum rules in general Asplund spaces. Asplund spaces form the appropriate setting for such approximate rules and for extremal principles. The proof of that result is rather sophisticated and long and can be omitted on a first reading. It relies on a primal characterization of the Fréchet subdifferentiability of a function at some point. The latter stems from the characterization of subdifferentiability of a convex function by calmness given in Lemma 3.29.

We also need the following extension result. In the sequel, given a function f on X and a subspace W of X, we denote by $f|_W$ the restriction of f to W.

Lemma 4.105. *Let W be a linear subspace of a normed space X and let $g : X \to \mathbb{R}$ be a convex continuous function. Given $z \in W$ and $z_W^* \in \partial(g|_W)(z)$, there exists some $z^* \in \partial g(z)$ such that $z^*|_W = z_W^*$.*

Proof. By the Hahn–Banach theorem, there exists some $y^* \in X^*$ that extends z_W^*. Let ι_W be the indicator function of W. Then we clearly have $y^* \in \partial(g + \iota_W)(z)$. Now, the definition of the Moreau–Rockafellar subdifferential shows that $\partial \iota_W(z) = W^\perp := \{x^* \in X^* : x^*|_W = 0\}$, and since g is convex continuous,

$$\partial(g + \iota_W)(z) = \partial g(z) + \partial \iota_W(z) = \partial g(z) + W^\perp.$$

Thus we can find $z^* \in \partial g(z)$ satisfying $y^* - z^* \in W^\perp$, hence $z^*|_W = z_W^*$. $\qquad\square$

4.6.1 A Characterization of Fréchet Subdifferentiability

We devote the present subsubsection to a characterization of the nonemptiness of the Fréchet subdifferential $\partial_F f(\bar{x})$ of a (nonconvex, nonsmooth) function f at some point $\bar{x}$ of its domain, $\mathrm{dom} f$. We will use the following notion of approximate Fréchet subdifferential. Given an arbitrary function $f : X \to \overline{\mathbb{R}}$ and $\varepsilon > 0$, the ε-*Fréchet subdifferential* $\partial_F^\varepsilon f(\bar{x})$ of f at $\bar{x} \in f^{-1}(\mathbb{R})$ is the set

$$\partial_F^\varepsilon f(x) := \left\{ x^* \in X^* : \liminf_{\|w\| \to 0_+} \frac{f(x+w) - f(x) - \langle x^*, w \rangle}{\|w\|} \geq -\varepsilon \right\}.$$

Equivalently,

$$\partial_F^\varepsilon f(\bar{x}) = \bigcap_{\eta > \varepsilon} \partial^\eta f(\bar{x}),$$

where $x^* \in X^*$ is in $\partial^\eta f(\bar{x})$ iff there exists some $\rho > 0$ such that

$$\forall x \in \rho B_X, \qquad f(\bar{x}+x) - f(\bar{x}) - \langle \bar{x}^*, x \rangle \geq -\eta \|x\|. \tag{4.61}$$

Clearly one has $\partial_F f(\bar{x}) = \partial_F^0 f(\bar{x})$. It is often simpler to use $\partial^\varepsilon f(\bar{x})$ rather than $\partial_F^\varepsilon f(\bar{x})$. In particular, when f is convex, one has $\partial^\varepsilon f(\bar{x}) = \partial(f + \varepsilon \| \cdot - \bar{x}\|)(\bar{x})$.

One can give a characterization of the nonemptiness of the sets $\partial^\varepsilon f(\bar{x})$ and $\partial_F f(\bar{x})$ that parallels the one we gave in the convex case. For such a purpose, given $\bar{x} \in X$, a function $f : X \to \overline{\mathbb{R}}$ finite at $\bar{x}$, and $\varepsilon, \rho > 0$, denoting by $B := B_X$ the closed unit ball of X, we introduce the function $f^{\varepsilon,\rho}$ given by

$$f^{\varepsilon,\rho}(x) := f(\bar{x}+x) - f(\bar{x}) + \varepsilon \|x\| \quad \text{for } x \in \rho B, \qquad f^{\varepsilon,\rho}(x) := +\infty \text{ for } x \in X \setminus \rho B.$$

Lemma 4.106. *Given an arbitrary function $f : X \to \mathbb{R}_\infty$, $\bar{x} \in \mathrm{dom} f$, and $\varepsilon > 0$, the simplified ε-subdifferential $\partial^\varepsilon f(\bar{x})$ of f at $\bar{x}$ contains some element of norm at most $c \in \mathbb{R}_+$ if and only if there exists $\rho > 0$ such that $\mathrm{co}(f^{\varepsilon,\rho})(\cdot) \geq -c \| \cdot \|$.*

The relation $\mathrm{co}(f^{\varepsilon,\rho})(\cdot) \geq -c \| \cdot \|$ can be rephrased more explicitly as follows: for all $m \geq 1$, $x_1, \ldots, x_m \in \rho B$, $(t_1, \ldots, t_m) \in \Delta_m := \{(t_1, \ldots, t_m) \in \mathbb{R}_+^m : t_1 + \cdots + t_m = 1\}$,

$$\sum_{i=1}^m t_i (f(\bar{x}+x_i) + \varepsilon \|x_i\|) \geq f(\bar{x}) - c \left\| \sum_{i=1}^m t_i x_i \right\|. \tag{4.62}$$

Proof. Let $\bar{x}^* \in \partial^\varepsilon f(\bar{x})$ and let $\rho > 0$ be as in the definition of this set. Then for $m \geq 1$ and $(t_1, \ldots, t_m) \in \Delta_m$, $x_1, \ldots, x_m \in \rho B$ we have

$$f(\bar{x}+x_i) + \varepsilon \|x_i\| \geq f(\bar{x}) + \langle \bar{x}^*, x_i \rangle, \quad i \in \mathbb{N} := \{1, \ldots, m\}.$$

Multiplying both sides by t_i and summing, we obtain (4.62) by taking $c := \|\bar{x}^*\|$.

Conversely, suppose $\mathrm{co}(f^{\varepsilon,\rho})(\cdot) \geq -c\|\cdot\|$ for some $\rho > 0$, $c \geq 0$. Then, since $-c\|0\| \leq \mathrm{co}(f^{\varepsilon,\rho})(0) \leq f^{\varepsilon,\rho}(0) = 0$, we have $\mathrm{co}(f^{\varepsilon,\rho})(0) = 0$ and $\mathrm{co}(f^{\varepsilon,\rho})(x) \geq \mathrm{co}(f^{\varepsilon,\rho})(0) - c\|x\|$. Thus, Lemma 3.29 yields some $\overline{x}^* \in \partial(\mathrm{co}(f^{\varepsilon,\rho}))(0) \cap cB_{X^*}$, so that for $x \in \rho B$, we get

$$f^{\varepsilon,\rho}(x) \geq \mathrm{co}(f^{\varepsilon,\rho})(x) \geq \mathrm{co}(f^{\varepsilon,\rho})(0) + \langle \overline{x}^*, x \rangle = \langle \overline{x}^*, x \rangle.$$

That ensures that $\overline{x}^* \in \partial^{\varepsilon} f(\overline{x})$. □

Now let us characterize the nonemptiness of the set $\partial_F f(\overline{x})$. In the sequel, (ε_n) is a fixed sequence of $(0,1)$ with limit 0.

Lemma 4.107. *Given $c \in \mathbb{R}_+$, $(\varepsilon_n) \to 0_+$, and an arbitrary function $f : X \to \overline{\mathbb{R}}$ finite at $\overline{x}$, one has $\partial_F f(\overline{x}) \cap cB_{X^*} \neq \varnothing$ if and only if there exists a sequence (ρ_n) of positive numbers such that, for $f^{\varepsilon,\rho}$ given as in the preceding lemma, one has*

$$\forall x \in X, \qquad (\mathrm{co}(\inf_n f^{\varepsilon_n,\rho_n}))(x) \geq -c\|x\|. \tag{4.63}$$

It can be added that $\overline{x}^* \in \partial_F f(\overline{x}) \cap cB_{X^*}$ if and only if there exists a sequence (ρ_n) of positive numbers such that $\mathrm{co}(\inf_n f^{\varepsilon_n,\rho_n})(0) = 0$ and $\overline{x}^* \in \partial_F \mathrm{co}(\inf_n f^{\varepsilon_n,\rho_n})(0)$. However, our aim is to obtain a condition that involves just an estimate about this function and no element of the dual space.

Proof. Let $\overline{x}^* \in \partial_F f(\overline{x}) \cap cB_{X^*}$. Let $\rho_n > 0$ be such that

$$\forall x \in \rho_n B, \quad f(\overline{x} + x) - f(\overline{x}) - \langle \overline{x}^*, x \rangle \geq -\varepsilon_n \|x\|$$

and let $g := \mathrm{co}(\inf_n f^{\varepsilon_n,\rho_n})$. Then for all n and $x \in X$, one has $f^{\varepsilon_n,\rho_n}(x) \geq \langle \overline{x}^*, x \rangle$, hence $g(x) \geq \langle \overline{x}^*, x \rangle$, in particular $g(0) \geq 0$, and in fact, $g(0) = 0$, since $f^{\varepsilon_n,\rho_n}(0) = 0$ for all n. Thus $\overline{x}^* \in \partial g(0)$. Using Lemma 3.29, we get (4.63). Conversely, this last relation ensures that $\partial g(0) \cap cB_{X^*} \neq \varnothing$. Now, for every $\overline{x}^* \in \partial g(0)$ we have $\overline{x}^* \in \partial_F f(\overline{x})$, since for all $n \in \mathbb{N}$ and all $x \in X$ we have

$$f^{\varepsilon_n,\rho_n}(x) \geq g(x) \geq \langle \overline{x}^*, x \rangle,$$

hence $\overline{x}^* \in \partial^{\varepsilon_n} f(\overline{x})$ (since $f^{\varepsilon_n,\rho_n}(0) = 0$). □

Applying Lemmas 3.29 and 1.54 to $h_n := f^{\varepsilon_n,\rho_n}$, we see that relation (4.63) holds if and only if for all $p \in \mathbb{N} \setminus \{0\}$ one has

$$\forall x \in X, \qquad \mathrm{co}(h_1, \ldots, h_p)(x) \geq -c\|x\|. \tag{4.64}$$

4.6.2 Separable Reduction

The following striking result is of independent interest, but it is not the final aim of this section.

Theorem 4.108. *Let $f : X \to \mathbb{R}_\infty$ be a lower semicontinuous function on a Banach space X and let $c \in \mathbb{R}_+$. Then for every separable subspace W_0 of X there exists a separable subspace W containing W_0 such that for all $w \in W$ the relation $\partial_F f(w) \cap cB_{X^*} \neq \varnothing$ holds if and only if $\partial_F(f|_W)(w) \cap cB_{W^*} \neq \varnothing$ holds.*

Proof. Without loss of generality, changing f into $c^{-1}f$ if necessary, we assume $c \leq 1$. Since f is lower semicontinuous, for all $x \in X$, we have

$$\eta(x) := \sup\{r \in \mathbb{R}_+ : \inf f(B(x,r)) > -\infty\} > 0.$$

Given a sequence $(\varepsilon_n) \to 0_+$ in $(0,1)$, we construct an increasing sequence $(W_k)_k$ of separable subspaces of X and a sequence $(A_k)_k$, where A_k is a dense countable subset of W_k. We start with W_0 and take for A_0 any dense countable subset of W_0.

Assume we have constructed (W_n, A_n) for $n = 0, \ldots, k$. Given $a \in X$, $k, m, p \in \mathbb{N} \setminus \{0\}$, $q > 0$, $\rho := (\rho_1, \ldots, \rho_p) \in (0, \eta(a)/2)^p$, $j \in J_{m,p} = (\mathbb{N}_p)^m$ (so that j includes the data of m and p), $t := (t_1, \ldots, t_m) \in \Delta_m$, let

$$X(a, \rho, j, q, t) := \left\{ x = (x_1, \ldots, x_m) \in \prod_{i=1}^{m} 2\rho_{j(i)} B, \; \|t_1 x_1 + \cdots + t_m x_m\| < q \right\} \quad (4.65)$$

and let $(w_1, \ldots, w_m) := w(a, \rho, j, k, q, t) \in X(a, \rho, j, q, t)$ be such that

$$\inf_{x \in X(a, \rho, j, q, t)} \left\{ \sum_{i=1}^{m} t_i(f(a + x_i) + \varepsilon_{j(i)} \|x_i\|) \right\} \geq \sum_{i=1}^{m} t_i(f(a + w_i) + \varepsilon_{j(i)} \|w_i\|) - \frac{1}{k}.$$

Let W_{k+1} be the closed linear space generated by W_k and the vectors $w_i(a, \rho, j, k, q, t)$ for $a \in A_k$, $m, p \in \mathbb{N} \setminus \{0\}$, $i \in \mathbb{N}_m$, $j \in J_{m,p} := \{j : \mathbb{N}_m \to \mathbb{N}_p\} = (\mathbb{N}_p)^m$, $q \in \mathbb{Q}$, $\rho \in (0, \eta(a)/2)^m \cap \mathbb{Q}^m$, $t := (t_1, \ldots, t_m) \in \Delta_m \cap \mathbb{Q}^m$. Clearly W_{k+1} is separable, so that we can take a dense countable subset A_{k+1} of W_{k+1} containing A_k.

Let $W := \mathrm{cl}(\bigcup_k W_k)$. Let $w \in W$ be such that $\partial_F(f|_W)(w) \cap cB_{W^*} \neq \varnothing$. Thus, there exists some decreasing sequence $(\rho_n)_n$ of rational numbers in $(0, \eta(w)/4) \cap (0,1)$ such that, setting

$$f_{n,w}^W(v) := f(w+v) - f(w) + \varepsilon_n \|v\| + \iota_{3\rho_n B}(v), \qquad v \in W,$$

$$f_{n,w}(x) := f(w+x) - f(w) + \varepsilon_n \|x\| + \iota_{\rho_n B}(x), \qquad x \in X,$$

where ι_{rB} denotes the indicator function of the ball $rB := rB_X$, one has

$$\forall p \geq 1, \forall v \in W, \qquad \mathrm{co}\left(f_{1,w}^W, \ldots, f_{p,w}^W\right)(v) \geq -c\|v\|. \quad (4.66)$$

Let us prove that for all $\alpha > 0$, we have

$$\forall p \geq 1, \forall x \in X, \qquad \mathrm{co}(f_{1,w}, \ldots, f_{p,w})(x) \geq -c\|x\| - \alpha. \quad (4.67)$$

This relation means that for all $m, p \in \mathbb{N} \setminus \{0\}$, all $j : \mathbb{N}_m \to \mathbb{N}_p$, all $t := (t_1, \ldots, t_m) \in \Delta_m$, and all $x_1, \ldots, x_m \in X$ with $x := t_1 x_1 + \cdots + t_m x_m$, we have

$$t_1 f_{j(1),w}(x_1) + \cdots + t_m f_{j(m),w}(x_m) \geq -c \|x\| - \alpha. \tag{4.68}$$

We may suppose that $t_1, \ldots, t_m$ are rational numbers. It also suffices to consider the case in which $x_i \in \rho_{j(i)} B$ for all $i \in \mathbb{N}_m$. Let q be a rational number satisfying $\|x\| < q < \|x\| + \alpha/5$, and let $\beta \in (0, \alpha/5)$ be such that $\beta < \eta(w)/2$, $\|x\| + \beta < q$. We take $k \geq \beta^{-1}$ with $d(w, W_k) < \min(\beta, \rho_p)$ and $a \in A_k$ such that $\|a - w\| \leq \min(\beta, \rho_p)$. Then $B(a, \eta(w)/2) \subset B(w, \eta(w))$, so that $\eta(a) \geq \eta(w)/2$. For $i \in \mathbb{N}_m$, setting $v_i := x_i + w - a$, we note that we have $\|v_i - x_i\| \leq \beta$, $\|v_i\| \leq \rho_{j(i)} + \rho_p \leq 2\rho_{j(i)}$, $\rho_i < \eta(w)/4 \leq \eta(a)/2$ and

$$\|t_1 v_1 + \cdots + t_m v_m\| \leq \|t_1 x_1 + \cdots + t_m x_m\| + \max_i \|v_i - x_i\| \leq \|x\| + \beta < q.$$

Thus $v_i \in X(a, \rho, j, q, t)$. The relation $w + x_i = a + v_i$ and the triangular inequality justify the first of the following string of inequalities, while the second one stems from the definition of w_i as an approximate minimizer on $X(a, \rho, j, q, t)$; the triangle inequality and the relations $a + w_i = w + z_i$, for $z_i := a - w + w_i$, explain the next ones:

$$\sum_{i=1}^{m} t_i(f(w + x_i) + \varepsilon_{j(i)} \|x_i\|) \geq \sum_{i=1}^{m} t_i(f(a + v_i) + \varepsilon_{j(i)} \|v_i\| - \varepsilon_{j(i)} \|x_i - v_i\|)$$

$$\geq \sum_{i=1}^{m} t_i(f(a + w_i) + \varepsilon_{j(i)} \|w_i\|) - 1/k - \beta$$

$$\geq \sum_{i=1}^{m} t_i(f(w + z_i) + \varepsilon_{j(i)}(\|a - w + w_i\| - \|a - w\|)) - 2\beta$$

$$\geq \sum_{i=1}^{m} t_i(f(w + z_i) + \varepsilon_{j(i)} \|z_i\|) - 3\beta.$$

Setting $z := \sum_{i=1}^{m} t_i z_i$, with $z_i := a - w + w_i \in W \cap 3\rho_{j(i)} B$, from (4.66), we get

$$\sum_{i=1}^{m} t_i f_{j(i),w}(x_i) \geq \sum_{i=1}^{m} t_i \left(f(w + z_i) - f(w) + \varepsilon_{j(i)} \|z_i\| \right) - 3\beta$$

$$= \sum_{i=1}^{m} t_i f_{j(i),w}^{W}(z_i) - 3\beta \geq -c \|z\| - 3\beta.$$

Now, since $z_i - w_i = a - w$ and $t_1 + \cdots + t_m = 1$, by (4.65) we have the estimate

$$\|z\| \leq \|t_1 w_i + \cdots + t_m w_m\| + \|a - w\| \leq q + \beta \leq \|x\| + \alpha/5 + \beta.$$

Since $c \leq 1$ and $\beta \leq \alpha/5$, relation (4.68) is established. Then for all $\alpha > 0$, (4.67) holds, so that (4.67) holds with $\alpha = 0$ and (4.63) is satisfied. $\square$

4.6.3   Application to Fuzzy Calculus

Using the previous results, one can prove the following crucial separable reduction property for sums.

Theorem 4.109. *Let W_0 be a separable subspace of a Banach space X, let $f \in \mathscr{F}(X)$, and let $g : X \to \mathbb{R}$ be convex continuous. Then there exists a separable subspace W of X containing W_0 such that for all $w,z \in W$ the relation $(\partial_F f(w) + \partial g(z)) \cap cB_{X^*} \neq \varnothing$ holds if and only if the relation $(\partial_F(f \mid_W)(w) + \partial(g \mid_W)(z)) \cap cB_{W^*} \neq \varnothing$ holds.*

Proof. Let us first consider the case that g is a continuous linear form. Then since $\partial_F(f + g)(w) = \partial_F f(w) + g$, with a similar relation for the restrictions to W, the result follows from Theorem 4.108 applied to $f + g$. An examination of its proof shows that the construction of W for f remains valid for $f + g$, since the same sequence (ρ_n) can be used for $f + g$ and for all $n, p \in \mathbb{N} \setminus \{0\}$ and $w \in W$,

$$(f + g)_{n,w}^W = f_{n,w}^W + g \mid_W, \quad (f + g)_{n,w} = f_{n,w} + g,$$

$$\mathrm{co}(f_{1,w} + g, \ldots, f_{p,w} + g) = \mathrm{co}(f_{1,w}, \ldots, f_{p,w}) + g,$$

and a similar relation with $\mathrm{co}(f_{1,w}^W + g \mid_W, \ldots, f_{p,w}^W + g \mid_W)$.

Now let us consider the general case. Given f and a separable subspace W_0 of X, let W be a separable subspace containing W_0 associated with f as in Theorem 4.108. Let $w, z \in W$ and let $w_W^* \in \partial_F(f \mid_W)(w)$, $z_W^* \in \partial(g \mid_W)(z)$ be such that $\|w_W^* + z_W^*\| \leq c$. Using Lemma 4.105, we can find some $z^* \in \partial g(z)$ such that $z^* \mid_W = z_W^*$. Then by the preceding special case, we can find $w^* \in \partial_F f(w)$ such that $\|w^* + z^*\| \leq c$, i.e., $w^* + z^* \in (\partial_F f(w) + z^*) \cap cB_{X^*} \subset (\partial_F f(w) + \partial g(z)) \cap cB_{X^*}$. $\square$

Recall that X is said to be a *∂_F-subdifferentiability space* if for every element f of the set $\mathscr{F}(X)$ of proper lower semicontinuous functions on X the set G of points $(w, f(w))$ such that $\partial_F f(w)$ is nonempty is dense in the graph of f. We say that X is a *reliable space* for ∂_F if whenever $f \in \mathscr{F}(X)$, $g : X \to \mathbb{R}$ is a convex continuous function, and $f + g$ attains its minimum on X at $\bar{x} \in X$, then for every $\varepsilon > 0$ there exist $w, z \in B(\bar{x}, \varepsilon)$ with $|f(w) - f(\bar{x})| \leq \varepsilon$ such that the relation $(\partial_F f(w) + \partial g(z)) \cap \varepsilon B_{X^*} \neq \varnothing$ holds. The space X is said to be *trustworthy* for ∂_F if the same holds when g is a Lipschitzian function and $\partial g(z)$ is replaced with $\partial_F g(z)$.

Theorem 4.110. *For a Banach space X, the following properties are equivalent:*

(a) X is an Asplund space
(b) For all $n \in \mathbb{N} \setminus \{0\}$, X^n is a reliable space for the Fréchet subdifferential ∂_F

(c) For all $n \in \mathbb{N} \setminus \{0\}$, X^n is trustworthy for the Fréchet subdifferential ∂_F
(d) X is a trustworthy space for the Fréchet subdifferential ∂_F
(e) X is a subdifferentiability space for the Fréchet subdifferential ∂_F

Proof. (a)$\Rightarrow$(b) Since X^n is an Asplund space by Corollary 3.101, it suffices to show that X is reliable whenever X is Asplund. Let $\varepsilon > 0$, $f \in \mathscr{F}(X)$, and a convex continuous function $g : X \to \mathbb{R}$ be given such that $f + g$ attains its minimum on X at $\overline{x} \in X$. Let $W_0 := \mathbb{R}\overline{x}$ and let W be a separable subspace of X containing W_0 associated to f as in Theorem 4.109. Since X is Asplund, the separable subspace W has a separable dual, hence a Fréchet smooth bump function. Then the fuzzy minimization rule shows that there exist $w, z \in B(\overline{x}, \varepsilon) \cap W$ such that $|f(w) - f(\overline{x})| \leq \varepsilon$ and $(\partial_F(f|_W)(w) + \partial(g|_W)(z)) \cap \varepsilon B_{W^*} \neq \varnothing$. Then we conclude from Theorem 4.109 that $(\partial_F f(w) + \partial g(z)) \cap \varepsilon B_{X^*} \neq \varnothing$.

In the next implications we may also suppose $n = 1$.
(b)$\Rightarrow$(c) Let $f \in \mathscr{F}(X)$, let $g : X \to \mathbb{R}$ be Lipschitzian with rate k such that $f + g$ attains its minimum on X at $\overline{x} \in X$. Then for all $(x, y) \in X^2$ one has

$$f(x) + g(y) + k\|x - y\| \geq f(x) + g(x) \geq f(\overline{x}) + g(\overline{x}).$$

Since X^2 is reliable, since h given by $h(x, y) := k\|x - y\|$ is convex continuous, and since $(x, y) \mapsto f(x) + g(y) + h(x, y)$ attains its minimum at $(\overline{x}, \overline{x})$, for all $\varepsilon > 0$ one can find $u \in B(0, \varepsilon)$, $w, z \in B(\overline{x}, \varepsilon, f)$, $w^* \in \partial_F f(w)$, $z^* \in \partial_F g(z)$, and $u^* \in B_{X^*}$ such that $\|(w^*, z^*) + k(u^*, -u^*)\| \leq \varepsilon$. It follows that $\|w^* + z^*\| \leq \|w^* + ku^*\| + \|z^* - ku^*\| \leq \varepsilon$ (if we endow X^2 with the sup norm and its dual with the sum norm).
(c)$\Rightarrow$(d) is obvious.
(d)$\Rightarrow$(e) The proof was given in Ekeland–Lebourg's theorem (Theorem 4.65).
(e)$\Rightarrow$(a) Let W be an open convex subset of X and let $f : W \to \mathbb{R}$ be a continuous convex function and let $\overline{x} \in W$, $\varepsilon > 0$ be given. Let $g := -f$. Since X is a subdifferentiability space for ∂_F, there exists some $x \in B(\overline{x}, \varepsilon) \cap W$ such that $\partial_F g(x)$ is nonempty. Then f is Fréchet differentiable at x. Thus f is densely Fréchet differentiable and X is an Asplund space. $\square$

Corollary 4.111. *The fuzzy calculus rules given for ∂_F are valid in Asplund spaces.*

4.7 Applications

In this section we just exhibit some direct and short applications.

4.7.1 Subdifferentials of Value Functions

Optimal value functions of optimization problems depending on parameters are of excruciating importance in analysis and optimization. Distance functions are of this type, and many results of game theory and optimal control theory rely on their study. Moreover, they play an important role in bilevel programming and in the study of Hamilton–Jacobi equations. Value functions are seldom differentiable. Providing conditions ensuring differentiability or subdifferentiability is not an easy task. They are of two types: *marginal functions* and *performance functions* respectively obtained as

$$m(w) := \sup_{x \in X} F(w,x), \qquad p(w) := \inf_{x \in X} F(w,x), \qquad w \in W.$$

Since subdifferentials are one-sided concepts, these two types should be distinguished. Hereinafter, the decision variable x belongs to an arbitrary set X, the parameter variable w belongs to a normed space W, and $F : W \times X \to \overline{\mathbb{R}}$ is a function called the *perturbation* function. We consider a nominal point $u \in W$ at which p is supposed to be finite. We do not assume attainment, a hypothesis that would much simplify the question, but is not always satisfied.

Changing F into $-F$, one deduces subdifferentiability results for m from superdifferentiability results about p. Thus, in order to avoid ambiguities about the sets of solutions or approximate solutions, we limit our study to the performance function p. For $\alpha > 0$, $w \in W$, we set

$$S(w, \alpha) := \{x \in X : F(w,x) < \inf F(\{w\} \times X) + \alpha\}. \tag{4.69}$$

Here we extend the addition to $\overline{\mathbb{R}} \times (0, +\infty)$ by setting $r + \alpha := -1/\alpha$ for $r = -\infty$, $r + \alpha := +\infty$ for $r = +\infty$, so that $S(w, \alpha)$ is always nonempty. Moreover, the set $S(w)$ of minimizers of F_w satisfies $S(w) = \bigcap_{\alpha > 0} S(w, \alpha)$.

In order to get some results, we impose a control of the behavior of F on some approximate solution set. In fact, instead of controlling the functions $F_x := F(\cdot, x)$ for all $x \in S(w, \alpha)$, it would suffice to control these functions for x in a sufficiently representative subset of $S(w, \alpha)$. In order to give a precise meaning to this idea, we introduce the *minimizing grill* of F_w as the family

$$\mathscr{M}_w := \{M \subset X : \inf F(\{w\} \times M) = p(w)\} = \{M \subset X : \forall \alpha > 0 \; M \cap S(w, \alpha) \neq \varnothing\}.$$

Of course, every member of the family $\mathscr{A}_w := \{S(w, \alpha) : \alpha > 0\}$ of approximate solution sets of F_w is a member of $\mathscr{M}_w$, but $\mathscr{M}_w$ is a much larger family, so that several assumptions below are less stringent than assumptions formulated in terms of the family $\mathscr{A}_w$. Both families play a natural role in minimization problems: M belongs to $\mathscr{M}_w$ iff M contains a minimizing sequence of F_w. In making assumptions about a family $(F_x)_{x \in M}$, one is willing to take $M \in \mathscr{M}_u$ as small as possible. The best case occurs when the set $S(u)$ of minimizers of F_u is nonempty: then one can take for M a singleton $\{x\}$, where $x \in S(u)$, or any subset of $S(u)$. However, we endeavor to avoid the assumption that $S(u)$ is nonempty.

4.7.1.1 Subdifferentiability of Marginal Functions

As mentioned above, we reduce the subdifferentiability of the marginal function m to the superdifferentiability of p (changing F into $-F$). Recall that the (Fréchet) *superdifferential* at u of a function $h : W \to \overline{\mathbb{R}}$ finite at u is the set

$$\widetilde{\partial}_F h(u) := -\partial_F(-h)(u).$$

Obviously, h is (Fréchet) differentiable at u iff it is subdifferentiable and superdifferentiable at u. Then one has $\partial_F h(u) = \widetilde{\partial}_F h(u) = \{Dh(u)\}$.

Superdifferentiability of p is easy to obtain. When the set $S(u)$ of minimizers of $F_u := F(u, \cdot)$ is nonempty and when for some $x \in S(u)$ the function $F(\cdot, x)$ is superdifferentiable at u, the performance function is clearly superdifferentiable at u. The following proposition does not require such assumptions.

Proposition 4.112. *The following condition on $u^* \in W^*$ ensures that $u^* \in \widetilde{\partial}_F p(u)$:*

(p^+) *for every $\varepsilon > 0$ there exists $\delta > 0$ such that for all $v \in B(u, \delta)$, $\alpha > 0$, there are $x \in S(u, \alpha)$ and $w^* \in B(u^*, \varepsilon)$ satisfying*

$$F(v, x) \leq F(u, x) + \langle w^*, v - u \rangle + \varepsilon \|v - u\|. \tag{4.70}$$

Proof. Given $\varepsilon > 0$, let $\delta > 0$ be as in condition (p^+). Then for every $v \in B(u, \delta)$ and every $\alpha > 0$, we pick $x \in S(u, \alpha)$ and $w^* \in B(u^*, \varepsilon)$ such that (4.70) is satisfied. Then we deduce from (4.70) and from the inequalities $p(v) \leq F(v, x)$, $F(u, x) \leq p(u) + \alpha$, $\langle w^*, v - u \rangle \leq \langle u^*, v - u \rangle + \varepsilon \|v - u\|$ that

$$p(v) \leq p(u) + \alpha + \langle u^*, v - u \rangle + 2\varepsilon \|v - u\| \qquad \forall v \in B(u, \delta),$$
$$p(v) \leq p(u) + \langle u^*, v - u \rangle + 2\varepsilon \|v - u\| \qquad \forall v \in B(u, \delta),$$

since $\alpha > 0$ is arbitrarily small, and we get that $u^* \in \widetilde{\partial} p(u)$. □

Given some fixed $M \in \mathcal{M}_u$, for $\alpha > 0$ we introduce the subset D_α^+ of W^* by

$$D_\alpha^+ := D_{\alpha, M}^+ := \{w^* \in W^* : \exists x \in S(u, \alpha) \cap M, \ w^* \in \widetilde{\partial}_F F_x(u)\}.$$

Corollary 4.113. *Suppose the following conditions bearing on some $M \in \mathcal{M}_u$ hold:*

(a^+) *For all $x \in M$ the function F_x is superdifferentiable at u*
(b^+) $\limsup_{\alpha \to 0_+} D_\alpha^+$ *is nonempty*
(e^+) *The family $(F_x)_{x \in M}$ is eventually equi-superdifferentiable at u in the following sense: for every $\varepsilon > 0$ there exist $\alpha, \delta > 0$ such that for all $x \in M \cap S(u, \alpha)$, $w^* \in \widetilde{\partial}_F F_x(u)$ (4.70) holds*

Then p is superdifferentiable at u and $\limsup_{\alpha \to 0_+} D_\alpha^+ \subset \widetilde{\partial}_F p(u)$.

Condition (e$^+$) is obviously satisfied when F_x is concave for each $x \in M$. This condition is a weakened form of the following assumption (which can be called *equi-superdifferentiability* at u of the family $(F_x)_{x \in M}$):

(e$'^+$)   For every $\varepsilon > 0$ there exists $\delta > 0$ such that for all $x \in M$, $w^* \in \widetilde{\partial}_F F_x(u)$, (4.70) holds.

Proof. Given $\varepsilon > 0$, we take $\delta > 0$, $\alpha > 0$ as in condition (e$^+$). Then we use condition (b$^+$) to pick $u^* \in \limsup_{\alpha \to 0_+} D_\alpha^+$, so that there exist $w^* \in B(u^*, \varepsilon)$, $x \in M \cap S(u, \alpha)$ satisfying $w^* \in \widetilde{\partial} F_x(u)$ and (4.70); this entails condition (p$^+$).   □

Of course, usual differentiability and equi-differentiability can be substituted for their one-sided counterparts used in the preceding corollary. These assumptions are satisfied in the next example.

Example. Let $W = X^*$, $u = 0$ and let $\varphi : X \to \mathbb{R}_\infty$ be an arbitrary function such that $\inf_X \varphi$ is finite. Taking $F(w, x) = \varphi(x) - \langle w, x \rangle$, so that the Fenchel transform of φ is $\varphi^* = -p$, and observing that the family $(F_x)_{x \in X}$, being composed of affine continuous functions, is equi-differentiable at 0, we get that $\partial \varphi^*(0) = -\widetilde{\partial}_F p(0)$ contains $-\overline{co}^* (\limsup_{\alpha \to 0_+} D_\alpha^+)$, where $D_\alpha^+ := \{-x : \varphi(x) \leq \inf_X \varphi + \alpha\}$. In particular, the set S of minimizers of φ satisfies $-\overline{co}^* S \subset \partial \varphi^*(0)$, a well-known fact.

4.7.1.2   Subdifferentiability of Performance Functions

Simple examples show that F may be smooth while p is not subdifferentiable. Such a case occurs when $W = \mathbb{R}$, $X := [-1, 1]$, and $F(w, x) := wx$, so that $p(w) = -|w|$. Thus, we need assumptions about the behavior of F ensuring some stability.

Definition 4.114. The perturbation F is said to be *compliant* at $u \in W$ with respect to some subset M of X if for every $\alpha > 0$ there exist $V \in \mathcal{N}(u)$ and $\beta > 0$ such that

$$\forall v \in V, \qquad\qquad S(v, \beta) \cap M \subset S(u, \alpha).$$

It is said that F is *docile* at u with respect to some subset M of X if for every $\alpha > 0$ there exists $V \in \mathcal{N}(u)$ such that $S(u, \alpha) \cap M \in \mathcal{M}_v$ for all $v \in V$, or in other words,

$$\forall \alpha > 0, \ \exists V \in \mathcal{N}(u), \ \forall v \in V, \ \forall \beta > 0, \qquad S(v, \beta) \cap S(u, \alpha) \cap M \neq \varnothing.$$

When one can take $M = W$, we just speak of compliant (resp. docile) perturbations. Thus a compliant perturbation is docile. On the other hand, when the set $S(u)$ of minimizers of F_u is nonempty, taking for M any subset of $S(u)$, we get that F is compliant with respect to M, but not necessarily docile with respect to M. Before quoting a compliance criterion, let us present examples.

Example. Let X be a closed nonempty subset of a normed space W and let F be given by $F(w,x) := \|w - x\|$. For every $u \in W$ and every $\alpha, \beta, \rho > 0$ such that $\beta + 2\rho \leq \alpha$ one has $S(v, \beta) \subset S(u, \alpha)$ for all $v \in B(u, \rho)$. Thus F is compliant.

Example. More generally, if X is an arbitrary set and if for all $x \in X$ the function F_x is Lipschitzian with rate c, then F is compliant at each point u of W, since for all $\alpha, \beta, \rho > 0$ such that $\beta + 2c\rho \leq \alpha$ one has $S(v, \beta) \subset S(u, \alpha)$ for all $v \in B(u, \rho)$.

Proposition 4.115 ([352, 820]). *Suppose that for all $x \in X$ the function F_x is lower semicontinuous and bounded below on W. Suppose there exist some $\lambda > p(u)$, $c \in \mathbb{R}_+$, and $V \in \mathcal{N}(u)$ such that for all $v \in V$, $x \in [F_v \leq \lambda]$ there exists $\rho > 0$ for which*

$$\forall w \in B(v, \rho), \qquad F_x(w) \leq F_x(v) + c\|v - w\|. \tag{4.71}$$

Then F is compliant at u and p is Lipschitzian with rate c on some $U \in \mathcal{N}(u)$.

Another compliance criterion can be deduced from Proposition 4.115. It uses the set $\widetilde{\partial}_F^\varepsilon F_x(v) := -\partial_F^\varepsilon(-F_x)(v)$ of $v^* \in W^*$ such that for all $\varepsilon' > \varepsilon$ the function $w \mapsto F_x(w) - \langle v^*, w \rangle - \varepsilon' \|w - v\|$ attains a local maximum at v. Of course, in that criterion, $\widetilde{\partial}^\varepsilon F_x(v)$ can be replaced with $\widetilde{\partial}_F F_x(v)$, or when the derivative exists, with $DF_x(v)$.

Corollary 4.116. *Suppose that for all $x \in X$ the function F_x is lower semicontinuous and bounded below on W and there exist some $\varepsilon > 0$, $\lambda > p(u)$, $c > 0$, and $V \in \mathcal{N}(u)$ such that for all $v \in V$ and all $x \in [F_v \leq \lambda]$ the function F_x is finite at v and $\widetilde{\partial}_F^\varepsilon F_x(v) \cap B(0, c) \neq \varnothing$. Then F is compliant at u and p is Lipschitzian with rate $c + \varepsilon$ around u.*

Proof. For all $v \in V$, $x \in [F_v \leq \lambda]$, picking $v^* \in \widetilde{\partial}_F^\varepsilon F_x(v) \cap B(0, c)$, one can find $\rho > 0$ such that

$$\forall w \in B(v, \rho), \qquad F_x(w) \leq F_x(v) + \langle v^*, w - v \rangle + \varepsilon \|w - v\| \leq F_x(v) + (c + \varepsilon) \|v - w\|.$$

Thus Proposition 4.115 applies. $\qquad\qquad\qquad\qquad\qquad\qquad\qquad\qquad\qquad\qquad\qquad \square$

The following assumptions ensuring subdifferentiability of p are rather stringent and complex. In particular, assumption (b$^-$) is not satisfied for $W = X = \mathbb{R}$, $F(w,x) = |w - x|$, $u = 0$, although $p = 0$ is differentiable with derivative 0. However, it is satisfied when $F(w,x) = (w - x)^2$. Assumption (e$^-$) is a weakened form of the following *equi-subdifferentiability* condition (in which M is a subset of X):

(e$_M^-$) for every $\varepsilon > 0$ there exists $\delta > 0$ such that for all $x \in M$ and $w^* \in \partial_F F_x(u)$, one has

$$\forall v \in B(u, \delta), \quad F_x(v) - F_x(u) - \langle w^*, v - u \rangle \geq -\varepsilon \|v - u\|. \tag{4.72}$$

This condition is satisfied if M is a singleton $\{x\}$ and F_x is equi-subdifferentiable at u or if for all $x \in M$, the function F_x is convex. When F is compliant (or just docile) at u, a natural choice for M is $M = S(u, \theta)$ for some $\theta > 0$.

A versatile subdifferentiability result follows.

Proposition 4.117. *If the following conditions hold for some $u^* \in W^*$, $M \subset X$, then p is subdifferentiable at u and $u^* \in \partial_F p(u)$.*

(a^-) F_x *is subdifferentiable at u for all $x \in M$*
(b^-) *For all $\varepsilon > 0$, $\alpha > 0$ there exists some $V \in \mathcal{N}(u)$ such that for all $v \in V$ one can find some $M_v \in \mathcal{M}_v$ contained in $S(u, \alpha) \cap M$ for which $\partial_F F_x(u) \cap B(u^*, \varepsilon) \neq \varnothing$ for all $x \in M_v$*
(e^-) *The family $(F_x)_{x \in M}$ is eventually equi-subdifferentiable at u in the following sense: for every $\varepsilon > 0$ there exist $\alpha, \delta > 0$ such that for all $x \in S(u, \alpha) \cap M$, $w^* \in \partial_F F_x(u)$, (4.72) holds*

Proof. Given $\varepsilon > 0$, condition (e^-) yields some $\alpha, \delta > 0$ such that (4.72) holds for all $x \in M$, $w^* \in \partial_F F_x(u)$. Taking a smaller δ if necessary, we may assume $V = B(u, \delta)$ in (b^-). Then, for every $v \in V$ and every $x \in M_v$, we can find $w^* \in \partial_F F_x(u) \cap B(u^*, \varepsilon)$. Since $M_v \subset S(u, \alpha) \cap M$, using (4.72) and the inequality $|\langle u^* - w^*, v - u \rangle| \leq \varepsilon \|v - u\|$, we get

$$F_x(v) \geq F_x(u) + \langle u^*, v - u \rangle - 2\varepsilon \|v - u\|.$$

Now, by definition of $\mathcal{M}_v$, we have $\inf\{F_x(v) : x \in M_v\} = p(v)$. Therefore

$$p(v) \geq p(u) + \langle u^*, v - u \rangle - 2\varepsilon \|v - u\|.$$

Since $\varepsilon > 0$ is arbitrary, this shows that $u^* \in \partial_F p(u)$. $\square$

The following two consequences are simpler than the preceding statement. We use the notion of limit inferior of a family $(A(t))_{t \in T}$ of subsets of W^* parameterized by a set T as $e(t) \to 0$, where $e : T \to P$ is a map with values in a topological space P and $0 \in P$: $u^* \in \liminf_{e(t) \to 0} A(t)$ iff $d(u^*, A(t)) \to 0$ as $e(t) \to 0$. This notion is a variant of the concept of limit inferior when T is a subset of a topological space P and $0 \in P$, which correspond to the case that e is the canonical injection of T into P. Here we take $T := M$, $P = \mathbb{R}_+$, $e(x) := F(u, x) - p(u)$, and $A(x) := D^-(x) := \partial_F F_x(u)$. Then $e^{-1}([0, \beta)) = S(u, \beta) \cap M$ and $u^* \in \liminf_{e(x) \to 0} D^-(x)$ iff for every $\varepsilon > 0$ there exists some $\beta > 0$ such that for all $x \in S(u, \beta) \cap M$ one has $\partial_F F_x(u) \cap B(u^*, \varepsilon) \neq \varnothing$.

Proposition 4.118. *Suppose that for some $M \in \mathcal{M}_u$ the conditions (a^-), (e^-) of Proposition 4.117 hold and*

(d) *The perturbation F is docile with respect to M*
(d^-) $\liminf_{e(x) \to 0} D^-(x)$ *is nonempty*

Then p is subdifferentiable at u and $\liminf_{e(x) \to 0} D^-(x) \subset \partial_F p(u)$.

Proof. Given $u^* \in \liminf_{e(x)\to 0} D^-(x)$, let us prove that condition (b$^-$) of Proposition 4.117 is satisfied. Given $\alpha, \varepsilon > 0$, condition (d$^-$) yields some $\beta \in (0, \alpha)$ such that for all $x \in S(u, \beta) \cap M$ there exists some $w^* \in \partial_F F_x(u) \cap B(u^*, \varepsilon)$. Since F is docile with respect to M, one can find $V \in \mathscr{N}(u)$ such that $S(u, \beta) \cap M \in \mathscr{M}_v$ for all $v \in V$. Then $M_v := S(u, \beta) \cap M$ is the required set. $\qquad\square$

In the next statement, we replace the docility condition by a compliance requirement and we use the classical inner limit as $t := (v, \beta) \to (u, 0)$ of the family $(A(t))_{t \in T}$ of subsets of W^* given, for some fixed $M \subset X$ and $t \in T := W \times (0, +\infty) \subset P := W \times \mathbb{R}_+$, by

$$A(v, \beta) := \{v^* \in W^* : \exists x \in S(v, \beta) \cap M, \ v^* \in \partial_F F_x(u)\} = D^-(S(v, \beta) \cap M).$$

Thus, $u^* \in \liminf_{(v, \beta)\to(u,0)} A(v, \beta)$ iff $d(u^*, A(v, \beta)) \to 0$ as $(v, \beta) \to (u, 0)$ in T. Since $A(v, \beta) \subset A(v, \gamma)$ for $\beta < \gamma$, one has $u^* \in \liminf_{(v, \beta)\to(u,0_+)} A(v, \beta)$ iff

$$\forall \varepsilon > 0 \ \exists \delta > 0 : \forall v \in B(u, \delta), \ \forall \beta > 0, \quad A(v, \beta) \cap B(u^*, \varepsilon) \neq \varnothing.$$

Proposition 4.119. *Suppose that for some $M \subset X$ the conditions (a$^-$), (e$^-$) of Proposition 4.117 hold and*

(c) The perturbation F is compliant with respect to M
(c$^-$) The set $A := \liminf_{(v, \beta)\to(u,0_+)} A(v, \beta)$ is nonempty

Then p is subdifferentiable at u and $A \subset \partial_F p(u)$.

Proof. Given $u^* \in A$, let us prove that condition (b$^-$) of Proposition 4.117 is satisfied. Given $\alpha, \varepsilon > 0$, the compliance assumption yields some $\beta > 0$ and $V \in \mathscr{N}(u)$ such that $S(v, \beta) \cap M \subset S(u, \alpha)$ for all $v \in V$. By condition (c$^-$), for some $\gamma > 0$ and some $\delta > 0$ with $V' := B(u, \delta) \subset V$, for all $v \in V'$ and all $\beta' \in (0, \gamma)$ we can find $x \in S(v, \beta') \cap M$ and $w^* \in \partial_F F_x(u) \cap B(u^*, \varepsilon)$. Taking a sequence $(\beta_n) \to 0_+$ in $(0, \inf(\beta, \gamma))$, we get some $x_n \in S(v, \beta_n) \cap M$ and $w_n^* \in \partial_F F_{x_n}(u) \cap B(u^*, \varepsilon)$. Let $M_v := \{x_n : n \in \mathbb{N}\}$. Then $M_v \in \mathscr{M}_v$ is the required set for condition (b$^-$). $\qquad\square$

Example. Let X be a nonempty closed subset of a normed space W and let $u \in W \setminus X$ be such that u has a best approximation $z \in X$ and such that $(x_n) \to z$ whenever $x_n \in X$ and $(d(u, x_n)) \to d_X(u) := \inf_{x \in X} d(u, x)$. Then if the norm of W is Fréchet differentiable at $z - u$, with derivative $j(z - u)$, one has $j(z - u) \in \partial_F d_X(u)$. In fact, when the norm is Fréchet differentiable at $z - u$, by Proposition 3.32, the multifunction $\partial \|\cdot\|$ is lower semicontinuous at $z - u$. Thus, for $F(w, x) := \|w - x\|$, $M := W$, our well-posedness assumption ensures that $j(z - u) \in \liminf_{e(x)\to 0} D^-(x)$. Since F is convex continuous, assumptions (a$^-$) and (e$^-$) are satisfied with $e(x) := \|x - u\| - d_X(u)$. Moreover, F is compliant by Proposition 4.115. Let us note that since the conditions of Corollary 4.113 are satisfied with $M = \{z\}$, we get that $j(z - u) \in \partial_F d_X(u) \cap \widetilde{\partial}_F d_X(u)$, so that d_X is Fréchet differentiable at u. $\qquad\square$

4.7.1.3   Differentiability Properties

Gathering the previous results, we obtain a differentiability property.

Proposition 4.120. *Suppose the following conditions hold for some $u^* \in W^*$ and $M \in \mathcal{M}_u$:*

(a) For all $x \in M$, F_x is differentiable at u and $DF_x \to u^*$ as $F(u,x) \to_M p(u)$
(d) F is docile at u with respect to M
(e) The family $(F_x)_{x \in M}$ is eventually equi-differentiable at u in the sense that for all $\varepsilon > 0$ there exist $\alpha > 0$, $\delta > 0$ such that

$$\forall v \in B(u,\delta), x \in S(u,\alpha) \cap M, \quad |F(v,x) - F(u,x) - \langle DF_x(u), v - u \rangle| \le \varepsilon \|v - u\|.$$

(4.73)

Then p is Fréchet differentiable at u and $Dp(u) = u^*$.

Proof. We note that for $D_\alpha := \{DF_x(u) : x \in S(\alpha,u) \cap M\}$, (a) implies that $\{u^*\} = \lim_{\alpha \to 0_+} D_\alpha$ in the sense that for every $\varepsilon > 0$ there exists $\beta > 0$ such that $D_\alpha \subset B(u^*, \varepsilon)$ for all $\alpha \in (0,\beta)$. Thus, the result follows from Corollary 4.113 and Proposition 4.118.   □

Using the methods of the previous proofs, one can obtain a circa-differentiability result.

Proposition 4.121 ([821]). *Suppose the following conditions hold for $M := S(u,\theta)$ with $\theta > 0$ and some $u^* \in W^*$:*

(a) For all $x \in M$, F_x is differentiable at u and $DF_x(u) \to u^*$ as $e(x) \to_M 0$
(d) F is docile at u with respect to $M := S(u,\theta)$
(e_C) The family $(F_x)_{x \in M}$ is eventually equi-circa-differentiable at u: for every $\varepsilon > 0$ there exist $\alpha > 0$, $\delta > 0$ such that for all $x \in S(u,\alpha)$ one has

$$|F_x(v) - F_x(w) - \langle DF_x(u), v - w \rangle| \le \varepsilon \|v - w\| \quad \forall v, w \in B(u,\delta).$$

Then p is circa-differentiable at u and $Dp(u) = u^*$.

Let us note that assumption (e_C) is a weakened form of the assumption that the family $(F_x)_{x \in M}$ is *equi-circa-differentiable* at u in the following sense: for every $\varepsilon > 0$ there exists $\delta > 0$ such that for every $x \in M$ one has

$$|F_x(v) - F_x(w) - \langle DF_x(u), v - w \rangle| \le \varepsilon \|v - w\| \quad \forall v, w \in B(u,\delta).$$

Now let us give a criterion for equi-circa-differentiability.

Lemma 4.122. *The following assumptions ensure that condition (e_C) holds:*

(a') There exists $\theta > 0$ such that for all $x \in S(u,\theta)$, F_x is differentiable on $B(u,\theta)$
(e_C') For every $\varepsilon > 0$ there exist $\alpha, \delta > 0$ such that for $v \in B(u,\delta)$, $x \in S(u,\alpha)$, one has $\|DF_x(v) - DF_x(u)\| \le \varepsilon$

Proof. Given $\varepsilon > 0$, we take $\alpha, \delta > 0$ as in (e'_C). Then for $x \in S(u, \alpha)$ one has

$$\forall v, w \in B(u, \delta), \qquad |F_x(v) - F_x(w) - \langle DF_x(u), v - w \rangle| \leq \varepsilon \|v - w\|,$$

by the mean value theorem applied to the functions $w \mapsto F_x(w) - DF_x(u)(w)$. □

Taking into account Corollary 4.116 and Lemma 4.122, we get the following corollary.

Corollary 4.123. *Suppose that for all $x \in X$ the function F_x is lower semicontinuous and bounded below on W and that for some $\theta > 0$ the following condition holds along with (a'), (e'_C):*

(f) There exist $c \in \mathbb{R}_+$, $V \in \mathcal{N}(u)$ such that for all $v \in V$ and all $x \in [F_v \leq p(u) + \theta]$ the function F_x is differentiable at v with $\|DF_x(v)\| \leq c$.

Then F is compliant at u and p is circa-differentiable at u.

Exercises

1. Suppose X is a topological space, F is lower semicontinuous at (u, x), and F_u is continuous at x for all $x \in X_0$, where X_0 is a subset of X such that for all sequences $(\varepsilon_n) \to 0_+$, $(u_n) \to u$ in W, (x_n) in X with $x_n \in S(u_n, \varepsilon_n)$ for each n, the sequence (x_n) has a cluster point in X_0. Show that if p is upper semicontinuous at u, then F is compliant at u. (See [820].)

2. Using a continuation method, prove Proposition 4.115.

3. Show that $\liminf_{e(x) \to 0} D^-(x) \subset \liminf_{(v, \alpha) \to (u, 0_+)} A_{v, \alpha}$ when F is docile.

4. Prove Proposition 4.121.

4.7.2 *Application to Regularization*

There are various ways of approximating a nonsmooth function by a more regular one. In finite dimensions, regularization by means of integral convolution with mollifier functions is especially useful for the study of partial differential operators. For optimization problems, infimal regularization is better adapted: it is valid even if the underlying space X is an infinite-dimensional Banach space, for it preserves infimal values and minimizers. Its general form is as follows: given a function $f : X \to \mathbb{R}_\infty$ and a "regularization kernel" $K_t : X^2 \to \overline{\mathbb{R}}_+$, one sets

$$f_t(w) := \inf\{f(x) + K_t(w, x) : x \in X\}, \qquad w \in W := X, \ t > 0.$$

The regularized function f_t thus appears as a special performance function. A simple form of the regularization kernel consists in taking $K_t(w,x) = tk(w-x)$ for $t > 0$, $w,x \in X$ with $k : X \to \overline{\mathbb{R}}_+$. In particular, the classical example of the *Baire* (or *Pasch–Hausdorff*) *regularization* is obtained with $k(x) = \|x\|$. Another form of the regularization kernel consists in taking $K_t(w,x) = tj(t^{-1}(w-x))$ for $t > 0$, $w,x \in X$, where $j : X \to \overline{\mathbb{R}}_+$ is a given function. The most popular processes are the *Moreau regularization* obtained with $j(\cdot) = (1/2)\|\cdot\|^2$ (or $k(\cdot) = (1/2)\|\cdot\|^2$) and the *rolling ball* regularization obtained with $j(x) = 1 - \sqrt{1 - \|x\|^2}$ for $\|x\| \in [-1,1]$, $+\infty$ otherwise. When $f_t := f \,\square\, j_t$ with $j_t(x) := tj(t^{-1}x)$, one has $f_t^* = f^* + tj^*$ and if $f_t^{**} = f_t$ one gets smoothness properties of f_t from rotundity properties of f_t^*. Moreover, the study of the convergence of f_t^* as $t \to 0$ is simple, so that continuity properties of the Fenchel transform may yield convergence results, at least under convexity assumptions. But we do not consider convergence issues. We rather deal with the case that t is fixed, so that the two different forms described above coincide.

Here our purpose is to show that the differentiability results of the preceding section can be applied to the case $K_t(w,x) = tk(w-x)$, where k is a function $k : X \to \mathbb{R}_+$ with the following properties:

(r1) k is coercive, Lipschitzian on bounded subsets, and $k(0) = 0$;
(r2) For every $c \in (0,1)$, $r > 0$ there exists some $m \in \mathbb{R}$ such that

$$k(w-x) \geq ck(x) - m \quad \forall w \in B(0,r), \forall x \in X;$$

(r3) k is continuously differentiable on X;
(r4) k is uniformly convex on bounded subsets: for every $r > 0$ there exists some nondecreasing function $\gamma : [0,2r] \to \mathbb{R}$ such that $\gamma(t) > 0$ for $t > 0$ and

$$\frac{1}{2}k(x) + \frac{1}{2}k(x') - k\left(\frac{1}{2}x + \frac{1}{2}x'\right) \geq \gamma(\|x - x'\|) \quad \forall x,x' \in B(0,r).$$

These conditions are satisfied for $k(\cdot) := s^{-1}\|\cdot\|^s$ with $s > 1$ when $(X, \|\cdot\|)$ is uniformly convex and uniformly smooth, in particular when X is a Hilbert space (see [984] for instance).

Since here the parameter $t \in (0,+\infty)$ is considered fixed, we do not mention it in the expression for F, so that the relationships with what precedes are clearer; however, the value function $p := f \,\square\, tk$ is now denoted by f_t.

Theorem 4.124. *Assume conditions (r1)–(r4). Suppose $f : X \to \mathbb{R}_\infty$ has a nonempty domain and is such that for some $s \in \mathbb{R}_+$ and all $u \in X$ the function $f(\cdot) + sk(u - \cdot)$ is convex and bounded below. Then, for $t > s$, the regularized function f_t of f given by*

$$f_t(w) = \inf_{x \in X} F(w,x), \text{ where } F(w,x) := f(x) + tk(w-x), \quad w \in X,$$

is of class C^1 on $W = X$ and $f_t \leq f$.

Proof. Clearly, assumption (r1) ensures that $f_t \leq f$. Taking $x_0 \in f^{-1}(\mathbb{R})$, for all $w \in W$ we have $f_t(w) \leq f(x_0) + tk(w - x_0) < +\infty$. Let $r > 0$ and $w \in B(0, r)$. Taking $c \in (0, 1)$ such that $ct > s$ and taking m associated with c and r as in assumption (r2), for $x \in X$ and $b := \inf(f(\cdot) + sk(\cdot))$ we get the estimate

$$F(w, x) \geq b - sk(x) + ctk(x) - mt \geq b - mt.$$

Then $f_t(w) \geq b - mt$ and for every $\beta \in (0, 1]$, $x \in S(w, \beta)$, we have

$$(ct - s)k(x) + b - mt \leq F(w, x) \leq f_t(w) + \beta \leq f(x_0) + tk(w - x_0) + 1,$$

so that the coercivity of k entails the existence of some $r_1 > 0$ such that $S(w, \beta) \subset S(w, 1) \subset M := B(0, r_1)$ for all $w \in B(0, r)$. The first of the preceding inequalities also shows that setting $\theta := 1$, $\lambda := p(u) + \theta$, increasing r_1 if necessary, we may suppose that $[F_w \leq \lambda] \subset B(0, r_1)$ for all $w \in B(0, r)$. Now for all $x \in X$, F_x is differentiable and $DF_x(w) = Dk(w - x)$, which is bounded for $w \in B(0, r)$, $x \in [F_w \leq \lambda]$, since k is Lipschitzian on the bounded set $B(0, r + r_1)$: condition (f) of Corollary 4.123 is satisfied. Let us show that $DF_x(w)$ converges as $w \to u$, $e(x) \to 0$, where $e(x) := f(x) + tk(w - x) - f_t(x)$.

For $u \in X$, $\alpha > 0$, and $x, x' \in S(u, \alpha)$ we have, with $x'' := \frac{1}{2}x + \frac{1}{2}x'$,

$$\frac{1}{2}(f(x) + tk(u - x)) + \frac{1}{2}(f(x') + tk(u - x')) \leq \frac{1}{2}(f_t(u) + \alpha) + \frac{1}{2}(f_t(u) + \alpha)$$
$$\leq f(x'') + tk(u - x'') + \alpha;$$

hence, using (r4) and the convexity of $f + sk(u - \cdot)$, we have

$$(t - s)\gamma(\|x - x'\|) \leq (t - s)\left(\frac{1}{2}k(u - x) + \frac{1}{2}k(u - x') - k(u - x'')\right)$$
$$\leq f(x'') + \alpha - \frac{1}{2}(f(x) + f(x')) + sk(u - x'') - \frac{s}{2}k(u - x) - \frac{s}{2}k(u - x') \leq \alpha.$$

It follows that the diameter of $S(u, \alpha)$ tends to 0 as $\alpha \to 0$. Since X is complete, the family $(S(u, \alpha))_{\alpha > 0}$ converges to some point $J_f(u)$ in X. Then when $w \to u$, $\alpha \to 0_+$, we have $\sup\{\|(w - x) - (u - J_f(u))\| : x \in S(u, \alpha)\} \to 0$. Since $DF_x(w) = -tDk(x - w)$ is continuous in (w, x) at $(u, J_f(u))$, we get that $\{DF_x(w) : x \in S(u, \alpha)\} \to u^* := -tDk(J_f(u) - u)$ as $w \to u$, $\alpha \to 0_+$, and condition (a') and (e'$_C$) of Corollary 4.122 are satisfied. The fact that f_t is of class C^1 ensues from its circa-differentiability at every point u of W. $\square$

Exercises

1. Let $f : X \to \mathbb{R}_\infty$, $t > 0$ and let f_t be the Baire regularized function of f. Show that either f_t is the greatest t-Lipschitzian function $g \le f$ or there is no such function and then $f_t = -\infty^X$.

2. Suppose k satisfies $tk(x) + (1-t)k(x') - k(tx + (1-t)x') \ge t(1-t)k(x-x')$ for all $t \in [0,1]$, $x, x' \in X$. Show that for $r, s > 0$ and every function f, one has $(f_r)_s \ge f_{r+s}$. Check that for $k := (1/2)\|\cdot\|^2$ in a Hilbert space this inequality is an equality.

3. Let $(X, \|\cdot\|)$ be a Hilbert space and let $k := (1/2)\|\cdot\|^2$.
(a) Show that for a closed proper convex function f the infimum in the definition of $f_t(w)$ is attained at a unique point $P_t(w)$.
(b) Prove that f_t is differentiable at every point $w \in X$ and $\nabla f_t(w) = (1/t)(w - P_t(w))$.
(c) Deduce that the distance function d_C to a closed convex subset C of X is differentiable and $\nabla d_C^2 = 2(I_X - P_C)$, where P_C is the projection from X to C.

4.7.3 Mathematical Programming Problems and Sensitivity

Let us consider the mathematical programming problem

$$(\mathscr{P}) \qquad \text{minimize } f(x) \text{ subject to } g(x) \in C,$$

where $f : X \to \mathbb{R}_\infty$ is finite at $\bar{x} \in g^{-1}(C)$ and $g : X \to Z$ is H-differentiable at $\bar{x}$, with X and Z Banach spaces, C a closed convex subset of Z. Let $Y := Z^*$. A first-order necessary optimality condition for an element $\bar{x} \in g^{-1}(C)$ can be expressed using the *Lagrangian* ℓ given by

$$\ell_y(x) := \ell(x, y) := f(x) + \langle y, g(x) \rangle, \qquad\qquad (x, y) \in X \times Y,$$

and the set

$$K(\bar{x}) := \{ y \in N(C, g(\bar{x})) : 0 \in \partial_D \ell_y(\bar{x}) = \partial_D f(\bar{x}) + y \circ g'(\bar{x}) \}$$

of *Karush–Kuhn–Tucker multipliers* at $\bar{x}$. Such a condition is usually obtained under a technical assumption called a constraint qualification condition. When C is a singleton, one usually requires that $g'(\bar{x})(X) = Z$; when C is a closed convex cone, a classical *qualification condition* is the Robinson condition

$$g'(\bar{x})(X) - \mathbb{R}_+(C - g(\bar{x})) = Z,$$

which is equivalent to the Mangasarian–Fromovitz condition when $Z = \mathbb{R}^m$, $C = \mathbb{R}^m_-$ (see Exercise 2 below).

In general, the *performance* (or value) function p associated to ($\mathscr{P}$) by

$$p(z) := \inf\{f(x) : g(x) + z \in C\}$$

is not differentiable. When p is differentiable, its derivative is a multiplier; more generally, as in the convex case, the following statement points out a link between the subdifferential of p and multipliers that is illuminating, since it shows that multipliers are not artificial tools but are naturally associated with the problem.

Proposition 4.125 ([809]). *For every solution $\bar{x}$ of ($\mathscr{P}$), one has $\partial_D p(0) \subset K(\bar{x})$.*

Proof. Let $\bar{y} \in \partial_D p(0)$, $\bar{z} := g(\bar{x}) \in C$. Let us set $F(z) := g^{-1}(C - z)$ for $z \in Z$. Let $w := r(z - \bar{z})$, with $z \in C$, $r > 0$. For $t \in [0, 1/r]$, one has $\bar{z} + tw = rtz + (1 - rt)\bar{z} \in C$, $g(\bar{x}) = (\bar{z} + tw) - tw$, hence $\bar{x} \in F(tw)$. Thus $p(tw) \le f(\bar{x}) = p(0)$. It follows that $\langle \bar{y}, w \rangle \le \liminf_{t \to 0_+} (1/t)(p(tw)) - p(0)) \le 0$. Since $z \in C$ and $r > 0$ are arbitrary, we get $\bar{y} \in (\mathbb{R}_+(C - \bar{z}))^0 = N(C, \bar{z})$.

Given $u, v \in X$, setting $w_{t,v} := t^{-1}(g(\bar{x}) - g(\bar{x} + tv))$, $w := -g'(\bar{x})u$ and noting that $(w_{t,v}) \to w$ as $(t, v) \to (0_+, u)$ and $g(\bar{x} + tv) + tw_{t,v} = g(\bar{x}) \in C$, hence $\bar{x} + tv \in F(tw_{t,v})$, we get $p(tw_{t,v}) \le f(\bar{x} + tv)$. It follows from the definition of $\partial_D p(0)$ that

$$\langle \bar{y}, -g'(\bar{x})u \rangle = \langle \bar{y}, w \rangle \le \liminf_{(t,v) \to (0_+, u)} \frac{1}{t}(p(tw_{t,v}) - p(0))$$

$$\le \liminf_{(t,v) \to (0_+, u)} \frac{1}{t}(f(\bar{x} + tv) - f(\bar{x})) = f^D(\bar{x}, u),$$

or $-\bar{y} \circ g'(\bar{x}) \in \partial_D f(\bar{x})$. Thus $0 \in \partial_D \ell_y(\bar{x})$ and $\bar{y} \in K(\bar{x})$. $\square$

Corollary 4.126. *If p is differentiable at 0, then for every solution $\bar{x}$ to ($\mathscr{P}$), one has $p'(0) \in K(\bar{x})$.*

One can get an optimality condition in the case that the constraint set is defined by inequalities without assuming differentiability of the map g. For that purpose, we need to express the normal cone to a sublevel set. We start with a special convex case.

Lemma 4.127. *Let $\bar{z}^* \in Z^*$ with $\|\bar{z}^*\| = 1$, $\gamma \in (0, 1)$ and let C be the Bishop–Phelps cone given by $C := \{z \in Z : \langle \bar{z}^*, z \rangle \ge \gamma\|z\|\}$. Then the polar cone of C is $C^0 = \mathbb{R}_+(\gamma B_{Z^*} - \bar{z}^*)$, and for $z \in C$ one has $N(C, z) = \{y \in C^0 : \langle y, z \rangle = 0\}$.*

Proof. The last relation is valid for every convex cone and is obtained by using the inclusion $C + z \subset C$, implying that $N(C, z) \subset N(C, 0) = C^0$ and by observing that for $y \in N(C, z)$ one has $\langle y, 2z - z \rangle \le 0$, $\langle y, 0 - z \rangle \le 0$, hence $N(C, z) \subset \{y \in C^0 : \langle y, z \rangle = 0\}$, while the reverse inclusion is obvious. To justify the expression of C^0 when C is the above Bishop–Phelps cone, one first observes that for all $y \in \gamma B_{Z^*} - \bar{z}^*$ and for all $r \in \mathbb{R}_+$, $z \in C$ one has $\langle y, z \rangle \le \gamma\|z\| + \langle -\bar{z}^*, z \rangle \le 0$, hence $ry \in C^0$. Conversely, since $Q := \mathbb{R}_+(\gamma B_{Z^*} - \bar{z}^*)$ is easily seen to be weak* closed, if $y \in Z^* \setminus \mathbb{R}_+(\gamma B_{Z^*} - \bar{z}^*)$, the Hahn–Banach theorem yields some $u \in X$ with $\|u\| = 1$

such that $\langle y, u \rangle > r \langle u^* - \bar{z}^*, u \rangle$ for all $r \in \mathbb{R}_+$, $u^* \in \gamma B_{Z^*}$. Since r is arbitrarily large, it follows that $\langle u^* - \bar{z}^*, u \rangle \leq 0$ for all $u^* \in \gamma B_{Z^*}$, and hence $\langle \bar{z}^*, u \rangle \geq \gamma \|u\|$ or $u \in C$. Since $\langle y, u \rangle > 0$, one has $y \notin C^0$. $\square$

Setting $g(z) := \gamma \|z\| - \langle \bar{z}^*, z \rangle$, so that $C = g^{-1}(\mathbb{R}_-)$, the formula for $N(C, z)$ is a special case of a general relation about the normal cone to the sublevel set

$$S_g(x) := \{w \in X : g(w) \leq g(x)\}.$$

Lemma 4.128. *Let $g \in \mathscr{F}(X)$. For $x \in \mathrm{dom}\, g$ one has $\mathbb{R}_+ \partial_D g(x) \subset N_D(S_g(x), x)$, $\mathbb{R}_+ \partial_F g(x) \subset N_F(S_g(x), x)$.*

Proof. Given $x^* \in \partial_D g(x)$ and $v \in T(S_g(x), x)$, taking sequences $(t_n) \to 0_+$, $(v_n) \to v$ such that $x + t_n v_n \in S_g(x)$, we have $x^* \in N_D(S_g(x), x)$, since

$$\langle x^*, v \rangle \leq \liminf_n \frac{1}{t_n}(g(x + t_n v_n) - g(x)) \leq 0.$$

If $x^* \in \partial_F g(x)$, given $\varepsilon > 0$ we can find $\delta > 0$ such that for $w \in S_g(x) \cap B(x, \delta)$ we have $\langle x^*, w - x \rangle \leq g(w) - g(x) + \varepsilon \|w - x\| \leq \varepsilon \|w - x\|$, so that $x^* \in N_F(S_g(x), x)$.
                                                                                      $\square$

In general, the inclusion $\mathbb{R}_+ \partial_D g(x) \subset N(S_g(x), x)$ is strict, as shown by the example of $g : \mathbb{R} \to \mathbb{R}$ given by $g(x) := \min(x, 2x)$, since $\partial_D g(0) = \varnothing$, whereas $N(S_g(0), 0) = \mathbb{R}_+$.

A reverse inclusion can be given in the convex case under a so-called qualification condition. In the nonconvex case, one may use a fuzzy inclusion.

Theorem 4.129 (Normal cone to a sublevel set). *Let g be a lower semicontinuous function on an F-smooth Banach space X and let $\bar{x} \in S := \{x \in X : g(x) \leq 0\}$, $\bar{x}^* \in N_F(S, \bar{x})$. Suppose $\liminf_{x \to \bar{x}} d(0, \partial_F g(x)) > 0$. Then for every $\varepsilon > 0$, there exist $x \in B(\bar{x}, \varepsilon, g)$, $x^* \in \partial_F g(x)$, and $r > 0$ such that $\|\bar{x}^* - r x^*\| < \varepsilon$.*

Proof. Without loss of generality, we may suppose $\bar{x} = 0$ and $g(\bar{x}) = 0$, replacing g by g' given by $g'(x) := g(x + \bar{x}) - g(\bar{x})$, so that $S' := (g')^{-1}(\mathbb{R}_-) \subset S - \bar{x}$ and observing that $N_F(S, \bar{x}) \subset N_F(S', 0)$. In view of the density of $\{(x, g(x)) : x \in \mathrm{dom}\, \partial_F g\}$ in the graph of g, the case $\bar{x}^* = 0$ is obvious. Thus, we may suppose $\|\bar{x}^*\| = 1$. Let $c < \liminf_{x \to \bar{x}} d(0, \partial_F g(x))$, $c > 0$. Given $\varepsilon \in (0, 1)$, $\alpha \in (0, \varepsilon/2)$, let $\rho \in (0, \varepsilon)$, $\rho < c/2$ be such that $2\alpha + 2\rho c^{-1}(2\alpha + 1) < \varepsilon$ and such that g is bounded below on $B := \rho B_X$, $d(0, \partial_F g(x)) > c$ for $x \in B$ and

$$\forall x \in S \cap (B \setminus \{0\}), \qquad\qquad \langle \bar{x}^*, x \rangle < \alpha \|x\|. \qquad\qquad (4.74)$$

Let $g_B := g + \iota_B$, and let C and C' be the Bishop–Phelps cones

$$C := \{x \in X : \langle \bar{x}^*, x \rangle \geq 2\alpha \|x\|\}, \qquad C' := \{x \in X : \langle \bar{x}^*, x \rangle \geq \alpha \|x\|\}.$$

In order to prove that $\bar{x}$ is a robust minimizer of g_B on C, let us first show that

$$\forall x \in X \setminus C', \qquad d_C(x) \geq \frac{\alpha}{2\alpha + 1} \|x\|. \tag{4.75}$$

Given $x \in X \setminus C'$, let $u \in B(0, \alpha(2\alpha + 1)^{-1} \|x\|)$, i.e., $\|u\| < \alpha \|x\| - 2\alpha \|u\|$. Then since $\langle \bar{x}^*, x \rangle < \alpha \|x\|$, we have

$$\langle \bar{x}^*, x + u \rangle \leq \alpha \|x\| + \|\bar{x}^*\| \cdot \|u\|$$
$$< 2\alpha \|x\| - 2\alpha \|u\| \leq 2\alpha \|x + u\|,$$

hence $x + u \in X \setminus C$, so that $B(x, \alpha(2\alpha + 1)^{-1} \|x\|) \subset X \setminus C$ and (4.75) holds.

Now, given $\delta > 0$ and $x \in B \setminus \{0\}$ satisfying $d_C(x) \leq \delta$, we have either $x \in C'$, hence $x \notin S$ (since $S \cap (B \setminus \{0\}) \cap C' = \varnothing$) and $g_B(x) > 0$, or $x \in X \setminus C'$ and $\|x\| \leq \alpha^{-1}(2\alpha + 1)d_C(x) \leq \alpha^{-1}(2\alpha + 1)\delta$. Both cases allow us to conclude that

$$\inf\{g_B(x) : x \in X, \, d_C(x) \leq \delta\} \geq \inf\{g_B(x) : x \in \alpha^{-1}(2\alpha + 1)\delta B_X\}.$$

Since g_B is lower semicontinuous at 0, the right-hand side converges to $g_B(0) = 0$ as $\delta \to 0$. Thus $\bar{x}$ is a robust minimizer of g_B on C.

Corollary 4.64 yields some $x \in B(\bar{x}, \rho)$, $y \in C$, $y^* \in N(C, y)$, $x^* \in \partial_F g_B(x)$ such that $\|x^* + y^*\| < \rho$. Then $x^* \in \partial_F g(x)$, since $x \in B(\bar{x}, \rho) = \mathrm{int} B$. Then Lemma 4.127 yields $s \in \mathbb{R}_+$ and $u^* \in 2\alpha B_{X^*}$ such that $y^* = s(u^* - \bar{x}^*)$. It follows that

$$c < \|x^*\| \leq \|x^* + y^*\| + \|-y^*\| < \rho + s(2\alpha + 1),$$

hence $s(2\alpha + 1) > c - \rho > c/2$ and $r := s^{-1} < 2c^{-1}(2\alpha + 1)$. Since $\|x^* + y^*\| < \rho$, we get

$$\|\bar{x}^* - rx^*\| \leq \|\bar{x}^* + ry^*\| + r\rho = \|u^*\| + r\rho \leq 2\alpha + r\rho < 2\alpha + 2\rho c^{-1}(2\alpha + 1) < \varepsilon.$$
$$\qquad\qquad\qquad\qquad\qquad\qquad\qquad\qquad\qquad\qquad\qquad\qquad\qquad\qquad\qquad\qquad \square$$

Let us note a consequence pertaining to the asymptotic subdifferential.

Corollary 4.130. *Let f be a lower semicontinuous function on an F-smooth Banach space X and let $\bar{x} \in \mathrm{dom} f$, $\bar{x}^* \in \partial_F^\infty f(\bar{x})$. Then for every $\varepsilon > 0$, there exist $x \in B(\bar{x}, \varepsilon, f)$, $x^* \in \partial_F f(x)$, and $t \in (0, \varepsilon)$ such that $\|\bar{x}^* - tx^*\| < \varepsilon$.*

Proof. The space $X \times \mathbb{R}$ is F-smooth and $g : X \times \mathbb{R} \to \mathbb{R}_\infty$ given by $g(x, r) := f(x) - r$ is lower semicontinuous with $\partial_F g(x, r) \subset \partial_F f(x) \times \{-1\}$, so that $\|z^*\| \geq 1$ for all $z \in X \times \mathbb{R}$ and $z^* \in \partial_F g(z)$. Then $E := \mathrm{epi} f = \{(x, r) \in X \times \mathbb{R} : g(x, r) \leq 0\}$. By definition of $\partial_F^\infty f(\bar{x})$, $(\bar{x}^*, 0) \in N_F(E, (\bar{x}, \bar{r}))$, where $\bar{r} := f(\bar{x})$. Given $\varepsilon > 0$, the preceding theorem yields some $t > 0$, $(x, r) \in B((\bar{x}, \bar{r}), \varepsilon/2)$, and $(x^*, r^*) \in \partial_F g(x, r)$ such that $|g(x, r)| < \varepsilon/2$, $\|(\bar{x}^*, 0) - t(x^*, r^*)\| < \varepsilon$. Thus $|f(x) - f(\bar{x})| < \varepsilon/2 + |r - f(\bar{x})| < \varepsilon$, $x^* \in \partial_F f(x)$, and $\|\bar{x}^* - tx^*\| < \varepsilon$. Moreover, since $r^* = -1$, we have $t < \varepsilon$. $\qquad \square$

Exercises

1. Check that the notion of multiplier adopted in this subsection coincides with the one considered in Sect. 3.5.4

2. Given a Banach space X and $g : X \to \mathbb{R}^{m+n}$, the *Mangasarian–Fromovitz qualification condition* at $\bar{x}$ is as follows:

(CQMF) $(g_i'(\bar{x}))_{1 \le i \le m}$ are linearly independent and there exists $u \in X$ such that $g_i'(\bar{x})u = 0$ for $i = 1, \ldots, m$, $g_j'(\bar{x})u < 0$ for $j = m+1, \ldots, m+n$.

Here g is assumed to be of class C^1 at $\bar{x}$ and one takes $C := \{0\} \times \mathbb{R}_-^n \subset Z := \mathbb{R}^m \times \mathbb{R}^n$. Show that condition (CQMF) is equivalent to the following dual condition:

(DCQMF) $y_1 g_1'(\bar{x}) + \cdots + y_{m+n} g_{m+n}'(\bar{x}) = 0$, $y_i \in \mathbb{R}$, for $i = 1, \ldots, m$, $y_j \in \mathbb{R}_+$ for $j = m+1, \ldots, m+n \implies y_h = 0$, $h = 1, \ldots, m+n$.

3. With the data of the preceding exercise, show that (CQMF) is equivalent to the Robinson constraint qualification condition

$$g'(\bar{x})(X) - \mathbb{R}_+(C - g(\bar{x})) = Z.$$

4.7.4 Openness and Metric Regularity Criteria

As observed above, since we have a decrease index, we can obtain various metric estimates. Here we focus on openness and metric regularity criteria for multimaps. In order to get such estimates, given a multimap $G : X \rightrightarrows Y$ between Asplund spaces, let us derive a description of the subdifferential of $f := d(0, G(\cdot))$.

Lemma 4.131. *Let $G : X \rightrightarrows Y$ be a multimap with closed graph between two Asplund spaces and let $f := d(0, G(\cdot))$, $S := G^{-1}(0)$. Given $x \in X \setminus S$ and $x^* \in \partial_F f(x)$, there exist sequences $(u_n) \to x$ in X, (v_n) in Y, (v_n^*) in Y^*, $(u_n^*) \to x^*$ in X^* such that $(\|v_n\|) \to f(x)$, $(\|v_n^*\|) \to 1$, $v_n \in G(x_n)$, $u_n^* \in D_F^* G(u_n, v_n)(v_n^*)$ for all $n \in \mathbb{N}$ and $(\langle v_n^*, v_n \rangle) \to f(x)$. Moreover, one can find a sequence $((z_n, z_n^*))$ in $\partial \|\cdot\|$ such that $(z_n - v_n) \to 0$, $(v_n^* - z_n^*) \to 0$.*

Proof. It suffices to apply Corollary 4.73 with (W, X, ε) changed into (X, Y, ε_n), where $(\varepsilon_n) \to 0_+$ and $j(x, v) := \|v\|$. Then in place of (u, v, w, u^*, v^*, x^*) we get $(u_n, v_n, z_n, u_n^*, v_n^*, z_n^*)$ with $v_n \in G(u_n)$, $\|v_n\| \le f(x) + \varepsilon_n$, $u_n^* \in D_F^* G(u_n, v_n)(v_n^*)$, $z_n^* \in \partial \|\cdot\|(z_n)$, $\|u_n^* - x^*\| \le \varepsilon_n$, $\|v_n^* - z_n^*\| \le \varepsilon_n$ for all $n \in \mathbb{N}$. Since G has a closed graph, we cannot have $\liminf_n \|v_n\| = 0$. We may take $\|v_n - z_n\| < \|v_n\|$, so that $\|z_n^*\| = 1$ and $\langle z_n^*, z_n \rangle = \|z_n\|$. Since $(\langle z_n^*, z_n \rangle) = (\|z_n\|) \to f(x)$ and $(\langle v_n^* - z_n^*, v_n \rangle) \to 0$, $(\langle z_n^*, v_n - z_n \rangle) \to 0$, since (v_n) and (z_n^*) are bounded, one gets $(\langle v_n^*, v_n \rangle) \to f(x)$. $\qquad \square$

Proposition 4.132. *Let $\bar{x} \in S := G^{-1}(0)$, where $G : X \rightrightarrows Y$ is a multimap with closed graph between two Asplund spaces. Suppose that for some $c, r > 0$ and for* *all* $x \in B(\overline{x},r) \setminus S$ *with* $f(x) := d(0,G(x)) < cr$ *there exist some* $\varepsilon > 0$ *such that one of the following conditions is satisfied:*

(a) *For all* $u \in B(x,\varepsilon)$, $v \in G(u)$ *with* $|\|v\| - f(x)| < \varepsilon$, $z \in B(v,\varepsilon)$, $z^* \in \partial \|\cdot\| (z)$, $v^* \in B(z^*,\varepsilon)$, $u^* \in D_F^* G(u,v)(v^*)$, *one has* $\|u^*\| \geq c^{-1}$;
(b) *For all* $u \in B(x,\varepsilon)$, $v \in G(u)$ *with* $|\|v\| - f(x)| < \varepsilon$, $v^* \in S_{Y^*}$, $u^* \in D_F^* G(u,v)(v^*)$ *satisfying* $|\langle v^*, v \rangle - f(x)| < \varepsilon$, *one has* $\|u^*\| \geq c^{-1}$.

Then for all $x \in B(\overline{x}, r/2)$ *one has*

$$d(x,S) \leq c^{-1} d(0,G(x)). \tag{4.76}$$

Proof. It suffices to prove that for every $b \in (0, c^{-1})$, every $x \in B(\overline{x}, r) \setminus S$ with $f(x) := d(0, G(x)) < cr$ and every $x^* \in \partial_F f(x)$ one has $\|x^*\| \geq b$. Suppose, to the contrary, that there exist $x \in B(\overline{x}, r) \setminus S$, with $f(x) < cr$ and $x^* \in \partial_F f(x)$ such that $\|x^*\| < b$. Let $\varepsilon > 0$ be as in the assumption, and let (u_n), (v_n), (z_n), (u_n^*), (v_n^*), (z_n^*) be the sequences given in the preceding lemma. For n large enough one has $u := u_n \in B(x,\varepsilon)$, $v := v_n \in G(u)$ with $|\|v\| - f(x)| < \varepsilon$, $z := z_n \in B(v,\varepsilon)$, $z^* := z_n^* \in \partial \|\cdot\| (z)$, $v^* = v_n^* \in B(z^*,\varepsilon)$, $u^* = u_n^* \in D_F^* G(u,v)(v^*)$, and $\|u^*\| < b$, a contradiction to assumption (a). The result also holds in case (b), since assumption (b) is stronger than assumption (a), the existence of $(z,z^*) \in \partial \|\cdot\| \cap (B(v,\varepsilon) \times B(v^*,\varepsilon))$ implying that $|\|v^*\| - 1| < \varepsilon$ and that $\langle v^*, v \rangle$ is close to $\langle z^*, z \rangle$, hence is close to $\|z\|$, $\|v\|$, $f(x)$. $\qquad \square$

Theorem 4.133. *Let* $\overline{x} \in S := G^{-1}(0)$, *where* $G : X \rightrightarrows Y$ *is a multimap with closed graph between two Asplund spaces. Suppose that for some* $c > 0$ *and some open neighborhoods* U *of* $\overline{x}$, V *of* 0 *and for all* $u \in U \setminus S$, $v \in G(u) \cap V$, $v^* \in S_{Y^*}$, $u^* \in D_F^* G(u,v)(v^*)$ *one has* $\|u^*\| \geq c^{-1}$. *Then for all* $x \in U$, *relation (4.76) holds.*

Proof. Let $r > 0$ be such that $B(\overline{x}, r) \subset U$, $B(0, cr) \subset V$. Given $x \in B(\overline{x}, r) \setminus S$ with $f(x) := d(0, G(x)) \geq cr$, relation (4.76) obviously holds. When $f(x) < cr$, taking $\varepsilon = \min(r - \|x - \overline{x}\|, cr - f(x))$, for $u \in B(x,\varepsilon)$, $v \in G(u) \cap (f(x) + \varepsilon)B_Y$, we have $u \in U$, $v \in V$, so that for every $u^* \in D_F^* G(u,v)(v^*)$, with $v^* \in S_{Y^*}$, one has $\|u^*\| \geq c^{-1}$, and Proposition 4.132 applies. $\qquad \square$

Let us pass to metric regularity results.

Theorem 4.134. *Let* X *and* Y *be Asplund spaces, let* U *and* V *be open subsets of* X *and* Y *respectively, let* $F : X \rightrightarrows Y$ *be a multimap with closed graph, and let* $c > 0$.

(a) F *is metrically regular on* $U \times V$ *with rate* c *if and only if for all* $u \in U$, $v \in F(u) \cap V$, $v^* \in S_{Y^*}$, $u^* \in D_F^* F(u,v)(v^*)$ *one has* $\|u^*\| \geq c^{-1}$.
(b) F *is open on* $U \times V$ *with rate* $a := 1/c$ *if and only if for all* $u \in U$, $v \in F(u) \cap V$, $v^* \in S_{Y^*}$, $u^* \in D_F^* F(u,v)(v^*)$ *one has* $\|u^*\| \geq c^{-1}$.
(c) $M := F^{-1}$ *is pseudo-Lipschitzian on* $V \times U$ *with rate* c *if and only if for all* $v \in V$, $u \in M(v) \cap U$, $u^* \in X^*$, $v^* \in D_F^* M(v,u)(u^*)$ *one has* $\|v^*\| \leq c\|u^*\|$.

Proof. Assuming that

$$\inf\{\|u^*\| : u^* \in D_F^* F(u,v)(v^*),\ u \in U,\ v \in F(u) \cap V,\ v^* \in S_{Y^*}\} \geq c,$$

let us prove that for all $(x,y) \in U \times V$ we have $d\left(x, F^{-1}(y)\right) \leq c\,d\left(y, F(x)\right)$. This follows from the preceding proposition, in which we set $G(x) := F(x) - y$ and replace V by $V - y$, so that $D_F^* G(u,v)(v^*) = D_F^* F(u, v+y)(v^*)$.

The sufficiency parts of (b) and (c) stem from what precedes and the equivalences of Theorem 1.139. The necessity part of (c) has been established in Proposition 4.26; by the mentioned equivalences, it entails the necessity parts of assertions (a) and (b).

□

4.7.5 Stability of Dynamical Systems and Lyapunov Functions

Let $f : X_0 \to X$ be a vector field on an open subset X_0 of a Banach space X and the associated differential equation

$$x'(t) = f(x(t)), \quad x(0) = z,$$

where $z \in X_0$ is the initial condition. For the sake of simplicity, we suppose that for all $z \in X_0$ this equation has a solution $x_z(\cdot) := x(\cdot, z)$ defined on $[0, +\infty)$. Such a property can be ensured by assuming a growth condition and some regularity on f (a local Lipschitz property, or, if X is finite-dimensional, just continuity). The following concepts are classical in mechanics and in the study of dynamical systems.

Definition 4.135. A closed subset $S \subset X_0$ of X is said to be *stable* for f if for every $\varepsilon > 0$ there exists some $\delta > 0$ such that $x(t, z) \in B(S, \varepsilon) := \{w \in X : d(w, S) < \varepsilon\}$ for all $(t, z) \in \mathbb{R}_+ \times (B(S, \delta) \cap X_0)$.

It is said to be *attractive* or *asymptotically stable* for f if it is stable and if there exists some $\alpha > 0$ such that for all $z \in B(S, \alpha) \cap X_0$, $d(x(t, z), S) \to 0$ as $t \to \infty$.

Note that when S is stable for f, for every $z \in S$ one has $f(z) \in T^D(S, z)$ (and even $f(z) \in T^I(S, z)$), since S is *invariant*, i.e., $x_z(t) \in S$ for all $t \in \mathbb{R}_+$ and all $z \in S$. In particular, when S is a singleton $\{a\}$, we have $f(a) = 0$.

Lyapunov introduced a method to ensure these stability properties. It extends the simple observation that S is stable for f if one can find a differentiable function $q : X_0 \to \mathbb{R}_+$ satisfying $q'(x) \cdot f(x) \leq 0$ for all $x \in X_0$ and the condition

$$q(x) \to 0 \iff d(x, S) \to 0. \tag{4.77}$$

In fact, in that case, given $\varepsilon > 0$, let $\gamma > 0$ and $\delta > 0$ be such that $q^{-1}([0, \gamma]) \subset B(S, \varepsilon)$ and $B(S, \delta) \subset q^{-1}([0, \gamma])$. Then for every $z \in B(S, \delta)$, one has $x(t, z) \in B(S, \varepsilon)$ for all $t \in \mathbb{R}_+$, since the function $t \mapsto q(x_z(t))$ is nonincreasing because we have

$$(q \circ x_z)'(t) = q'(x_z(t)) \cdot f(x_z(t)) \leq 0.$$

A similar argument holds if q is a differentiable *Lyapunov function for S*, i.e., a differentiable function $q : X_0 \to \mathbb{R}_+$ satisfying (4.77) and for some $c \in \mathbb{R}_+$,

$$\forall x \in X, \qquad q'(x) \cdot f(x) + cq(x) \leq 0.$$

In such a case the function $t \mapsto e^{ct}q(x_z(t))$ is nonincreasing, so that one has $q(x_z(t)) \leq e^{-ct}q(x_z(0)) = e^{-ct}q(z)$ for all $t \in \mathbb{R}_+$. Moreover, we note that when c is positive, S is attractive.

As shown by examples, it is of interest to extend the preceding method to nonsmooth Lyapunov functions. Let us say that a lower semicontinuous function $q : X_0 \to \overline{\mathbb{R}}_+$ is a *Lyapunov function* for S if it satisfies condition (4.77) and $\langle x^*, f(x) \rangle + cq(x) \leq 0$ for all $x \in X$. We also say that (p,q,c) is a *Lyapunov triple* for S if $c \in \mathbb{R}_+$, $p, q : X_0 \to \mathbb{R}_+$, p is continuous, q is lower semicontinuous and satisfies (4.77), and p, q, c satisfy the following conditions:

(L1) $\langle x^*, f(x) \rangle + p(x) + cq(x) \leq 0$ for all $x \in X_0$ and all $x^* \in \partial_F q(x)$;
(L2) when $c = 0$, then p is Lipschitzian on bounded subsets, q is unbounded on unbounded subsets, and $p(x) \to 0 \implies d(x,S) \to 0$.

Clearly, if (p,q,c) is a Lyapunov triple, then q is a Lyapunov function.

Theorem 4.136. *Suppose X is F-smooth. If f is locally Lipschitzian and q is a Lyapunov function for S, then S is stable.*

If (p,q,c) is a Lyapunov triple for S, then S is attractive, provided that in the case $c = 0$, f is bounded on bounded subsets.

Proof. In order to show the stability of S in both cases, it suffices to prove that for all $z \in X$, the function $t \mapsto e^{ct}q(x_z(t)) + \int_0^t e^{cs}p(x_z(s))ds$ is nonincreasing on $\mathbb{R}_+$. Since x_z and p are continuous, the last term is differentiable. To study the first one, let us use a special case of the Leibniz rule for ∂_F: if $g, h : W \to \mathbb{R}$ are two lower semicontinuous functions on an open subset W of a normed space, h being positive and differentiable, then

$$\forall x \in W, \qquad \partial(gh)(x) = h(x)\partial g(x) + g(x)h'(x).$$

Taking $W := (0, +\infty)$, $g(t) := q(x_z(t))$, $h(t) := e^{ct}$ and using Corollary 4.98, we are led to check that (L1) implies that for all $t \in \mathbb{R}_+$, $t^* \in \partial g(t)$ one has

$$e^{ct}t^* + ce^{ct}q(x_z(t)) + e^{ct}p(x_z(t)) \leq 0, \tag{4.78}$$

$\mathbb{R}$ being identified with its dual. Since $t^* \in \partial_F(q \circ x_z)(t)$ and since q is inf-compact on the image under x_z of a compact interval, Theorem 4.70 yields sequences $(t_n) \to t$, $(t_n^*) \to t^*$, $(y_n) \to_q x_z(t)$, (y_n^*), (v_n^*) such that $(\|y_n^* - v_n^*\|) \to 0$, $(\|v_n^*\| \cdot \|x_z(t_n) - y_n\|) \to 0$, with $y_n^* \in \partial q(y_n)$, $t_n^* \in D_F^* x_z(t_n)(v_n^*)$ for all $n \in \mathbb{N}$. The last relation means that $t_n^* = \langle v_n^*, x_z'(t_n) \rangle = \langle v_n^*, f(x_z(t_n)) \rangle$. Since f is locally Lipschitzian and $(x_z(t_n)) \to x_z(t)$, $(\|v_n^*\| \cdot \|x_z(t_n) - y_n\|) \to 0$, from (L1) we get

$$t^* = \lim_n \langle v_n^*, f(y_n) \rangle = \lim_n \langle y_n^*, f(y_n) \rangle \leq \limsup_n (-p(y_n) - cq(y_n)).$$

Since p and q are lower semicontinuous, we obtain $t^* \leq -p(x_z(t)) - cq(x_z(t))$ and (4.78), and hence the stability of S.

Let us turn to the second assertion. When $c > 0$, condition (4.77) and the fact that $t \mapsto e^{ct} q(x_z(t))$ is bounded ensure that $d(x_z(t), S) \to 0$ as $t \to +\infty$. When $c = 0$, since $q \circ x_z$ is bounded on $\mathbb{R}_+$, $x_z(\mathbb{R}_+)$ is bounded. By (L2) and our assumption on f, $p \circ x_z$ is Lipschitzian. Since $\int_0^\infty p(x_z(s))ds$ is finite, it is easy to see that $p(x_z(s)) \to 0$ when $s \to \infty$ (exercise). Then (L2) ensures that $d(x_z(s), S) \to 0$ as $s \to \infty$.   □

Exercises

1. Prove the fact used in the proof of Theorem 4.136 that if $r : \mathbb{R}_+ \to \mathbb{R}_+$ is Lipschitzian and integrable on $\mathbb{R}_+$, then $r(t) \to 0$ as $t \to +\infty$.

2. Given a function $g : \mathbb{R}^2 \to \mathbb{R}_+$ of class C^1, let $f : \mathbb{R}^2 \to \mathbb{R}^2$ be defined by

$$f(x,y) := (-y - xg(x,y), x - yg(x,y)).$$

(a) Check that q given by $q(x,y) := x^2 + y^2$ is a Lyapunov function for $S := \{(0,0)\}$.
(b) Check that S is attractive if for some $c > 0$ one has $g(x,y) \geq c$ for all $(x,y) \in \mathbb{R}^2$.

3. **(a)** Let X be a Hilbert space and let $A : X \to X$ be a symmetric continuous linear map that is positive semidefinite, i.e., such that $q(x) := (Ax \mid x) \geq 0$ for all $x \in X$. Show that $S := \{0\}$ is stable for the vector field f given by $f(x) := -Ax$. Show that S is attractive if A is positive definite, i.e., if for some $c > 0$ one has $q(\cdot) \geq c\|\cdot\|^2$.
(b) Let f be a vector field of class C^1 on a Hilbert space X such that $A := -f'(0)$ is positive definite. Show that $S := \{0\}$ is attractive. [Hint: Use q as in (a).]

4. Use a Lyapunov function to show that $S := \{(x,y) \in \mathbb{R}^2 : x^2 + y^2 = 1\}$ is a stable set for the vector field f given by $f(x,y) := (y, -x)$. [Hint: Take $q(x,y) := |x^2 + y^2 - 1|$.]

5. Consider the differential equation $x'(t) = -(1/4)x(t)(x(t) + 4)(x(t) - 2)$, which occurs in population models. Let $q : \mathbb{R} \to \overline{\mathbb{R}}_+$ be given by $q(x) := (x+4)^2$ for $x \in (-\infty, -1]$, $q(x) := +\infty$ for $x \in (-1,1)$, $q(x) = (x-2)^2$ for $x \in [1, +\infty)$.
(a) Show that $S := \{-4, 2\}$ and q are such that $q(x) \to 0$ if and only if $d(x, S) \to 0$.
(b) Check that q is a Lyapunov function for S. Conclude that S is attractive. (See [996].)

6. Show that the function $q : \mathbb{R}^2 \to \overline{\mathbb{R}}_+$ given by $q(x,y) := |x^2 + y^2 - 1|$ when $x^2 + y^2 \geq 1/2$, $+\infty$ otherwise, is a Lyapunov function for the unit circle S in $\mathbb{R}^2$ when one considers the vector field f given by

$$f(x,y) := \left(-y + x(1 - x^2 - y^2), x + y(1 - x^2 - y^2)\right).$$

7. Let C be the *Cantor set* consisting of the set of $x \in [0,1]$ whose ternary expansion $x = \sum_{n \geq 1} 3^{-n} x_n$ with $x_n \in \{0,1,2\}$ is such that $x_n \neq 1$ for all n. Since C is closed, $[0,1] \setminus C$ is the union of a countable family of open intervals (a_k, b_k). Define $q : C \to \mathbb{R}$ by $q(x) := \sum_{n \geq 1} 2^{-n-1} x_n$ for $x := \sum_{n \geq 1} 3^{-n} x_n \in C$.
(**a**) Show that $q(a_k) = q(b_k)$ for all k. [Hint: Observe that $a_k = 0.x_1 \ldots x_k 0222\ldots$, while $b_k = 0.x_1 \ldots x_k 2000\ldots$, so that q can be extended by continuity by setting $q(x) = q(a_k) = q(b_k)$ for $x \in (a_k, b_k)$. Extend further q to $\mathbb{R}$ by requiring it to be even and by setting $q(x) = 1$ for $x > 1$.]
(**b**) Show that $q(x) \to 0 \Leftrightarrow d(x,S) \to 0$ for $S := \{0\}$ and that the set of differentiability points of q is $\mathbb{R} \setminus (C \cup (-C))$ and for such points x one has $q'(x) = 0$. Conclude that the condition $q'(x) \cdot f(x) \leq 0$ at all points of differentiability of q does not suffice to ensure that S is stable (otherwise, for every vector field, S would be stable). (See [136, 137].)

4.8 Notes and Remarks

The definition of a derivative with the help of a convergence on the space of functions from X to Y, as given in the supplement, is due to G. de Lamadrid [610]; see also [49, 307, 946, 947] for supplements and historical information. Although the generalization of Fréchet and Hadamard subdifferentials to bornological subdifferentials is alluring, we resisted the attraction of such a systematic generalization.

The origins of Fréchet subdifferentiation are not easy to detect. They were used early on, in [244, 352, 363, 602, 603]. For Hadamard subdifferentiation, see [78, 780, 782], which were written independently (the manuscript of the last paper remained four years in the hands of the referee or of the editor). It is not known to the author whether $\partial_H = \partial_D$ in Hadamard smooth spaces, even in the case that $S_j := j^{-1}(1)$ is a smooth submanifold of X and $(u,r) \mapsto ru$ realizes a diffeomorphism from $S_j \times (0, +\infty)$ onto $X \setminus \{0\}$, as is the case when j is a smooth norm (off 0).

The passages from analytic notions to geometric concepts is one of the key features of nonsmooth analysis. Corollary 4.17 is taken from [413] for the directional case. Proposition 4.22 appeared in [119, Lemma 6]. We give some attention to order properties, which have often been neglected up to now, a surprising fact, since nonsmooth analysis pertains to one-sided analysis. Proposition 4.49 is new.

Tangent cones in the nonregular case were introduced by Bouligand [158, 159]; see also Severi [891]. Their study was included in a general framework by Choquet [209]. The use of tangent cones in optimization was initiated by Dubovitskii and Miljutin [327, 328]; see also [619]. Their viewpoint was only partially dualized. For analytical concepts, Pshenichnii [853] was the pioneer; then Clarke [211, 214] and numerous authors made other proposals. They will be considered in the next chapter. A general comparison was given by Ioffe [519].

The fuzzy viewpoint is of fundamental importance. It was initiated in [513, 715, and others] and developed in [362, 363, 516–528, 603, 718]. Theorem 4.110 can be found in [528, 995]. The weak fuzzy rules are typical of the work of Ioffe.

The concept of linearly coherent family is due to Ioffe [529]. The sum rule under a linear metric qualification condition was devised by Ioffe for his subdifferential in [530, 531]. Linear coherence is called local linear regularity or bounded linear regularity in several papers [67, 76, 77, 992]. In the convex case, a global version can be given to this property that is quite unexpected in the nonconvex case.

The subsection about softness is adapted from [823].

The subsection about value functions incorporates results from [820] and gives a simplified version of [821]. The latter paper was prompted by genericity results about the existence of optimal solutions to the problem of minimizing $F_w := F(w, \cdot)$. Such results require circa (or strict) differentiability of the performance function p; differentiability properties would not suffice. These results are in the line of the work by Ekeland and Lebourg [352]; see also [37, 341, 565, 623, 809, 877, 878, 999]. In order to get genericity properties, one has to rely on deep results of Preiss [850].

For the applications of these strict or circa differentiability results to existence and genericity properties, we refer to [821], and for previous results of this kind, to [37, 72, 108, 352, 623, 999, 1000] and their bibliographies; they might also be relevant to the methods of [551]. The last example of Sect. 4.7.2 giving a differentiability property of the distance function to a subset of a normed space is a variant of results of [396, Corollary 3.5], [784, Proposition 1.5], [810, Corollary 2.10], [999, Corollary 2] given under additional smoothness properties of the norm or additional assumptions on X. The terminology " marginal function" is used by economists. Some mathematicians use it both for supremum and infimum functions. In order to avoid confusion, we keep the terminology " marginal function" for supremal value functions and we strive to propagate the terminology " performance function" for infimal value functions. That allows one to make a clear distinction.

The regularization we consider is of Moreau type; for simplicity, we do not take a general regularization kernel as in [157, 204, 205] but limit our illustration to an infimal convolution process as in [25, 810]. For other results about regularization processes in Banach spaces, see [157, 204, 205, 904]. In these references X is complete and f is assumed to be bounded below; on the other hand, as in the Lasry–Lions method for Hilbert spaces [613], an iteration of the regularization process enables one to get rid of the convexity condition made above.

Chapter 5
Circa-Subdifferentials, Clarke Subdifferentials

The days of our youth are the days of our glory.

—Lord Byron

We devote the present chapter to one of the most famous attempts to generalize the concept of derivative. When limited to the class of locally Lipschitzian functions, it is of simple use, a fact that explains its success. The general case requires a more sophisticated approach. We choose a geometrical route to it involving the concept of normal cone. It makes easy the proofs of calculus rules. In fact, in this theory, a complete primal–dual picture is available: besides a normal cone concept, one has a notion of tangent cone to a set, and besides a subdifferential for a function one has a notion of directional derivative. Moreover, inherent convexity properties ensure a full duality between these notions. Furthermore, the geometrical notions are related to the analytical notions in the same way as those that have been obtained for elementary subdifferentials. These facts represent great theoretical and practical advantages.

However, some drawbacks are experienced. The main one concerns the lack of precision of the approximations to sets given by tangent cones and directional derivatives for functions: the convexification process that comes on top of the limiting process that yields the notions of the present chapter mostly explains this lack of accuracy. As a result, the Clarke subdifferential is often too large, and an inclusion of the form $0 \in \partial f(x)$ may not bring much information; correspondingly, the normal cone of Clarke is often too big to produce every usable information: for instance, an inclusion of the form $-f'(x) \in N_C(S,x)$ is of no use if the normal cone $N_C(S,x)$ to S at x (in the sense of Clarke) is the whole space.

Nonetheless, the tools we present in this chapter can serve as surrogate smooth concepts, especially in a first stage or when no other result is available. Many articles and books have adopted such a strategy. We devote the last section of the chapter to slightly more precise variants. They keep the main features of Clarke's concepts.

J.-P. Penot, *Calculus Without Derivatives*, Graduate Texts in Mathematics 266,
DOI 10.1007/978-1-4614-4538-8_5, © Springer Science+Business Media New York 2013

5.1   The Locally Lipschitzian Case

In this chapter X is a Banach space, W is an open subset of X, and $\mathscr{L}(W)$ denotes the set of locally Lipschitzian functions on W. It forms an important class of functions on W enjoying nice stability properties. The possibility of extending differential calculus to such a class in a simple way makes Clarke's approach alluring.

5.1.1   Definitions and First Properties

The definition of the Clarke derivative of a locally Lipschitzian function is simple.

Definition 5.1. The *Clarke derivate* (or *circa-derivate*) $f^C(x, \cdot)$ of $f \in \mathscr{L}(W)$ at $x \in W$ is defined by

$$f^C(x, u) := \limsup_{(t,w) \to (0_+, x)} \frac{f(w + tu) - f(w)}{t}, \qquad u \in X.$$

The *Clarke subdifferential* of f at x is the support set of $f^C(x, \cdot)$ given by

$$\partial_C f(x) := \{x^* \in X^* : \langle x^*, \cdot \rangle \leq f^C(x, \cdot)\}.$$

Simple consequences of this definition can be drawn, and simple calculus rules can be derived, both for the subdifferential and for the derivate. We start with the latter.

Proposition 5.2. *Let $f \in \mathscr{L}(W)$ and let $x \in W$. Then*

(a) *The function $f^C(x, \cdot)$ is finite and sublinear;*
(b) *If k is the Lipschitz rate of f near x then $|f^C(x, \cdot)| \leq k\|\cdot\|$;*
(c) *$f^C(\cdot, \cdot)$ is upper semicontinuous on $W \times X$;*
(d) *$f^C(x, -u) = (-f)^C(x, u)$ for all $(x, u) \in W \times X$.*

Proof. (a) The finiteness of $f^C(x, \cdot)$ and assertion (b) stem from the inequality $t^{-1}|f(w + tu) - f(w)| \leq k\|u\|$, where k is the Lipschitz rate of f on a neighborhood of x. Since $f^C(x, \cdot)$ is obviously positively homogeneous, let us prove that it is subadditive. Given $u, v \in X$, let us write

$$\frac{f(w + t(u + v)) - f(w)}{t} = \frac{f(w + tu + tv) - f(w + tu)}{t} + \frac{f(w + tu) - f(w)}{t}$$

and use the fact that the upper limit of a sum is not greater than the sum of the upper limits of the summands. Here the upper limit is taken over $(t, w) \to (0_+, x)$, so that $w + tu \to x$, and the upper limit of the first term of the right-hand side is bounded above by $f^C(x, v)$, while the upper limit of the second term is exactly $f^C(x, u)$. Thus

$$f^C(x, u+v) \leq f^C(x,v) + f^C(x,u).$$

(c) Let $(x,u) \in W \times X$ and let $((x_n, u_n)) \to (x,u)$. For each $n \in \mathbb{N}$ we can find $(t_n, w_n) \in (0, 2^{-n}) \times B(x_n, 2^{-n})$ such that $t_n^{-1}(f(w_n + t_n u_n) - f(w_n)) \geq f^C(x_n, u_n) - 2^{-n}$. Then $(w_n) \to x$ and if k is the Lipschitz rate of f on some $V \in \mathcal{N}(x)$, we get

$$\limsup_n f^C(x_n, u_n) \leq \limsup_n t_n^{-1}(f(w_n + t_n u_n) - f(w_n))$$

$$\leq \limsup_n t_n^{-1}(f(w_n + t_n u) - f(w_n)) + k \|t_n u_n - t_n u\|) \leq f^C(x,u).$$

(d) Setting $z := w - tu$, we get that $(t,z) \to (0_+, x)$ iff $(t,w) \to (0_+, x)$, hence

$$f^C(x, -u) = \limsup_{(t,w) \to (0_+, x)} t^{-1}(f(w - tu) - f(w))$$

$$\leq \limsup_{(t,z) \to (0_+, x)} t^{-1}((-f)(z + tu) - (-f)(z)) = (-f)^C(x,u).$$

Similarly, for $v := -u$, one has $(-f)^C(x, -v) \leq f^C(x,v)$, hence $(-f)^C(x,u) = f^C(x, -u)$. □

Using duality, corresponding properties can be derived for the subdifferential.

Proposition 5.3. *Let $f \in \mathscr{L}(W)$ and let $x \in W$. Then*

(a) The set $\partial_C f(x)$ is a nonempty weak* compact convex subset of X^*;
(b) If f is Lipschitzian with rate k near x then $\|x^*\| \leq k$ for each $x^* \in \partial_C f(x)$;
(c) If $((x_n, x_n^*))_n$ is a sequence in $W \times X^*$ such that $(x_n) \to x$, $x_n^* \in \partial_C f(x_n)$ for all n and (x_n^*) has a weak* cluster point x^*, then x^* belongs to $\partial_C f(x)$;
(d) $\partial_C(-f)(x) = -\partial_C f(x)$.

Proof. (a), (b) The nonemptiness of $\partial_C f(x) := \partial f^C(x, \cdot)(0)$ results from the Hahn–Banach theorem and the fact that $f^C(x, \cdot)$ is sublinear and continuous. If f is Lipschitzian with rate k near x, then for $x^* \in \partial_C f(x)$, one has $|\langle x^*, \cdot \rangle| \leq |f^C(x, \cdot)| \leq k \|\cdot\|$, so that $\|x^*\| \leq k$. Since $\partial_C f(x)$ is clearly weak* closed, it is weak* compact.

(c) Let $((x_n, x_n^*))_n$ be a sequence in $W \times X^*$ such that $(x_n) \to x$, $x_n^* \in \partial_C f(x_n)$ for all n and (x_n^*) has a weak* cluster point x^*. Then for all $u \in X$ one has

$$\langle x^*, u \rangle \leq \limsup_n \langle x_n^*, u \rangle \leq \limsup_n f^C(x_n, u) \leq f^C(x, u)$$

by assertion (c) of Proposition 5.2, so that $x^* \in \partial_C f(x)$.

(d) One has $x^* \in \partial_C(-f)(x)$ iff for all $u \in X$, setting $v := -u$, one has $\langle x^*, v \rangle \leq (-f)^C(x,v) = f^C(x, -v)$ or $\langle -x^*, u \rangle = \langle x^*, v \rangle \leq f^C(x,u)$ iff $-x^* \in \partial_C f(x)$. □

Corollary 5.4. *If $f \in \mathscr{L}(W)$, then the multimap $\partial_C f(.)$ is upper semicontinuous on W for the weak* topology on X^*.*

Proof. Suppose, to the contrary, that for some weak* open subset V of X^* containing $\partial_C f(x)$ there exist sequences $(x_n) \to x$ and (x_n^*) with $x_n^* \in \partial_C f(x_n) \backslash V$ for all $n \in \mathbb{N}$. If f is Lipschitzian with rate k near x, then $\|x_n^*\| \leq k$ for all n large enough. Thus (x_n^*) has some weak* cluster point $x^* \in kB_{X^*}$. By assertion (c) of the preceding statement, we have $x^* \in \partial_C f(x)$. Then for some $n \in \mathbb{N}$, we have $x_n^* \in V$, a contradiction. Thus one has $\partial_C f(x') \subset V$ for x' close to x.   □

The correspondence between f^C and $\partial_C f$ can be inverted, thanks to the symmetry of the Minkowski–Hörmander duality between closed convex subsets of X^* and their support functions.

Proposition 5.5. *Let $f : W \to \mathbb{R}$ be a locally Lipschitzian function. Then*

$$\forall (x, v) \in W \times X, \qquad f^C(x, v) = \max\{x^*(v) : x^* \in \partial_C f(x)\}.$$

Later on, we will make a comparison with the subdifferential of convex analysis for f convex. Let us now make a comparison with the directional subdifferential.

Proposition 5.6. *Let $f : W \to \mathbb{R}$ be a locally Lipschitzian function. Then*

$$\forall (x, v) \in W \times X, \qquad f^C(x, v) \geq f^D(x, v), \quad \partial_D f(x) \subset \partial_C f(x).$$

In particular, if $f \in \mathscr{L}(W)$ is Gâteaux differentiable at x, one has $Df(x) \in \partial_C f(x)$. If f is circa-differentiable (=strictly differentiable) at x, then $\partial_C f(x) = \{Df(x)\}$.

Proof. The inequalities $f^C(x, v) \geq \limsup_{t \to 0_+} t^{-1}(f(x+tv) - f(x)) \geq f^D(x, v)$ are obvious and yield the announced inclusion. If f is Hadamard differentiable at x, in particular if f is a locally Lipschitzian function and is Gâteaux differentiable at x, we have $\partial_D f(x) = \{Df(x)\}$, hence $Df(x) \in \partial_C f(x)$.

When f is circa-differentiable at x, it is Lipschitzian around x, and from the definition of f^C one sees that $f^C(x, \cdot) = Df(x)$. It follows that $\partial_C f(x) = \{Df(x)\}$.   □

Exercise. When f is just differentiable at x, one may have $\partial_C f(x) \neq \{Df(x)\}$. Find an example. [Hint: Use a construction as in Exercise 7 of Sect. 5.2.]

Corollary 5.7. *For $f \in \mathscr{L}(W)$ and $x \in W$ one has $\partial_\ell^D f(x) \subset \partial_C f(x)$, where $\partial_\ell^D f(x)$ is the set of weak* cluster points of sequences (x_n^*) with $x_n^* \in \partial_D f(x_n)$ for some sequence (x_n) with limit x.*

Corollary 5.8. *If a locally Lipschitzian function $f : W \to \mathbb{R}$ attains its minimum on W at x, then $0 \in \partial_C f(x)$.*

Proof. In such a case one has $f^D(x, \cdot) \geq 0$, $0 \in \partial_D f(x)$, hence $0 \in \partial_C f(x)$.   □

Because $\partial_C f(x)$ may be large, such a conclusion is not always particularly informative. However, it can be exploited because the Clarke subdifferential satisfies good calculus rules, as we will show in the next subsection.

5.1.2 Calculus Rules in the Locally Lipschitzian Case

These rules are direct consequences of Proposition 5.3 and the rules for computing upper limits.

Proposition 5.9. *Let $f : W \to \mathbb{R}$ be a locally Lipschitz function on an open subset W of X. Then $\partial_C(-f)(x) = -\partial_C f(x)$ and for all $r \in \mathbb{R}$ one has $\partial_C(rf)(x) = r\partial_C f(x)$.*

Theorem 5.10. *Let $f, g : W \to \mathbb{R}$ be locally Lipschitz functions. Then*

$$\partial_C(f+g)(x) \subset \partial_C f(x) + \partial_C g(x),$$
$$\partial_C(f \vee g)(x) \subset \operatorname{co}(\partial_C f(x) \cup \partial_C g(x)).$$

Proof. The first relation is a consequence (via the equality $\partial_C f(x) = \partial f^C(x, \cdot)(0)$ and the corresponding rule of convex analysis) of the inequality

$$(f+g)^C(x, \cdot) \leq f^C(x, \cdot) + g^C(x, \cdot),$$

following from the fact that the upper limit of a sum is bounded above by the sum of the upper limits of the summands.

In order to prove the second relation it suffices to compare the support functions of both sides. Let $h := f \vee g := \max(f, g)$, let $u \in X$, and let $(x_n, t_n) \to (x, 0_+)$ be such that $h^C(x, u) = \lim_n t_n^{-1}(h(x_n + t_n u) - h(x_n))$. Taking a subsequence and exchanging the roles of f and g if necessary, we may suppose $h(x_n + t_n u) = f(x_n + t_n u)$ for all $n \in \mathbb{N}$. Then since $h(x_n) \geq f(x_n)$, we have

$$h^C(x, u) \leq \limsup_n t_n^{-1}(f(x_n + t_n u) - f(x_n)) \leq f^C(x, u) \leq f^C(x, u) \vee g^C(x, u).$$

Since $f^C(x, \cdot) \vee g^C(x, \cdot)$ is the support function of the weak* compact convex set $\operatorname{co}(\partial_C f(x) \cup \partial_C g(x))$, we get the result. □

These rules can be extended by induction to every finite families of $\mathscr{L}(W)$. More precisely, if $(f_1, \ldots, f_k)$ is a finite family of $\mathscr{L}(W)$, setting

$$f(x) := \max_{1 \leq i \leq k} f_i(x), \qquad I(x) := \{i \in \{1, \ldots, k\} : f_i(x) = f(x)\},$$

using the fact that $f = \max\{f_i : i \in I(x)\}$ on a neighborhood of x, one has

$$\partial_C f(x) \subset \operatorname{co}(\cup_{i \in I(x)} \partial_C f_i(x)).$$

More rules can be obtained with the help of an appropriate mean value theorem. We deduce this mean value theorem from the following special composition rule.

Lemma 5.11. *Suppose $f : W \to \mathbb{R}$ is locally Lipschitzian and W contains the segment $[x, y]$. Then the function h given by $h(r) := f(x_r)$ with $x_r := x + r(y - x)$ is Lipschitzian on $[0, 1]$ and for all $r \in [0, 1]$ one has*

$$\partial_C h(r) \subset \langle \partial_C f(x_r), y - x \rangle := \{ \langle x^*, y - x \rangle : x^* \in \partial_C f(x_r) \}. \tag{5.1}$$

Proof. The fact that h is Lipschitzian stems from the compactness of $[0,1]$. Since the two closed convex sets appearing in relation (5.1) are compact intervals of $\mathbb{R}$, it suffices to prove that for $v = -1, +1$ one has

$$\max\{ r^* v : r^* \in \partial_C h(r) \} \le \max\{ \langle x^*, y - x \rangle v : x^* \in \partial_C f(x_r) \}.$$

The left-hand side is $h^C(r, v)$. Since $x + s(y - x) \to x_r$ as $s \to r$, the right-hand side is

$$f^C(x_r, v(y - x)) \ge \limsup_{(t,s) \to (0_+, r)} \frac{1}{t} \left(f(x + s(y - x) + tv(y - x)) - f(x + s(y - x)) \right)$$

$$= \limsup_{(t,s) \to (0_+, r)} \frac{1}{t} \left(h(s + tv) - h(s) \right) = h^C(r, v),$$

and the inequality justifying the statement is proved.   □

Theorem 5.12 (Lebourg's mean value theorem [621]). *Let $f : W \to \mathbb{R}$ be locally Lipschitzian on an open subset W of X containing $[x, y]$. Then there exist some $w \in \,]x, y[:= \{ (1 - t)x + ty : t \in (0, 1) \}$, $w^* \in \partial_C f(w)$ such that*

$$f(y) - f(x) = \langle w^*, y - x \rangle.$$

Proof. Let $h : \mathbb{R} \to \mathbb{R}$ be given by $h(r) := f(x_r)$ for $x_r := x + r(y - x)$ and let k be given by $k(r) = h(r) + r[f(x) - f(y)]$. Since k is continuous and $k(0) = k(1) = f(x)$, there is some $r \in (0, 1)$ such that k attains either its minimum or its maximum on $[0, 1]$ at r. Let $w := x_r$. By Corollary 5.8 one has $0 \in \partial_C k(r)$. Using Theorem 5.10 and Lemma 5.11, one gets $0 \in \langle \partial_C f(w), y - x \rangle + f(x) - f(y)$.   □

We are in a position to derive chain rules and other calculus rules.

Theorem 5.13 (Chain rule). *Let X and Y be Banach spaces, let W (resp. Z) be an open subset of X (resp. Y), let $g : W \to Y$, $h : Z \to \mathbb{R}$ be locally Lipschitz maps such that $g(W) \subset Z$, and let $f := h \circ g$.*

(a) *If $Y = \mathbb{R}^n$, setting $g := (g_1, \ldots, g_n)$ and denoting by $\overline{co}^*(A)$ the w^*-closed convex hull of a subset A of X^*, for all $x \in W$ one has*

$$\partial_C f(x) \subset \overline{co}^* \{ \partial_C (y^* \circ g)(x) : y^* \in \partial_C h(g(x)) \} \tag{5.2}$$

$$\subset \overline{co}^* \left\{ \sum_{i=1}^{n} y_i^* \partial_C g_i(x) : y^* = (y_1^*, \ldots, y_n^*) \in \partial_C h(g(x)) \right\}. \tag{5.3}$$

(b) *If g is circa-differentiable (=strictly differentiable) at $x \in W$, then*

$$\partial_C f(x) \subset \partial_C h(g(x)) \circ g'(x).$$

(c) *If, moreover, $g'(x)(X) = Y$, this inclusion is an equality.*

Proof. The local Lipschitz property of f is straightforward. Let $x \in W$, $y := g(x)$.

(a) The inclusion $\partial_C f(x) \subset \overline{co}^*(A)$, with $A := \{\partial_C(y^* \circ g)(x) : y^* \in \partial_C h(y)\}$, is equivalent to the inequality $f^C(x, \cdot) \le \sup\langle A, \cdot \rangle$ between the respective support functions. Given $u \in X$, we pick a sequence $((t_n, x_n))_n \to (0_+, x)$ such that $\lim_n (1/t_n)(f(x_n + t_n u) - f(x_n)) = f^C(x, u)$. Let $y_n \in [g(x_n), g(x_n + t_n u)]$ and $y_n^* \in \partial_C h(y_n)$ be given by the Lebourg's theorem (Theorem 5.12):

$$f(x_n + t_n u) - f(x_n) h(g(x_n + t_n u)) - h(g(x_n)) = \langle y_n^*, g(x_n + t_n u) - g(x_n) \rangle. \quad (5.4)$$

Since g is continuous, we have $(y_n) \to y := g(x)$, and since $\partial_C h$ is locally bounded, we may assume that (y_n^*) has a weak* cluster point y^* that belongs to $\partial_C h(y)$ by assertion (c) of Proposition 5.3.

Since $(t_n^{-1}(g(x_n + t_n u)) - g(x_n)))_n$ is bounded and $(y_n^*) \to y^*$ in norm, since Y is finite-dimensional, we get $(\langle y_n^* - y^*, t_n^{-1}(g(x_n + t_n u) - g(x_n)) \rangle)_n \to 0$, whence

$$f^C(x, u) = \lim_n \frac{\langle y^*, g(x_n + t_n u) - g(x_n) \rangle}{t_n} \le (y^* \circ g)^C(x, u).$$

Now, there exists some $x^* \in \partial_C(y^* \circ g)(x)$ such that $(y^* \circ g)^C(x, u) = \langle x^*, u \rangle$. Thus $x^* \in A$ and $f^C(x, u) \le \langle x^*, u \rangle \le \sup\langle A, u \rangle$. Relation (5.3) follows from relation (5.2), Theorem 5.10, and Proposition 5.9, since $y^* \circ g = y_1^* g_1 + \cdots + y_n^* g_n$.

(b) Let us suppose now that g is circa-differentiable at $x \in W$. Let $u \in X$ and let $((t_n, x_n))_n \to (0_+, x)$ be chosen as above. Applying again the mean value theorem, we get some $y_n \in [g(x_n), g(x_n + t_n u)]$ and some $y_n^* \in \partial_C h(y_n)$ such that relation (5.4) holds. Let $x^* \in \partial_C f(x)$ and let y^* be a weak* cluster point of (y_n^*). By circa-differentiability of g, we have $(t_n^{-1}(g(x_n + t_n u) - g(x_n)))_n \to g'(x)(u)$, so that

$$\lim_n \frac{1}{t_n}(f(x_n + t_n u) - f(x_n)) = \lim_n \frac{1}{t_n}\langle y_n^*, g(x_n + t_n u) - g(x_n) \rangle = \langle y^*, g'(x)(u) \rangle.$$

It follows that $\langle x^*, u \rangle \le f^C(x, u) \le \sup\{\langle y^* \circ g'(x), u \rangle : y^* \in \partial_C h(y)\}$. Since $\partial_C h(y)$ is weak* compact, $\partial_C h(y) \circ g'(x) = g'(x)^\mathsf{T}(\partial_C h(y))$ is weak* compact, hence weak* closed (and convex). Since u is arbitrary in X, it follows that $x^* \in g'(x)^\mathsf{T}(\partial_C h(y))$.

(c) Finally, suppose that g is circa-differentiable at $x \in W$ and $g'(x)(X) = Y$. Then the Graves–Lyusternik theorem ensures that g is open at x. Thus, for every sequence $(y_n) \to y$, one can find a sequence $(x_n) \to x$ such that $g(x_n) = y_n$ for n large. Given $u \in X$, $v := g'(x)(u)$, let us take $(y_n) \to y$, $(t_n) \to 0_+$ such that $h^C(y, v) = \lim_n (1/t_n)(h(y_n + t_n v) - h(y_n))$. Now, since g is circa-differentiable at x, we have $y_n + t_n v = g(x_n) + t_n g'(x)u = g(x_n + t_n u) + t_n z_n$, where $(z_n) \to 0$. Since h is locally Lipschitzian, we get $h^C(y, v) = \lim_n (1/t_n)(h(g(x_n + t_n u))) - h(g(x_n)) \le f^C(x, u)$. Thus, for all $y^* \in \partial_C h(y)$, we have $y^* \circ g'(x) \le f^C(x, \cdot)$ and $y^* \circ g'(x) \in \partial_C f(x)$. $\square$

Results involving order are scarce for Clarke subdifferentials. In particular, the homotonicity property $\partial f(\bar{x}) \subset \partial g(\bar{x})$ when $f \le g$ and $f(\bar{x}) = g(\bar{x})$ is not satisfied for $\partial = \partial_C$, as the example of $f := -|\cdot|$, $g = 0$ on $X := \mathbb{R}$, with $\bar{x} := 0$ shows. Thus, the next results are noteworthy.

Proposition 5.14. *Let V and W be two Banach spaces, let $A \in L(V,W)$ with $W = A(V)$, $\overline{v} \in V$, $\overline{w} := A\overline{v}$ and let j and p be locally Lipschitzian functions on V and W respectively such that $p \circ A \le j$ and such that for all sequences $(w_n) \to \overline{w}$, $(\alpha_n) \to 0_+$ one can find a sequence $(v_n) \to \overline{v}$ satisfying $A(v_n) = w_n$ and $j(v_n) \le p(w_n) + \alpha_n$ for all $n \in \mathbb{N}$. Then one has*

$$A^{\mathsf{T}}(\partial_C p(\overline{w})) \subset \partial_C j(\overline{v}). \tag{5.5}$$

When p is the distance function d_S to some closed subset S of W, the requirement can be restricted to sequences $(w_n) \to \overline{w}$ in S.

Proof. Let $\overline{w}^* \in \partial_C p(\overline{w})$ and let $v \in V$, $w := Av$. Let us pick sequences $(t_n) \to 0_+$, $(w_n) \to \overline{w}$ such that $(1/t_n)(p(w_n + t_n w) - p(w_n)) \to p^C(\overline{w}, w)$. By assumption, we can find a sequence $(v_n) \to \overline{v}$ such that $A(v_n) = w_n$ and $j(v_n) \le p(w_n) + t_n^2$ for all $n \in \mathbb{N}$. Since $p(w_n + t_n w) \le j(v_n + t_n v)$ and $-p(w_n) \le -j(v_n) + t_n^2$, we have

$$\langle A^{\mathsf{T}} \overline{w}^*, v \rangle = \langle \overline{w}^*, w \rangle \le p^C(\overline{w}, w) \le \limsup_n \frac{1}{t_n}(j(v_n + t_n v) - j(v_n) + t_n^2) \le j^C(\overline{v}, v).$$

Since v is an arbitrary element of V, this means that $A^{\mathsf{T}} \overline{w}^* \in \partial_C j(\overline{v})$.

When $p := d_S$ for a closed subset S of W, given $\overline{w}^* \in \partial_C p(\overline{w})$, $v \in V$, $w := Av$, $(t_n) \to 0_+$, $(w_n) \to \overline{w}$ such that $(1/t_n)(d_S(w_n + t_n w) - d_S(w_n)) \to d_S^C(\overline{w}, w)$, we pick $w_n' \in S$ such that $\|w_n' - w_n\| \le d_S(w_n) + t_n^2$. Then we can find a sequence $(v_n) \to \overline{v}$ such that $A(v_n) = w_n'$ and $j(v_n) \le t_n^2$ for all $n \in \mathbb{N}$. Since $d_S(w_n + t_n w) \le d_S(w_n' + t_n w) + \|w_n - w_n'\|$, we have

$$d_S^C(\overline{w}, w) \le \limsup_n t_n^{-1} \left[(d_S(w_n' + t_n w) + \|w_n - w_n'\|) - (\|w_n' - w_n\| - t_n^2) \right]$$

$$\le \limsup_n t_n^{-1} \left[j(v_n + t_n v) + t_n^2 \right] \le \limsup_n t_n^{-1} \left[j(v_n + t_n v) - j(v_n) + 2t_n^2 \right]$$

$$\le j^C(\overline{v}, v).$$

Thus $\langle A^{\mathsf{T}} \overline{w}^*, v \rangle = \langle \overline{w}^*, w \rangle \le d_S^C(\overline{w}, w) \le j^C(\overline{v}, v)$ for all $v \in X$ and $A^{\mathsf{T}} \overline{w}^* \in \partial_C j(\overline{v})$. $\square$

Exercises

1. For $f, g \in \mathscr{L}(W)$, show that $\partial_C(f \wedge g)(x) \subset \mathrm{co}(\partial_C f(x) \cup \partial_C g(x))$ with $f \wedge g := \min(f, g)$.

2. (Danskin's theorem) Let X be a Banach space and let S be a compact metric space. Given a function $g : S \times X \to \mathbb{R}$ that is jointly continuous and whose derivative with respect to the second variable exists and is jointly continuous, let f be the *marginal function* given by

$$f(x) := \max_{s \in S} g(s,x).$$

Show that f is locally Lipschitzian and that for all $x, v \in X$ one has

$$f^C(x,v) = f^D(x,v) = \max\{D_2 g(s,x).v : s \in S(x)\},$$

where $S(x) := \{s \in S : g(s,x) = f(x)\}$. Show that $\partial_C f(x) = \mathrm{co}\{D_2 g(s,x) : s \in S(x)\}$.

3. (a) With the assumptions of the preceding exercise, suppose that for some $\bar{x} \in X$ one has $D_2 g(s',\bar{x}) \neq D_2 g(s'',\bar{x})$ whenever $s' \neq s''$ in S. Assuming that f is differentiable at $\bar{x}$, show that $S(\bar{x})$ is a singleton.
(b) Let E be a closed subset of a Euclidean space X and let $\bar{x} \in X \backslash E$ be such that d_E is differentiable at $\bar{x}$. Show that there is a unique best approximation $\bar{e}$ of $\bar{x}$ in E and that $\nabla d_E(\bar{x}) = (\bar{x} - \bar{e})/\|\bar{x} - \bar{e}\|$. [Hint: Take $S := E \cap B(\bar{x},r)$ for $r > d_E(\bar{x})$ and define g by $g(s,x) := -\|s - x\|^2$.]

4. With the notation of Exercise 2, suppose that S is a compact subset of the dual Y^* of a Banach space Y and that $g(s,x) = \langle s, h(x) \rangle$ for $(s,x) \in S \times X$, where $h : X \to Y$ is a locally Lipschitzian map. Check that f is locally Lipschitzian and show that when for some $\bar{x} \in X$, the set $S(\bar{x})$ is a singleton $\{\bar{y}^*\}$, then $\partial_C f(\bar{x}) \subset \partial_C(\bar{y}^* \circ h)(\bar{x})$.

5. For $f \in \mathscr{L}(W)$, where W is an open subset of X and $x \in W, v \in X$, show that $f^C(x,v) = \limsup_{w \to x} f^D(w,v)$, where $f^D(w,\cdot)$ is the lower derivate of f at w.

6. (Leibniz rule) Let $g, h \in \mathscr{L}(W)$, where W is an open subset of X. Show that

$$\partial_C(gh)(x) \subset \partial_C(g(x)h(\cdot) + h(x)g(\cdot))(x) \subset g(x)\partial_C h(x) + h(x)\partial_C g(x).$$

Suppose $h(W) \subset \mathbb{P}$. Show that $\partial_C(g/h)(x) \subset h(x)^{-2}(h(x)\partial_C g(x) - g(x)\partial_C h(x))$.

7. Let X be a Hilbert space identified with its dual X^* and let $f \in \mathscr{L}(X)$. Given $\bar{x} \in X$, let $\bar{x}^*$ be the element of least norm of $\partial_C f(\bar{x})$. Show that $-\bar{x}^*$ is a direction of descent for f in the sense that for $t > 0$ small enough, one has $f(\bar{x} - t\bar{x}^*) \leq f(\bar{x}) - (1/2)t\|\bar{x}^*\|^2$.

5.1.3 The Clarke Jacobian and the Clarke Subdifferential in Finite Dimensions

In finite dimensions, a characterization of the Clarke subdifferential can be given using the famous Rademacher's theorem, several nice proofs of which are available (see [218, 360], for instance).

Theorem 5.15 (Rademacher). *A locally Lipschitzian function f on $\mathbb{R}^d$ is differentiable on a set whose complement has (Lebesgue) measure 0.*

We can deduce from this result an important representation of $\partial_C f$.

Theorem 5.16. *Let* $f : W \to \mathbb{R}$ *be a Lipschitzian function on an open subset* W *of* $\mathbb{R}^d$. *Let* N *be a set of measure zero in* W *and let* N_f *be the set of points of* W *at which* f *is not differentiable. Then, for all* $\bar{x} \in W$, $\partial_C f(\bar{x})$ *is the convex hull* $C(\bar{x})$ *of the set* $A(\bar{x})$ *of limits of sequences* (x_n^*) *such that for some sequence* $(x_n) \to \bar{x}$ *in* $Z := W \backslash (N \cup N_f)$ *one has* $x_n^* = f'(x_n)$ *for all* n:

$$\partial_C f(\bar{x}) = \mathrm{co}\{x^* : \exists (x_n) \to_Z \bar{x}, \exists (x_n^*) \to x^*, \forall n \in \mathbb{N} \; x_n^* = f'(x_n)\}. \tag{5.6}$$

Proof. Let us first observe that the set $A(\bar{x})$ between curly braces in the above formula is nonempty, since Z is dense in W and f' is bounded on Z. Thus $A(\bar{x})$ is compact and (by Carathéodory's theorem) $C(\bar{x})$ is compact too. Proposition 5.3 (c) ensures that $A(\bar{x})$ is contained in $\partial_C f(\bar{x})$, so that by the convexity of $\partial_C f(\bar{x})$, the convex hull $C(\bar{x})$ of $A(\bar{x})$ is also contained in $\partial_C f(\bar{x})$. In order to prove that this inclusion is an equality, it remains to prove that the support function of $C(\bar{x})$ (or $A(\bar{x})$) is not less than the support function $f^C(\bar{x}, \cdot)$ of $\partial_C f(\bar{x})$ or that for every unit vector v of $\mathbb{R}^d$,

$$f^C(\bar{x}, v) \leq \sup\{\langle x^*, v \rangle : x^* \in A(\bar{x})\}. \tag{5.7}$$

Let $r > \sup\{\langle x^*, v \rangle : x^* \in A(\bar{x})\}$. A compactness argument using the boundedness of $\partial_C f$ around $\bar{x}$ shows that there exists some $\delta > 0$ such that for all $z \in Z \cap B(\bar{x}, 2\delta)$ one has $f'(z).v < r$. Since $(N \cup N_f) \cap B(\bar{x}, 2\delta)$ has measure 0, it follows from Fubini's theorem that for almost all $z \in B(\bar{x}, \delta)$ the line segment $L_z := \{z + tv : t \in (0, \delta)\}$ meets $(N \cup N_f) \cap B(\bar{x}, 2\delta)$ in a set of one-dimensional measure zero. For such a z and $t \in (0, \delta)$ one has $z + [0, \delta]v \subset B(\bar{x}, 2\delta)$, hence $f'(z + sv) \cdot v \leq r$ and

$$f(z + tv) - f(z) = \int_0^t f'(z + sv) \cdot v \, \mathrm{d}s \leq rt. \tag{5.8}$$

Since f is continuous, this inequality is in fact valid for every $z \in B(\bar{x}, \delta)$ and every $t \in (0, \delta)$. It follows from the definition of f^C that $f^C(\bar{x}, v) \leq r$. $\square$

Let us give an outline of a notion that is related to the preceding result.

Definition 5.17. Given a locally Lipschitzian map $g : W \to \mathbb{R}^p$, where W is an open subset of $\mathbb{R}^n$, and a set N of measure zero in W, the *Clarke Jacobian* or *circa-Jacobian* of g at $\bar{x} \in \mathbb{R}^n$ is given by the following formula, in which $Z := W \backslash (N \cup N_g)$, N_g being the set of points of nondifferentiability of g:

$$\partial_C g(\bar{x}) = \mathrm{co} \partial_L g(\bar{x}), \quad \text{where} \quad \partial_L g(\bar{x}) := \{A : \exists (x_k) \to_Z \bar{x}, (g'(x_k)) \to A\}.$$

A version of Rademacher's theorem asserts that N_g has measure zero. Thus, we see that $\partial_C g(\bar{x})$ is a nonempty compact convex subset of $L(\mathbb{R}^n, \mathbb{R}^p)$ that is obviously contained in the product of the subdifferentials at $\bar{x}$ of the components of g. We admit the fact that $\partial_C g(\bar{x})$ does not depend on the choice of the measure-zero set N [956]. Theorem 5.16 ensures that for $p = 1$ this definition of $\partial_C g(\bar{x})$ coincides with the one given earlier.

Proposition 5.18 (Vectorial mean value theorem). *Let* $g : W \to \mathbb{R}^p$ *be locally Lipschitzian, where W is an open convex subset of* $\mathbb{R}^n$. *Then for all* $x, y \in W$ *one has*

$$g(y) - g(x) \in \operatorname{co}(\partial_C g([x,y]) \cdot (y - x)).$$

Proof. Let us first consider the case in which $[x,y] \cap N_g$ is a set of one-dimensional measure zero. Then since $\partial_C g([x,y])$ is compact,

$$g(y) - g(x) = \int_0^1 g'(x + t(y - x)) \cdot (y - x) \mathrm{d}t \subset \operatorname{co}(\partial_C g([x,y]) \cdot (y - x)).$$

The general case is obtained by a passage to the limit, using a decomposition of $\mathbb{R}^n$ as $(\mathbb{R}(y - x)) \oplus (\mathbb{R}(y - x))^\perp$ and Fubini's theorem. $\qquad\square$

Let us admit the next result, a chain rule involving the circa-Jacobian.

Theorem 5.19 ([214, Theorem 2.6.6]). *Let* $f := h \circ g$, *where* $g : W \to \mathbb{R}^p$ *and* $h : \mathbb{R}^p \to \mathbb{R}$ *are locally Lipschitzian, W being an open subset of* $\mathbb{R}^n$. *Then*

$$\forall x \in W, \qquad \partial_C f(x) \subset \operatorname{co}(\partial_C h(g(x)) \circ \partial_C g(x)).$$

More attention has been given to the Clarke Jacobian of g than to the notion we describe now. Given a map $g : X \to Y$ between two normed spaces and $x \in X$, let us introduce the multimap $\Delta_C g(x) : Y^* \rightrightarrows X^*$ with closed convex values defined by

$$\Delta_C g(x)(y^*) := \partial_C(y^* \circ g)(x), \qquad y^* \in Y^*.$$

When f is Lipschitzian around x, $\Delta_C g(x)$ is a bounded odd *fan* in the sense that there exists $\kappa > 0$ such that for all $y^*, z^* \in Y^*$, $r, s \in \mathbb{R}$ one has $\Delta_C g(x)(y^*) \subset \kappa \|y^*\| B_{X^*}$,

$$\Delta_C g(x)(ry^* + sz^*) \subset r\Delta_C g(x)(y^*) + s\Delta_C g(x)(z^*).$$

This fan can be related to the circa-Jacobian of g.

Corollary 5.20. *Let* $g : W \to Y := \mathbb{R}^p$ *be locally Lipschitzian, W being an open subset of* $\mathbb{R}^n$. *Then*

$$\forall x \in W, y^* \in Y^*, \qquad \Delta_C g(x)(y^*) = y^* \circ (\partial_C g(x)).$$

Proof. For every $x \in W$, $y^* \in Y^*$, the preceding theorem with $h := y^*$ ensures that

$$\Delta_C g(x)(y^*) := \partial_C(y^* \circ g)(x) \subset \operatorname{co}(y^* \circ \partial_C g(x)) = y^* \circ \partial_C g(x),$$

since $\partial_C g(x)$ is convex. Conversely, given $A \in \partial_C g(x)$, using Carathéodory's theorem in $\mathbb{R}^q$ with $q := np$, one can find some $t_i \in [0,1]$ with $t_0 + \cdots + t_q = 1$ and some $A_i \in \partial_L g(x) \subset L(\mathbb{R}^n, \mathbb{R}^q)$ for $i = 1, \ldots, q$ such that $A = t_0 A_0 + \cdots + t_q A_q$. By definition of $\partial_L g(x)$ there exist sequences $(x_{i,n})_n \to_Z x$, for $i = 0, \ldots, q$ such that $(Dg(x_{i,n}))_n \to A_i$ for all $i \in \{0, \ldots, q\}$. Then by upper semicontinuity of $\partial_C(y^* \circ g)$ and convexity of $\partial_C(y^* \circ g)(x)$ one has

$$y^* \circ A = \sum_{i=0}^{q} t_i y^* \circ A_i = \lim_n \left(\sum_{i=0}^{q} t_i D(y^* \circ g)(x_{i,n}) \right) \in \partial_C(y^* \circ g)(x). \qquad \square$$

Exercises

1. Use Theorem 5.16 to compute $\partial_C f(0,0)$, where $f : \mathbb{R}^2 \to \mathbb{R}$ is defined by $f(x,y) := \max(\min(x-y, -x-y), y)$. [Hint: Identify the set N_f of points of nondifferentiability of f as

$$N_f := \{(x,y) : x < 0, \ x = 2y\} \cup \{(x,y) : x > 0, \ x = -2y\} \cup \{0\} \times \mathbb{R}_-$$

and consider the three open connected components of $\mathbb{R}^2 \setminus N_f$ on which f takes the values $y, x-y, -x-y$ respectively; conclude that $\partial_C f(0,0)$ is the convex hull of the points $(0,1), (1,-1), (-1,-1)$.]

2. (a) Use an example similar to the one in Exercise 1 to show that for a locally Lipschitzian function f on the product $X := X_1 \times X_2$ of two normed spaces, neither of the sets $\partial_C f(x_1, x_2)$, $\partial_{1,C} f(x_1, x_2) \times \partial_{2,C} f(x_1, x_2)$ is larger than the other one. Here $\partial_{1,C} f(x_1, x_2) := \partial_C f(\cdot, x_2)(x_1)$, and a similar notation is used for $\partial_{2,C} f(x_1, x_2)$.
(b) Prove that $\partial_{1,C} f(x_1, x_2) \subset p_1(\partial_C f(x_1, x_2))$, where $p_1 : X^* \to X_1^*$ is the canonical projection (the transpose of $x_1 \mapsto (x_1, 0)$).

3. Let X_1, X_2 be finite-dimensional Banach spaces and let $f \in \mathcal{L}(X_1 \times X_2)$ be such that for all $x_2 \in X$, $f(\cdot, x_2)$ is convex. Using the notation of the preceding exercise, show that for every $x_1 \in X_1$, $x_2 \in X_2$ one has $p_1(\partial_C f(x_1, x_2)) \subset \partial_{1,C} f(x_1, x_2)$. [Hint: Use Theorem 5.16.]

4. Given a bounded measurable function g on an interval I of $\mathbb{R}$, for $x \in \text{int}(I)$ let

$$a(x) := \text{ess}\sup g(x) := \sup\{r : \forall \delta > 0, \mu(\{x' \in [x - \delta, x + \delta] : g(x') > r\}) > 0\}$$

and let $b(x) := \text{ess}\inf g(x) := -\text{ess}\sup(-g)(x)$, μ being the Lebesgue measure on I. Given $c \in I$, let $f(x) := \int_c^x g(s) d\mu(s)$ for $x \in I$. Show that $\partial_C f(x) = [b(x), a(x)]$.

5.2 Circa-Normal and Circa-Tangent Cones

A geometric approach parallels the analytical approach we have followed up to now. As in the case of the directional subdifferential, it has strong links with the analytical approach and is available in both a primal form and a dual form. It is a key to a simple way of extending the preceding concepts and rules to non-Lipschitzian functions, a question we will deal with in the next section.

We start with the primal concept of tangent cone. Here and elsewhere, $x \to_E a$ means that x converges to a while remaining in the subset E.

Definition 5.21. Given a subset E of a normed space X and $a \in cl(E)$, the *Clarke tangent cone* (or *circa-tangent cone*) to E at a is the set $T^C(E,a)$ of $v \in X$ such that for all sequences $(e_n) \to_E a$, $(t_n) \to 0_+$ there exists a sequence (v_n) with limit v such that $e_n + t_n v_n \in E$ for all $n \in \mathbb{N}$. Thus

$$T^C(E,a) := \liminf_{(e,t) \to_{E \times \mathbb{P}} (a,0)} \frac{1}{t}(E - e).$$

A connection with the Clarke derivate is given in the following statement.

Lemma 5.22 (Hiriart–Urruty [481]). *A vector v belongs to $T^C(E,a)$ if and only if $d_E^C(a,v) \le 0$, where d_E is the distance function associated with E.*

Proof. Let $v \in T^C(E,a)$ and let $(t_n, a_n) \to (0_+, a)$ be such that

$$d_E^C(a,v) = \lim_n \frac{1}{t_n} \left(d_E(a_n + t_n v) - d_E(a_n) \right).$$

Let us pick $e_n \in E$ such that $\|e_n - a_n\| < d_E(a_n) + t_n^2$. Since $(e_n) \to_E a$, the definition of $T^C(E,a)$ yields a sequence $(v_n) \to v$ such that $e_n + t_n v_n \in E$ for all $n \in \mathbb{N}$. Then

$$d_E(a_n + t_n v) \le \|(a_n + t_n v) - (e_n + t_n v_n)\| \le d_E(a_n) + t_n^2 + t_n \|v - v_n\|,$$

so that $\lim_n t_n^{-1} \left(d_E(a_n + t_n v) - d_E(a_n) \right) \le \lim_n (t_n + \|v - v_n\|) = 0$ and $d_E^C(a,v) \le 0$.

Conversely, let $v \in X$ be such that $d_E^C(a,v) \le 0$. Then for all sequences $(t_n) \to 0_+$, $(e_n) \to_E a$, we have $\limsup_n t_n^{-1} \left(d_E(e_n + t_n v) - d_E(e_n) \right) \le 0$. Since $e_n \in E$, we have $d_E(e_n) = 0$. Let $e_n' \in E$ be such that $\|e_n + t_n v - e_n'\| \le d_E(e_n + t_n v) + t_n^2$ and let $v_n := t_n^{-1}(e_n' - e_n)$. Then we have $e_n + t_n v_n = e_n' \in E$ for each $n \in \mathbb{N}$ and

$$\limsup_n \|v - v_n\| = \limsup_n t_n^{-1} \|e_n + t_n v - e_n'\| \le \limsup_n t_n^{-1} (d_E(e_n + t_n v) + t_n^2) \le 0.$$

Thus $(v_n) \to v$ and $v \in T^C(E,a)$. □

Exercise. Check that for every subset E of a normed space X, every $a \in cl(E)$, and every $v \in X$ one has $d_E^C(a,v) \ge 0$. Deduce from this the relation $T^C(E,a) = \{v \in$

$X : d_E^C(a,v) = 0\}$. [Hint: For all sequences $(e_n) \to_E a$, $(t_n) \to 0_+$ one has $d_E^C(a,v) \geq \limsup_n t_n^{-1}(d_E(e_n + t_n v) - d_E(e_n)) \geq 0$.]

The main feature of the Clarke tangent cone is given in the following proposition.

Proposition 5.23. *The Clarke tangent cone $T^C(E,a)$ to E at $a \in \mathrm{cl}(E)$ is a closed convex cone contained in the tangent cone $T(E,a)$ to E at a (and is even contained in the incident tangent cone $T^I(E,a) := \liminf_{t \to 0_+} t^{-1}(E - a)$). Moreover, one has*

$$T(E,a) + T^C(E,a) \subset T(E,a),$$

$$T^I(E,a) + T^C(E,a) \subset T^I(E,a),$$

and for every subset F of X such that $E \subset F \subset \mathrm{cl}(E)$ one has $T^C(F,a) = T^C(E,a)$.

These inclusions may help in computing $T^C(E,a)$, as the next examples below show.

Proof. The closedness of $T^C(E,a)$ is a general property of inner limits. The stability of $T^C(E,a)$ by homotheties is obvious. The stability of $T^C(E,a)$ under addition stems from the preceding lemma and the sublinearity of $d_E^C(a,\cdot)$; we also encourage the reader to give a direct proof.

Now let $u \in T(E,a)$ and $v \in T^C(E,a)$. There exists a sequence $((t_n, u_n)) \to (0_+, u)$ such that $e_n := a + t_n u_n \in E$ for all $n \in \mathbb{N}$. Then one can find a sequence $(v_n) \to v$ such that $e_n + t_n v_n \in E$ for all $n \in \mathbb{N}$. Thus $a + t_n(u_n + v_n) \in E$ for all $n \in \mathbb{N}$, and since $(u_n + v_n) \to u + v$, one gets $u + v \in T(E,a)$. The proof with $T(E,a)$ replaced with $T^I(E,a)$ is similar.

Since $0 \in T^I(E,a)$, we deduce from the inclusion $T^I(E,a) + T^C(E,a) \subset T^I(E,a)$ that $T^C(E,a) \subset T^I(E,a) \subset T(E,a)$.

Now, if $E \subset F \subset \mathrm{cl}(E)$ and $v \in T^C(E,a)$, for all sequences $(t_n) \to 0_+$, $(f_n) \to_F a$, we can find a sequence (e_n) in E such that $\|e_n - f_n\| \leq t_n^2$ for all n. Then $(e_n) \to_E a$, and if $(v_n) \to v$ is such that $e_n + t_n v_n \in E$ for all n, one has $f_n + t_n w_n \in E \subset F$ for $w_n := v_n + t_n^{-1}(e_n - f_n)$ and $(w_n) \to v$, so that $v \in T^C(F,a)$. The inclusion $T^C(F,a) \subset T^C(E,a)$ is proved similarly. $\square$

The definition of $T^C(E,a)$ shows that this cone may be very small and may fail to give a local approximation to the subset E of X.

Example 5.1. Let $E := (\mathbb{R} \times \{0\}) \cup (\{0\} \times \mathbb{R}) \subset X := \mathbb{R}^2$, $a := (0,0)$. Then $T^C(E,a) = \{(0,0)\}$, whereas $T^I(E,a) = T(E,a) = E$.

Example 5.2 (The pie test). Let $E := \{(r,s) \in \mathbb{R}^2 : s \geq -|r|\}$, $a = (0,0)$. Then $T^C(E,a) = \{(r,s) \in \mathbb{R}^2 : s \geq |r|\}$, whereas $T(E,a) = E$.

Example 5.3 (Rockafellar [880]). Let X, Y be normed spaces, let W be an open subset of X, and let $g : W \to Y$ be a Lipschitzian map with rate c. Then the Clarke tangent cone $T^C(G,(a,b))$ to the graph G of g at $(a,b) \in G$ is not just a convex cone but a linear subspace. Moreover, $(u,v) \in T^C(G,(a,b))$ iff g is *directionally circa-differentiable* at a in the direction u in the sense that for all sequences $(a_n) \to a$,

$(t_n) \to 0_+$, $(u_n) \to u$, the sequence $((g(a_n + t_n u_n) - g(a_n)/t_n)_n$ converges to some vector $y \in Y$ and $y = v$. In order to prove these assertions, let $(u,v) \in T^C(G,(a,b))$, so that for all sequences $(a_n) \to a$, $(t_n) \to 0_+$, since $(b_n) := (g(a_n)) \to b := g(a)$, one can find sequences $(u_n) \to u$, $(v_n) \to v$ such that $(a_n, b_n) + t_n(u_n, v_n) \in G$ for all n. Then $g(a_n + t_n u_n) - g(a_n) = t_n v_n$ and g is circa-differentiable at a in the direction u with derivative $g'(a)u = v$. Conversely, when this property occurs, for all sequences $((a_n, b_n)) \to_G (a,b)$, $(t_n) \to 0_+$, $(u_n) \to u$, for $v_n := (1/t_n)(g(a_n + t_n u_n) - g(a_n))$, one has $(v_n) \to v := g'(a)(u)$ and $(a_n, b_n) + t_n(u_n, v_n) \in G$ for all n, so that $(u,v) \in T^C(G,(a,b))$. Since $g'(a)(-u) = -g'(a)(u)$ by the above definition, the cone $T^C(G,(a,b))$ is a linear subspace.

Example 5.4. If G is the graph of the function $(x,r) \longmapsto \|x\|$ on $X \times \mathbb{R}$, then $T^C(G,(0,0,0)) = \{0\} \times \mathbb{R} \times \{0\}$ although G is a Lipschitzian submanifold of $X \times \mathbb{R} \times \mathbb{R}$ whose dimension is the dimension of X plus 1. In this illustration of the preceding example we see that the local behavior of G around $(a,b) = ((0,0),0)$ is not accurately reflected by $T^C(G,(a,b))$.

Definition 5.24. The *Clarke normal cone* or *circa-normal cone* $N_C(E,a)$ to a subset E of X at $a \in \mathrm{cl}(E)$ is the polar cone of $T^C(E,a)$:

$$N_C(E,a) := \left(T^C(E,a)\right)^0 := \{x^* \in X^* : \forall v \in T^C(E,a) \langle x^*, v \rangle \le 0\}.$$

Since $T^C(E,a)$ may be small, $N_C(E,a)$ may be correspondingly large, and a relation of the form $x^* \in -N_C(E,a)$ may be poorly informative. In Example 5.1 above, for instance, one has $N_C(E,a) = X^*$, whereas $N(E,a) = \{0\}$.

Since $T^C(E,a)$ is a closed convex cone, the first part of the following proposition is a consequence of the bipolar theorem.

Proposition 5.25. *(a) The Clarke normal cone $N_C(E,a)$ to a subset E of X at $a \in \mathrm{cl}(E)$ is a weak* closed convex cone. The Clarke tangent cone $T^C(E,a)$ is in turn the polar cone to $N_C(E,a)$.*
(b) $N_C(E,a) = \mathrm{cl}^*(\mathbb{R}_+ \partial_C d_E(a))$.

Proof. It remains to prove assertion (b). Let $r \in \mathbb{R}_+$ and $x^* \in \partial_C d_E(a)$. Lemma 5.22 asserts that for every $v \in T^C(E,a)$ one has $d_E^C(a,v) \le 0$, hence $\langle rx^*, v \rangle \le 0$ and $rx^* \in N_C(E,a)$. Since $N_C(E,a)$ is weak* closed, we get $\mathrm{cl}^*(\mathbb{R}_+ \partial_C d_E(a)) \subset N_C(E,a)$.

Now let $a^* \in X^* \setminus \mathrm{cl}^*(\mathbb{R}_+ \partial_C d_E(a))$. The separation theorem yields some $v \in X$ such that $\langle a^*, v \rangle > \sup\{\langle rx^*, v \rangle : r \in \mathbb{R}_+, x^* \in \partial_C d_E(a)\}$. Since r is arbitrarily large, one has $\langle x^*, v \rangle \le 0$ for all $x^* \in \partial_C d_E(a)$, hence $d_E^C(a,v) \le 0$ and $v \in T^C(E,a)$. Picking $x^* \in \partial_C d_E(a)$, one has $\langle a^*, v \rangle > \langle 0x^*, v \rangle = 0$, hence $a^* \notin \left(T^C(E,a)\right)^0 := N_C(E,a)$, and the proof is complete. $\square$

The map $E \mapsto N_C(E,a)$ is not antitone (i.e., order-reversing) and the map $E \mapsto T^C(E,a)$ is not homotone (i.e., order-preserving), as the following examples show.

Example 5.5. Let $X = \mathbb{R}^2$, $E := (\mathbb{R} \times \{0\}) \cup (\{0\} \times \mathbb{R})$, $F = \mathbb{R} \times \{0\}$, $a := (0,0)$. Then $T^C(E,a) = \{(0,0)\}$, $T^C(F,a) = F$ and $N_C(E,a) = \mathbb{R}^2$, $N_C(F,a) = \{0\} \times \mathbb{R}$.

Example 5.6. Let $X = \mathbb{R}^2$, $E := \{(r,s) : s \le |r|\}$, $F := \mathbb{R} \times \mathbb{R}_-$, $a := (0,0)$. Then $T^C(E,a) = \{(r,s) : s \le -|r|\}$, $T^C(F,a) = \mathbb{R} \times \mathbb{R}_-$ and $N_C(E,a) = \{(r,s) : s \ge |r|\}$, $N_C(F,a) = \{0\} \times \mathbb{R}_+$.

In spite of this major drawback, the Clarke tangent cones and normal cones coincide with the usual concepts in some important cases.

Proposition 5.26. (a) If E is a convex subset of X and $a \in \text{cl}(E)$, then $T^C(E,a) = T(E,a)$ and $N_C(E,a) = N(E,a) = \{x^* \in X^* : \forall x \in E \ \langle x^*, x-a \rangle \le 0\}$.
(b) If E is a submanifold of class C^1 of X, then $T^C(E,a) = T(E,a)$ and $N_C(E,a) = N(E,a)$.

Proof. (a) Let $v \in \mathbb{R}_+(E-a)$, $v = r(e-a)$ with $r \in \mathbb{R}_+$, $e \in E$. For all sequences $(t_n) \to 0_+$, $(e_n) \to a$ with $e_n \in E$ for all $n \in \mathbb{N}$, we have $e_n + t_n r(e-e_n) \in E$ for $n \in \mathbb{N}$ so large that $t_n r \in [0,1]$ and $(r(e-e_n))_n \to v$, so that $v \in T^C(E,a)$. Since $T^C(E,a)$ is closed, we get $T(E,a) = \text{cl}(\mathbb{R}_+(E-a)) \subset T^C(E,a)$. The reverse inclusion is a general fact observed above.
(b) By definition of a submanifold of class C^1 of X, there exist a closed subspace Y of X and a C^1-diffeomorphism $\varphi : U \to V$ of an open neighborhood U of a onto an open neighborhood V of 0 such that $\varphi(a) = 0$ and $\varphi(E \cap U) = Y \cap V$. Then the following result shows that $T^C(E,a) = D\varphi^{-1}(0)(T^C(Y,0)) = D\varphi^{-1}(0)(Y) = T(E,a)$. By polarity, $N_C(E,a) = N(E,a)$. $\qquad\square$

Let us give some calculus rules for tangent and normal cones. We start with images. Here we recall that a map $h : X \to Y$ is *open* at $e \in E \subset X$ from E onto $F \subset Y$ if for every sequence $(y_n) \to_F h(e)$ there exists a sequence $(x_n) \to_E e$ such that $h(x_n) = y_n$ for all n large enough.

Proposition 5.27. Let X, Y be normed spaces, let W be an open subset of X, and let $h : W \to Y$ be a mapping that is circa-differentiable at some point e of a subset E of W. Let F be a subset of Y such that $h(E) \subset F$. Suppose h is open at e from E onto F. Then $h'(e)(T^C(E,e)) \subset T^C(F,h(e))$ and $h'(e)^\mathsf{T}(N_C(F,h(e))) \subset N_C(E,e)$.
In particular, if h is a bijection of class C^1 at e, with inverse of class C^1 at $h(e)$ and $h(E) = F$, one has $h'(e)(T^C(E,e)) = T^C(F,h(e))$ and $N_C(E,e) = (h'(e)^\mathsf{T})^{-1}(N_C(F,h(e)))$.

Proof. Let $u \in T^C(E,e)$, let $v := h'(e)(u)$, and let $((t_n,y_n))$ be a sequence of $\mathbb{P} \times F$ with limit $(0,h(e))$. Since h is open at e from E onto F, there exists a sequence (e_n) in E such that $(e_n) \to e$ and $h(e_n) = y_n$ for n large. Since $u \in T^C(E,e)$, there exists a sequence $(u_n) \to u$ such that $e_n + t_n u_n \in E$ for all $n \in \mathbb{N}$. Then $h(e_n + t_n u_n) \in F$ and since h is circa-differentiable at e, $v_n := t_n^{-1}(h(e_n + t_n u_n) - h(e_n)) \to h'(e)(u) = v$. Since $y_n + t_n v_n = h(e_n) + t_n v_n = h(e_n + t_n u_n) \in F$ for all $n \in \mathbb{N}$, one gets that $v \in T^C(F,h(e))$. The inclusion $h'(e)^\mathsf{T}(N_C(F,h(e))) \subset N_C(E,e)$ follows by polarity.
When h^{-1} is also circa-differentiable at $h(e)$ and $h(E) = F$, interchanging the roles of h and h^{-1}, we also get $Dh^{-1}(h(e))(T^C(F,h(e))) \subset T^C(E,e)$, and since $Dh^{-1}(h(e)) = (Dh(e))^{-1}$, equality holds. $\qquad\square$

Proposition 5.28. *Let X, Y be normed spaces, let $A \subset X$, $B \subset Y$, and let $(x,y) \in A \times B$. Then*

$$T^C(A \times B, (x,y)) = T^C(A,x) \times T^C(B,y),$$

$$N_C(A \times B, (x,y)) = N_C(A,x) \times N_C(B,y).$$

Proof. Since the projections $p_X : X \times Y \to X$, $p_Y : X \times Y \to Y$ are continuous and open, the inclusion $T^C(A \times B, (x,y)) \subset T^C(A,x) \times T^C(B,y)$ is a consequence of the preceding proposition (or of the definition). The proof of the reverse inclusion is also a direct application of the definition. Equality for normal cones follows by polarity. $\square$

A crucial relationship between normal cones and subdifferentials is revealed in the next result.

Theorem 5.29. *Given a function $f : X \to \overline{\mathbb{R}}$ finite at $x \in X$ and Lipschitzian around x, let $E := \mathrm{epi}\, f$, $e := (x, f(x))$. Then*

$$\begin{aligned}
T^C(E,e) &= \quad \mathrm{epi}\, f^C(x,\cdot), \\
x^* \in \partial_C f(x) &\Leftrightarrow (x^*, -1) \in N_C(E,e).
\end{aligned}$$

Proof. Let $(u,r) \in T^C(E,e)$. Let $((x_n, t_n))_n \to (x, 0_+)$ be such that

$$f^C(x,u) = \lim(1/t_n)\left(f(x_n + t_n u) - f(x_n)\right).$$

Since f is continuous around x, it follows that $e_n := (x_n, f(x_n)) \to_E e$. By definition of $T^C(E,e)$ there exists a sequence $((u_n, r_n))_n \to (u,r)$ such that $e_n + t_n(u_n, r_n) \in E$ for $n \in \mathbb{N}$ large enough. This inclusion can be written $f(x_n) + t_n r_n \geq f(x_n + t_n u_n)$, and one gets $f^C(x,u) \leq \lim_n r_n = r$, so that $(u,r) \in \mathrm{epi}\, f^C(x,\cdot)$.

Conversely, let $(u,r) \in \mathrm{epi}\, f^C(x,\cdot)$. Let $(t_n) \to 0_+$, $(e_n) := (x_n, s_n) \to_E e$. By definition of f^C, one has $\limsup_n (1/t_n)\left(f(x_n + t_n u) - f(x_n)\right) \leq f^C(x,u) \leq r$. It follows that there exists a sequence $(r_n) \to r$ such that $r_n \geq (1/t_n)\left(f(x_n + t_n u) - f(x_n)\right)$ for all large $n \in \mathbb{N}$. Then $(x_n + t_n u, f(x_n) + t_n r_n) \in E$, whence $(x_n + t_n u, s_n + t_n r_n) \in E$ for such n's. Therefore $(u,r) \in T^C(E,e)$.

Now $(x^*, -1) \in N_C(E,e)$ means that $\langle x^*, u \rangle - r \leq 0$ for all $(u,r) \in T^C(E,e)$. Since $T^C(E,e) = \mathrm{epi}\, f^C(x,\cdot)$, this property is equivalent to $x^* \leq f^C(x,\cdot)$ or $x^* \in \partial_C f(x)$. $\square$

Corollary 5.30. *If $f : X \to \mathbb{R}$ is locally Lipschitzian, then for all $x \in X$ one has*

$$\partial_C f(x) = \{x^* \in (X^* : (x^*, -1) \in \mathrm{cl}^*)\mathbb{R}_+ \partial_C d_{\mathrm{epi}\, f}(x, f(x))\}.$$

The last result of this section is a key ingredient for optimality conditions in mathematical programming problems. Here $g : X \to \overline{\mathbb{R}}$ is Lipschitzian around $\overline{x} \in X$.

Proposition 5.31. *Let $S := \{x \in X : g(x) \leq g(\overline{x})\}$. If $0 \notin \partial_C g(\overline{x})$, then one has*

$$\{v : g^C(\bar{x}, v) \leq 0\} \subset T^C(S, \bar{x}), \tag{5.9}$$

$$N_C(S, \bar{x}) \subset \mathbb{R}_+ \partial_C g(\bar{x}). \tag{5.10}$$

Proof. Since $g^C(\bar{x}, \cdot)$ is the support function of the weak* closed convex set $\partial_C g(\bar{x})$, which does not contain 0, there exists some $u \in X$ such that $g^C(\bar{x}, u) < 0$. Let us show first that such a vector belongs to $T^C(S, \bar{x})$. By definition of $g^C(\bar{x}, u)$, for every $\alpha > 0$ satisfying $g^C(\bar{x}, u) < -\alpha$ and for all sequences $(t_n) \to 0_+$, $(x_n) \to_S \bar{x}$, for n large enough we have $g(x_n + t_n u) - g(x_n) \leq -t_n \alpha$, hence $g(x_n + t_n u) \leq g(\bar{x})$ and $x_n + t_n u \in S$, since $x \in S$ means that $g(x) \leq g(\bar{x})$. This shows that $u \in T^C(S, \bar{x})$.

Now let $v \in X$ be such that $g^C(\bar{x}, v) \leq 0$. Taking u satisfying $g^C(\bar{x}, u) < 0$, since $g^C(\bar{x}, \cdot)$ is sublinear, for all $t > 0$ we have $g^C(\bar{x}, v + tu) < 0$, hence $v + tu \in T^C(S, \bar{x})$. Since this cone is closed, taking the limit as $t \to 0_+$, we get $v \in T^C(S, \bar{x})$.

Since the weak* compact set $\partial_C g(\bar{x})$ does not contain 0, the cone $\mathbb{R}_+ \partial_C g(\bar{x})$ is easily seen to be weak* closed. Its polar cone is $\{v : g^C(\bar{x}, v) \leq 0\}$. Since polarity reverses inclusions and since $Q^{00} = Q$ when Q is a weak* closed convex cone, relation (5.10) follows from relation (5.9). $\square$

Exercises

1. Let $f : \mathbb{R}^n \to \mathbb{R}$ be Lipschitzian around $\bar{x} \in \mathbb{R}^n$, let E (resp. G) be the epigraph (resp. graph) of f, and let $\bar{x}_f := (\bar{x}, f(\bar{x}))$.

(a) Prove that f is differentiable at $\bar{x}$ iff $T(G, \bar{x}_f) := T^D(G, \bar{x}_f)$ is a hyperplane.
(b) Prove that f is circa-differentiable at $\bar{x}$ iff $T^C(G, \bar{x}_f)$ is a hyperplane.
(c) Prove that f is circa-differentiable at $\bar{x}$ iff $T^C(E, \bar{x}_f)$ is a half-space.
(d) Show that $T^D(E, \bar{x}_f)$ may be a half-space while f is not differentiable at $\bar{x}$.

2. Let $E := \mathbb{R}^d \backslash \text{int } \mathbb{R}^d_+$. Compute the sets $T^C(E, a)$ and $N_C(E, a)$ for $a \in E$.

3. Compute the sets $T^C(E, a)$ and $N_C(E, a)$ for $a := (0, 0)$ and

(a) $E := \{(r, s) \in \mathbb{R}^2 : s \geq -|r|^\alpha\}$ with $\alpha > 0$;
(b) $E := \{(r, s) \in \mathbb{R}^2 : s \geq |r|^\alpha\}$ with $\alpha > 0$.
(c) $E := (\mathbb{R} \times \mathbb{R}_+) \cup \{0\} \times \mathbb{R}$.

4. Given $E \subset X$ and $v \in X$, let $F := E + v$. Show that for all $a \in \text{cl}(E)$ one has $T^C(F, a + v) = T^C(E, a)$ and $N_C(F, a + v) = N_C(E, a)$.

5. Deduce from Proposition 5.27 that the notion of Clarke tangent cone can be defined for subsets of manifolds of class C^1.

6. The *paratingent cone* to a subset E of a normed space X at $a \in \text{cl} E$ is the set $T^P(E, a)$ of $u \in X$ such that there exist sequences $(t_n) \to 0_+$, $(a_n) \to a$, $(u_n) \to u$ such that $a_n \in E$, $a_n + t_n u_n \in E$ for all $n \in \mathbb{N}$. Show that $T^P(E, a) + T^C(E, a) \subset T^P(E, a)$.

Consider the case in which E is the graph or the epigraph of a function. Define concepts of paratingent derivative and coderivative of a multimap.

Show by some examples that $T^P(E,a)$ may be very large.

7. Give an example of a subset E of $\mathbb{R}^2$ containing $a := (0,0)$ such that $T^C(E,a) \neq T^D(E,a) = T^I(E,a)$ even though this last cone is convex. [Hint: Given a decreasing sequence $(r_n) \to 0_+$, define an even function f that is affine on each interval $[r_{n+1}, r_n]$ and such that $f(r_{2n}) = 0$, $f(r_{2n+1}) = r_{2n+1}^2$, with $\limsup_n r_{2n+1}^{-2}(r_{2n+1} - r_{2n+2}) < \infty$, $\limsup_n r_{2n+1}^{-2}(r_{2n} - r_{2n+1}) < \infty$.]

5.3   Subdifferentials of Arbitrary Functions

The concepts of Clarke derivate and Clarke subdifferential can be extended to every function either via an analytical approach or via a geometrical approach. We adopt the second approach, which is more natural, but we also present analytical formulas.

5.3.1   Definitions and First Properties

The geometric approach relies on the relationship disclosed by Theorem 5.29. It shows that the definition we adopt now is compatible with the one we used for Lipschitzian functions.

Definition 5.32. Given $f : X \to \overline{\mathbb{R}}$ finite at $x \in X$, let $E := \operatorname{epi} f$, $e := (x, f(x))$. The *Clarke derivate* and the *Clarke subdifferential* of f at x are defined respectively by

$$f^C(x,u) := \min\{r \in \mathbb{R} : (u,r) \in T^C(E,e)\}, \tag{5.11}$$

$$\partial_C f(x) := \{x^* \in X^* : (x^*, -1) \in N_C(E,e)\}. \tag{5.12}$$

Since $T^C(E,e)$ is closed, the above infimum is attained when it is finite, so that we write min, according to a classical convention. Since for $x^* \in X^*$ one has $(x^*, -1) \in N_C(E,e)$ iff $\langle x^*, u \rangle - r \leq 0$ for all $(u,r) \in T^C(E,e)$, iff $\langle x^*, u \rangle \leq f^C(x,u)$, the definition of $\partial_C f(x)$ can be reformulated as follows:

$$\partial_C f(x) = \{x^* \in X^* : \langle x^*, \cdot \rangle \leq f^C(x, \cdot)\}. \tag{5.13}$$

Let us note that the normal cone to a subset S at $x \in S$ can be interpreted as the subdifferential at x of the indicator function ι_S of S:

$$N_C(S,x) = \partial_C \iota_S(x). \tag{5.14}$$

In fact, since the epigraph E of ι_S is $S \times \mathbb{R}_+$, by Proposition 5.28 and the fact that the normal cone to the convex set $\mathbb{R}_+$ is the normal cone of convex analysis, one has $x^* \in \partial_C \iota_S(x)$ iff $(x^*, -1) \in N_C(E, (x,0)) = N_C(S,x) \times \mathbb{R}_-$ iff $x^* \in N_C(S,x)$.

Let us prove an extension of Proposition 5.5.

Proposition 5.33. *For a function $f : X \to \overline{\mathbb{R}}$ finite at $x \in X$, the following assertions are equivalent:*

(a) $f^C(x,0) > -\infty$;
(b) $f^C(x,0) = 0$;
(c) $\partial_C f(x) \neq \varnothing$.

Moreover, under each of these conditions, for all $u \in X$, one has

$$f^C(x,u) = \sup\{\langle x^*, u \rangle : x^* \in \partial_C f(x)\}. \tag{5.15}$$

Proof. Let $E := \text{epi} f$ and let $e := (x, f(x))$. Since $T^C(E, e)$ is the epigraph of $f^C(x, \cdot)$ and $(0,0) \in T^C(E, e)$, one has $f^C(x,0) \leq 0$, and by sublinearity, either $f^C(x,0) = -\infty$ or $f^C(x,0) = 0$. Thus (a)$\Leftrightarrow$(b). Clearly (c)$\Rightarrow$(a). Suppose (b) holds. Then since $f^C(x, \cdot)$ is lower semicontinuous and sublinear, its epigraph $T^C(E, e)$ being a closed convex cone, for all $u \in X$ one has $f^C(x, u) > -\infty$. Then $f^C(x, \cdot)$ is the supremum of the continuous linear forms bounded above by $f^C(x, \cdot)$: (5.15) holds and (c) is satisfied. $\qquad\square$

Proposition 5.34. *If $f : X \to \overline{\mathbb{R}}$ is convex and finite at $x \in X$, then $\partial_C f(x) = \partial_{MR} f(x)$, where $\partial_{MR} f(x)$ is the subdifferential of f at x in the sense of convex analysis.*

Proof. In view of Definition 5.32, for $E := \text{epi} f$, $x^* \in \partial_C f(x)$ if and only if $(x^*, -1) \in N_C(E, e) = N(E, e)$ (Proposition 5.26), if and only if $x^* \in \partial_{MR} f(x)$. $\qquad\square$

Proposition 5.35. *If X, Y are normed spaces, if $f \in \mathscr{F}(X)$, $g \in \mathscr{F}(Y)$, and if h is defined by $h(x,y) := f(x) + g(y)$, then for all $(x,y) \in \text{dom} h$, $(u,v) \in X \times Y$ one has*

$$h^C((x,y),(u,v)) \leq f^C(x,u) + g^C(y,v), \tag{5.16}$$

$$\partial_C h(x,y) \subset \partial_C f(x) \times \partial_C g(y). \tag{5.17}$$

Proof. Let $q : X \times \mathbb{R} \times Y \times \mathbb{R} \to X \times Y \times \mathbb{R}$ be given by $q(x', r', y', s') = (x', y', r' + s')$ and let $x_f := (x, f(x))$, $y_g := (y, g(y))$, $z_h := (x, y, h(x,y))$. Since q is a continuous linear map that is open from $\text{epi} f \times \text{epi} g$ onto $\text{epi} h$ as is easily seen, Propositions 5.27 and 5.28 ensure that one has

$$q(T^C(\text{epi} f, x_f) \times T^C(\text{epi} g, y_g)) = q(T^C(\text{epi} f \times \text{epi} g, (x_f, y_g)) \subset T^C(\text{epi} h, z_h).$$

Thus, for all $(u, r) \in T^C(\text{epi} f, x_f)$, $(v, s) \in T^C(\text{epi} g, y_g)$ one has $(u, v, r + s) \in T^C(\text{epi} h, z_h)$, hence $h^C((x,y),(u,v)) \leq r + s$. Taking the infimum over $r \geq f^C(x,u)$, $s \geq g^C(y,v)$, one gets inequality (5.16).

Now let $(x^*, y^*) \in \partial_C h(x, y)$. Since $(0,0) \in T^C(\text{epi} g, y_g)$, for every $(u, r) \in T^C(\text{epi} f, x_f)$, the inclusion of the first part of the proof yields $(u, 0, r) \in T^C(\text{epi} h, z_h)$, hence $h^C((x, y), (u, 0)) \leq f^C(x, u)$, so that $\langle x^*, u \rangle = \langle (x^*, y^*), (u, 0) \rangle \leq f^C(x, u)$ and $x^* \in \partial_C f(x)$. The relation $y^* \in \partial_C g(y)$ is obtained similarly.   □

A precise analysis of the normal cone to the epigraph E of a function f will be useful. With this aim, let us introduce the *Clarke singular (or asymptotic) subdifferential* of f at x as the set

$$\partial_C^\infty f(x) := \{x^* \in X^* : (x^*, 0) \in N_C(E, e)\}. \tag{5.18}$$

Proposition 5.36. *Let $f : X \to \overline{\mathbb{R}}$ be finite at $x \in X$ and let $e := (x, f(x))$, $E := \text{epi} f$. Then $\partial_C^\infty f(x)$ is a weak* closed convex cone and one has the decomposition*

$$N_C(E, e) = (\mathbb{P}(\partial_C f(x) \times \{-1\})) \cup (\partial_C^\infty f(x) \times \{0\}). \tag{5.19}$$

Moreover, $\partial_C f(x) + \partial_C^\infty f(x) = \partial_C f(x)$. If $\partial_C f(x)$ is nonempty, then $\partial_C^\infty f(x)$ is the recession cone of $\partial_C f(x)$ and $N_C(E, e) = \mathbb{R}_+ (\partial_C f(x) \times \{-1\}) + \partial_C^\infty f(x) \times \{0\}$.

Proof. The right-hand side of relation (5.19) is clearly contained in the left-hand side. Let $(x^*, r^*) \in N_C(E, e)$. Then for every $(v, r) \in T^C(E, e)$, one has $\langle x^*, v \rangle + r^* r \leq 0$. Since $T^C(E, e)$ is an epigraph and contains $(0, 0)$, one has $r^* r \leq 0$ for all $r \in \mathbb{R}_+$, hence $r^* \leq 0$. If $r^* = 0$, one has $x^* \in \partial_C^\infty f(x)$; if $r^* < 0$, by (5.12), one can write $(x^*, r^*) = |r^*| (x^*/|r^*|, -1) \in \mathbb{P}(\partial_C f(x) \times \{-1\})$, so that in both cases (x^*, r^*) belongs to the right-hand side. When $\partial_C f(x)$ is nonempty, the last equality is an easy consequence of the first one and of the convexity of $N_C(E, e)$.

Given $y^* \in \partial_C f(x)$ and $z^* \in \partial_C^\infty f(x)$, one has $(y^* + z^*, -1) = (y^*, -1) + (z^*, 0) \in N_C(E, e)$, hence $y^* + z^* \in \partial_C f(x)$. In fact, since $0 \in \partial_C^\infty f(x)$, the inclusion $\partial_C f(x) + \partial_C^\infty f(x) \subset \partial_C f(x)$ is an equality. Finally, the last assertion corresponds to a general fact in convex analysis: the preceding inclusion shows that $\partial_C^\infty f(x)$ is contained in the recession cone of $\partial_C f(x)$; on the other hand, given $a^* \in \partial_C f(x)$, if x^* belongs to the recession cone of $\partial_C f(x)$, for every $r \in \mathbb{R}_+$ one has $(a^* + rx^*, -1) \in N_C(E, e)$, hence $(x^*, 0) = \lim_{r \to +\infty} r^{-1}(a^* + rx^*, -1) \in N_C(E, e)$, so that $x^* \in \partial_C^\infty f(x)$.   □

Corollary 5.37. *If $f : X \to \mathbb{R}$ is Lipschitzian around x, then $\partial_C^\infty f(x) = \{0\}$.*

Proof. When f is Lipschitzian around $x \in X$, the set $\partial_C f(x)$ is bounded and nonempty, so that its recession cone $\partial_C^\infty f(x)$ is $\{0\}$.   □

The calculus rules we have in view in the next subsection will require "qualification conditions," i.e., additional assumptions. We take a geometric approach to dealing with them in introducing a notion of locally uniform feasible direction.

Definition 5.38. The *(circa-)hypertangent cone* $H^C(E, a)$ (or $H(E, a)$ for simplicity) to a subset E of X at $a \in \text{cl}(E)$ is the set of vectors $u \in X$ for which there exists $\varepsilon > 0$ such that $e + tv \in E$ for all $e \in E \cap B(a, \varepsilon)$, $t \in (0, \varepsilon)$, $v \in B(u, \varepsilon)$. Equivalently,

$$u \in H(E,a) \Leftrightarrow \forall (e_n) \to_E a, \ (t_n) \to 0_+, \ (u_n) \to u, \ \exists m \in \mathbb{N} : \forall n \geq m, \ e_n + t_n u_n \in E.$$

When the set $H(E,a)$ is nonempty, we say that E has the *cone property around a*, or that E is *epi-Lipschitzian at a*. This property plays an important role in the study of elliptic partial differential equations and in shape optimization.

Example 5.1. Let E be the epigraph of a Lipschitzian function $f : W \to \mathbb{R}$ on some open subset W of a normed space X. Then $u := (0,1) \in H(E,a)$ for all $a \in E$, as is easily checked. This partially explains the terminology.

Example 5.2. Let E be a convex subset with nonempty interior. Then E has the cone property around all $a \in \mathrm{cl}(E)$. In fact, given $a \in \mathrm{cl}(E)$, $u \in X$ such that $a + u \in \mathrm{int}E$, one can find $\varepsilon \in (0,1)$ such that $B(a+u,2\varepsilon) \subset E$, whence for $e \in E \cap B(a,\varepsilon)$, $t \in (0,\varepsilon)$, $v \in B(u,\varepsilon)$ one has $e + v \in B(a+u,2\varepsilon)$, hence $e + tv = (1-t)e + t(e+v) \in E$ by convexity.

Example 5.3. Suppose X is finite-dimensional. Then one can show that a subset E of X has the cone property around $a \in E$ if and only if $T^C(E,a)$ has nonempty interior (Exercise 5). The necessity part of this assertion is proved in the next theorem.

Theorem 5.39. *The set $H(E,a)$ of hypertangent vectors to a subset E of X at $a \in \mathrm{cl}(E)$ is an open convex cone contained in $T^C(E,a)$. Moreover one has*

$$T^C(E,a) + H(E,a) = H(E,a).$$

If E has the cone property around a (i.e., if $H(E,a) \neq \varnothing$), then one has

$$H(E,a) = \mathrm{int}T^C(E,a), \quad T^C(E,a) = \mathrm{cl}(H(E,a)) \quad N_C(E,a) = (H(E,a))^0.$$

Proof. Since $0 \in T^C(E,a)$, the inclusion $H(E,a) \subset T^C(E,a) + H(E,a)$ holds. Let $u \in H(E,a)$ and $v \in T^C(E,a)$. Given sequences $(e_n) \to_E a$, $(t_n) \to 0_+$, $(w_n) \to w := u + v$, one can find a sequence $(v_n) \to v$ such that $x_n := e_n + t_n v_n \in E$ for all $n \in \mathbb{N}$. Since $(u_n) := (w_n - v_n) \to u$, one has $x_n + t_n u_n \in E$ for n large, hence $e_n + t_n w_n \in E$ for such n's. Thus $w = u + v \in H(E,a)$. The convexity of $H(E,a)$ is a consequence of the inclusions $H(E,a) \subset T^C(E,a)$ and $T^C(E,a) + H(E,a) \subset H(E,a)$.

Since $H(E,a)$ is open and contained in $T^C(E,a)$, one has $H(E,a) \subset \mathrm{int}T^C(E,a)$. Conversely, given $v \in \mathrm{int}T^C(E,a)$, assuming one has some $u \in H(E,a)$, for $r > 0$ small enough, one can write $v = (v - ru) + ru \in T^C(E,a) + H(E,a) \subset H(E,a)$. Therefore $H(E,a) = \mathrm{int}T^C(E,a)$ in such a case. Moreover, for every $w \in T^C(E,a)$ one has $w = \lim_n(w + 2^{-n}u)$, with $w + 2^{-n}u \in H(E,a)$ for all $n \in \mathbb{N}$, hence $w \in \mathrm{cl}(H(E,a))$. Since $T^C(E,a)$ is closed and contains $H(E,a)$, the equality $T^C(E,a) = \mathrm{cl}(H(E,a))$ holds whenever E has the cone property around a.

Given $a^* \in (H(E,a))^0$, $u \in H(E,a)$, and $w \in T^C(E,a)$, taking $w_n := w + 2^{-n}u$, we see that $\langle a^*, w \rangle = \lim_n \langle a^*, w_n \rangle \leq 0$, hence $a^* \in N^C(E,a)$ and $(H(E,a))^0 \subset N_C(E,a)$. The reverse inclusion follows from the containment $H(E,a) \subset T^C(E,a)$. $\square$

Exercise. Show that if E has the cone property around a, then $H(E,a) = H(\mathrm{cl}(E),a)$. This fact is used in the proof of the next corollary.

Corollary 5.40. *Suppose that E has the cone property around $a \in \mathrm{cl}(E)$. Then the multimap $N_C(E,\cdot)$ is closed at a on $\mathrm{cl}(E)$: if $(x_n) \to a$ in $\mathrm{cl}(E)$, (x_n^*) has a weak* limit point x^* and $x_n^* \in N_C(E,x_n)$ for all $n \in \mathbb{N}$, then $x^* \in N_C(E,a)$.*

Proof. Since E has the cone property around a, one has $T^C(E,a) = \mathrm{cl}(H(E,a))$, so that it suffices to show that $\langle x^*, u \rangle \leq 0$ for all $u \in H(E,a)$ when x^* is a weak* cluster point of a sequence (x_n^*) as in the statement. The definition of $H(E,a)$ shows that $u \in H(E,x) \subset T^C(E,x)$ for $x \in \mathrm{cl}\,E$ close enough to a; thus one has $\langle x_n^*, u \rangle \leq 0$ for n large enough. Taking a weak* converging subnet, one sees that $\langle x^*, u \rangle \leq 0$. $\qquad\square$

Let us give an example of the use of the hypertangent cone.

Proposition 5.41. *Let E and F be two subsets of X and let $a \in \mathrm{cl}(E \cap F)$ be such that $T^C(E,a) \cap H(F,a) \neq \varnothing$. Then $T^C(E,a) \cap T^C(F,a) \subset T^C(E \cap F,a)$.*

Proof. Let $u \in T^C(E,a) \cap H(F,a)$. For all sequences $(a_n) \to_{E \cap F} a$, $(t_n) \to 0_+$ one can find a sequence $(u_n) \to u$ such that $a_n + t_n u_n \in E$ for all $n \in \mathbb{N}$. Since $u \in H(F,a)$, one has $a_n + t_n u_n \in F$ for n large, hence $a_n + t_n u_n \in E \cap F$ for n large and $u \in T^C(E \cap F,a)$.

Now let $v \in T^C(E,a) \cap T^C(F,a)$ and let $v_k = v + 2^{-k}u$ for $k \in \mathbb{N}$. Then $v_k \in T^C(E,a) \cap H(F,a)$ by Theorem 5.39 and the convexity of $T^C(E,a)$. The first part of the proof shows that $v_k \in T^C(E \cap F,a)$. Since this cone is closed and since $(v_k) \to v$, one gets $v \in T^C(E \cap F,a)$. $\qquad\square$

The cone property will be used for functions through the following definition.

Definition 5.42. *A function $f : X \to \overline{\mathbb{R}}$ finite at $x \in X$ is said to have the cone property around x if its epigraph E has the cone property around $e := (x, f(x))$. More precisely, f is said to have the cone property around x (or to be directionally Lipschitzian at x) in the direction $u \in X$ if $(u,r) \in H(E,e)$ for some $r \in \mathbb{R}$.*

Let us present some concrete criteria for the cone property for $f : X \to \overline{\mathbb{R}}$.

Proposition 5.43. *Under each of the following conditions f has the cone property around $x \in f^{-1}(\mathbb{R})$:*

(a) f is Lipschitzian around x.
(b) f is convex and bounded above on some neighborhood of some point y.
(c) f is the indicator function ι_S of a subset S having the cone property around x.
(d) f is nondecreasing with respect to the order induced by a convex cone K with nonempty interior.
(e) X is finite-dimensional and $\mathrm{int}\,\mathrm{dom}\,f^C(x,\cdot)$ is nonempty.

Proof. (a) If f is Lipschitzian around x, then one has $(0,1) \in H(\mathrm{epi}\,f,(x,f(x)))$.

(b) Suppose f is convex and bounded above by m on some neighborhood of some point y. Then $(y, m+1)$ belongs to the interior of the epigraph E of f, so that E has the cone property around $(x, f(x))$ by Example 5.2 above.

(c) If $f = \iota_S$, $u \in H(S,x)$, then $(u,r) \in H(S \times \mathbb{R}_+, (x,0))$ for all $r > 0$.

(d) Suppose f is *homotone* with respect to the order induced by K, i.e., that $f(x') \le f(x'')$ when $x',x'' \in X$ with $x'' - x' \in K$. Given $u \in -\mathrm{int}K$, let us show that $(u,1) \in H(\mathrm{epi}\, f, (x,f(x)))$. Let us pick $\varepsilon \in (0,1)$ such that $B(u,\varepsilon) \subset -K$. Then for every $(w,s) \in \mathrm{epi}\, f$, $t \in (0,1)$, $v \in B(u,\varepsilon)$, $r \in [1-\varepsilon, 1+\varepsilon]$, we have $tv \in -K$, hence $f(w+tv) \le f(w) \le s \le s + tr$ and $(w+tv, s+tr) \in \mathrm{epi}\, f$; thus $(u,1) \in H(\mathrm{epi}\, f, (x,f(x)))$.

(e) Suppose X is finite-dimensional and $\mathrm{int\, dom}\, f^C(x,\cdot)$ is nonempty. Then the convex function $f^C(x,\cdot)$ is continuous on $\mathrm{int\, dom}\, f^C(x,\cdot)$, so that $T^C(\mathrm{epi}\, f, (x,f(x)))$ has nonempty interior. Since X is finite-dimensional, we get that $\mathrm{epi}\, f$ has the cone property by Example 5.3 above. $\square$

It can be shown that f has the cone property around $x \in f^{-1}(\mathbb{R})$ in the direction u if and only if

$$f^0(x,u) := \inf_{\varepsilon > 0} \sup \left\{ \frac{f(w+tv) - s}{t} : (t,w,v) \in T_\varepsilon \times B(x,\varepsilon) \times B(u,\varepsilon), s \in I_f(x,w,\varepsilon) \right\}$$

(5.20)

is finite, where $T_\varepsilon := (0,\varepsilon)$, $I_f(x,w,\varepsilon) := \{s \in (f(x) - \varepsilon, f(x) + \varepsilon) : s \ge f(w)\}$. Moreover,

$$(u,r) \in H(E,e) \Leftrightarrow f^0(x,u) < r, \qquad (5.21)$$

so that $f^0(x,u) = \inf\{r : (u,r) \in H(E,e)\}$. When f is continuous at x it can be checked (exercise) that the complicated expression for $f^0(x,u)$ boils down to

$$f^0(x,u) = \limsup_{(t,v,w) \to (0_+, u, x)} \frac{1}{t}(f(w+tv) - f(w)).$$

When f is Lipschitzian around x, one has $f^0(x,\cdot) = f^C(x,\cdot)$.

Proposition 5.44 ([875]). *If the function $f : X \to \overline{\mathbb{R}}$ is finite at $x \in X$ and has the cone property around x, then $f^0(x,\cdot) = f^C(x,\cdot)$ on $\mathrm{dom}\, f^0(x,\cdot)$, and $f^0(x,\cdot)$ is continuous there. Moreover, the following equivalence holds, and if X is complete, then $\mathrm{dom}\, f^0(x,\cdot) = \mathrm{int\, dom}\, f^C(x,\cdot)$:*

$$x^* \in \partial_C f(x) \Leftrightarrow x^* \le f^0(x,\cdot). \qquad (5.22)$$

Proof. Suppose f has the cone property around x. Let E be its epigraph and $e := (x, f(x))$. Then $\mathrm{dom}\, f^0(x,\cdot)$ is the projection $p_X(H(E,e))$ of $H(E,e)$ on X. Since the canonical projection $p_X : X \times \mathbb{R} \to X$ is open, $p_X(H(E,e))$ is open. Thus, $\mathrm{dom}\, f^0(x,\cdot) \subset \mathrm{int\, dom}\, f^C(x,\cdot)$. Moreover, $f^0(x,\cdot) \ge f^C(x,\cdot)$, since the strict epigraph of $f^0(x,\cdot)$ is $H(E,e) \subset T^C(E,e) = \mathrm{epi}\, f^C(x,\cdot)$. Now $f^0(x,\cdot)$ is convex and locally bounded above at each point of $\mathrm{dom}\, f^0(x,\cdot) = p_X(H(E,e))$. Thus $f^0(x,\cdot)$ is continuous on $\mathrm{dom}\, f^0(x,\cdot)$. Moreover, for each $u \in \mathrm{dom}\, f^0(x,\cdot)$ and each $s \ge f^C(x,u)$, there exists a sequence $((u_n, s_n)) \to (u,s)$ with $(u_n, s_n) \in H(E,e)$, since

$H(E,e)$ is dense in $T^C(E,e) = \text{epi}\, f^C(x,\cdot)$. Therefore $f^0(x,u) = \lim_n f^0(x,u_n) \leq s$, whence $f^0(x,u) \leq f^C(x,u)$ and equality holds.

Relation (5.22) is a consequence of the following string of equivalences, which stems from Theorem 5.39: $x^* \leq f^0(x,\cdot)$ iff $\langle x^*, u \rangle \leq r$ for all $(u,r) \in H(E,e)$ iff $(x^*,-1) \in (H(E,e))^0 = (T^C(E,e))^0 = N_C(E,e)$ iff $x^* \in \partial_C f(x)$.

If X is complete, the lower semicontinuous function $f^C(x,\cdot)$ is continuous on the interior of its domain, so that for all $u \in \text{int dom}\, f^C(x,\cdot)$ and all $s > f^C(x,u)$ one has $(u,s) \in \text{int}(T^C(E,e)) = H(E,e)$, hence $f^0(x,u) \leq s$ and $u \in \text{dom}\, f^0(x,\cdot)$.   □

Corollary 5.45. *If $f : X \to \overline{\mathbb{R}}$ is finite and continuous at $x \in X$ and has the cone property around x, then $-f$ has the cone property and*

$$\partial_C(-f)(x) = -\partial_C f(x). \tag{5.23}$$

Proof. Let E (resp. F) be the epigraph of f (resp. $-f$) and let $e := (x, f(x))$, $e' := (x, -f(x))$. Given $x^* \in \partial_C(-f)(x)$, let us show that $-x^* \leq f^0(x,\cdot)$ or that $\langle -x^*, u \rangle \leq r$ for all $(u,r) \in H(E,e)$. It suffices to prove that for all $(u,r) \in H(E,e)$ we have $(-u,r) \in H(F,e')$. Let $((x_n, s_n)) \to_F e'$, $(t_n) \to 0_+$, $((-u_n, r_n)) \to (-u,r)$. Since f is continuous, we have $(f(x_n - t_n u_n)) \to f(x)$, so that for n large enough, $(x_n - t_n u_n, f(x_n - t_n u_n)) + t_n(u_n, r_n) \in E$, or $f(x_n) \leq f(x_n - t_n u_n) + t_n r_n$, hence $(x_n - t_n u_n, -f(x_n) + t_n r_n) \in F$ and also $(x_n - t_n u_n, s_n + t_n r_n) \in F$. Thus $(-u,r) \in H(F,e')$.   □

Exercises

1. Let S be a subset of a normed space X, let $a \in \text{cl}\, S$, let ι_S be the indicator function of S, and let $E := S \times \mathbb{R}_+$. Check that $T^C(E,(a,0)) = T^C(S,a) \times \mathbb{R}_+$ and

$$(\iota_S)^C (a,\cdot) = \iota_{T^C(S,a)}(\cdot).$$

2. (a) Show that for a function $f : X \to \overline{\mathbb{R}}$ finite at $x \in X$ and $u \in X$ one has

$$f^C(x,u) = \inf_{\varepsilon > 0} \sup_{(t,w,s) \in (0,\varepsilon) \times B(x,\varepsilon) \times I_f(x,w,\varepsilon)} \inf_{v \in B(u,\varepsilon)} \frac{1}{t}(f(w+tv) - s),$$

where $I_f(x,w,\varepsilon) := \{s \in (f(x) - \varepsilon, f(x) + \varepsilon) : s \geq f(w)\}$. Prove that when f is lower semicontinuous at x, this expression can be slightly simplified into

$$f^C(x,u) = \inf_{\varepsilon > 0} \sup_{(t,w) \in (0,\varepsilon) \times B(x,\varepsilon,f)} \inf_{v \in B(u,\varepsilon)} \frac{1}{t}(f(w+tv) - f(w)),$$

where $B(x,\varepsilon,f) := \{w \in B(x,\varepsilon) : |f(w) - f(x)| < \varepsilon\}$. When f is continuous at x,

$$f^C(x,u) = \inf_{\varepsilon>0} \sup_{(t,w)\in(0,\varepsilon)\times B(x,\varepsilon)} \inf_{v\in B(u,\varepsilon)} \frac{1}{t}\left(f(w+tv) - f(w)\right).$$

These expressions show that $f^C(x,\cdot)$ is the upper epi-limit as $(t,w,s) \to (0,x,f(x))$ in $\mathbb{P} \times \mathrm{epi} f$ of the functions $f_{t,w,s}$ given by $f_{t,w,s}(v) = t^{-1}(f(w+tv) - s)$ for $(t,w,s) \in \mathbb{P} \times \mathrm{epi} f$. Although this connection with variational convergences is important, one may prefer to use a geometrical approach rather than an analytical approach relying on these formulas when dealing with Clarke derivates.

(b) Prove relation (5.20) when f has the cone property around $x \in f^{-1}(\mathbb{R})$ in the direction u.

3. Show that a function $f : X \to \overline{\mathbb{R}}$ may be finite at $x \in X$ and have the cone property around x and satisfy $\partial_C^\infty f(x) \neq \{0\}$. [Hint: Take for f the indicator function of a subset having the cone property around x, and observe that $\partial_C^\infty f(x) = \partial_C f(x)$ is a cone.]

4. Show that the inclusion $\partial_C h(x,y) \subset \partial_C f(x) \times \partial_C g(y)$ for h given by $h(x,y) := f(x) + g(y)$ may be strict.

5. Show that when X is finite-dimensional, for $e \in E \subset X$ one has $H(E,e) = \mathrm{int} T^C(E,e)$. (See [873].)

6. Let E,F be subsets of a normed space X and let $x \in E \cap F$ be such that $T^D(E,x) \cap T^C(F,x)$ is convex and $T^D(E,x) \cap H(F,x) \neq \varnothing$. Show that $T^D(E,x) \cap T^C(F,x) \subset T^D(E \cap F,x)$. [Hint: Use Theorem 5.39 and the inclusion $T^D(E,x) \cap H(F,x) \subset T^D(E \cap F,x)$ to get that for every $v \in T^D(E,x) \cap T^C(F,x)$ and all $u \in T^D(E,x) \cap H(F,x)$, $t \in (0,1)$ one has $v_t := (1-t)v + tu \in T^D(E,x) \cap H(F,x)$.]

7. With the notation and assumptions of the preceding exercise, suppose that $T^C(F,x) = T^D(F,x)$. Then show that $T^D(E,x) \cap T^D(F,x) = T^D(E \cap F,x)$.

5.3.2 Regularity

Calculus rules can be improved when one assumes that the functions are regular enough. A precise meaning can be given to the word "regular." Let us observe that the results below could be given variants by assuming coincidence of the Clarke notions with the incident ones (instead of the contingent ones), a requirement weaker than the one in the next definition.

Definition 5.46. A subset E of a normed space X is said to be circa-regular, or Clarke-regular, or simply *regular* if there is no risk of confusion, at $a \in \mathrm{cl}(E)$ if $T^C(E,a) = T^D(E,a)$.

A function $f : X \to \overline{\mathbb{R}}$ finite at $x \in X$ is said to be circa-regular, or simply *regular*, at x if its epigraph is regular at $(x,f(x))$, or equivalently, if $f^C(x,\cdot) = f^D(x,\cdot)$, the lower directional derivate of f at x.

The dual requirements are equivalent to the respective primal properties:

Proposition 5.47. *A set E is regular at $a \in \mathrm{cl}(E)$ if and only if $N_C(E,a) = N_D(E,a)$. A function $f : X \to \overline{\mathbb{R}}$ is regular at $x \in f^{-1}(\mathbb{R})$ if and only if $\partial_C f(x) = \partial_D f(x)$ and $\partial_C^\infty f(x) = \partial_D^\infty f(x)$.*

Proof. If $T^C(E,a) = T^D(E,a)$, then $N_C(E,a) := \left(T^C(E,a)\right)^0 = \left(T^D(E,a)\right)^0 =: N_D(E,a)$. Conversely, suppose $N_C(E,a) = N_D(E,a)$. Then by the bipolar theorem,

$$T^C(E,a) = (N_C(E,a))^0 = (N_D(E,a))^0 = \left(T^D(E,a)\right)^{00} \supset T^D(E,a),$$

and since $T^C(E,a) \subset T^D(E,a)$, equality holds.

If f is regular at $x \in f^{-1}(\mathbb{R})$ then, denoting by E its epigraph and setting $e := (x, f(x))$, one has $N_C(E,e) = N_D(E,e)$, so that $\partial_C f(x) = \partial_D f(x)$ and $\partial_C^\infty f(x) = \partial_D^\infty f(x)$. Conversely, suppose $\partial_C f(x) = \partial_D f(x)$ and $\partial_C^\infty f(x) = \partial_D^\infty f(x)$. Then by Propositions 4.18 and 5.36, one has $N_C(E,e) = N_D(E,e)$. Thus E is regular at e and f is regular at x. $\qquad\square$

Every convex set is regular at every point of its closure, and every convex function is regular at every point of its domain. If a function $f : X \to \overline{\mathbb{R}}$ is finite at $x \in X$ and is circa-differentiable at x, then one sees that f is regular at x and $\partial_C f(x) = \{f'(x)\}$, $\partial_C^\infty f(x) = \partial_D^\infty f(x) = \varnothing$. If $G \subset X \times Y$ is the graph of a map $g : X \to Y$ that is circa-differentiable at x, then G is regular at $(x, g(x))$. Other examples arise from the calculus rules we present in the next subsection. Let us complete Proposition 5.31 with the following simple observation.

Proposition 5.48. *Let g be Lipschitzian around $\overline{x}$ and let $S := \{x \in X : g(x) \le g(\overline{x})\}$. If $0 \notin \partial_C g(\overline{x})$ and if g is regular at $\overline{x}$, then S is regular at $\overline{x}$ and one has $T^C(S, \overline{x}) = \{v : g^C(\overline{x}, v) \le 0\} = T^D(S, \overline{x})$.*

Proof. Let $v \in T^D(S, \overline{x})$: there exist sequences $(t_n) \to 0_+$, $(v_n) \to v$ such that $\overline{x} + t_n v_n \in S$ or $g(\overline{x} + t_n v_n) \le g(\overline{x})$ for all $n \in \mathbb{N}$. Then one has $g^D(\overline{x}, v) \le 0$, hence $g^C(\overline{x}, v) \le 0$ by regularity of g at $\overline{x}$ and $v \in T^C(S, \overline{x})$ by Proposition 5.31. Since $T^C(S, \overline{x}) \subset T^D(S, \overline{x})$, the announced double equality ensues. $\qquad\square$

5.3.3 Calculus Rules

The following result generalizes both Propositions 5.27, 5.41 and will be a key to some calculus rules.

Proposition 5.49. *Let X, Y be normed spaces, let W be an open subset of X, let F (resp. G) be a subset of X (resp. Y), and let $g : W \to Y$ be a mapping that is circa-differentiable at some point a of $E := F \cap g^{-1}(G)$. Suppose $A(T^C(F,a)) \cap H(G,b) \ne \varnothing$, where $A := g'(a)$, $b := g(a)$. Then $T^C(F,a) \cap A^{-1}(T^C(G,b)) \subset$*

$T^C(E,a)$. *If F and G are regular at a and b respectively, then equality holds and E is regular at a.*

Proof. Let us first show that if $u \in T^C(F,a) \cap A^{-1}(H(G,b))$, then $u \in T^C(E,a)$. Let $(t_n) \to 0_+$, $(a_n) \to_E a$. Since $u \in T^C(F,a)$, there exists a sequence $(u_n) \to u$ such that $a_n + t_n u_n \in F$ for all $n \in \mathbb{N}$. Then since g is circa-differentiable at a,

$$w_n := \frac{1}{t_n}(g(a_n + t_n u_n) - g(a_n)) \to w := A(u).$$

Since $w \in H(G,b)$, there exists $m \in \mathbb{N}$ such that $g(a_n + t_n u_n) = g(a_n) + t_n w_n \in G$ for all $n \geq m$. Thus $a_n + t_n u_n \in E$ for $n \geq m$ and one gets that $u \in T^C(E,a)$.

Now let $v \in T^C(F,a) \cap A^{-1}(T^C(G,b))$ and let $v_k = v + 2^{-k}u$ for $k \in \mathbb{N}$, where u is as above. Then $v_k \in T^C(F,a) \cap A^{-1}(H(G,b))$ by Theorem 5.39 and the convexity of $T^C(F,a)$. The first part of the proof shows that $v_k \in T^C(E,a)$. Since this cone is closed and since $(v_k) \to v$, one gets $v \in T^C(E,a)$.

Suppose F (resp. G) is regular at a (resp. b). Then since $T^D(E,a) \subset T^D(F,a) \cap A^{-1}(T^D(G,b))$, we get $T^D(E,a) \subset T^C(E,a)$, and these inclusions are equalities. $\square$

One can easily derive a chain rule from the preceding proposition.

Theorem 5.50. *Let X,Y be normed spaces, let W be an open subset of X, let $g : W \to Y$ be circa-differentiable at $\overline{x} \in X$ of X, and let $h : Y \to \overline{\mathbb{R}}$ be finite at $\overline{y} := g(\overline{x})$. Let $f := h \circ g$. Suppose there exists some $u \in X$ such that $h^0(\overline{y}, g'(\overline{x})(u)) < +\infty$. Then*

$$f^C(\overline{x}, \cdot) \leq h^C(\overline{y}, \cdot) \circ g'(\overline{x}), \tag{5.24}$$

$$\partial_C f(\overline{x}) \subset (g'(\overline{x}))^{\mathsf{T}}(\partial_C h(\overline{y})) := \partial_C h(\overline{y}) \circ g'(\overline{x}). \tag{5.25}$$

If h is regular at $\overline{y}$, (5.24) and (5.25) are equalities.

Proof. Let $k : X \times \mathbb{R} \to Y \times \mathbb{R}$ be given by $k(x,r) = (g(x),r)$ and let E (resp. G) be the epigraph of f (resp. h), so that $E = k^{-1}(G)$. Let $a := (\overline{x}, f(\overline{x}))$, $b := k(a)$, $A := k'(a)$. Since k is circa-differentiable at a and since our assumptions ensure that $k'(a)(u,r) = (g'(\overline{x})(u),r) \in H(G,b)$ for $r > h^0(\overline{y}, g'(\overline{x})(u))$, taking $F := X \times \mathbb{R}$, and replacing g by k in the preceding proposition, we get $A^{-1}(T^C(G,b)) \subset T^C(E,a)$, or equivalently, relation (5.24).

Let $\overline{x}^* \in \partial_C f(\overline{x})$. Then by relation (5.24), one has $\overline{x}^* \in \partial(h^C(\overline{y}, \cdot) \circ g'(\overline{x}))(0)$ in the sense of convex analysis, and by our qualification assumption and Corollary 5.44, $h^C(\overline{y}, \cdot)$ is continuous at some point of $g'(\overline{x})(X)$. Thus, the chain rule of convex analysis ensures that $\overline{x}^* \in \partial_C h(\overline{y}) \circ g'(\overline{x})$.

When $h^C(\overline{y}, \cdot) = h^D(\overline{y}, \cdot)$, (5.24) is an equality, since $h^D(\overline{y}, \cdot) \circ g'(\overline{x}) \leq f^D(\overline{x}, \cdot) \leq f^C(\overline{x}, \cdot)$. Moreover, since $h^C(\overline{y}, \cdot)$ is convex and continuous at $g'(\overline{x})(u)$, (5.25) is an equality by a composition rule of convex analysis. $\square$

Now let us turn to a rule for the sum of two functions.

Theorem 5.51. *Let* $f, g : X \to \overline{\mathbb{R}}$ *be two functions finite at* $\overline{x} \in X$, *lower semicontinuous at* $\overline{x}$, *and such that* $\operatorname{dom} f^C(\overline{x}, \cdot) \cap \operatorname{dom} g^0(\overline{x}, \cdot) \neq \varnothing$. *Then the following relations hold; they are equalities when* f *and* g *are regular at* $\overline{x}$:

$$(f + g)^C(\overline{x}, \cdot) \leq f^C(\overline{x}, \cdot) + g^C(\overline{x}, \cdot), \tag{5.26}$$

$$\partial_C(f + g)(\overline{x}) \subset \partial_C f(\overline{x}) + \partial_C g(\overline{x}). \tag{5.27}$$

Proof. Let $F := \operatorname{epi} f \times \mathbb{R} \subset X \times \mathbb{R}^2$, $G := \{(x, r, s) \in X \times \mathbb{R}^2 : (x, s) \in \operatorname{epi} g\}$ and let $E := F \cap G$, $e := (\overline{x}, f(\overline{x}), g(\overline{x}))$. Then $T^C(F, e) = T^C(\operatorname{epi} f, (\overline{x}, f(\overline{x}))) \times \mathbb{R}$ and

$$H(G, e) = \{(x, r, s) \in X \times \mathbb{R}^2 : (x, s) \in H(\operatorname{epi} g, (\overline{x}, g(\overline{x})))\}.$$

Thus, we have $(u, r, s) \in T^C(F, e) \cap H(G, e)$ for every $u \in \operatorname{dom} f^C(\overline{x}, \cdot) \cap \operatorname{dom} g^0(\overline{x}, \cdot)$, $r \geq f^C(\overline{x}, u)$, $s > g^0(\overline{x}, u)$. It follows from Proposition 5.41 or Proposition 5.49 that

$$T^C(F, e) \cap T^C(G, e) \subset T^C(F \cap G, e) = T^C(E, e).$$

Now the epigraph H of $f + g$ is the image of E by the continuous linear mapping $h : X \times \mathbb{R}^2 \to X \times \mathbb{R}$ given by $h(x, r, s) = (x, r + s)$. The map h is open at e from E onto $H = h(E)$, since for every sequence $((x_n, q_n)) \to_H f(\overline{x}) + g(\overline{x})$, setting $s_n := g(x_n)$, $r_n := q_n - s_n$, one has, by the lower semicontinuity of f and g at $\overline{x}$,

$$\liminf_n r_n \geq \liminf_n f(x_n) \geq f(\overline{x}), \quad \liminf_n s_n \geq \liminf_n g(x_n) \geq g(\overline{x}),$$

hence $\limsup_n r_n \leq \lim_n (r_n + s_n) - \liminf_n s_n \leq (f(\overline{x}) + g(\overline{x})) - g(\overline{x}) = f(\overline{x})$ and similarly $\limsup_n s_n \leq g(\overline{x})$, so that $((x_n, r_n, s_n)) \to (\overline{x}, f(\overline{x}), g(\overline{x}))$. We conclude from Proposition 5.27 that $h'(e)(T^C(E, e)) \subset T^C(H, h(e))$. Therefore, since $h'(e) = h$, we have

$$h(T^C(F, e) \cap T^C(G, e)) \subset h'(e)(T^C(E, e)) \subset T^C(H, h(e)),$$

and by definition of h, $(f + g)^C(\overline{x}, \cdot) \leq f^C(\overline{x}, \cdot) + g^C(\overline{x}, \cdot)$.

Since $g^C(\overline{x}, \cdot)$ is finite and continuous at some $u \in \operatorname{dom} f^C(\overline{x}, \cdot)$, relation (5.27) follows from the classical sum rule for subdifferentials of convex functions.

When f and g are regular at $\overline{x}$, taking into account the relations $f^D(\overline{x}, \cdot) + g^D(\overline{x}, \cdot) \leq (f + g)^D(\overline{x}, \cdot) \leq (f + g)^C(\overline{x}, \cdot)$, we get equality in (5.26). Then (5.27) is also an equality by the rule of convex analysis we just used. $\square$

Corollary 5.52. *Let* $f : X \to \overline{\mathbb{R}}$ *be lower semicontinuous and finite at* $\overline{x} \in X$ *and let* g *be locally Lipschitzian around* $\overline{x}$. *Then relations (5.26), (5.27) hold.*

Proof. This follows from the relations $\operatorname{dom} g^0(\overline{x}, \cdot) = X$, $0 \in \operatorname{dom} f^C(\overline{x}, \cdot)$. $\square$

Corollary 5.53. *Let* S *be a closed subset of* X *and let* $\overline{x} \in S$ *be a local minimizer of a locally Lipschitzian function* j. *Then* $0 \in \partial_C j(\overline{x}) + N_C(S, \overline{x})$.

Proof. Taking for f the indicator function of S and $g = j$, the result is a consequence of the preceding corollary and of the rule $0 \in \partial_C(j + \iota_S)(\bar{x})$. $\qquad\square$

Corollary 5.54. *Let f, $g_1, \ldots, g_k$ be locally Lipschitzian functions and let $\bar{x}$ be a minimizer of f on the set $S := \{x \in X : g_i(x) \leq 0 \; i \in \mathbb{N}_k\}$. Let $I := \{i \in \mathbb{N}_k : g_i(\bar{x}) = 0\}$. Suppose*

$$(t_i) \in \mathbb{R}_+^k, \; t_i = 0 \; \forall i \in \mathbb{N}_k \backslash I, \; 0 \in t_1 \partial_C g_1(\bar{x}) + \cdots + t_k \partial_C g_k(\bar{x}) \Longrightarrow t_i = 0 \; \forall i \in \mathbb{N}_k.$$

Then there exist $x^* \in \partial_C f(\bar{x})$, $y_i \in \mathbb{R}_+$, $x_i^* \in \partial_C g_i(\bar{x})$ for $i \in \mathbb{N}_k$ such that

$$x^* + y_1 x_1^* + \cdots + y_k x_k^* = 0, \qquad y_i g_i(\bar{x}) = 0 \quad \text{for all } i \in \mathbb{N}_k. \qquad (5.28)$$

Proof. Since $\partial_C g_i(\bar{x})$ is nonempty for all $i \in \mathbb{N}_k$ and $y_i = 0$ for $i \in \mathbb{N}_k \setminus I$, we may suppose $I = \mathbb{N}_k$. Let $g := \max_{i \in I} g_i$. We cannot have $0 \in \partial_C g(\bar{x})$, since otherwise, we could find some $t_i \in \mathbb{R}_+$ with sum 1 such that $0 \in t_1 \partial_C g_1(\bar{x}) + \cdots + t_k \partial_C g_k(\bar{x})$, a contradiction to our qualification condition. Now, by Theorem 5.10, $\partial_C g(\bar{x})$ is the convex hull of the sets $\partial_C g_i(\bar{x})$ for $i \in I$. The Karush-Kuhn-Tucker relation (5.28) ensues. Proposition 5.31 asserts that $N_C(S, \bar{x}) \subset \mathbb{R}_+ \partial_C g(\bar{x})$. $\qquad\square$

There is a special result for separable functions that does not require any qualification condition. However, it is just an inclusion, not an equality as in the case of the Fréchet subdifferential.

Proposition 5.55. *Let X and Y be normed spaces and let $g : X \to \overline{\mathbb{R}}$, $h : Y \to \overline{\mathbb{R}}$ be finite and lower semicontinuous at $\bar{x} \in X$ and $\bar{y} \in Y$ respectively. Then for f given by $f(x, y) = g(x) + h(y)$, one has $\partial_C f(\bar{x}, \bar{y}) \subset \partial_C g(\bar{x}) \times \partial_C h(\bar{y})$.*

Proof. The roles of g and h being symmetric, it suffices to show that for every $(x^*, y^*) \in \partial_C f(\bar{x}, \bar{y})$ one has $x^* \in \partial_C g(\bar{x})$. This will be a consequence of the fact that setting $\bar{p} := f(\bar{x}, \bar{y})$, $\bar{q} := g(\bar{x})$, for every $(u, r) \in T^C(\text{epi } g, (\bar{x}, \bar{q}))$ one has $(u, 0, r) \in T^C(\text{epi } f, (\bar{x}, \bar{y}, \bar{p}))$, since then one has $\langle x^*, u \rangle - r = \langle (x^*, y^*), (u, 0) \rangle - r \leq 0$. Now, for $(u, r) \in T^C(\text{epi } g, (\bar{x}, \bar{q}))$ and sequences $(t_n) \to 0_+$, $((x_n, y_n, p_n)) \to (\bar{x}, \bar{y}, \bar{p})$ in epi f, one has $((x_n, q_n)) \to_{\text{epi } g} (\bar{x}, \bar{q})$ for $q_n := g(x_n) + p_n - f(x_n, y_n) = p_n - h(y_n)$: since g and h are lower semicontinuous at $\bar{x}$ and $\bar{y}$ respectively, $\liminf_n q_n \geq \liminf_n g(x_n) \geq g(\bar{x}) := \bar{q}$, while $\limsup_n q_n \leq \lim_n p_n - \liminf_n h(y_n) \leq \bar{p} - h(\bar{y}) = g(\bar{x})$. By definition of $T^C(\text{epi } g, (\bar{x}, \bar{q}))$, there exists a sequence $((u_n, r_n)) \to (u, r)$ such that $(x_n, q_n) + t_n(u_n, r_n) \in \text{epi } g$ for all $n \in \mathbb{N}$. Then

$$g(x_n + t_n u_n) + h(y_n) \leq q_n + t_n r_n + h(y_n) = p_n + t_n r_n$$

and $(x_n + t_n u_n, y_n, p_n + t_n r_n) \in \text{epi } f$ for all $n \in \mathbb{N}$. This shows that $(u, 0, r) \in T^C(\text{epi } f, (\bar{x}, \bar{y}, \bar{p}))$. $\qquad\square$

Let us turn to properties involving order.

Theorem 5.56. *Let $f, g : X \to \overline{\mathbb{R}}$ be two functions finite and lower semicontinuous at $\bar{x} \in X$ and such that $\text{dom} f^C(\bar{x}, \cdot) \cap \text{dom} g^0(\bar{x}, \cdot) \neq \varnothing$ and $f(\bar{x}) = g(\bar{x})$. Then for*

$h := f \vee g := \max(f,g)$, *the following relations hold; they are equalities when f and g are regular at* $\overline{x}$:

$$h^C(\overline{x},\cdot) \leq f^C(\overline{x},\cdot) \vee g^C(\overline{x},\cdot), \tag{5.29}$$

$$\partial_C h(\overline{x}) \subset \mathrm{co}(\partial_C f(\overline{x}) \cup \partial_C g(\overline{x})) \cup (\partial_C f(\overline{x}) + \partial_C^{\infty} g(\overline{x})) \cup (\partial_C^{\infty} f(\overline{x}) + \partial_C g(\overline{x})). \tag{5.30}$$

Proof. Let $F := \mathrm{epi}\, f$, $G := \mathrm{epi}\, g$ and let $e := (\overline{x}, h(\overline{x})) \in H := \mathrm{epi}\, h$. Since $H = F \cap G$, Proposition 5.49 shows that $T^C(F,e) \cap T^C(G,e) \subset T^C(H,e)$, with equality when F and G are regular at e. Then, using the indicator functions of these cones and a sum rule of convex analysis, one gets $N^C(H,e) \subset N^C(F,e) + N^C(G,e)$ with equality when F and G are regular at e. Given $z^* \in \partial_C h(\overline{x})$, writing $(z^*, -1) = (x^*, -r) + (y^*, -s) \in N^C(F,e) + N^C(G,e)$ with $r,s \in \mathbb{R}_+$ and considering separately the three cases $(r,s) \in \mathbb{P} \times \mathbb{P}$, $(r,s) = (1,0)$, and $(r,s) = (0,1)$, we get inclusion (5.30). $\qquad\square$

Proposition 5.57. *Let* $f = h \circ g$, *where* $g : W \to \mathbb{R}$ *is lower semicontinuous on an open subset* W *of* X *and* $h : S \to \mathbb{R}$ *is defined and continuous on an open interval* S *of* $\mathbb{R}$ *containing* $g(W)$. *If* h *is circa-differentiable at* $\overline{s} := g(\overline{x})$ *with* $h'(\overline{s}) > 0$, *one has* $\partial_C f(\overline{x}) = h'(\overline{s}) \partial_C g(\overline{x})$.

Proof. The inverse function theorem ensures that there exist open intervals $J \subset S$, $I \subset \mathbb{R}$ containing $\overline{s}$ and $\overline{r} := h(\overline{s})$ respectively such that h induces a homeomorphism from J onto I. Then, denoting by F (resp. G) the epigraph of f (resp. g), one has $(x,r) \in F \cap (W \times I)$ if and only if $(x,s) \in G \cap (W \times J)$ and $r = h(s)$. Since $\widehat{h} : (x,s) \mapsto (x, h(s))$ is a homeomorphism from $W \times J$ onto $W \times I$ that is circa-differentiable at $(\overline{x}, \overline{s})$, $\widehat{h}^{-1}$ being circa-differentiable at $(\overline{x}, \overline{r})$, one has $(\overline{x}^*, -1) \in N_C(F, (\overline{x}, \overline{r}))$ if and only if $(\overline{x}^*, -h'(\overline{s})) \in N_C(G, (\overline{x}, \overline{s}))$ or $\overline{x}^* \in \partial_C f(\overline{x})$ if and only if $\overline{y}^* := \overline{x}^*/h'(\overline{s}) \in \partial_C g(\overline{x})$. $\qquad\square$

Now let us consider extensions of Proposition 5.14 about performance functions.

Proposition 5.58. *Let* $A \in L(V,W)$ *be a surjective, continuous, linear map between two normed spaces, let* $j : V \to \overline{\mathbb{R}}$, $p : W \to \overline{\mathbb{R}}$ *be lower semicontinuous and such that* $p \circ A \leq j$. *Suppose that for some* $\overline{v} \in V$ *and all sequences* $(w_n) \to_p \overline{w} := A\overline{v}$, $(\alpha_n) \to 0_+$ *one can find a sequence* $(v_n) \to \overline{v}$ *such that* $j(v_n) \leq p(w_n) + \alpha_n$ *and* $A(v_n) = w_n$ *for all* $n \in \mathbb{N}$. *Then one has*

$$A^{\mathsf{T}}(\partial_C p(\overline{w})) \subset \partial_C j(\overline{v}). \tag{5.31}$$

When p *is the distance function to a subset* S *of* W *and* $\overline{w} \in \mathrm{cl}S$, *the assumption can be restricted to sequences* $(w_n) \to \overline{w}$ *in* S.

This result takes a simpler form when $W = V$, A is the identity map, and j is such that $j \geq p := d_S$ and $j = 0$ on S. Then one has $\partial_C d_S(\overline{x}) \subset \partial_C j(\overline{x})$ for all $\overline{x} \in S$.

Proof. Let us denote by E (resp. P) the epigraph of j (resp. p), let us set $\overline{x} := (\overline{v}, j(\overline{v}))$, $\overline{y} := (\overline{w}, p(\overline{w}))$, and let us show that for all $(u,r) \in T^C(E, \overline{x})$ we have $(Au, r) \in T^C(P, \overline{y})$. Given sequences $(t_n) \to 0_+$, $((w_n, s_n)) \to_P \overline{y}$, we use our assumption to pick a sequence $(v_n) \to \bar{v}$ such that $j(v_n) \le p(w_n) + t_n^2 \le s_n + t_n^2$ and $A(v_n) = w_n$ for all $n \in \mathbb{N}$. Then $j(\bar{v}) \le \liminf_n j(v_n) \le \limsup_n j(v_n) \le \lim_n s_n = p(\bar{w})$, and since $p \circ A \le j$, equality holds and $(j(v_n)) \to j(\bar{v})$. Moreover, $((v_n, s_n + t_n^2)) \to_E (\bar{v}, j(\bar{v})) := \bar{x}$, so that there exists a sequence $((u_n, r_n)) \to (u, r)$ such that $(v_n, s_n + t_n^2) + t_n(u_n, r_n) \in E$ for all n. Then since $p \circ A \le j$, we get $(Av_n, s_n) + t_n(Au_n, r_n + t_n) \in P$ for all n. Since $(Au_n) \to Au$ and $(t_n + r_n) \to r$, this proves that $(Au, r) \in T^C(P, \bar{y})$. Taking polars, we get that for all $(\bar{w}^*, \bar{r}^*) \in N_C(P, \bar{y})$ we have $(A^\top \bar{w}^*, \bar{r}^*) \in N_C(E, \bar{x})$. Taking $\bar{r}^* = -1$, we obtain (5.31).

Now let us suppose $p = d_S$, where S is a subset of W and $\bar{w} \in \mathrm{cl}S$. Again we have $j(\bar{v}) = p(\bar{w})$, since we can find a sequence $(w_n) \to_S \bar{w}$. Let $\bar{x} := (\bar{v}, 0)$, $\bar{y} := (\bar{w}, 0)$. Let us prove again that $(Au, r) \in T^C(P, \bar{y})$ for all $(u, r) \in T^C(E, \bar{x})$. Given sequences $(t_n) \to 0_+$, $((w_n, s_n)) \to_P \bar{y}$, we pick sequences (w'_n) in S, $(v'_n) \to \bar{v}$ such that $d(w'_n, w_n) \le d_S(w_n) + t_n^2 \le s_n + t_n^2$, $j(v'_n) \le t_n^2$ and $A(v'_n) = w'_n$ for all $n \in \mathbb{N}$. Then $((v'_n, t_n^2)) \to_E (\bar{v}, j(\bar{v})) := \bar{x}$, so that there exists a sequence $((u_n, r_n)) \to (u, r)$ such that $(v'_n, t_n^2) + t_n(u_n, r_n) \in E$ for all n. Then since $p \circ A \le j$, we get $(Av'_n, t_n^2) + t_n(Au_n, r_n) \in P$ for all n. Since d_S is Lipschitzian with rate 1 and $d(w'_n, w_n) \le s_n + t_n^2$, we get $(w_n, s_n + 2t_n^2) + t_n(Au_n, r_n) \in P$ for all n. Since $(Au_n) \to Au$ and $(r_n + 2t_n) \to r$, this proves that $(Au, r) \in T^C(P, \bar{y})$. Then the proof can be finished as above. $\qquad\square$

An analogue of Proposition 4.49 is valid for the Clarke subdifferential.

Proposition 5.59. *Let $f : X \to \overline{\mathbb{R}}$, $g : Y \to \overline{\mathbb{R}}$ be lower semicontinuous functions on normed spaces X and Y respectively, and let $h : X \times Y \to \overline{\mathbb{R}}$ be given by $h(x, y) := \max(f(x), g(y))$. Suppose that for some $(\bar{x}, \bar{y}) \in X \times Y$ one has $f(\bar{x}) = g(\bar{y}) \in \mathbb{R}$. Then*

$$\partial_C h(\bar{x}, \bar{y}) \subset \mathrm{co}\left((\partial_C f(\bar{x}) \times \{0\}) \cup (\{0\} \times \partial_C g(\bar{y}))\right) = \bigcup_{\lambda \in [0,1]} (1 - \lambda)\partial_C f(\bar{x}) \times \lambda \partial_C g(\bar{y}).$$

Proof. Using support functions, it suffices to show that for all $(u, v) \in X \times Y$ one has

$$h^C((\bar{x}, \bar{y}), (u, v)) \le \max(f^C(\bar{x}, u), g^C(\bar{y}, v)).$$

By definition of h^C, given $s \ge \max(f^C(\bar{x}, u), g^C(\bar{y}, v))$ we have to prove that for any sequence $(t_n) \to 0_+$ and any sequence $((x_n, y_n, r_n))$ in epi h with limit $(\bar{x}, \bar{y}, h(\bar{x}, \bar{y}))$ one can find a sequence $((u_n, v_n, s_n)) \to (u, v, s)$ such that $(x_n, y_n, r_n) + t_n(u_n, v_n, s_n) \in$ epi h for all n. Since $((x_n, r_n)) \to (\bar{x}, f(\bar{x}))$ in epi f, we can find a sequence $((u_n, p_n)) \to (u, s)$ such that $(x_n, r_n) + t_n(u_n, p_n) \in$ epi f for all n. Similarly, one can find a sequence $((v_n, q_n)) \to (v, s)$ such that $(y_n, r_n) + t_n(v_n, q_n) \in$ epi g for all n. Then, taking $s_n := \max(p_n, q_n)$, we get the required sequence. $\qquad\square$

The proof of the following result is similar to the proof of Corollary 4.80. It is also a consequence of the fact that δ_f^C is bounded above by the similar index build with $\partial_F f$ or $\partial_D f$.

Proposition 5.60. *For every Banach space X and for every $f \in \mathcal{F}(X)$, the function $\delta_f^C : X \to \overline{\mathbb{R}}_+$ given by $\delta_f^C(x) := \inf\{\|x^*\| : x^* \in \partial_C f(x)\}$ is a decrease index for f.*

Lebourg's mean value theorem can be extended to non-Lipschitzian functions. Its statement is similar to the one for the Fréchet mean value theorem, but it is valid in every Banach space. Its proof being identical to the one for that subdifferential, in order to avoid repetition, it will not be presented here.

Exercises

1. Given $f : X \to \overline{\mathbb{R}}$ and $x \in f^{-1}(\mathbb{R})$, show that for every $r \in \mathbb{R}_+$, $c \in \mathbb{R}$, $\overline{x} \in X$, for g given by $g(u) := f(u+\overline{x})+c$, one has $\partial_C(rf)(x) = r\partial_C f(x)$, $\partial_C g(x-\overline{x}) = \partial_C f(x)$.

2. Given $f : X \to \overline{\mathbb{R}}$, $x \in f^{-1}(\mathbb{R})$ and $\ell \in X^*$, show that for $h := f + \ell$ one has

$$h^C(x,u) = f^C(x,u) + \langle \ell, u \rangle, \quad \partial_C h(x) = \partial_C f(x) + \ell.$$

More generally, show that if $g : X \to \overline{\mathbb{R}}$ is circa-differentiable at x, $h := f + g$, then

$$h^C(x,u) = f^C(x,u) + \langle g'(x), u \rangle, \quad \partial_C h(x) = \partial_C f(x) + g'(x).$$

3. Give an example of two functions $g : X \to \overline{\mathbb{R}}$, $h : Y \to \overline{\mathbb{R}}$ such that for f given by $f(x,y) := g(x)+h(y)$ the inclusion $\partial_C f(\overline{x},\overline{y}) \subset \partial_C g(\overline{x}) \times \partial_C h(\overline{y})$ is strict.

4. (a) Give an analytic proof to Theorem 5.51. [See [214, pp. 102–105] or [874].]
(b) Give an analytic proof to Theorem 5.50. [See [214, pp. 106–108] or [874].]

5. Let $A : V \to W$ be a surjective continuous linear map between two normed spaces, let $j : V \to \overline{\mathbb{R}}$ be lower semicontinuous, and let $p : W \to \mathbb{R}$ be Lipschitzian with rate c near some $\overline{w} \in W$. Suppose that $j \geq p \circ A$ and that for some $\overline{v} \in A^{-1}(\overline{w})$ and all sequences $(w_n) \to \overline{w}$, $(t_n) \to 0_+$ one can find a sequence $(v_n) \to \overline{v}$ such that $(t_n^{-1}(j(v_n) - p(w_n)) + ct_n^{-1} \|A(v_n) - w_n\|) \to 0$. Then adapting the proof of Proposition 5.58, show that

$$A^\top(\partial_C p(\overline{w})) \subset \partial_C j(\overline{v}).$$

6. Combine Theorems 5.50 and 5.51 to deal with the subdifferential of a function of the form $f + h \circ g$. Give also a direct proof using Proposition 5.49.

7. Define the *radial hypertangent cone* $H^r(E,a)$ to a subset E of X at $a \in \mathrm{cl}(E)$ as the set of vectors $u \in X$ for which there exists $\varepsilon > 0$ such that $e + tu \in E$ for all $e \in E \cap B(a,\varepsilon)$, $t \in (0,\varepsilon)$. Note that $H(E,a) \subset H^r(E,a) \subset T^C(E,a)$ and that $H^r(E,a)$ is a convex cone.

5.4   Limits of Tangent and Normal Cones

We have seen the interest in assuming regularity in order to transform the inclusions of calculus rules into equalities. Now we intend to relate the concepts of this chapter to limiting notions and to find links with regularity.

We shall show that in finite dimensions, the Clarke tangent cone can be considered the persistent parts of the tangent cones at nearby points, i.e., the limit inferior of the tangent cones at neighboring points as they converge to the nominal point. We also investigate generalizations of this characterization to infinite-dimensional spaces. A first step is the following inclusion.

Theorem 5.61 ([784]). *For every subset E of a Banach space X and every $a \in$ cl(E) one has*

$$\liminf_{x \to_E a} T^D(E,x) \subset T^C(E,a). \qquad (5.32)$$

Proof. Given $v \in X \setminus T^C(E,a)$, let us show that there exists $\alpha > 0$ such that

$$\forall \delta > 0 \; \exists e \in B(a,\delta) \cap E : \qquad B(v,\alpha) \cap T^D(E,e) = \varnothing, \qquad (5.33)$$

i.e., $v \notin \liminf_{x \to_E a} T^D(E,x)$. By definition of $T^C(E,a)$ there exists $\beta > 0$ such that

$$\forall \rho > 0 \; \exists x \in B(a,\rho) \cap E, \; \exists t \in (0,\rho) : \qquad (x + t B(v,\beta)) \cap E = \varnothing. \qquad (5.34)$$

Pick $\alpha \in (0,\beta)$. Given $\delta > 0$, let $\rho \in (0,1)$ be such that $\rho(\|v\| + \alpha + 1) < \delta$. Then taking x and t as in (5.34), for all e in the drop $D(x,B) := \mathrm{co}(\{x\} \cup B)$ with $B := B[x + tv, t\alpha]$, one has $\|e - a\| \leq \|e - x\| + \|x - a\| \leq t\|v\| + t\alpha + \rho < \delta$. Setting $C := \mathbb{R}_+(B - x) = \mathbb{R}_+ B[v,\alpha]$, the truncated drop theorem (Lemma 1.99) yields some $e \in E \cap D(x,B)$ such that $E \cap (e + C) \cap B(e, \beta - \alpha) = \{e\}$. It follows that $T^D(E,e) \cap B(v,\alpha) = \varnothing$. $\qquad \square$

In order to state the following consequence, let us say that E is *sleek* at $a \in \mathrm{cl}(E)$ if the multimap $T^D(E,\cdot)$ is lower semicontinuous at a on $E \cup \{a\}$.

Corollary 5.62. *If a subset E of X is sleek at $a \in \mathrm{cl}(E)$, then it is regular at a.*

Proof. Since sleekness at a means that $T^D(E,a) \subset \liminf_{x \to_E a} T^D(E,x)$, the regularity of E at a stems from the inclusions (5.32) and $T^C(E,a) \subset T^D(E,a)$. $\qquad \square$

Theorem 5.61 has interesting consequences concerning continuous tangent vector fields. In essence, it implies that for continuous vector fields, it is equivalent to taking the tangency condition in the sense of the tangent cone or in the sense of the Clarke tangent cone. This fact, expounded in the next statement, is just a consequence of the characterization of lower semicontinuity of a multimap in terms of selections.

Corollary 5.63. *Given a subset E of a Banach space X and $a \in \mathrm{cl}(E)$, let $v : E \cup \{a\} \to X$ be continuous at a and such that $v(x) \in T^D(E,x)$ for all $x \in E$. Then $v(a) \in T^C(E,a)$.*

An inclusion opposite to the one in Theorem 5.61 can be obtained, provided one replaces the tangent cones $T^D(E,x)$ with the *weak** tangent cones $T^{**}(E,x)$, where $T^{**}(E,x)$ is the set of limit points of bounded families $(t^{-1}(e_t - x))_{t>0}$ in X^{**} for the weak** topology $\sigma(X^{**}, X^*)$, where $e_t \in E$ for all $t > 0$. Here we consider that X (hence E) is embedded in the second dual space X^{**} and we set $T^\sigma(E,x) := T^{**}(E,x) \cap X$. Thus, the definition of $T^\sigma(E,x)$ differs from that of $T^D(E,x)$ by the use of the weak topology instead of the strong topology. When X is reflexive, $T^\sigma(E,x) = T^{**}(E,x)$. Of course, when X is finite-dimensional, one has $T^\sigma(E,x) = T^D(E,x)$.

Theorem 5.64. *Let E be a subset of a Banach space X and let $a \in \mathrm{cl}(E)$. Denote by $N_L^{\mathrm{cl}}(E,a)$ the set of weak* cluster points of bounded sequences (x_n^*) such that for some sequence $(x_n) \to_E a$, one has $x_n^* \in N_F(E,x_n)$ for all n. Then one has*

$$\liminf_{x \to_E a} T^D(E,x) \subset T^C(E,a) \subset \liminf_{x \to_E a} T^{**}(E,x), \tag{5.35}$$

$$N_L^{\mathrm{cl}}(E,a) \subset N_C(E,a). \tag{5.36}$$

In particular, if E has the cone property at a or if X is finite-dimensional, one has $T^C(E,a) = \liminf_{x \to_E a} T^D(E,x)$ and E is sleek at a if and only if E is regular at a.

Proof. To prove (5.35) it remains to show that given $v \in T^C(E,a)$ and $\varepsilon > 0$, one can find $\delta > 0$ such that for all $x \in E \cap B(a,\delta)$ one has $(v + \varepsilon B_{X^{**}}) \cap T^{**}(E,x) \neq \varnothing$, where $B_{X^{**}}$ is the closed unit ball in X^{**}. Since $d_E^C(a,v) \leq 0$, let $\delta > 0$ be such that $t^{-1}(d_E(x+tv) - d_E(x)) < \varepsilon$ for all $(t,x) \in (0,\delta) \times B(a,\delta)$. Then for all $t \in (0,\delta)$, $x \in E \cap B(a,\delta)$ there exists some $e_{t,x} \in E$ such that $t^{-1}\|x+tv - e_{t,x}\| < \varepsilon$. For $x \in E \cap B(a,\delta)$, since the family $(t^{-1}(e_{t,x} - x))_{t \in (0,\delta)}$ is contained in $B(v,\varepsilon)$, hence in the closed ball $v + \varepsilon B_{X^{**}}$, which is $\sigma(X^{**}, X^*)$-compact, it has a weak** cluster point $u_x \in v + \varepsilon B_{X^{**}}$ as $t \to 0_+$. By definition, we have $u_x \in T^{**}(E,x)$. Thus (5.35) holds.

In order to prove (5.36), let $a^* \in N_L^{\mathrm{cl}}(E,a)$, $v \in T^C(E,a)$ and let us show that $\langle a^*, v \rangle \leq 0$. By definition of $N_L^{\mathrm{cl}}(E,a)$, there exist a bounded sequence (x_n^*) and a sequence $(x_n) \to_E a$ such that a^* is a weak* cluster point of (x_n^*) and $x_n^* \in N_F(E,x_n)$ for all $n \in \mathbb{N}$. Using (5.35), we can find a sequence (v_n^{**}) in X^{**} such that $(\|v_n^{**} - v\|) \to 0$ and $v_n^{**} \in T^{**}(E,x_n)$ for all $n \in \mathbb{N}$. Since $N_F(E,x_n)$ is the polar cone of $T^{**}(E,x_n)$ by Exercise 7 of Sect. 4.1.3, we have $\langle x_n^*, v_n^{**} \rangle \leq 0$ for all $n \in \mathbb{N}$, $\liminf_n \langle a^* - x_n^*, v \rangle \leq 0$, whence, as required,

$$\langle a^*, v \rangle \leq \liminf_n (\langle a^* - x_n^*, v \rangle + \langle x_n^*, v - v_n^{**} \rangle + \langle x_n^*, v_n^{**} \rangle) \leq \lim_n \|x_n^*\| \cdot \|v - v_n^{**}\| = 0.$$

When E has the cone property at a, one has $T^C(E,a) = \mathrm{cl}(H(E,a))$, $H(E,a) \subset \liminf_{x \to_E a} T^D(E,x)$, which is closed, so that $T^C(E,a) \subset \liminf_{x \to_E a} T^D(E,x)$ and equality holds. When X is finite-dimensional, this equality stems from (5.35). $\square$

In reflexive Banach spaces the preceding results can be made more precise.

Theorem 5.65 (Borwein–Strojwas [131]). *Let E be a closed subset of a reflexive Banach space X and let $a \in E$. Then denoting by $\overline{\mathrm{co}}(S)$ the closed convex hull of a subset S of X, one has*

$$T^C(E,a) = \liminf_{x \to_E a} T^\sigma(E,x) = \liminf_{x \to_E a} \overline{\mathrm{co}}(T^\sigma(E,x)). \qquad (5.37)$$

In view of the inclusions of the preceding theorem, it suffices to prove that the right-hand side of these equalities is contained in $T^C(E,a)$. We give two proofs in special cases, referring to [131] for the general case. The first one is valid in finite-dimensional spaces; it gives the scheme of the second one, which is more technical.

Proof when X has a Fréchet differentiable norm and E is proximinal (i.e., all points of X have a best approximation in E). Let $u \in S_X \cap \liminf_{x \to_E a} \overline{\mathrm{co}}(T^\sigma(E,x))$. Let us show that $u \in T^C(E,a)$ by proving that for all $\varepsilon > 0$ we have $d_E^C(a,u) \leq \varepsilon$. The definition of a limit inferior yields some $\delta > 0$ such that

$$\forall x \in E \cap B(a,4\delta), \qquad B(u,\varepsilon) \cap \overline{\mathrm{co}}(T^\sigma(E,x)) \neq \varnothing. \qquad (5.38)$$

Given $(t,w) \in (0,\delta) \times B(a,\delta)$, let us set $f(s) := d_E(w+su)$ for $s \in [0,t]$. We intend to show that whenever $f(s)$ is positive (i.e., $w + su \notin E$), the right upper Dini derivative $D^+f(s)$ of f at s is bounded above by ε. Using the mean value theorem between $r := \sup\{r' \in [0,t] : f(r') = 0\}$ with $r = 0$ when $f(r') > 0$ for all $r' \in [0,t]$ and t, this estimate will give, by continuity of f,

$$f(t) - f(0) \leq f(t) = f(t) - f(r) \leq (t-r)\sup\{D^+f(s) : s \in (r,t)\} \leq (t-r)\varepsilon \leq \varepsilon t.$$

Since (t,w) is taken arbitrarily in $(0,\delta) \times B(a,\delta)$, we obtain

$$d_E^C(a,u) \leq \sup\{t^{-1}(d_E(w+tu) - d_E(w)) : (t,w) \in (0,\delta) \times B(a,\delta)\} \leq \varepsilon.$$

By our choice of (t,w), for $s \in [0,t]$ we have $w + su \in B(a,2\delta)$, hence $d_E(w+su) \leq 2\delta$. By assumption, there is a best approximation x of $w+su$ in E: $\|w+su-x\| = d_E(w+su)$. Thus $\|x-a\| \leq \|x-(w+su)\| + \|(w+su)-a\| < 4\delta$, and (5.38) ensures that we can find $v \in B(u,\varepsilon) \cap \overline{\mathrm{co}}(T^\sigma(E,x))$. Let $s \in [0,t]$ with $f(s) > 0$ and let $S(w+su-x)$ be the derivative of the norm at $w+su-x \neq 0$. By Fermat's rule and Exercise 7 of Sect. 4.1.3, we have $\langle S(w+su-x),v\rangle \leq 0$ and

$$D^+f(s) := \limsup_{s' \to s_+} \frac{1}{s'-s}\left(d_E(w+s'u) - d_E(w+su)\right)$$

$$\leq \limsup_{s' \to s_+} \frac{1}{s'-s}\left(\|(w+s'u) - x\| - \|(w+su) - x\|\right) = \langle S(w+su-x),u\rangle$$

$$\leq \|S(w+su-x)\|\,\|u-v\| + \langle S(w+su-x),v\rangle \leq \varepsilon,$$

since $S(\cdot)$ takes its value in the closed unit ball of X^*. $\square$

Proof when X has a Fréchet differentiable and uniformly Gâteaux differentiable norm satisfying the Kadec–Klee property. The Kadec–Klee property is the requirement that the weak topology and the strong topology induce the same topology on the unit sphere S_X. The norm is uniformly Gâteaux differentiable if it is Gâteaux differentiable on $X\setminus\{0\}$ and if for all $u \in S_X$ the limit

$$\lim_{t\to 0}\frac{1}{t}(\|z+tu\| - \|z\|) = \langle S(z), u\rangle$$

is uniform for $z \in S_X$. In such a case, for every $r > 0$, all sequences $(t_n) \to 0_+$, (z_n) in X with $(\|z_n\|) \to r$ one has $t_n^{-1}(\|z_n + t_n u\| - \|z_n\| - t_n\langle S(z_n), u\rangle) \to 0$.

We follow the steps of the preceding proof, starting with a unit vector u in $\liminf_{x\to_E a}\overline{\mathrm{co}}(T^\sigma(E,x))$. Again, given $\varepsilon > 0$, our aim is to show that $d_E^C(a,u) \le \varepsilon$ by proving that $D^+ f(s) \le \varepsilon$ for all $(t,w) \in (0,\delta) \times B(a,\delta)$, $s \in [0,t]$ satisfying $f(s) > 0$, where $f(s) := d_E(w + su)$ and where $\delta > 0$ is associated with $\varepsilon > 0$ as in (5.38).

By a theorem of Lau [618], the set F of points of $X\setminus E$ having a best approximation in E is dense in $X\setminus E$. Given a sequence $(s_n) \to 0_+$ such that $D^+ f(s) = \lim_n (1/s_n)(f(s+s_n) - f(s))$, let $y_n \in B(w + su, s_n^2) \cap F$ and let $x_n \in E$ be a best approximation of y_n in E: $\|x_n - y_n\| = d_E(y_n)$. Then for some $m \in \mathbb{N}$ we have $\|y_n - a\| < 2\delta$ for $n \ge m$, since $B(a,2\delta)$ is a neighborhood of $w + su$. Thus, as in the preceding proof, we have $x_n \in B(a,4\delta)$ and we can find $v_n \in B(u,\varepsilon) \cap \overline{\mathrm{co}}(T^\sigma(E,x_n))$. If $S(y_n - x_n)$ is the Fréchet derivative of the norm at $y_n - x_n$, we have $\langle S(y_n - x_n), v_n\rangle \le 0$. Thus, since d_E is Lipschitzian with rate 1 and since the norm is uniformly Gâteaux differentiable and $(\|y_n - x_n\|) \to d_E(w + su)$, we have

$$D^+ f(s) = \limsup_n \frac{d_E(w + su + s_n u) - d_E(w + su)}{s_n}$$

$$\le \limsup_n \frac{\|y_n + s_n u - x_n\| + s_n^2 - \|y_n - x_n\| + s_n^2}{s_n} = \limsup_n \langle S(y_n - x_n), u\rangle$$

$$\le \limsup_n (\|S(y_n - x_n)\|\,\|u - v_n\| + \langle S(y_n - x_n), v_n\rangle) \le \varepsilon,$$

and we conclude the proof as above. $\square$

We deduce from the preceding result a dual characterization. A similar representation can be given in terms of proximal normals; we refer to [119, 525, 926].

Corollary 5.66. *For a closed subset E of a reflexive Banach space X and $a \in E$, setting $N_L(E,a) := \mathrm{w}^* - \mathrm{seq} - \limsup_{x\to_E a} N_F(E,x)$, one has*

$$N_C(E,a) = \overline{\mathrm{co}}^*(N_L(E,a)).$$

Proof. Since by Exercise 7 of Sect. 4.1.3, for all $x \in E$ one has $N_F(E,x) = (T^\sigma(E,x))^0 = (\overline{\mathrm{co}}(T^\sigma(E,x)))^0$, taking polars in relation (5.37) and applying Theorem 1.84, endowing X^* with the weak topology, one gets

$$N_C(E,a) = \left(w^* - \text{seq} - \limsup_{x \to_E a} N_F(E,x) \right)^{00} = \overline{\text{co}}^*(N_L(E,x)). \qquad \square$$

Exercise. Prove that the inclusion $N_C(E,a) \subset w^* - \limsup_{x \to_E a} N_C(E,x)$ may be strict, even if the space is finite-dimensional. (See [873].)

5.5   Moderate Subdifferentials

In the present section we deal with constructions that have several similarities with the Clarke concepts. In particular, convexity properties are automatically fulfilled. However, the new subdifferential is smaller, hence more precise; it even coincides with the singleton formed by the derivative when the latter exists, a feature that is not satisfied by the Clarke subdifferential. On the other hand, the continuity (or robustness) property of Proposition 5.3 (c) is not automatically satisfied by the moderate subdifferential.

For brevity and simplicity we adopt a geometric viewpoint.

5.5.1   Moderate Tangent Cones

In order to get a clear insight into the following constructions, let us point out that for every cone Q of a linear space there exists a canonical convex cone C contained in Q, its star. In general, the *star* $\text{st}(Q)$ of a subset Q of a linear space X is the set of $x \in Q$ such that for all $q \in Q$ and all $t \in [0,1]$ one has $x + t(q - x) \in Q$. Thus $\text{st}(Q)$ is the set of points x such that Q is star-shaped at x. Clearly, $\text{st}(Q)$ is convex. When Q is a cone, it is easy to check that $\text{st}(Q) = Q \boxminus Q$, where the operation $\boxminus$ is given by

$$A \boxminus B := \{x \in X : B + x \subset A\}.$$

Exercise. When Q is the epigraph of a positively homogeneous function, $\text{st}(Q)$ is the epigraph of a sublinear function.

The concepts we introduce now are closely related to such constructions.

Definition 5.67. The *moderate tangent cone* to a subset E of a normed space X at $a \in \text{cl}E$ is the set $T^M(E,a)$ of $w \in X$ such that for every $z \in X$ and all sequences $(t_n) \to 0_+$, $(z_n) \to z$ satisfying $a + t_n z_n \in E$ for all $n \in \mathbb{N}$, there exists a sequence $(w_n) \to w$ such that $a + t_n z_n + t_n w_n \in E$ for all $n \in \mathbb{N}$.

The *moderate normal cone* to E at a is the polar cone $N_M(E,a)$ of $T^M(E,a)$.

Let us make a comparison with cones we introduced previously, in particular with the circa-tangent cone and the *incident cone* to E at a, i.e., the set $T^I(E,a)$ of $v \in X$ such that for all $(t_n) \to 0_+$ there exists some $(v_n) \to v$ satisfying $a + t_n v_n \in E$ for all n.

Proposition 5.68. *For every subset E of a normed space X and every $a \in \mathrm{cl}E$ the set $T^M(E,a)$ is a closed convex cone. Moreover, one has*

$$T^C(E,a) \subset T^M(E,a) \subset T^I(E,a) \subset T^D(E,a). \tag{5.39}$$

These inclusions may be strict (Exercises 3 and 4).

Proof. The set $T^M(E,a)$ is closed, since it appears as the intersection of the family of sets $\liminf_n t_n^{-1}(E - a - t_n z_n)$ indexed by the triples $((t_n),z,(z_n))$ satisfying $a + t_n z_n \in E$ for all $n \in \mathbb{N}$. It is clearly stable under multiplication by positive real numbers. Let us prove that it is convex. Let $v,w \in T^M(E,a)$ and let $z \in X$ and sequences $(t_n) \to 0_+$, $(z_n) \to z$ satisfying $a + t_n z_n \in E$ for all $n \in \mathbb{N}$ be given. Since $w \in T^M(E,a)$, there exists a sequence $(w_n) \to w$ such that $a + t_n z_n + t_n w_n \in E$ for all $n \in \mathbb{N}$. Since $v \in T^M(E,a)$ and $(w_n + z_n) \to w + z$, there exists a sequence $(v_n) \to v$ such that $a + t_n(z_n + w_n) + t_n v_n \in E$ for all $n \in \mathbb{N}$. Since $(w_n + v_n) \to w + v$, we get that $w + v \in T^M(E,a)$.

The inclusion $T^M(E,a) \subset T^I(E,a)$ is obvious (take $z = 0$, $z_n = 0$ for all n). Let $w \in T^C(E,a)$ and let $z \in X$ and sequences $(t_n) \to 0_+$, $(z_n) \to z$ satisfying $a + t_n z_n \in E$ for all $n \in \mathbb{N}$. Then $(a_n) := (a + t_n z_n) \to_E a$, so that there exists a sequence $(w_n) \to w$ such that $a + t_n z_n + t_n w_n = a_n + t_n w_n \in E$ for all $n \in \mathbb{N}$. Thus $w \in T^M(E,a)$. $\qquad \square$

One can easily check the inclusions

$$T^I(E,a) \boxminus T^D(E,a) \subset T^M(E,a) \subset T^I(E,a) \boxminus T^I(E,a). \tag{5.40}$$

In fact, if $w \in T^I(E,a) \boxminus T^D(E,a)$, given $z \in X$ and sequences $(t_n) \to 0_+$, $(z_n) \to z$ satisfying $a + t_n z_n \in E$ for all $n \in \mathbb{N}$, we have $z \in T^D(E,a)$, hence $v := w + z \in T^I(E,a)$, so that there exists a sequence $(v_n) \to v$ such that $a + t_n v_n \in E$ for all n; then $(w_n) := (v_n - z_n) \to w$ and $a + t_n z_n + t_n w_n \in E$ for all $n \in \mathbb{N}$, whence $w \in T^M(E,a)$. The second inclusion means that for all $v \in T^I(E,a)$, $w \in T^M(E,a)$ one has $v + w \in T^I(E,a)$. Such is the case because given a sequence $(t_n) \to 0_+$, we can find a sequence $(v_n) \to v$ such that $a + t_n v_n \in E$ for all n, and the definition of $T^M(E,a)$ yields a sequence $(w_n) \to w$ satisfying $a + t_n v_n + t_n w_n \in E$ for all n, so that $v + w \in T^I(E,a)$.

Thus, when E is *tangentable* (or derivable) at a in the sense that $T^I(E,a) = T^D(E,a)$, one has

$$T^M(E,a) = T^D(E,a) \boxminus T^D(E,a) = T^I(E,a) \boxminus T^I(E,a).$$

Since tangentability is a rather mild assumption, this equality offers an interpretation of $T^M(E,a)$ that is often satisfied. The preceding observations show that the stars $T^{MD}(E,a) := T^D(E,a) \boxminus T^D(E,a)$ and $T^{MI}(E,a) := T^I(E,a) \boxminus T^I(E,a)$ of $T^D(E,a)$ and $T^I(E,a)$ respectively are tangent cones of interest. This assertion is reinforced by the next result, which stems from the fact that for every convex cone C containing 0 one has $C \boxminus C = C$.

Proposition 5.69. *If E is a convex subset of X and $a \in \mathrm{cl}E$, then $T^M(E,a)$ coincides with the usual tangent cone $T(E,a) := \mathrm{cl}(\mathbb{R}_+(E-a))$ to E at a. More generally, if E is tangentable at a and if $T^D(E,a)$ is convex, then $T^M(E,a) = T^D(E,a)$.*

The last assertion also yields the following result (which can also be established via invariance under diffeomorphisms).

Proposition 5.70. *If E is a submanifold of class C^1 of X and $e \in E$, then $T^M(E,e)$ coincides with the classical tangent cone to E at e.*

The invariance under diffeomorphism we alluded to above can be made more precise, as in the following statement. Note that here h is just Hadamard differentiable at a, and not necessarily circa-differentiable at a as in Proposition 5.27.

Proposition 5.71. *Let X, Y be normed spaces and let E (resp. F) be a subset of an open subset W of X (resp. of Y). Let $h : W \to Y$ be Hadamard differentiable at $a \in \mathrm{cl}E \cap W$ and directionally open at a on E with respect to F in the following sense: for every $z \in T^D(F,b)$ with $h(E) \subset F$, $b := h(a)$ and for all sequences $(z_n) \to z$, $(t_n) \to 0_+$ with $b + t_n z_n \in F$ for all n there exists a convergent sequence (w_n) such that $h(a + t_n w_n) = b + t_n z_n$ and $a + t_n w_n \in E$ for n large enough. Then*

$$h'(a)(T^M(E,a)) \subset T^M(F,b), \qquad h'(a)^\mathsf{T}(N_M(F,b)) \subset N_M(E,a). \qquad (5.41)$$

The directional openness assumption is satisfied if h has a right inverse k that is Hadamard differentiable at $b := h(a)$ and such that $k(F) \subset E$.

Proof. Let $u \in T^M(E,a)$ and let $v := h'(a)(u)$. Given $z \in T^D(F,b)$ and sequences $(z_n) \to z$, $(t_n) \to 0_+$ with $b + t_n z_n \in F$ for all n, our assumption yields $w \in X$ and a sequence $(w_n) \to w$ such that $h(a + t_n w_n) = b + t_n z_n$ and $a + t_n w_n \in E$ for n large enough. Then since $u \in T^M(E,a)$, there exists a sequence $(u_n) \to u$ such that $e_n := a + t_n w_n + t_n u_n \in E$ for all n. Then $h(e_n) \in F$, and since h is Hadamard differentiable at a, one has $h(a + t_n w_n + t_n u_n) = b + t_n z_n + t_n v_n$ for some sequence $(v_n) \to v$. Thus $v \in T^M(F,b)$. The inclusion for normal cones follows by polarity. $\qquad \square$

An adapted notion of hypertangent cone will be useful for calculus rules.

Definition 5.72. The *moderate hypertangent cone* $H^M(E,a)$ to a subset E of X at $a \in \mathrm{cl}(E)$ is the set of vectors $w \in X$ such that for all $z \in X$ and all sequences $(t_n) \to 0_+$, $(w_n) \to w$, $(z_n) \to z$ satisfying $a + t_n z_n \in E$ for all $n \in \mathbb{N}$, one has $a + t_n z_n + t_n w_n \in E$ for all n large enough.

It is easy to check that $H^M(E,a)$ is a convex cone such that

$$H^C(E,a) \subset H^M(E,a) \subset T^M(E,a), \qquad (5.42)$$

$$T^M(E,a) + H^M(E,a) = H^M(E,a). \qquad (5.43)$$

Proposition 5.73. *If $H^M(E,a)$ is nonempty, one has*

$$T^M(E,a) = \mathrm{cl}(H^M(E,a)), \quad \mathrm{int}T^M(E,a) \subset H^M(E,a), \quad N_M(E,a) = (H^M(E,a))^0.$$

Proof. The inclusion $\mathrm{cl}(H^M(E,a)) \subset T^M(E,a)$ stems from the closedness of $T^M(E,a)$. Let $u \in H^M(E,a)$. Then by (5.43), for every $w \in T^M(E,a)$ and every $t > 0$ one has $w_t := w + tu \in H^M(E,a)$. Since $(w_t)_t \to w$ as $t \to 0_+$, we get that $w \in \mathrm{cl}(H^M(E,a))$. Thus $T^M(E,a) = \mathrm{cl}(H^M(E,a))$, and the relation $N_M(E,a) = (H^M(E,a))^0$ ensues.

Let $w \in \mathrm{int}T^M(E,a)$. Let us pick $u \in H^M(E,a)$ and observe that for $t > 0$ small enough we have $w = (w - tu) + tu \in T^M(E,a) + H^M(E,a) = H^M(E,a)$. $\qquad\square$

Exercises

1. *The Treiman's tangent cone* $T^B(E,a)$ *to* E *at* $a \in \mathrm{cl}E$ *is the set of* $v \in X$ *such that for all bounded sequences* (w_n) *in* X *and* $(t_n) \to 0_+$ *satisfying* $a + t_n w_n \in E$ *for all* $n \in \mathbb{N}$, *there exists a sequence* $(v_n) \to v$ *such that* $a + t_n w_n + t_n v_n \in E$ *for all* $n \in \mathbb{N}$. Check that $T^B(E,a)$ is convex and that $T^C(E,a) \subset T^B(E,a) \subset T^M(E,a)$. Write inclusions for the associated normal cones. Prove that these inclusions may be strict.

2. Let X,Y be normed spaces, let $A \subset X$, $B \subset Y$, and let $(x,y) \in \mathrm{cl}A \times \mathrm{cl}B$. Then

$$T^M(A \times B, (x,y)) = T^M(A,x) \times T^M(B,y),$$

$$N_M(A \times B, (x,y)) = N_M(A,x) \times N_M(B,y).$$

3. Show that for $E := \mathrm{epi}\, f$, with $f(r) := -|r|$ for $r \in \mathbb{R}$ and $a := (0,0)$, one has $T^M(E,a) = \mathrm{epi}\,|\cdot| \neq T^D(E,a)$.

4. Show that if $f : X \to \mathbb{R}$ is differentiable at x, but not circa-differentiable there, then for $E := \mathrm{epi}\, f$, $a := (x, f(x))$, one has $T^M(E,a) \neq T^C(E,a)$. [Hint: Show that $T^M(E,a) = \mathrm{epi}\, f'(a,\cdot)$.]

5. (a) Give a direct proof of the first assertion of Proposition 5.69.
(b) Prove the assertion preceding Proposition 5.71.

6. Define the *directional hypertangent cone* to a subset E of a normed space X at $a \in \mathrm{cl}(E)$, or *cone of feasible directions* to E at a, as the set $H^D(E,a)$ of $u \in X$ such that for all sequences $(t_n) \to 0_+$, $(u_n) \to u$ one has $a + t_n u_n \in E$ for n large enough.

(a) Show that $H^D(E,a) = X \backslash T^D(X \backslash E, a)$.
(b) Prove that $H^M(E,a) = H^D(E,a) \boxminus T^D(E,a)$.

7. (a) Check the relations $H^C(E,a) \subset H^M(E,a) \subset H^D(E,a)$.
(b) Check the equality $T^M(E,a) + H^M(E,a) = H^M(E,a)$.

5.5.2 Moderate Subdifferentials

Let us turn to moderate derivates of functions. Again, we have a primal and a dual approach.

It is easy to show that if E is the epigraph of a function, then for every $a \in \mathrm{cl}E$, the moderate tangent cone $T^M(E,a)$ is an epigraph. It is natural to introduce the associated function.

Definition 5.74. The *moderate derivate* of a function $f : X \to \overline{\mathbb{R}}$ at $x \in f^{-1}(\mathbb{R})$ in the direction $u \in X$ is the function $f^M(x,\cdot)$ whose epigraph is $T^M(E,e)$, where E is the epigraph of f and $e := (x,f(x))$: $\mathrm{epi}f^M(x,\cdot) = T^M(E,e)$. Equivalently,

$$f^M(x,u) := \min\{r \in \mathbb{R} : (u,r) \in T^M(E,e)\}, \qquad\qquad u \in X.$$

The *moderate subdifferential* of $f : X \to \overline{\mathbb{R}}$ at $x \in f^{-1}(\mathbb{R})$ is the set

$$\partial_M f(x) := \{x^* \in X^* : x^*(\cdot) \le f^M(x,\cdot)\}.$$

Since $T^C(E,e) \subset T^M(E,e) \subset T^D(E,e)$, one has $f^D(x,\cdot) \le f^M(x,\cdot) \le f^C(x,\cdot)$ and $\partial_D f(x) \subset \partial_M f(x) \subset \partial_C f(x)$. These inequalities and inclusions may be strict.

Example. For f given by $f(0) := 0$, $f(x) := x\sin(1/x)$ for $x \in \mathbb{R}\setminus\{0\}$, one has $\partial_M f(0) = \partial_C f(0) = [-1,1]$, $\partial_D f(0) = \varnothing$. For $g : \mathbb{R} \to \mathbb{R}$ given by $g(x) = xf(x)$, one has $\partial_M g(0) = \{g'(0)\} = \{0\}$, $\partial_C g(0) = [-1,1]$.

The convexity of $T^M(E,e)$ can be translated into an analytical property.

Proposition 5.75. *For every function f finite at x, $f^M(x,\cdot)$ is sublinear. Moreover,*

$$x^* \in \partial_M f(x) \iff (x^*,-1) \in N_M(\mathrm{epi}f,(x,f(x))).$$

This equivalence follows from the definitions, with $E := \mathrm{epi}f$, $e := (x,f(x))$:

$$x^* \in \partial_M f(x) \Leftrightarrow \forall(u,r) \in T^M(E,e)\ \langle(x^*,-1),(u,r)\rangle = \langle x^*,u\rangle - r \le 0.$$

Under some mild assumptions, analytical expressions can be given for the sub-derivate f^M.

Proposition 5.76. *Suppose $f : X \to \overline{\mathbb{R}}$ is finite at x and epi-derivable at x in the sense that $T^D(E,e) = T^I(E,e)$ for $E := \mathrm{epi}f$, $e := (x,f(x))$. Then for all $u \in X$, one has*

$$f^M(x,u) = \sup_{v \in \mathrm{dom}f^D(x,\cdot)} \left(f^D(x,u+v) - f^D(x,v)\right).$$

If $f^D(x,\cdot) = f^I(x,\cdot)$ is sublinear, then $f^M(x,\cdot) = f^D(x,\cdot)$ and $\partial_M f(x) = \partial_D f(x)$. In particular, if f is Hadamard differentiable at x, then $\partial_M f(x) = \{f'(x)\}$.

Proof. We have seen that $T^M(E,e) = T^D(E,e) \boxminus T^D(E,e)$ when E is tangentable at e. Then it remains to check that when $A := B \boxminus C$ and A, B, C are the epigraphs of functions g, h, k respectively, one has $g = h \boxminus k$, where

$$(h \boxminus k)(u) := \sup_{v \in \text{dom}\,k} (h(u+v) - k(v)).$$

In fact, one has

$$(u,r) \in A \Leftrightarrow \forall (v,s) \in C \ (u+v, r+s) \in B \Leftrightarrow \forall v \in \text{dom}\,k, \forall s \geq k(v), \ r+s \geq h(u+v)$$

$$\Leftrightarrow \forall v \in \text{dom}\,k \ \ r \geq h(u+v) - k(v) \Leftrightarrow r \geq (h \boxminus k)(u).$$

If $f^D(x,\cdot) = f^l(x,\cdot)$ is sublinear, then $T^M(E,e) = T^D(E,e)$ and $f^M(x,\cdot) = f^D(x,\cdot)$. $\square$

Exercise. Suppose f is finite and epi-derivable at x and $f^D(x,\cdot)$ is superlinear (i.e., $-f^D(x,\cdot)$ is sublinear). Show that $f^M(x,\cdot) = -f^D(x,\cdot)$.

Another analytical expression of f^M can be given when f is Lipschitzian around x. It is akin to the one for the circa-derivate (Definition 5.1). In fact, it suffices to suppose that $f(x)$ is finite and f is *directionally stable* at x, i.e., for all $v \in X \setminus \{0\}$ there exist $r, c > 0$ such that $|f(x+tw) - f(x)| \leq c \|tw\|$ for all $w \in B(v,r), t \in [0,r]$. This condition is obviously satisfied when f is Lipschitzian around x.

Proposition 5.77. *For every $f : X \to \mathbb{R}$ finite and continuous at $x \in X$ one has*

$$f^M(x,u) \leq \sup_{v \in X} \limsup_{(t,v',u') \to (0,v,u)} \frac{1}{t} \left(f(x+tv'+tu') - f(x+tv') \right). \tag{5.44}$$

If, moreover, f is directionally stable at x, for $f^\diamond(x,u) := \inf\{q : (u,q) \in H^M(E,e)\}$, one has

$$f^M(x,u) \leq f^\diamond(x,u) = \sup_{v \in X} \limsup_{t \to 0_+} \frac{1}{t} \left(f(x+tv+tu) - f(x+tv) \right).$$

For $u \in \text{dom}\, f^\diamond(x,\cdot)$ equality holds.

Proof. More generally, we first prove that for every f finite at x, setting for $u, v \in X$,

$$f^\diamond_v(x,u) := \limsup_{\substack{(t,v',u',r) \to (0,v,u,f(x)) \\ t>0, \ r \geq f(x+tv')}} \frac{1}{t} \left(f(x+tv'+tu') - r \right), \tag{5.45}$$

$E := \text{epi}\,f, e := (x, f(x))$, one has

$$f^M(x,u) \leq f^\diamond(x,u) := \inf\{q : (u,q) \in H^M(E,e)\} \leq \sup_{v \in X} f^\diamond_v(x,u). \tag{5.46}$$

These estimates are more general than the first assertion: when f is continuous at x, the right-hand sides in (5.44) and (5.46) coincide, since then

$$f_v^\diamond(x,u) = \limsup_{(t,u',v')\to(0_+,u,v)} \frac{1}{t}\left(f(x+tv'+tu') - f(x+tv')\right),$$

as is easily seen, using the fact that for every $\varepsilon > 0$, $v \in X$ there exists $\delta > 0$ such that $f(x+tv') \in (f(x) - \varepsilon, f(x) + \varepsilon)$ whenever $t \in (0,\delta)$, $v' \in B(v,\delta)$.

The first inequality in (5.46) stems from the inclusion $H^M(E,e) \subset T^M(E,e)$. In order to prove the other one, given $w := (u,q) \in X \times \mathbb{R}$ satisfying $q > \sup_{v\in X} f_v^\diamond(x,u)$, let us show that $(u,q) \in H^M(E,e)$. Given sequences $(t_n) \to 0_+$, $(w_n) := ((u_n,q_n)) \to w$, $(z_n) := ((v_n,s_n)) \to z := (v,s)$ satisfying $e + t_n z_n \in E$ for all $n \in \mathbb{N}$, let us show that $e + t_n z_n + t_n w_n \in E$ for all $n \in \mathbb{N}$ large enough. We have $f(x+t_n v_n) \le r_n := f(x) + t_n s_n$, and $(r_n) \to f(x)$. Since $q > f_v^\diamond(x,u)$, we have

$$q > \limsup_n t_n^{-1}(f(x+t_n v_n + t_n u_n) - r_n).$$

Thus, $q_n \ge t_n^{-1}(f(x+t_n v_n + t_n u_n) - r_n)$ for all $n \in \mathbb{N}$ large enough. Then for such n's, we get $(x+t_n v_n + t_n u_n, f(x) + t_n s_n + t_n q_n) \in E$, so that $(u,q) \in H^M(E,e)$.

Now suppose that f is directionally stable at x and let $w := (u,q) \in H^M(E,e)$. Let us show that for every $v \in X$ we have $f_v^\diamond(x,u) \le q$; this will prove that the second inequality in (5.46) is an equality. Given sequences $(t_n) \to 0_+$, $(u_n) \to u$, $(v_n) \to v$, $(r_n) \to f(x)$ such that $(t_n^{-1}[f(x+t_n v_n + t_n u_n) - r_n]) \to f_v^\diamond(x,u)$ and $r_n \ge r_n' := f(x+t_n v_n)$ for all $n \in \mathbb{N}$, using the assumption that f is directionally stable at x, taking a subsequence if necessary, we may assume that the bounded sequence $(s_n) := (t_n^{-1}[r_n' - f(x)])$ has a limit $s \in \mathbb{R}$. Then $(x+t_n v_n, f(x) + t_n s_n) \in E$ for all n and $(z_n) := ((v_n,s_n)) \to z := (v,s)$. Since $w \in H^M(E,e)$, we have $e + t_n z_n + t_n w_n \in E$ for the sequence $(w_n) := ((u_n,q_n)) \to (u,q) = w$ with $q_n := q$ for all n. Thus $f(x+t_n v_n + t_n u_n) \le f(x) + t_n s_n + t_n q$ for all n, whence $t_n^{-1}(f(x+t_n v_n + t_n u_n) - r_n') \le q$. Since $r_n \ge r_n'$, we get $f_v^\diamond(x,u) \le q$.

Now let us prove the last assertion. Let $u \in \mathrm{dom} f^\circ(x,\cdot)$, so that there exists some $q \in \mathbb{R}$ such that $(u,q) \in H^M(E,e)$. Then for all $t > 0$ and all $r \in \mathbb{R}$ such that $(u,r) \in T^M(E,e)$ one has $(u,r)+t(u,q) \in H^M(E,e)$, hence $(u,(1+t)^{-1}(r+tq)) \in H^M(E,e)$ and $f^\circ(x,u) \le (1+t)^{-1}(r+tq)$. Taking the limit as $t \to 0_+$, we get $f^\circ(x,u) \le r$. Therefore $f^\circ(x,u) \le \inf\{r : (u,r) \in T^M(E,e)\} = f^M(x,u)$, and equality holds. $\square$

The following definition will be convenient.

Definition 5.78. A subset E of X is said to have the *moderate cone property* around $a \in \mathrm{cl}(E)$ if $H^M(E,a)$ is nonempty. A function $f : X \to \overline{\mathbb{R}}$ finite at $x \in X$ has the moderate cone property around x if its epigraph E has the moderate cone property around $e := (x,f(x))$.

Obviously these notions are less demanding than the corresponding notions for the Clarke concepts.

Let us introduce the *singular moderate subdifferential* of f at x as the set

$$\partial_M^\infty f(x) := \{x^* \in X^* : (x^*, 0) \in N_M(\text{epi}\, f, e)\}, \qquad \text{where } e := (x, f(x)).$$

The decomposition of $N_M(\text{epi}\, f, e)$ one gets is similar to that of Proposition 5.36.

Proposition 5.79. *Let $f : X \to \overline{\mathbb{R}}$ be finite at $x \in X$ and let $e := (x, f(x))$, $E := \text{epi}\, f$. Then $\partial_M^\infty f(x)$ is a weak* closed convex cone, and one has the decomposition*

$$N_M(E, e) = (\mathbb{P}(\partial_M f(x) \times \{-1\})) \cup (\partial_M^\infty f(x) \times \{0\}).$$

Moreover $\partial_M f(x) + \partial_M^\infty f(x) = \partial_M f(x)$, and if $\partial_M f(x)$ is nonempty, then $\partial_M^\infty f(x)$ is the recession cone of $\partial_M f(x)$ and $N_M(E, e) = \mathbb{R}_+(\partial_M f(x) \times \{-1\}) + \partial_M^\infty f(x) \times \{0\}$.

Since $N_M(E, e) \subset N_C(E, e)$, hence $\partial_M^\infty f(x) \subset \partial_C^\infty f(x)$, we get that $\partial_M^\infty f(x) = \{0\}$ when f is Lipschitzian around x.

The proof of the next result is a simple adaptation of the proof of Corollary 5.45.

Corollary 5.80. *If $f : X \to \overline{\mathbb{R}}$ is finite and continuous at $x \in X$ and has the moderate cone property around x, then*

$$\partial_M(-f)(x) = -\partial_M f(x). \qquad (5.47)$$

5.5.3 Calculus Rules for Moderate Subdifferentials

As in the case of Clarke subdifferentials, calculus rules can be improved when one assumes that the functions are regular enough. A corresponding concept of regularity is as follows. Clearly, it is less stringent than Clarke regularity.

Definition 5.81. A subset E of a normed space X is said to be *moderately regular*, or simply *M-regular*, at $a \in \text{cl}(E)$ if $T^M(E, a) = T^D(E, a)$. A function $f : X \to \overline{\mathbb{R}}$ finite at $x \in X$ is said to be *moderately regular*, or simply M-regular, at x if $f^M(x, \cdot) = f^D(x, \cdot)$.

The dual requirements are equivalent to the respective primal properties, as one can show as for the case of circa-regularity.

Proposition 5.82. *A set E is M-regular at $a \in \text{cl}(E)$ if and only if $N_M(E, a) = N_D(E, a)$. A function f is M-regular at $x \in f^{-1}(\mathbb{R})$ iff $\partial_M f(x) = \partial_D f(x)$ and $\partial_M^\infty f(x) = \partial_D^\infty f(x)$.*

A convex set is M-regular at every point of its closure and a convex function is M-regular at every point of its domain. If a function $f : X \to \overline{\mathbb{R}}$ is finite at $x \in X$ and is Hadamard differentiable at x, then one sees that f is M-regular at x and $\partial_M f(x) = \{f'(x)\}$, $\partial_M^\infty f(x) = \partial_D^\infty f(x) = \varnothing$. Other examples arise from the calculus rules we present next. Their proofs, being similar to those given for the Clarke subdifferential, are left as exercises. Again, we take a geometric starting point.

Proposition 5.83. *Let X, Y be normed spaces, let W be an open subset of X, let F (resp. G) be a subset of X (resp. Y), and let $g : W \to Y$ be Hadamard differentiable at a point a of $E := F \cap g^{-1}(G)$. Suppose $A(T^M(F, a)) \cap H^M(G, b) \neq \varnothing$, where $A := g'(a)$, $b := g(a)$. Then $T^M(F, a) \cap A^{-1}(T^M(G, b)) \subset T^M(E, a)$. If F and G are M-regular at a and b respectively, then equality holds and E is M-regular at a.*

A chain rule can be derived from Proposition 5.83 as for Clarke subdifferentials.

Theorem 5.84. *Let X, Y be normed spaces, let W be an open subset of X, let $h : Y \to \overline{\mathbb{R}}$ be finite at $\overline{y} \in Y$, and let $g : W \to Y$ be a mapping that is Hadamard differentiable at some point $\overline{x}$ of X such that $\overline{y} = g(\overline{x})$. Let $f := h \circ g$. Suppose there exists some $u \in X$ such that $h^\circ(\overline{y}, g'(\overline{x})(u)) < +\infty$. Then one has*

$$f^M(\overline{x}, \cdot) \leq h^M(\overline{y}, \cdot) \circ g'(\overline{x}), \tag{5.48}$$

$$\partial_M f(\overline{x}) \subset g'(\overline{x})^\intercal(\partial_M h(\overline{y})) := \partial_M h(\overline{y}) \circ g'(\overline{x}). \tag{5.49}$$

If h is M-regular at $\overline{y}$, these relations are equalities.

Theorem 5.85. *Let $f, g : X \to \overline{\mathbb{R}}$ be two lower semicontinuous functions finite at $\overline{x} \in X$ such that there exists some $u \in \mathrm{dom} f^M(\overline{x}, \cdot) \cap \mathrm{dom} g^\circ(\overline{x}, \cdot)$. Then*

$$(f + g)^M(\overline{x}, \cdot) \leq f^M(\overline{x}, \cdot) + g^M(\overline{x}, \cdot), \tag{5.50}$$

$$\partial_M(f + g)(\overline{x}) \subset \partial_M f(\overline{x}) + \partial_M g(\overline{x}). \tag{5.51}$$

If f and g are M-regular at $\overline{x}$, these relations are equalities.

For separable functions, a sum rule does not require additional assumptions.

Proposition 5.86. *If X, Y are normed spaces, $f : X \to \mathbb{R}_\infty$, $g : Y \to \mathbb{R}_\infty$, and if h is defined by $h(x, y) := f(x) + g(y)$, then for every $(x, y) \in \mathrm{dom} h$, $(u, v) \in X \times Y$, one has*

$$h^M((x, y), (u, v)) \leq f^M(x, u) + g^M(y, v), \tag{5.52}$$

$$\partial_M h(x, y) \subset \partial_M f(x) \times \partial_M g(y). \tag{5.53}$$

If f and g are M-regular at x and y respectively, these relations are equalities.

Now let us state a moderate version of Proposition 5.58; its proof is similar.

Proposition 5.87. *Let $A \in L(V, W)$ be a surjective continuous linear map between two normed spaces, let $j : V \to \overline{\mathbb{R}}$, $p : W \to \overline{\mathbb{R}}$ be lower semicontinuous and such that $p \circ A \leq j$. Suppose that for some $\overline{v} \in j^{-1}(\mathbb{R})$, $\overline{w} := A\overline{v} \in p^{-1}(\mathbb{R})$ and every sequence $(\alpha_n) \to 0_+$, one of the following assumptions is satisfied:*

(a) *For every sequence* $(w_n) \to \overline{w}$ *one can find a sequence* $(v_n) \to \overline{v}$ *such that* $j(v_n) \leq p(w_n) + \alpha_n$ *and* $A(v_n) = w_n$ *for all* $n \in \mathbb{N}$.

(b) $p = d_S$ *for some closed subset* S *of* W *containing* $\overline{w}$ *and for every sequence* $(w_n) \to_S \overline{w}$ *one can find a sequence* $(v_n) \to \overline{v}$ *such that* $j(v_n) \leq p(w_n) + \alpha_n$ *and* $A(v_n) = w_n$ *for all* $n \in \mathbb{N}$.

Then one has

$$A^\mathsf{T}(\partial_M p(\overline{w})) \subset \partial_M j(\overline{v}).$$

Exercises

1. Prove Theorem 5.84.

2. Prove Theorem 5.85.

3. Prove Proposition 5.86.

4. The *Treiman subderivate* of a function $f : X \to \overline{\mathbb{R}}$ at $x \in f^{-1}(\mathbb{R})$ in the direction $u \in X$ is the function $f^B(x, \cdot)$ whose epigraph is $T^B(E, e)$, where E is the epigraph of f and $e := (x, f(x))$: $\mathrm{epi} f^B(x, \cdot) = T^B(E, e)$. The *Treiman subdifferential* of $f : X \to \overline{\mathbb{R}}$ at $x \in f^{-1}(\mathbb{R})$ is the set

$$\partial_B f(x) := \{x^* \in X^* : x^*(\cdot) \leq f^B(x, \cdot)\}.$$

Prove for this concept results similar to those of this section. (See [931, 934].)

5. Suppose that for every subset E of a normed space X and all $a \in \mathrm{cl}E$ one is given a set $\gamma(E, a)$ of sequences $((t_n, e_n))_n$ converging to $(0_+, a)$. Then define $T^\gamma(E, a)$ as the set of $v \in X$ such that for every $((t_n, e_n))_n$ in $\gamma(E, a)$ there exists a sequence $(v_n) \to v$ in X with $e_n + t_n v_n \in E$ for every n.

(a) Check that $T^\gamma(E, a) = T^I(E, a)$ when $((t_n, e_n))_n \in \gamma(E, a)$ iff $e_n = a$ for all n.

(b) Check that $T^\gamma(E, a) = T^C(E, a)$ when $((t_n, e_n))_n \in \gamma(E, a)$ iff $e_n \in E$ for all n.

(c) Give interpretations of $T^B(E, a)$ and $T^M(E, a)$ in terms of appropriate convergences γ.

(d) Suppose that for every $((t_n, e_n))_n \in \gamma(E, a)$, every $u \in X$, and every sequence $(u_n) \to u$ satisfying $e_n + t_n u_n \in E$ for all n one has $((t_n, e_n + t_n u_n))_n \in \gamma(E, a)$. Show that $T^\gamma(E, a)$ is convex in such a case. (See [790].)

6. Define a hypertangent cone associated with a convergence γ as in the preceding exercise and use it to display some calculus rules.

7. Let S be a subset of a Banach space X and let $f : X \to \mathbb{R}$ be a Lipschitzian function that attains its minimum over S at $\overline{x} \in S$. Show that $0 \in \partial_M f(\overline{x}) + N_M(S, \overline{x})$. Compare this condition with the necessary condition using the Clarke subdifferential. [Hint: Use the penalization lemma and the sum rule of Theorem 5.85.]

5.6   Notes and Remarks

The present chapter deals with notions that have given a strong impetus to nonsmooth analysis, thanks to the landmark contributions of Clarke [211, 212, 214, 218] supplemented with some simplifications or complements due to Hiriart–Urruty [480, 481], Ioffe [514, 525], Penot [816], Rockafellar [873–875], Thibault [908–913], Treiman [926–936], and others. Here analytical and geometrical primal notions are available. They even enjoy pleasant convexity and robustness properties. It follows that they can be recovered from dual notions, a distinct feature of this theory. Again, the passages from analytical notions to geometrical notions are multiple and useful. However, the theory cannot be expected to be miraculous: these notions do not reflect accurately the behaviors of the involved sets or functions and they cannot be considered real approximations. Still the theory shows the way to what can be expected in terms of optimality conditions and other aims. Moreover, at least in the Lipschitzian case, the apparatus is simple enough, and good calculus rules exist, two facts that explain its wide success. It is a specific feature of our presentation that we prefer a geometrical approach in the non-Lipschitzian case. It relies on the fundamental idea of Clarke to use the subdifferential of the distance function to a set and its relationship to the normal cone.

In favorable spaces, the notions presented in this chapter can be considered convexified objects made out of limiting subdifferentials and limiting normal cones. This fact illustrates the unity of the field. While it appears to some authors a hodgepodge of disparate creatures, it can be considered an integrated biome, with its natural evolution and kindred species. Moreover, when some continuity property (or "sleekness" property) is available, one gets coincidence with the concepts previously considered.

The definition of the Clarke tangent cone we have adopted appeared in [481]; see also [908]. The definition of the Clarke hypertangent cone we have chosen is the one in [218] and is more restrictive than the original one in [875]; it is more convenient for a number of statements and proofs, as for instance in calculus rules.

It seems that Corollary 5.44 and relation (5.23) for a directionally Lipschitzian function are new. Also, the properties related to order have not appeared elsewhere (but see [816] for what concerns the metric form of Proposition 5.58).

The moderate tangent cone has been introduced in [790, 829] under the name "prototangent cone" or "pseudo-strict tangent cone," along with a full range of tangent cones in a unified way; see also [566]. Moderate subdifferentials appeared in [707, 708]; see also [103, 141]. For another proposal of spectra of tangent cones see [933]. Calculus rules for the moderate subdifferential are given in [103, 141] and [707, 708]. J. Treiman [926–928, 931–936] has made remarkable efforts to promote normal cones and subdifferentials as small as possible, an aim shared with the preceding references and [526, 566]. He also proposed the first correct proof of the link between the Clarke tangent cone and the limit inferior of the nearby directional tangent cones [926] in infinite dimensions, the proof in [784] presenting a flaw in infinite dimensions; for other proofs, see [39, 791]. The way to such a result has been paved by papers dealing with semicontinuity results of tangent cones in view of applications to dynamical systems [230–232].

The number of papers dealing with optimality conditions, optimal control problems, or analysis problems in terms of Clarke's generalized gradients (an expression we prefer to avoid when no scalar product is available) is huge. As explained above, this fact is due to the simplicity and the qualities of the theory and to the success of the books [214, 218].

Chapter 6
Limiting Subdifferentials

Two is my limit and I've already exceeded it.

—Ring Lardner

The fuzzy character of the rules devised in Chap. 4 incites us to pass to the limit. Such a process is simple enough in finite-dimensional Banach spaces. However, since a number of problems are set in functional spaces, one is led to examine what can be done in infinite-dimensional spaces. It appears that the situation may depend on the nature of the space. For that reason, we mainly limit the study of this chapter to the framework of Asplund spaces (Sects. 6.1–6.5). As seen in Chap. 4, all the rules concerning Fréchet subdifferentials in Fréchet smooth Banach spaces are valid in the framework of Asplund spaces. Passing to the limit gives a particularly simple and striking character to the rules concerning sums and compositions. However, in Sect. 6.6, we give some attention to a limiting procedure involving directional subdifferentials. Such a construction is particularly adapted to the wide class of weakly compactly generated (WCG) spaces that encompasses separable spaces and reflexive spaces.

Because multimaps play an important role in a number of problems, we give much attention to limiting coderivatives. This choice stems from the fact that the limiting constructions we present for subdifferentials of functions involve their epigraphs. Thus, we are naturally led to the associated epigraph multimaps. Moreover, we are also interested in the analysis of maps or multimaps between Banach spaces without smoothness assumptions. Therefore, we must accept some complexity of the limiting processes in product spaces, since an interplay between weak* convergence and strong convergence occurs in such a case. The reader who cherishes simplicity will get an easier first reading on assuming that all spaces are finite-dimensional. While some developments concerning compactness in particular would become useless, the constructions and many results would remain of interest.

J.-P. Penot, *Calculus Without Derivatives*, Graduate Texts in Mathematics 266,
DOI 10.1007/978-1-4614-4538-8_6, © Springer Science+Business Media New York 2013

The calculus rules one gets with the limiting constructions of the present chapter are alluring. Still, these constructions are not without weaknesses. In particular, the precision of approximations is often lost and inclusions or order properties are not preserved by these limiting processes.

Throughout the chapter, weak* convergence in the dual X^* of a Banach space is denoted by $\xrightarrow{*}$. The notation $(x_n^*) \to^* x^*$ means that (x_n^*) is bounded and has x^* as a weak* cluster point.

6.1   Limiting Constructions with Firm Subdifferentials

In Asplund spaces, the limiting constructions we present are rather simple. We recall that closed balls of the dual of an Asplund space are sequentially compact for the weak* topology, i.e., bounded sequences of such a dual space have weak* convergent subsequences.

6.1.1   Limiting Subdifferentials and Limiting Normals

Definition 6.1. Given a function $f : X \to \overline{\mathbb{R}}$ on an Asplund space X and $\overline{x} \in f^{-1}(\mathbb{R})$, the (firm) *limiting subdifferential* of f at $\overline{x}$ is the set $\partial_L f(\overline{x})$ of $\overline{x}^* \in X^*$ such that there is a sequence of pairs (x_n, x_n^*) in the graph of $\partial_F f$ such that $(x_n^*) \xrightarrow{*} \overline{x}^*$ and $(x_n) \to_f \overline{x}$, that is, $(\|x_n - \overline{x}\|) \to 0$ and $(f(x_n)) \to f(\overline{x})$.

For a Lipschitzian function f, $\partial_L f(\overline{x})$ can be seen as the projection on X^* of the intersection of $\{\overline{x}\} \times X^* \times \{f(\overline{x})\}$ with the sequential closure of the subjet of f:

$$J_F^1 f := \{(w, w^*, f(w)) : w \in \mathrm{dom} f, \ w^* \in \partial_F f(w)\},$$

when $X \times X^* \times \mathbb{R}$ is endowed with the product topology of the strong topology on X, $\mathbb{R}$ and the weak* topology on X^*. This definition and the representation of the Clarke subdifferential $\partial_C f$ we have seen in Chap. 4 yield the following inclusions:

$$\partial_F f(\overline{x}) \subset \partial_L f(\overline{x}) \subset \partial_C f(\overline{x}).$$

They may be strict and $\partial_L f(\overline{x})$ may be nonconvex, as the next examples show.

Example. Let $f : \mathbb{R} \to \mathbb{R}$ be defined by $f(x) := -|x|$. Then $\partial_F f(0) = \varnothing$, $\partial_L f(0) = \{-1, 1\}$, $\partial_C f(0) = [-1, 1]$.

Example. Let $f : \mathbb{R}^2 \to \mathbb{R}$ be defined by $f(x_1, x_2) := |x_1| - |x_2|$. Then $\partial_F f(0,0) = \varnothing$, $\partial_L f(0,0) = [-1, 1] \times \{-1, 1\}$, $\partial_C f(0,0) = [-1, 1]^2$.

Example. Let $f : \mathbb{R}^2 \to \mathbb{R}$ be given by $f(x_1, x_2) := ||x_1| + x_2|$. Then $\partial_F f(0, 0) = \{(x_1^*, x_2^*) : |x_1^*| \leq x_2^* \leq 1\}$, $\partial_C f(0, 0) = [-1, 1] \times [-1, 1]$, $\partial_L f(0, 0) = \{(x_1^*, x_2^*) : |x_1^*| \leq x_2^* \leq 1\} \cup \{(x_1^*, x_2^*) : x_1^* \in [-1, 1], \ |x_1^*| = -x_2^* \leq 1\}$.

In general Banach spaces one has to use a more sophisticated limiting process.

Definition 6.2. When X is a general Banach space one defines $\partial_L f(\overline{x})$ as the set of $\overline{x}^* \in X^*$ such that there exist sequences $(\varepsilon_n) \to 0_+$, $(x_n) \to_f \overline{x}$, $(x_n^*) \to^* \overline{x}^*$ such that $x_n^* \in \partial_F^{\varepsilon_n} f(x_n)$ for all $n \in \mathbb{N}$, where $\partial_F^\varepsilon f(x) := \partial_F(f + \varepsilon q_x)(x)$ with $q_x(u) := \|u - x\|$:

$$\partial_F^\varepsilon f(x) := \left\{ x^* \in X^* : \liminf_{\|w\| \to 0_+} \frac{f(x + w) - f(x) - \langle x^*, w \rangle}{\|w\|} \geq -\varepsilon \right\}.$$

Proposition 6.3. *When X is an Asplund space, Definitions 6.1 and 6.2 coincide.*

Proof. It suffices to prove that $\overline{x}^* \in \partial_L f(\overline{x})$ in the sense of Definition 6.1 whenever $\overline{x}^*$ is a weak* cluster point of a bounded sequence (x_n^*) such that $x_n^* \in \partial_F^{\varepsilon_n} f(x_n)$ for some sequences $(\varepsilon_n) \to 0_+$, $(x_n) \to_f \overline{x}$. By definition of $\partial_F^{\varepsilon_n} f(x_n)$ there exists $\delta_n > 0$ such that for all $w \in \delta_n B_X$ one has $f(x_n + w) - f(x_n) - \langle x_n^*, w \rangle \geq -2\varepsilon_n \|w\|$. Thus x_n is a local minimizer of $u \mapsto f(u) - \langle x_n^*, u - x_n \rangle + 2\varepsilon_n \|u - x_n\|$. The fuzzy minimization rule yields $u_n \in B(x_n, \varepsilon_n, f)$, $u_n^* \in \partial_F f(u_n)$ such that $\|u_n^* - x_n^*\| \leq 3\varepsilon_n$. Then $(u_n) \to_f \overline{x}$ and $(u_n^*) \to^* \overline{x}^*$, so that $\overline{x}^* \in \partial_L f(\overline{x})$ by the weak* sequential compactness of closed balls of X^*. $\qquad\square$

A first advantage of this limiting process is that it produces nonempty subdifferentials under reasonable assumptions.

Proposition 6.4 (Limiting subdifferentiability of Lipschitz functions). *Let X be an Asplund space. Let f be a function on X that is Lipschitzian around $\overline{x} \in f^{-1}(\mathbb{R})$. Then $\partial_L f(\overline{x}) \neq \varnothing$. Moreover, if c is the Lipschitz rate of f at $\overline{x}$, then $\|\overline{x}^*\| \leq c$ for every $\overline{x}^* \in \partial_L f(\overline{x})$.*

If X is a WCG space and if f is Lipschitzian around $\overline{x} \in f^{-1}(\mathbb{R})$, then $\partial_L f(\overline{x})$ is a weak* compact set.

Proof. By Theorem 4.65 there is a sequence $(x_n) \to \overline{x}$ such that $\partial_F f(x_n) \neq \varnothing$. Let $x_n^* \in \partial_F f(x_n)$. Then for every $c' > c$, there is a number m such that $\|x_n^*\| \leq c'$ for $n \geq m$. Since closed balls are weak* sequentially compact in X^*, (x_n^*) has a subsequence weak* converging to some $\overline{x}^* \in c' B_{X^*}$. By definition, $\overline{x}^* \in \partial_L f(\overline{x})$. Since every element $\overline{x}^* \in \partial_L f(\overline{x})$ is obtained as the weak* limit of such a sequence (x_n^*), and since c' is arbitrarily close to c, every element $\overline{x}^*$ of $\partial_L f(\overline{x})$ must be in $c B_{X^*}$.

When X is a WCG Banach space, taking a decreasing sequence $(r_n) \to 0_+$ such that f is c'-Lipschitzian on $B(\overline{x}, r_1)$ and setting $F_n := \partial_F f(B(\overline{x}, r_n))$, Theorem 3.109 ensures that $\partial_L f(\overline{x})$ is the intersection over $n \geq 1$ of the sets $\mathrm{cl}^*(F_n)$, hence is a weak* closed subset of $c' B_{X^*}$, hence is weak* compact. $\qquad\square$

Let us turn to a geometrical counterpart.

Definition 6.5. The *(firm) limiting normal cone* to a subset S of an Asplund space X at $\bar{x} \in S$ is the set $N_L(S,\bar{x})$ of weak* limits of sequences (x_n^*) for which there exists a sequence (x_n) of S converging to $\bar{x}$ such that $x_n^* \in N_F(S,x_n)$ for all $n \in \mathbb{N}$.

This cone can be interpreted in terms of the indicator function ι_S to S:

$$N_L(S,\bar{x}) = \partial_L \iota_S(\bar{x}).$$

When X is a general Banach space, keeping this relation as the definition of the *limiting normal cone* to S, one is led to introduce the *ε-approximate normal cone*

$$N_F^\varepsilon(S,x) := \left\{ x^* \in X^* : \limsup_{w \to x,\ w \in S \setminus \{x\}} \frac{\langle x^*, w - x \rangle}{\|w - x\|} \le \varepsilon \right\}$$

and to declare that $\bar{x}^* \in N_L(S,\bar{x})$ if and only if there exist sequences $(\varepsilon_n) \to 0_+$, $(x_n) \to_S \bar{x}$, $(x_n^*) \to^* \bar{x}^*$ such that $x_n^* \in N_F^{\varepsilon_n}(S,x_n)$ for all $n \in \mathbb{N}$. Proposition 6.3 ensures that when X is an Asplund space both definitions coincide.

Proposition 6.6. *If C is a convex subset of an arbitrary normed space X and if $\bar{x} \in C$, then one has $N_L(C,\bar{x}) = N(C,\bar{x})$, the normal cone in the sense of convex analysis:*

$$N_L(C,\bar{x}) = \{\bar{x}^* \in X^* : \forall x \in C, \ \langle \bar{x}^*, x - \bar{x} \rangle \le 0\}.$$

Proof. Let us note that for every $\varepsilon > 0$ one has $x^* \in N_F^\varepsilon(C,\bar{x})$ if and only if for all $\gamma > \varepsilon$ there exists $\delta > 0$ such that $\langle x^*, w - \bar{x} \rangle \le \gamma \|w - \bar{x}\|$ for all $w \in C \cap B(\bar{x}, \delta)$. Since C is convex, this inequality entails that $\langle x^*, x - \bar{x} \rangle \le \gamma \|x - \bar{x}\|$ for all $x \in C$, the convex function $x \mapsto \gamma \|x - \bar{x}\| - \langle x^*, x - \bar{x} \rangle$ having $\bar{x}$ as a local minimizer on C. Therefore, given $x^* \in N_L(S,\bar{x})$, so that there exist sequences $(\varepsilon_n) \to 0_+$, $(x_n) \to_C \bar{x}$, $(x_n^*) \to^* \bar{x}^*$, and $x_n^* \in N_F^{\varepsilon_n}(C,x_n)$ for all $n \in \mathbb{N}$, one has $\langle x_n^*, x - x_n \rangle \le 2\varepsilon_n \|x - x_n\|$ for all $x \in C$, and passing to the limit, $\langle x^*, x - \bar{x} \rangle \le 0$ for all $x \in C$. The reverse inclusion $N(C,\bar{x}) \subset N_L(C,\bar{x})$ is obvious. $\square$

It is useful to employ a characterization of the limiting normal cone to a set S in terms of the distance function to S. We start with a characterization of $\partial_L d_S$.

Lemma 6.7. *Let S be a closed subset of an Asplund space X and let $\bar{x} \in S$. Then $\bar{x}^* \in \partial_L d_S(\bar{x})$ if and only if there exist sequences $(x_n) \to \bar{x}$, $(x_n^*) \to^* \bar{x}^*$ such that $x_n \in S$, $x_n^* \in \partial_F d_S(x_n)$ for all $n \in \mathbb{N}$.*

Proof. Let $\bar{x}^* \in \partial_L d_S(\bar{x})$. By definition, there are sequences $(w_n) \to \bar{x}$, $(w_n^*) \to^* \bar{x}^*$ such that $w_n^* \in \partial_F d_S(w_n)$ for all $n \in \mathbb{N}$. When $P := \{p \in \mathbb{N} : w_p \in S\}$ is infinite, the conclusion holds. Thus we suppose P is finite, and using the Asplund space version of Theorem 4.74, for $n \in N := \mathbb{N} \setminus P$, we pick $x_n \in S$ and $x_n^* \in \partial_F d_S(x_n)$ such that $\|x_n - w_n\| \le d_S(w_n) + 2^{-n}$, $\|x_n^* - w_n^*\| < 2^{-n}$. Then $(x_n) \to_S \bar{x}$, $(x_n^*) \to^* \bar{x}^*$. The converse is obvious. $\square$

Proposition 6.8. *The limiting normal cone to a subset S of an Asplund space X satisfies*

$$N_L(S,\overline{x}) = \mathbb{R}_+ \partial_L d_S(\overline{x}), \qquad\qquad \overline{x} \in S.$$

Proof. Let $\overline{x}^* \in N_L(S,\overline{x})$: there exists a sequence $((x_n, x_n^*))$ in $S \times X^*$ such that $(x_n) \to \overline{x}$, $(x_n^*) \overset{*}{\to} \overline{x}^*$ and $x_n^* \in N_F(S, x_n)$ for all $n \in \mathbb{N}$. Since $(r_n) := (\|x_n^*\|)$ is bounded, taking a subsequence, we may suppose $(r_n) \to r$ for some $r \in \mathbb{R}_+$ and $(u_n^*) \overset{*}{\to} u^*$ in B_{X^*} for some u_n^*, $u^* \in B_{X^*}$ with $x_n^* = r_n u_n^*$ for all n. If $r = 0$, we have $\overline{x}^* = 0 \in \mathbb{R}_+ \partial_L d_S(\overline{x})$. If $r \neq 0$, since $u_n^* \in N_F(S, x_n) \cap B_{X^*} = \partial_F d_S(x_n)$ for all $n \in \mathbb{N}$, we get $u^* \in \partial_L d_S(\overline{x})$ and $\overline{x}^* = $w*-$\lim_n r_n u_n^* = r u^* \in \mathbb{R}_+ \partial_L d_S(\overline{x})$.

Now let $\overline{x}^* \in \mathbb{R}_+ \partial_L d_S(\overline{x})$, i.e., $\overline{x}^* = r u^*$ for some $r \in \mathbb{R}_+$ and $u^* \in \partial_L d_S(\overline{x})$. Lemma 6.7 yields $(u_n) \to_S \overline{x}$, $(u_n^*) \overset{*}{\to} u^*$ such that $u_n^* \in \partial_F d_S(u_n)$ for all $n \in \mathbb{N}$. Then we have $r u_n^* \in N_F(S, u_n)$ for all $n \in \mathbb{N}$ and $(r x_n^*) \overset{*}{\to} \overline{x}^*$, so that $\overline{x}^* \in N_L(S, \overline{x})$. ☐

In general, $N_L(S,\overline{x})$ is nonconvex, hence cannot be the polar of some tangent cone (see Exercise 1). It may even happen that $N_L(S,\overline{x})$ is not closed in the norm topology (see Exercise 2).

Let us note that the relationship between the Fréchet subdifferential of a function and the normal cone to its epigraph can be extended to the limiting concepts.

Proposition 6.9. *Let $f : X \to \overline{\mathbb{R}}$ be a lower semicontinuous function on an Asplund space and let $x \in f^{-1}(\mathbb{R})$. Then, denoting by E the epigraph of f and setting $e := (x, f(x))$ one has*

$$x^* \in \partial_L f(x) \iff (x^*, -1) \in N_L(E, e).$$

Proof. Given $x^* \in \partial_L f(x)$, let $(x_n) \to_f x$, $(x_n^*) \overset{*}{\to} x^*$ such that $w_n^* \in \partial_F f(x_n)$ for all $n \in \mathbb{N}$. Then $(e_n) := ((x_n, f(x_n))) \to e$, $((x_n^*, -1)) \overset{*}{\to} (x^*, -1)$, and $(x_n^*, -1) \in N_F(E, e_n)$ for all $n \in \mathbb{N}$. Thus $(x^*, -1) \in N_L(E, e)$.

Conversely, let $x^* \in X^*$ be such that $(x^*, -1) \in N_L(E, e)$. Then there exist sequences $(e_n) \to_E e$, $((x_n^*, r_n)) \overset{*}{\to} (x^*, 1)$ such that $(x_n^*, -r_n) \in N_F(E, e_n)$. Setting $e_n := (x_n, s_n)$, since f is lower semicontinuous, we have $f(x) \leq \liminf_n f(x_n) \leq \limsup_n f(x_n) \leq \lim_n s_n = f(x)$, so that $(x_n) \to_f x$. Dropping a finite number of terms, we may suppose $r_n > 0$ for all $n \in \mathbb{N}$; then $r_n^{-1} x_n^* \in \partial_F f(x_n)$ and $(r_n^{-1} x_n^*) \overset{*}{\to} x^*$, so that $x^* \in \partial_L f(x)$. ☐

The simplest relationship between the Clarke subdifferential and the limiting subdifferential concerns the case of a Lipschitzian function.

Theorem 6.10. *Let W be an open subset of an Asplund space X and let $f : W \to \mathbb{R}$ be a locally Lipschitzian function. Then for all $a \in W$ one has*

$$\partial_C f(a) = \overline{co}^*(\partial_L f(a)).$$

Proof. It suffices to prove that the support functions of these sets coincide, i.e.,

$$f^C(a,u) = \sup\{\langle a^*, u\rangle : \exists (x_n) \to a, \exists (x_n^*) \overset{*}{\to} a^*,\; x_n^* \in \partial_F f(x_n) \forall n\}, \qquad (6.1)$$

for all $u \in X$. Since $\partial_C f$ is weak* sequentially upper semicontinuous by Proposition 5.3 and since its graph contains the graph of $\partial_F f$, the right-hand side is not greater than the left-hand side. In order to prove the reverse inequality, let us pick sequences $(a_n) \to a$, $(t_n) \to 0_+$ such that $f^C(a,u) = \lim_n (1/t_n)(f(a_n + t_n u) - f(a_n))$. Given a sequence $(\varepsilon_n) \to 0_+$, the mean value theorem in Asplund spaces yields some $w_n \in [a_n, a_n + t_n u]$, some $x_n \in B(w_n, \varepsilon_n t_n)$, and $x_n^* \in \partial_F f(x_n)$ such that

$$\langle x_n^*, t_n u\rangle \geq f(a_n + t_n u) - f(a_n) - \varepsilon_n t_n.$$

Since f is Lipschitzian around a, the sequence (x_n^*) is bounded, hence has a subsequence that weak* converges to some $a^* \in \partial_L f(a)$. Dividing both sides of the preceding inequality by t_n and passing to the limit, we get $\langle a^*, u\rangle \geq f^C(a,u)$, hence the opposite inequality we were looking for.   □

Corollary 6.11. *For a closed subset E of an Asplund space X and $a \in E$ one has*

$$N_C(E,a) = \overline{co}^*(N_L(E,a))).$$

Proof. When X is an Asplund space one has $\partial_C d_E(a) = \overline{co}^*(\partial_L d_E(a))$, hence by Propositions 5.25, 4.13, using the notation lim sup for the sequential weak* lim sup,

$$N_C(E,a) = cl^*(\mathbb{R}_+ \partial_C d_E(a)) = cl^*(\mathbb{R}_+ \overline{co}^*(\partial_L d_E(a)))$$
$$\subset \overline{co}^*(\mathbb{R}_+ \partial_L d_E(a)) \subset \overline{co}^*(\limsup_{x \to_E a} \mathbb{R}_+ \partial_F d_E(x)) \subset \overline{co}^*(\limsup_{x \to_E a} N_F(E,x)).$$

On the other hand, since $N_C(E,a)$ is a weak* closed convex cone, relation (5.36) implies that $\overline{co}^*(N_L(E,a)) \subset N_C(E,a)$. Thus, equality holds.   □

Using the representation of the normal cone to an epigraph and an Asplund space version of Corollary 4.130, we get the following result.

Theorem 6.12. *Let X be an Asplund space, $f \in \mathscr{F}(X)$, and $a \in \mathrm{dom} f$. Then*

$$\partial_C f(a) = \overline{co}^*(\partial_L f(a)) + \partial_C^\infty f(a),$$
$$\partial_C^\infty f(a) = \overline{co}^*(\{w^* - \lim_n r_n x_n^*,\; x_n^* \in \partial_F f(x_n),\; (x_n) \to_f a,\; (r_n) \to 0_+\}).$$

Exercise. Prove these two relations.

Exercises

1. Let $S := \text{epi} f$, with $f(x) := -|x|$ for $x \in \mathbb{R}$ and let $\bar{x} := (0,0)$. Then $N_L(S,\bar{x}) = (\mathbb{R}_+ e) \cup (\mathbb{R}_+ e')$ for $e := (-1,-1)$, $e' := (1,-1)$.

2. (Fitzpatrick) Let X be a separable Hilbert space with an orthonormal basis $\{e_n : n \in \mathbb{N}\}$ and let S be the cone

$$S := \mathbb{R}_+ e_0 \cup \{r(e_0 - ne_n) + s(e_0 - se_p) : r, s \in \mathbb{R}_+, \ p > n > 0\}.$$

Check that S is closed but that $N_L(S, \cdot)$ is not closed at 0. [Hint: Identifying X^* with X, show that $e_0 + (1/n)e_n \in N_L(S,0)$ but $e_0 \notin N_L(S,0)$. See [146], [718, p. 11].]

3. Let X be an arbitrary normed space, let $f : X \to \mathbb{R}$ be a lower semicontinuous function on X, and let $x \in f^{-1}(\mathbb{R})$. Denote by E the epigraph of f and set $e := (x, f(x))$. Show that for every given $\varepsilon > 0$ one can find $\alpha > 0$, $\beta > 0$ such that

$$x^* \in \partial_F^\alpha f(x) \Longrightarrow (x^*, -1) \in N_F^\varepsilon(E,e),$$

$$(x^*, -1) \in N_F^\beta(E,e) \Longrightarrow x^* \in \partial_F^\varepsilon f(x).$$

(See [525].)

4. Deduce from the preceding exercise that Proposition 6.9 is valid in every normed space X.

5. Check that for a closed subset E of a Banach space and for a boundary point $\bar{x}$ of E one has the inclusion $N_L(E,\bar{x}) \subset \mathbb{R}_+ \partial_L d_E(\bar{x})$.

6. Check that for a closed subset E of a Banach space and for a boundary point e of E one has $N_L(E,e) \subset N_L(\text{bdry}E,e)$, where $\text{bdry}E$ denotes the boundary of E.

7. Show that for a closed subset E of a Banach space X, for a boundary point e of E, and for all $\varepsilon \geq 0$ one has $N_F^\varepsilon(E,e) \subset \{x^* \in X^* : \forall v \in T(E,e) \ \langle x^*, v \rangle \leq \varepsilon \|v\|\}$.

6.1.2 Limiting Coderivatives

The concept of *limiting coderivative* of a multimap defined as follows will be instrumental for the study of correspondences.

Definition 6.13. Let $F : X \rightrightarrows Y$ be a multimap between two normed spaces X, Y. The *limiting coderivative* of F at $(\bar{x}, \bar{y}) \in F$ is the multimap $D_L^* F(\bar{x}, \bar{y}) : Y^* \rightrightarrows X^*$ defined by

$$D_L^* F(\bar{x}, \bar{y})(y^*) := \{x^* \in X^* : (x^*, -y^*) \in N_L(F, (\bar{x}, \bar{y}))\}, \qquad y^* \in Y^*.$$

The *mixed (limiting) coderivative* of F at $(\bar{x},\bar{y})$ is the multimap $D_M^* F(\bar{x},\bar{y}) : Y^* \rightrightarrows X^*$ defined by $x^* \in D_M^* F(\bar{x},\bar{y})(y^*)$ if and only if there exist sequences $(\varepsilon_n) \to 0_+$, $((x_n,y_n)) \to_F (\bar{x},\bar{y})$, $(x_n^*) \to^* x^*$, $(y_n^*) \to y^*$ such that $(x_n^*, -y_n^*) \in N_F^{\varepsilon_n}(F,(x_n,y_n))$ for all $n \in \mathbb{N}$.

When $F(\bar{x})$ is a singleton $\{\bar{y}\}$, we just write $D_L^* F(\bar{x})$ instead of $D_L^* F(\bar{x},\bar{y})$ and $D_M^* F(\bar{x})$ instead of $D_M^* F(\bar{x},\bar{y})$.

Since strong convergence entails weak* convergence, one has

$$\forall y^* \in Y^*, \qquad D_M^* F(\bar{x},\bar{y})(y^*) \subset D_L^* F(\bar{x},\bar{y})(y^*),$$

and this inclusion is an equality when Y is finite-dimensional.

In infinite dimensions the inclusion may be strict, even when F is single-valued.

Example–Exercise. Let Y be a separable Hilbert space with an orthonormal basis (e_n) and let $f : \mathbb{R} \to Y$ be even and defined by $f(x) = (2x - 2^{-n})e_n + (2^{-n} - x)e_{n+1}$ for $x \in [2^{-n-1}, 2^{-n}]$, $f(0) = 0$. For all $y^* \in Y^*$, since $(\langle y^*, e_n \rangle) \to 0$, one has $D_F^* f(0)(y^*) = \{0\}$, $D_F^* f(2^{-n})(y^*) = \varnothing$. For $x \in (2^{-n-1}, 2^{-n})$, f is of class C^1 at x; hence $D_F^* f(x)(y^*) = \langle y^*, 2e_n - e_{n+1} \rangle$. It follows that $D_M^* f(0)(y^*) = \{0\}$. In order to prove that $D_L^* f(0)(y^*) \neq \{0\}$, let $x_n = (1/2)(2^{-n-1} + 2^{-n})$, $v_n^* := 2e_n - e_{n+1}$, $u_n := v_n / \|v_n\|$, $y_n^* := y^* + u_n^*$, $x_n^* := \langle y_n^*, 2e_n - e_{n+1} \rangle$. Check that $\|u_n^*\| = 1$, $(y_n^*) \overset{*}{\to} y^*$, $(x_n^*) \to \sqrt{5}$, $x_n^* \in D_F^* f(x_n)(y_n^*)$, and $\sqrt{5} \in D_L^* f(0)(y^*)$. □

Proposition 6.14. *(a) If X and Y are arbitrary normed spaces and if $g : X \to Y$ is circa-differentiable at $\bar{x} \in X$, then $D_L^* g(\bar{x}) = D_M^* g(\bar{x}) = g'(\bar{x})^\mathsf{T}$.*
(b) If $F : X \rightrightarrows Y$ has a convex graph, then $D_L^* F(\bar{x},\bar{y}) = D_M^* F(\bar{x},\bar{y}) = D_F^* F(\bar{x},\bar{y})$.

Proof. (a) We have seen that $D_F^* g(\bar{x}) = A^\mathsf{T}$, with $A := g'(\bar{x})$. It remains to show that for all $y^* \in Y$, $x^* \in D_L^* g(\bar{x})(y^*)$ one has $x^* = A^\mathsf{T}(y^*)$. Let $((x_n^*, y_n^*)) \to^* (x^*, y^*)$ be such that $(x_n^*, -y_n^*) \in N_F^{\varepsilon_n}(\mathrm{gph}(g),(x_n,y_n))$ for some sequences $(\varepsilon_n) \to 0_+$ in $(0,1]$, $(x_n) \to \bar{x}$, $(y_n) := (g(x_n)) \to g(\bar{x})$. Since g is circa-differentiable at $\bar{x}$, we can find a sequence $(\delta_n) \to 0_+$ such that

$$\forall x \in B(x_n,\delta_n), \qquad \|g(x) - g(x_n) - A(x - x_n)\| \leq \varepsilon_n \|x - x_n\|.$$

Hence for all $x \in B(x_n,\delta_n)$, $\|g(x) - g(x_n)\| \leq \ell \|x - x_n\|$, where $\ell := \|A\| + 1$. Since $(x_n^*, -y_n^*) \in N_F^{\varepsilon_n}(\mathrm{gph}(g),(x_n,y_n))$, taking a smaller δ_n if necessary, for all $(x,y) \in B((x_n,y_n),\delta_n)$ with $y = g(x)$, taking into account these estimates, we get

$$\langle x_n^*, x - x_n \rangle - \langle y_n^*, g(x) - g(x_n) \rangle \leq 2\varepsilon_n(\|x - x_n\| + \|g(x) - g(x_n)\|),$$

$$\langle x_n^*, x - x_n \rangle - \langle y_n^*, A(x - x_n) \rangle \leq 2\varepsilon_n(\ell + 1)\|x - x_n\| + \varepsilon_n \|y_n^*\| \|x - x_n\|.$$

Since x can be taken arbitrarily in $B(x_n,(1+\ell)^{-1}\delta_n)$, we obtain $\|x_n^* - A^\mathsf{T}(y_n^*)\| \leq \varepsilon_n(\|y_n^*\| + 2\ell + 2)$. Since $(x_n^*, A^\mathsf{T}(y_n^*)) \to^* (x^*, A^\mathsf{T}(y^*))$, we get $x^* - A^\mathsf{T}(y^*) = 0$.

Assertion (b) stems from the coincidence of normal cones to a convex subset (Proposition 6.6). □

The geometric characterization of the limiting subdifferential given in Proposition 6.9 can be interpreted in terms of coderivatives using Exercise 4 of Sect. 6.1.1.

Proposition 6.15. *If $f : X \to \overline{\mathbb{R}}$ is lower semicontinuous and if $E_f : X \rightrightarrows \mathbb{R}$ is the multimap with graph $\mathrm{epi}\, f$, the limiting subdifferential of f at $\overline{x} \in f^{-1}(\mathbb{R})$ satisfies*

$$\partial_L f(\overline{x}) = D_L^* E_f(\overline{x}, f(\overline{x}))(1) := \{\overline{x}^* : (\overline{x}^*, -1) \in N_L(E_f, (\overline{x}, f(\overline{x})))\}.$$

Since $N_F(E_f, e) \subset X^* \times \mathbb{R}_-$ for all $e \in E_f$, we have $N_L(E_f, e) \subset X^* \times \mathbb{R}_-$. This inclusion incites us to introduce the *singular limiting subdifferential* of f at $\overline{x}$ by

$$\partial_L^\infty f(\overline{x}) := \{\overline{x}^* \in X^* : (\overline{x}^*, 0) \in N_L(E_f, \overline{x}_f)\},$$

where $\overline{x}_f := (\overline{x}, f(\overline{x}))$, so that by homogeneity, one has the decomposition

$$N_L(E_f, \overline{x}_f) = \mathbb{P}(\partial_L f(\overline{x}) \times \{-1\}) \cup (\partial_L^\infty f(\overline{x}) \times \{0\}).$$

Proposition 6.16 (Scalarization). *Let X and Y be Asplund spaces and let $g : X \to Y$ be continuous at $\overline{x} \in X$. Then, for all $y^* \in Y^*$, one has the following inclusion, which is an equality when g is Lipschitzian around $\overline{x}$:*

$$\partial_L(y^* \circ g)(\overline{x}) \subset D_M^* g(\overline{x})(y^*).$$

Proof. Given $y^* \in Y^*$, let us set $f := y^* \circ g$. For all $x \in X$, Proposition 4.25 shows that $\partial_F f(x) \subset D_F^* g(x)(y^*)$. Given $\overline{x}^* \in \partial_L f(\overline{x})$ and sequences $(x_n) \to_f \overline{x}$, $(x_n^*) \xrightarrow{*} \overline{x}^*$ with $x_n^* \in \partial_F f(x_n)$ for all $n \in \mathbb{R}$, by continuity of g at $\overline{x}$ one has $((x_n, g(x_n))) \to (\overline{x}, g(\overline{x}))$. Since $(x_n^*, -y^*) \in N_F(\mathrm{gph}\, g, (x_n, g(x_n)))$, it follows that $\overline{x}^* \in D_M^* g(\overline{x})(y^*)$.

Now let us prove the opposite inclusion when g is Lipschitzian around $\overline{x}$ with rate ℓ. Let $x^* \in D_M^* g(\overline{x})(y^*)$ and let $(x_n) \to \overline{x}$, $(x_n^*) \xrightarrow{*} \overline{x}^*$, $(y_n^*) \to y^*$ with $x_n^* \in D_F^* g(x_n)(y_n^*)$ for all $n \in \mathbb{N}$. For n large enough g is stable at x_n, so that Proposition 4.25 ensures that $x_n^* \in \partial_F(y_n^* \circ g)(x_n)$. Since $y_n^* \circ g - y^* \circ g$ is Lipschitzian with rate $\varepsilon_n := \ell \|y_n^* - y^*\|$, we have $x_n^* \in \partial_F^{\varepsilon_n}(y^* \circ g)(x_n)$. Then $x^* \in \partial_L(y^* \circ g)(\overline{x})$ by Proposition 6.3. $\qquad\square$

Exercises

1. Let $f : X \to \mathbb{R}$ be a function finite at $\overline{x}$. Show that $D_L^* f(\overline{x})(1) \subset \partial_L f(\overline{x})$.

2. Suppose f is continuous around $\overline{x}$. Deduce from Exercise 6 of Sect. 6.1.1 that $\partial_L f(\overline{x}) \subset D_L^* f(\overline{x})(1)$. Recover this inclusion from the scalarization result above.

3. Deduce from Exercise 1 that $\partial_L^\infty f(\overline{x}) \subset D_L^* f(\overline{x})(0)$ if f is continuous around $\overline{x}$.

6.1.3 Some Elementary Properties

Let us give some elementary properties of the limiting subdifferential on the class
$\mathscr{F}(X)$ of proper lower semicontinuous functions on a normed space X.

Proposition 6.17. *(a) If $f, g \in \mathscr{F}(X)$ coincide around x, then $\partial_L f(x) = \partial_L g(x)$.*
(b) If $f \in \mathscr{F}(X)$ is convex, then $\partial_L f(x) = \{x^* \in X^* : f \geq x^* + f(x) - x^*(x)\}$.
(c) If $f \in \mathscr{F}(X)$ attains a local minimum at $x \in \mathrm{dom}\, f$, then $0 \in \partial_L f(x)$.
*(d) If $\lambda > 0$, $A \in L(X,Y)$ with $A(X) = Y$, $b \in Y$, $g \in \mathscr{F}(X)$, $\ell \in X^*$, $c \in \mathbb{R}$ and
$f(x) = \lambda g(Ax+b) + \langle \ell, x \rangle + c$, then $\partial_L f(x) = \lambda A^{\mathsf{T}}(\partial_L g(Ax+b)) + \ell$.*
*(e) If $f \in \mathscr{F}(X \times Y)$ is given by $f(x,y) = g(x) + h(y)$ with $g \in \mathscr{F}(X)$, $h \in \mathscr{F}(Y)$,
then $\partial_L f(x,y) = \partial_L g(x) \times \partial_L h(y)$.*
*(f) If $f = h \circ g$, where $h \in \mathscr{F}(Y)$, $g : X \to Y$ is circa-differentiable at $\bar{x} \in X$ and
open at $\bar{x}$, then $g'(\bar{x})^{\mathsf{T}}(\partial_L h(g(\bar{x}))) \subset \partial_L f(\bar{x})$.*
*(g) If $f = g + h$, where $g : X \to \mathbb{R}$ is circa-differentiable at $\bar{x} \in X$ and $h \in \mathscr{F}(X)$,
then $\partial_L f(\bar{x}) = g'(\bar{x}) + \partial_L h(\bar{x})$.*
(h) If $g : X \to \mathbb{R}$ is circa-differentiable at $\bar{x} \in X$, then $\partial_L g(\bar{x}) = \{g'(\bar{x})\}$.

Proof. (a) is obvious. (b) stems from the equality $\partial^\varepsilon f(x) = \partial_{MR} f(x) + \varepsilon B_{X^*}$
when f is convex and from the closedness property of $\partial_{MR} f$. One can also use
the characterizations of the subdifferentials in terms of the normal cones to the
epigraphs. (c) is a consequence of the inclusion $\partial_F f(x) \subset \partial_L f(x)$. When X and Y are
Asplund, (d) and (e) follow from similar properties with the Fréchet subdifferential
by a passage to the limit. In the general case one deduces (d) from (f) and elementary
computations and one proves (e) by using approximate Fréchet subdifferentials
using the fact that for every $\varepsilon > 0$ one has $\partial_F^\varepsilon f(x,y) = (\partial_F^\varepsilon g(x), \partial_F^\varepsilon h(y))$ if one
endows $X \times Y$ with the sum norm.

(f) Let $f = h \circ g$ as in (f), let $\bar{y} := g(\bar{x})$, $A := g'(\bar{x})$, and let $\bar{y}^* \in \partial_L h(\bar{y})$. There exist
sequences $(\varepsilon_n) \to 0_+$, $(y_n) \to_h \bar{y}$, $(y_n^*) \to^* \bar{y}^*$ satisfying $y_n^* \in \partial_F^{\varepsilon_n} h(y_n)$ for all $n \in \mathbb{N}$.
Since g is open at $\bar{x}$, we can find a sequence $(x_n) \to \bar{x}$ such that $y_n = g(x_n)$ for n large
enough. Let $(\alpha_n), (\beta_n), (\gamma_n) \to 0_+$ in $(0,1)$ and $m \in \mathbb{N}$ be such that for all $n \geq m$ one
has $\gamma_n > \varepsilon_n$, $\alpha_n(\|A\| + \gamma_n) \leq \beta_n$ and

$$h(y_n + y) - h(y_n) - \langle y_n^*, y \rangle \geq -\gamma_n \|y\| \qquad \forall y \in \beta_n B_Y,$$

$$g(x_n + x) - g(x_n) - A(x) \in \gamma_n \|x\| B_Y \qquad \forall n \geq m, \forall x \in \alpha_n B_X.$$

Plugging $y = g(x_n + x) - g(x_n) \in (\|A\| + \gamma_n) \|x\| B_Y$ in the first relation, we get

$$f(x_n + x) - f(x_n) - \langle y_n^*, A(x) \rangle \geq -\gamma_n(\|A\| + \gamma_n) \|x\| - \gamma_n \|y_n^*\| \|x\|.$$

Since (y_n^*) is bounded, we see that $(\eta_n) \to 0$ for $\eta_n := \gamma_n(\|A\| + \gamma_n + \|y_n^*\|)$. Since
$(A^{\mathsf{T}} y_n^*) \to^* A^{\mathsf{T}} \bar{y}^*$ and $(f(x_n)) = (h(y_n)) \to h(\bar{y}) = f(\bar{x})$, we get $A^{\mathsf{T}} \bar{y}^* \in \partial_L f(\bar{x})$.

Assertion (g) can be deduced from (e) and (f) by introducing $G : X \to X \times \mathbb{R}$,
$H : X \times \mathbb{R} \to \mathbb{R}$ defined by $G(x) := (x, h(x))$ and $H(x,r) := g(x) + r$ and observing
that one has a double inclusion, since $h = f - g$. A direct proof is easy. Note that

(g) implies (h) by taking $h = 0$ and that (g) can be deduced from (h) and the sum rule below when X is Asplund. However, (g) and (h) hold in an arbitrary Banach space X. □

The following geometrical property is an easy consequence of Proposition 6.17.

Corollary 6.18. *If A and B are closed subsets of Banach spaces X and Y respectively, and $(\bar{x}, \bar{y}) \in A \times B$, then $N_L(A \times B, (\bar{x}, \bar{y})) = N_L(A, \bar{x}) \times N_L(B, \bar{y})$.*

Proof. Since $\iota_{A \times B}(x, y) = \iota_A(x) + \iota_B(y)$, the result follows from Proposition 6.17 (e). □

Now let us turn to properties involving order.

Proposition 6.19. *Let X be an Asplund space, let $g := (g_1, \ldots, g_m) : X \to \mathbb{R}^m$ with $g_i \in \mathscr{L}(X)$, and let $h : \mathbb{R}^m \to \mathbb{R}$ be of class C^1 around $\bar{y} := g(\bar{x})$ and nondecreasing in each of its m arguments near $\bar{y}$, with $h'(\bar{y}) \neq 0$. Let $f := h \circ g$. Then*

$$\partial_L f(\bar{x}) \subset h'(\bar{y}) \circ (\partial_L g_1(\bar{x}), \ldots, \partial_L g_m(\bar{x})). \tag{6.2}$$

Proof. The result follows from Proposition 4.48 by a passage to the limit. □

Proposition 6.20. *Let X, Y be Asplund spaces, let $f : X \to \overline{\mathbb{R}}$ be finite and Lipschitzian around $\bar{x} \in X$, let $g : Y \to \mathbb{R}$ be of class C^1 at $\bar{y} \in Y$ with $g'(\bar{y}) \neq 0$, and let h be given by $h(x, y) := \max(f(x), g(y))$. Then, if $f(\bar{x}) = g(\bar{y})$, for $(\bar{x}^*, \bar{y}^*) \in \partial_L h(\bar{x}, \bar{y})$ with $\bar{y}^* \neq g'(\bar{y})$ there exists $\lambda \in [0, 1]$ such that*

$$(\bar{x}^*, \bar{y}^*) \in (1 - \lambda) \partial_L f(\bar{x}) \times \lambda \partial_L g(\bar{y}).$$

Proof. Let $((x_n, y_n)) \to (\bar{x}, \bar{y})$ and $((x_n^*, y_n^*)) \xrightarrow{*} (\bar{x}^*, \bar{y}^*)$ be such that $(x_n^*, y_n^*) \in \partial_F h(x_n, y_n)$ for all $n \in \mathbb{N}$. Let us first show that the set $P := \{p \in \mathbb{N} : g(y_p) > f(x_p)\}$ is finite. Otherwise, we would have $y_p^* \in \partial_F g(y_p)$ for all $p \in P$, as easily seen, hence $\bar{y}^* \in \partial_L g(\bar{y}) = \{g'(\bar{y})\}$, a contradiction with $\bar{y}^* \neq g'(\bar{y})$. If $Q := \{q \in \mathbb{N} : f(x_q) > g(y_q)\}$ is infinite, for all $q \in Q$ we have $h(x, y) = f(x)$ for (x, y) near (x_q, y_q), hence $x_q^* \in \partial_F f(x_q)$, $y_q^* = 0$ and $\bar{x}^* \in \partial_L f(\bar{x})$, $\bar{y}^* = 0$, so that the result holds with $\lambda = 0$. It remains the case when $N := \{n \in \mathbb{N} : f(x_n) = g(y_n)\}$ is infinite. Then, observing that for $n \in N$ large enough we have $g'(y_n) \neq 0$ and $y_n^* \neq g'(y_n)$, Proposition 4.49 yields some $\lambda_n \in (0, 1]$ such that $(x_n^*, y_n^*) \in \lambda_n \partial_F f(x_n) \times (1 - \lambda_n) g'(y_n)$. Since $\bar{y}^* \neq g'(\bar{y})$, 0 cannot be a limit point of (λ_n). Thus, (λ_n) has a non null limit point λ and we get $(\bar{x}^*, \bar{y}^*) \in \lambda \partial_L f(\bar{x}) \times (1 - \lambda) \partial_L g(\bar{y})$. □

Proposition 6.21. *Let V, W be Banach spaces, let $A \in L(V, W)$ with $W = A(V)$, $\bar{w} \in W$, $B \subset p^{-1}(\bar{w})$, $j \in \mathscr{F}(V)$, $p \in \mathscr{F}(W)$ be such that $p \circ A \leq j$.*
(a) Suppose that for every sequence $(w_n) \to_p \bar{w}$ one can find $\bar{v} \in B$ and a sequence $(v_n) \to \bar{v}$ such that $A(v_n) = w_n$ and $j(v_n) = p(w_n)$ for all $n \in \mathbb{N}$ large enough. Then one has

$$A^\top(\partial_L p(\bar{w})) \subset \bigcup_{\bar{v} \in B} \partial_L j(\bar{v}). \tag{6.3}$$

(b) If V is an Asplund space, the same inclusion holds when for all sequences $(w_n) \to_p \overline{w}$, $(\alpha_n) \to 0_+$ one can find $\overline{v} \in B$ and a sequence $(v_n) \to \overline{v}$ such that $A(v_n) = w_n$ and $j(v_n) \leq p(w_n) + \alpha_n$ for all $n \in \mathbb{N}$ large enough.

(c) If V and W are Asplund spaces and $p := d_S$ for a closed subset S of W, to get inclusion (6.3) it suffices to assume that for all sequences $(w_n) \to \overline{w}$, $(\alpha_n) \to 0_+$ with $w_n \in S$ for all $n \in \mathbb{N}$ one can find $\overline{v} \in B$ and a sequence $(v_n) \to \overline{v}$ such that $A(v_n) = w_n$ and $j(v_n) \leq p(w_n) + \alpha_n$ for all $n \in \mathbb{N}$ large enough.

For B one can take a singleton $\{\overline{v}\}$, $p^{-1}(\overline{w})$ or any intermediate choice.

Proof. (a) Let $\overline{w}^* \in \partial_L p(\overline{w})$. There exist sequences $(w_n) \to_p \overline{w}$, $(w_n^*) \to^* \overline{w}^*$, $(\varepsilon_n) \to 0_+$ such that $w_n^* \in \partial_F^{\varepsilon_n} p(w_n)$ for all $n \in \mathbb{N}$. Let $(\delta_n) \to 0_+$ be such that

$$\forall w \in B[w_n, \delta_n], \qquad p(w_n) \leq p(w) - \langle w_n^*, w - w_n \rangle + 2\varepsilon_n \|w - w_n\|.$$

Let $c := \|A\|$, let $\overline{v} \in B$, and let $(v_n) \to \overline{v}$ be such that $A(v_n) = w_n$ and $j(v_n) = p(w_n)$ for all $n \in \mathbb{N}$ large enough. Then since $p(Av) \leq j(v)$ and $p(w_n) = j(v_n)$, we have

$$\forall v \in B[v_n, \delta_n/c], \qquad j(v_n) \leq j(v) - \langle w_n^*, A(v - v_n) \rangle + 2c\varepsilon_n \|v - v_n\|.$$

Thus $A^\mathsf{T}(w_n^*) \in \partial_F^{2c\varepsilon_n} j(v_n)$, and since $(A^\mathsf{T}(w_n^*)) \to^* A^\mathsf{T}(\overline{w}^*)$, $(v_n) \to \overline{v}$, and $(j(v_n)) \to j(\overline{v})$, we get $A^\mathsf{T}(\overline{w}^*) \in \partial_L j(\overline{v})$.

(b) Suppose V is an Asplund space and the assumption is relaxed as in part (b) of the statement. Taking $(\varepsilon_n) \to 0_+$, (δ_n), $(w_n) \to_p \overline{w}$, $(w_n^*) \to^* \overline{w}^*$ as above and $\alpha_n := \delta_n \varepsilon_n$, we can find $\overline{v} \in B$ and a sequence $(v_n) \to \overline{v}$ such that $A(v_n) = w_n$ and $j(v_n) \leq p(w_n) + \alpha_n$ for all $n \in \mathbb{N}$ large enough. Then for all n large enough, we have

$$\forall v \in B[v_n, \delta_n/c], \qquad j(v_n) \leq j(v) - \langle w_n^*, A(v - v_n) \rangle + 2c\varepsilon_n \|v - v_n\| + \alpha_n.$$

The Ekeland principle yields some $z_n \in B[v_n, \delta_n/2c]$ such that z_n is a minimizer of

$$v \mapsto j(v) - \langle w_n^*, A(v - v_n) \rangle + 2c\varepsilon_n \|v - v_n\| + 2c\varepsilon_n \|v - z_n\|$$

on $B[v_n, \delta_n/c]$. Then z_n is a local minimizer of this function, so that the fuzzy sum rule yields some $u_n \in B(z_n, \delta_n/2c)$ and $u_n^* \in \partial_F j(u_n)$ such that $u_n^* - A^\mathsf{T}(w_n^*) \in 5c\varepsilon_n B_{V^*}$. Since $(A^\mathsf{T}(w_n^*)) \to^* A^\mathsf{T}(\overline{w}^*)$, we obtain $A^\mathsf{T}(\overline{w}^*) \in \partial_L j(\overline{v})$.

(c) When V and W are Asplund spaces and $p := d_S$ for a closed subset S of W, given $\overline{w}^* \in \partial_L p(\overline{w})$, using Lemma 6.7, one can find sequences $(w_n) \to \overline{w}$ in S and $(w_n^*) \to^* \overline{w}^*$ such that $w_n^* \in \partial_F d_S(w_n)$ for all $n \in \mathbb{N}$ and finish the proof as above. $\square$

6.1.4 Calculus Rules Under Lipschitz Assumptions

Calculus rules with limiting subdifferentials in Asplund spaces take attractive forms, especially in the Lipschitzian case.

Theorem 6.22 (Sum rule for limiting subdifferentials). *Let X be an Asplund space and let $f = f_1 + \cdots + f_k$, where $f_1, \ldots, f_{k-1}$ are Lipschitzian around $\bar{x}$ and f_k is lower semicontinuous on X and finite at $\bar{x}$. Then*

$$\partial_L f(\bar{x}) \subset \partial_L f_1(\bar{x}) + \cdots + \partial_L f_k(\bar{x}). \tag{6.4}$$

Proof. Let $\bar{x}^* \in \partial_L f(\bar{x})$ and let $(x_n) \to_f \bar{x}$, $(x_n^*) \xrightarrow{*} \bar{x}^*$ with $x_n^* \in \partial_F f(x_n)$ for all n. Given a sequence $(\varepsilon_n) \to 0_+$, by the fuzzy sum rule for Fréchet subdifferentials (Theorem 4.69) there are sequences of pairs $(x_{i,n}, x_{i,n}^*) \in \partial_F f_i$, for $i \in \mathbb{N}_k$, such that $\|x_{i,n} - x_n\| \le \varepsilon_n$, $|f_i(x_{i,n}) - f_i(x_n)| \le \varepsilon_n$, $\|x_n^* - (x_{1,n}^* + \cdots + x_{k,n}^*)\| \le \varepsilon_n$.

Since $f_1, \ldots, f_{k-1}$ are Lipschitzian near $\bar{x}$, there is an r such that $\|x_{i,n}^*\| \le r$ for $i \in \mathbb{N}_{k-1}$, $n \in \mathbb{N}$. But then $(\|x_{k,n}^*\|)$ is also bounded (e.g., by $\|x^*\| + kr + 1$). Since balls in X^* are sequentially weak* compact, we may assume that every sequence $(x_{i,n}^*)$ weak* converges to some $\bar{x}_i^*$, and since $(f_i(x_{i,n})) \to f_i(\bar{x})$, then $\bar{x}^* = \bar{x}_1^* + \cdots + \bar{x}_k^*$: (6.4) holds. □

Theorem 6.23 (Chain rule for limiting subdifferentials). *Let X and Y be Asplund spaces, let $g : X \to Y$ be a map with closed graph that is continuous at $\bar{x} \in X$, and let $h : Y \to \mathbb{R}_\infty$ be Lipschitzian around $\bar{y} := g(\bar{x})$. Then*

$$\partial_L(h \circ g)(\bar{x}) \subset D_L^* g(\bar{x})(\partial_L h(\bar{y})).$$

Proof. Let $f := h \circ g$ and let $\bar{x}^* \in \partial_L f(\bar{x})$. There exists a sequence of pairs (x_n, x_n^*) in the graph of $\partial_F f$ such that $(x_n) \to \bar{x}$, $(x_n^*) \xrightarrow{*} \bar{x}^*$. Given a sequence $(\varepsilon_n) \to 0_+$, Theorem 4.70 yields sequences (u_n) in X, (y_n) in Y, (u_n^*) in X^*, (v_n^*), (y_n^*) in Y^* such that $y_n^* \in \partial_F h(y_n)$, $\|y_n^* - v_n^*\| \le \varepsilon_n$, $\|x_n^* - u_n^*\| \le \varepsilon_n$, $u_n^* \in D_F^* g(u_n)(v_n^*)$, $\|g(u_n) - g(x_n)\| \le \varepsilon_n$, $\|y_n - g(x_n)\| \le \varepsilon_n$ for all $n \in \mathbb{N}$. Since h is Lipschitzian around $\bar{y}$, the sequences (y_n^*) and (v_n^*) are bounded. We may assume they weak* converge to some $\bar{y}^* \in \partial_L h(\bar{y})$, since $(y_n) \to \bar{y}$, g being continuous at $\bar{x}$. Then we get $\bar{x}^* \in D_L^* g(\bar{x})(\bar{y}^*)$. □

Remark. If $\bar{x}^* \in \partial_F f(\bar{x})$ one has a slightly more precise conclusion: there exists some $\bar{y}^* \in \partial_L h(\bar{y})$ such that $-\bar{y}^* \in D_M^* g^{-1}(\bar{y}, \bar{x})(-\bar{x}^*)$. This observation stems from the fact that $(\|(\bar{x}^*, 0) - (0, y_n^*) - (u_n^*, -v_n^*)\|) \to 0$.

Exercises

1. Show that for $f, g \in \mathcal{L}(X)$ the inclusion $\partial_L(f+g)(\bar{x}) \subset \partial_L f(\bar{x}) + \partial_L g(\bar{x})$ may be strict. [Hint: Take $X := \mathbb{R}$, $f := |\cdot|$, $g := -|\cdot|$.]

2. Deduce from the preceding exercise that inclusion (6.2) may be strict. [Hint: Take $m = 2$, and h given by $h(y_1, y_2) = y_1 + y_2$.]

3. Show that for $f \in \mathcal{F}(X)$, $g \in \mathcal{L}(X)$, $\bar{x} \in \mathrm{dom} f$, one has $\partial_L^\infty(f+g)(\bar{x}) = \partial_L^\infty f(\bar{x})$.

6.2 Some Compactness Properties

The following result is representative of the kind of properties we will consider in the present section in view of applications to general calculus rules.

Proposition 6.24. *Let Q be a weak* locally compact cone of the dual X^* of a Banach space. Then a net $(x_i^*)_{i \in I}$ of Q, that weak* converges to 0 converges in norm to 0.*

Proof. Let V be a neighborhood of 0 in the weak* topology such that $Q \cap V$ is weak* compact. Since weak* compact subsets are bounded, we can find $s > 0$ such that $Q \cap V \subset s B_{X^*}$. Then for all $\varepsilon > 0$ there exists $i_\varepsilon \in I$ such that $x_i^* \in \varepsilon s^{-1} V$ for $i \geq i_\varepsilon$. Thus, $x_i^* \in \varepsilon B_{X^*}$ for $i \geq i_\varepsilon$ and $(x_i^*)_{i \in I} \to 0$. □

In order to present a characterization of weak* locally compact cones of the dual of a Banach space X, let us say that a cone Q of X^* is a *Loewen cone* if there exist $\gamma > 0$ and a compact subset K of X (for the strong topology) such that Q coincides with the set

$$Q_\gamma(K) := \left\{ x^* \in X^* : h_K(x^*) := \max_{x \in K} \langle x^*, x \rangle \geq \gamma \|x^*\| \right\}.$$

Let us note that one may restrict one's attention to the case $\gamma = 1$, replacing K by $\gamma^{-1} K$; then we write $Q(K)$ instead of $Q_1(K)$. However, it is convenient to make use of the scalar γ.

This class contains the class of *Bishop–Phelps cones* for which K is a singleton.

If X is finite-dimensional, every closed cone of X^* is contained in a Loewen cone (take for K the unit ball of X). Another case of interest is that in which Q is the positive polar cone $P^\oplus := -P^0$ of a closed cone P with nonempty interior. In fact, given $v \in \mathrm{int} P$ and $r > 0$ such that $B(v, r) \subset P$, for all $x^* \in Q$ we have $\langle x^*, v - ru \rangle \geq 0$ for all $u \in B_X$, hence $\langle x^*, v \rangle \geq r \|x^*\|$, and Q is even contained in a Bishop–Phelps cone. Such an inclusion is a special case of the following lemma, in which one takes for B the ball $r B_X$ and for C a singleton. Loewen cones correspond to the choice of such a ball for B and for C a compact subset.

Lemma 6.25. *If B,C are subsets of a normed space X and if P is a cone of X, one has the following implication; the reverse implication holds whenever $C + P$ is closed, convex:*

$$B \subset C + P \Longrightarrow P^0 \subset Q(B,C) := \{x^* \in X^* : h_B(x^*) \leq h_C(x^*)\}.$$

Proof. Let $x^* \in P^0$. For all $b \in B$ one can find $c \in C$, $p \in P$ such that $b = c + p$, so that $\langle x^*, b - c \rangle \leq 0$ and $\langle x^*, b \rangle \leq \sup_{c \in C}\langle x^*, c \rangle =: h_C(x^*)$. Thus, $h_B(x^*) \leq h_C(x^*)$.

Conversely, suppose $C + P$ is closed, convex, and there exists $b \in B \setminus (C + P)$. Then the Hahn–Banach theorem yields some $x^* \in X^*$ such that $\sup\{\langle x^*, c + p \rangle : c \in C, p \in P\} < \langle x^*, b \rangle$. Then one has $x^* \in P^0$ and $h_B(x^*) \geq \langle x^*, b \rangle > h_C(x^*)$, hence $x^* \notin Q(B,C)$. $\qquad\qquad\square$

Lemma 6.26. *Every Loewen cone $Q_\gamma(K)$ is weak* closed, weak* locally compact, and contained in a finite union of Bishop–Phelps cones.*

Proof. Let us first note that for $\beta \in (0, \gamma)$ there exists a finite set $F \subset K$ such that $Q_\gamma(K) \subset Q_\beta(F)$, which is the union of the cones $Q_\beta(\{w\})$ for $w \in F$. In fact, from the open covering $\{B(x, \gamma - \beta) : x \in K\}$ of the compact set K, we can extract a finite covering $\{B(x, \gamma - \beta) : x \in F\}$ of K. Then for every $x^* \in Q_\gamma(K)$, we have

$$\gamma \|x^*\| \leq \max_{w \in F} \sup_{u \in B_X} \langle x^*, w + (\gamma - \beta)u \rangle \leq \max_{w \in F}\langle x^*, w \rangle + (\gamma - \beta)\|x^*\|;$$

hence $\beta \|x^*\| \leq \max_{w \in F}\langle x^*, w \rangle$ and $x^* \in Q_\beta(F)$.

Now let us show that $Q_\gamma(K)$ is weak* closed. This is obvious when K is a singleton $\{w\}$, since then $Q_\gamma(K)$ is the sublevel at the level 0 of the weak* lower semicontinuous convex function $x^* \mapsto \gamma \|x^*\| - \langle x^*, w \rangle$. Thus, if F is a finite set, $Q_\gamma(F)$ is weak* closed too. Now let us consider the general case. Let x^* be in the weak* closure of $Q_\gamma(K)$. Given $\beta \in (0, \gamma)$, let F be a finite subset of K such that $Q_\gamma(K) \subset Q_\beta(F)$. Since $Q_\beta(F)$ is weak* closed, we have $x^* \in Q_\beta(F)$, hence $\beta \|x^*\| \leq h_F(x^*) \leq h_K(x^*)$. Since β is arbitrarily close to γ, we get $\gamma \|x^*\| \leq h_K(x^*)$ and $x^* \in Q_\gamma(K)$.

Finally, let us show that $Q := Q_\gamma(K)$ is weak* locally compact. We know (Exercise 5, Sect. 1.3) that it suffices to exhibit a weak* compact neighborhood of 0 in Q. We pick $\beta \in (0, \gamma)$ and a finite subset F of K such that $Q \subset Q_\beta(F)$. Then $V := F^0$ is a neighborhood of 0 in X^* endowed with the weak* topology, and for $x^* \in Q \cap F^0 \subset Q_\beta(F) \cap F^0$, we have $\beta \|x^*\| \leq 1$. Thus $Q_\gamma(K) \cap V$ is contained in $\beta^{-1} B_{X^*}$, hence is compact in the weak* topology. $\qquad\square$

We are ready to present the announced characterization.

Proposition 6.27. *For a weak* closed and convex cone Q of X^* the following assertions are equivalent:*

(a) Q is weak* locally compact;
(b) There exists a weak* neighborhood V of 0 such that $Q \cap V$ is bounded;

(c) Q is contained in a Loewen cone.
(d) If $(x_i^*)_{i\in I}$ is a bounded net of Q that weak converges to 0, then $(x_i^*) \to 0$.*

Proof. (a)$\Rightarrow$(b) is obvious, since every weak* compact set is bounded.
(b)$\Rightarrow$(c) We may suppose $V = F^0$, where F is a finite set $F := \{x_1,\dots,x_m\}$. Let $r > 0$ be such that $Q \cap V \subset rB_{X^*}$. Let $x^* \in Q$. When $h_F(x^*) \le 0$, for all $t \in \mathbb{R}_+$ we have $tx^* \in Q \cap V$, hence $\|tx^*\| \le r$ and $x^* = 0$. When $h_F(x^*) > 0$, we have $x^*/h_F(x^*) \in Q \cap V$, hence $\|x^*\| \le rh_F(x^*)$ and $x^* \in Q_{1/r}(F)$.
(c)$\Rightarrow$(a) This follows from the preceding lemma, since Q is weak* closed.

Since we have seen that (a)$\Rightarrow$(d), it remains to prove the converse or that (d)$\Rightarrow$(b). If (b) does not hold, for every finite set F of X, the set $Q \cap F^0$ is unbounded. Let $z_{F,n}^* \in Q \cap F^0$ be such that $r_{F,n} := \|z_{F,n}^*\| \ge n$ and let $x_{F,n}^* := z_{F,n}^*/r_{F,n}$. Let $\mathscr{P}_f(X)$ be the set of finite subsets of X and let $I := \mathscr{P}_f(X) \times \mathbb{N}$. The product order of the inclusion on $\mathscr{P}_f(X)$ and of the usual order on $\mathbb{N}$ is directed. The net $(x_{F,n}^*)_{(F,n)\in I}$ weak* converges to 0 and is bounded, but $\|x_{F,n}^*\| = 1$ for all $(F,n) \in I$, so that (d) does not hold. $\square$

The following condition generalizes the cone property.

Definition 6.28. A closed subset S of a normed space X is said to have the *cone property up to a compact set* around $\bar{x} \in S$, or to be *compactly epi-Lipschitzian* around $\bar{x}$, if there exist a neighborhood V of $\bar{x}$, a neighborhood U of 0, a compact subset K of X, and $\tau > 0$ such that

$$\forall t \in [0,\tau], \qquad S \cap V + tU \subset S + tK. \tag{6.5}$$

We note that in (6.5) we may assume that U is the unit ball: if $\gamma > 0$ is such that $\gamma B_X \subset U$, setting $K' := \gamma^{-1}K$, for all $s := \gamma t \in [0, \gamma\tau]$ we have $S \cap V + sB_X \subset S + sK'$.

The cone property corresponds to the case in which K is a singleton $\{-w\}$: then taking $U := \gamma B_X$, relation (6.5) can be written $S \cap V + [0,\tau]B[w,\gamma] \subset S$.

Relation (6.5) can be viewed as a concrete condition ensuring the inclusion $N(S,x) \subset Q_\gamma(K)$ for $x \in S \cap V$:

Proposition 6.29. *Suppose S has the cone property up to a compact set K around $\bar{x}$, as in (6.5). Then there exist $\gamma > 0$ and a neighborhood V of $\bar{x}$ such that*

$$\forall x \in S \cap V, \qquad \gamma B_X \subset K + T(S,x),$$

$$\forall x \in S \cap V, \qquad N(S,x) \subset Q_\gamma(K).$$

Proof. Let $\gamma, \delta, \tau > 0$ be such that (6.5) holds for $U := \gamma B_X$, $V := B(\bar{x}, \delta)$. Given $x \in S \cap V$, $u \in U$, for $t \in (0,\tau)$ let $z_t \in K$ be such that $x + tu - tz_t \in S$. Taking a sequence $(t_n) \to 0$ in $(0,\tau)$ such that (z_{t_n}) has a limit $\bar{z}$ in K, we get $u - \bar{z} \in T(S,x)$, hence $\gamma B_X \subset K + T(S,x)$. This inclusion entails the second one, as Lemma 6.25 shows with $B := \gamma B_X$, $C := K$, $P := T(S,x)$. $\square$

The following notion will play a central role. It is always satisfied in finite-dimensional spaces.

Definition 6.30 ([804, 807]). A subset S of a normed space X is said to be *sequentially normally compact* at $\bar{x} \in S$, in short *normally compact* at $\bar{x} \in S$, if for all sequences $(x_n) \to_S \bar{x}$, $(x_n^*) \overset{*}{\to} 0$ with $x_n^* \in N_F(S, x_n)$ for all $n \in \mathbb{N}$, one has $(x_n^*) \to 0$.

In the sequel we omit the word "sequentially" because we use sequences only. The terminology is explained by the following equivalence.

Lemma 6.31. *For a subset S of X the following assertions are equivalent:*

(a) S is normally compact at $\bar{x} \in S$;
(b) If $(x_n) \to_S \bar{x}$, every sequence (x_n^*) in the unit sphere S_{X^*} of X satisfying $x_n^* \in N_F(S, x_n)$ for all $n \in \mathbb{N}$ has a nonnull weak* cluster point.
If B_{X^*} is sequentially weak* compact (in particular when X is an Asplund space) the preceding assertions are equivalent to the following one:
(c) If $(x_n) \to_S \bar{x}$, every sequence (x_n^*) in S_{X^*} satisfying $x_n^* \in N_F(S, x_n)$ for all $n \in \mathbb{N}$ has a subsequence that weak* converges to a nonnull limit.

The proof is similar to the proof of Proposition 3.72 and is even simpler, since one can use sequences only. One uses the fact that a sequence of the weak* compact set B_{X^*} either weak* converges to 0 or has a nonnull weak* cluster point.

Let us give some criteria for normal compactness.

Proposition 6.32. *Let S be a closed subset of a Banach space X such that B_{X^*} is sequentially weak* compact and let $\bar{x} \in S$. Then each of the following conditions implies that S is normally compact at $\bar{x}$:*
(a) S is a closed convex set with nonempty interior;
(b) There exist $V \in \mathcal{N}(\bar{x})$ and a Loewen cone $Q_\gamma(K)$ such that $N(S, x) \subset Q_\gamma(K)$ for all $x \in S \cap V$;
(c) S is compactly tangentially determined around $\bar{x}$, i.e., there exist $\gamma > 0$, a compact subset K of X, and $V \in \mathcal{N}(\bar{x})$ such that $\gamma B_X \subset T(S, x) + K$ for all $x \in S \cap V$;
(d) S satisfies the cone property up to a compact set around $\bar{x}$;
(e) S satisfies the cone property around $\bar{x}$.

Proof. (a) The result has been proved in Lemma 3.71 (a).
(b) It is a consequence of Propositions 6.24 and 6.27.
(c) and (d) are a consequences of (b) and Proposition 6.29 (or of its proof).
(e) is a special case of (d). $\square$

The next definition is an adaptation to multimaps of the preceding notion.

Definition 6.33. A multimap $F : X \rightrightarrows Y$ is said to be *coderivatively compact* (resp. *strongly coderivatively compact*) at $(\bar{x}, \bar{y})$ if for all sequences $((x_n, y_n)) \to (\bar{x}, \bar{y})$ in F, (x_n^*), (y_n^*) with $x_n^* \in D_F^* F(x_n, y_n)(y_n^*)$ for all $n \in \mathbb{N}$, one has $(x_n^*) \to 0$ whenever $(x_n^*) \overset{*}{\to} 0$ and $(y_n^*) \to 0$ (resp. $(y_n^*) \overset{*}{\to} 0$).

Clearly, F is coderivatively compact at $(\bar{x}, \bar{y})$ when F is *coderivatively bounded* at $(\bar{x}, \bar{y})$ in the following sense: for all sequences $((x_n, y_n)) \to (\bar{x}, \bar{y})$ in F, (x_n^*), (y_n^*) with $x_n^* \in D_F^* F(x_n, y_n)(y_n^*)$ for all $n \in \mathbb{N}$ one has $(x_n^*) \to 0$ whenever $(y_n^*) \to 0$. The terminology is justified by the next exercise.

Exercise. Recall that the norm of a *process* $H : X \rightrightarrows Y$ is defined by

$$\|H\| := \sup\{\|y\| : y \in H(B_X)\}.$$

(a) Check that for two processes $H, K : X \rightrightarrows Y$ one has $\|H + K\| \leq \|H\| + \|K\|$ and $\|rH\| = r\|H\|$ for all $r > 0$.
(b) Show that $F : X \rightrightarrows Y$ is coderivatively bounded around $(\overline{x}, \overline{y})$ if and only if there exist $c > 0$ and $W \in \mathcal{N}(\overline{x}, \overline{y})$ such that $\|D_F^* F(x, y)\| \leq c$ for all $(x, y) \in W$.   □

A multimap $F : X \rightrightarrows Y$ is strongly coderivatively compact at $(\overline{x}, \overline{y}) \in F$ when its graph is normally compact at $(\overline{x}, \overline{y})$. A subset S of X is normally compact at $\overline{x} \in S$ if and only if for every Banach space Y and $\overline{y} \in Y$, the multimap $F : X \rightrightarrows Y$ with graph $S \times Y$ is coderivatively compact at $(\overline{x}, \overline{y})$. When Y is finite-dimensional, strong coderivative compactness coincides with coderivative compactness. Coderivative compactness is obviously satisfied when X is finite-dimensional. In view of the following lemma, it is also satisfied when F is pseudo-Lipschitzian around $(\overline{x}, \overline{y})$.

Lemma 6.34. *If $F : X \rightrightarrows Y$ is pseudo-Lipschitzian around $(\overline{x}, \overline{y})$, then F is coderivatively bounded at $(\overline{x}, \overline{y})$, hence is coderivatively compact at $(\overline{x}, \overline{y})$.*

In fact, there exists $c > 0$ such that $\|x^*\| \leq c\|y^*\|$ for all $y^* \in Y^*$, all (x, y) near $(\overline{x}, \overline{y})$, and all $x^* \in D_F^* F(x, y)(y^*)$, hence for all $x^* \in D_M^* F(x, y)(y^*)$.

If Y is finite-dimensional, the same inequality holds when $x^* \in D_L^* F(\overline{x}, \overline{y})(y^*)$.

Proof. The first assertion is the content of Proposition 4.26. A passage to the limit yields the estimate for $D_M^* F(x, y)$: given $y^* \in Y^*$, $x^* \in D_M^* F(x, y)(y^*)$, taking sequences $((x_n, y_n)) \to (x, y)$ in F, $(x_n^*) \xrightarrow{*} x^*$, $(y_n^*) \to y^*$ with $x_n^* \in D_F^* F(x_n, y_n)(y_n^*)$, for all n we get $\|x^*\| \leq \liminf_n \|x_n^*\| \leq c \liminf_n \|y_n^*\| = c \|y^*\|$.

When $\dim Y < \infty$, one has $D_L^* F(\overline{x}, \overline{y}) = D_M^* F(\overline{x}, \overline{y})$.   □

Corollary 6.35. *If $M : W \rightrightarrows Z$ is metrically regular around $(\overline{w}, \overline{z})$, then M^{-1} is coderivatively compact at $(\overline{z}, \overline{w})$.*

Proof. The result stems from the fact that M^{-1} is pseudo-Lipschitzian around $(\overline{z}, \overline{w})$.   □

Exercise. Give a direct proof that a convex multimap $F : X \rightrightarrows Y$ such that $B(\overline{y}, s) \subset F(B(\overline{x}, r))$ for some $r, s > 0$ is such that F^{-1} is coderivatively compact at $(\overline{y}, \overline{x})$.

Let us say that $F : X \rightrightarrows Y$ has the *partial cone property up to a compact set* around $(\overline{x}, \overline{y})$ if there exist $\alpha, \beta, \tau > 0$, a neighborhood W of $(\overline{x}, \overline{y})$, and a compact subset K of X such that

$$\forall t \in [0, \tau], \qquad F \cap W + t\alpha B_X \times \{0\} \subset F + t(K \times \beta B_Y). \qquad (6.6)$$

Proposition 6.36. *If F has the partial cone property up to a compact set around $(\overline{x}, \overline{y})$ then there exist $\alpha, \beta > 0$, a neighborhood W of $(\overline{x}, \overline{y})$, and a compact subset K of X such that for all $(x, y) \in F \cap W$, $y^* \in Y^*$, $x^* \in D_F^* F(x, y)(y^*)$ one has*

$$\alpha \|x^*\| \leq \beta \|y^*\| + h_K(x^*). \tag{6.7}$$

In turn, this property ensures that F is coderivatively compact at $(\bar{x}, \bar{y})$.

Proof. Let $\alpha, \beta, \tau > 0$, $W \in \mathcal{N}(\bar{x}, \bar{y})$, and let K be a compact subset of X such that (6.6) holds. Let $\gamma > 0$ be such that $K \subset \gamma B_X$. Then for all $(x, y) \in F \cap W$, $u \in B_X$, and all $t \in [0, \tau]$ one can find $(x_t, y_t) \in F$, $z_t \in K$, and $v_t \in B_Y$ such that

$$(x, y) + t\alpha(u, 0) = (x_t, y_t) + t(z_t, \beta v_t),$$

so that $\|(x_t, y_t) - (x, y)\| \leq t(\alpha + \beta + \gamma)$. Hence, given $y^* \in Y^*$, $x^* \in D_F^* F(x, y)(y^*)$, for some modulus $\varepsilon(\cdot)$ one has

$$t\alpha\langle x^*, u \rangle = \langle (x^*, -y^*), (x_t - x, y_t - y) \rangle + t \langle (x^*, -y^*), (z_t, \beta v_t) \rangle$$
$$\leq t\varepsilon(t) + th_K(x^*) + t\beta \|y^*\|.$$

Dividing both sides by t, passing to the limit as $t \to 0_+$, and taking the supremum over $u \in B_X$, one gets (6.7).

Now let us assume this estimate. Given sequences $((x_n, y_n)) \to (\bar{x}, \bar{y})$ in F, $(x_n^*) \xrightarrow{*} 0$, $(y_n^*) \to 0$ with $x_n^* \in D_F^* F(x_n, y_n)(y_n^*)$ for all $n \in \mathbb{N}$, one has $(x_n^*) \to 0$ since (using subsequences) we may assume that a sequence (z_n) in K such that $\langle x_n^*, z_n \rangle = h_K(x_n^*)$ has a limit $z^* \in K$, and hence $(h_K(x_n^*)) \to 0$. $\square$

The proofs of the following results are left as exercises.

Proposition 6.37. *Let X and Y be two Asplund spaces, let $D \subset Y$, and let $f : X \to Y$ be circa-differentiable at $\bar{x} \in C := f^{-1}(D)$ with $f'(\bar{x})(X) = Y$. Then C is normally compact at $\bar{x}$ if and only if D is normally compact at $f(\bar{x})$.*

Proposition 6.38. *Let $F : X \rightrightarrows Y$ with a closed graph and let $g : X \to Y$ be circa-differentiable at $\bar{x} \in X$. Then $G := F + g$ is coderivatively compact (resp. strongly coderivatively compact, resp. normally compact, resp. coderivatively bounded) at $(\bar{x}, \bar{y} + g(\bar{x}))$ if and only if F has the same property at $(\bar{x}, \bar{y})$.*

It is also of interest to introduce compactness properties for subdifferentials.

Definition 6.39. A function $f : X \to \overline{\mathbb{R}}$ on a normed space X is said to be *subdifferentially compact* at a point $\bar{x}$ where it is finite if for all sequences $(x_n) \to_f \bar{x}$, $(t_n) \to 0_+$, $(w_n^*) \xrightarrow{*} 0$ such that $w_n^* \in t_n \partial_F f(x_n)$ for all $n \in \mathbb{N}$, one has $(w_n^*) \to 0$.

Such a notion is related to coderivative compactness via the epigraph multimap.

Proposition 6.40. *A lower semicontinuous function $f : X \to \overline{\mathbb{R}}$ on an Asplund space is subdifferentially compact at a point $\bar{x}$ where it is finite if and only if its epigraph multimap $E := E_f$ is coderivatively compact at $\bar{x}_f := (\bar{x}, f(\bar{x}))$.*

Proof. Suppose E is coderivatively compact at $\bar{x}_f$. Given sequences $(t_n) \to 0_+$, $(x_n) \to_f \bar{x}$, $(w_n^*) \xrightarrow{*} 0$ such that $w_n^* \in t_n \partial_F f(x_n)$ for all $n \in \mathbb{N}$, one has $(w_n^*, -t_n) \in N_F(E, (x_n, f(x_n)))$, hence $(w_n^*) \to 0$, and f is subdifferentially compact at $\bar{x}$.

Conversely, suppose f is subdifferentially compact at $\bar{x}$ and let $((w_n, r_n)) \to \bar{x}_f$ in E, $((w_n^*, r_n^*)) \xrightarrow{*} (0,0)$ with $(w_n^*, -r_n^*) \in N_F(E, (w_n, r_n))$ for all n. Let $N := \{n \in \mathbb{N} : r_n^* = 0\}$, so that $w_n^* \in r_n^* \partial_F f(w_n)$ for all $n \in \mathbb{N} \setminus N$ and $(w_n^*) \to 0$ if N is finite, since f is subdifferentially compact at $\bar{x}$. It remains to consider the case in which N is infinite. Using Corollary 4.130 and a sequence $(\varepsilon_n) \to 0_+$, for all $n \in N$ we can find $t_n \in (0, \varepsilon_n)$, $x_n \in B(w_n, \varepsilon_n, f)$, $x_n^* \in \partial_F f(x_n)$ such that $\|w_n^* - t_n x_n^*\| < \varepsilon_n$. Then $(t_n x_n^*) \xrightarrow{*} 0$, hence $(t_n x_n^*) \to 0$ along N, since f is subdifferentially compact at $\bar{x}$. Therefore $(w_n^*) \to 0$ and E_f is coderivatively compact at $\bar{x}_f$.   □

Exercises

1. Check that a subset S of X is normally compact at $\bar{x} \in S$ if and only if for every Banach space Y and $\bar{y} \in Y$, the multimap $F : X \rightrightarrows Y$ with graph $S \times Y$ is coderivatively compact at $(\bar{x}, \bar{y})$.

2. Suppose $F : X \rightrightarrows Y$ has the *strong partial cone property up to a compact set* around $(\bar{x}, \bar{y})$ in the following sense: there exist $\alpha, \tau > 0$, a neighborhood W of $(\bar{x}, \bar{y})$, and compact subsets K of X, L of Y such that

$$\forall t \in [0, \tau], \qquad F \cap W + t\alpha B_X \times \{0\} \subset F + t(K \times L). \tag{6.8}$$

(a) Show that for all $(x, y) \in F \cap W$, $y^* \in Y^*$, $x^* \in D_F^* F(x, y)(y^*)$ one has $\alpha \|x^*\| \leq h_K(x^*) + h_L(y^*)$.
(b) Prove that the latter property ensures that F is strongly coderivatively compact at $(\bar{x}, \bar{y})$.

3. Check that if the graph of $F : X \rightrightarrows Y$ has the cone property up to a compact set around $(\bar{x}, \bar{y})$, then $F : X \rightrightarrows Y$ has the strong partial cone property up to a compact set around $(\bar{x}, \bar{y})$.

4. Check that a subset S of X has the cone property up to a compact set at $\bar{x} \in S$ if and only if for every Banach space Y and $\bar{y} \in Y$, the multimap $F : X \rightrightarrows Y$ with graph $S \times Y$ has the strong partial cone property up to a compact set around $(\bar{x}, \bar{y})$.

6.3 Calculus Rules for Coderivatives and Normal Cones

Since limiting subdifferentials are related to limiting coderivatives and limiting normal cones, it is sensible to deduce calculus rules for subdifferentials from calculus rules for normal cones and coderivatives under operations on sets or multimaps such as intersections and direct and inverse images. We start with intersections.

6.3.1 Normal Cone to an Intersection

Unions and intersections are such basic operations with sets that they deserve priority. Simple examples show that the simple rule $N(F \cup G, x) \subset N(F, x) \cap N(G, x)$ for two subsets F, G of a Banach space, $x \in F \cap G$ is satisfied for the firm and the directional normal cones but not for the limiting normal cone. Thus, we focus our attention on intersections. We start with the observation that metric estimates yield a rule for the normal cone to an intersection.

Theorem 6.41 (Normal cone to an intersection). *Let $(S_1, \ldots, S_k)$ be a family of closed subsets of an Asplund space satisfying the following linear coherence condition at $\bar{x} \in S := S_1 \cap \cdots \cap S_k$: for some $c > 0$, $\rho > 0$,*

$$\forall x \in B(\bar{x}, \rho), \qquad d(x, S) \le c \, d(x, S_1) + \cdots + c \, d(x, S_k). \tag{6.9}$$

Then one has

$$N_L(S, \bar{x}) \subset N_L(S_1, \bar{x}) + \cdots + N_L(S_k, \bar{x}). \tag{6.10}$$

The result follows from a passage to the limit in Theorem 4.75; but we present another proof.

Proof. Let $\bar{x}^* \in N_L(S, \bar{x})$, so that by Proposition 6.8, $\bar{x}^* = r\bar{u}^*$ for some $r \in \mathbb{R}_+$, $\bar{u}^* \in \partial_L d_S(\bar{x})$. Let $f := c \, d(\cdot, S_1) + \cdots + c \, d(\cdot, S_k)$, so that $d_S \le f$ and $f \mid_S = 0$. Proposition 6.21 ensures that $\bar{u}^* \in \partial_L f(\bar{x})$. The sum rule yields $\bar{u}_i^* \in c \partial_L d_{S_i}(\bar{x})$ such that $\bar{u}^* = \bar{u}_1^* + \cdots + \bar{u}_k^*$. Then $\bar{x}_i^* := r \bar{u}_i^* \in N_L(S_i, \bar{x})$ and $\bar{x}^* = \bar{x}_1^* + \cdots + \bar{x}_k^*$. $\qquad \square$

The study of the limiting normal cone to an intersection we undertake now makes use of the alliedness property that appeared in Chap. 4. It generalizes the notion of direct sum of linear spaces.

Definition 6.42 ([813]). A finite family $(S_i)_{i \in I}$ ($I := \mathbb{N}_k$) of closed subsets of a normed space X is said to be *allied* at $\bar{x} \in S := S_1 \cap \cdots \cap S_k$ if whenever $x_{n,i}^* \in N_F(S_i, x_{n,i})$ with $(x_{n,i})_n \in S_i$ for $(n, i) \in \mathbb{N} \times I$, $(x_{n,i})_n \to \bar{x}$,

$$(\|x_{n,1}^* + \cdots + x_{n,k}^*\|)_n \to 0 \implies \forall i \in I \quad (\|x_{n,i}^*\|)_n \to 0.$$

This property can be reformulated as follows: there exist $\rho > 0$, $c > 0$ such that

$$\forall x_i \in S_i \cap B(\bar{x}, \rho), \ x_i^* \in N_F(S_i, x_i), \qquad c \max_{i \in I} \|x_i^*\| \le \|x_1^* + \cdots + x_k^*\|. \tag{6.11}$$

This reformulation follows by homogeneity from the fact that one can find $\rho > 0$ and $c > 0$ such that for $x_i \in S_i \cap B(\bar{x}, \rho)$, $x_i^* \in N_F(S_i, x_i)$ with $\|x_1^* + \cdots + x_k^*\| < c$ one has $\max_{i \in I} \|x_i^*\| < 1$ or, equivalently, $\max_{i \in I} \|x_i^*\| \ge 1 \implies \|x_1^* + \cdots + x_k^*\| \ge c$.

The result that follows reduces alliedness to an easier requirement.

Proposition 6.43. *A finite family $(S_i)_{i \in I}$ $(I := \mathbb{N}_k)$ of closed subsets of a normed space X is allied at $\bar{x} \in S := S_1 \cap \cdots \cap S_k$ if and only if given $x_{n,i} \in S_i$, $x_{n,i}^* \in \partial_F d_{S_i}(x_{n,i})$ for $(n, i) \in \mathbb{N} \times I$, with $(x_{n,i})_n \to \bar{x}$, one has*

$$\left(\left\| x_{n,1}^* + \cdots + x_{n,k}^* \right\| \right)_n \to 0 \Longrightarrow \left(\max_{i \in I} \left\| x_{n,i}^* \right\| \right)_n \to 0. \tag{6.12}$$

Proof. Since $\partial_F d_{S_i}(x_{n,i}) \subset N_F(S_i, x_{n,i})$ for all $x_{n,i} \in S_i$ and all (n, i), condition (6.12) follows from alliedness. Conversely, suppose condition (6.12) is satisfied. Let $(x_{n,i})_n \to \bar{x}$ in S_i, $(x_{n,i}^*)_n$ in X^* be sequences satisfying $(\| x_{n,1}^* + \cdots + x_{n,k}^* \|)_n \to 0$ and $x_{n,i}^* \in N_F(S_i, x_{n,i})$ for all $(n, i) \in \mathbb{N} \times I$. Let $r_n := \max_{i \in I}(\| x_{n,i}^* \|)$. If (r_n) is bounded, setting $w_{n,i}^* := x_{n,i}^*/r \in N(S_i, x_{n,i}) \cap B_{X^*} = \partial_F d_{S_i}(x_{n,i})$ with $r > \sup_n r_n$, we get that $(w_{n,i}^*)_n \to 0$, hence $(x_{n,i}^*)_n \to 0$. It remains to discard the case in which (r_n) is unbounded. Taking a subsequence, we may suppose $(r_n) \to +\infty$. Setting $u_{n,i}^* := x_{n,i}^*/r_n$, so that $(\| u_{n,1}^* + \cdots + u_{n,k}^* \|)_n \to 0$, we obtain from our assumption that $(\| u_{n,i}^* \|)_n \to 0$ for all $i \in I$, a contradiction to $\max_{i \in I} \| u_{n,i}^* \| = 1$ for all $n \in \mathbb{N}$. $\square$

We have seen in Proposition 4.81 that alliedness implies linear coherence in Fréchet smooth spaces or in Asplund spaces. Let us give another proof.

Theorem 6.44. *Let $(S_i)_{i \in I}$ $(I := \mathbb{N}_k)$ be a finite family of closed subsets of an Asplund space X that is allied at $\bar{x} \in S := S_1 \cap \cdots \cap S_k$. Then there exist $c, \rho > 0$ such that the linear coherence condition*

$$\forall x \in B(\bar{x}, \rho), \qquad d(x, S) \le cd(x, S_1) + \cdots + cd(x, S_k), \tag{6.13}$$

is satisfied, whence

$$N_L(S, \bar{x}) \subset N_L(S_1, \bar{x}) + \cdots + N_L(S_k, \bar{x}). \tag{6.14}$$

Proof. Relation (6.11) yields some $\gamma, \rho \in (0, 1)$ such that for all $x_i \in S_i \cap B(\bar{x}, 5\rho)$, $x_i^* \in N_F(S_i, x_i)$ for $i \in I$ satisfying $\| x_1^* + \cdots + x_k^* \| < 3\gamma$ one has $\sup_{i \in I} \| x_i^* \| < 1/2$. It follows from Lemma 6.7 that for all $w_i \in B(\bar{x}, 2\rho)$, $w_i^* \in \partial_F d(\cdot, S_i)(w_i)$ for $i \in I$ satisfying $\| w_1^* + \cdots + w_k^* \| < 2\gamma$ one has $\sup_{i \in I} \| w_i^* \| < 1$, since we can find $x_i \in S_i$, $x_i^* \in N_F(S_i, x_i)$ such that $\| x_i^* - w_i^* \| < \gamma/k < 1/2$, and $\| x_i - w_i \| < d(w_i, S_i) + \rho \le 3\rho$, and hence $x_i \in B(\bar{x}, 5\rho)$. Let $f : X \to \mathbb{R}$ be given by $f(x) := d(x, S_1) + \cdots + d(x, S_k)$. Let $x \in B(\bar{x}, \rho) \setminus S$ and $x^* \in \partial_F f(x)$. Since the S_i's are closed, we have $\delta_j := d(x, S_j) > 0$ for some $j \in I$. Let $\delta \in (0, \delta_j) \cap (0, \rho)$. The fuzzy sum rule yields $w_i \in B(x, \delta)$ and $w_i^* \in \partial_F d(\cdot, S_i)$ for $i \in I$ such that $\| w_1^* + \cdots + w_k^* - x^* \| < \gamma$. Since $\delta < \delta_j$, we have $w_j \in X \setminus S_j$, hence $\| w_j^* \| = 1$. Thus $\| w_1^* + \cdots + w_k^* \| \ge 2\gamma$ and $\| x^* \| \ge \gamma$. It follows from Theorems 1.114, 4.80 that $d(x, S) \le (1/\gamma) f(x)$ for all $x \in B(\bar{x}, \rho)$. $\square$

A weaker notion of nice joint behavior can be given (it is weaker because a weak* convergence assumption is added). It is always satisfied in finite-dimensional spaces.

Definition 6.45 ([813, Definition 3.2]). A finite family $(S_i)_{i \in I}$ of closed subsets of a normed space X with $I := \mathbb{N}_k$ is said to be *synergetic* at $\bar{x} \in S := S_1 \cap \cdots \cap S_k$ if $(x_{n,i}) \to \bar{x}$, $(x_{n,i}^*) \overset{*}{\to} 0$ are such that $x_{n,i} \in S_i$, $x_{n,i}^* \in N_F(S_i, x_{n,i})$ for all $(n,i) \in \mathbb{N} \times I$ and $(x_{n,1}^* + \cdots + x_{n,k}^*) \to 0$ implies that for all $i \in I$, one has $(x_{n,i}^*) \to 0$.

Two subsets are synergetic at some point $\bar{z}$ of their intersection whenever one of them is normally compact at $\bar{z}$. However, it may happen that they are synergetic at $\bar{z}$ while none of them is normally compact at $\bar{z}$. This happens for $A \times B$ and $C \times D$ with $\bar{z} := (\bar{x}, \bar{y})$, A (resp. D) being normally compact at $\bar{x}$ (resp. $\bar{y}$) while B and C are arbitrary (for instance singletons in infinite-dimensional spaces).

The preceding notion can be related to alliedness with the help of the following *normal qualification condition* (NQC):

$$x_i^* \in N_L(S_i, \bar{x}), \; x_1^* + \cdots + x_k^* = 0 \Longrightarrow x_1^* = \cdots = x_k^* = 0. \tag{6.15}$$

Proposition 6.46. *A finite family $(S_i)_{i \in I}$ $(I := \mathbb{N}_k)$ of closed subsets of an Asplund space X is allied at $\bar{x} \in S := S_1 \cap \cdots \cap S_k$ if and only if it is synergetic at $\bar{x}$ and the normal qualification condition (6.15) holds.*

In particular, if X is finite-dimensional, (6.15) implies alliedness and (6.13), (6.14).

Proof. The necessity condition ("only if" assertion) is obvious (see (6.11)). Conversely, suppose $(S_i)_{i \in I}$ is synergetic and (NQC) holds. Let $x_{n,i} \in S_i$ and let $x_{n,i}^* \in N_F(S_i, x_{n,i})$ for $(n,i) \in \mathbb{N} \times I$ be such that $(x_{n,i})_n \to \bar{x}$, $(\|x_{n,1}^* + \cdots + x_{n,k}^*\|)_n \to 0$. We may assume that $r_n := \max(\|x_{n,1}^*\|, \ldots, \|x_{n,k}^*\|)$ is positive for all n. Let $w_{n,i}^* := x_{n,i}^* / r_n$. Let r be a limit point of (r_n) in $\overline{\mathbb{R}}_+ := [0, +\infty]$. Taking subsequences, we may assume that (r_n) converges to r and that $(w_{n,i}^*)_n$ weak* converges to some $w_i^* \in B_{X^*}$ for all $i \in I$. Then $w_i^* \in N_L(S_i, \bar{x})$, and if $r \neq 0$, one has $w_1^* + \cdots + w_k^* = 0$. The (NQC) condition implies that $w_i^* = 0$, a contradiction to the synergy of the family $(S_i)_{i \in I}$ and the fact that there is some $j \in I$ such that $\|w_{n,j}^*\| = 1$ for infinitely many $n \in \mathbb{N}$. Thus $r = 0$. Since r is an arbitrary limit point of (r_n), one gets $(r_n) \to 0$. $\qquad \square$

Exercises

1. Check that the inclusion $N_L(F \cup G, \bar{x}) \subset N_L(F, \bar{x}) \cap N_L(G, \bar{x})$ for $F, G \subset X, \bar{x} \in F \cap G$ is not satisfied for $X := \mathbb{R}^2$, $F := \mathbb{R} \times \{0\}, G := \{0\} \times \mathbb{R}, \bar{x} := (0,0)$.

2. A family $(S_i)_{i \in I}$ of closed subsets of a normed space X, with $I := \mathbb{N}_k$, is said to satisfy the *limiting qualification condition* (LQC) at $\bar{x} \in S$ if whenever $(x_{n,i}) \to \bar{x}$, $(x_{n,i}^*) \overset{*}{\to} x_i^*$ with $x_{n,i} \in S_i$, $x_{n,i}^* \in N_F(S_i, x_{n,i})$ for all $(n,i) \in \mathbb{N} \times I$ and $(x_{n,1}^* + \cdots + x_{n,k}^*) \to 0$, one has $x_i^* = 0$ for all $i \in I$. Note that this condition is a consequence of the normal qualification condition (NQC), hence is also a consequence of alliedness. Show that $(S_i)_{i \in I}$ is allied at $\bar{x}$ if and only if it is synergetic at $\bar{x}$ and (LQC) holds.

3. For $\varepsilon > 0$, the ε-*plastering* of a cone P of a normed space Z is the set

$$P_\varepsilon := \{z \in Z : d(z,P) < \varepsilon \|z\|\} \cup \{0\}.$$

Two cones P,Q of Z are said to be *apart* if $\mathrm{gap}(P \cap S_Z, Q \cap S_Z) > 0$, where S_Z is the unit sphere in Z and for two subsets C,D of Z, the *gap* between C and D is defined by $\mathrm{gap}(C,D) := \inf\{\|x - y\| : x \in C, y \in D\}$.

Show that P,Q are apart if and only if for some $\varepsilon > 0$ one has $P_\varepsilon \cap Q_\varepsilon = \{0\}$, if and only if for some $\alpha > 0$ one has $P_\alpha \cap Q = \{0\}$.

4. Show that a pair (F,G) of closed subsets of a normed space Z is allied at $\bar{x} \in F \cap G$ if and only if it satisfies the following *local uniform alliedness* (LUA) property: there exists $\varepsilon > 0$ such that for all $y \in F \cap B(\bar{x},\varepsilon)$, $z \in G \cap B(\bar{x},\varepsilon)$ the cones $N_F(F,y)$ and $N_F(G,z)$ are apart.

5. Show that the (LUA) property at $\bar{x} \in F \cap G$ is equivalent to the *fuzzy qualification condition* (FQC) at $\bar{x}$: there exists $\gamma \in (0,1)$ such that for all $y \in F \cap B(\bar{x},\gamma)$, $z \in G \cap B(\bar{x},\gamma)$ one has

$$(N_F(F,y) + \gamma B_{Z^*}) \cap (-N_F(G,z) + \gamma B_{Z^*}) \cap B_{Z^*} \subset (1 - \gamma)B_{Z^*}. \qquad (6.16)$$

6. Let $F := \mathbb{R}_+ \times \mathbb{R}_+$, $G := \mathbb{R}_+ \times \mathbb{R}_-$ in $X := \mathbb{R}^2$, $\bar{x} = (0,0)$. Then $\{0\} \times \mathbb{R}_- \subset N_L(F,\bar{x}) \cap (-N_L(G,\bar{x}))$, so that conditions (6.15), (6.16) are not satisfied, whereas for all $x \in X \setminus (F \cap G)$, y, z close enough to $\bar{x}$ and $y^* \in \partial_F d_F(y)$, $z^* \in \partial_F d_G(z)$ one has $\|y^* + z^*\| \geq 1$ and relation (6.11) holds.

7. Check with an example that the metric estimate (6.13) of the linear coherence condition is a more general property than alliedness or synergy. [Hint: Take an infinite-dimensional Banach space W, endow $X := W \times \mathbb{R}$ with the sum norm, consider $F := \{0\} \times \mathbb{R}_-$, $G := \{0\} \times \mathbb{R}_+$, and show that $d(\cdot,F \cap G) \leq d(\cdot,F) + d(\cdot,G)$ but that F,G are not allied at $(0,0)$ and that conditions (6.15), (6.16) are not satisfied.]

6.3.2 Coderivative to an Intersection of Multimaps

Now let us pass to multimaps. Since the graph of a multimap is a subset of a product space, the preceding concepts can be adapted to such a product structure in order to get refined conditions. For the sake of simplicity of notation, we limit our study to families of two members and we identify a multimap with its graph.

Definition 6.47. Two multimaps $F,G : X \rightrightarrows Y$ are said to be *range-allied* (resp. *source-allied*) at $\bar{z} \in F \cap G$ if $(w_n) \to \bar{z}$ in F, $(z_n) \to \bar{z}$ in G, (w_n^*), (z_n^*) in $X^* \times Y^*$ are such that $w_n^* := (u_n^*, v_n^*) \in N_F(F,w_n)$, $z_n^* = (x_n^*, y_n^*) \in N_F(G,z_n)$ for all $n \in \mathbb{N}$ and $(w_n^* + z_n^*) \to 0$ implies that one has $(v_n^*) \to 0$ (resp. $(u_n^*) \to 0$).

Clearly, if $F := B \times C$, $G := D \times E$, where C and E are allied at $\bar{y} \in C \cap E$, then F and G are range-allied at $\bar{z} := (\bar{x}, \bar{y})$ for all $\bar{x} \in B \cap D$. A similar assertion holds for source-alliedness, since F, G are source-allied if and only if F^{-1}, G^{-1} are range-allied.

It is also easy to see that when F, G are range-allied at $\bar{z}$ and F or G is coderivatively compact at $\bar{z}$, then the graphs of F, G are synergetic at $\bar{z}$. Similarly, if F, G are source-allied at $\bar{z}$ and F^{-1} or G^{-1} is coderivatively compact at $(\bar{y}, \bar{x})$, then F, G are synergetic at $\bar{z}$.

Calculus rules for the intersection of two multimaps are given in the next statement. Here, for two multimaps $P, Q : X \rightrightarrows Y$ between two linear spaces the multimap $P \square Q : X \rightrightarrows Y$ is the multimap whose graph is the sum of the graphs of F and G, i.e., is defined by

$$(P \square Q)(x) := \{P(u) + Q(v) : u, v \in X, \ u + v = x\}, \quad x \in X.$$

Note that if P and Q are the epigraph multimaps associated with functions f, g respectively, then the vertical closure of $P \square Q$ is the epigraph multimap associated with the infimal convolution $f \square g$.

Proposition 6.48. *Let $F, G : X \rightrightarrows Y$ and let $\bar{z} := (\bar{x}, \bar{y}) \in F \cap G$. Then*

$$D_D^* F(\bar{x}, \bar{y}) \square D_D^* G(\bar{x}, \bar{y}) \subset D_D^* (F \cap G)(\bar{x}, \bar{y}), \tag{6.17}$$

$$D_F^* F(\bar{x}, \bar{y}) \square D_F^* G(\bar{x}, \bar{y}) \subset D_F^* (F \cap G)(\bar{x}, \bar{y}). \tag{6.18}$$

Suppose X and Y are Asplund spaces and the graphs of F and G are closed. Then in order that the inclusion

$$D_L^* (F \cap G)(\bar{x}, \bar{y}) \subset D_L^* F(\bar{x}, \bar{y}) \square D_L^* G(\bar{x}, \bar{y}) \tag{6.19}$$

hold, it suffices that one of the following assumptions be satisfied:

(a) *The graphs of F and G are allied at $\bar{z}$;*
(b) *They are synergetic at $\bar{z}$ and satisfy the (NQC) condition (6.15) at $\bar{z}$;*
(c) *They are synergetic at $\bar{z}$ and satisfy the following condition:*

$$u^* \in (-D_L^* F(\bar{x}, \bar{y})(v^*)) \cap D_L^* G(\bar{x}, \bar{y})(-v^*) \Longrightarrow u^* = 0, \ v^* = 0; \tag{6.20}$$

(d) *F is strongly coderivatively compact at $(\bar{x}, \bar{y})$, G^{-1} is coderivatively compact at $(\bar{y}, \bar{x})$, and (6.20) holds;*
(e) *F^{-1} is coderivatively compact at $(\bar{y}, \bar{x})$, G is strongly coderivatively compact at $(\bar{x}, \bar{y})$, and (6.20) holds;*
(f) *F and G are range-allied at $\bar{z}$, either F or G is coderivatively compact at $(\bar{x}, \bar{y})$, and the following condition holds:*

$$(-D_M^* F(\bar{x}, \bar{y})(0)) \cap D_M^* G(\bar{x}, \bar{y})(0) = \{0\}; \tag{6.21}$$

(g) F and G are source-allied at $\bar{z}$, either F^{-1} or G^{-1} is coderivatively compact at $(\bar{y},\bar{x})$, and the following condition holds:

$$(-D_M^* F^{-1}(\bar{y},\bar{x})(0)) \cap D_M^* G^{-1}(\bar{y},\bar{x})(0) = \{0\}. \qquad (6.22)$$

Proof. The first assertion is an immediate consequence of an inclusion for the normal cone to an intersection; here one uses the facts that the normal cones are convex and that the passage to the normal cone is antitone.

Under the assumptions (a), (b), (c), inclusion (6.19) is a consequence of Proposition 6.46 and of the preceding analysis, observing that condition (6.20) is equivalent to (6.15) for $I := \{1,2\}$, $S_1 := F$, $S_2 := G$ and that $(y^*,x^*) \in D_L^* F(\bar{x},\bar{y}) \square D_L^* G(\bar{x},\bar{y})$ if and only if $(x^*,-y^*) \in N_L(F,\bar{z}) + N_L(G,\bar{z})$.

Let us prove case (d) by showing that the graphs of F and G are synergetic at $\bar{z} := (\bar{x},\bar{y})$ whenever F is strongly coderivatively compact at $\bar{z}$ and G^{-1} is coderivatively compact at $(\bar{y},\bar{x})$. In fact, if $(w_n) \to \bar{z}$ in F, $(z_n) \to \bar{z}$ in G, $(w_n^*) \overset{*}{\to} 0$, $(z_n^*) \overset{*}{\to} 0$ are such that $w_n^* := (u_n^*,v_n^*) \in N_F(F,w_n)$, $z_n^* := (x_n^*,y_n^*) \in N_F(G,z_n)$ for all $n \in \mathbb{N}$ and $(w_n^* + z_n^*) \to 0$, we have $(u_n^*) \to 0$, since F is strongly coderivatively compact at $(\bar{x},\bar{y})$, whence $(x_n^*) \to 0$ and $(y_n^*) \to 0$, $(v_n^*) \to 0$, since G^{-1} is coderivatively compact at $(\bar{y},\bar{x})$. Case (e) is similar.

Let us prove case (f). Suppose F and G are range-allied at $\bar{z}$, F is coderivatively compact at $\bar{z}$, and relation (6.21) holds. Let us prove that F,G are allied at $\bar{z}$. Let $(w_n) \to \bar{z}$ in F, $(z_n) \to \bar{z}$ in G, (w_n^*), (z_n^*) be sequences such that $(w_n^* + z_n^*) \to 0$, and $w_n^* := (u_n^*,v_n^*) \in N_F(F,w_n)$, $z_n^* := (x_n^*,y_n^*) \in N_F(G,z_n)$ for all $n \in \mathbb{N}$. Taking subsequences, we may assume that $r_n := \|w_n^*\|$ is positive for all large n and that (r_n) converges to some $r \in \overline{\mathbb{R}}_+$ and that (w_n^*/r_n) weak* converges to some $\bar{w}^* := (\bar{u}^*,\bar{v}^*)$. Let us prove that $r = 0$. If r is a positive number or $+\infty$, since F and G are range-allied, we have $(v_n^*) \to 0$, hence $\bar{v}^* = \lim_n(v_n^*/r_n) = 0$. Since $(w_n^* + z_n^*) \to 0$, we also have $(y_n^*/r_n) \to 0$. Then $\bar{u}^* \in (D_M^* F(\bar{x},\bar{y})(0)) \cap (-D_M^* G(\bar{x},\bar{y})(0))$, so that $\bar{u}^* = 0$ by condition (6.21). Now, since F is coderivatively compact at $\bar{z}$ and $(u_n^*/r_n) \overset{*}{\to} 0$, $(v_n^*/r_n) \to 0$, we have $(u_n^*/r_n) \to 0$, a contradiction to $\|w_n^*\|/r_n = 1$.

Case (g) is similar, changing F and G into F^{-1} and G^{-1} respectively. □

The following corollary stems from (6.17)–(6.19).

Corollary 6.49. *Let X and Y be Asplund spaces, let $F,G : X \rightrightarrows Y$ be two closed multimaps that are soft (resp. F-soft) at $(\bar{x},\bar{y}) \in F \cap G$ and satisfy one of the assumptions (a)–(g) of Proposition 6.48. Then $F \cap G$ is soft (resp. F-soft) at $(\bar{x},\bar{y})$ and*

$$D_L^*(F \cap G)(\bar{x},\bar{y}) = D_L^* F(\bar{x},\bar{y}) \square D_L^* G(\bar{x},\bar{y}).$$

Given normed spaces X, Y_i, for $i \in \mathbb{N}_k$ and multimaps $F_i : X \rightrightarrows Y_i$, in order to estimate the coderivative of the multimap $F := (F_1,\ldots,F_k) : X \rightrightarrows Y := Y_1 \times \cdots \times Y_k$ defined by $F(x) := F_1(x) \times \cdots \times F_k(x)$, let us introduce the following definition, in which $\bar{x} \in X$, $\bar{y}_i \in F_i(\bar{x})$ for $i \in \mathbb{N}_k$.

Definition 6.50. The multimaps $F_i : X \rightrightarrows Y_i$ are said to be *cooperative* (resp. *coordinated*) at $(\bar{x}, \bar{y}_1, \dots, \bar{y}_k)$ if the graphs of the multimaps $M_i : X \rightrightarrows Y := Y_1 \times \cdots \times Y_k$ given by $M_1(x) := F_1(x) \times Y_2 \times \cdots \times Y_k$, $M_i(x) := Y_1 \times \cdots \times Y_{i-1} \times F_i(x) \times Y_{i+1} \times \cdots \times Y_k$ for $i = 2, \dots, k-1$, $M_k(x) := Y_1 \times \cdots \times Y_{k-1} \times F_k(x)$ are *allied* (resp. *synergetic*) at $(\bar{x}, \bar{y}_1, \dots, \bar{y}_k)$.

It is easy to see that $F_1, \dots, F_k$ are cooperative (resp. coordinated) whenever all but one of the F_i's are coderivatively bounded around $(\bar{x}, \bar{y}_i, 0)$ (resp. coderivatively compact at $(\bar{x}, \bar{y}_i)$).

Corollary 6.51. *Let X, Y_1, Y_2 be Asplund spaces and let the multimaps $F_1 : X \rightrightarrows Y_1$, $F_2 : X \rightrightarrows Y_2$ have closed graphs. If they are cooperative at $(\bar{x}, \bar{y}_1, \bar{y}_2)$, then for every $(\bar{y}_1^*, \bar{y}_2^*) \in Y_1^* \times Y_2^*$, one has*

$$D_L^*(F_1, F_2)(\bar{x}, \bar{y}_1, \bar{y}_2)(\bar{y}_1^*, \bar{y}_2^*) \subset D_L^* F_1(\bar{x}, \bar{y}_1)(\bar{y}_1^*) + D_L^* F_2(\bar{x}, \bar{y}_2)(\bar{y}_2^*). \tag{6.23}$$

The same relation holds if

$$(-D_M^* F_1(\bar{x}, \bar{y}_1)(0)) \cap D_M^* F_2(\bar{x}, \bar{y}_2)(0) = \{0\} \tag{6.24}$$

and if F_1, F_2 are coordinated at $(\bar{x}, \bar{y}_1, \bar{y}_2)$, in particular, if either F_1 or F_2 is coderivatively compact at $(\bar{x}, \bar{y}_1)$ or $(\bar{x}, \bar{y}_2)$ respectively.

Proof. Let $F := (F_1, F_2)$ and let M_1 and M_2 be defined as above, so that $F = M_1 \cap M_2$ and one has the relations

$$D_L^* F_1(\bar{x}, \bar{y}_1)(\bar{y}_1^*) = D_L^* M_1(\bar{x}, \bar{y}_1, \bar{y}_2)(\bar{y}_1^*, 0),$$

$$D_L^* F_2(\bar{x}, \bar{y}_2)(\bar{y}_2^*) = D_L^* M_2(\bar{x}, \bar{y}_1, \bar{y}_2)(0, \bar{y}_2^*),$$

and similar ones in which the limiting coderivatives are replaced with mixed coderivatives. Expressing $D_L^* M_1(\bar{x}, \bar{y}_1, \bar{y}_2) \square D_L^* M_2(\bar{x}, \bar{y}_1, \bar{y}_2)$, the first assertion is a consequence of Theorem 6.44.

The proof of the second one is similar to the proof of case (f) of Proposition 6.48, observing that here we can dispense with the condition that M_1 or M_2 is coderivatively compact at $\bar{z} := (\bar{x}, \bar{y}_1, \bar{y}_2) \in M_1 \cap M_2$. The details are left as an exercise. $\qquad \square$

Corollary 6.51 can be generalized to a finite family of multimaps in an obvious way. We just state an application to the case of a map with values in $\mathbb{R}^k$.

Corollary 6.52. *Let X be an Asplund space and let $f := (f_1, \dots, f_k) : X \to \mathbb{R}^k$. Suppose $(\mathrm{epi} f_1, \dots, \mathrm{epi} f_k)$ is cooperative at $(\bar{x}, \bar{y}) := (\bar{x}, \bar{y}_1, \dots, \bar{y}_k) := (\bar{x}, f_1(\bar{x}), \dots, f_k(\bar{x}))$. Then for all $(\bar{y}_1^*, \dots, \bar{y}_k^*) \in \mathbb{R}^k$ one has*

$$D_L^* f(\bar{x}, \bar{y})(\bar{y}_1^*, \dots, \bar{y}_k^*) \subset D_L^* f_1(\bar{x}, \bar{y}_1)(\bar{y}_1^*) + \cdots + D_L^* f(\bar{x}, \bar{y}_k)(\bar{y}_k^*).$$

The versatility of set-valued analysis can be experienced through the following statement whose proof consists in taking inverses.

Corollary 6.53. *Let X_1, X_2, Y be Asplund spaces, let $G_1 : X_1 \rightrightarrows Y$, $G_2 : X_2 \rightrightarrows Y$ be multimaps with closed graphs and let $G : X_1 \times X_2 \rightrightarrows Y$ be defined by $G(x_1,x_2) := G_1(x_1) \cap G_2(x_2)$ for $(x_1,x_2) \in X := X_1 \times X_2$. If G_1^{-1} and G_2^{-1} are cooperative at $(\bar{y},\bar{x}_1,\bar{x}_2)$ then for every $y^* \in Y^*$ one has*

$$D_L^* G(\bar{x}_1,\bar{x}_2,\bar{y})(y^*) \subset D_L^* G_1(\bar{x}_1,\bar{y})(y^*) \times D_L^* G_2(\bar{x}_2,\bar{y})(y^*).$$

The same conclusion holds when G_1^{-1} and G_2^{-1} are coordinated at $(\bar{y},\bar{x}_1,\bar{x}_2)$ and

$$(-D_M^* G_1^{-1}(\bar{y},\bar{x}_1)(0)) \cap D_M^* G_2^{-1}(\bar{y},\bar{x}_2)(0) = \{0\}. \tag{6.25}$$

Proof. One has $y \in G(x_1,x_2)$ if and only if $(x_1,x_2) \in F_1(y) \times F_2(y)$ for $F_1 := G_1^{-1}$, $F_2 := G_2^{-1}$. Thus, the result stems from Corollary 6.51 when coderivatives are rewritten in terms of normal cones (exercise). ☐

Exercises

1. Show that the multimaps $F' : X \rightrightarrows Y'$, $F'' : X \rightrightarrows Y''$ are cooperative (resp. coordinated) at $(\bar{x},\bar{y}',\bar{y}'')$ iff for all sequences $((x_n',y_n')) \to (\bar{x},\bar{y}')$ in F', $((x_n'',y_n'')) \to (\bar{x},\bar{y}'')$ in F'', $(x_n'^*),(x_n''^*)$ in X^* (resp. $(x_n'^*),(x_n''^*) \overset{*}{\to} 0$), $(y_n'^*),(y_n''^*) \to 0$ such that $(x_n'^* + x_n''^*) \to 0$ and $x_n'^* \in D_F^* F'(x_n',y_n')(y_n'^*)$, $x_n''^* \in D_F^* F''(x_n'',y_n'')(y_n''^*)$ for all n, one has $(x_n'^*) \to 0$ (and $(x_n''^*) \to 0$).

2. (**a**) Check that if F_1 is coderivatively bounded around $(\bar{x},\bar{y}_1)$ or if F_2 is coderivatively bounded around $(\bar{x},\bar{y}_2)$, then F_1 and F_2 are cooperative at $(\bar{x},\bar{y}_1,\bar{y}_2)$.
(**b**) Check that if F_1 is coderivatively compact at $(\bar{x},\bar{y}_1)$ (or if F_2 is coderivatively compact at $(\bar{x},\bar{y}_2)$), then F_1 and F_2 are coordinated at $(\bar{x},\bar{y}_1,\bar{y}_2)$.

3. Show that two subsets S_1, S_2 of a normed space X are allied (resp. synergetic) at $\bar{x} \in S_1 \cap S_2$ if and only if the multimaps $F_1, F_2 : X \rightrightarrows Y := \{0\}$ with graphs $S_1 \times \{0\}$ and $S_2 \times \{0\}$ respectively are cooperative (resp. coordinated) at $(\bar{x},0,0)$.

4. With the notation of Definition 6.50, prove that F_1 and F_2 are cooperative at $(\bar{x},\bar{y}_1,\bar{y}_2)$ if and only if M_1 and M_2 are source-allied at $(\bar{x},(\bar{y}_1,\bar{y}_2))$.

5. (**Restriction of a multimap**) Let $F : X \rightrightarrows Y$ be a multimap and let $C \subset X$. Denote by F_C the multimap defined by $F_C(x) := F(x)$ for $x \in C$, $F_C(x) = \varnothing$ for $x \in X \setminus C$. Check that $F_C = F \cap G$ with $G := C \times Y$. Describe alliedness of (F,G) in terms of $D_F^* F$ and $N(C,\cdot)$. Derive from that an inclusion for $D_L^* F_C$.

6. (**Restriction of a multimap**) Let $F : X \rightrightarrows Y$, C and G be as in the preceding exercise. Describe synergy of (F,G) in terms of $D_F^* F$ and $N(C,\cdot)$. Show that if C is normally compact at $\bar{x} \in C$ or if F is coderivatively compact at $(\bar{x},\bar{y})$, then (F,G) is synergetic at $(\bar{x},\bar{y})$.

7. (Restriction of a multimap) Let $F : X \rightrightarrows Y$ be a multimap and let $C \subset X$. Define $0_C : X \rightrightarrows Y$ by $0_C(x) := \{0\}$ for $x \in C$, $0_C(x) := \varnothing$ for $x \in X \setminus C$, so that 0_C is the restriction to C of the null multimap. Check that $F_C = F + 0_C$. Derive from the results about sums of multimaps (Sect. 6.3.6) an inclusion for $D_L^* F_C$ and compare the required assumptions with those in the preceding two exercises.

6.3.3 Normal Cone to a Direct Image

Now let us evaluate the limiting normal cone to a direct image. Here we take the image under a continuous linear map p; the case of a map of class C^1 is left as an exercise. We need some topological notions. A multimap $M : W \rightrightarrows Z$ between two metric spaces is said to be *lower semicontinuous* or *inner semicontinuous* at $(\overline{w}, B)$ (on $E \subset W$), where B is some subset of Z, if for every sequence (w_n) (of E) converging to $\overline{w}$ there exist some $\overline{z} \in B$ and a sequence $(z_n) \to \overline{z}$ such that $z_n \in M(w_n)$ for n in an infinite subset N of $\mathbb{N}$. Three special cases are of interest in the preceding definition: the case that B is a singleton $\{\overline{z}\}$, the case that $B = M(\overline{w})$, and the case $B := Z$. The first case coincides with the usual notion of lower semicontinuity at $(\overline{w}, \overline{z})$. In the last case, this property has been renamed *semicompactness* of M at $\overline{w}$.

Let us recall that given a map $p : V \to W$ between two metric spaces $V, W, A \subset V$, $e \in E \subset W$, p is said to be *proper* at (A, e) with respect to E if for every sequence $(e_n) \to_E e$ there exist $a \in A$ and a sequence $(v_n) \to a$ such that $p(v_n) = e_n$ for all n in an infinite subset of $\mathbb{N}$. The case $A := \{a\}$ corresponds to openness of p at a. The case $A := V$ is usual properness of p at e. Clearly, $p : V \to W$ is proper at (A, e) with respect to E if and only if $M := p^{-1} : W \rightrightarrows Z := V$ is lower semicontinuous at (e, A) on E. On the other hand, $M : W \rightrightarrows Z$ is lower semicontinuous at $(\overline{w}, B)$ (on $E \subset W$) if and only if the restriction p_M of the canonical projection $W \times Z \to W$ to the graph of M is proper at $(A, \overline{w})$ with $A := \{\overline{w}\} \times B$.

Proposition 6.54. *Let U, V, W be normed spaces, W being an Asplund space, let $C \subset V$, $E \subset W$, and let $p : V \to W$ be linear and continuous and such that $p(C) \subset E$. Let $e \in E$, $A \subset p^{-1}(e) \cap C$.*
(a) If $p \mid_C$ is proper at (A, e) with respect to E, then one has

$$N_L(E, e) \subset \bigcup_{c \in A} (p^{\mathsf{T}})^{-1} (N_L(C, c)).$$

In particular, if p is open from C to E at $c \in A$, one has $N_L(E, e) \subset (p^{\mathsf{T}})^{-1} (N_L(C, c))$.
(b) If $V = W \times U$ and C is (the graph of) a multimap from W to U with domain E that is lower semicontinuous at (e, B) on E for some $B \subset U$, one has $N_L(E, e) \subset \bigcup_{b \in B} D_M^* C(e, b)(0)$.

Proof. (a) Let us first recall from Proposition 2.108 that for $e^* \in N_F(E, e)$ we have $p^{\mathsf{T}}(e^*) \in N_F(C, c)$ for all $c \in p^{-1}(e)$. Here, since p is linear and continuous, a direct proof is even easier than in the case that p is differentiable (exercise). Now let $e^* \in$

$N_L(E,e)$. There exist sequences $(e_n) \to_E e$, $(e_n^*) \overset{*}{\to} e^*$ such that $e_n^* \in N_F(E,e_n)$ for all n. When $p|_C$ is proper at (A,e) with respect to E there exist $c \in A$ and a sequence $(c_n) \to_C c$ such that $p(c_n) = e_n$ for n in an infinite subset N of $\mathbb{N}$. By what precedes, we have $p^\mathsf{T}(e_n^*) \in N_F(C,c_n)$. Since $(p^\mathsf{T}(e_n^*)) \overset{*}{\to} p^\mathsf{T}(e^*)$, we conclude that $p^\mathsf{T}(e^*) \in N_L(C,c)$. Taking $A := \{c\}$, we get the second assertion of (a).

(b) Suppose $V = W \times U$ and C is (the graph of) a multimap from W to U with domain E that is lower semicontinuous at (e,B). Since for $p := p_W$, $p^\mathsf{T}(e_n^*) = (e_n^*,0) \in N_F(C,c_n)$, we see that $e^* \in D_M^* C(e,b)(0)$, where $c := (e,b) \in A := \{e\} \times B$.   □

A study of the direct image $E := F(C)$ of a subset C of a normed space X by a multimap $F : X \rightrightarrows Y$ with values in another normed space Y can be derived from the preceding proposition and from results about intersections of sets.

Proposition 6.55. *Let $F : X \rightrightarrows Y$ be a multimap with closed graph between two Asplund spaces and let C be a closed subset of X, $\bar{y} \in E := F(C)$, $B \subset F^{-1}(\bar{y}) \cap C$. Suppose that the multimap $y \mapsto F^{-1}(y) \cap C$ is lower semicontinuous at $(\bar{y}, B)$ on E. (a) If for all $\bar{x} \in B$, the sets F and $C \times Y$ are (linearly) coherent around $(\bar{x}, \bar{y})$ in the sense that for some $c > 0$, $\rho > 0$ one has*

$$\forall(x,y) \in B((\bar{x},\bar{y}),\rho), \quad d((x,y),F \cap (C \times Y)) \leq cd((x,y),F) + cd(x,C), \quad (6.26)$$

then the following inclusion holds:

$$N_L(F(C),\bar{y}) \subset \bigcup_{\bar{x} \in B} D_L^* F^{-1}(\bar{y},\bar{x})(N_L(C,\bar{x})). \quad (6.27)$$

(b) Suppose that for $\bar{x} \in B$, F and $C \times Y$ are allied at $(\bar{x},\bar{y})$. Then F and $C \times Y$ are coherent around $(\bar{x},\bar{y})$.

Proof. (a) Since for all $(x,y) \in X \times Y$ one has $d(x,C) = d((x,y),C \times Y)$, one sees that (6.26) means that F and $C \times Y$ are (linearly) coherent around $(\bar{x},\bar{y})$ in the sense of Theorem 6.41, so that

$$N_L(F \cap (C \times Y),(\bar{x},\bar{y})) \subset N_L(F,(\bar{x},\bar{y})) + N_L(C \times Y,(\bar{x},\bar{y})).$$

Since $F(C) = p_Y(F \cap (C \times Y))$ and $N_L(C \times Y,(\bar{x},\bar{y})) = N_L(C,\bar{x}) \times \{0\}$, applying Proposition 6.54, for all $y^* \in N_L(F(C),\bar{y})$ one gets some $(u^*,v^*) \in N_L(F,(\bar{x},\bar{y}))$, $w^* \in N_L(C,\bar{x})$ such that $(0,y^*) = (u^*,v^*) + (w^*,0)$ or $y^* = v^* \in D_L^* F^{-1}(\bar{y},\bar{x})(w^*)$, so that (6.27) holds. One can also call upon Proposition 6.21 with j given by $j(x,y) := cd((x,y),F) + cd(x,C)$ (exercise: follow the proof of Proposition 6.56 in the next subsection).

(b) Theorem 6.41 and the relation $d((x,y),C \times Y) = d(x,C)$ entail (6.26).   □

Exercises

1. With the data of Proposition 6.54 (b), show that if C is pseudo-Lipschitzian around $c := (e, u)$ on E, then $N_L(E, e) = D_M^* C(e, u)(0)$.

2. Using Proposition 2.108, derive the following variant of Proposition 6.54. Let V, W be normed spaces, $C \subset V$, $E \subset W$, let $p : V \to W$ be linear and continuous and such that $p(C) \subset E$. Let $c \in C$, $e := p(c)$.
(a) Check that $N(E, e) \subset (p^\mathsf{T})^{-1}(N(C, c))$. If $T(E, e) \subset p(T(C, c))$, in particular if there exists a map $q : W \to V$ that is Hadamard differentiable at e and such that $q(e) = c$, $q(E) \subset C$, $p \circ q \mid_E = I_E$, show that $N(E, e) = (p^\mathsf{T})^{-1}(N(C, c))$.
(b) Prove that $N_F(E, e) \subset (p^\mathsf{T})^{-1}(N_F(C, c))$ and $N_F^\varepsilon(E, e) \subset (p^\mathsf{T})^{-1}(N_F^{\|p\|\varepsilon}(C, c))$ for all $\varepsilon \in \mathbb{R}_+$.
(c) Show that if p is open at c from C to E with a linear rate κ, then one has $(p^\mathsf{T})^{-1}(N_F^\gamma(C, c)) \subset N_F^{\kappa\gamma}(E, e)$, in particular $N_F(E, e) = (p^\mathsf{T})^{-1}(N_F(C, c))$.

3. Show that the first assertion of Proposition 6.21 is a consequence of assertion (a) of Proposition 6.54. Show that conversely, the particular case of this assertion is a consequence of the first assertion of Proposition 6.21.

6.3.4 Normal Cone to an Inverse Image

Inverse images are closely linked to intersections and direct images. First, we note that the inverse image $P := F^{-1}(Q)$ of $Q \subset Y$ by a multimap $F : X \rightrightarrows Y$ can be considered the direct image by $F^{-1} : Y \rightrightarrows X$ of Q. This observation explains the analogy between the next proposition and the preceding one. Also, denoting by $d : x \mapsto (x, x)$ the diagonal map, for $F, G \subset X$ one has $F \cap G = d^{-1}(F \times G)$; conversely, the inverse image $P := F^{-1}(Q)$ of $Q \subset Y$ by a multimap $F : X \rightrightarrows Y$ is such that $F \cap (P \times Q) = F \cap (X \times Q)$ and $P = p_X(F \cap (X \times Q))$, where $p_X : X \times Y \to X$ is the first projection. An estimate of the limiting normal cone to an inverse image can be derived from this observation.

Proposition 6.56. *Let $F : X \rightrightarrows Y$ be a multimap with closed graph between two Asplund spaces and let Q be a closed subset of Y, $P := F^{-1}(Q)$, $\bar{x} \in P$, $B \subset F(\bar{x}) \cap Q$. Suppose that the multimap $x \mapsto F(x) \cap Q$ is lower semicontinuous at $(\bar{x}, B)$ on P.*
(a) If for all $\bar{y} \in B$, F and Q are (linearly) coherent around $(\bar{x}, \bar{y})$ in the sense that for some $c > 0$, $\rho > 0$ one has

$$\forall (x, y) \in B((\bar{x}, \bar{y}), \rho), \qquad d(x, P) \leq cd((x, y), F) + cd(y, Q), \qquad (6.28)$$

then the following inclusion holds:

$$N_L(P, \bar{x}) \subset \bigcup_{\bar{y} \in B} D_L^* F(\bar{x}, \bar{y})(N_L(Q, \bar{y})). \qquad (6.29)$$

(b) Suppose that for all $\bar{y} \in B$, F and Q are allied at $(\bar{x}, \bar{y}) \in F \cap (P \times Q)$ in the sense that whenever $((x_n, y_n)) \to (\bar{x}, \bar{y})$ in F, $(z_n) \to \bar{y}$ in Q, $(x_n^*) \to 0$ in X^*, (y_n^*), (z_n^*) are sequences in Y^* such that $(y_n^* - z_n^*) \to 0$, $x_n^* \in D_F^* F(x_n, y_n)(y_n^*)$, $z_n^* \in N_F(Q, z_n)$ for all $n \in \mathbb{N}$, one has $(y_n^*) \to 0$, $(z_n^*) \to 0$. Then F and Q are coherent around $(\bar{x}, \bar{y})$.

Proof. (a) Let $\bar{x}^* \in N_L(P, \bar{x})$, so that by Proposition 6.8, $\bar{x}^* = r\bar{u}^*$ for some $r \in \mathbb{R}_+$, $\bar{u}^* \in \partial_L d_P(\bar{x})$. Let $j : X \times Y \to \mathbb{R}$ be given by $j(x, y) := cd((x, y), F) + cd(y, Q)$, so that $d_P(x) \le j(x, y)$ for all $(x, y) \in B(\bar{x}, \rho) \times B(\bar{y}, \rho)$. By our semicontinuity assumption, for every sequence $(x_n) \to \bar{x}$ in P there exist $\bar{y} \in B$ and a sequence $(y_n) \to \bar{y}$ such that $j(x_n, y_n) = 0$ for all n in an infinite subset of $\mathbb{N}$. Proposition 6.21 ensures that $(\bar{u}^*, 0) \in \partial_L j(\bar{x}, \bar{y})$ (replacing $\bar{y}$ by another point of B if necessary). The sum rule yields $(\bar{u}^*, \bar{v}^*) \in c\partial_L d_F(\bar{x}, \bar{y})$, $\bar{w}^* \in c\partial_L d_Q(\bar{y})$ such that $(\bar{u}^*, 0) = (\bar{u}^*, \bar{v}^*) + (0, \bar{w}^*)$. Then $\bar{z}^* := r\bar{w}^* \in N_L(Q, \bar{y})$, and for $\bar{y}^* := r\bar{w}^* = -r v^*$ one has $\bar{x}^* \in D_L^* F(\bar{x}, \bar{y})(\bar{y}^*)$.

(b) Given $\bar{y} \in B$, the alliedness assumption on F and Q is easily seen to be equivalent to the alliedness of $\mathrm{gph} F$ and $X \times Q$. Then by Theorem 6.41, there exist $c > 0$, $\rho > 0$ such that for all $x \in B(\bar{x}, \rho)$, $y \in B(\bar{y}, \rho)$ one has

$$d((x, y), F \cap (X \times Q)) \le cd((x, y), F) + cd((x, y), X \times Q). \qquad (6.30)$$

Now, for all $x \in X$ one has

$$\inf_{u \in P} d(x, u) \le \inf\{d(x, u) + d(y, v) : y \in Y, \ (u, v) \in F \cap (P \times Q)\},$$

$$d(x, P) \le \inf\{d((x, y), F \cap (P \times Q)) : y \in Y\}.$$

Since $F \cap (P \times Q) = F \cap (X \times Q)$, $d((x, y), X \times Q) = d(y, Q)$, we get (6.28).   □

Using the sum rule, the preceding result yields a Lagrange multiplier rule.

Corollary 6.57. *Let X and Y be Asplund spaces, let Q be a closed subset of Y, let $F : X \rightrightarrows Y$ be a multimap with closed graph, and let $\bar{x} \in P := F^{-1}(Q)$ be a minimizer of a Lipschitzian function $f : X \to \mathbb{R}$ on P. Assume that for some $\bar{y} \in F(\bar{x}) \cap Q$, the multimap $x \mapsto F(x) \cap Q$ is lower semicontinuous at $(\bar{x}, \bar{y})$ on P and that F and Q are allied at $(\bar{x}, \bar{y})$. Then there exists some $y^* \in N_L(Q, \bar{y})$ such that*

$$0 \in \partial_L f(\bar{x}) + D_L^* F(\bar{x}, \bar{y})(y^*).$$

Exercises

1. Relate Proposition 6.56 to Theorem 6.44 and Proposition 6.54, using the fact that with the notation of Proposition 6.56, P is the image of $F \cap (X \times Q)$ by the canonical projection $p_X : X \times Y \to X$.

2. Suppose F and Q are *synergetic* at $(\bar{x},\bar{y}) \in F \cap (P \times Q)$ in the sense that whenever $((x_n,y_n)) \to (\bar{x},\bar{y})$ in F, $(z_n) \to \bar{y}$ in Q, $(x_n^*) \stackrel{*}{\to} 0$ in X^*, $(y_n^*) \stackrel{*}{\to} 0$, $(z_n^*) \stackrel{*}{\to} 0$ in Y^* are such that $(y_n^* - z_n^*) \to 0$, $x_n^* \in D_F^* F(x_n,y_n)(y_n^*)$, $z_n^* \in N_F(Q,z_n)$ for all $n \in \mathbb{N}$, one has $(y_n^*) \to 0$, $(z_n^*) \to 0$. Show that (6.30) holds, provided the following condition is satisfied:

$$y^* \in N_L(Q,\bar{y}), \quad 0 \in D_L^* F(\bar{x},\bar{y})(y^*) \Longrightarrow y^* = 0.$$

3. Show that Theorem 6.41 is a consequence of Proposition 6.56. [Hint: $F \cap G = d^{-1}(F \times G)$, where $d : X \to X \times X$ is the diagonal map $x \mapsto (x,x)$.]

4. Establish the results of the next subsection using Proposition 6.56. [Hint: Observe that for $E := G \circ F, H : X \times Z \rightrightarrows X \times Y \times Z$ given by $H(x,z) := \{x\} \times Y \times \{z\}$ one has $E = H^{-1}((F \times Z) \cap (X \times G))$.]

6.3.5 Coderivatives of Compositions

Now let us study the coderivatives of $E := G \circ F$, where $F : X \rightrightarrows Y, G : Y \rightrightarrows Z$ are multimaps between Asplund spaces. We set

$$F_Z := \{((x,z),y) : (x,y) \in F, \, z \in Z\}, \quad G_X := \{(y,(x,z)) : x \in X, (y,z) \in G\},$$

$$C := \{((x,z),y) : (x,y) \in F, (y,z) \in G\},$$

considered as a multimap $C : X \times Z \rightrightarrows Y$, so that

$$C = F_Z \cap G_X^{-1}, \tag{6.31}$$

with $G_X^{-1} = \{((x,z),y) : x \in X, \, (y,z) \in G\} = X \times G^{-1}$, and denoting by $p_{X \times Z}$ the canonical projection $X \times Z \times Y \to X \times Z$, one has

$$E := G \circ F = p_{X \times Z}(C).$$

Thus, an estimate of the coderivative of E can be derived from an inclusion for an intersection and an inclusion for a projection. We first deal with the projection process. As in Proposition 6.54, in order to get some versatility, we introduce a subset B of $C(\bar{x},\bar{z})$. The extreme cases $B = C(\bar{x},\bar{z})$ and B a singleton, $B = \{\bar{y}\}$, are the most remarkable cases, but intermediate situations may occur.

Lemma 6.58. *Suppose C is lower semicontinuous at $((\bar{x},\bar{z}),B)$ on E for some subset B of $C(\bar{x},\bar{z})$ and that*

$$D_M^* C((\bar{x},\bar{z}),\bar{y})(0) \subset \bigcup_{y^* \in Y^*} D_L^* F(\bar{x},\bar{y})(y^*) \times D_L^* G^{-1}(\bar{z},\bar{y})(-y^*) \tag{6.32}$$

for all $\bar{y} \in B$, *or, more generally,*

$$\bigcup_{\bar{y} \in B} D_M^* C((\bar{x}, \bar{z}), \bar{y})(0) \subset \bigcup_{\bar{y} \in B} \bigcup_{y^* \in Y^*} D_L^* F(\bar{x}, \bar{y})(y^*) \times D_L^* G^{-1}(\bar{z}, \bar{y})(-y^*). \qquad (6.33)$$

Then for $E := G \circ F$, *one has*

$$D_L^* E(\bar{x}, \bar{z}) \subset \bigcup_{\bar{y} \in B} D_L^* F(\bar{x}, \bar{y}) \circ D_L^* G(\bar{y}, \bar{z}). \qquad (6.34)$$

Proof. Let $z^* \in Z^*$ and let $x^* \in D_L^* E(\bar{x}, \bar{z})(z^*)$. Then $(x^*, -z^*) \in N_L(E, (\bar{x}, \bar{z}))$, and since C is lower semicontinuous at $(\bar{x}, \bar{z}, B)$ on E, Proposition 6.54(b) yields some $\bar{y}_0 \in B$ such that $(x^*, -z^*) \in D_M^* C((\bar{x}, \bar{z}), \bar{y}_0)(0)$; hence by (6.33), there exist some $\bar{y} \in B$, $y^* \in Y^*$ such that $x^* \in D_L^* F(\bar{x}, \bar{y})(y^*)$, $-z^* \in D_L^* G^{-1}(\bar{z}, \bar{y})(-y^*)$ or $y^* \in D_L^* G(\bar{y}, \bar{z})(z^*)$, and (6.34) holds. $\qquad \square$

Remark. Let us observe that (6.32) is a consequence of the condition

$$\forall \bar{y} \in B, \qquad N_L(C, (\bar{x}, \bar{z}, \bar{y})) \subset N_L(F_Z, (\bar{x}, \bar{z}, \bar{y})) + N_L(G_X^{-1}, (\bar{x}, \bar{z}, \bar{y})). \qquad (6.35)$$

In fact, given $(x^*, z^*) \in D_L^* C(\bar{x}, \bar{z}, \bar{y})(0)$, i.e., $(x^*, z^*, 0) \in N_L(C, (\bar{x}, \bar{z}, \bar{y}))$, relation (6.35) asserts that one can find $(u^*, y^*, w^*) \in X^* \times Y^* \times Z^*$ such that

$$(u^*, 0, -y^*) \in N_L(F_Z, (\bar{x}, \bar{z}, \bar{y})), \quad (0, w^*, y^*) \in N_L(G_X^{-1}, (\bar{x}, \bar{z}, \bar{y})),$$

and

$$(x^*, z^*, 0) = (u^*, 0, -y^*) + (0, w^*, y^*),$$

whence $u^* = x^*$, $w^* = z^*$. One easily deduces from the preceding relations that $(x^*, -y^*) \in N_L(F, (\bar{x}, \bar{y}))$, $(z^*, y^*) \in N_L(G^{-1}, (\bar{z}, \bar{y}))$, or $(x^*, z^*) \in D_L^* F(\bar{x}, \bar{y})(y^*) \times D_L^* G^{-1}(\bar{z}, \bar{y})(-y^*)$. $\qquad \square$

Example. Suppose F is a single-valued map that is continuous at $\bar{x}$. Then for every multimap G, C is lower semicontinuous on E at $((\bar{x}, \bar{z}), B)$ for $B := \{\bar{y}\}$ with $\bar{y} := F(\bar{x}), \bar{z} \in G(\bar{y})$.

Example. Suppose $G := g^{-1}$, where $g : Z \to Y$ is continuous at $\bar{z}$ and $g(\bar{z}) \in F(\bar{x})$. Then C is lower semicontinuous on E at $((\bar{x}, \bar{z}), B)$ for $B := \{g(\bar{z})\}$.

Now let us use sufficient conditions for (6.33) derived from rules for intersections. We first point out links with metric estimates.

Proposition 6.59. *Suppose* X, Y, Z *are Asplund spaces,* F *and* G *have closed graphs,* C *is lower semicontinuous at* $((\bar{x}, \bar{z}), B)$ *on* E, *and for every* $\bar{y} \in B$ *there are* $c > 0$ *and a neighborhood* U *of* $(\bar{x}, \bar{z}, \bar{y})$ *such that for all* $(x, z, y) \in U$ *one has*

$$d((x, z, y), C) \leq cd((x, y), F) + cd((y, z), G). \qquad (6.36)$$

Then (6.33) and (6.34) hold.

Proof. Let us set

$$j(x,z,y) := cd((x,y),F) + cd((y,z),G) = cd((x,z,y),F_Z) + cd((x,z,y),G_X^{-1}),$$

so that (6.36) can be rewritten $d_C \leq j$. Since for every closed subset S of an Asplund space and every $s \in S$ one has $N_L(S,s) = \mathbb{R}_+ \partial_L d_S(s)$, Proposition 6.21 ensures that for all $\bar{y} \in B$ condition (6.35) is satisfied, so that (6.33) and (6.34) hold. $\qquad\square$

Now let us introduce conditions in terms of coderivatives. Since $C = F_Z \cap G_X^{-1}$, in order to get relation (6.35), it is natural to use Theorem 6.44 and Proposition 6.46. Now we observe that the sets F_Z and G_X^{-1} are allied (resp. synergetic) at $(\bar{x},\bar{z},\bar{y})$ if and only if the multimaps $F_Z^{-1} : Y \rightrightarrows X \times Z$ and $G_X : Y \rightrightarrows X \times Z$ are allied (resp. synergetic) at $(\bar{y},\bar{x},\bar{z})$. In view of Definition 6.50, the latter means that the multimaps $F^{-1} : Y \rightrightarrows X$ and $G : Y \rightrightarrows X$ are cooperative (resp. coordinated) at $(\bar{y},\bar{x},\bar{z})$. This condition can be explicitly expressed thus: *for all sequences* $((x_n,y_n)) \to (\bar{x},\bar{y})$ *in* F, $((w_n,z_n)) \to (\bar{y},\bar{z})$ *in* G, $(w_n^*),(y_n^*)$ *in* Y^*, $(x_n^*) \to 0$ *in* X^*, $(z_n^*) \to 0$ *in* Z^* *with* $(w_n^* - y_n^*) \to 0$, $x_n^* \in D_F^* F(x_n,y_n)(y_n^*)$, $w_n^* \in D_F^* G(w_n,z_n)(z_n^*)$, *one has* $(y_n^*) \to 0$ *(resp.* $(y_n^*) \to 0$ *whenever* $(y_n^*) \xrightarrow{*} 0$).

Obviously, F^{-1} and G are coordinated at $(\bar{y},\bar{x},\bar{z})$ if Y is finite-dimensional. It is also the case if F^{-1} is coderivatively compact at $(\bar{y},\bar{x})$ or if G is coderivatively compact at $(\bar{y},\bar{z})$. On the other hand, when F^{-1} (resp. G) is coderivatively bounded around $(\bar{y},\bar{x})$ (resp. $(\bar{y},\bar{z})$), F^{-1} and G are cooperative at $(\bar{y},\bar{x},\bar{z})$.

Theorem 6.60. *Suppose X,Y,Z are Asplund spaces, F and G have closed graphs. If for some subset B of $C(\bar{x},\bar{z})$, C is lower semicontinuous at $((\bar{x},\bar{z}),B)$ on E and if F^{-1} and G are cooperative at $(\bar{y},\bar{x},\bar{z})$ for all $\bar{y} \in B$, then (6.34) holds.*

In particular, if C is lower semicontinuous at $(\bar{x},\bar{z},\bar{y})$ and if F^{-1} and G are cooperative at $(\bar{y},\bar{x},\bar{z})$, then (6.32) and (6.34) with $B := \{\bar{y}\}$ hold.

We can also use Proposition 6.48. However, the qualification condition we present in the next corollary is weaker than the condition obtained from (6.15) or (6.20) for F_Z and G_X^{-1}.

Corollary 6.61. *Suppose X,Y,Z are Asplund spaces, F^{-1} and G have closed graphs and are coordinated at $(\bar{y},\bar{x},\bar{z})$ for all $\bar{y} \in B \subset C(\bar{x},\bar{z})$. Suppose C is lower semicontinuous at $((\bar{x},\bar{z}),B)$ on $E := G \circ F$. Then (6.33) holds whenever the following condition is satisfied for all $\bar{y} \in B$:*

$$(-D_M^* F^{-1}(\bar{y},\bar{x})(0)) \cap D_M^* G(\bar{y},\bar{z})(0) = \{0\}. \tag{6.37}$$

Of course, when Y is finite-dimensional, condition (6.37) is equivalent to

$$(D_L^* F(\bar{x},\bar{y}))^{-1}(0) \cap D_L^* G(\bar{y},\bar{z})(0) = \{0\}, \tag{6.38}$$

but when Y is infinite-dimensional (6.37) is less restrictive.

Proof. We apply Corollary 6.53 with $X_1 := X$, $X_2 := Z$, $G_1 := F$, $G_2 := G^{-1}$, since $C(x,z) = F(x) \cap G^{-1}(z)$ for all $(x,z) \in X \times Z$ and since (6.37) coincides with (6.25). $\square$

Using Lemma 6.34, we get the following consequence.

Corollary 6.62. *Suppose X, Y, Z are Asplund spaces, F and G have closed graphs, C is lower semicontinuous at $((\bar{x}, \bar{z}), B)$ on $E := G \circ F$, and for every $\bar{y} \in B$, either G is pseudo-Lipschitzian around $(\bar{y}, \bar{z})$ or F^{-1} is pseudo-Lipschitzian around $(\bar{y}, \bar{x})$. Then (6.33) holds.*

A simple case in which relation (6.34) holds is given in the next corollary.

Corollary 6.63. *If G is a single-valued map that is circa-differentiable at $\bar{y}$, then*

$$D_L^*(G \circ F)(\bar{x}, \bar{z}) \subset D_L^* F(\bar{x}, \bar{y}) \circ (G'(\bar{y}))^\top.$$

Proof. It is easy to see that C is lower semicontinuous at $((\bar{x}, \bar{z}), \bar{y})$ on $E := G \circ F$ for $z := G(\bar{y})$. Since G is circa-differentiable at $\bar{y}$, it is Lipschitzian around $\bar{y}$, hence pseudo-Lipschitzian around $(\bar{y}, \bar{z})$ with $\bar{z} = g(\bar{y})$. Taking into account the relation $D_L^* G(\bar{y}) = (G'(\bar{y}))^\top$, the result follows from the preceding corollary. $\square$

Replacing Y with $Y \times Y$ and taking for G a continuously differentiable operation, in particular a continuous bilinear operation, one gets several calculus rules. In the next section we consider the case of the sum.

Exercises

1. Deduce from Exercise 2 of Sect. 6.3.3 the following implications.
(a) If the multimap $C : X \times Z \rightrightarrows Y$ of (6.31) is lower semicontinuous at $((\bar{x}, \bar{z}), \bar{y})$ on E with a linear rate in the sense that there exists some $k > 0$ such that $d(\bar{y}, C(x,z)) \leq k \|(x,z) - (\bar{x}, \bar{z})\|$ for (x,z) near $(\bar{x}, \bar{z})$, and if Y is finite-dimensional, then

$$D_D^* F(\bar{x}, \bar{y}) \circ D_D^* G(\bar{y}, \bar{z}) \subset D_D^* E(\bar{x}, \bar{z}). \tag{6.39}$$

(b) If $C : X \times Z \rightrightarrows Y$ is lower semicontinuous at $((\bar{x}, \bar{z}), \bar{y})$ on E with a linear rate, then

$$D_F^* F(\bar{x}, \bar{y}) \circ D_F^* G(\bar{y}, \bar{z}) \subset D_F^* E(\bar{x}, \bar{z}). \tag{6.40}$$

Hint: Use the inclusion (and its directional variant)

$$N_F(F_Z, (\bar{x}, \bar{z}, \bar{y})) + N_F(G_X^{-1}, (\bar{x}, \bar{z}, \bar{y})) \subset N_F(C, (\bar{x}, \bar{z}, \bar{y})). \tag{6.41}$$

(c) Combine these results with Lemma 6.58 to get an exact expression for the coderivatives of the composition $E := G \circ F$.

(**d**) Deduce from what precedes conditions ensuring that E is soft (resp. F-soft) at $(\bar{x},\bar{z})$ when F is soft (resp. F-soft) at $(\bar{x},\bar{y})$ and G is soft (resp. F-soft) at $(\bar{y},\bar{z})$.

2. Assuming that $C : X \times Z \rightrightarrows Y$ is lower semicontinuous at $((\bar{x},\bar{z}),\bar{y})$ on E with a linear rate and

$$\forall y^* \in Y^*, \qquad D_L^* F(\bar{x},\bar{y})(y^*) \times D_L^* G^{-1}(\bar{z},\bar{y})(-y^*) \subset D_F^* C(\bar{x},\bar{z},\bar{y})(0), \qquad (6.42)$$

or C is pseudo-Lipschitzian around $((\bar{x},\bar{z}),\bar{y})$ on E and

$$\forall y^* \in Y^*, \qquad D_L^* F(\bar{x},\bar{y})(y^*) \times D_L^* G^{-1}(\bar{z},\bar{y})(-y^*) \subset D_M^* C(\bar{x},\bar{z},\bar{y})(0), \qquad (6.43)$$

show that $D_L^* F(\bar{x},\bar{y}) \circ D_L^* G(\bar{y},\bar{z}) \subset D_L^* E(\bar{x},\bar{z})$.

3. (**a**) Assuming that for some subset B of $C(\bar{x},\bar{z})$ one has

$$\bigcap_{\bar{y} \in C(\bar{x},\bar{z})} D_D^* C(\bar{x},\bar{z},\bar{y})(0) \subset \bigcup_{\bar{y} \in B} \bigcup_{y^* \in Y^*} D_D^* F(\bar{x},\bar{y})(y^*) \times D_D^* G^{-1}(\bar{z},\bar{y})(-y^*), \qquad (6.44)$$

show that

$$D_D^*(G \circ F)(\bar{x},\bar{z}) \subset \bigcup_{\bar{y} \in B} D_D^* F(\bar{x},\bar{y}) \circ D_D^* G(\bar{y},\bar{z}). \qquad (6.45)$$

(**b**) Assuming that for some subset B of $C(\bar{x},\bar{z})$ one has

$$\bigcap_{\bar{y} \in C(\bar{x},\bar{z})} D_F^* C(\bar{x},\bar{z},\bar{y})(0) \subset \bigcup_{\bar{y} \in B} \bigcup_{y^* \in Y^*} D_F^* F(\bar{x},\bar{y})(y^*) \times D_F^* G^{-1}(\bar{z},\bar{y})(-y^*), \qquad (6.46)$$

show that

$$D_F^*(G \circ F)(\bar{x},\bar{z}) \subset \bigcup_{\bar{y} \in B} D_F^* F(\bar{x},\bar{y}) \circ D_F^* G(\bar{y},\bar{z}). \qquad (6.47)$$

4. Suppose (6.46) (resp. (6.44)) holds and F is a single-valued map that is stable at $\bar{x}$ or G^{-1} is a single-valued map that is stable at $\bar{z}$. Show that (6.47) holds (resp. (6.45) holds if Y is finite-dimensional).

In particular, when (6.46) (resp. (6.44)) holds and F is a single-valued map that is Fréchet differentiable (resp. Hadamard differentiable and stable) at $\bar{x}$, one has $D_F^* E(\bar{x},\bar{z}) = F'(\bar{x})^\mathsf{T} \circ D_F^* G(\bar{y},\bar{z})$ (resp. $D^* E(\bar{x},\bar{z}) = F'(\bar{x})^\mathsf{T} \circ D^* G(\bar{y},\bar{z})$), and when G^{-1} is a single-valued map that is Fréchet differentiable (resp. Hadamard differentiable and stable) at $\bar{z}$, one has $D_F^* E(\bar{x},\bar{z}) = D_F^* F(\bar{x},\bar{y}) \circ ((G^{-1})'(\bar{z})^\mathsf{T})^{-1}$ (resp. $D_F^* E(\bar{x},\bar{z}) = D_F^* F(\bar{x},\bar{y}) \circ ((G^{-1})'(\bar{z})^\mathsf{T})^{-1})$.

Get a similar assertion for $D_L^* E(\bar{x},\bar{z})$ under circa-differentiability of F or G^{-1}.

5. Let $X = Y = Z = \mathbb{R}$ and let F,G be defined by $F(x) = \{0,x\}$ for $x \in X$ and $G(y) = \{|y-1|\}$ for $y \in Y$. Check that $E(x) := G(F(x)) = \{1,|x-1|\}$ for $x \in X$ and $C = \mathbb{R} \times \{(1,0)\} \cup \{(x,|x-1|,x) : x \in \mathbb{R}\}$. Let $(\bar{x},\bar{z}) = (2,1)$, $\bar{y}_1 = 0$, and $\bar{y}_2 = 2$, so that $C(\bar{x},\bar{z}) = \{\bar{y}_1,\bar{y}_2\}$. If D^* stands for D_D^*, D_F^*, D_L^*, check that for every $y^* \in \mathbb{R}$, $z^* \in \mathbb{R}$, $D^*F(\bar{x},\bar{y}_1)(y^*) = \{0\}$, $D^*F(\bar{x},\bar{y}_2)(y^*) = \{y^*\}$, $D^*G(\bar{y}_1,\bar{z})(z^*) = \{-z^*\}$, $D^*G(\bar{y}_2,\bar{z})(z^*) = \{z^*\}$. Prove that $D^*C(\bar{x},\bar{z},\bar{y}_1)(0) = \{0\} \times \mathbb{R}$ and $D^*C(\bar{x},\bar{z},\bar{y}_2)(0) = \{(x^*,z^*) : x^* + z^* = 0\}$. Therefore, (6.44), (6.46), and (6.33) are satisfied, so that (6.45), (6.47), and (6.34) hold.

Check directly these inclusions by showing that $D^*E(\bar{x},\bar{z}) = D_F^*E(\bar{x},\bar{z}) = \{(0,0)\}$ and $D_L^*E(\bar{x},\bar{z})(z^*) = \{0,z^*\}$ for every $z^* \in \mathbb{R}$.

6. The purposes of this exercise and of the next one are to show that conditions (6.32) and (6.33) are slightly more general than condition (6.37).

Let $X = Y = Z = \mathbb{R}$ and let F,G be given by $F(x) = \{0\}$ for $x \in X$, $G(y) := [0,y]$ for $y \in \mathbb{R}_+$, $G(y) := \varnothing$ otherwise, so that $E := G \circ F = X \times \{0\}$. For $\bar{x} = 0$, $\bar{y} = 0$, $\bar{z} = 0$, show that the qualification condition (6.37) does not hold, since

$$(D_M^*F(\bar{x},\bar{y}))^{-1}(0) \cap D_M^*G(\bar{y},\bar{z})(0) = (-\infty,0].$$

[Hint: Check that $D^*F(\bar{x},\bar{y})(y^*) = \{0\}$ for all $y^* \in \mathbb{R}$, $D^*G(\bar{y},\bar{z})(z^*) = (-\infty,0]$ for $z^* \in \mathbb{R}_+$, $D^*G(\bar{y},\bar{z})(z^*) = (-\infty,-z^*]$ else and $D^*E(\bar{x},\bar{z})(z^*) = \{0\}$, similar relations holding for the Fréchet, the limiting, and the mixed coderivatives.]

On the other hand, check that $C = (\mathbb{R} \times \{0\}) \times \{0\}$, so that $N_L(C,(\bar{x},\bar{z},\bar{y})) = \{0\} \times \mathbb{R}^2$, while $N_L(F_Z,(0,0,0)) = \{0\} \times \{0\} \times \mathbb{R}$, $N_L(G_X^{-1},(0,0,0)) = \{(0,z^*,y^*) : y^* \leq 0, z^* \leq -y^*\}$ and (6.32) and (6.33) are satisfied.

7. Let $X = Y = Z = \mathbb{R}$, $F(x) = \{x\}$ for $x \in \{0\} \cup \{a_n : n \in \mathbb{N}\}$, $F(x) = \varnothing$ otherwise, where (a_n) is a decreasing sequence with limit 0 and $G(y) = \mathbb{R}_+$ for $x \in \mathbb{R}_+$, $G(y) = \varnothing$ otherwise. Let $\bar{x} = 0$, $\bar{y} = 0$, $\bar{z} = 0$. Check that $D_L^*F(\bar{x},\bar{y})(y^*) = \mathbb{R}$ for all $y^* \in \mathbb{R}$, $D_L^*G(\bar{y},\bar{z})(z^*) = \mathbb{R}_-$ for $z^* \in \mathbb{R}_+$, $D_L^*G(\bar{y},\bar{z})(z^*) = \varnothing$ otherwise, so that $D_M^*F(\bar{x},\bar{y})^{-1}(0) \cap D_M^*G(\bar{y},\bar{z})(0) = \mathbb{R}_- \neq \{0\}$. On the other hand, check that $C(x,z) = F(x)$ for $(x,z) \in \mathbb{R} \times \mathbb{R}_+$, $C(x,z) = \varnothing$ otherwise, and $N_L(C,(\bar{x},\bar{z},\bar{y})) = \mathbb{R} \times \mathbb{R}_- \times \mathbb{R}$, so that $D_L^*C(\bar{x},\bar{z},\bar{y})(0) = \mathbb{R} \times \mathbb{R}_-$ and thus condition (6.33) holds with $B := \{0\}$.

6.3.6 Coderivatives of Sums

Now we turn to the case of the sum $S := F_1 + F_2$ of two multimaps $F_1,F_2 : X \rightrightarrows Y$. There are several ways of reducing the computation of the coderivatives of a sum to the case of a composition (Exercise 1). We use the decomposition $S = G \circ F$, where

$$F := (F_1,F_2) : x \rightrightarrows F_1(x) \times F_2(x), \quad G : (y_1,y_2) \mapsto y_1 + y_2.$$

We introduce the corresponding resultant $C : X \times Y \rightrightarrows Y^2$ by

$$C(x,z) := \{(y_1,y_2) \in F_1(x) \times F_2(x) : y_1 + y_2 = z\}, \qquad (x,z) \in X \times Y.$$

Here and in the sequel, $\bar{x} \in X$, $\bar{y}_1 \in F_1(\bar{x})$, $\bar{y}_2 \in F_2(\bar{x})$, $\bar{z} := \bar{y}_1 + \bar{y}_2$.

An assumption about $F := (F_1, F_2)$ yields a useful inclusion for $D_L^*(F_1 + F_2)$.

Theorem 6.64. *Suppose that F_1 and F_2 have closed graphs, C is lower semicontinuous at $((\bar{x}, \bar{z}), B)$ on S for some subset B of $C(\bar{x}, \bar{z})$, and for every $\bar{y} := (\bar{y}_1, \bar{y}_2) \in B$, one has*

$$D_L^*(F_1, F_2)(\bar{x}, (\bar{y}_1, \bar{y}_2))(y_1^*, y_2^*) \subset D_L^* F_1(\bar{x}, \bar{y}_1)(y_1^*) + D_L^* F_2(\bar{x}, \bar{y}_2)(y_2^*). \tag{6.48}$$

Then for all $z^* \in Y^*$, one has the inclusion

$$D_L^*(F_1 + F_2)(\bar{x}, \bar{z})(z^*) \subset \bigcup_{(\bar{y}_1, \bar{y}_2) \in B} D_L^* F_1(\bar{x}, \bar{y}_1)(z^*) + D_L^* F_2(\bar{x}, \bar{y}_2)(z^*). \tag{6.49}$$

Proof. This is a consequence of Corollary 6.63, G being linear continuous, hence circa-differentiable, with $(G'(\bar{y}))^\mathsf{T}(z^*) = (z^*, z^*)$, so that Corollary 6.63 yields

$$\forall z^* \in Y^* \quad D_L^* S(\bar{x}, \bar{z})(z^*) \subset \bigcup_{\bar{y} \in B} \left(D_L^* F(\bar{x}, \bar{y}) \circ (G'(\bar{y}))^\mathsf{T} \right)(z^*) = \bigcup_{\bar{y} \in B} D_L^* F(\bar{x}, \bar{y})(z^*, z^*).$$

Then (6.48) entails (6.49). $\qquad\square$

Example. It is easy to see that when F_1 and F_2 are the epigraph multimaps associated with lower semicontinuous functions f_1 and f_2 respectively, then C is lower semicontinuous on S at $((\bar{x}, \bar{z}), B)$ for every $B \subset C(\bar{x}, \bar{z})$ for $\bar{z} \ge f_1(\bar{x}) + f_2(\bar{x})$. $\qquad\square$

Relation (6.48) can be replaced with the following one, in which we set $M_1(x) := F_1(x) \times Y$, $M_2(x) := Y \times F_2(x)$: for all $(\bar{y}_1, \bar{y}_2) \in B$,

$$N_L((F_1, F_2), (\bar{x}, \bar{y}_1, \bar{y}_2)) \subset N_L(M_1, (\bar{x}, \bar{y}_1, \bar{y}_2)) + N_L(M_2, (\bar{x}, \bar{y}_1, \bar{y}_2)).$$

Taking into account Corollary 6.51, we get the following corollary.

Corollary 6.65. *Let X, Y_1, Y_2 be Asplund spaces and let the multimaps $F_1 : X \rightrightarrows Y_1$, $F_2 : X \rightrightarrows Y_2$ have closed graphs. If C is lower semicontinuous at $((\bar{x}, \bar{z}), B)$ on S for some subset B of $C(\bar{x}, \bar{z})$ and if for every $\bar{y} := (\bar{y}_1, \bar{y}_2) \in B$, F_1 and F_2 are cooperative at $(\bar{x}, \bar{y}_1, \bar{y}_2)$, then for all $z^* \in Y^*$, relation (6.49) holds. In particular, (6.49) holds whenever F_1 and F_2 are coordinated at $(\bar{x}, \bar{y}_1, \bar{y}_2)$ for all $\bar{y} := (\bar{y}_1, \bar{y}_2) \in B$ and*

$$(-D_M^* F_1(\bar{x}, \bar{y}_1))(0) \cap D_M^* F_2(\bar{x}, \bar{y}_2)(0) = \{0\}. \tag{6.50}$$

Along with Exercise 2 below, Theorem 6.64 yields sum rules in equality form by combining the assumptions. We leave this task to the reader, but we focus on a case of special interest.

Corollary 6.66. *Suppose F_2 is a single-valued map that is circa-differentiable at $\bar{x}$. Let $y_2 \in Y$ be such that $F_2(\bar{x}) = \{y_2\}$. Then for all $\bar{y}_1 \in F_1(\bar{x})$ and all $y^* \in Y^*$,*

$$D_L^* S(\bar{x}, \bar{y}_1 + \bar{y}_2)(y^*) = D_L^* F_1(\bar{x}, \bar{y}_1)(y^*) + F_2'(\bar{x})^\mathsf{T}(y^*).$$

Proof. It is easy to see that the multimaps F_1, F_2 are coordinated at $(\bar{x}, \bar{y}_1, \bar{y}_2)$ and (6.50) holds. Moreover, C is lower semicontinuous at $((\bar{x}, \bar{z}), (\bar{y}_1, \bar{y}_2))$ on S. Theorem 6.64 yields

$$D_L^*(F_1 + F_2)(\bar{x}, \bar{z})(z^*) \subset D_L^* F_1(\bar{x}, \bar{y}_1)(z^*) + F_2'(\bar{x})^\mathsf{T}(z^*).$$

Since $F_1(x) = S(x) - F_2(x)$, a similar argument proves the reverse inclusion.   □

Exercises

1. Show that for $F_1, F_2 : X \rightrightarrows Y$, where X and Y are normed spaces, the sum $S := F_1 + F_2$ can be written $F_1 + F_2 = Q \circ P$, where $P : x \rightrightarrows \{x\} \times F_1(x)$ and $Q : (x, y) \rightrightarrows y + F_2(x)$. Check that the corresponding resultant R is given by $R(x, z) = \{x\} \times R_1(x, z)$ with $R_1(x, z) := F_1(x) \cap (z - F_2(x))$. Give results similar to Theorem 6.64 and its Corollaries using the decomposition $F_1 + F_2 = Q \circ P$. (See [658].)

2. (a) Let $F_1, F_2 : X \rightrightarrows Y$, $S := F_1 + F_2$. Assuming that Y is finite-dimensional and the multimap $C : X \times Y \rightrightarrows Y \times Y$ is lower semicontinuous at $((\bar{x}, \bar{z}), (\bar{y}_1, \bar{y}_2))$ on S with a linear rate, show that

$$D_D^* F_1(\bar{x}, \bar{y}_1) + D_D^* F_2(\bar{x}, \bar{y}_2) \subset D_D^* S(\bar{x}, \bar{y}_1 + \bar{y}_2). \tag{6.51}$$

(b) Assuming that C is lower semicontinuous at $((\bar{x}, \bar{z}), (\bar{y}_1, \bar{y}_2))$ on S with a linear rate, show that

$$D_F^* F_1(\bar{x}, \bar{y}_1) + D_F^* F_2(\bar{x}, \bar{y}_2) \subset D_F^* S(\bar{x}, \bar{y}_1 + \bar{y}_2). \tag{6.52}$$

(c) Assuming that C is lower semicontinuous at $((\bar{x}, \bar{z}), (\bar{y}_1, \bar{y}_2))$ on S with a linear rate and that

$$\forall y^* \in Y^*, \quad (D_L^* F_1(\bar{x}, \bar{y}_1)(y^*) + D_L^* F_2(\bar{x}, \bar{y}_2)(y^*), -y^*) \subset D_F^* R_1((\bar{x}, \bar{z}), \bar{y}_1)(0),$$
$$\tag{6.53}$$

where R_1 is defined in Exercise 1, show that

$$D_L^* F_1(\bar{x}, \bar{y}_1) + D_L^* F_2(\bar{x}, \bar{y}_2) \subset D_L^* S(\bar{x}, \bar{y}_1 + \bar{y}_2). \tag{6.54}$$

3. Show by the following example that condition (6.48) may be weaker than condition (6.50). Let $X = Y = Z = \mathbb{R}$, $F_1 = \mathbb{R}_- \times \mathbb{R}_-$, $F_2 = \mathbb{R}_+ \times \mathbb{R}_+$, and $(\bar{x}, \bar{z}) = (0,0)$, $(\bar{y}_1, \bar{y}_2) = (0,0)$. Check that $D_L^* F_1(\bar{x}, \bar{y}_1) = \mathbb{R}_- \times \mathbb{R}_+$ and $D_L^* F_2(\bar{x}, \bar{y}_2) = \mathbb{R}_+ \times \mathbb{R}_-$, whence $(-D_M^* F_1(\bar{x}, \bar{y}_1)(0)) \cap D_M^* F_2(\bar{x}, \bar{y}_2)(0) = \mathbb{R}_- \neq \{0\}$. On the other hand, since $(F_1, F_2) = \{0\} \times \mathbb{R}_- \times \mathbb{R}_+$, check that $D_L^*(F_1, F_2)(\bar{x}, \bar{y}_1, \bar{y}_2) = \mathbb{R}_- \times \mathbb{R}_+ \times \mathbb{R}$, and condition (6.48) holds.

6.4 General Subdifferential Calculus

It is easy to give examples showing that without Lipschitzian assumptions or compactness assumptions, the sum rule or the chain rule of Sect. 6.1.4 may fail.

Example. Let $f, g : \mathbb{R} \to \mathbb{R}$ be given by $f(x) := x^{1/3}$, $g := (-f)^+$. Then $\partial_L f(0) = \varnothing$, $\partial_L g(0) = \mathbb{R}_-$, and $\partial_L(f+g)(0) = \mathbb{R}_+$.

Happily, under appropriate assumptions, the rules for coderivatives in the preceding subsections entail rules for functions in the class $\mathscr{F}$ of proper lower semicontinuous functions. We start with metric estimates.

Theorem 6.67 (Chain rule for limiting subdifferentials). *Let X and Y be Asplund spaces, let $g : X \to Y$ be continuous around $\bar{x} \in X$, let $h \in \mathscr{F}(Y)$, let $f := h \circ g$, and let $\bar{y} := g(\bar{x})$, $\bar{z} := h(\bar{y})$. Suppose g and h are coherent around $(\bar{x}, \bar{y}, \bar{z})$ in the sense that there exist some $c > 0$, $\rho > 0$ such that for all $(x, y, r) \in B((\bar{x}, \bar{y}, \bar{z}), \rho)$ one has*

$$d((x, r), \operatorname{epi} f) \leq cd((y, r), \operatorname{epi} h) + cd((x, y), \operatorname{gph} g). \tag{6.55}$$

Then one has the inclusions

$$\partial_L(h \circ g)(\bar{x}) \subset D_L^* g(\bar{x})(\partial_L h(\bar{y})), \tag{6.56}$$

$$\partial_L^\infty(h \circ g)(\bar{x}) \subset D_L^* g(\bar{x})(\partial_L^\infty h(\bar{y})). \tag{6.57}$$

Proof. Let $S := \{(x, y, r) \in X \times Y \times \mathbb{R} : (x, r) \in \operatorname{epi} f\}$ and let $j : X \times Y \times \mathbb{R} \to \mathbb{R}$ be given by $j(x, y, r) := cd((y, r), \operatorname{epi} h) + cd((x, y), \operatorname{gph} g)$, so that $d(\cdot, S) \leq j$ around $(\bar{x}, \bar{y}, \bar{z})$. The conclusion is a consequence of Proposition 6.21 with $V = W := X \times Y \times \mathbb{R}$, the sum rule, and the relations

$$\partial_L d_{\operatorname{epi} f}(\bar{x}, \bar{z}) = \{(x^*, z^*) : ((x^*, 0, z^*) \in \partial_L d_S(\bar{x}, \bar{z})\},$$

$$\partial_L d_{\operatorname{epi} h}(\bar{y}, \bar{z}) = \{(y^*, z^*) : ((0, y^*, z^*) \in \partial_L d_{X \times \operatorname{epi} h}(\bar{x}, \bar{y}, \bar{z})\},$$

$$\partial_L d_{\operatorname{gph} g}(\bar{x}, \bar{y}) = \{(x^*, y^*) : ((x^*, y^*, 0) \in \partial_L d_{\operatorname{gph} g \times \mathbb{R}}(\bar{x}, \bar{y}, \bar{z})\}.$$

One can also use Proposition 6.21 with $V := X \times Y \times \mathbb{R}$, $W := X \times \mathbb{R}$, A being the canonical projection, observing that $j(x_n, g(x_n), r_n) = 0$ whenever $(x_n, r_n) \in \operatorname{epi} f$. $\square$

Now let us consider infinitesimal assumptions.

Theorem 6.68 (Chain rule for limiting subdifferentials). *Let X and Y be Asplund spaces, let $g : X \to Y$ be continuous around $\bar{x} \in X$, and let $h \in \mathcal{F}(Y)$ be such that g and the epigraph multimap E_h associated with h are cooperative at $(\bar{x}, \bar{y}, \bar{z})$ for $\bar{y} := g(\bar{x})$, $\bar{z} := h(\bar{y})$. Then inclusions (6.56) and (6.57) hold.*

In particular, if g and E_h are coordinated at $(\bar{x}, \bar{y}, \bar{z})$ and if the following qualification condition is satisfied, then these inclusions hold:

$$(-D_M^* g^{-1}(\bar{y}, \bar{x})(0)) \cap \partial_L^\infty h(\bar{y}) = \{0\}. \tag{6.58}$$

Proof. Let us apply Theorem 6.60 and Corollary 6.61 with $Z := \mathbb{R}$, $F := g$, $G := E_h$, $B := \{\bar{y}\}$. Since g is continuous, the resultant multimap C is lower semicontinuous at $((\bar{x}, \bar{z}), \bar{y})$ on the epigraph of $h \circ g$. Condition (6.58) transcribes (6.37). Inclusion (6.34) yields (6.56) and (6.57) when applied to 1 and 0 respectively. $\qquad\square$

Using Corollary 4.130, the cooperative property can be made explicit as follows: whenever $(x_n) \to \bar{x}$, $(x_n^*) \to 0$, $(y_n) \to \bar{y}$, $(s_n) \to 0_+$, (v_n^*), (y_n^*) in Y^* are such that $(v_n^* - y_n^*) \to 0$, $x_n^* \in D_F^* g(x_n)(v_n^*)$, $y_n^* \in s_n \partial_F h(y_n)$ for all $n \in \mathbb{N}$, one has $(y_n^*) \to 0$. For the coordination property one adds $(y_n^*) \overset{*}{\to} y_n^*$ in the assumptions.

Using Proposition 6.40, we get the following consequence.

Corollary 6.69. *Let X and Y be Asplund spaces, let $g : X \to Y$ be continuous around $\bar{x} \in X$, and let $h \in \mathcal{F}(Y)$ be such that g and the epigraph multimap E_h associated to h satisfy condition (6.58) for $\bar{y} := g(\bar{x})$, $\bar{z} := h(\bar{y})$. Suppose that either g^{-1} is coderivatively compact at $(\bar{x}, \bar{y})$ or h is subdifferentially compact at $\bar{y}$. Then inclusions (6.56) and (6.57) hold.*

Let us turn to sums of functions. Let us say that a family $(f_1, \dots, f_k)$ of functions on X is *coordinated at $\bar{x}$* if the family $(F_1, \dots, F_k)$ is coordinated at $(\bar{x}, f_1(\bar{x}), \dots, f_k(\bar{x}))$, where F_i is the epigraph multimap associated with f_i. Observing that the epigraph multimap F associated with $f := f_1 + \cdots + f_k$ is the sum of the multimaps $F_1, \dots, F_k$, we derive the next statement from the last corollary and the example following Theorem 6.64; we also use Proposition 6.40.

Theorem 6.70. *Let X be an Asplund space and let $f_1, f_2 \in \mathcal{F}(X)$ be finite at $\bar{x}$ and such that*

$$\partial_L^\infty f_1(\bar{x}) \cap (-\partial_L^\infty f_2(\bar{x})) = \{0\}. \tag{6.59}$$

Then if f_1 and f_2 are coordinated at $\bar{x}$, in particular if f_1 or f_2 is subdifferentially compact at $\bar{x}$, for $f := f_1 + f_2$ one has

$$\partial_L f(\bar{x}) \subset \partial_L f_1(\bar{x}) + \partial_L f_2(\bar{x}), \tag{6.60}$$

$$\partial_L^\infty f(\bar{x}) \subset \partial_L^\infty f_1(\bar{x}) + \partial_L^\infty f_2(\bar{x}). \tag{6.61}$$

Proof. It suffices to observe that condition (6.59) amounts to (6.50) and to apply (6.49) to $z^* = 1$ and $z^* = 0$, with $B := \{(f_1(\bar{x}), f_2(\bar{x}))\}$. $\qquad\square$

Corollary 6.71. *Let X be an Asplund space and let $f = f_1 + \cdots + f_k$, where $f_i \in \mathscr{F}(X)$ is finite at $\bar{x}$ for $i \in \mathbb{N}_k$ and such that*

$$x_1^* + \cdots + x_k^* = 0, \ x_i^* \in \partial_L^\infty f_i(\bar{x}) \implies x_i^* = 0, \ i \in \mathbb{N}_k. \tag{6.62}$$

If all f_i's but one are subdifferentially compact at $\bar{x}$, then

$$\partial_L f(\bar{x}) \subset \partial_L f_1(\bar{x}) + \cdots + \partial_L f_k(\bar{x}), \tag{6.63}$$

$$\partial_L^\infty f(\bar{x}) \subset \partial_L^\infty f_1(\bar{x}) + \cdots + \partial_L^\infty f_k(\bar{x}). \tag{6.64}$$

Proof. The epigraph multimap E_f of f is the sum of the epigraph multimaps $F_i := E_{f_i}$ for $i \in \mathbb{N}_k$. The case $k = 2$ stems from the theorem, since (6.62) reduces to (6.59). Then assuming that $f_2, \ldots, f_k$ are subdifferentially compact at $\bar{x}$, an induction on k yields the result, since the family $f_1, \ldots, f_{k-1}$ satisfies (6.62) with k changed into $k-1$, so that by (6.64), the functions $g := f_1 + \cdots + f_{k-1}$ and f_k satisfy (6.59). □

Exercise. Find connections between Theorem 6.68 and Theorem 6.70.

Exercise. Deduce from a sum rule for functions a rule for the coderivative of a composition $H := G \circ F$ using the indicator functions of F, G, H. [Hint: Use relation (4.18).]

Exercise. Let X and Y be Asplund spaces, let $g : X \to Y$ be continuous around $\bar{x} \in X$, let $h \in \mathscr{F}(X \times Y)$, and let $f \in \mathscr{F}(X)$ be given by $f(x) := h(x, g(x))$. Set $\bar{y} := g(\bar{x})$. Give conditions ensuring that

$$\partial_L f(\bar{x}) \subset \{x^* + \partial_L(y^* \circ g)(\bar{x}) : (x^*, y^*) \in \partial_L h(\bar{x}, \bar{y})\},$$

$$\partial_L^\infty f(\bar{x}) \subset \{x^* + \partial_L^\infty(y^* \circ g)(\bar{x}) : (x^*, y^*) \in \partial_L^\infty h(\bar{x}, \bar{y})\}.$$

6.5   Error Bounds and Metric Estimates

The limiting concepts presented above enable one to give nice statements concerning metric estimates. These concepts can be slightly refined. We devote the next subsection to such a refinement.

6.5.1   *Upper Limiting Subdifferentials and Conditioning*

We saw in Chap. 1 that for a function $f : X \to \overline{\mathbb{R}}_+$ and $\bar{x} \in S := f^{-1}(\{0\})$ the conditioning rate $\gamma_f(\bar{x}) := \liminf_{x \to \bar{x}, \ x \in X \setminus S} f(x)/d_S(x)$ of f at $\bar{x}$ can be estimated using a decrease index and the decrease principle. Since $\delta_{f,F}$ given by $\delta_{f,F}(x) := \inf\{\|x^*\| : x^* \in \partial_F f(x)\}$ is a decrease index and since $\partial_F f(x) \subset \partial_L f(x)$, it follows that $\delta_{f,L}$ given by $\delta_{f,L}(x) := \inf\{\|x^*\| : x^* \in \partial_L f(x)\}$ is also a decrease index. The following notion is a more refined tool.

Definition 6.72. The *upper limiting subdifferential* of $f : X \to \overline{\mathbb{R}}$ at $\overline{x} \in f^{-1}(\mathbb{R})$ is the set

$$\partial_L^> f(\overline{x}) := \{\overline{x}^* : \exists\, (x_n) \to_f \overline{x},\ (x_n^*) \overset{*}{\to} \overline{x}^*,\ x_n^* \in \partial_F f(x_n),\ f(x_n) > f(\overline{x})\}.$$

Let us note that $\partial_L^> f(\overline{x})$ may be much smaller than $\partial_L f(\overline{x})$, as the next example shows, and is more appropriate to the study of conditioning than $\partial_L f(\overline{x})$.

Example. Let S be a closed subset of a finite-dimensional space X and let $\overline{x} \in S$, $f := d_S$. Then for every $\overline{x}^* \in \partial_L^> f(\overline{x})$ one has $\|\overline{x}^*\| = 1$, while $0 \in \partial_L f(\overline{x})$. Thus, the replacement of $\partial_L^> f$ by $\partial_L f$ in the next theorem would not be possible.

However, $\partial_L^> f(\overline{x})$ cannot be taken as a substitute for $\partial_L f(\overline{x})$ in every respect. It may be empty, even for a Lipschitzian function f on $\mathbb{R}$ such as $-|\cdot|$, with $\overline{x} := 0$. Its calculus rules are poorer than those for ∂_L.

Exercise. Consider rules for $\partial_L^>(h \circ g)$, $\partial_L^>(f_1 + f_2)$, $\partial_L^> \max(f_1, f_2)$.

When X is an Asplund space and $f := d_S$, one can give a refined form of Lemma 6.7, showing that elements of $\partial_L^> d_S(\overline{x})$ are weak* limits of approximate normals.

Lemma 6.73. *Let S be a closed subset of an Asplund space X and let $\overline{x}^* \in \partial_L^> d_S(\overline{x})$ with $\overline{x} \in S$. Then there exist sequences $(x_n) \to \overline{x}$ in S, (u_n) in S_X, $(t_n) \to 0_+$, $(x_n^*) \overset{*}{\to} \overline{x}^*$ such that $(\langle x_n^*, u_n \rangle) \to 1$, $(t_n^{-1} d_S(x_n + t_n u_n)) \to 1$, $x_n^* \in \partial_F d_S(x_n)$ for all $n \in \mathbb{N}$.*

Proof. Given $\overline{x}^* \in \partial_L^> d_S(\overline{x})$, let $(w_n) \to_{X\backslash S} \overline{x}$, $(w_n^*) \overset{*}{\to} \overline{x}^*$ with $w_n^* \in \partial_F d_S(w_n)$ for all n. Theorem 4.74 yields some $x_n \in S$, $x_n^* \in \partial_F d_S(x_n)$ such that $\|x_n^* - w_n^*\| \le 2^{-n}$, $t_n := \|x_n - w_n\| \le d_S(w_n)(1 + 2^{-n})$, $|\langle x_n^*, w_n - x_n \rangle - d_S(w_n)| \le t_n 2^{-n}$ for all n. Setting $u_n := t_n^{-1}(w_n - x_n)$, we have $(t_n) \to 0_+$, $(t_n^{-1} d_S(w_n)) \to 1$, $(\langle x_n^*, u_n \rangle) \to 1$. $\qquad\square$

A direct proof. Given $\overline{x}^* \in \partial_L^> d_S(\overline{x})$, let $(w_n) \to x$ in $X \setminus S$, $(w_n^*) \overset{*}{\to} \overline{x}^*$ with $w_n^* \in \partial_F d_S(w_n)$ for all $n \in \mathbb{N}$. Given a sequence $(\varepsilon_n) \to 0$ in $(0, 1/3)$, let $\delta_n \in (0, \varepsilon_n)$, $\delta_n \le \varepsilon_n d_S(w_n)$ be such that

$$\forall w \in B[w_n, 2\delta_n],\quad \langle w_n^*, w - w_n \rangle \le d_S(w) - d_S(w_n) + \varepsilon_n \|w - w_n\|. \tag{6.65}$$

Let us pick $y_n \in S$ such that $\|y_n - w_n\| \le d_S(w_n) + \delta_n^2$. Then for all $x \in S \cap B[y_n, 2\delta_n]$, taking $w := w_n + x - y_n$, we have

$$d_S(w_n + x - y_n) \le \|(w_n + x - y_n) - x\| = \|w_n - y_n\| \le d_S(w_n) + \delta_n^2,$$

$$\langle w_n^*, x - y_n \rangle = \langle w_n^*, w - w_n \rangle \le \delta_n^2 + \varepsilon_n \|x - y_n\|.$$

Setting $f_n(x) := \langle w_n^*, y_n - x \rangle + \varepsilon_n \|x - y_n\|$, so that $f_n \geq -\delta_n^2$ on $S \cap B[w_n, 2\delta_n]$ and $f_n(y_n) = 0$, the Ekeland variational principle yields some $v_n \in S \cap B[y_n, \delta_n]$ such that for all $x \in S \cap B[y_n, 2\delta_n]$,

$$\langle w_n^*, y_n - v_n \rangle + \varepsilon_n \|v_n - y_n\| \leq \langle w_n^*, y_n - x \rangle + \varepsilon_n \|x - y_n\| + \delta_n \|x - v_n\|,$$

whence, simplifying both sides and using the triangle inequality, we obtain

$$\forall x \in S \cap B[y_n, 2\delta_n], \qquad \langle w_n^*, x - v_n \rangle \leq \varepsilon_n \|x - v_n\| + \delta_n \|x - v_n\| \leq 2\varepsilon_n \|x - v_n\|.$$

In particular, for all $x \in S \cap B[v_n, \delta_n]$, we have $\langle w_n^*, x - v_n \rangle \leq 2\varepsilon_n \|x - v_n\|$. Thus, the function g_n defined by $g_n(x) := 2\varepsilon_n \|x - v_n\| - \langle w_n^*, x - v_n \rangle + \iota_S(x)$ attains a local minimum at v_n. The fuzzy sum rule yields some $x_n \in S \cap B(v_n, \delta_n)$, $z_n^* \in N_F(S, x_n)$, $u_n^* \in B_{X^*}$ such that $\|2\varepsilon_n u_n^* - w_n^* + z_n^*\| \leq \varepsilon_n$. Then we have $\|z_n^* - w_n^*\| \leq 3\varepsilon_n$, and setting $t_n := \|y_n - w_n\|$, $u_n := t_n^{-1}(w_n - y_n)$, taking $w := w_n - \delta_n u_n$ in relation (6.65), we get, since $-d_S(w_n) \leq -t_n + \delta_n^2$ and $t_n \geq d_S(w_n) \geq \delta_n/\varepsilon_n > \delta_n$,

$$\delta_n \langle w_n^*, -u_n \rangle \leq \|w_n - \delta_n u_n - y_n\| - d_S(w_n) + \varepsilon_n \|\delta_n u_n\|$$
$$\leq t_n - \delta_n + (-t_n + \delta_n^2) + \varepsilon_n \delta_n = \delta_n(-1 + \delta_n + \varepsilon_n),$$

hence $\langle w_n^*, u_n \rangle \geq 1 - 2\varepsilon_n$. Observing that $\|z_n^*\| \leq \|w_n^*\| + 3\varepsilon_n = 1 + 3\varepsilon_n$ and $\|z_n^*\| \geq \|w_n^*\| - 3\varepsilon_n = 1 - 3\varepsilon_n > 0$, setting $x_n^* := z_n^*/\|z_n^*\|$, so that $x_n^* \in N_F(S, x_n) \cap B_{X^*} = \partial_F d_S(x_n)$, $\|x_n^* - z_n^*\| \leq 3\varepsilon_n$, we obtain $\langle x_n^*, u_n \rangle \geq \langle w_n^*, u_n \rangle - \|w_n^* - z_n^*\| - \|x_n^* - z_n^*\| \geq 1 - 8\varepsilon_n$ and $\langle x_n^*, u_n \rangle \leq 1$. Let $r_n := d_S(w_n)$, so that $r_n \leq t_n \leq r_n + \delta_n^2 \leq r_n + \varepsilon_n r_n$. Since $d_S(x_n + t_n u_n) \leq \|t_n u_n\| = t_n$ and

$$d_S(x_n + t_n u_n) = d_S(w_n + x_n - y_n) \geq d_S(w_n) - \|x_n - y_n\| \geq r_n - 2\delta_n \geq r_n - 2\varepsilon_n r_n,$$

we get $(t_n^{-1} r_n) \to 1$ and $(t_n^{-1} d_S(x_n + t_n u_n)) \to 1$. $\qquad\square$

Our main criterion for conditioning is displayed in the next result.

Theorem 6.74. *Let X be an Asplund space and let $f \in \mathscr{F}(X)$ be nonnegative. Then the conditioning rate $\gamma_f(\bar{x}) := \liminf_{x \to \bar{x}, x \in X \setminus S} f(x)/d_S(x)$ of f at $\bar{x} \in S := f^{-1}(0)$ satisfies $\gamma_f(\bar{x}) \geq d(0, \partial_L^> f(\bar{x}))$.*

Proof. By Theorems 1.114 and 4.80, setting $\bar{c} := d(0, \partial_L^> f(\bar{x}))$, it suffices to prove that for all $c \in (0, \bar{c})$ there exists $r > 0$ such that $\|x^*\| > c$ whenever $x^* \in \partial_F f(x)$ for some $x \in B(\bar{x}, 2r) \setminus S$ with $f(x) < cr$. If it is not the case, there exist some $c \in (0, \bar{c})$ and some sequences $(x_n) \to \bar{x}$ in $X \setminus S$, (x_n^*) in X^* such that $(f(x_n)) \to 0$ and $x_n^* \in \partial_F f(x_n) \cap cB_{X^*}$ for all $n \in \mathbb{N}$. Since cB_{X^*} is weak* sequentially compact, the sequence (x_n^*) has a subsequence that weak* converges to some $\bar{x}^*$. Then $\bar{x}^* \in \partial_L^> f(\bar{x}) \cap cB_{X^*}$, a contradiction to $c < d(0, \partial_L^> f(\bar{x}))$. $\qquad\square$

If one is not interested in a quantitative result, a qualitative variant may suffice.

Proposition 6.75. *Let X be an Asplund space and let $f \in \mathscr{F}(X)$ be nonnegative and such that for some $\bar{x} \in S := f^{-1}(0)$, one has $0 \notin \partial_L^> f(\bar{x})$. Then f is linearly conditioned around $\bar{x}$: there exist $c > 0$, $r > 0$ such that $d(x,S) \le f(x)/c$ for all $x \in B(\bar{x}, r)$.*

Proof. Again it suffices to prove that there exist $r, c > 0$ such that $\|x^*\| > c$ whenever $x^* \in \partial_F f(x)$ for some $x \in B(\bar{x}, 2r) \setminus S$ satisfying $f(x) < cr$. If it is not the case, one can find sequences $(x_n) \to \bar{x}$, $(x_n^*) \to 0$ such that $(f(x_n)) \to 0_+$ and $x_n^* \in \partial_F f(x_n)$ for all $n \in \mathbb{N}$. Then one gets $0 \in \partial_L^> f(\bar{x})$, contrary to our assumption. $\square$

Remark. We note that in the preceding proposition, one can replace $\partial_L^> f(\bar{x})$ by its variant $\partial_{L,s}^> f(\bar{x})$ in which the weak* convergence is replaced by strong convergence.

Remark. The pointwise character of the preceding results is remarkable. However, one must realize that it cannot replace the local result of Theorems 1.114, 4.80. An example showing the difference is provided by the choice for f of the norm on a separable Hilbert space X: then $0 \in \partial_L^> f(0)$, whereas $\|x^*\| = 1$ for all $x \in X \setminus S$ and $x^* \in \partial_F f(x)$. However, $0 \notin \partial_{L,s}^> f(0)$.

As a sample of the applications one can derive from Theorem 6.74, let us give the next two corollaries.

Corollary 6.76. *Let $\bar{x} \in S := g^{-1}(C)$, $\bar{y} := g(\bar{x})$, where $g : X \to Y$ is a continuous map between two Asplund spaces and C is a closed subset of Y. Suppose there exists $\bar{c} > 0$ such that $\|\bar{x}^*\| \ge \bar{c}$ for all $\bar{x}^* \in D_L^* g(\bar{x})(\partial_L^> d_C(\bar{y}))$. Then for all $c \in (0, \bar{c})$ there exists a neighborhood V of $\bar{x}$ such that for all $x \in V$ one has*

$$d(x,S) \le c^{-1} d(g(x),C), \tag{6.66}$$

$$N_L(S,\bar{x}) \subset D_L^* g(\bar{x})(N_L(C,\bar{y})). \tag{6.67}$$

Proof. Let $f := d(g(\cdot),C)$. By Theorem 6.74, to prove inequality (6.66), it suffices to show that $\inf\{\|x^*\| : x^* \in \partial_L^> f(\bar{x})\} \ge \bar{c}$, or, by our assumption, that $\partial_L^> f(\bar{x}) \subset D_L^* g(\bar{x})(\partial_L^> d_C(\bar{y}))$. Given $\bar{x}^* \in \partial_L^> f(\bar{x})$, let $(x_n) \to_f \bar{x}$, $(x_n^*) \overset{*}{\to} \bar{x}^*$ with $f(x_n) > 0$, $x_n^* \in \partial_F f(x_n)$ for all $n \in \mathbb{N}$. Then $g(x_n) \in Y \setminus C$ and we can find $((u_n, v_n)) \to (\bar{x}, g(\bar{x}))$, $(y_n) \to g(\bar{x})$, $((u_n^*, v_n^*))$, (y_n^*) satisfying $v_n = g(u_n)$, $y_n \in Y \setminus C$, $y_n^* \in \partial_F d_C(y_n)$, $u_n^* \in D_F^* g(u_n)$ for all n and $(\|u_n^* - x_n^*\|) \to 0$, $(\|v_n^* - y_n^*\|) \to 0$. Since (y_n^*) is contained in B_{Y^*}, taking subsequences, we may assume that (y_n^*) weak* converges to some $\bar{y}^* \in \partial_L^> d_C(\bar{y})$. Then $(v_n^*) \overset{*}{\to} \bar{y}^*$, $(u_n^*) \overset{*}{\to} \bar{x}^*$, so that $\bar{x}^* \in D_L^* g(\bar{x})(\bar{y}^*) \subset D_L^* g(\bar{x})(\partial_L^> d_C(\bar{y}))$.
   A similar proof shows that $\partial_L f(\bar{x}) \subset D_L^* g(\bar{x})(\partial_L d_C(\bar{y}))$. Then by Proposition 6.8 we get (6.67). $\square$

Corollary 6.77. *Let F and G be two closed subsets of an Asplund space and let $\bar{x} \in E := F \cap G$. Suppose there exists $\bar{c} > 0$ such that $\|\bar{y}^* + \bar{z}^*\| \ge \bar{c}$ for all $(\bar{y}^*, \bar{z}^*) \in (\partial_L^> d_F(\bar{x}) \times \partial_L d_G(\bar{x})) \cup (\partial_L d_F(\bar{x}) \times \partial_L^> d_G(\bar{x}))$. Then for all $c < \bar{c}$ there exists a neighborhood V of $\bar{x}$ such that for all $x \in V$ one has*

$$d(x, F \cap G) \leq c\, d(x, F) + c\, d(x, G),$$

$$N_L(F \cap G, \overline{x}) \subset N_L(F, \overline{x}) + N_L(G, \overline{x}).$$

Proof. Let $g : x \mapsto (x, x)$ be the diagonal map, so that $E = g^{-1}(F \times G)$. It is easy to see that $D_L^* g(\overline{x})(y^*, z^*) = y^* + z^*$ for all $y^*, z^* \in X^*$ and $\partial_L^> d_{F \times G}(\overline{x}, \overline{x}) \subset (\partial_L^> d_F(\overline{x}) \times \partial_L d_G(\overline{x})) \cup (\partial_L d_F(\overline{x}) \times \partial_L^> d_G(\overline{x}))$. Thus, the result is a consequence of the preceding corollary. $\square$

Exercise. Let X be an Asplund space and for $f : X \to \overline{\mathbb{R}}$ finite at x let

$$\partial_L^> f(x) := \{x^* : \exists (x_n) \subset S_f^>(x),\ x_n^* \in \partial_F f(x_n),\ (x_n) \to_f x,\ (x_n^*) \overset{*}{\to} x^*\},$$

where $S_f^>(x) := \{w \in X : f(w) \geq f(x)\}$. Note that $\partial_F f(x) \subset \partial_L^> f(x)$, $\partial_L^> f(x) \subset \partial_L^> f(x)$ and that when x is a local minimizer of f one has $\partial_L^> f(x) = \partial_L f(x)$. Check that $\partial_L^> f(x)$ is the sequential weak* closure of $\partial_F f(x)$ when x is a local strict maximizer of f.

Exercise. Let S be a nonempty closed subset of an Asplund space X and for $r > 0$ let $S_r := \{x \in X : d(x, S) \leq r\}$, $w \in X \setminus S$ such that $d(w, S) = r$.
(a) Show that $\mathbb{R}_+ \partial_L^> d_S(w) \subset N_L(S_r, w)$. [Hint: Use the facts that $d_S(x) = d_{S_r}(x) + r$ when $d_S(x) \geq r$ and that $\partial_L^> d_{S_r}(w) = \partial_L d_{S_r}(w)$.]
(b) Show that this inclusion is an equality when w is not a local maximizer of d_S.
(c) Check that when $S = S_X$ and $w = 0$ one has $\mathbb{R}_+ \partial_L^> d_S(w) = \varnothing$, $N_L(S_r, w) = \{0\}$.
(d) Show that $\partial_L^> d_S(w) \subset N_L(S_r, w) \setminus \{0\}$ when S_r is normally compact at w.

6.5.2   Application to Regularity and Openness

We devote the present subsection to pointwise criteria for openness, regularity, and pseudo-Lipschitz properties of multimaps.

Theorem 6.78. *Let Y and Z be Asplund spaces and let $M : Y \rightrightarrows Z$ be a multimap with closed graph. Then the following assertions are equivalent:*
 (a) M is pseudo-Lipschitzian around $(\overline{y}, \overline{z}) \in M$;
 (b) M is coderivatively compact at $(\overline{y}, \overline{z})$ and $\|D_M^* M(\overline{y}, \overline{z})\| < \infty$;
 (c) M is coderivatively compact at $(\overline{y}, \overline{z})$ and $D_M^* M(\overline{y}, \overline{z})(0) = \{0\}$.
 Moreover, under these conditions, the exact pseudo-Lipschitz rate $\mathrm{lip}(M, (\overline{y}, \overline{z}))$ of M at $(\overline{y}, \overline{z})$ satisfies the estimate $\|D_M^* M(\overline{y}, \overline{z})\| \leq \mathrm{lip}(M, (\overline{y}, \overline{z}))$.

Proof. (a)$\Rightarrow$(b) and the estimates of the exact pseudo-Lipschitz rate of M are given in Proposition 6.34. (b)$\Rightarrow$(c) is obvious from the relation $\|y^*\| \leq c\|z^*\|$ for $c :=$ $\|D_M^* M(\overline{y}, \overline{z})\|$, $z^* \in Z^*$, $y^* \in D_M^* M(\overline{y}, \overline{z})(z^*)$.
(c)$\Rightarrow$(a) In view of Theorem 4.134, it suffices to find some $c > 0$, $V \in \mathcal{N}(\overline{y})$, $W \in \mathcal{N}(\overline{z})$ such that for all $v \in V$, $w \in M(v) \cap W$, $w^* \in S_{Z^*}$, $v^* \in D_F^* M(v, w)(w^*)$

one has $\|v^*\| \le c$. If that is not possible, there exist sequences $(y_n) \to \bar{y}$, $(z_n) \to \bar{z}$, (w_n^*) in S_{Z^*}, (v_n^*) in Y^* such that $z_n \in M(y_n)$, $v_n^* \in D_F^* M(y_n, z_n)(w_n^*)$, $r_n := \|v_n^*\| \ge n$ for all n. Taking a subsequence if necessary, we may assume that $(y_n^*) := (r_n^{-1} v_n^*)$ has a weak* limit $y^* \in D_M^* M(\bar{y}, \bar{z})(0)$. Our assumption yields $y^* = 0$, and since M is coderivatively compact at $(\bar{y}, \bar{z})$ and $(w_n^*/r_n) \to 0$, we obtain $(y_n^*) \to 0$, a contradiction to $\|y_n^*\| = 1$. $\square$

From the equivalences of Theorem 1.139 one deduces the next criteria.

Theorem 6.79. *Let X and Y be Asplund spaces. For a multimap $F : X \rightrightarrows Y$ with closed graph and $(\bar{x}, \bar{y}) \in F$, the following assertions are equivalent.*

(a) *F is open around $(\bar{x}, \bar{y})$ with a linear rate .*
(b) *F^{-1} is coderivatively compact at $(\bar{y}, \bar{x})$ and $\|D_M^* F^{-1}(\bar{y}, \bar{x})\| < \infty$.*
(c) *F^{-1} is coderivatively compact at $(\bar{y}, \bar{x})$ and $D_M^* F^{-1}(\bar{y}, \bar{x})(0) = \{0\}$.*
(d) *F is metrically regular around $(\bar{x}, \bar{y})$.*

The following example shows that the coderivative compactness assumption in the preceding criteria cannot be dropped, even if one replaces the condition $D_M^* F^{-1}(\bar{y}, \bar{x})(0) = \{0\}$ by the stronger condition $\ker D_L^* F(\bar{x}, \bar{y}) = \{0\}$.

Example–Exercise. Let $X := Y$ be a separable Hilbert space with an orthonormal basis $\{e_n : n \in \mathbb{N}\}$, let $C := \{x = \sum_{n \ge 1} x_n e_n : \forall n \in \mathbb{N} \ |x_n| \le 2^{-n}\}$, $a := \sum_{n \ge 1} e_n/n$, $b := \sum_{n \ge 1} e_n/n^2$, $A := [-1, 1]a/\|a\|$, and let $F : X \rightrightarrows Y$ be given by $F(x) = x + C$ for $x \in A$, $F(x) = \varnothing$ for $x \in X \setminus A$. Then F has a closed convex graph, and since $\mathrm{span}\, C$ is dense in X, for $(\bar{x}, \bar{y}) := (0, 0)$ one has

$$N_L(F, (\bar{x}, \bar{y})) \subset N_L(\{0\} \times C, (0, 0)) = X^* \times \{0\},$$

whence $\ker D_L^* F(\bar{x}, \bar{y}) = \{0\}$. However, F is not open at $(\bar{x}, \bar{y})$: for all $r > 0$ one has $0 \notin \mathrm{int}\, F(r B_X)$, since for $s \in [-r/\|a\|, r/\|a\|]$, $t > 0$ the relation $tb \in sa + C$ entails $|t/n^2 - s/n| \le 2^{-n}$ for all n, which is impossible. $\square$

6.6   Limiting Directional Subdifferentials

In Hadamard smooth spaces, a limiting construction that mimics the one for limiting firm subdifferentials can be given. In view of the strong analogies with the proofs for limiting firm subdifferentials, our treatment is more concise.

Definition 6.80. Given a function $f : X \to \overline{\mathbb{R}}$ on a normed space X and $x \in f^{-1}(\mathbb{R})$, the *limiting directional subdifferential* of f at x is the set $\partial_\ell f(x)$ of $x^* \in X^*$ that are weak* cluster points of some bounded sequence (x_n^*) such that for some sequence $(x_n) \to_f x$, $x_n^* \in \partial_D f(x_n)$ for all $n \in \mathbb{N}$. Recall that we write $(x_n^*) \to^* x^*$ to mean that the sequence (x_n^*) is bounded and that x^* is a weak* cluster point of it.

For Lipschitzian functions on smooth spaces, one gets nonempty subdifferentials. The proof is the same as that for the firm case, using Theorem 4.65.

Proposition 6.81 (Subdifferentiability of Lipschitzian functions). *If f is Lipschitzian with rate c at $\overline{x}$, then $\partial_\ell f(\overline{x})$ is contained in cB_{X^*}. If, moreover, X is an H-smooth Banach space, then $\partial_\ell f(\overline{x}) \neq \varnothing$.*

When X is a WCG space and f is Lipschitzian around x, the definition can be simplified thanks to the Borwein–Fitzpatrick theorem (Theorem 3.109):

$$\partial_\ell f(x) := \mathrm{w}^* - \mathrm{seq} - \limsup_{w \to x} \partial_D f(w) = \mathrm{w}^* - \limsup_{w \to x} \partial_D f(w).$$

Moreover, we have some coincidence results.

Theorem 6.82. *(a) Let X be a WCG space and let f be an element of the set $\mathscr{L}(X)$ of locally Lipschitzian functions on X. Then for all $x \in X$, one has*

$$\partial_\ell f(x) = \mathrm{w}^* - \limsup_{w \to x} \partial_H f(w) = \partial_h f(x) := \mathrm{w}^* - \mathrm{seq} - \limsup_{w \to x} \partial_H f(w).$$

(b) If X is a WCG Asplund space, then for all $x \in X$, one has $\partial_\ell f(x) = \partial_L f(x)$.

Proof. (a) Let $x^* \in \partial_\ell f(x)$: there exist sequences $(x_n) \to x$, $(x_n^*) \to^* x^*$ such that $x_n^* \in \partial_D f(x_n)$ for all $n \in \mathbb{N}$. Since X is a subdifferentiability space for ∂_H, given a weak* closed neighborhood V of 0 in X^*, one can find $y_n \in B(x_n, 2^{-n})$, $y_n^* \in x_n^* + V$ such that $y_n^* \in \partial_H f(y_n)$. Since (y_n^*) is bounded, one can find $y^* \in x^* + V$ such that $(y_n^*) \to^* y^*$. Then $y^* \in \partial_h f(x)$, and since the family of closed neighborhoods of 0 is a base of neighborhoods of 0, $x^* \in \mathrm{cl}^* \partial_h f(x)$. Since X is a WCG space, the set $\partial_h f(x)$ is weak* closed. Thus $x^* \in \partial_h f(x)$. The opposite inclusion $\partial_h f(x) \subset \partial_\ell f(x)$ stems from the inclusion $\partial_H f(x) \subset \partial_D f(x)$ by a passage to the limit.
(b) When X is a WCG Asplund space it is a subdifferentiability space for ∂_F, and the same proof with ∂_F instead of ∂_H is valid. □

Definition 6.83. The *limiting directional normal cone* to a subset E of a normed space X at $\overline{x} \in E$ is

$$N_\ell(E, \overline{x}) := \partial_\ell \iota_E(\overline{x}).$$

Thus, $\overline{x}^* \in N_\ell(E, \overline{x})$ iff $\overline{x}^*$ is a weak* cluster point of a bounded sequence (x_n^*) for which there exists a sequence $(x_n) \to_E \overline{x}$ such that $x_n^* \in N_D(E, x_n)$ for all $n \in \mathbb{N}$.
Let us give a comparison with the subdifferential of the distance function to E.

Proposition 6.84. *For every subset E of a WCG space X and every $\overline{x} \in E$ one has*

$$\partial_\ell d_E(\overline{x}) \subset N_\ell(E, \overline{x}).$$

Proof. Let $\overline{x}^* \in \partial_\ell d_E(\overline{x})$. By the preceding theorem, there are sequences $(w_n) \to \overline{x}$, $(w_n^*) \to^* \overline{x}^*$ such that $w_n^* \in \partial_H d_E(w_n)$ for all $n \in \mathbb{N}$. When $P := \{p \in \mathbb{N} : w_p \in E\}$ is infinite, the conclusion holds. Thus we suppose P is finite, and given a closed neighborhood V of 0 in the weak* topology, for $n \in N := \mathbb{N} \setminus P$, using the approximate projection theorem (Theorem 4.74), we pick $e_n \in E$, $e_n^* \in N_H(E, e_n) \cap$ S_{X^*} such that $\|e_n - w_n\| \le d_E(w_n) + 2^{-n}$, $e_n^* - w_n^* \in V$. Then $(e_n) \to \bar{x}$ in E, (e_n^*) has a weak* cluster point $e^* \in x^* + V$. Thus $e^* \in \mathrm{w}^* - \limsup_{y(\in E) \to \bar{x}} N_H(E, y) \cap B_{X^*}$ and $\bar{x}^*$ belongs to the weak* closure of this set that is weak* closed and is a sequential limsup by the Borwein–Fitzpatrick theorem. $\square$

A concept of limiting directional coderivative of a multimap can be defined as in the case of the limiting coderivative.

Definition 6.85. The *limiting directional coderivative* of a multimap $F : X \rightrightarrows Y$ between two normed spaces at $(\bar{x}, \bar{y}) \in F$ is the multimap $D_\ell^* F(\bar{x}, \bar{y}) : Y^* \rightrightarrows X^*$ defined by

$$D_\ell^* F(\bar{x}, \bar{y})(y^*) := \{x^* \in X^* : (x^*, -y^*) \in N_\ell(F, (\bar{x}, \bar{y}))\}, \qquad y^* \in Y^*.$$

The *mixed limiting directional coderivative* of F at $(\bar{x}, \bar{y})$ is the multimap $D_m^* F(\bar{x}, \bar{y}) :$ $Y^* \rightrightarrows X^*$ defined by $x^* \in D_m^* F(\bar{x}, \bar{y})(y^*)$ iff there exist sequences $((x_n, y_n)) \to_F (\bar{x}, \bar{y})$, $(x_n^*) \to^* x^*$, $(y_n^*) \to y^*$ such that $(x_n^*, -y_n^*) \in N_D(F, (x_n, y_n))$ for all $n \in \mathbb{N}$.

Clearly, for all $y^* \in Y^*$, one has the inclusion $D_m^* F(\bar{x}, \bar{y})(y^*) \subset D_\ell^* F(\bar{x}, \bar{y})(y^*)$, and this inclusion is an equality when Y is finite-dimensional.

Proposition 6.86. *If* $g : X \to Y$ *is of class* D^1 *at* $\bar{x} \in X$, *then* $D_\ell^* g(\bar{x}) = D_m^* g(\bar{x}) = g'(\bar{x})^\mathsf{T}$.
If $F : X \rightrightarrows Y$ *has a convex graph, then* $D_\ell^* F(\bar{x}, \bar{y}) = D_m^* F(\bar{x}, \bar{y}) = D^* F(\bar{x}, \bar{y})$.

Proof. Let $A := g'(\bar{x})$, so that $D_D^* g(\bar{x}) = A^\mathsf{T}$. Since $D_D^* g(\bar{x}) \subset D_m^* g(\bar{x}) \subset D_\ell^* g(\bar{x})$, it remains to show that $x^* = A^\mathsf{T}(y^*)$ for all $y^* \in Y$ and all $x^* \in D_\ell^* g(\bar{x})(y^*)$. Let $((x_n^*, y_n^*)) \to^* (x^*, y^*)$ be such that $(x_n^*, -y_n^*) \in N_D(\mathrm{gphg}, (x_n, y_n))$ for some sequences $(x_n) \to \bar{x}$, $(y_n) := (g(x_n)) \to g(\bar{x})$. Since g is of class D^1 at $\bar{x}$, we have $x_n^* = y_n^* \circ g'(x_n)$ for n large enough. Then for all $u \in X$, taking a net $((x_{n(i)}^*, y_{n(i)}^*))_{i \in I} \xrightarrow{*} (x^*, y^*)$, so that $((y_{n(i)}^*, g'(x_{n(i)})u))_i \to \langle y^*, g'(\bar{x})u \rangle$, we get $\langle y^*, Au \rangle = \langle x^*, u \rangle$, hence $x^* = A^\mathsf{T}(y^*)$.
The second assertion stems from a property of normal cones to a convex set. $\square$

Let us give a characterization of the limiting directional subdifferential.

Proposition 6.87. *If* $f : X \to \overline{\mathbb{R}}$ *is lower semicontinuous and finite at* $\bar{x} \in X$ *and if* $E : X \rightrightarrows \mathbb{R}$ *is the multimap with graph* epi$f$, *the limiting directional subdifferential of* f *at* $\bar{x}$ *satisfies*

$$\partial_\ell f(\bar{x}) = D_\ell^* E(\bar{x}, f(\bar{x}))(1) := \{\bar{x}^* : (\bar{x}^*, -1) \in N_\ell(\mathrm{epi}f, (\bar{x}, f(\bar{x})))\}.$$

Proof. Given $\bar{x}^* \in \partial_\ell f(\bar{x})$, we can find sequences $(x_n) \to_f \bar{x}$, $(x_n^*) \to^* \bar{x}^*$ such that $x_n^* \in \partial_D f(x_n)$ for all $n \in \mathbb{N}$. Then by Corollary 4.15, $(x_n^*, -1) \in N_D(E, (x_n, f(x_n)))$, and since $((x_n, f(x_n))) \to_E (\bar{x}, f(\bar{x})) = \bar{x}_f$, we get $(\bar{x}^*, -1) \in N_\ell(E, \bar{x}_f)$.
Conversely, let $(\bar{x}^*, -1) \in N_\ell(E, \bar{x}_f)$. There exist sequences $((x_n, r_n)) \to (\bar{x}, f(\bar{x}))$ in E, $((x_n^*, -r_n^*)) \to^* (\bar{x}^*, -1)$ such that $(x_n^*, -r_n^*) \in N_D(E, (x_n, r_n))$ for all $n \in \mathbb{N}$. For n large we have $r_n^* > 0$, so that $u_n^* := x_n^* / r_n^* \in \partial_D f(x_n)$ by Corollary 4.15. Then

taking a subnet $((x^*_{n(i)}, -r^*_{n(i)}))_{i \in I} \xrightarrow{*} (\overline{x}^*, -1)$, we see that $(u^*_n) \to^* \overline{x}^*$ and since $f(\overline{x}) \leq \liminf_n f(x_n) \leq \limsup_n f(x_n) \leq \lim_n r_n = f(\overline{x})$, hence $(f(x_n)) \to f(\overline{x})$, we get $\overline{x}^* \in \partial_\ell f(\overline{x})$. $\qquad \square$

Proposition 6.88 (Scalarization). *Let $g : X \to Y$ be continuous at $\overline{x} \in X$. Then for all $y^* \in Y^*$, one has the following inclusion; it is an equality when g is Lipschitzian near $\overline{x}$, X is a WCG Banach space, and Y is finite-dimensional:*

$$\partial_\ell (y^* \circ g)(\overline{x}) \subset D^*_m g(\overline{x})(y^*).$$

Proof. The proof of the inclusion is similar to that for the subdifferential ∂_L.

Let us prove the opposite inclusion when X is a WCG space, Y is finite-dimensional, and g is Lipschitzian near $\overline{x}$. Let $x^* \in D^*_m g(\overline{x})(y^*)$ and let $(x_n) \to \overline{x}$, $(x^*_n) \to^* x^*$, $(y^*_n) \to y^*$ with $x^*_n \in D^*_D g(x_n)(y^*_n)$ for all $n \in \mathbb{N}$. For n large enough g is stable at x_n, hence tangentially compact, so that Proposition 4.25 ensures that $x^*_n \in \partial_D (y^*_n \circ g)(x_n)$. Since $y^*_n \circ g - y^* \circ g$ is Lipschitzian with rate $\varepsilon_n := \ell \|y^*_n - y^*\|$, where ℓ is a Lipschitz rate of g near $\overline{x}$, we see that $x^*_n \in \partial_D (y^* \circ g + \varepsilon_n \|\cdot - x_n\|)(x_n)$. Then for every weak* closed neighborhood V of 0, Theorem 4.83 yields some $w_n \in B(x_n, \varepsilon_n)$, some $w^*_n \in \partial_D (y^* \circ g)(w_n)$ such that $w^*_n \in x^*_n + \varepsilon_n B_{X^*} + V$. Since (w^*_n) is bounded, it has a weak* cluster point $w^* \in x^* + V$. Since $\partial_\ell (y^* \circ g)(\overline{x})$ is weak* closed, one gets $x^* \in \partial_\ell (y^* \circ g)(\overline{x})$. $\qquad \square$

Exercise. Let E_f be the epigraph of $f \in \mathscr{F}(X)$ finite at $\overline{x} \in X$ and let $\overline{x}_f := (\overline{x}, f(\overline{x}))$.
(a) Check that $N_\ell(E_f, \overline{x}_f)$ is included in $X^* \times \mathbb{R}_-$.
(b) Introduce the *singular limiting directional subdifferential* of f at $\overline{x}$ by $\overline{x}^* \in \partial^\infty_\ell f(\overline{x})$ iff $(\overline{x}^*, 0) \in N_\ell(E_f, \overline{x}_f)$ and prove the decomposition

$$N_\ell(E_f, \overline{x}_f) = (\partial^\infty_\ell f(\overline{x}) \times \{0\}) \cup \mathbb{R}_+ (\partial_\ell f(\overline{x}) \times \{-1\}).$$

6.6.1 Some Elementary Properties

Let us give some properties of the limiting directional subdifferential.

Proposition 6.89. *(a) If $f, g \in \mathscr{F}(X)$ coincide around x, then $\partial_\ell f(x) = \partial_\ell g(x)$.*
(b) If $f \in \mathscr{F}(X)$ is convex, then $\partial_\ell f(x) = \partial_{MR} f(x)$.
(c) If $f \in \mathscr{F}(X)$ attains a local minimum at $x \in \mathrm{dom}\, f$, then $0 \in \partial_\ell f(x)$.
(d) If $\lambda > 0$, $A \in L(X,Y)$ with $A(X) = Y$, $b \in Y$, $g \in \mathscr{F}(X)$, $\ell \in X^*$, $c \in \mathbb{R}$ and $f(x) = \lambda g(Ax + b) + \langle \ell, x \rangle + c$, then $\partial_\ell f(x) = \lambda A^\mathsf{T}(\partial_\ell g(Ax + b)) + \ell$.
(e) If $g \in \mathscr{F}(X)$, $h \in \mathscr{F}(Y)$ and f is given by $f(x,y) = g(x) + h(y)$, then $\partial_\ell f(x,y) \subset \partial_\ell g(x) \times \partial_\ell h(y)$. If X or Y is a WCG space, equality holds.
(f) If $f = g + h$, where $g \in \mathscr{F}(X)$, $h : X \to \mathbb{R}$ is of class D^1 at $\overline{x} \in X$, then $\partial_\ell f(\overline{x}) = \partial_\ell g(\overline{x}) + h'(\overline{x})$.
(g) If $h : X \to \mathbb{R}$ is of class D^1 at $\overline{x} \in X$, then $\partial_\ell h(\overline{x}) = \{h'(\overline{x})\}$.

(h) If $f = h \circ g$, where $h \in \mathscr{F}(Y)$, $g : X \to Y$ is of class D^1 at $\bar{x} \in X$ and open at $\bar{x}$, then $g'(\bar{x})^{\mathsf{T}}(\partial_\ell h(g(\bar{x}))) \subset \partial_\ell f(\bar{x})$. If $g'(\bar{x})(X) = Y$ and g is of class C^1 or Y is finite-dimensional, this inclusion is an equality.

Proof. (a) is obvious. (b) stems from the equality $\partial_D f = \partial_{MR} f$ when f is convex and the closedness of $\partial_{MR} f$. (c) is a consequence of the inclusion $\partial_D f(x) \subset \partial_\ell f(x)$. (d), (e), (f), and (g) follow from similar properties with the directional subdifferential by a passage to the limit. When in (e) X or Y is a WCG space, equality holds in view of Theorem 6.82. (h) Let $f = h \circ g$, let $\bar{y} := g(\bar{x})$, $A := g'(\bar{x})$, and let $\bar{y}^* \in \partial_\ell h(\bar{y})$. There exist sequences $(y_n) \to_h \bar{y}$, $(y_n^*) \to^* \bar{y}^*$ satisfying $y_n^* \in \partial_D h(y_n)$ for all $n \in \mathbb{N}$. Since g is open at $\bar{x}$, we can find a sequence $(x_n) \to \bar{x}$ such that $y_n = g(x_n)$ for n large. By Proposition 4.40, $y_n^* \circ g'(x_n) \in \partial_D f(x_n)$ and $(y_n^* \circ g'(x_n)) \to^* \bar{y}^* \circ g'(\bar{x})$, so that $g'(\bar{x})^{\mathsf{T}}(\bar{y}^*) = \bar{y}^* \circ g'(\bar{x}) \in \partial_\ell f(\bar{x})$. The last assertion stems from Proposition 4.42. $\quad\square$

Now let us turn to properties involving order. A first one can be derived from Proposition 4.48 by a passage to the limit.

Proposition 6.90. *Let $f := h \circ g$, where $g := (g_1, \ldots, g_m) : X \to \mathbb{R}^m$, $g_i \in \mathscr{L}(X)$ for $i \in \mathbb{N}_m$, $h : \mathbb{R}^m \to \mathbb{R}$ is of class C^1 around $\bar{y} := g(\bar{x})$ and nondecreasing in each of its m arguments near $\bar{y}$, with $h'(\bar{y}) \neq 0$. Then*

$$\partial_\ell f(\bar{x}) \subset h'(\bar{y}) \circ (\partial_\ell g_1(\bar{x}), \ldots, \partial_\ell g_m(\bar{x})). \tag{6.68}$$

The next property is analogous to Proposition 6.20.

Proposition 6.91. *Let X and Y be Banach spaces, let $f : X \to \bar{\mathbb{R}}$ be finite and Lipschitzian around $\bar{x} \in X$, let $g : Y \to \mathbb{R}$ be of class D^1 at $\bar{y} \in Y$ with $g'(\bar{y}) \neq 0$ and let h be given by $h(x, y) = \max(f(x), g(y))$. Then, if $f(\bar{x}) = g(\bar{y})$, for $(\bar{x}^*, \bar{y}^*) \in \partial_\ell h(\bar{x}, \bar{y})$ with $\bar{y}^* \neq g'(\bar{y})$ there exists $\lambda \in [0, 1]$ such that*

$$(\bar{x}^*, \bar{y}^*) \in (1 - \lambda)\partial_\ell f(\bar{x}) \times \lambda \partial_\ell g(\bar{y}).$$

Proposition 6.92. *Let V, W be Banach spaces, let $A \in L(V, W)$ with $W = A(V)$, $\bar{v} \in V$, $\bar{w} := A\bar{v}$, $j \in \mathscr{F}(V)$, $p \in \mathscr{F}(W)$ such that $p \circ A \leq j$. Suppose that for every sequence $(w_n) \to_p \bar{w}$ one can find a sequence $(v_n) \to \bar{v}$ such that $A(v_n) = w_n$ and $j(v_n) = p(w_n)$ for all $n \in \mathbb{N}$ large enough. Then one has*

$$A^{\mathsf{T}}(\partial_\ell p(\bar{w})) \subset \partial_\ell j(\bar{v}). \tag{6.69}$$

Proof. Let $\bar{w}^* \in \partial_\ell p(\bar{w})$. There exist sequences $(w_n) \to_p \bar{w}$, $(w_n^*) \to^* \bar{w}^*$ such that $w_n^* \in \partial_D p(w_n)$ for all $n \in \mathbb{N}$. Let $(v_n) \to \bar{v}$ be such that $A(v_n) = w_n$ and $j(v_n) = p(w_n)$ for all $n \in \mathbb{N}$ large enough. Then $A^{\mathsf{T}}(w_n^*) \in \partial_D j(v_n)$, and since $(A^{\mathsf{T}}(w_n^*)) \to^* A^{\mathsf{T}}(\bar{w}^*)$ and $(v_n) \to_j \bar{v}$, we get $A^{\mathsf{T}}(\bar{w}^*) \in \partial_\ell j(\bar{v})$. $\quad\square$

The following geometrical properties are easy consequences of Proposition 6.89.

Proposition 6.93. *(a) If A and B are closed subsets of Banach spaces X and Y respectively, and $(\bar{x}, \bar{y}) \in A \times B$, then $N_\ell(A \times B, (\bar{x}, \bar{y})) \subset N_\ell(A, \bar{x}) \times N_\ell(B, \bar{y})$ and equality holds if X or Y is a WCG space.*
(b) If C is a closed convex subset of a Banach space X, then $N_\ell(C, \bar{x})$ coincides with the normal cone in the sense of convex analysis: $N_\ell(C, \bar{x}) = N_{MR}(C, \bar{x}) := \{\bar{x}^* \in X^* : \forall x \in C, \langle \bar{x}^*, x - \bar{x} \rangle \leq 0\}$.

Proof. (a) The assertion follows from Proposition 6.89 (e) with $g := \iota_A$, $h := \iota_B$.
(b) The assertion stems from Proposition 6.89 (b), with $f := \iota_C$. $\qquad\qquad\square$

6.6.2 Calculus Rules Under Lipschitz Assumptions

Calculus rules with limiting directional subdifferentials in WCG spaces keep some attractive features.

Theorem 6.94 (Sum rule for limiting directional subdifferentials). *Let X be a WCG space, and let $f = f_1 + \cdots + f_k$, where $f_1, \ldots, f_k \in \mathscr{L}(X)$. Then*

$$\partial_\ell f(\bar{x}) \subset \partial_\ell f_1(\bar{x}) + \cdots + \partial_\ell f_k(\bar{x}). \tag{6.70}$$

Proof. We know that X is H-smooth. Let $\bar{x}^* \in \partial_\ell f(\bar{x})$, and let $(x_n) \to \bar{x}$, $(x_n^*) \overset{*}{\to} \bar{x}^*$ with $x_n^* \in \partial_H f(x_n)$ for all n. Given a weak* closed neighborhood V of 0 in X^* and a sequence $(\varepsilon_n) \to 0_+$, by the fuzzy sum rule for Hadamard subdifferentials (Theorem 4.69) there are sequences $((x_{i,n}, x_{i,n}^*)) \in \partial_H f_i$, for $i \in \mathbb{N}_k$, such that $d(x_{i,n}, x_n) \leq \varepsilon_n$, $x_n^* \in x_{1,n}^* + \cdots + x_{k,n}^* + V$. Since for some $r > 0$ one has $x_{i,n}^* \in r B_{X^*}$ for all $(i,n) \in \mathbb{N}_k \times \mathbb{N}$, one can find $y_i^* \in \partial_\ell f_i(\bar{x})$ such that $(x_{i,n}^*) \to^* y_i^*$ for $i \in \mathbb{N}_k$ and $\bar{x}^* \in y_1^* + \cdots + y_k^* + V$. Since $S := \partial_\ell f_1(\bar{x}) + \cdots + \partial_\ell f_k(\bar{x})$ is weak* compact and $\bar{x}^* \in S + V$ for every weak* closed neighborhood V of 0, one gets $\bar{x}^* \in S$. $\quad\square$

Theorem 6.95 (Chain rule for limiting directional subdifferentials). *Let X and Y be WCG spaces, let $g : X \to Y$ be a map that is Lipschitzian around $\bar{x} \in X$, and let $h : Y \to \mathbb{R}_\infty$ be Lipschitzian around $\bar{y} := g(\bar{x})$. Then*

$$\partial_\ell(h \circ g)(\bar{x}) \subset D_\ell^* g(\bar{x})(\partial_\ell h(\bar{y})).$$

Proof. Let $f := h \circ g$ and let $\bar{x}^* \in \partial_\ell f(\bar{x})$. Let r (resp. s) be the Lipschitz rate of f (resp. h) on a neighborhood of $\bar{x}$ (resp. $\bar{y}$) and let G be the graph of g. Then, by the penalization lemma, for x near $\bar{x}$ one has

$$f(x) = \inf\{f(w) + r \|w - x\| : w \in X\} = \inf\{h(y) + r \|w - x\| : (w, y) \in G\}$$

$$= \inf\{h(y) + r \|w - x\| + (r + s)d_G(w, y) : (w, y) \in X \times Y\}.$$

Let $j: X \times X \times Y \to \mathbb{R}$ be defined by $j(x,w,y) := h(y) + r\|x-w\| + (r+s)d_G(w,y)$. For every sequence $(x_n) \to \bar{x}$, one has $((x_n, x_n, g(x_n))) \to (\bar{x}, \bar{x}, \bar{y})$ and $j(x_n, x_n, g(x_n))$ $= f(x_n)$ for all n. Proposition 6.92 implies that $(\bar{x}^*, 0, 0) \in \partial_\ell j(\bar{x}, \bar{x}, \bar{y})$. Theorem 6.94 yields $(\bar{w}^*, \bar{v}^*) \in \partial_\ell d_G(\bar{x}, \bar{y})$, $\bar{y}^* \in \partial_\ell h(\bar{y})$, $\bar{z}^* \in B_{X^*}$ such that

$$(\bar{x}^*, 0, 0) = (0, 0, \bar{y}^*) + r(\bar{z}^*, -\bar{z}^*, 0) + (r+s)(0, \bar{w}^*, \bar{v}^*).$$

Then one has $\bar{x}^* = r\bar{z}^* = (r+s)\bar{w}^* \in (r+s)D_\ell g(\bar{x})(-\bar{v}^*) = D_\ell g(\bar{x})(\bar{y}^*)$. □

Exercises

1. Show that for $f, g \in \mathscr{L}(X)$ the inclusion $\partial_\ell(f+g)(\bar{x}) \subset \partial_\ell f(\bar{x}) + \partial_\ell g(\bar{x})$ may be strict. [Hint: Take $X := \mathbb{R}$, $f := |\cdot|$, $g := -|\cdot|$.]

2. Deduce from the preceding exercise that inclusion (6.68) may be strict. [Hint: Take $m = 2$ and h given by $h(y_1, y_2) = y_1 + y_2$.]

3. **(a)** Declare that a subset S of a normed space X is *directionally normally compact* at $\bar{x} \in S$ if for all sequences $(x_n) \to_S \bar{x}$, $(x_n^*) \xrightarrow{*} 0$ with $x_n^* \in N_D(S, x_n)$ for all $n \in \mathbb{N}$, one has $(x_n^*) \to 0$. Compare this property with normal compactness at $\bar{x}$.
(b) Prove that S is directionally normally compact at $\bar{x} \in S$ if and only if every sequence (x_n^*) in S_{X^*} has a nonnull weak* cluster value whenever, for some sequence $(x_n) \to_S \bar{x}$, it satisfies $x_n^* \in N_D(S, x_n)$ for all $n \in \mathbb{N}$.
(c) Give criteria for directional normal compactness.

6.7   Notes and Remarks

Most of the ingredients of the first five sections of this chapter are inspired by pioneer notes and papers by Kruger [594–599], the numerous papers of Mordukhovich starting with [713, 714], and his monograph [718]. For what concerns calculus rules, we rely on the idea in [547, 658, 806, 813] that a good collective behavior gives better results than the assumption that most factors are nice (or one factor in the case of a pair). The reader may find difficulties in this chapter due to the fact that it presents several versions of this idea of good collective behavior. Thus, it may be advisable to skip all but one of them on a first reading, for instance alliedness. On the other hand, the reader may notice that almost all results could be deduced from results pertaining to multimaps. Such a direct route would make the presentation much shorter. However, we have preferred a slower pace that starts with sets rather than multimaps. In such a way, our starting point is less complex, and the reader who just needs a result about sets has simpler access to it.

The basic idea of passing to the limit enables one to gather precious information about the behavior of the function or the set around the specific point of interest. The accuracy of the elementary normal cones and subdifferentials may be lost, but to a lesser extent than with Clarke's notions. This advantage is due to the fact that no automatic convexification occurs. On the other hand, in such an approach, one cannot expect the beautiful duality relationships of Clarke's concepts. The idea of taking limits appeared as early as 1976 in Mordukhovich's pioneer paper [713] and in his book [716], where it is used for the needs of optimal control theory. The reader will find interesting developments about the history of the birth of limiting concepts in [531] and in the commentary of Chap. 1 of the monograph [718].

A decisive advantage of the notions of the present chapter lies in the calculus rules that are precise inclusions rather than approximate rules, at least under some qualification conditions. These rules and constructs are particularly striking in finite-dimensional spaces. Such rules for the case in which all the functions but one are Lipschitzian were first proved in the mimeographed paper [513] and announced in [512]. The finite-dimensional calculus for lower semicontinuous functions was first presented in [520]. The qualification conditions there were more restrictive than those announced in the note [715], but the proofs with the latter are identical to those with Ioffe's qualification conditions. Similar conditions appeared in Rockafellar's paper [879]. The calculus in the infinite-dimensional case was first presented in Kruger's paper [595]. The Asplund space theory appeared only in 1996 [721, 722], although the possibility of an extension to Asplund spaces was indicated by Fabian [362, 363, 373], who was the first to apply separable reduction but did not work with limiting constructions.

Normal compactness of a subset appeared in [804] in the convex case and in [807] in the nonconvex case. Its discovery was influenced by some methods of Brézis [170] and Browder [176] in nonlinear functional analysis and by the general views of [787] about compactness properties. The work of Loewen [671] was also decisive. The latter elaborated upon the notion of compactly epi-Lipschitzian set due to Borwein and Strojwas [131], which generalizes the notion of epi-Lipschitzian set as explained in Sect. 6.2. The latter notion was introduced by Rockafellar [874, 875] as a convenient qualification condition. Comprehensive characterizations of compactly epi-Lipschitzian closed convex subsets of normed spaces are presented in [143]. See also [358, 530]. Proposition 6.36 is inspired by [530, Theorem 2] and [577].

The notion of coderivatively compact multimap is a natural adaptation of normal compactness to the case of a product structure. It appeared in [807] with the aim of getting openness results. It is also considered in [530, 577, 718] and in numerous papers of Mordukhovich and his coauthors under the name partial sequential normal compactness (PSNC). Strong coderivative compactness of a multimap is called there strong partial sequential normal compactness. The initial contribution of Mordukhovich and Shao [723] was of local character rather than pointwise character, but they quickly changed their presentations after reading the preprints of [807] and [577]. The subtleties of sequential versus nonsequential compactness are explored in [116, 530, 723, 727]. The numerous equivalences of fuzzy calculus,

limiting calculus, and the Asplund property have confirmed the idea that the class of
Asplund spaces is the appropriate framework for limiting constructions with Fréchet
subdifferentials [528, 722, 995].

The approach through metric estimates of Theorem 6.13 and Propositions 6.41
and 6.59 is close in spirit to [524, 531, 547, 802] and [813].

Numerous authors have tackled the questions of conditioning and error bounds,
after the pioneer paper of Hoffman [505]. Among them are Azé [53, 54],
Azé–Corvellec [55], Azé–Hiriart–Urruty [57], Bolte–Daniilidis–Lewis–Shiota [106,
107], Burke [181, 182], Burke–Ferris [184], Burke–Deng [183] , Cominetti [225],
Cornejo–Jourani–Zălinescu   [229],   Coulibaly–Crouzeix   [242],   Dontchev–
Rockafellar [318, 321], Henrion–Outrata [473], [474], Henrion–Jourani–Outrata
[476], Ioffe [511, 531–533, 535, 536, 538, 539, 541], Ioffe–Outrata [546], Jourani–
Thibault [575], Klatte–Kummer [590], Kummer [606], Łojasiewicz [678], Ng–Yang
[747], Ngai–Théra [750–752, 752, 753], Pang [777], Penot [808, 825], Wu–Ye
[962, 963, 965], Zhang-Treiman [990], and many more. The convex case is especially
rich and has been treated by Auslender–Crouzeix [43], Burke–Tseng [187], Ioffe
[537], Ioffe–Sekiguchi [548], Klatte–Kummer [590], Lewis–Pang [649], Li–Singer
[656], Ngai–Théra [752], Robinson [868], Song [899], Zălinescu [983], among
others.

Some concepts similar to the upper limiting subdifferential of [546, 825] were
used previously in [214, 718, 728], but they are different.

The section dealing with limiting directional objects is rather new, although it
has some connections with proposals by Ioffe. For the sake of brevity this material
is not fully developed here.

Chapter 7
Graded Subdifferentials, Ioffe Subdifferentials

The time has come for the epoch of world literature, and everyone must endeavor to hasten this epoch.

—Goethe, *Conversations*, 1827

In this last chapter we present an approach valid in every Banach space. The key idea, due to A.D. Ioffe, that yields such a universal theory consists in reducing the study to a convenient class of linear subspaces. Initially, Ioffe used the class of finite-dimensional subspaces of X [512, 513, 515, 516]; then he turned to the class of closed separable subspaces, which has some convenient permanence properties [527]. Since such an approach presents some analogy with the notion of inductive limit of topological linear spaces, one could call the obtained subdifferentials inductive subdifferentials rather than geometric subdifferentials. However, we adopt a different terminology that is evocative of this restriction process, and we keep Ioffe's notation ∂_G.

The price to pay to get such a general theory is a certain complexity of concepts and proofs. The latter can be skipped by the user on a first reading. The reader will be glad to find statements presenting rather familiar properties.

7.1 The Lipschitzian Case

In the sequel, X is an arbitrary Banach space and $\mathscr{S}(X)$ denotes the family of closed separable linear subspaces of X. Given a linear subspace W of X and a function $f : X \to \mathbb{R}_\infty$, we denote by $j_W : W \to X$ the canonical injection and by f_W the restriction to W of the function f, i.e., $f_W := f \circ j_W$. Also, we denote by $r_W : X^* \to W^*$ the restriction map defined by $r_W = j_W^\mathsf{T}$, so that for all $x^* \in X^*$, $w \in W$,

$$\langle r_W(x^*), w \rangle = \langle x^*, j_W(w) \rangle = x^*(j_W(w)) = \langle x_W^*, w \rangle.$$

For convenience, again we use the notation $(x_n^*) \to^* x^*$ to mean that the sequence (x_n^*) is bounded and has x^* as a weak* cluster point.

7.1.1 Some Uses of Separable Subspaces

The following lemmas show convenient properties of the family $\mathscr{S}(X)$.

Lemma 7.1. *Let $A : X \to Y$ be a surjective linear continuous map between two Banach spaces. Then for every $Z \in \mathscr{S}(Y)$ there exists some $W \in \mathscr{S}(X)$ such that $A(W) = Z$.*

Proof. By Michael's selection theorem (Theorem 1.40) there exists a continuous right inverse $B : Y \to X$ of A. Given $Z \in \mathscr{S}(Y)$, let D be a countable dense subset of Z and let W be the closure of the set E of rational combinations of elements of $B(D)$, i.e., the closed linear subspace generated by $B(D)$. Then $W \in \mathscr{S}(X)$, and since A and B are continuous, we have $A(W) = A(\mathrm{cl}(E)) \subset \mathrm{cl}(A(E)) \subset Z$ and $B(Z) = B(\mathrm{cl}(D)) \subset \mathrm{cl}(B(D)) \subset W$, hence $Z = A(B(Z)) \subset A(W)$. Thus $A(W) = Z$. $\square$

Another proof. Let $\{z_n\}$ be a dense countable subset of Z, with $z_0 = 0$. The open mapping theorem ensures that there exists some $c > 0$ such that for all $y \in Y$ one can find some $x \in A^{-1}(y)$ satisfying $\|x\| \le c\|y\|$. Thus, for all $(m,n) \in \mathbb{N}^2$ one can find $w_{m,n} \in X$ such that $A(w_{m,n}) = z_n - z_m$ and $\|w_{m,n}\| \le c\|z_n - z_m\|$. Let W be the closure of the linear subspace W_0 spanned by $\{w_{m,n} : (m,n) \in \mathbb{N}^2\}$. It is a separable subspace of X, and since $z_n = A(w_{0,n})$, one has $A(W) = A(\mathrm{cl}(W_0)) \subset \mathrm{cl}(A(W_0)) \subset Z$. In fact, these inclusions are equalities because for all $z \in Z$ one can find a sequence $(z_{k(n)})_n \to z$ for some map $k : n \mapsto k(n)$ from $\mathbb{N}$ to $\mathbb{N}$ with $\|z_{k(n+1)} - z_{k(n)}\| \le 2^{-n}\|z\|$, $\|z_{k(n)} - z\| \le 2^{-n}\|z\|$, $k(0) = 0$. Then the series with general term $w_{k(n+1),k(n)}$ is absolutely convergent, and its sum $w \in W$ satisfies $A(w) = z$. $\square$

Remark. Moreover, one may observe that the set $\mathscr{S}_A(X)$ of $W \in \mathscr{S}(X)$ such that $A(W) = Z$ for some $Z \in \mathscr{S}(Y)$ is cofinal in $\mathscr{S}(X)$: for every $W_0 \in \mathscr{S}(X)$, setting $Z := \mathrm{cl}(A(W_0)) \in \mathscr{S}(Y)$, there exists $W \in \mathscr{S}(X)$ such that $A(W) = Z$. Then $W_0 + W \in \mathscr{S}_A(X)$, since $A(W_0 + W) = Z$ and W_0 is contained in $W_0 + W$. $\square$

Lemma 7.2. *Let S be a closed subset of a Banach space X and let $W_0 \in \mathscr{S}(X)$. Then there exists $W \in \mathscr{S}(X)$ containing W_0 such that $d(x,S) = d(x,S \cap W)$ for all $x \in W$.*

Proof. Starting with W_0, we define inductively an increasing sequence $(W_n)_{n \ge 1}$ of $\mathscr{S}(X)$ such that $d(\cdot, S) = d(\cdot, S \cap W_n)$ on W_{n-1}. Assuming that $W_1, \ldots, W_n$ satisfying this property have been defined, in order to define W_{n+1} we take a countable dense subset $D_n := \{w_p : p \in \mathbb{N}\}$ of W_n; here, to avoid heavy notation we do not write the dependence on n of the elements of D_n. For all $p \in \mathbb{N}$ we pick sequences $(x_{k,p})_{k \ge 0}$ of

S such that $d(w_p, x_{k,p}) \leq d(w_p, S) + 2^{-k}$ for all $k, p \in \mathbb{N}$. Let W_{n+1} be the closure of the linear span of the set $\{x_{k,p} : (k,p) \in \mathbb{N}^2\} \cup D_n$. Then $W_n \subset W_{n+1}$, $W_{n+1} \in \mathscr{S}(X)$, and for all $w_p \in D_n$ we have $d(w_p, S) = \inf_k d(w_p, x_{k,p}) = d(w_p, S \cap W_{n+1})$. Since $d(\cdot, S)$ and $d(\cdot, S \cap W_{n+1})$ are continuous and since D_n is dense in W_n, these two functions coincide on W_n. Thus, our inductive construction is achieved.

Finally, we take for W the closure of the space spanned by the union of the family (W_n) and we use the inequalities $d(x, S) = d(x, S \cap W_{n+1}) \geq d(x, S \cap W) \geq d(x, S)$ for $x \in W_n$, and a density argument as above extends the relation $d(\cdot, S) = d(\cdot, S \cap W)$ from the union of the W_n's to W. $\qquad\square$

Remark. If $(S_n)_{n \in \mathbb{N}}$ is a countable collection of closed subsets and if $W_0 \in \mathscr{S}(X)$, then there exists $W \in \mathscr{S}(X)$ containing W_0 such that $d(x, S_n) = d(x, S_n \cap W)$ for all $x \in W$ and all $n \in \mathbb{N}$. In order to prove this, we use the construction of the lemma to get an increasing sequence $(W_n)_{n \geq 1}$ of $\mathscr{S}(X)$ containing W_0 such that $d(x, S_i) = d(x, S_i \cap W_n)$ for all $x \in W_{n-1}$ and all $i = 0, \ldots, n-1$. Then we take for W the closed linear span of the union of the family (W_n).

7.1.2 The Graded Subdifferential and the Graded Normal Cone

The following notion has some analogy with the scheme of Galerkin approximations in numerical analysis and could be called the Galerkin–Ioffe subdifferential. It has been called the *geometric approximate subdifferential* by A. Ioffe. For f in the set $\mathscr{L}(X)$ of locally Lipschitzian functions on X, $x \in X$, and $W \in \mathscr{S}(X)$, it involves the sets

$$\partial_D^W f(x) := \left\{ w^* \in W^* : \forall w \in W \ \langle w^*, w \rangle \leq df(x, w) := \liminf_{t \to 0_+} \frac{1}{t}[f(x + tw) - f(x)] \right\}$$

and

$$\partial_\ell^W f(\bar{x}) := \{ \overline{w}^* \in W^* : \exists (x_n) \to \bar{x}, \exists (w_n^*) \xrightarrow{*} \overline{w}^*, \ w_n^* \in \partial_D^W f(x_n) \ \forall n \}.$$

Thus, $\partial_\ell^W f(\bar{x}) := w^* - \text{seq} - \limsup_{x \to \bar{x}} \partial_D^W f(x)$. Introducing the function $f_{x,W} \in \mathscr{L}(W)$ given by $f_{x,W}(w) := f(x + w)$ for $w \in W$, one has $df(x, w) = f_{x,W}^D(0, w)$, hence $\partial_D^W f(x) = \partial_D f_{x,W}(0)$. Although the set $\partial_\ell^W f(\bar{x})$ stands in W^*, it is distinct from $\partial_\ell f_{\bar{x},W}(0)$, since the sequence (x_n) may be out of W. When $W \in \mathscr{S}(X, \bar{x}) := \{ W \in \mathscr{S}(X) : \bar{x} \in W \}$, one has $\partial_\ell f_{\bar{x},W}(0) = \partial_\ell f_W(\bar{x})$, where f_W denotes the restriction of f to W; but still $\partial_\ell^W f(\bar{x})$ may differ from $\partial_\ell f_W(\bar{x})$. Since $\partial_\ell^W f(\bar{x})$ is a subset of W^* and not a subset of X^*, we use its inverse image by the restriction map $r_W := j_W^\mathsf{T} : X^* \to W^*$ given by $r_W(x^*) = x^* \mid_W$ to get a subset of X^*.

Definition 7.3. The *graded subdifferential* of $f \in \mathscr{L}(X)$ at $\bar{x} \in X$ is the set

$$\partial_G f(\bar{x}) = \bigcap_{W \in \mathscr{S}(X)} r_W^{-1}(\partial_\ell^W f(\bar{x})). \tag{7.1}$$

Given $W, Z \in \mathscr{S}(X)$ with $W \subset Z$, one has $r_Z^{-1}(\partial_\ell^Z f(\bar{x})) \subset r_W^{-1}(\partial_\ell^W f(\bar{x}))$. It follows that in relation (7.1), instead of taking the intersection over the whole family $\mathscr{S}(X)$ one may take the intersection over a *cofinal subfamily* $\mathscr{Z}$ in the sense that for all $W \in \mathscr{S}(X)$ one can find some $Z \in \mathscr{Z}$ such that $W \subset Z$.

Proposition 7.4. *If c is the Lipschitz rate of $f \in \mathscr{L}(X)$ at $\bar{x} \in X$, then $\partial_G f(\bar{x})$ is a nonempty weak* compact subset of cB_{X^*} and one has*

$$\partial_G f(\bar{x}) = \bigcap_{W \in \mathscr{S}(X)} r_W^{-1}(\partial_\ell^W f(\bar{x})) \cap cB_{X^*}. \tag{7.2}$$

Moreover, the multimap $\partial_G f$ is upper semicontinuous from X endowed with the topology associated with the norm to X^* endowed with the weak* topology.

Proof. For every $\bar{x}^* \in \partial_G f(\bar{x})$ and every $W \in \mathscr{S}(X)$ one has $r_W(\bar{x}^*) \in \partial_\ell^W f(\bar{x})$, hence $\|r_W(\bar{x}^*)\|_{W^*} \le c$, since for some sequences $(x_n) \to \bar{x}$, $(c_n) \to c$ one has $df(x_n, \cdot) \le c_n \|\cdot\|$ for all n. It follows that $\|\bar{x}^*\| \le c$. Thus $\partial_G f(\bar{x})$ is the intersection of the family $(r_W^{-1}(\partial_\ell^W f(\bar{x})) \cap cB_{X^*})_{W \in \mathscr{S}(X)}$. Moreover, the sets of this family are nonempty, since $\partial_\ell^W f(\bar{x})$ contains $\partial_\ell f_{\bar{x},W}(0)$, and for all $w^* \in \partial_\ell^W f(\bar{x})$, the Hahn–Banach theorem yields some $x^* \in r_W^{-1}(w^*)$ such that $\|x^*\| = \|w^*\| \le c$. Given $W, Z \in \mathscr{S}(X)$ satisfying $W \subset Z$, one obviously has $\partial_D^Z f(x) |_W \subset \partial_D^W f(x)$, hence $\partial_\ell^Z f(x) |_W \subset \partial_\ell^W f(x)$ and $r_Z^{-1}(\partial_\ell^Z f(\bar{x})) \cap cB_{X^*} \subset r_W^{-1}(\partial_\ell^W f(\bar{x})) \cap cB_{X^*}$. Thus, the directed family $(r_W^{-1}(\partial_\ell^W f(\bar{x})) \cap cB_{X^*})_{W \in \mathscr{S}(X)}$ of weak* compact subsets of X^* has nonempty intersection.

Given an open neighborhood V of 0 in X^* for the weak* topology, we can find $\delta > 0$ such that $\partial_G f(x) \subset \partial_G f(\bar{x}) + V$ for all $x \in B(\bar{x}, \delta)$: otherwise, there would exist sequences $(x_n) \to \bar{x}$, (x_n^*) in $X^* \backslash (\partial_G f(\bar{x}) + V)$ with $x_n^* \in \partial_G f(x_n)$ for all n, and since (x_n^*) is bounded, it would have a weak* cluster point x^* and for all $W \in \mathscr{S}(X)$, using the weak* sequential compactness of B_{W^*}, we would have $r_W(x^*) \in \partial_\ell^W f(\bar{x})$, hence $x^* \in \partial_G f(\bar{x})$, a contradiction to $x^* \in X^* \backslash (\partial_G f(\bar{x}) + V)$, which is weak* closed. $\square$

Another approach to $\partial_G f(\bar{x})$ can be given. It consists in taking the inverse image under r_W before passing to the lim sup. For the sake of clarity, let us set

$$\partial_X^W f(x) := r_W^{-1}(\partial_D^W f(x)) = \{x^* \in X^* : \forall w \in W \langle x^*, w \rangle \le df(x, w)\}$$

and let us define the *nuclear subdifferential* of $f \in \mathscr{L}(X)$ at $x \in X$ as the set

$$\partial_N f(\bar{x}) = \bigcap_{W \in \mathscr{S}(X)} \bigcup_{r > 0} w^* - \limsup_{x \to \bar{x}} (\partial_X^W f(x) \cap rB_{X^*}). \tag{7.3}$$

The original terminology given by Ioffe was the "nucleus of the approximate subdifferential." The interest of this approach is that the sets $\partial_X^W f(x)$ are subsets of X^* for all $W \in \mathscr{S}(X)$.

Remark. Using the function $f_{x,W} \in \mathcal{L}(W)$ given by $f_{x,W}(w) := f(x+w)$ for $w \in W$, we observed that $df(x,w) = f^D_{x,W}(0,w)$, hence $\partial^W_X f(x) = r^{-1}_W(\partial_D f_{x,W}(0))$. When W belongs to the family $\mathcal{S}(X,x)$ of $W \in \mathcal{S}(X)$ containing x, one has $\partial^W_D f(x) = r^{-1}_W(\partial_D f_W(x))$, where $f_W := f_{0,W} := f \circ j_W$ is the restriction of f to W.

Exercise. Check that $\partial^W_X f(x) = \partial_D f^{x+W}(x)$, where for $S \subset X$, $f^S := f + \iota_S$.

Let $\partial^W_{\ell,X} f(\overline{x})$ be the set of weak* cluster points of bounded sequences (x^*_n) such that $x^*_n \in \partial^W_X f(x_n)$ for some sequence $(x_n) \to \overline{x}$ in X. With the notation $\to^*$, we have

$$\partial^W_{\ell,X} f(\overline{x}) := \{\overline{x}^* \in X^* : \exists r > 0, \ \exists (x_n) \to \overline{x}, \ \exists (x^*_n) \to^* \overline{x}^*, \ x^*_n \in \partial^W_X f(x_n) \cap rB_{X^*} \ \forall n\}.$$

Proposition 7.5. *For $f \in \mathcal{L}(X)$, $\overline{x} \in X$ one has*

$$\partial_N f(\overline{x}) = \bigcap_{W \in \mathcal{S}(X)} \partial^W_{\ell,X} f(\overline{x}) = \partial_G f(\overline{x}).$$

Proof. Let us first show that $\partial_G f(\overline{x}) \subset \partial_N f(\overline{x})$ by observing that for all $W \in \mathcal{S}(X)$,

$$r^{-1}_W\left(\partial^W_\ell f(\overline{x})\right) \subset \partial^W_{\ell,X} f(\overline{x}) \subset \bigcup_{r>0} w^* - \limsup_{x \to \overline{x}} \left(\partial^W_X f(x) \cap rB_{X^*}\right). \tag{7.4}$$

In fact, given $\overline{x}^* \in r^{-1}_W(\partial^W_\ell f(\overline{x}))$ there exist sequences $(x_n) \to \overline{x}$ in X, $(w^*_n) \overset{*}{\to} \overline{x}^*_W := r_W(\overline{x}^*)$. Since f is Lipschitzian on a neighborhood of $\overline{x}$, one can find $r > 0$ such that $w^*_n \in rB_{W^*}$ for all n. The Hahn–Banach theorem yields some $y^*_n \in rB_{X^*}$ such that $r_W(y^*_n) = w^*_n$. Let $\overline{y}^*$ be a weak* cluster point of (y^*_n). Since r_W is weak* continuous, one has $r_W(\overline{y}^*) = w^* - \lim_n w^*_n = r_W(\overline{x}^*)$. Let $x^*_n := y^*_n + \overline{x}^* - \overline{y}^* \in sB_{X^*}$ with $s := r + \|\overline{x}^* - \overline{y}^*\|$. Then $r_W(x^*_n) = w^*_n$ and $(x^*_n) \to^* \overline{x}^*$, so that $\overline{x}^* \in \partial^W_{\ell,X} f(\overline{x})$. The second inclusion is obvious.

In order to prove that the inclusions (7.4) are equalities, it suffices to show that given $r > c$, the Lipschitz rate of f at $\overline{x}$, a decreasing sequence $(\varepsilon_n) \to 0_+$, $\overline{x}^* \in \bigcap_n \mathrm{cl}^*(F^W_n)$ with $F^W_n := \partial^W_X f(B(\overline{x}, \varepsilon_n)) \cap rB_{X^*}$, one has $\overline{x}^* \in r^{-1}_W(\partial^W_\ell f(\overline{x}))$. For all $n \in \mathbb{N}$, one has $r_W(F^W_n) \subset \partial^W_D f(B(\overline{x}, \varepsilon_n))$, and since r_W is weak* continuous and nonexpansive, $r_W(\overline{x}^*) \in r_W(\mathrm{cl}^*(F^W_n)) \subset \mathrm{cl}^* r_W(F^W_n) \subset rB_{W^*}$. Since W is a WCG space, the Borwein–Fitzpatrick theorem yields some (x_n), $(w^*_n) \overset{*}{\to} r_W(\overline{x}^*)$ such that $x_n \in B(\overline{x}, \varepsilon_n)$, $w^*_n \in \partial^W_D f(x_n)$. Thus $r_W(\overline{x}^*) \in \partial^W_\ell f(\overline{x})$ and $\overline{x}^* \in r^{-1}_W(\partial^W_\ell f(\overline{x}))$. □

The following exercise draws attention to the dangers of making inappropriate extensions.

Exercise. Given $W \in \mathcal{S}(X)$, let $E : W^* \rightrightarrows X^*$ be the normalized extension given by $E(w^*) := \{x^* \in X^* : r_W(x^*) = w^*, \ \|x^*\| = \|w^*\|\}$. Let $f \in \mathcal{L}(X)$, with $X := \mathbb{R}^2$, be the function $(r,s) \mapsto s$. Check that for $W := \mathbb{R} \times \{0\}$, $Z = \mathbb{R}^2$ one has $E(\partial_D f_W(0)) \cap E(\partial_D f_Z(0)) = \varnothing$ although $\partial^Z_D f(x) \subset \partial^W_D f(x)$ are both nonempty.

The following lemma will be useful.

Lemma 7.6. *Let I be a directed set and let $(W_i)_{i \in I}$ be a cofinal subfamily of $\mathscr{S}(X)$ such that $W_i \subset W_j$ for $i \leq j$ in I. Given a function $f : X \to \mathbb{R}_\infty$ that is Lipschitzian with rate r around $\bar{x} \in f^{-1}(\mathbb{R})$, for $i \in I$, let $x_i^* \in C_i := \mathrm{w}^* - \limsup_{x \to \bar{x}}(\partial_D^{W_i} f(x) \cap rB_{X^*})$. Then $(x_i^*)_{i \in I}$ has a weak* cluster point $\bar{x}^* \in \partial_N f(\bar{x}) = \partial_G f(\bar{x})$.*

Proof. We have $C_j \subset C_i \subset rB_{X^*}$ for $i, j \in I$ satisfying $i \leq j$. Let $\bar{x}^*$ be a weak* cluster point of $(x_j^*)_{j \in I}$ in the weak* compact set rB_{X^*}. Since for all $i \in I$ the set C_i is weak* closed, we have $\bar{x}^* \in C_i$. Thus $\bar{x}^* \in \cap_{i \in I} C_i$. Since $(W_i)_{i \in I}$ is cofinal, $\bar{x}^* \in \partial_N f(\bar{x})$. □

Let us turn to a geometric concept.

Definition 7.7. The *graded normal cone* to a subset S of X at $x \in S$ is the cone $N_G(S, x)$ generated by $\partial_G d_S(x)$:

$$N_G(S, x) := \mathbb{R}_+ \partial_G d_S(x).$$

7.1.3 Relationships with Other Subdifferentials

Let us compare the graded subdifferential with other subdifferentials.

Proposition 7.8. *For a locally Lipschitzian function f on an Asplund space X one has*

$$\partial_L f(\bar{x}) \subset \partial_\ell f(\bar{x}) \subset \partial_G f(\bar{x}) \subset \overline{\mathrm{co}}^*(\partial_G f(\bar{x})) = \partial_C f(\bar{x}).$$

Proof. The inclusion $\partial_L f(\bar{x}) \subset \partial_\ell f(\bar{x})$ (resp. $\partial_\ell f(\bar{x}) \subset \partial_G f(\bar{x})$) stems from the obvious inclusion $\partial_F f(x) \subset \partial_D f(x)$ (resp. $\partial_D f(x) \subset \partial_X^W f(x)$) for all $x \in X$, $W \in \mathscr{S}(X)$. Given $\bar{x}^* \in \partial_G f(\bar{x})$ and $v \in X$, for every $W \in \mathscr{S}(X)$ one can find sequences $(x_n) \to \bar{x}$, $(x_n^*) \to^* \bar{x}^*$ with $\langle x_n^*, v \rangle \leq f^D(x_n, v) \leq f^C(x_n, v)$. Thus $\langle \bar{x}^*, v \rangle \leq \limsup_n \langle x_n^*, v \rangle \leq \limsup_n f^C(x_n, v) \leq f^C(x, v)$ by Proposition 5.2 (c). Since v is arbitrary, we get $\bar{x}^* \in \partial_C f(\bar{x})$.

To prove that $\partial_C f(\bar{x}) \subset \overline{\mathrm{co}}^*(\partial_G f(\bar{x}))$ it suffices to show that for all $v \in X$ one has

$$f^C(\bar{x}, v) \leq \sup\{\langle \bar{x}^*, v \rangle : \bar{x}^* \in \partial_G f(\bar{x})\}. \tag{7.5}$$

Let $r > c$, the Lipschitz rate of f at $\bar{x}$. The definition of $f^C(\bar{x}, v)$ yields $(w_n) \to \bar{x}$, $(t_n) \to 0_+$ such that $t_n^{-1}(f(w_n + t_n v) - f(w_n)) \to f^C(\bar{x}, v)$. For all W in the set $\mathscr{W}$ of $W \in \mathscr{S}(X)$ containing v and $\{w_n : n \in \mathbb{N}\}$, the mean value theorem in the H-smooth space W yields some $u_n \in [w_n, w_n + t_n v] + 2^{-n} B_W$ and some $u_n^* \in \partial_D f_W(u_n)$ such that $\langle u_n^*, t_n v \rangle \geq f(w_n + t_n v) - f(w_n) - 2^{-n} t_n$. Taking a sequence (x_n^*) in rB_{X^*} such that $r_W(x_n^*) = u_n^*$ for all n and a weak* cluster point $x^*(W)$ of (x_n^*) and then a weak* cluster point of $(x^*(W))_{W \in \mathscr{W}}$, one gets an element $\bar{x}^*$ of $\partial_N f(\bar{x}) = \partial_G f(\bar{x})$ by Lemma 7.6 with $\langle \bar{x}^*, v \rangle \geq f^C(\bar{x}, v)$, since $\langle x^*(W), v \rangle \geq f^C(\bar{x}, v)$ for all $W \in \mathscr{W}$ and (7.5) holds. □

Theorem 7.9 (Coincidence in WCG spaces). *Let X be a WCG space and let $f :$
$X \to \mathbb{R}_\infty$ be Lipschitzian around $\bar{x} \in X$. Let $\partial_h f(x) := \mathrm{w}^* - \mathrm{seq} - \limsup_{w \to x} \partial_H f(w)$
and $\partial_\ell f(x) := \mathrm{w}^* - \mathrm{seq} - \limsup_{w \to x} \partial_D f(w)$ as in Theorem 6.82. Then*

$$\partial_G f(\bar{x}) = \partial_h f(\bar{x}) = \partial_\ell f(\bar{x}). \tag{7.6}$$

If X is a WCG Asplund space, then one has

$$\partial_G f(\bar{x}) = \partial_L f(\bar{x}).$$

Proof. Let $U \in \mathcal{N}(\bar{x})$ and $c > 0$ be such that f is c-Lipschitzian on U. Since for all
$u \in U$ and all $W \in \mathscr{S}(X)$ one has $\partial_H f(u) \subset \partial_D f(u) \subset \partial_X^W f(u)$, one gets $\partial_h f(\bar{x}) \subset$
$\partial_\ell f(\bar{x}) \subset \partial_{\ell,X}^W f(\bar{x})$, hence $\partial_h f(\bar{x}) \subset \partial_\ell f(\bar{x}) \subset \partial_G f(\bar{x})$. Since $\partial_h f(\bar{x})$ is weak* closed,
to prove (7.6) it suffices to show that $\partial_G f(\bar{x}) \subset \mathrm{cl}^*(\partial_h f(\bar{x}))$. Let $\bar{x}^* \in \partial_G f(\bar{x})$ and let
$\varepsilon > 0$, $V := W^\perp + \varepsilon B_{X^*}$, where W is a finite-dimensional subspace of X containing
$\bar{x}$. Since the sets like V form a base of neighborhoods of 0 in the weak* topology, it
suffices to show that $(\bar{x}^* + V) \cap \partial_h f(\bar{x})$ is nonempty. Since X is a WCG space and
$\bar{x}^* \in \mathrm{w}^* - \limsup_{x \to \bar{x}} \partial_X^W f(x)$, given a sequence $(\varepsilon_n) \to 0_+$ in $(0, \varepsilon/3)$, one can find
sequences $(x_n) \to \bar{x}$, $(x_n^*) \overset{*}{\to} \bar{x}^*$ in X^* such that $x_n^* \in \partial_X^W f(x_n)$ and $|\langle x_n^* - \bar{x}^*, w \rangle| \le$
$\varepsilon_n \|w\|$ for all n and all $w \in W$. Thus x_n is a robust local minimizer on X of

$$u \mapsto f(u) + \iota_W(u) - \langle \bar{x}^*, u \rangle + 2\varepsilon_n \|u\|.$$

The fuzzy minimization rule (Corollary 4.64) yields sequences (u_n), (u_n^*), (v_n^*) such
that $u_n \in B(x_n, \varepsilon_n)$, $u_n^* \in \partial_H f(u_n)$, $v_n^* \in W^\perp$ and $\|u_n^* + v_n^* - \bar{x}^*\| \le 3\varepsilon_n$ for all n. Thus
$u_n^* \in \bar{x}^* + W^\perp + 3\varepsilon_n B_{X^*} \subset \bar{x}^* + V$. Let u^* be the weak* limit of a subsequence of (u_n^*)
in $c B_{X^*}$. Then $u^* \in \partial_h f(\bar{x})$, and since V is weak* closed, $u^* \in \bar{x}^* + V$.

If X is both a WCG space and an Asplund space, we know that $\partial_\ell f(\bar{x}) = \partial_L f(\bar{x})$. $\square$

In separable Banach spaces, one has a simple criterion at one's disposal.

Lemma 7.10. *Let X be a separable Banach space, let (X_n) be an increasing
sequence of linear subspaces whose union is dense in X, and let $f \in \mathscr{L}(X)$, $\bar{x} \in X$,
$\bar{x}^* \in X^*$. Let $(x_n) \to \bar{x}$ in X, $(x_n^*) \overset{*}{\to} \bar{x}^*$ in X^* be such that $\langle x_n^*, x \rangle \le df(x_n, x)$ for all
$x \in X_n$. Then $\bar{x}^* \in \partial_\ell f(\bar{x}) = \partial_G f(\bar{x})$.*

Proof. We first observe that we may assume that for all $n \in \mathbb{N}$, X_n is finite-
dimensional. If it is not the case, taking a countable subset $\{e_k : k \in \mathbb{N}\}$ and for
$n \in \mathbb{N}$ some $p(n) \in \mathbb{N}$ and $x_{k,n} \in X_{p(n)}$ with $\|e_k - x_{k,n}\| \le 2^{-n}$ for $k = 0, \dots, n$ with
$p(n+1) \ge p(n)$, we can replace (X_n) with the sequence (W_n) and x_n^* with $x_{p(n)}^*$,
where W_n is the linear hull of $\{x_{k,n} : k = 0, \dots, n\}$. By assumption, for all n we have
$r_{X_n}(x_n^*) \in \partial_D f_{x_n, X_n}(0)$, where $f_{x_n, X_n} \in \mathscr{L}(X_n)$ is given by $f_{x_n, X_n}(x) := f(x_n + x)$ for
$x \in X_n$. Let r be the Lipschitz rate of f on a neighborhood of $\bar{x}$ and let $(\varepsilon_n) \to 0_+$ in
$(0, 1]$. Then for $x \in X_n$ small enough we have $f(x_n + x) - \langle x_n^*, x \rangle + \varepsilon_n \|x\| \ge f(x_n)$.
The penalization lemma ensures that the function

$$x \mapsto f(x_n + x) - \langle x_n^*, x \rangle + \varepsilon_n \|x\| + (2r + 1)d(x, X_n)$$

attains a local minimum at 0. The sum rule in the H-smooth space X yields $w_n \in B(x_n, \varepsilon_n)$, $w_n^* \in \partial_D f(w_n)$, $u_n^* \in 2\varepsilon_n B_{X^*}$, $z_n^* \in X_n^\perp \cap (2r+1)B_{X^*}$ such that $w_n^* - x_n^* = z_n^* + u_n^*$. Then $(w_n) \to \bar{x}$, and since $(w_n^* - x_n^*)$ is bounded and $(\langle w_n^* - x_n^*, x \rangle) \to 0$ for all $x \in \cup_n X_n$, we get $(w_n^*) \xrightarrow{*} \bar{x}^*$. Thus $\bar{x}^* \in \partial_\ell f(\bar{x})$. $\square$

One is led to the following question: in the definition of $\partial_G f(\bar{x})$ can one substitute for $\mathscr{S}(X)$ some other family of subspaces of X? We consider the important case of the family of finite-dimensional subspaces of X.

Proposition 7.11. *Given a Banach space X, $\bar{x} \in X$, and $f \in \mathscr{L}(X)$, denoting by $\mathscr{D}(X)$ the family of finite-dimensional subspaces of X, one has*

$$\partial_G f(\bar{x}) = \bigcap_{W \in \mathscr{D}(X)} \partial_{\ell, X}^W f(\bar{x}).$$

Proof. Since $\mathscr{D}(X) \subset \mathscr{S}(X)$, it suffices to show that for all $\bar{x}^* \in X^*$ and for all $W \in \mathscr{S}(X)$ one has $\bar{x}^* \in \partial_{\ell, X}^W f(\bar{x})$ whenever $\bar{x}^* \in \partial_{\ell, X}^Z f(\bar{x})$ for all $Z \in \mathscr{D}(X)$. Given such an $\bar{x}^*$ and $W \in \mathscr{S}(X)$, picking $\rho > 0$ such that f is Lipschitzian with rate r on $B(\bar{x}, \rho)$, by Proposition 7.5 it is enough to prove that $\bar{x}^* \in K := \mathrm{w}^* - \limsup_{x \to \bar{x}} (\partial_X^W f(x) \cap r B_{X^*})$ or even that for every finite subset F of X one has $\bar{x}^* \in K + F^0$ (we use the fact that K is weak* compact and that F^0 is an element of a basis of neighborhoods of 0 for the weak* topology). Let Z be the linear span of F and let $(x_n) \to \bar{x}$ in $B(\bar{x}, \rho)$, $(x_n^*) \to^* \bar{x}^*$ be such that $x_n^* \in \partial_X^Z f(x_n) \cap r B_{X^*}$ for all n. Since Z is finite-dimensional, 0 is a local minimizer of the restriction to Z of the $(2r + \varepsilon_n)$-Lipschitzian function $g_n : z \mapsto f(x_n + z) - \langle x_n^*, z \rangle + \varepsilon_n \|z\|$. The penalization lemma ensures that 0 is a local minimizer of $g_n + (2r + \varepsilon_n) d_Z$ on X, hence on W. Since W is an H-smooth space, one can find $v_n, w_n \in \varepsilon_n B_W$ such that $0 \in \partial_D g_n(w_n) + (2r + \varepsilon_n) \partial_D d_Z(v_n) + \varepsilon_n B_{W^*}$ in W^*. Let $y_n^* \in \partial_X^W f(x_n + w_n) \cap r B_{X^*}$ be such that

$$0 \in r_W(y_n^* - x_n^*) + (2r + \varepsilon_n) \partial_D(d_Z \mid w)(v_n) + 2\varepsilon_n B_{W^*}.$$

One may suppose that $x_n^* - y_n^* \in 2\varepsilon_n B_{X^*} + Z^\perp \subset F^0$ and that $(x_n^* - y_n^*) \to^* \bar{x}^* - \bar{y}^* \in Z^\perp$ for some $\bar{y}^* \in r B_{X^*}$. Then $\bar{y}^* \in K$ and $x_n^* \in \partial_X^W f(x_n + w_n) + 2\varepsilon_n B_{X^*} + Z^\perp$. For all $m \in \mathbb{N}$ there exists $n \geq m$ with $\bar{x}^* - \bar{y}^* \in x_n^* - y_n^* + \varepsilon_n B_{X^*} + Z^\perp \subset 3\varepsilon_n B_{X^*} + Z^\perp$; hence $\bar{x}^* \in \bar{y}^* + Z^\perp \subset K + F^0$.

7.1.4 Elementary Properties in the Lipschitzian Case

Let us give some elementary properties of the graded subdifferential on the class of locally Lipschitzian functions. We start with a simple composition property.

Proposition 7.12. *Let X, Y be two Banach spaces, let $A : X \to Y$ be a surjective linear continuous map, let $f \in \mathscr{L}(X)$, $g \in \mathscr{L}(Y)$ be such that $f \geq g \circ A$. Suppose* *that for some $\bar{x} \in X$ and every sequence $(y_n) \to \bar{y} := A\bar{x}$ there exists a sequence $(x_n) \to \bar{x}$ such that $Ax_n = y_n$ and $f(x_n) = g(y_n)$ for all n large enough. Then one has $A^{\mathsf{T}}(\partial_G g(\bar{y})) \subset \partial_G f(\bar{x})$.*

If $f = g \circ A$, then one has $\partial_G f(\bar{x}) = A^{\mathsf{T}}(\partial_G g(\bar{y}))$.

Proof. Let $\bar{y}^* \in \partial_G g(\bar{y})$, let $\bar{x}^* := A^{\mathsf{T}}(\bar{y}^*)$, and let $W \in \mathscr{S}(X)$. One has $Z := \mathrm{cl}\,A(W) \in \mathscr{S}(Y)$, so that by definition, there exist sequences $(y_n) \to \bar{y}$ in Y, $(z_n^*) \xrightarrow{*} r_Z(\bar{y}^*)$ satisfying $z_n^* \in \partial_D^Z g(y_n)$ for all n. Taking a sequence $(x_n) \to \bar{x}$ satisfying $A(x_n) = y_n$, $f(x_n) = g(y_n)$ for all large n, setting $w_n^* := A_W^{\mathsf{T}}(z_n^*)$, where A_W is the restriction of A to W and Z, for all $w \in W$ one has $\langle w_n^*, w \rangle = \langle z_n^*, A_W(w) \rangle \leq dg(y_n, Aw) \leq df(x_n, w)$, hence $w_n^* \in \partial_D^W f(x_n)$. Since A_W^{T} is weak* continuous, $(w_n^*) \xrightarrow{*} A_W^{\mathsf{T}}(r_Z(\bar{y}^*)) = r_W(A^{\mathsf{T}}(\bar{y}^*)) = r_W(\bar{x}^*)$, one gets $r_W(\bar{x}^*) \in \partial_\ell^W f(\bar{x})$. Thus $\bar{x}^* \in \partial_G f(\bar{x})$.

Now suppose $f = g \circ A$ and let $\bar{x}^* \in \partial_G f(\bar{x})$. Let $w \in N := A^{-1}(0)$, $V := \mathbb{R}w$. For all $x \in X$, $v \in V$ we have $df(x, v) = 0$, $df(x, -v) = 0$, hence $\langle \bar{x}^*, v \rangle = 0$, since $r_V(\bar{x}^*) \in \partial_\ell^V f(\bar{x})$. Thus there exists $\bar{y}^* \in Y^*$ such that $\bar{x}^* = A^{\mathsf{T}}(\bar{y}^*)$. Let us show that $\bar{y}^* \in \partial_G g(\bar{y})$. Given $Z \in \mathscr{S}(Y)$, Lemma 7.1 yields some $W \in \mathscr{S}(X)$ such that $A(W) = Z$. Let $A_W \in L(W, Z)$ be the restriction of A to W and Z. Since $r_W(\bar{x}^*) \in \partial_\ell^W f(\bar{x})$ there exist sequences $(x_n) \to \bar{x}$ in X, $(w_n^*) \xrightarrow{*} r_W(\bar{x}^*)$ in W^* such that $w_n^* \in \partial_D^W f(x_n)$ for all n. For $w \in W$, $x \in X$, $y := Ax$, $z \in Z$, let us set $f_{x,W}(w) = f(x + w)$, $g_{y,Z}(z) = g(y + z)$, so that $f_{x,W} = g_{y,Z} \circ A_W$ and $\partial_D f_{x,W}(0) = A_W^{\mathsf{T}}(\partial_D g_{y,Z}(0))$. Thus, for all $n \in \mathbb{N}$, we have $w_n^* \in \partial_D f_{x_n,W}(0) = A_W^{\mathsf{T}}(\partial_D g_{y_n,Z}(0))$ with $y_n := Ax_n$. Let $z_n^* \in \partial_D g_{y_n,Z}(0)$ be such that $w_n^* = A_W^{\mathsf{T}}(z_n^*)$. The Banach–Schauder theorem ensures that $\omega(A_W)\|z_n^*\| \leq \|w_n^*\|$, where $\omega(A_W)$ is the openness rate of A_W. Thus (z_n^*) is bounded, and since B_{W^*} is weak* sequentially compact, we can find a weak* convergent subsequence. Let $z^* \in Z^*$ be the limit of this subsequence. Since A_W^{T} is weak* continuous, we have $A_W^{\mathsf{T}}(z^*) = \lim_n w_n^* = r_W(\bar{x}^*)$. On the other hand,

$$r_W(\bar{x}^*) = r_W(A^{\mathsf{T}}(\bar{y}^*)) = \bar{y}^* \circ A \circ j_W = \bar{y}^* \circ j_Z \circ A_W = r_Z(\bar{y}^*) \circ A_W = A_W^{\mathsf{T}}(r_Z(\bar{y}^*)),$$

so that $A_W^{\mathsf{T}}(z^*) = A_W^{\mathsf{T}}(r_Z(\bar{y}^*))$. Since A_W^{T} is injective, one gets $z^* = r_Z(\bar{y}^*)$ and $r_Z(\bar{y}^*) = \mathrm{w}^* - \lim_n z_n^* \in \partial_\ell^Z g(\bar{y})$. Therefore $\bar{y}^* \in \partial_G g(\bar{y})$. $\square$

Exercise. Prove this proposition using the alternative characterizations of $\partial_G f(\bar{x})$ and $\partial_G g(\bar{y})$.

For the next proposition similar to Proposition 7.12 but involving a distance function, we need a technical result of independent interest.

Lemma 7.13. *Let E be a nonempty subset of a Banach space X, let $g := d_E$, $\bar{x} \in X$, $\bar{x}^* \in \partial_G g(\bar{x})$, and let $W_0 \in \mathscr{S}(X)$. Then there exist some $W \in \mathscr{S}(X)$ containing $\bar{x}$ and W_0 and sequences $(w_n) \to \bar{x}$ in W, $(w_n^*) \xrightarrow{*} r_W(\bar{x}^*)$ in W^* such that $w_n^* \in \partial_D^W g(w_n)$ and $d(\cdot, E) = d(\cdot, E \cap W)$ on W. In particular, $r_W(\bar{x}^*) \in \partial_\ell g_W(\bar{x})$, where g_W is the restriction of g to W. Moreover, if $\bar{x} \in E$, then one can take $w_n \in E \cap W$ for all n.*

Proof. We construct inductively a sequence (W_n) in $\mathscr{S}(X)$ such that $W_n \subset W_{n+1}$, $d(\cdot, E) = d(\cdot, E \cap W_{n+1})$ on W_{n+1}, $\bar{x} \in W_{n+1}$ for all n. For $n = 0$, W_1 is given by Lemma 7.2. Suppose W_k has been defined for $k = 1, \ldots, n$. Since $r_{W_n}(\bar{x}^*) \in$

$\partial_\ell^{W_n} d_E(\overline{x})$, there exist sequences $(x_{n,p})_p \to \overline{x}$ in X, $(w^*_{n,p}) \xrightarrow{*} r_{W_n}(\overline{x}^*)$ in W_n^* satisfying $w^*_{n,p} \in \partial_D^{W_n} d_E(x_{n,p})$ for all $p \in \mathbb{N}$. Using again Lemma 7.2, we pick $W_{n+1} \in \mathscr{S}(X)$ such that $W_n \subset W_{n+1}$, $x_{n,p} \in W_{n+1}$ for all $p \in \mathbb{N}$ and $d(\cdot, E) = d(\cdot, E \cap W_{n+1})$ on W_{n+1}. Let W be the closure of $\cup_n W_n$. Then by density, we have $d(\cdot, E) = d(\cdot, E \cap W)$ on W. Given a sequence $(\varepsilon_p) \to 0_+$, we may suppose $x_{n,p} \in B(\overline{x}, \varepsilon_p)$ for all n, p and we can pick $x^*_{n,p} \in r_{W_n}^{-1}(w^*_{n,p}) \cap B_{X^*}$ for all n, p with $(x^*_{n,p})_p \to_* \overline{x}^*$, as we already observed. Let d be a metric on W^* inducing the weak* topology on B_{W^*}. Since $(r_W(x^*_{n,p}))_p \to_* r_W(\overline{x}^*)$, we can find some $p(n) \geq n$ in $\mathbb{N}$ such that $d(r_W(x^*_{n,p(n)}), r_W(\overline{x}^*)) \leq \varepsilon_n$. Replacing X, X_n, x_n, x_n^* with W, W_n, $x_{n,p(n)}$, $r_W(x_{n,p(n)})$ in Lemma 7.10, we get $r_W(\overline{x}^*) \in \partial_\ell g_W(\overline{x})$ and sequences $(w_n) \to \overline{x}$ in W, $(w_n^*) \xrightarrow{*} r_W(\overline{x}^*)$ in W^* such that $w_n^* \in \partial_D^W g(w_n)$. If $\overline{x} \in E$, since W is H-smooth and B_{W^*} is metrizable for the weak* topology, invoking Theorem 4.87, we may require that $w_n \in E \cap W$ for all n. $\square$

Proposition 7.14. *Let X, Y be two Banach spaces, let $A : X \to Y$ be a surjective linear continuous map, let E be a closed subset of Y, and let $f \in \mathscr{L}(X)$ be such that $f \geq g \circ A$ for $g := d_E$. Let $\overline{x} \in A^{-1}(E)$. Suppose that for every sequence $(y_n) \to \overline{y} := A(\overline{x})$ in E one can find a sequence $(x_n) \to \overline{x}$ satisfying $A(x_n) = y_n$ and $f(x_n) = 0$ for all $n \in \mathbb{N}$ large enough. Then one has $A^\top(\partial_G g(\overline{y})) \subset \partial_G f(\overline{x})$.*

Proof. Let $\overline{y}^* \in \partial_G g(\overline{y})$, let $\overline{x}^* := A^\top(\overline{y}^*)$, and let $W \in \mathscr{S}(X)$ containing $\overline{x}$. Let us show that $r_W(\overline{x}^*) \in \partial_\ell^W f(\overline{x})$ to obtain that $\overline{x}^* \in \partial_G f(\overline{x})$. Let $Z_0 := \mathrm{cl}(A(W))$. Replacing X, W_0, $\overline{x}$, $\overline{x}^*$ with Y, Z_0, $\overline{y}$, $\overline{y}^*$ in the preceding lemma, we can find $Z \in \mathscr{S}(Y)$ containing $\overline{y}$ and Z_0 and sequences $(z_n) \to \overline{y}$ in $E \cap Z$, $(z_n^*) \xrightarrow{*} r_Z(\overline{y}^*)$ in Z^* such that $z_n^* \in \partial_D^Z g(z_n)$ for all $n \in \mathbb{N}$ and $d(y, E \cap Z) = d(y, E)$ for all $y \in Z$. By assumption there exists a sequence $(x_n) \to \overline{x}$ satisfying $A(x_n) = z_n$ and $f(x_n) = 0$ for all n large enough. Let $A_W \in L(W, Z)$ be the restriction of A to W and Z. Setting $w_n^* := A_W^\top(z_n^*)$, for all $w \in W$ one has $(w_n^*) \xrightarrow{*} A_W^\top(r_Z(\overline{y}^*)) = r_W(A^\top(\overline{y}^*)) = r_W(\overline{x}^*)$ and

$$\langle w_n^*, w \rangle = \langle z_n^*, A_W(w) \rangle \leq dg(z_n, Aw) \leq df(x_n, w),$$

hence $w_n^* \in \partial_D^W f(x_n)$. This shows that $r_W(\overline{x}^*) \in \partial_\ell^W f(\overline{x})$. Since $W \in \mathscr{S}(X)$ is arbitrary, one gets $\overline{x}^* \in \partial_G f(\overline{x})$. $\square$

We are ready to give a list of properties.

Proposition 7.15. *(a) If f, $g \in \mathscr{L}(X)$ coincide around x, then $\partial_G f(x) = \partial_G g(x)$.*
   (b) If $f \in \mathscr{L}(X)$ is convex, then $\partial_G f(x) = \partial_{MR} f(x)$.
   (c) If $f \in \mathscr{L}(X)$ attains a local minimum at $x \in X$, then $0 \in \partial_G f(x)$.
   (d) If $\lambda > 0$, $A \in L(X, Y)$ with $A(X) = Y$, $b \in Y$, $\ell \in X^*$, $c \in \mathbb{R}$, and $f(x) = \lambda g(Ax + b) + \langle \ell, x \rangle + c$, then $\partial_G f(x) = \lambda A^\top(\partial_G g(Ax + b)) + \ell$.
   (e) If $g \in \mathscr{L}(X)$, $h \in \mathscr{L}(Y)$, $f(x, y) = g(x) + h(y)$, then $\partial_G f(x, y) \subset \partial_G g(x) \times \partial_G h(y)$, with equality when h is convex or just radially differentiable.
   (f) If $f = g + h$, where $g \in \mathscr{L}(X)$, $h : X \to \mathbb{R}$ is circa-differentiable at $\overline{x} \in X$, then $\partial_G f(\overline{x}) = \partial_G g(\overline{x}) + h'(\overline{x})$.
   (g) If $h : X \to \mathbb{R}$ is circa-differentiable at $\overline{x} \in X$, then $\partial_G h(\overline{x}) = \{h'(\overline{x})\}$.

(h) If $f = g \circ h$, where $g \in \mathscr{L}(Y)$, $h : X \to Y$ is circa-differentiable at $\bar{x} \in X$ and $h'(\bar{x})(X) = Y$, then $h'(\bar{x})^{\mathsf{T}}(\partial_G g(h(\bar{x}))) = \partial_G f(\bar{x})$.

Proof. (a) is obvious. (b) stems from the equalities $\partial_D^W f(u) = \partial_{MR} f_{u,W}(0)$, $\partial_\ell^W f(x) = \partial_{MR} f_{x,W}(0)$ when f is convex, $W \in \mathscr{S}(X)$, $u \in W$ with $f_{u,W}(w) := f(u+w)$. (c) is a consequence of the inclusion $\partial_D f(x) \subset \partial_G f(x)$. (d) follows from Proposition 7.12. The inclusion of (e) stems from the fact that the set $\mathscr{S}(X) \times \mathscr{S}(Y)$ is cofinal in $\mathscr{S}(X \times Y)$. When h is radially differentiable one has $df((x,y),(u,v)) = dg(x,u) + dh(y,v)$ and $\partial_D^{W \times Z} f(x,y) = \partial_D^W g(x) \times \partial_D^Z h(y)$. (f) is a direct consequence of the definitions. (g) corresponds to the special case $g = 0$ in (f). (h) can be deduced from (d) and (f) by taking $g = 0$. The function k given by $k(x) := g(h(x)) - g(A(x - \bar{x}) + h(\bar{x}))$, where $A := h'(\bar{x})$, since the Lipschitz rate of k at $\bar{x}$ is 0, so that $k'(\bar{x}) = 0$. $\square$

Theorem 7.16. *Let f, g be two locally Lipschitzian functions on a Banach space X. Then for all $x \in X$,*

$$\partial_G(f + g)(x) \subset \partial_G f(x) + \partial_G g(x).$$

Proof. Let $r > 0$ be such that f and g are r-Lipschitzian on some neighborhood of x. Given $x^* \in \partial_G(f + g)(x)$, $W \in \mathscr{S}(X)$, let $(x_n) \to x$, $(x_n^*) \to^* x^*$ be such that $x_n^* \in \partial_X^W(f + g)(x_n)$ for all n and let V be a weak* closed neighborhood of 0 in W^*. Theorem 4.83 and the Hahn–Banach theorem yield sequences (y_n), (z_n) in X, (y_n^*), (z_n^*) in rB_{X^*} such that $y_n, z_n \in (x_n + W) \cap B(x_n, 2^{-n})$, $r_W(y_n^*) \in \partial_D f_{x_n,W}(y_n - x_n)$, $r_W(z_n^*) \in \partial_D g_{x_n,W}(z_n - x_n)$, $r_W(y_n^*) + r_W(z_n^*) \in r_W(x_n^*) + V$ for all $n \in \mathbb{N}$. Thus $y_n^* \in \partial_X^W f(y_n)$, $z_n^* \in \partial_X^W g(z_n)$, and one may assume that (y_n^*) and (z_n^*) have weak* cluster points $y_{V,W}^*$, $z_{V,W}^*$ respectively such that $r_W(y_{V,W}^* + z_{V,W}^* - x^*) \in V$. Let

$$A_W := w^* - \limsup_{y \to x}(\partial_X^W f(y) \cap rB_{X^*}), \qquad B_W := w^* - \limsup_{z \to x}(\partial_X^W g(z) \cap rB_{X^*}),$$

$C_W := A_W + B_W$. Since C_W is weak* compact and $r_W(x^*) \in r_W(C_W) - V$ for every closed neighborhood V of 0 in W^*, one has $r_W(x^*) \in r_W(C_W)$. Let $y^*(W) \in A_W$, $z^*(W) \in B_W$ be such that $r_W(y^*(W) + z^*(W)) = r_W(x^*)$ and let y^*, z^* be cluster points of the nets $(y^*(W))_{W \in \mathscr{S}(X)}$, $(z^*(W))_{W \in \mathscr{S}(X)}$ respectively. Since $A_W \subset A_Z$, $B_W \subset B_Z$ for $Z \subset W$, one has $y^* \in \bigcap_{Z \in \mathscr{S}(X)} A_Z = \partial_G f(x)$, $z^* \in \bigcap_{Z \in \mathscr{S}(X)} B_Z = \partial_G g(x)$ and $r_Z(y^* + z^*) = r_Z(x^*)$ for all $Z \in \mathscr{S}(X)$, whence $x^* = y^* + z^*$. $\square$

Now let us turn to properties involving order.

Proposition 7.17. *Let X, Y be Banach spaces, let $f : X \to \overline{\mathbb{R}}$ be finite and Lipschitzian around $\bar{x}$, let $g : Y \to \mathbb{R}$ be of class D^1 around $\bar{y}$ with $g'(\bar{y}) \neq 0$, let $(\bar{x}, \bar{y}) \in X \times Y$ be such that $f(\bar{x}) = g(\bar{y})$, and let h be given by $h(x,y) := \max(f(x), g(y))$. Then $(\bar{x}^*, 0) \in \partial_G h(\bar{x}, \bar{y}) \Longrightarrow \bar{x}^* \in \partial_G f(\bar{x})$.*

Proof. Let $(\bar{x}^*, 0) \in \partial_G h(\bar{x}, \bar{y})$. For all $W \in \mathscr{S}(X)$, $Z \in \mathscr{S}(Y)$ we have $(\bar{x}^*, 0) \in \partial_{\ell, X \times Y}^{W \times Z} h(\bar{x}, \bar{y})$. Let $((x_n, y_n)) \to (\bar{x}, \bar{y})$ and $((x_n^*, y_n^*)) \to^* (\bar{x}^*, 0)$ be such that $(x_n^*, y_n^*) \in \partial_{X \times Y}^{W \times Z} h(x_n, y_n)$. Considering three cases as in the proof of Proposition 6.91, we can show that $\bar{x}^* \in \partial_{\ell, X}^W f(\bar{x})$. Since W is arbitrary in $\mathscr{S}(X)$, we get $\bar{x}^* \in \partial_G f(\bar{x})$. $\square$

Some properties of normal cones can be deduced from Propositions 7.14, 7.15.

Proposition 7.18. *Let S be a closed subset S of a Banach space X and let $\bar{x} \in S$.*

(a) *The normal cone $N_G(S,\bar{x})$ to S at $\bar{x}$ does not depend on the choice of the norm on X among those inducing the topology of X.*

(b) *One has $N_G(S,\bar{x}) = [1,+\infty)\partial_G d_S(\bar{x})$.*

(c) *If S is convex, then $N_G(S,\bar{x})$ coincides with the normal cone in the sense of convex analysis: $N_G(S,\bar{x}) = \{\bar{x}^* \in X^* : \forall x \in S, \langle \bar{x}^*, x - \bar{x} \rangle \le 0\}$.*

(d) *If A and B are closed subsets of Banach spaces X and Y respectively and $(\bar{x},\bar{y}) \in A \times B$, then $N_G(A \times B, (\bar{x},\bar{y})) \subset N_G(A,\bar{x}) \times N_G(B,\bar{y})$, and equality holds when d_B is radially differentiable around $\bar{y}$, in particular when B is convex.*

Proof. (a) The result follows from Proposition 7.14, taking for A the identity map and $f = c d'_S$ where $c > 0$ and d'_S is the distance associated with an equivalent norm $\|\cdot\|'$.

(b) For all $r \in (0,1)$ one has $d_S \le r^{-1} d_S$ and hence $r\partial_G d_S(\bar{x}) \subset \partial_G d_S(\bar{x})$ by Proposition 7.14 and $0\partial_G d_S(\bar{x}) \subset \partial_G d_S(\bar{x})$ by Proposition 7.15 (c). Thus $\mathbb{R}_+ \partial_G d_S(\bar{x}) = [1,+\infty)\partial_G d_S(\bar{x})$.

(c) The assertion stems from Proposition 7.15 (b), with $f := d_C$.

(d) Since $d_{A \times B}(x,y) = d_A(x) + d_B(y)$ when one takes the sum norm on $X \times Y$, Proposition 7.15 (e) implies that $\mathbb{R}_+ \partial_G d_{A \times B}(\bar{x},\bar{y}) \subset \mathbb{R}_+ \partial_G d_A(\bar{x}) \times \mathbb{R}_+ \partial_G d_B(\bar{y})$ and the inclusion of normal cones. It also ensures the equality case.   □

Proposition 7.19. *Let X, Y be Banach spaces, let $\bar{x} \in E \subset X$, $F \subset Y$ and let $A \in L(X,Y)$ be such that $A(E) \subset F$. Suppose that for every sequence $(y_n) \to \bar{y} := A\bar{x}$ in F there exists a sequence $(x_n) \to \bar{x}$ in E such that $Ax_n = y_n$ for all $n \in \mathbb{N}$ large enough. Then*

$$A^\top(N_G(F,\bar{y})) \subset N_G(E,\bar{x}). \tag{7.7}$$

In particular, this inclusion holds when $E = A^{-1}(F)$ and $Y = A(X)$.

This inclusion is an equality when A is an isomorphism and $F = A(E)$.

Proof. Setting $f := \|A\| d_E$, $g := d_F$, let us observe that for all $x \in X$, we have

$$g(Ax) \le \inf_{u \in E} \|Ax - Au\| \le \inf_{u \in E} \|A\| \|u - x\| = f(x).$$

Moreover, for every sequence $(y_n) \to_F \bar{y}$, there exists a sequence $(x_n) \to_E \bar{x}$ such that $Ax_n = y_n$ for all n large enough. Then $f(x_n) = 0$ for all n, and Proposition 7.14 ensures that for all $\bar{y}^* \in \partial_G g(\bar{y})$ one has $A^\top(\bar{y}^*) \in \|A\| \partial_G d_E(\bar{x})$. Relation (7.7) ensues. Interchanging E and F and changing A into A^{-1} yields the last assertion.   □

The next proposition will justify the extension given in the next section.

Proposition 7.20. *For every locally Lipschitzian function f on a Banach space X and $\bar{x} \in X$, $\bar{r} := f(\bar{x})$, $\bar{e} := (\bar{x}, f(\bar{x}))$, one has*

$$\partial_G f(\bar{x}) = \{x^* \in X^* : (x^*, -1) \in N_G(\text{epi} f, \bar{e})\}.$$

Proof. Let $f \in \mathscr{L}(X), \bar{x} \in X$ and let E be the epigraph of f. Without loss of generality we may assume that f is globally Lipschitzian with rate $c > 0$ and even that $c = 1$ (since we can change the norm of X to the norm $c \|\cdot\|$). Then endowing $X \times \mathbb{R}$ with the sum norm, we have $d_E(x,r) = (f(x) - r)^+$ for all $(x,r) \in X \times \mathbb{R}$, hence

$$\forall (x,r) \in X \times \mathbb{R}, \qquad f(x) \leq j(x,r) := d_E(x,r) + r.$$

Moreover, for every sequence $(x_n) \to \bar{x}$, we have $j(x_n, r_n) = f(x_n)$ for $r_n := f(x_n)$ and $((x_n, r_n)) \to (\bar{x}, \bar{r})$. Thus, Proposition 7.12 implies that for all $\bar{x}^* \in \partial_G f(\bar{x})$ one has $(\bar{x}^*, 0) \in \partial_G j(\bar{x}, \bar{r}) = \partial_G d_E(\bar{x}, \bar{r}) + (0, 1)$ or $(\bar{x}^*, -1) \in \partial_G d_E(\bar{x}, \bar{r})$, since $j = d_E + \ell$, where ℓ is the linear form $(x,r) \mapsto r$. Conversely, given $\bar{x}^* \in X^*$ satisfying $(\bar{x}^*, -1) \in \partial_G d_E(\bar{x}, \bar{r})$, one has $(\bar{x}^*, 0) \in \partial_G j(\bar{x}, \bar{r})$, hence $\bar{x}^* \in \partial_G f(\bar{x})$ by Proposition 7.17. $\qquad\square$

7.2 Subdifferentials of Lower Semicontinuous Functions

This section is devoted to the extension of the graded subdifferential to the class of lower semicontinuous functions. Throughout, X is an arbitrary Banach space.

Definition 7.21. Let f be a member of the family $\mathscr{F}(X)$ of proper lower semicontinuous functions on X. The *graded subdifferential* of f at $\bar{x} \in \mathrm{dom} f$ is the set

$$\partial_G f(\bar{x}) := \{ x^* \in X^* : (x^*, -1) \in N_G(\mathrm{epi} f, (\bar{x}, \bar{r})) \} \qquad \text{for } r := f(\bar{x}).$$

This definition is justified by Proposition 7.20. Moreover, when applied to the indicator function of a set S, this definition turns out to be compatible with the definition we gave of the normal cone to S.

Proposition 7.22. *For every closed subset S of X and for every $\bar{x} \in S$ one has*

$$N_G(S, \bar{x}) = \partial_G \iota_S(\bar{x}).$$

Proof. The epigraph E of ι_S is just $S \times \mathbb{R}$, so that taking the sum norm on $X \times \mathbb{R}$, one has $d_E(x,r) = d_S(x) + r^-$, where $r^- := \max(-r, 0)$. Since d_S and $r \mapsto r^-$ are both Lipschitzian, Proposition 7.15 (b), (e) gives $\partial_G d_E(\bar{x}, 0) = \partial_G d_S(\bar{x}) \times [[-1, 0]]$. Thus, $\bar{x}^* \in \partial_G \iota_S(\bar{x})$ if and only if there exist $r \geq 0$, $s^* \in [-1, 0]$, $w^* \in \partial_G d_S(\bar{x})$ such that $(\bar{x}^*, -1) = r(w^*, s^*)$, i.e., $\bar{x}^* \in r \partial_G d_S(\bar{x})$ for some $r \geq 1$, or, by assertion (b) of Proposition 7.18, $\bar{x}^* \in N_G(S, \bar{x})$. $\qquad\square$

Now let us give a crucial sum rule.

Theorem 7.23. *Let $g \in \mathscr{F}(X)$, $h \in \mathscr{L}(X)$. Then for all $\bar{x} \in \mathrm{dom} g$,*

$$\partial_G(g + h)(\bar{x}) \subset \partial_G g(\bar{x}) + \partial_G h(\bar{x}).$$

Proof. Without loss of generality we may assume that h is globally Lipschitzian with rate 1. Let F, G, H be the epigraphs of $f := g + h$, g, h respectively. Observing that for a, $r \in \mathbb{R}$ one has

$$\inf\{|q - r| : q \geq a\} = (a - r)^+$$

and that $r \mapsto r^+ := \max(r, 0)$ is sublinear, one gets, for $(x, y, r) \in X \times Y \times \mathbb{R}$,

$$d_F(x, r) = \inf\{\|x - u\| + (s + h(u) - r)^+ : u \in X, \ s \geq g(u)\}$$
$$\leq \inf\{\|x - u\| + (s - p)^+ + (h(u) - q)^+ : (u, s) \in G, \ p + q = r\}$$
$$\leq \inf\{2\|x - u\| + (s - p)^+ + (h(x) - q)^+ : (u, s) \in G, \ p + q = r\}$$
$$\leq \inf\{2d_G(x, p) + d_H(x, q) : p, q \in \mathbb{R}, \ p + q = r\}.$$

Setting $j(x, p, q) := 2d_G(x, p) + d_H(x, q)$, defining $A \in L(X \times \mathbb{R}^2, X \times \mathbb{R})$ by $A(x, r, s)$ $:= (x, r + s)$, for every sequence $((x_n, r_n)) \to (\bar{x}, \bar{r}) := (\bar{x}, f(\bar{x}))$ in F the sequence $((x_n, r_n - h(x_n), h(x_n)))$ converges to $(\bar{x}, \bar{p}, \bar{q}) := (\bar{x}, g(\bar{x}), h(\bar{x}))$ and satisfies $A(x_n, r_n - h(x_n), h(x_n)) = (x_n, r_n)$, $j((x_n, r_n - h(x_n), h(x_n)) = 0$ for all $n \in \mathbb{N}$. Since $A^{\mathsf{T}}(x^*, r^*) = (x^*, r^*, r^*)$, Proposition 7.14 and Theorem 7.16 ensure that for all $(\bar{u}^*, -\bar{r}^*) \in \partial_G d_F(\bar{x}, \bar{r})$ there exist $(\bar{v}^*, -\bar{p}^*) \in 2\partial_G d_G(\bar{x}, \bar{p})$ and $(\bar{w}^*, -\bar{q}^*) \in \partial_G d_H(\bar{x}, \bar{q})$ such that

$$(\bar{u}^*, -\bar{r}^*, -\bar{r}^*) = (\bar{v}^*, -\bar{p}^*, 0) + (\bar{w}^*, 0, -\bar{q}^*),$$

so that $\bar{u}^* = \bar{v}^* + \bar{w}^*$, $\bar{p}^* = \bar{r}^* = \bar{q}^*$. Thus, given $\bar{x}^* \in \partial_G f(\bar{x})$ and $c \geq 1$ such that $(\bar{x}^*, -1) = c(\bar{u}^*, -\bar{r}^*) \in c\partial_G d_F(\bar{x}, \bar{r})$, since $c\bar{r}^* = 1$, setting $\bar{y}^* := c\bar{v}^* \in \partial_G g(\bar{x})$, $\bar{z}^* := c\bar{w}^* \in \partial_G h(\bar{x})$, one has $\bar{x}^* = \bar{y}^* + \bar{z}^*$. $\qquad \square$

Corollary 7.24. *Let* g, $h \in \mathscr{F}(X)$, *h being circa-differentiable at* $\bar{x} \in \mathrm{dom}\, g \cap \mathrm{dom}\, h$. *Then*

$$\partial_G(g + h)(\bar{x}) = \partial_G g(\bar{x}) + h'(\bar{x}).$$

Proof. Since h is Lipschitzian around $\bar{x}$ with $\partial_G h(\bar{x}) = \{h'(\bar{x})\}$ and $g = f - h$ for $f := g + h$, one has $\partial_G f(\bar{x}) \subset \partial_G g(\bar{x}) + h'(\bar{x})$ and $\partial_G g(\bar{x}) \subset \partial_G f(\bar{x}) - h'(\bar{x})$. $\qquad \square$

Theorem 7.25. *The assertions of Proposition 7.15 are valid on the class of lower semicontinuous functions.*

Proof. Assertion (a) is obvious. Assertion (b) is a consequence of the fact that when $f \in \mathscr{F}(X)$ is convex, $d(\cdot, \mathrm{epi}\, f)$ is convex. (c) If $f \in \mathscr{F}(X)$ attains a local minimum at $\bar{x}$, modifying f outside some closed ball and subtracting $f(\bar{x})$, we may assume that $\bar{x}$ is a global minimizer of f and $f(\bar{x}) = 0$. Then the epigraph E_f of f is contained in $X \times \mathbb{R}_+$, so that for all $(x, r) \in X \times \mathbb{R}$, one has

$$g(x, r) := d((x, r), E_f) + r \geq d((x, r), X \times \mathbb{R}_+) + r = (-r)^+ + r \geq 0 = g(\bar{x}, f(\bar{x})).$$

Since g attains its minimum at $\overline{x}_f := (\overline{x}, f(\overline{x}))$, by Proposition 7.15 (c) and (f) we get $(0,0) \in \partial_G d(\cdot, E_f)(\overline{x}_f) + (0,1)$ or $(0,-1) \in \partial_G d_{E_f}(\overline{x}_f)$; hence $0 \in \partial_G f(\overline{x})$.

(d) Let $f(x) = g(x+b) + c$, with $f, g \in \mathscr{F}(X)$. The epigraph of f being a translate of the epigraph of g, one has $\partial_G f(x) = \partial_G g(x+b)$. Now suppose $f = \lambda g$. Setting $A(x,r) := (x, \lambda^{-1}r)$, $\|(x,t)\|' = \|x\| + \lambda |t|$ for $(x,r) \in X \times \mathbb{R}$, the distances to the epigraphs F and G of f and g for the sum norm and the norm $\|\cdot\|'$ respectively are related by $d_F(x,t) = d'_G(A(x,t))$. Since A is onto, we get $\partial_G d_F(x,t) = A^{\mathsf{T}} \partial_G d'_G(A(x,t))$. Thus $(x^*, -1) \in r\partial_G d_F(x, f(x))$ for some $r \in \mathbb{R}_+$ if and only if $(x^*, -\lambda) \in r\partial_G d'_G(x, g(x))$ or $\lambda^{-1}x^* \in \partial_G g(x)$ and $\partial_G f(x) = \lambda \partial_G g(x)$.

Now let us suppose $f = g \circ A$ for some surjective $A \in L(X,Y)$ and $g \in \mathscr{F}(Y)$. We may suppose the norm on Y satisfies $\|y\| = \inf\{\|x\| : x \in A^{-1}(y)\}$. Then denoting again by F and G the epigraphs of f and g respectively, for $(x,t) \in X \times \mathbb{R}$ one has

$$d_G(Ax,t) = \inf\{\|Ax - Aw\| + |t - r| : (w,r) \in F\} \leq d_F(x,t).$$

In fact, $d_G(Ax,t) = d_F(x,t)$ since for $(w,r) \in F$, $u \in \ker A$, one has $(w+u, r) \in F$,

$$\|Ax - Aw\| + |t - r| = \inf\{\|x - w - u\| : u \in \ker A\} + |t - r| \geq d_F(x,t).$$

Thus, for $t = g(Ax)$, one has $\partial_G d_F(x,t) = (A \times I)^{\mathsf{T}}(\partial_G d_G(Ax,t))$ and $x^* \in \partial_G f(x)$ if and only if for some $r \in \mathbb{R}_+$ and some $(z^*, -s^*) \in \partial_G d_G(Ax,t)$, one has $(x^*, -1) = r(A^{\mathsf{T}}z^*, -s^*)$ or $rs^* = 1$, $x^* = A^{\mathsf{T}}(z^*/s^*) := A^{\mathsf{T}}(y^*)$ with $y^* := z^*/s^* \in \partial_G g(Ax)$. Thus $\partial_G f(x) = A^{\mathsf{T}}(\partial_G g(Ax))$.

(e) Suppose $f(x,y) = g(x) + h(y)$ with $g \in \mathscr{F}(X)$, $h \in \mathscr{F}(Y)$ and let $\overline{x} \in X$, $\overline{y} \in Y$, $\overline{r} := g(\overline{x})$, $\overline{s} := h(\overline{y})$, $\overline{t} := \overline{r} + \overline{s}$. Noting that for all $(u,r,v,s) \in X \times \mathbb{R} \times Y \times \mathbb{R}$ one has

$$\inf\{|r+s-t| : t \geq g(u) + h(v)\} \leq \inf\{|r-r'| + |s-s'| : r' \geq g(u), \, s' \geq h(v)\},$$

one gets $d((x,y,t), \text{epi}f) \leq \inf\{d((x,r), \text{epi}g) + d((y,s), \text{epi}h) : r + s = t\}$ for all $(x,y,t) \in X \times Y \times \mathbb{R}$. Let $A : (X \times \mathbb{R}) \times (Y \times \mathbb{R}) \to X \times Y \times \mathbb{R}$ be the surjective linear map defined by $A(x,r,y,s) := (x,y,r+s)$, so that $A^{\mathsf{T}}(u^*,v^*,t^*) = (u^*,t^*,v^*,t^*)$ for all $(u^*,v^*,t^*) \in X^* \times Y^* \times \mathbb{R}$. For every sequence $((x_n,y_n,t_n)) \to (\overline{x},\overline{y},\overline{t})$ in $\text{epi}f$, the lower semicontinuity of g and h ensures that $(g(x_n)) \to \overline{r}$, $(h(y_n)) \to \overline{s}$. Thus, by Propositions 7.14 and 7.15(e), given $(\overline{x}^*,\overline{y}^*) \in \partial_G f(\overline{x},\overline{y})$ and $c \geq 1$ such that $(c^{-1}\overline{x}^*, c^{-1}\overline{y}^*, -c^{-1}) \in \partial_G d((\cdot,\cdot,\cdot), \text{epi}f)(\overline{x},\overline{y},\overline{r}+\overline{s})$, one has $(c^{-1}\overline{x}^*, -c^{-1}) \in \partial_G d((\cdot,\cdot), \text{epi}g)(\overline{x},\overline{r})$, $(c^{-1}\overline{y}^*, -c^{-1}) \in \partial_G d((\cdot,\cdot), \text{epi}h)(\overline{y},\overline{s})$ and $\overline{x}^* \in \partial_G f(\overline{x})$, $\overline{y}^* \in \partial_G h(\overline{y})$. The preceding corollary is the corresponding version of assertion (f) with $g \in \mathscr{F}(X)$. Assertion (g) of Proposition 7.15 is unchanged, since a circa-differentiable function is locally Lipschitzian. □

Let us extend the comparison with the Clarke subdifferential we gave above.

Proposition 7.26. *For every Banach space X, for every closed subset S of X, every $\overline{x} \in S$, and every lower semicontinuous function f on X one has*

$$N_C(S,\bar{x}) = \overline{\mathrm{co}}^*(N_G(S,\bar{x})), \qquad \overline{\mathrm{co}}^*(\partial_G f(\bar{x})) \subset \partial_C f(\bar{x}).$$

If f is Lipschitzian around $\bar{x}$, this inclusion is an equality.

Proof. Let $\bar{x}^* \in N_G(S,\bar{x})$, so that there exists some $r \in [1,+\infty)$ such that $r^{-1}\bar{x}^* \in \partial_G d_S(\bar{x}) \subset \partial_C d_S(\bar{x})$, whence $\bar{x}^* \in \mathbb{R}_+ \partial_C d_S(\bar{x}) \subset N_C(S,\bar{x})$. Since $N_C(S,\bar{x})$ is weak* closed and convex, one gets $\overline{\mathrm{co}}^*(N_G(S,\bar{x})) \subset N_C(S,\bar{x})$. Conversely, since

$$\mathbb{R}_+ \partial_C d_S(\bar{x}) = \mathbb{R}_+ \overline{\mathrm{co}}^*(\partial_G d_S(\bar{x})) \subset \overline{\mathrm{co}}^*(\mathbb{R}_+ \partial_G d_S(\bar{x})) = \overline{\mathrm{co}}^*(N_G(S,\bar{x}))$$

by Proposition 7.8 and since $N_C(S,\bar{x}) = \mathrm{cl}^*(\mathbb{R}_+ \partial_C d_S(\bar{x}))$ by relation (5.25), one has $N_C(S,\bar{x}) \subset \overline{\mathrm{co}}^*(N_G(S,\bar{x}))$ and equality holds.

Let E be the epigraph of f and let $\bar{x}_f := (\bar{x}, f(\bar{x}))$. For all $\bar{x}^* \in \partial_G f(\bar{x})$ one has $(\bar{x}^*, -1) \in N_G(E, \bar{x}_f) \subset N_C(E, \bar{x}_f)$, and hence $\bar{x}^* \in \partial_C f(\bar{x})$. Since $\partial_C f(x)$ is convex and weak* closed, one gets the inclusion $\overline{\mathrm{co}}^*(\partial_G f(\bar{x})) \subset \partial_C f(\bar{x})$. Proposition 7.8 ensures that this inclusion is an equality when f is Lipschitzian at $\bar{x}$ with rate c. □

7.3　Notes and Remarks

A.D. Ioffe has published many articles presenting his subdifferentials, beginning with [512–516]; see also [517, 520–522, 524, 529, 530]. In spite of their attractive character, not many researchers have made use of them, though there are notable exceptions, among whom are L. Thibault and his coauthors. The author of this book hopes that the present chapter will contribute to the dissemination of his approach. Its writing has been inspired by [527] and [541]. Proposition 7.11 clarifies the links of the present definition of $\partial_G f$ with the approximate subdifferential $\partial_A f$ as introduced by Ioffe.

We have abandoned the terminology "geometric subdifferential" chosen by A.D. Ioffe because we consider that the concept is not more geometric than the other ones and also because it is not used by all other authors. We could have associated the name of Galerkin, in view of the similarities with the classical approximation scheme of numerical analysis. Of course, such a terminology would not be justified by a personal involvement; but that is also the case with the Fréchet and the Hadamard subdifferentials.

References

1. Abraham, R., Robbin, J.: Transversal Mappings and Flows. Benjamin, New York (1967)
2. Adly, S., Buttazzo, G., Théra, M.: Critical points for nonsmooth energy functions and applications. Nonlinear Anal. **32**(6), 711–718 (1998)
3. Adly, S., Ernst, E., Théra, M.: On the Dieudonné theorem in reflexive Banach spaces. Cybern. Syst. Anal. **38**(3), 339–343 (2002)
4. Alberti, G., Ambrosio, L., Cannarsa, P.: On the singularities of convex functions. Manuscripta Math. **76**(3–4), 421–435 (1992)
5. Aliprantis, C.D., Border, K.C.: Infinite Dimensional Analysis. Springer, New York (1999)
6. Amahroq, T., Jourani, A., Thibault, L.: A general metric regularity in Asplund Banach spaces. Numer. Funct. Anal. Optim. **19**(3–4), 215–226 (1998)
7. Amara, C., Ciligot-Travain, M.: Lower CS-closed sets and functions. J. Math. Anal. Appl. **239**(2), 371–389 (1999)
8. Ambrosio, L.: On some properties of convex functions. (Italian) Atti Accad. Naz. Lincei Cl. Sci. Fis. Mat. Natur. Rend. Lincei (9) Mat. Appl. **3**(3), 195–202 (1992)
9. Amir, D., Lindenstrauss, J.: The structure of weakly compact sets in Banach spaces. Ann. Math. **88**, 35–44 (1968)
10. Anh, L.Q., Khanh, P.Q.: Semicontinuity of the solution set of parametric multivalued vector quasiequilibrium problems. J. Math. Anal. Appl. **294**(2), 699–711 (2004)
11. Aragón Artacho, F.J., Dontchev, A.L.: On the inner and outer norms of sublinear mappings. Set-Valued Anal. **15**(1), 61–65 (2007)
12. Aragón Artacho, F.J., Geoffroy, M.H.: Characterization of metric regularity of subdifferentials. J. Convex Anal. **15**(2), 365–380 (2008)
13. Aragón Artacho, F.J., Mordukhovich, B.S.: Metric regularity and Lipschitzian stability of parametric variational inequalities. Nonlinear Anal. **72**, 1149–1170 (2010)
14. Artstein-Avidan, S., Milman, V.: The concept of duality in convex analysis, and the characterization of the Legendre transform. Ann. Math. **169**(2), 661–674 (2009)
15. Arutyunov, A.V.: Optimality Conditions: Abnormal and Degenerate Problems. Kluwer, Dordrecht (2000)
16. Arutyunov, A.V.: Covering mappings in metric spaces and fixed points (in Russian). Russ. Math. Dokl. **76**(2), 665–668 (2007)
17. Arutyunov, A.V., Izmailov, A.F.: Directional stability theorem and directional metric regularity. Math. Oper. Res. **31**, 526–543 (2006)
18. Arutyunov, A., Avakov, E., Dmitruk, A., Gelman, B., Obukhovskii, V.V.: Locally covering maps in metric spaces and coincidence points. J. Fixed Point Theor. Appl. **5**, 106–127 (2009)
19. Arutyunov, A.V., Akharov, E.R., Izmailov, A.F.: Directional regularity and metric regularity. SIAM J. Optim. **18**, 810–833 (2007)
20. Asplund, E.: Fréchet differentiability of convex functions. Acta Math. **121**, 31–47 (1968)

21. Asplund, E.: Chebyshev sets in Hilbert spaces. Trans. Am. Math. Soc. **144**, 235–240 (1969)
22. Asplund, E.: Differentiability of the metric projection in finite-dimensional Euclidean space. Proc. Am. Math. Soc. **38**, 218–219 (1973)
23. Asplund, E., Rockafellar, R.T.: Gradients of convex functions. Trans. Am. Math. Soc. **139**, 443–467 (1969)
24. Attouch, H.: Variational Convergence for Functions and Operators. Applicable Mathematics Series. Pitman, Boston (1984)
25. Attouch, H., Azé, D.: Approximation and regularization of arbitrary functions in Hilbert spaces by the Lasry-Lions method. Ann. Inst. H. Poincaré Anal. Non Linéaire **10**, 289–312, (1993)
26. Attouch, H., Théra, M.: A general duality principle for the sum of two operators. J. Convex Anal. **3**(1), 1–24 (1996)
27. Attouch, H., Wets, R.J.-B.: Quantitative stability of variational systems: I. The epigraphical distance. Trans. Am. Math. Soc. **328**(2), 695–729 (1992)
28. Attouch, H., Lucchetti, R., Wets, R.J.-B.: The topology of the ρ-Hausdorff distance. Ann. Mat. Pura Appl. **160**(4), 303–320 (1992)
29. Attouch, H., Buttazzo, G., Michaille, G.: Variational Analysis in Sobolev and BV Spaces. MPS-SIAM Series in Optimization, vol. 6. SIAM, Philadelphia (2006)
30. Aubin, J.-P.: Gradients généralisés de Clarke. Micro-cours, CRM, University of Montréal (1977)
31. Aubin, J.-P.: Mathematical Methods of Game and Economic Theory. Studies in Mathematics and Its Applications, vol. 7. North Holland, Amsterdam (1979)
32. Aubin, J.-P.: Contingent derivatives of set-valued maps and existence of solutions to nonlinear inclusions and differential inclusions. In: Nachbin, L. (ed.) Advances in Mathematics, Supplementary Study, pp. 160–232 (1981)
33. Aubin, J.-P.: Lipschitz behavior of solutions to convex minimization problems. Math. Oper. Res. **9**, 87–111 (1984)
34. Aubin, J.-P.: L'Analyse Non Linéaire et ses Motivations Economiques. Masson, Paris (1984)
35. Aubin, J.-P.: Mutational and Morphological Analysis. Tools for Shape Evolution and Morphogenesis. Systems and Control: Foundations and Applications. Birkhäuser, Boston (1999)
36. Aubin, J.-P., Ekeland, I.: Estimates of the duality gap in nonconvex programming. Math. Oper. Res. **1**, 225–245 (1976)
37. Aubin, J.-P., Ekeland, I.: Applied Nonlinear Analysis. Pure and Applied Mathematics. Wiley-Interscience, New York (1984)
38. Aubin, J.-P., Frankowska, H.: On inverse functions theorems for set-valued maps. J. Math. Pures Appl. **66**, 71–89 (1987)
39. Aubin, J.-P., Frankowska, H.: Set-Valued Analysis. Systems and Control: Foundations and Applications, vol. 2. Birkhäuser, Boston (1990)
40. Auslender, A.: Differential stability in nonconvex and nondifferentiable programming. Math. Prog. Study. **10**, 29–41 (1976)
41. Auslender, A.: Stability in mathematical programming with nondifferentiable data. SIAM J. Contr. Optim. **22**, 239–254 (1984)
42. Auslender, A., Cominetti, R.: A comparative study of multifunction differentiability with applications in mathematical programming. Math. Oper. Res. **16**, 240–258 (1991)
43. Auslender, A., Crouzeix, J.-P.: Global regularity theorems. Math. Oper. Res. **13**, 243–253 (1988)
44. Auslender, A., Teboulle, M.: Asymptotic Cones and Functions in Optimization and Variational Inequalities. Springer, New York (2003)
45. Aussel, D., Corvellec, J.-N., Lassonde, M.: Mean value property and subdifferential criteria for lower semicontinuous functions. Trans. Am. Math. Soc. **347**, 4147–4161 (1995)
46. Aussel, D., Corvellec, J.-N., Lassonde, M.: Nonsmooth constrained optimization and multidirectional mean value inequalities. SIAM J. Optim. **9**, 690–706 (1999)

47. Aussel, D., Daniilidis, A., Thibault, L.: Subsmooth sets: functional characterizations and related concepts. Trans. Am. Math. Soc. **357**, 1275–1302 (2004)

48. Avakov, E.R., Agrachevand, A.A., Arutyunov, A.V.: The level set of a smooth mapping in a neighborhood of a singular point. Math. Sbornik. **73**, 455–466 (1992)

49. Averbukh, V.I., Smolyanov, O.G.: The theory of differentiation in linear topological spaces. Russ. Math. Surv. **22**(6), 201–258 (1967)

50. Averbukh, V.I., Smolyanov, O.G.: The various definitions of the derivative in linear topological spaces. Uspehi Mat. Nauk **23**(4)(162), 67–116 (1968); English translation: Russ. Math. Surv. **23**(4), 67–113 (1968)

51. Avez, A.: Calcul Différentiel. Masson, Paris (1983)

52. Azé, D.: Eléments d'Analyse Convexe et Variationnelle. Ellipses, Paris (1997)

53. Azé, D.: A survey on error bounds for lower semicontinuous functions. ESAIM Proc. **13**, 1–17 (2003)

54. Azé, D.: A unified theory for metric regularity of multifunctions. J. Convex Anal. **13**(2), 225–252 (2006)

55. Azé, D., Corvellec, J.-N.: Variational methods in classical open mapping theorems. J. Convex Anal. **13**(3–4), 477–488 (2006)

56. Azé, D., Corvellec, J.-N.: A variational method in fixed point results with inwardness conditions. Proc. Am. Math. Soc. **134**(12), 3577–3583 (2006)

57. Azé, D., Hiriart-Urruty, J.-B.: Optimal Hoffman-type estimates in eigenvalue and semidefinite inequality constraints. J. Global Optim. **24**(2), 133–147 (2002)

58. Azé, D., Hiriart-Urruty, J.-B.: Analyse variationnelle et optimisation. Eléments de cours, exercices et problèmes corrigés. Cepadues, Toulouse (2010)

59. Azé, Penot, D., J.-P.: Operations on convergent families of sets and functions. Optimization **21**, 521–534 (1990)

60. Azé, D., Penot, J.-P.: Qualitative results about the convergence of convex sets and convex functions. In: Ioffe, A., et al. (eds.) Optimization and Nonlinear Analysis. Pitman Research in Mathematics Series, vol. 244, pp. 1–24. Longman, Harlow (1992)

61. Azé, D., Penot, J.-P.: Uniformly convex and uniformly smooth functions. Ann. Fac. Sci. Toulouse **6**(4), 705–730 (1995)

62. Azé, D., Poliquin, R.A.: Equicalmness and epiderivatives that are pointwise limits. J. Optim. Theor. Appl. **96**(3), 555–573 (1998)

63. Azé, D., Chou, C.C., Penot, J.-P.: Subtraction theorems and approximate openness for multifunctions: topological and infinitesimal viewpoints. J. Math. Anal. Appl. **221**, 33–58 (1998)

64. Azé, D., Corvellec, J.-N., Lucchetti, R.E.: Variational pairs and applications to stability in nonsmooth analysis. Nonlinear Anal. Theor. Meth. Appl. **49A**(5), 643–670 (2002)

65. Bachir, M.: A non-convex analogue to Fenchel duality. J. Funct. Anal. **181**(2), 300–312 (2001)

66. Bacciotti, A., Ceragioli, F., Mazzi, L.: Differential inclusions and monotonicity conditions for nonsmooth Lyapunov functions. Set-Valued Anal. **8**(3), 299–309 (2000)

67. Bakan, A., Deutch, F., Li, W.: Strong CHIP, normality and linear regularity of convex sets. Trans. Am. Math. Soc. **357**, 3831–3863 (2005)

68. Balakrishnan, A.V.: Applied Functional Analysis, 2nd edn. Springer, New York (1981)

69. Banach, S.: Théorie des Opérations Linéaires. Subw. Funduszu Narodowej, Warsaw (1932)

70. Bank, B., Guddat, J., Klatte, D., Kummer, B., Tammer, K.: Non-Linear Parametric Optimization. Akademie-Verlag, Berlin (1982)

71. Baranger, J.: Existence de solutions pour des problèmes d'optimisation non convexes. J. Math. Pures Appl. **52**, 377–406 (1973)

72. Baranger, J., Témam, R.: Nonconvex optimization problems depending on a parameter. SIAM J. Contr. **13**, 146–152 (1975)

73. Barbu, V., Precupanu, T.: Convexity and Optimization in Banach Spaces. D. Reidel, Dordrecht (1986)

74. Bardi, M., Capuzzo-Dolcetta, I.: Optimal Control and Viscosity Solutions of Hamilton-Jacobi-Bellman Equations. Birkhäuser, Basel (1997)
75. Bauschke, H.H., Combettes, P.-L.: Convex Analysis and Monotone Operator Theory in Hilbert Spaces. Springer, New York (2011)
76. Bauschke, H.H., Borwein, J.M., Li, W.: Strong conical hull intersection property, bounded linear regularity, Jameson's property (G) and error bounds in convex optimization. Math. Program. **86**, 135–160 (1999)
77. Bauschke, H.H., Borwein, J.M., Tseng, P.: Bounded linear regularity, strong CHIP and CHIP are distinct properties. J. Convex Anal. **7**, 395–413 (2000)
78. Bazaraa, M.S., Goode, J.J., Nashed, M.Z.: On the cones of tangents with applications to mathematical programming. J. Optim. Theor. Appl. **13**, 389–426 (1974)
79. Beauzamy, B.: Introduction to Banach Spaces and Their Geometry. Mathematics Studies, vol. 86. North Holland, Amsterdam (1982)
80. Bector, C.R., Chandra, S., Dutta, J.: Principles of Optimization Theory. Alpha Science International, London (2004)
81. Bednarczuk, E.: Sensitivity in mathematical programming: a review. Modelling, identification, sensitivity analysis and control of structures. Contr. Cybern. **23**(4), 589–604 (1994)
82. Bednarczuk, E., Penot, J.-P.: On the positions of the notions of well-posed minimization problems. Boll. Unione Mat. Ital. **6-B**(7), 665–683 (1992)
83. Bednarczuk, E., Penot, J.-P.: Metrically well-set minimization problems. Appl. Math. Optim. **26**, 273–285 (1992)
84. Bednarczuk, E.M., Song, W.: Contingent epiderivative and its applications to set-valued optimization. Contr. Cybern. **27**(3), 375–386 (1998)
85. Bednarczuk, E., Pierre, M., Rouy, E., Sokolowski, J.: Tangent sets in some functional spaces. Nonlinear Anal. Theor. Meth. Appl. **42**(5), 871–886 (2000)
86. Beer, G.: Topologies on Closed and Closed Convex Sets. Mathematics and Its Applications, vol. 268. Kluwer, Dordrecht (1993)
87. Beer, G.: On the compactness theorem for sequences of closed sets. Math. Balk. New Ser. **16**(1–4), 327–338 (2002)
88. Beer, G., Lucchetti, R.: Convex optimization and the epi-distance topology. Trans. Am. Math. Soc. **327** (1990), 795–813
89. Beer, G., Lucchetti, R.: The epi-distance topology: continuity and stability results with applications to convex optimization problems. Math. Oper. Res. **17**(3), 715–726 (1992)
90. Beer, G., Lucchetti, R.: Convergence of epigraphs and of sublevel sets. Set-Valued Anal. **1**(2), 159–183 (1993)
91. Beer, G., Lucchetti, R.: Weak topologies for the closed subsets of a metrizable space. Trans. Am. Math. Soc. **335**(2), 805–822 (1993)
92. Benahmed, S.: Sur les Méthodes Variationnelles en Analyse Multivoque. PhD thesis, University of Toulouse, November 2009
93. Benahmed, S., Azé, D.: On fixed points of generalized set-valued contractions. Bull. Aust. Math. Soc. **81**(1), 16–22 (2010)
94. Ben-Tal, A.B., Nemirovski, A.: Lectures on Modern Convex Optimization: Analysis, Algorithms, and Engineering Applications. SIAM-MPS, Philadelphia (2001)
95. Benoist, J.: The size of the Dini subdifferential. Proc. Am. Math. Soc. **129**(2), 525–530 (2001)
96. Benoist, J.: Intégration du sous-différentiel proximal: un contre exemple (Integration of the proximal subdifferential: a counterexample). Can. J. Math. **50**(2), 242–265 (1998)
97. Benoist, J.: Approximation and regularization of arbitrary sets in finite dimensions. Set-Valued Anal. **2**, 95–115 (1994)
98. Benyamini, Y., Lindenstrauss, J.: Geometric Nonlinear Functional Analysis, vol. 1. American Mathematical Society Colloquium Publications no. 48. American Mathematical Society, Providence (2000)
99. Berger, M.: Convexité dans le plan, dans l'espace et au-delà. De la puissance et de la complexité d'une notion simple, 1, 2. Ellipses, Paris (2006)

100. Bertsekas, D.P.: Nonlinear Programming. Athena Scientific, Boston (1999)
101. Bianchini, S., Bressan, A.: On a Lyapunov functional relating shortening curves and viscous conservation laws. Nonlinear Anal. Theor. Meth. Appl. **51**(4), 649–662 (2002)
102. Bierstone, E., Milman, P.D.: Subanalytic Geometry. Model Theory, Algebra, and Geometry. Math. Sci. Res. Inst. Publ., vol. 39, pp. 151–172. Cambridge University Press, Cambridge (2000)
103. Birge, J.R., Qi, L.: Semiregularity and generalized subdifferentials with applications to optimization. Math. Oper. Res. **18**(4), 982–1005 (1993)
104. Bishop, E., Phelps, R.R.: The support functional of convex sets. In: Klee, V. (ed.) Convexity. Proc. Symposia Pure Math. vol. VII, pp. 27–35. American Mathematical Society, Providence (1963)
105. Bishop, E., Phelps, R.R.: A proof that every Banach space is subreflexive. Bull. Am. Math. Soc. **67**, 97–98 (1961)
106. Bolte, J., Daniilidis, A., Lewis, A., Shiota, M.: Clarke critical values of subanalytic Lipschitz continuous functions. Ann. Polon. Math. **87**, 13–25 (2005)
107. Bolte, J., Daniilidis, A., Lewis, A., Shiota, M.: Clarke subgradients of stratifiable functions. SIAM J. Optim. **18**(2), 556–572 (2007)
108. Bonnans, J.F., Shapiro, A.: Perturbation Analysis of Optimization Problems. Springer, New York (2000)
109. Borwein, J.: Multivalued convexity and optimization: a unified approach to inequality and equality constraints. Math. Program. **13**(2), 183–199 (1977)
110. Borwein, J.: Weak tangent cones and optimization in a Banach space. SIAM J. Contr. Optim. **16**(3), 512–522 (1978)
111. Borwein, J.: Stability and regular points of inequality systems. J. Optim. Theor. Appl. **48**, 9–52 (1986)
112. Borwein, J.M.: Epi-Lipschitz-like sets in Banach space: theorems and examples. Nonlinear Anal. Theor. Appl. **11**, 1207–1217 (1987)
113. Borwein, J.M.: Minimal cuscos and subgradients of Lipschitz functions. In: Baillon, J.-B., Théra, M. (eds.) Fixed Point Theory and Its Applications. Pitman Research Notes in Mathematics Series, vol. 252, pp. 57–82. Longman, Essex (1991)
114. Borwein, J.: Differentiability properties of convex of Lipschitz, and of semicontinuous mappings in Banach spaces. In: Ioffe, A., et al. (eds.) Optimization and Nonlinear Analysis. Pitman Research in Mathematics Series vol. 244, pp. 39–52. Longman, Harlow (1992)
115. Borwein, J.M., Fabián, M.: A note on regularity of sets and of distance functions in Banach space. J. Math. Anal. Appl. **182**, 566–570 (1994)
116. Borwein, J.M., Fitzpatrick, S.: Weak* sequential compactness and bornological limit derivatives. J. Convex Anal. **2**(1,2), 59–67 (1995)
117. Borwein, J.M., Fitzpatrick, S.P.: Existence of nearest points in Banach spaces. Can. J. Math. **61**, 702–720 (1989)
118. Borwein, J.M., Fitzpatrick, S.: A weak Hadamard smooth renorming of $L_1(\Omega, \mu)$. Can. Math. Bull. **36**(4), 407–413 (1993)
119. Borwein, J.M., Giles, J.R.: The proximal normal formula in Banach space. Trans. Am. Math. Soc. **302**(1), 371–381 (1987)
120. Borwein, J., Goebel, R.: Notions of relative interior in Banach spaces. Optimization and related topics, 1. J. Math. Sci. (N.Y.) **115**(4), 2542–2553 (2003)
121. Borwein, J.M., Ioffe, A.D.: Proximal analysis in smooth spaces. Set-Valued Anal. **4**(1), 1–24 (1996)
122. Borwein, J.M., Lewis, A.S.: Duality relationships for entropy-like minimization problems. SIAM J. Contr. Optim. **29**(2), 325–338 (1991)
123. Borwein, J.M., Lewis, A.S.: Partially finite convex programming I. Quasi-relative interiors and duality theory. Math. Prog. B **57**(1), 15–48 (1992). II. Explicit lattice models. Math. Prog. B, **57**(1), 49–83 (1992)
124. Borwein, J.M., Lewis, A.S.: Partially-finite programming in L_1 and the existence of maximum entropy estimates. SIAM J. Optim. **3**(2), 248–267 (1993)

125. Borwein, J.M., Lewis, A.S.: Strong rotundity and optimization. SIAM J. Optim. **4**(1), 146–158 (1994)

126. Borwein, J.M., Lewis, A.S.: Convex Analysis and Nonlinear Optimization. Theory and Examples. Can. Math. Soc., Springer, New York (2000)

127. Borwein, J., O'Brien, R.: Tangent cones and convexity. Can. Math. Bull. **19**(3), 257–261 (1976)

128. Borwein, J.M., Preiss, D.: A smooth variational principle with applications to subdifferentiability and to differentiability of convex functions. Trans. Am. Math. Soc. **303**(2), 517–527 (1987)

129. Borwein, J.M., Strojwas, H.M.: Directionally Lipschitzian mappings on Baire spaces. Can. J. Math. **36**, 95–130 (1984)

130. Borwein, J.M., Strojwas, H.M.: Tangential approximations. Nonlinear Anal. **9**(12) 1347–1366 (1985)

131. Borwein, J.M., Strojwas, H.M.: Proximal analysis and boundaries of of closed sets in Banach space. I. Theory. Can. J. Math. **38**(2), 431–452 (1986)

132. Borwein, J.M., Strojwas, H.M.: Proximal analysis and boundaries of closed sets in Banach space. II. Applications. Can. J. Math. **39**, 428–472 (1987)

133. Borwein, J.M., Strojwas, H.M.: The hypertangent cone. Nonlinear Anal. **13**(2), 125–144 (1989)

134. Borwein, J.M., Vanderwerff, J.: Differentiability of conjugate functions and perturbed minimization principles. J. Convex Anal. **16**(9) 707–7011 (2009)

135. Borwein, J.M., Wang, X.: Distinct differentiable functions may share the same Clarke subdifferential at all points, Proc. Am. Math. Soc. **125**(3), 807–813 (1997)

136. Borwein, J.M., Zhu, Q.J.: A survey of subdifferential calculus with applications. Nonlinear Anal. **38**, 687–773 (1999)

137. Borwein, J.M., Zhu, Q.J.: Techniques of Variational Analysis. Canadian Books in Math., vol. 20. Can. Math. Soc., Springer, New York (2005)

138. Borwein, J.M., Zhu, Q.J.: Variational methods in convex analysis. J. Global Optim. **35**(2), 197–213 (2006)

139. Borwein, J.M., Zhuang, D.: On Fan's minimax theorem. Math. Program. **34**, 232–234 (1986)

140. Borwein, J.M., Zhuang, D.: Verifiable necessary and sufficient conditions for openness and regularity of set-valued and single-valued maps. J. Math. Anal. Appl. **134**, 441–459 (1988)

141. Borwein, J.M., Fitzpatrick, S., Giles, J.R.: The differentiability of real functions on normed linear space using generalized subgradients. J. Math. Anal. Appl. **128**, 512–534 (1987)

142. Borwein, D., Borwein, J.M., Wang, X.: Approximate subgradients and coderivatives in $\mathbb{R}^n$. Set-Valued Anal. **4**, 375–398 (1996)

143. Borwein, J.M., Lucet, Y., Mordukhovich, B.S.: Compactly epi-Lipschitzian convex sets and functions in normed spaces. J. Convex Anal. **7**, 375–393 (2000)

144. Borwein, J.M., Treiman, J.S., Zhu, Q.J.: Necessary conditions for constrained optimization problems with semicontinuous and continuous data. Trans. Am. Math. Soc. **350**, 2409–2429 (1998)

145. Borwein, J., Moors, W.B., Shao, Y.: Subgradient representation of multifunctions. J. Aust. Math. Soc. B **40**(3), 301–313 (1999)

146. Borwein, J., Fitzpatrick, S., Girgensohn, R.: Subdifferentials whose graphs are not norm×weak* closed. Can. Math. Bull. **46**(4), 538–545 (2003)

147. Borwein, J.M., Burke, J.V., Lewis, A.S.: Differentiability of cone-monotone functions on separable Banach space. Proc. Am. Math. Soc. **132**(4), 1067–1076 (2004)

148. Borwein, J., Cheng, L., Fabian, M., Revalski, J.P.: A one perturbation variational principle and applications. Set-Valued Anal. **12**(1–2), 49–60 (2004)

149. Bosch, P., Jourani, A., Henrion, R.: Sufficient conditions for error bounds and applications. Appl. Math. Optim. **50**(2), 161–181 (2004)

150. Boţ, R.I., Wanka, G.: Farkas results with conjugate functions. SIAM J. Optim. **15**, 540–554 (2005)

151. Boţ, R.I., Wanka, G.: An alternative formulation of a new closed cone constraint qualification. Nonlinear Anal. Theor. Meth. Appl. **64**, 1367–1381 (2006)

152. Boţ, R.I., Wanka, G.: A weaker regularity condition for subdifferential calculus and Fenchel duality in infinite dimensional spaces. Nonlinear Anal. **65**, 2787–2804 (2006)

153. Boţ, R.I., Grad, S.-M., Wanka, G.: New regularity conditions for strong and total Fenchel–Lagrange duality in infinite-dimensional spaces. Nonlinear Anal. **69**, 323–336 (2008)

154. Boţ, R.I., Grad, S.-M., Wanka, G.: On strong total Lagrange duality for convex optimization problems. J. Math. Anal. Appl. **337**, 1315–1325 (2008)

155. Boţ, R.I., Csetnek, E.R., Wanka, G.: Regularity conditions via quasi-relative interior in convex programming. SIAM J. Optim. **19**(1), 217–233 (2008)

156. Bouchitté, G., Buttazzo, G., Fragal, I.: Mean curvature of a measure and related variational problems. Dedicated to Ennio De Giorgi. Ann. Scuola Norm. Sup. Pisa Cl. Sci. **25**(4) (1997), (1–2), 179–196 (1998)

157. Bougeard, M., Penot, J.-P., Pommellet, A.: Towards minimal assumptions for the infimal convolution regularization. J. Math. Anal. Appl. **64**, 245–270 (1991)

158. Bouligand, G.: Sur les surfaces dépourvues de points hyperlimites. Ann. Soc. Polon. Math. **9**, 32–41 (1930)

159. Bouligand, G.: Introduction à la Géométrie Infinitésimale Directe. Gauthiers-Villars, Paris (1932)

160. Bounkhel, M., Al-Yusof, R.: Proximal analysis in reflexive smooth Banach spaces. Nonlinear Anal. **73**(7), 1921–1939 (2010)

161. Bounkhel, M., Jofré, A.: Subdifferential stability of the distance function and its applications to nonconvex economies and equilibrium. J. Nonlinear Convex Anal. **5**(3), 331–347 (2004)

162. Bounkhel, M., Thibault, L.: Nonconvex sweeping process and prox-regularity in Hilbert space. J. Nonlinear Convex Anal. **6**(2) 359–374 (2005)

163. Bourass, A., Giner, E.: Kuhn–Tucker conditions and integral functionals. J. Convex Anal. **8**(2), 533–553 (2001)

164. Bourbaki, N.: Elements of Mathematics. General Topology. Addison-Wesley, Reading, MA (1971). Translated from the French, Hermann, Paris, 1940

165. Bourbaki, N.: Variétés différentielles et analytiques. Fascicule de résultats. Hermann, Paris (1967)

166. Boyd, S., Vandenberghe, L.: Convex Optimization. Cambridge University Press, New York (2004)

167. Bressan, A.: Hamilton–Jacobi Equations and Optimal Control: An Illustrated Tutorial. NTNU, Trondheim (2001)

168. Bressan, A.: On the intersection of a Clarke cone with a Boltyanskii cone. SIAM J. Contr. Optim. **45**(6), 2054–2064 (2007)

169. Bressan, A., Piccoli, B.: Introduction to the mathematical theory of control. AIMS Series on Applied Mathematics, vol. 2. American Institute of Mathematical Sciences (AIMS), Springfield (2007)

170. Brézis, H.: Equations et inéquations nonlinéaires dans les espaces vectoriels en dualité. Annales Inst. Fourier **18**, 115–175 (1968)

171. Brézis, H., Browder, F.E.: A general principle on ordered sets in nonlinear functional analysis. Adv. Math. **21**(3), 355–364 (1976)

172. Bridson, M., Haefliger, A.: Metric Spaces of Non-positive Curvature. Springer, Berlin (1995)

173. Briec, W.: Minimum distance to the complement of a convex set: duality result. J. Optim. Theor. Appl. **93**(2), 301–319 (1997)

174. Brøndsted, A.: On a lemma of Bishop and Phelps. Pac. J. Math. **55**, 335–341 (1974)

175. Brøndsted, A., Rockafellar, R.T.: On the subdifferentiability of convex functions. Proc. Am. Math. Soc. **16**, 605–611 (1965)

176. Browder, F.E.: Nonlinear Operators and Nonlinear Equations of Evolution in Banach Spaces. Proc. Symposia Pure Math., vol. 18. American Mathematical Society, Providence (1976)

177. Bucur, D., Buttazzo, G.: Variational Methods in Some Shape Optimization Problems. Appunti dei Corsi Tenuti da Docenti della Scuola (Lecture Notes of a course in Scuola Normale Superiore), Pisa (2002)

178. Burachik, R.S., Fitzpatrick, S.: On a family of convex functions associated to subdifferentials. J. Nonlinear Convex Anal. **6**(1), 165–171 (2005); Erratum ibid. (3), 535 (2005)

179. Burachik, R.S., Jeyakumar, V.: A dual condition for the convex subdifferential sum formula with applications. J. Convex Anal. **12**, 279–290 (2005)

180. Burachik, R., Jeyakumar, V., Wu, Z.Y.: Necessary and sufficient conditions for stable conjugate duality. Nonlinear Anal. Theor. Meth. Appl. **64**, 1998–2006 (2006)

181. Burke, J.V.: Calmness and exact penalization. SIAM J. Contr. Optim. **29**, 493–497 (1991)

182. Burke, J.V.: An exact penalization viewpoint of constrained optimization. SIAM J. Contr. Optim. **29**, 968–998 (1991)

183. Burke, J.V., Deng, S.: Weak sharp minima revisited. I. Basic theory. Well-posedness in optimization and related topics (Warsaw, 2001). Contr. Cybern. **31**(3), 439–469 (2002)

184. Burke, J.V., Ferris, M.C.: Weak sharp minima in mathematical programming. SIAM J. Contr. Optim. **31**(5), 1340–1359 (1993)

185. Burke, J.V., Poliquin, R.A.: Optimality conditions for non-finite valued convex composite functions. Math. Program. B **57**(1), 103–120 (1992)

186. Burke, J.V., Qi, L.Q.: Weak directional closednesss and generalized subdifferentials. J. Math. Anal. Appl. **159**(2), 485–499 (1991)

187. Burke, J.V., Tseng, P.: A unified analysis of Hoffman's bound via Fenchel duality. SIAM J. Optim. **6**, 265–282 (1996)

188. Burke, J.V., Ferris, M.C., Qian, M.: On the Clarke subdifferential of the distance function of a closed set. J. Math. Anal. Appl. **166**, 199–213 (1992)

189. Burke, J.V., Lewis, A.S., Overton, M.L.: Approximating subdifferentials by random sampling of gradients. Math. Oper. Res. **27**(3), 567–584 (2002)

190. Bussotti, P.: On the genesis of the Lagrange multipliers. J. Optim. Theor. Appl. **117**, 453–459 (2003)

191. Bustos Valdebenito, M.: ε-gradients pour les fonctions localement lipschitziennes et applications. (French) [ε-gradients for locally Lipschitz functions and applications] Numer. Funct. Anal. Optim. **15**(3–4), 435–453 (1994)

192. Buttazzo, G., Giaquinta, M., Hildebrandt, S.: One-dimensional Variational Problems. An Introduction. Oxford Lecture Series in Mathematics and Its Applications, vol. 15. Clarendon Press, Oxford (1998)

193. Campa, I., Degiovanni, M.: Subdifferential calculus and nonsmooth critical point theory. SIAM J. Optim. **10**(4), 1020–1048 (2000)

194. Cánovas, M.J., Dontchev, A.L., López, M.A., Parra, J.: Metric regularity of semi-infinite constraint systems. Math. Program. B **104**(2–3), 329–346 (2005)

195. Carathéodory, C.: Calculus of variations and partial differential equations of the first order. Part I: Partial Differential Equations of the First Order. Holden-Day, San Francisco (1965). Part II: Calculus of Variations. Holden-Day, San Francisco (1967)

196. Caristi, J.: Fixed point theorems for mappings satisfying inwardness conditions. Trans. Am. Math. Soc. **215**, 241–251 (1976)

197. Cartan, H.: Cours de Calcul Différentiel. Hermann, Paris (1977)

198. Castaing, C., Valadier, M.: Convex Analysis and Measurable Multifunctions. Lecture Notes in Mathematics, vol. 580. Springer, Berlin (1977)

199. Cellina, A.: On the bounded slope condition and the validity of the Euler Lagrange equation. SIAM J. Contr. Optim. **40**(4), 1270–1279 (2001/2002)

200. Cellina, A.: The classical problem of the calculus of variations in the autonomous case: relaxation and Lipschitzianity of solutions. Trans. Am. Math. Soc. **356**(1), 415–426 (2004)

201. Cellina, A.: The Euler Lagrange equation and the Pontriagin maximum principle. Boll. Unione Mat. Ital. Sez. B Artic. Ric. Mat. **8**(2), 323–347 (2005)

202. Cellina, A.: Necessary conditions in the calculus of variations. Rend. Cl. Sci. Mat. Nat. **142**, 225–235 (2008/2009)
203. Cellina, A., Colombo, G., Fonda, A.: A continuous version of Liapunov's convexity theorem. Ann. Inst. H. Poincaré Anal. Non Linéaire **5**(1), 23–36 (1988)
204. Cepedello Boiso, M.: Approximation of Lipschitz functions by Δ-convex functions in Banach spaces. Isr. J. Math. **106**, 269–284 (1998)
205. Cepedello Boiso, M.: On regularization in superreflexive Banach spaces by infimal convolution formulas. Studia Math. **129**(3), 265–284 (1998)
206. Cesari, L.: Optimization - Theory and Applications. Problems with Ordinary Differential Equations. Applications of Mathematics, vol. 17. Springer, New York (1983)
207. Chen, X., Nashed, Z., Qi, L.Q.: Smoothing methods and semismooth methods for nondifferentiable operator equations. SIAM J. Numer. Anal. **38**, 1200–1216 (2000)
208. Chong Li, Ng, K.F., Pong, T.K.: Constraint qualifications for convex inequality systems with applications in constrained optimization. SIAM J. Optim. **19**(1), 163–187 (2008)
209. Choquet, G.: Convergences. Annales de l'Université de Grenoble **23**, 55–112 (1947)
210. Ciligot-Travain, M.: An intersection formula for the normal cone associated with the hypertangent cone. J. Appl. Anal. **5**(2), 239–247 (1999)
211. Clarke, F.H.: Necessary Conditions for Nonsmooth Problems in Optimal Control and the Calculus of Variations. PhD thesis, Department of Mathematics, University of Washington, Seattle (1973)
212. Clarke, F.H.: Generalized gradients and applications. Trans. Am. Math. Soc. **205**, 247–262 (1975)
213. Clarke, F.H.: A new approach to Lagrange multipliers. Math. Oper. Res. **1**, 97–102 (1976)
214. Clarke, F.H.: Optimization and Nonsmooth Analysis. Wiley, New York, 1983. Second edition: Classics in Applied Mathematics, vol. 5. Society for Industrial and Applied Mathematics (SIAM), Philadelphia (1990)
215. Clarke, F.H.: Methods of Dynamic and Nonsmooth Optimization. CBMS-NSF Regional Conference Series in Applied Mathematics, vol. 57. Society for Industrial and Applied Mathematics (SIAM), Philadelphia (1989)
216. Clarke, F.H., Ledyaev, Yu.S.: Mean value inequality in Hilbert space. Trans. Am. Math. Soc. **344**, 307–324 (1994)
217. Clarke, F.H., Ledyaev, Yu.S., Stern, R.J., Wolenski, P.R.: Proximal analysis and minimization principles. J. Math. Anal. Appl. **196**, 722–735 (1995)
218. Clarke, F.H., Ledyaev, Yu.S., Stern, R.J., Wolenski, P.R.: Nonsmooth Analysis and Control Theory. Graduate Texts in Mathematics, vol. 178. Springer, New York (1998)
219. Coban, M.M., Kenderov, P.S., Revalski, J.P.: Generic well-posedness of optimization problems in topological spaces. Mathematika **36**, 301–324 (1989)
220. Collier, J.B.: A class of strong differentiability spaces. Proc. Am. Math. Soc. **53**(2), 420–422 (1975)
221. Colombo, G., Goncharov, V.V.: Variational inequalities and regularity properties of closed sets in Hilbert spaces. J. Convex Anal. **8**(1), 197–221 (2001)
222. Combari, C., Laghdir, M., Thibault, L.: A note on subdifferentials of convex composite functionals. Arch. Math. **67**(3), 239–252 (1996)
223. Combari, C., Thibault, L.: On the graph convergence of subdifferentials of convex functions. Proc. Am. Math. Soc. **126**(8), 2231–2240 (1998)
224. Combari, C., Laghdir, M., Thibault, L.: On subdifferential calculus for convex functions defined on locally convex spaces. Ann. Sci. Math. Québec **23**(1), 23–36 (1999)
225. Cominetti, R.: Metric regularity, tangent sets, and second-order optimality conditions. Appl. Math. Optim. **21**(3), 265–287 (1990)
226. Cominetti, R., Penot, J.-P.: Tangent sets of order one and two to the positive cones of some functional spaces. Appl. Math. Optim. **36**(3), 291–312 (1997)
227. Contesse, L.: On the boundedness of certain point-to-set maps and its application in optimization. In: Recent Advances in System Modelling and Optimization (Santiago, 1984), pp. 51–68. Lecture Notes in Control and Inform. Sci., vol. 8. Springer, Berlin (1986)

228. Contesse, L., Penot, J.-P.: Continuity of the Fenchel correspondence and continuity of polarities. J. Math. Anal. Appl. **156**(2), 305–328 (1991)
229. Cornejo, O., Jourani, A., Zalinescu, C.: Conditioning and upper-Lipschitz inverse subdifferentials in nonsmooth optimization problems. J. Optim. Theor. Appl. **95**(1), 127–148 (1997)
230. Cornet, B.: A remark on tangent cones, working paper. Université Paris-Dauphine (1979)
231. Cornet, B.: Regular properties of tangent and normal cones. Cahiers Math de la Décision. University of Paris-Dauphine (1981)
232. Cornet, B.: An existence theorem of slow solutions for a class of differential inclusions. J. Math. Anal. Appl. **96**, 130–147 (1983)
233. Cornet, B.: Regularity properties of open tangent cones. Math. Program. Stud. **30**, 17–33 (1987)
234. Cornet, B., Laroque, G.: Lipschitz properties of solutions in mathematical programming. J. Optim. Theor. Appl. **53**(3), 407–427 (1987)
235. Cornet, B., Vial, J.-Ph.: Lipschitzian solutions of perturbed nonlinear programming problems. SIAM J. Contr. Optim. **24**(6), 1123–1137 (1986)
236. Correa, R., Gajardo, P.: Eigenvalues of set-valued operators in Banach spaces. Set-Valued Anal. **13**(1), 1–19 (2005)
237. Correa, R., Jofré, A.: Tangentially continuous directional derivatives in nonsmooth analysis. J. Optim. Theor. Appl. **61**(1), 1–21 (1989)
238. Correa, R., Hiriart-Urruty, J.-B., Penot, J.-P.: A note on connected set-valued mappings. Boll. Un. Mat. Ital. C **5**(6) (1986), **5**(1), 357–366 (1987)
239. Correa, R., Jofré, A., Thibault, L.: Characterization of lower semicontinuous convex functions. Proc. Am. Math. Soc. **116**(1), 67–72 (1992)
240. Correa, R., Gajardo, P., Thibault, L.: Links between directional derivatives through multidirectional mean value inequalities. Math. Program. B **116**(1–2), 57–77 (2009)
241. Correa, R., Gajardo, P., Thibault, L.: Various Lipschitz-like properties for functions and sets. I. Directional derivative and tangential characterizations. SIAM J. Optim. **20**(4), 1766–1785 (2010)
242. Coulibaly, A., Crouzeix, J.-P.: Condition numbers and error bounds in convex programming. Math. Program. B **116**(1–2), 79–113 (2009)
243. Covitz, H., Nadler, S.B. Jr.: Multi-valued contraction mappings in generalized metric spaces. Isr. J. Math. **8**, 5–11 (1970)
244. Crandall, M.G., Lions, P.-L.: Viscosity solutions to Hamilton–Jacobi equations. Trans. Am. Math. Soc. **277**, 1–42 (1983)
245. Craven, B.D.: Control and Optimization. Chapman and Hall Mathematics Series. Chapman and Hall, London (1995)
246. Craven, B.D.: Avoiding a constraint qualification. Optimization **41**(4), 291–302 (1997)
247. Craven, B.D., Ralph, D., Glover, B.M.: Small convex-valued subdifferential in mathematical programming. Optimization **32**, 1–21 (1995)
248. Czarnecki, M.-O., Gudovich, A.N.: Representations of epi-Lipschitzian sets. Nonlinear Anal. **73**(8), 2361–2367 (2010)
249. Czarnecki, M.O., Rifford, L.: Approximation and regularization of Lipschitz functions: convergence of the gradients. Trans. Am. Math. Soc. **358**(10), 4467–4520 (2006)
250. Dacorogna, B.: Introduction to the Calculus of Variations. Translated from the 1992 French original. Second edition. Imperial College Press, London (2009)
251. Dacorogna, B.: Direct methods in the calculus of variations, 2nd edn. Applied Mathematical Sciences, vol. 78. Springer, New York (2008)
252. Dal Maso, G.: An Introduction to Γ-Convergence. Birkhäuser, Basel (1993)
253. Daneš, J.: A geometric theorem useful in nonlinear functional analysis. Boll. Un. Mat. Ital. **6**(4), 369–375 (1972)
254. Daneš, J.: On local and global moduli of convexity. Comment. Math. Univ. Carolinae **17**(3), 413–420 (1976)
255. Daneš, J., Durdil, J.: A note on geometric characterization of Fréchet differentiability. Comment. Math. Univ. Carolinae **17**(1), 195–204 (1976)

256. Daniilidis, A., Georgiev, P.: Cyclic hypomonotonicity, cyclic submonotonicity, and integration. J. Optim. Theor. Appl. **122**(1), 19–39 (2004)

257. Daniilidis, A., Jules, F., Lassonde, M.: Subdifferential characterization of approximate convexity: the lower semicontinuous case. Math. Program. B **116**, 115–127 (2009)

258. Davis, W.J., Figiel, T., Johnson, W.B., Pelczynski, A.: Factoring weakly compact operators. J. Funct. Anal. **17**, 311–327 (1974)

259. de Barra, G., Fitzpatrick, S., Giles, J.R.: On generic differentiability of locally Lipschitz functions on a Banach space. Proc. CMA (ANU) **20**, 39–49 (1988)

260. De Blasi, F.S., Myjak, J.: Sur la convergence des approximations successives pour les contractions non linéaires dans un espace de Banach. C.R. Acad. Sci. Paris **283**, 185–187 (1976)

261. De Blasi, F.S., Myjak, J.: Some generic properties in convex and nonconvex optimization theory. Comment. Math. Prace Mat. **24**, 1–14 (1984)

262. de Figueiredo, D.G.: Topics in Nonlinear Analysis. Lecture Notes, vol. 48. University of Maryland, College Park (1967)

263. De Giorgi, E., Marino, A., Tosques, M.: Problemi di evoluzione in spaci metrici de massima pendenza, Atti Acad. Naz. Lincei Cl Sci. Fis. Mat. Natur. Rend. Lincei **68**(8), 180–187 (1980)

264. Degiovanni, M.: Nonsmooth critical point theory and applications. Proceedings of the Second World Congress of Nonlinear Analysts. Part 1 (Athens, 1996). Nonlinear Anal. **30**(1), 89–99 (1997)

265. Degiovanni, M.: A survey on nonsmooth critical point theory and applications. In: From Convexity to Nonconvexity, pp. 29–42. Nonconvex Optim. Appl., vol. 55. Kluwer, Dordrecht (2001)

266. Degiovanni, M., Marzocchi, M.: A critical point theory for nonsmooth functionals. Ann. Mat. Pura Appl. **167**(4), 73–100 (1994)

267. Deimling, K.: Nonlinear Functional Analysis. Springer, Berlin (1985)

268. Deimling, K.: Multivalued Differential Equations. De Gruyter, Berlin (1992)

269. Dempe, S., Vogel, S.: The generalized Jacobian of the optimal solution in parametric optimization. Optimization **50**(5–6), 387–405 (2001)

270. Dempe, S., Dutta, J., Lohse, S.: Optimality conditions for bilevel programming problems. Optimization **55**(5–6), 505–524 (2006)

271. Dempe, S., Dutta, J., Mordukhovich, B.S.: New necessary optimality conditions in optimistic bilevel programming. Optimization **56**(5–6), 577–604 (2007)

272. Demyanov, V.F.: The rise of nonsmooth analysis: its main tools. Kibernet. Sistem. Anal. **188**(4), 63–85 (2002); translation in Cybern. Syst. Anal. **38**(4), 527–547 (2002)

273. Demyanov, V.F., Jeyakumar, V.: Hunting for a smaller convex subdifferential. J. Global Optim. **10**(3), 305–326 (1997)

274. Dem'yanov, V.F., Malozëmov, V.N.: Introduction to Minimax. Reprint of the 1974 edition. Dover, New York (1990)

275. Demyanov, V.F., Roshchina, V.A.: Optimality conditions in terms of upper and lower exhausters. Optimization **55**(5–6), 525–540 (2006)

276. Demyanov, V.F., Roshchina, V.A.: Exhausters and subdifferentials in non-smooth analysis. Optimization **57**(1), 41–56 (2008)

277. Demyanov, V.F., Rubinov, A.M.: Constructive Nonsmooth Analysis. Approximation and Optimization, vol. 7. P. Lang, Frankfurt (1995)

278. Demyanov, V., Rubinov, A. (eds.): Quasidifferentiability and Related Topics. Nonconvex Optimization and Its Applications, vol. 43. Kluwer, Dordrecht (2000)

279. Dem'yanov, V.F., Vasil'ev, L.V.: Nondifferentiable Optimization. Optimization Software, New York (1985)

280. Dem'yanov, V.F., Lemaréchal, C., Zowe, J.: Approximation to a set-valued mapping. I. A proposal. Appl. Math. Optim. **14**(3), 203–214 (1986)

281. Dem'yanov, V.F., Stavroulakis, G.E., Polyakova, L.N., Panagiotopoulos, P.: Quasidifferentiability and Nonsmooth Modelling in Mechanics, Engineering and Economics. Nonconvex Optimization and Its Applications, vol. 10. Kluwer, Dordrecht (1996)

282. Deville, R.: Smooth variational principles and non-smooth analysis in Banach spaces. In: Clarke, F.H., Stern, R.J. (eds.) Nonlinear Analysis, Differential Equations and Control, pp. 369–405. Kluwer, Dordrecht (1999)

283. Deville, R., Ghoussoub, N.: Perturbed minimization principles and applications. In: Handbook of the Geometry of Banach spaces, vol. I, pp. 393–435. North Holland, Amsterdam (2001)

284. Deville, R., Ivanov, M.: Smooth variational principles with constraints. Arch. Math. **69**, 418–426 (1997)

285. Deville, R., Maaden, A.: Smooth variational principles in Radon–Nikodým spaces. Bull. Aust. Math. Soc. **60**(1), 109–118 (1999)

286. Deville, R., Revalski, J.P.: Porosity of ill-posed problems. Proc. Am. Math. Soc. **128**(4), 1117–1124 (2000)

287. Deville, R., Zizler, V.: Farthest points in w^*-compact sets. Bull. Aust. Math. Soc. **38**(3), 433–439 (1988)

288. Deville, R., Godefroy, G., Zizler, V.: A smooth variational principle with applications to Hamilton–Jacobi equations in infinite dimensions. J. Funct. Anal. **111**, 197–212 (1993)

289. Deville, R., Godefroy, G., Zizler, V.: Smoothness and Renormings in Banach Spaces. Pitman Monographs, vol. 64. Longman, London (1993)

290. Di, S., Poliquin, R.: Contingent cone to a set defined by equality and inequality constraints at a Fréchet differentiable point. J. Optim. Theor. Appl. **81**(3), 469–478 (1994)

291. Dien, P.H., Luc, D.T.: On the calculation of generalized gradients for a marginal function. Acta Math. Vietnam. **18**(2), 309–326 (1993)

292. Diestel, J.: Geometry of Banach Spaces – Selected Topics. Lecture Notes in Mathematics, vol. 485. Springer, New York (1975)

293. Diestel, J.: Sequences and Series in Banach Spaces. Graduate Texts in Mathematics, vol. 92. Springer, New York (1984)

294. Dieudonné, J.: Foundations of Modern Analysis. Academic Press, New York (1960)

295. Dinh, N., Lee, G.M., Tuan, L.A.: Generalized Lagrange multipliers for nonconvex directionally differentiable programs. In: Jeyakumar, V., Rubinov, A. (eds.) Continuous Optimization: Current Trends and Modern Applications, pp. 293–319. Springer, New York (2005)

296. Dinh, N., Goberna, M.A., López, M.A.: From linear to convex systems: consistency, Farkas' lemma and applications. J. Convex Anal. **13**, 279–290 (2006)

297. Dinh, N., Goberna, M.A., López, M.A., Son, T.Q.: New Farkas-type results with applications to convex infinite programming. ESAIM Contr. Optim. Cal. Var. **13**, 580–597 (2007)

298. Dinh, N., Mordukhovich, B.S., Nghia, T.T.A.: Subdifferentials of value functions and optimality conditions for some classes of DC and bilevel infinite and semi-infinite programs. Math. Program. B **123**(1), 101–138 (2010)

299. Dinh, N., Nghia, T.T.A., Vallet, G.: A closedness condition and its applications to DC programs with convex constraints. Optimization **59**(3–4), 541–560 (2010)

300. Dinh, N., Vallet, G., Nghia, T.T.A.: Farkas-type results and duality for DC programs with convex constraints. J. Convex Anal. **15**(2), 235–262 (2008)

301. Dmitruk, A.V., Kruger, A.Y.: Metric regularity and systems of generalized equations. J. Math. Anal. Appl. **342**(2), 864–873 (2008)

302. Dmitruk, A.V., Kruger, A.Y.: Extensions of metric regularity. Optimization **58**(5), 561–584 (2009)

303. Dmitruk, A.V., Miliutin, A.A., Osmolovskii, N.P.: Liusternik's theorem and the theory of extrema. Russ. Math. Surv. **35**, 11–51 (1981)

304. Dolecki, S.: A general theory of necessary conditions. J. Math. Anal. Appl. **78**, 267–308 (1980)

305. Dolecki, S.: Tangency and differentiation: some applications of convergence theories. Ann. Mat. Pura Appl. **130**, 223–255 (1982)

306. Dolecki, S.: Hypertangent cones for a special class of sets. In: Optimization: Theory and Algorithms (Confolant, 1981), pp. 3–11. Lecture Notes in Pure and Appl. Math., vol. 86. Dekker, New York (1983)

307. Dolecki, S., Greco, G.: Tangency vis-à-vis differentiability by Peano, Severi and Guareschi. J. Convex Anal. **18**(2), 301–339 (2011)

308. Dontchev, A.L.: Implicit function theorems for generalized equations. Math. Program. A **70**(1), 91–106 (1995)

309. Dontchev, A.L.: The Graves theorem revisited. J. Convex Anal. **3**, 45–53 (1996)

310. Dontchev, A.L.: A local selection theorem for metrically regular mappings. J. Convex Anal. **11**(1), 81–94 (2004)

311. Donchev, T., Dontchev, A.L.: Extensions of Clarke's proximal characterization for reachable mappings of differential inclusions. J. Math. Anal. Appl. **348**(1), 454–460 (2008)

312. Dontchev, A.L., Frankowska, H.: Lyusternik-Graves theorem and fixed points. Proc. Am. Math. Soc. **139**(2), 521–534 (2011)

313. Dontchev, A.L., Hager, W.W.: Lipschitzian stability in nonlinear control and optimization. SIAM J. Contr. Optim. **31**(3), 569–603 (1993)

314. Dontchev, A.L., Hager, W.W.: An inverse function theorem for set-valued maps. Proc. Am. Math. Soc. **121**, 481–489 (1994)

315. Dontchev, A.L., Hager, W.W.: Implicit functions, Lipschitz maps and stability in optimization. Math. Oper. Res. **19**, 753–768 (1994)

316. Dontchev, A.L., Lewis, A.S.: Perturbations and metric regularity. Set-Valued Anal. **13**(4), 417–438 (2005)

317. Dontchev, A.L., Rockafellar, R.T.: Characterizations of Lipschitzian stability in nonlinear programming. In: Mathematical Programming with Data Perturbations, pp. 65–82. Lecture Notes in Pure and Appl. Math., vol. 195. Dekker, New York (1998)

318. Dontchev, A.L., Rockafellar, R.T.: Regularity and conditioning of solution mappings in variational analysis. Set-Valued Anal. **12**(1–2), 79–109 (2004)

319. Dontchev, A.L., Rockafellar, R.T.: Parametrically robust optimality in nonlinear programming. Appl. Comput. Math. **5**(1), 59–65 (2006)

320. Dontchev, A.L., Rockafellar, R.T.: Robinson's implicit function theorem and its extensions. Math. Program. B **117**(1–2), 129–147 (2009)

321. Dontchev, A.L., Rockafellar, R.T.: Implicit Functions and Solution Mappings. A View from Variational Analysis. Springer, New York (2009)

322. Dontchev, A.L., Veliov, V.L. Metric regularity under approximations, Control Cybernet. **38**(4B), 1283–1303 (2009)

323. Dontchev, A.L., Zolezzi, T.: Well-posed Optimization Problems. Lectures Notes in Math., vol. 1543. Springer, Berlin (1993)

324. Dontchev, A.L., Lewis, A.S., Rockafellar, R.T.: The radius of metric regularity. Trans. Am. Math. Soc. **355**(2), 493–517 (2003)

325. Dontchev, A.L., Quincampoix, M., Zlateva, N.: Aubin criterion for metric regularity. J. Convex Anal. **13**(2), 281–297 (2006)

326. Dries, V.D., Miller, C.: Geometries, categories and o-minimal structures. Duke Math. J. **84**, 497–540 (1996)

327. Dubovitskii, A.Y., Milyiutin, A.A.: Extremum problems in the presence of constraints. Dokl. Akad. Nauk. SSSR **149**, 759–762 (1963)

328. Dubovitskii, A.Y., Milyiutin, A.A.: Extremum problems in the presence of restrictions. USSR Comput. Math. Phys. **5**, 1–80 (1965)

329. Dunford, N., Schwartz, J.: Linear Operators I. Wiley-Interscience, New York (1958)

330. Duong, P.C., Tuy, H.: Stability, surjectivity and local invertibility of non differentiable mappings. Acta Math. Vietnamica **3**, 89–105 (1978)

331. Durea, M., Strugariu, R.: Quantitative results on openness of set-valued mappings and implicit multifunctions, Pac. J. Optim. **6**(3), 533–549 (2010)

332. Dutta, J.: Generalized derivatives and nonsmooth optimization, a finite dimensional tour. With discussions and a rejoinder by the author. Top **13**(2), 185–314 (2005)

333. Eberhard, A., Wenczel, R.: Some sufficient optimality conditions in nonsmooth analysis. SIAM J. Optim. **20**(1), 251–296 (2009)

334. Edelstein, M.: Farthest points of sets in uniformly convex Banach spaces. Isr. J. Math. **4**, 171–176 (1966)

335. Edmond, J.F., Thibault, L.: Inclusions and integration of subdifferentials. J. Nonlinear Convex Anal. **3**(3), 411–434 (2002)

336. Edwards, D.A.: On the homeomorphic affine embedding of a locally compact cone into a Banach dual space endowed with the vague topology. Proc. Lond. Math. Soc. **14**, 399–414 (1964)

337. Edwards, R.E.: Functional Analysis. Theory and Applications. Holt, Rinehart and Winston, New York (1965). Reprint by Dover, New York (1995)

338. Eells, J., Jr.: A setting for global analysis. Bull. Am. Math. Soc. **72**, 751–807 (1966)

339. Egorov, Y.V.: Some necessary conditions for optimality in Banach spaces. Math. Sbornik **64**, 79–101 (1964)

340. Ekeland, I.: Sur les problèmes variationnels. C.R. Acad. Sci. Paris A-B **275**, A1057–A1059 (1972)

341. Ekeland, I.: On the variational principle. J. Math. Anal. Appl. **47**, 324–353 (1974)

342. Ekeland, I.: Legendre duality in nonconvex optimization and calculus of variations. SIAM J. Contr. Optim. **15**(6), 905–934 (1977)

343. Ekeland, I.: Nonconvex minimization problems. Bull. Am. Math. Soc. **1**(3), 443–474 (1979)

344. Ekeland, I.: Nonconvex duality. Analyse non convexe (Proc. Colloq., Pau, 1977). Bull. Soc. Math. France Mém. **60**, 45–55 (1979)

345. Ekeland, I.: Ioffe's mean value theorem. In: Convex Analysis and Optimization (London, 1980), pp. 35–42. Res. Notes in Math., vol. 57. Pitman, Boston (1982)

346. Ekeland, I.: Two results in convex analysis. In: Optimization and Related Fields (Erice, 1984), pp. 215–228. Lecture Notes in Mathematics, vol. 1190. Springer, Berlin (1986)

347. Ekeland, I.: Convexity Methods in Hamiltonian Mechanics. Ergebnisse der Mathematik und ihrer Grenzgebiete (3), vol. 19. Springer, Berlin (1990)

348. Ekeland, I.: Non-convex duality. In: Complementarity, Duality and Symmetry in Nonlinear Mechanics, pp. 13–19. Adv. Mech. Math., vol. 6. Kluwer, Boston (2004)

349. Ekeland, I.: An inverse function theorem in Fréchet spaces. Ann. Inst. H. Poincaré Anal. Non-Linéaire **28**(1), 91–105 (2011)

350. Ekeland, I., Ghoussoub, N.: Selected new aspects of the calculus of variations in the large. Bull. Am. Math. Soc. (N.S.) **39**(2), 207–265 (2002)

351. Ekeland, I., Lasry, J.-M.: Duality in nonconvex variational problems. In: Advances in Hamiltonian Systems (Rome, 1981), pp. 73–108. Ann. CEREMADE. Birkhäuser, Boston (1983)

352. Ekeland, I., Lebourg, G.: Generic Fréchet differentiability and perturbed optimization problems in Banach spaces. Trans. Am. Math. Soc. **224**(2), 193–216 (1976)

353. Ekeland, I., Témam, R.: Convex Analysis and Variational Problems. North Holland, Amsterdam (1976). Translated from the French, Dunod–Gauthier–Villars, Paris (1974)

354. Ekeland, I., Turnbull, T.: Infinite-Dimensional Optimization and Convexity. Chicago Lectures in Mathematics. University of Chicago Press, Chicago (1983)

355. Ekeland, I., Valadier, M.: Representation of set-valued mappings. J. Math. Anal. Appl. **35**, 621–629 (1971)

356. El Abdouni, B.: Thibault, L.: Quasi-interiorly e-tangent cones to multifunctions. Numer. Funct. Anal. Optim. **10**(7–8), 619–641 (1989)

357. Eremin, I.I.: The penalty method in convex programming. Sov. Math. Dokl. **8**, 458–462 (1966)

358. Ernst, E., Théra, M.: Boundary half-strips and the strong CHIP. SIAM J. Optim. **18**, 834–852 (2007)

359. Ernst, E., Théra, M., Volle, M.: Optimal boundedness criteria for extended-real-valued functions. Optimization **56**(3), 323–338 (2007)

360. Evans, L.C., Gariepy, R.F.: Measure Theory and Fine Properties of Functions. CRC, Boca Raton (1992)

361. Fabián, M.: On minimum principles. Acta Polytechnica **20** 109–118 (1983)

362. Fabian, M.: Subdifferentials, local ε-supports and Asplund spaces. J. Lond. Math. Soc. II **34**, 568–576 (1986)

363. Fabian, M.: Subdifferentiability and trustworthiness in the light of a new variational principle of Borwein and Preiss. Acta Univ. Carol. Math. Phys. **30**(2), 51–56 (1989)

364. Fabian, M.: Gâteaux Differentiability of Convex Functions and Topology, Weak Asplund Spaces. Canadian Math. Soc. Series of Monographs and Advanced Texts. Wiley, New York (1997)

365. Fabian, M., Loewen, P.D.: A generalized variational principle. Can. J. Math. **53**(6), 1174–1193 (2001)

366. Fabian, M., Mordukhovich, B.S.: Smooth variational principles and characterizations of Asplund spaces. Set-Valued Anal. **6**, 381–406 (1998)

367. Fabian, M., Mordukhovich, B.S.: Separable reduction and supporting properties of Fréchet-like normals in Banach spaces. Can. J. Math. **51**(1), 26–48 (1999)

368. Fabian, M., Mordukhovich, B.S.: Separable reduction and extremal principles in variational analysis. Nonlinear Anal. Theor. Meth. Appl. **49A**(2), 265–292 (2002)

369. Fabian, M., Mordukhovich, B.S.: Sequential normal compactness versus topological normal compactness in variational analysis. Nonlinear Anal. Theor. Meth. Appl. **54A**(6), 1057–1067 (2003)

370. Fabian, M., Preiss, D.: On intermediate differentiability of Lipschitz functions on certain Banach spaces. Proc. Am. Math. Soc. **113**(3), 733–740 (1991)

371. Fabian, M., Preiss, D.: On the Clarke generalized Jacobian. Rend. Circ. Mat. Palermo **52**(Suppl 14), 305–307 (1987)

372. Fabian, M., Revalski, J.: A variational principle in reflexive spaces with Kadec–Klee norm. J. Convex Anal. **16**(1), 211–226 (2009)

373. Fabián, M., Zhivkov, N.V.: A characterization of Asplund spaces with the help of local ε-supports of Ekeland and Lebourg. C.R. Acad. Bulgare Sci. **38**(6), 687–674 (1985)

374. Fabián, M., Hajek, P., Vandernerff, J.: On smooth variational principles in Banach spaces. J. Math. Anal. Appl. **197**(1), 153–172 (1996)

375. Fabian, M., Godefroy, G., Zizler, V.: A note on Asplund generated Banach spaces. Bull. Pol. Acad. Sci. Math. **47**(3), 221–230 (1999)

376. Fabián, M., et al.: Functional Analysis and Infinite Dimensional Geometry. CMS Books in Mathematics, vol. 8. Springer, New York (2001)

377. Fabian, M., Montesinos, V., Zizler, V.: Weakly compact sets and smooth norms in Banach spaces. Bull. Aust. Math. Soc. **65**(2), 223–230 (2002)

378. Fabian, M., Loewen, P.D., Mordukhovich, B.S.: Subdifferential calculus in Asplund generated spaces. J. Math. Anal. Appl. **322**(2), 787–795 (2006)

379. Fabian, M., Loewen, P.D., Wang, X.: ε-Fréchet differentiability of Lipschitz functions and applications. J. Convex Anal. **13**(3–4), 695–709 (2006)

380. Fabian, M.J., Henrion, R., Kruger, A.Y., Outrata, J.V.: Error bounds: necessary and sufficient conditions. Set-Valued Var. Anal. **18**(2), 121–149 (2010)

381. Facchinei, F., Pang, J.S.: Finite Dimensional Variational Inequalities and Complementary Problems I. Springer, New York (2003)

382. Facchinei, F., Pang, J.S.: Finite-Dimensional Variational Inequalities and Complementarity Problems II. Springer, New York (2003)

383. Fajardo, M.D., López, M.A.: Locally Farkas–Minkowski systems in convex semi-infinite programming. J. Optim. Theor. Appl. **103**, 313–335 (1999)

384. Fan, K.: Minimax theorems. Natl. Acad. Sci. Proc. USA **39**, 42–47 (1953)

385. Fang, D.H., Li, C., Ng, K.F.: Constraint qualifications for optimality conditions and total Lagrange dualities in convex infinite programming. Nonlinear Anal. **73**(5), 1143–1159 (2010)

386. Fang, J.X.: The variational principle and fixed point theorem in certain topological spaces. J. Math. Anal. Appl. **202**(2), 398–412 (1996)

387. Federer, H.: Curvature measures. Trans. Am. Math. Soc. **93**, 418–491 (1959)

388. Fenchel, W.: On conjugate convex functions. Can. J. Math. **1**, 73–77 (1949)
389. Fenchel, W.: Convex Cones, Sets and Functions. Logistic Project Report. Department of Mathematics, Princeton University, Princeton (1953)
390. Ferrer, J.: Rolle's theorem fails in ℓ_2. Am. Math. Monthly **103** 161–165 (1996)
391. Ferriero, A., Marchini, E.M.: On the validity of the Euler–Lagrange equation. J. Math. Anal. Appl. **304**, 356–369 (2005)
392. Fiacco, A.V.: Introduction to Sensitivity and Stability Analysis in Nonlinear Programming. Academic Press, New York (1983)
393. Fiacco, A.V., Mc Cormick, G.P.: Nonlinear Programming. Wiley, New York (1968)
394. Figiel, T.: On the moduli of convexity and smoothness. Studia Math. **56**, 121–155 (1976)
395. Filippov, A.F.: Classical solutions of differential inclusions with multivalued right-hand sides. SIAM J. Contr. **5** 609–621 (1967)
396. Fitzpatrick, S.: Metric projection and the differentiability of the distance functions. Bull. Aust. Math. Soc. **22**, 291–312 (1980)
397. Fitzpatrick, S.: Differentiation of real-valued functions and continuity of metric projections. Proc. Am. Math. Soc. **91**(4), 544–548 (1984)
398. Fitzpatrick, S., Lewis, A.S.: Weak-star convergence of convex sets. J. Convex Anal. **13**(3–4), 711–719 (2006)
399. Fitzpatrick, S., Phelps, R.R.: Differentiability of the metric projection in Hilbert space. Trans. Am. Math. Soc. **270**(2), 483–501 (1982)
400. Fitzpatrick, S.P., Simons, S.: The conjugates, compositions and marginals of convex functions. J. Convex Anal. **8**, 423–446 (2001)
401. Flåm, S.D.: Solving convex programs by means of ordinary differential equations. Math. Oper. Res. **17**(2), 290–302 (1992)
402. Flåm, S.D.: Upward slopes and inf-convolutions. Math. Oper. Res. **31**(1), 188–198 (2006)
403. Flåm, S.D., Seeger, A.: Solving cone-constrained convex programs by differential inclusions. Math. Program. A **65**(1), 107–121 (1994)
404. Flåm, S., Jongen, H.Th., Stein, O.: Slopes of shadow prices and Lagrange multipliers. Optim. Lett. **2**(2), 143–155 (2008)
405. Flåm, S.D., Jongen, H.Th., Stein, O.: Slopes of shadow prices and Lagrange multipliers. Optim. Lett. **2**(2), 143–155 (2008)
406. Flåm, S.D., Hiriart-Urruty, J.-B., Jourani, A.: Feasibility in finite time. J. Dyn. Contr. Syst. **15**(4), 537–555 (2009)
407. Fougères, A.: Coercivité des intégrandes convexes normales. Application à la minimisation des fonctionelles intégrales et du calcul des variations, Séminaire d'analyse convexe Montpellier, exposé no. 19 (1976)
408. Frankowska, H.: An open mapping principle for set-valued maps. J. Math. Anal. Appl. **127**, 172–180 (1987)
409. Frankowska, H.: High order inverse function theorems. Analyse non linéaire (Perpignan, 1987). Ann. Inst. H. Poincaré Anal. Non Linéaire Suppl. **6**, 283–303 (1989)
410. Frankowska, H.: Some inverse mapping theorems. Annales Institut H. Poincaré Anal. Non Linéaire **7**, 183–234 (1990)
411. Frankowska, H.: Conical inverse mapping theorems. Bull. Aust. Math. Soc. **45**(1), 53–60 (1992)
412. Frankowska, H, Quincampoix, M.: Hölder metric reegularity of set-valued maps. Math. Program. 132(1–2), 333–354 (2012)
413. Frankowska, H., Plaskacz, S., Rzezuchowski, T.: Set-valued approach to Hamilton–Jacobi–Bellman equations. In: Set-valued Analysis and Differential Inclusions (Pamporovo, 1990), pp. 105–118. Progr. Systems Control Theory, vol. 16. Birkhäuser, Boston (1993)
414. Friedman, A.: Variational Principles and Free Boundary Problems. Wiley, New York (1982)
415. Furi, M., Vignoli, A.: About well-posed optimization problems for functionals in metric spaces. J. Optim. Theor. Appl. **5**, 225–229 (1970)
416. Furi, M., Vignoli, A.: A characterization of well-posed minimum problems in a complete metric space. J. Optim. Theor. Appl. **5**, 452–461 (1970)

417. Fusek, P., Klatte, D., Kummer, B.: Examples and counterexamples in Lipschitz analysis. Well-posedness in optimization and related topics (Warsaw, 2001). Contr. Cybern. **31**(3), 471–492 (2002)

418. Gelfand, I.M., Fomin, S.V.: Calculus of Variations. Revised English edition. 3rd. printing. Prentice-Hall, Englewood Cliffs (1965)

419. Gautier, S.: Affine and eclipsing multifunctions. Numer. Funct. Anal. Optim. **11**(7/8), 679–699 (1990)

420. Gauvin, J.: The generalized gradient of a marginal function in mathematical programming. Math. Oper. Res. **4**(4), 458–463 (1979)

421. Gauvin, J.: Directional derivative for the value function in mathematical programming. In: Nonsmooth Optimization and Related Topics (Erice, 1988), pp. 167–183. Ettore Majorana Internat. Sci. Ser. Phys. Sci., vol. 43. Plenum, New York (1989)

422. Gauvin, J., Dubeau, F.: Differential properties of the marginal function in mathematical programming. Optimality and stability in mathematical programming. Math. Program. Stud. **19**, 101–119 (1982)

423. Gauvin, J., Dubeau, F.: Some examples and counterexamples for the stability analysis of nonlinear programming problems. Sensitivity, stability and parametric analysis. Math. Program. Stud. **21**, 69–78 (1984)

424. Gauvin, J., Janin, R.: Directional Lipschitzian optimal solutions and directional derivative for the optimal value function in nonlinear mathematical programming. Analyse non linéaire (Perpignan, 1987). Ann. Inst. H. Poincaré Anal. Non Linéaire **6**(suppl.), 305–324 (1989)

425. Gauvin, J., Janin, R.: Directional derivative of the value function in parametric optimization. Ann. Oper. Res. **27**(1–4), 237–252 (1990)

426. Gauvin, J., Tolle, J.W.: Differential stability in nonlinear programming. SIAM J. Contr. Optim. **15**(2), 294–311 (1977)

427. Geoffroy, M.H.: Approximation of fixed points of metrically regular mappings. Numer. Funct. Anal. Optim. **27**(5–6), 565–581 (2006)

428. Geoffroy, M., Lassonde, M.: On a convergence of lower semicontinuous functions linked with the graph convergence of their subdifferentials. In: Constructive, Experimental, and Nonlinear Analysis (Limoges, 1999), pp. 93–109. CMS Conf. Proc., vol. 27. American Mathematical Society, Providence (2000)

429. Geoffroy, M., Lassonde, M.: Stability of slopes and subdifferentials. Set-Valued Anal. **11**(3), 257–271 (2003)

430. Georgiev, P.: The strong Ekeland Variational Principle, the strong drop theorem and applications. J. Math. Anal. Appl. **131**, 1–21 (1988)

431. Georgiev, P., Zlateva, N.: Lasry–Lions regularizations and reconstruction of subdifferentials. C.R. Acad. Bulgare Sci. **51**(9–10), 9–12 (1998)

432. Georgiev, P., Kutzarova, D., Maaden, A.: On the smooth drop property. Nonlinear Anal. **26**(3), 595–602 (1996)

433. Geremew, W., Mordukhovich, B.S., Nam, N.M.: Coderivative calculus and metric regularity for constraint and variational systems. Nonlinear Anal. **70**(1), 529–552 (2009)

434. Giannessi, F.: Constrained Optimization and Image Space Analysis, vol. 1. Separation of Sets and Optimality Conditions. Mathematical Concepts and Methods in Science and Engineering, vol. 49. Springer, New York (2005)

435. Giles, J.R.: A subdifferential characterisation of Banach spaces with the Radon–Nikodym property. Bull. Aust. Math. Soc. **66**(2), 313–316 (2002)

436. Giles, J.R.: Convex Analysis with Applications in Differentiation of Convex Functions. Pitman Research Lecture Notes in Mathematics, vol. 58. Longman, London (1982)

437. Giles, J.: Generalising generic differentiability properties of convex functions. In: Lau, A.T.M., et al. (eds.) Topological Linear Spaces, Algebras and Related Areas. Pitman Research Lecture Notes in Mathematics, vol. 316, pp. 193–207. Longman, London (1994)

438. Giles, J., Vanderwerff, J.: Asplund spaces and a variant of weak uniform rotundity. Bull. Aust. Math. Soc. **61**(3), 451–454 (2000)

439. Giner, E.: On the Clarke subdifferential of an integral functional on L_p, $1 \leq p < \infty$. Can. Math. Bull. **41**(1), 41–48 (1998)

440. Giner, E.: Calmness properties and contingent subgradients of integral functionals on Lebesgue spaces L_p, $1 \leq p < \infty$. Set-Valued Anal. **17**(3), 223–243 (2009)

441. Giner, E.: Necessary and sufficient conditions for the interchange between infimum and the symbol of integration. Set-Valued Anal. **17**(4), 321–357 (2009)

442. Giner, E.: Lagrange multipliers and lower bounds for integral functionals. J. Convex Anal. **17**(1) 301–308 (2010)

443. Giner, E.: Subdifferential regularity and characterizations of Clarke subgradients of integral functionals. J. Nonlinear Convex Anal. **9**(1), 25–36 (2008)

444. Giner, E.: Michel–Penot subgradients and integral functionals. Preprint, University Paul Sabatier, Toulouse (2006)

445. Goberna, M.A., López, M.A.: Linear Semi-Infinite Optimization. Wiley, Chichester (1998)

446. Goberna, M.A., López, M.A., Pastor, J.: Farkas-Minkowski in semi-infinite programming. Appl. Math. Optim. **7**, 295–308 (1981)

447. Godefroy, G., Troyanski, S., Whitfield, J., Zizler, V.: Smoothness in weakly compactly generated Banach spaces. J. Funct. Anal. **52**, 344–352 (1983)

448. Goebel, K., Kirk, W.A.: Topics in Metric Fixed Point Theory. Cambridge Studies in Advanced Math., vol. 28. Cambridge University Press, Cambridge (1990)

449. Goebel, R., Rockafellar, R.T.: Local strong convexity and local Lipschitz continuity of the gradient of convex functions. J. Convex Anal. **15**(2), 263–270 (2008)

450. Goldstine, H.H.: A History of the Calculus of Variations from the 17th Through the 19th Century. Studies in the History of Mathematics and Physical Sciences, vol. 5. Springer, New York (1980)

451. Góra, P., Stern, R.J.: Subdifferential analysis of the Van der Waerden function. J. Convex Anal. **18**(3), 699–705 (2011)

452. Gowda, M.S., Teboulle, M.: A comparison of constraint qualification in infinite-dimensional convex programming. SIAM J. Contr. Optim. **28**, 925–935 (1990)

453. Granas, A., Lassonde, M.: Some elementary general principles of convex analysis. Topol. Meth. Nonlinear Anal. **5**(1), 23–37 (1995)

454. Graves, L.M.: Some mapping theorems. Duke Math. J. **17**, 111–114 (1950)

455. Gromov, M.: Metric Structures for Riemannian and Non-Riemannian Spaces. Prog. in Math., vol. 152. Birkhäuser, Boston (1999)

456. Gudovich, A., Kamenskii, M., Quincampoix, M.: Existence of equilibria of set-valued maps on bounded epi-Lipschitz domains in Hilbert spaces without invariance conditions. Nonlinear Anal. Theor. Meth. Appl. **72**(1), 262–276 (2010)

457. Guillemin, V., Pollack, A.: Differential Topology. Prentice Hall, Englewood Cliffs (1976)

458. Ha, T.X.D.: Some variants of the Ekeland Variational Principle for a set-valued map. J. Optim. Theor. Appl. **124**(1), 187–206 (2005)

459. Hadamard, J.: Cours d'Analyse. Ecole Polytechnique 2eme division, 1928–1929, Premier semestre, mimeographied notes

460. Hagler, J., Sullivan, F.E.: Smoothness and weak star sequential compactness. Proc. Am. Math. Soc. **78**, 497–503 (1980)

461. Halkin, H.: A satisfactory treatment of equality and operator constraints in the Dubovitskii–Milyutin optimization formalism. J. Optim. Theor. Appl. **6** 138–149 (1970)

462. Halkin, H.: Implicit functions and optimization problems without continuous differentiability of the data. Collection of articles dedicated to the memory of Lucien W. Neustadt. SIAM J. Contr. **12**, 229–236 (1974)

463. Halkin, H.: Interior mapping theorem with set-valued derivatives. J. Analyse Math. **30**, 200–207 (1976)

464. Halkin, H., Neustadt, L.W.: General necessary conditions for optimization problems. Proc. Natl. Acad. Sci. USA **56**, 1066–1071 (1966)

465. Hamel, A.H.: Equivalents to Ekeland's variational principle in uniform spaces. Nonlinear Anal. Theor. Meth. Appl. **62A**(5), 913–924 (2005)

466. Hamilton, R.S.: The inverse function theorem of Nash and Moser. Bull. Am. Math. Soc. **7**(1), 65–222 (1982)

467. Hanner, O.: On the uniform convexity of L^p and ℓ^p. Ark. Mat. **3**, 239–244 (1956)

468. Hantoute, A., López, M.A., Zălinescu, C.: Subdifferential calculus rules in convex analysis: a unifying approach via pointwise supremum functions. SIAM J. Optim. **19**, 863–882 (2008)

469. Hare, W.L., Lewis, A.S.: Estimating tangent and normal cones without calculus. Math. Oper. Res. **30**(4), 785–799 (2005)

470. Hare, W.L., Poliquin, R.A.: Prox-regularity and stability of the proximal mapping. J. Convex Anal. **14**(3), 589–606 (2007)

471. Haydon, R.: A counterexample to certain questions about scattered compact sets. Bull. Lond. Math. Soc. **22**, 261–268 (1990)

472. He, Y., Sun, J.: Error bounds for degenerate cone inclusion problems. Math. Oper. Res. **30**, 701–717 (2005)

473. Henrion, R., Outrata, J.: A subdifferential condition for calmness of multifunctions. J. Math. Anal. Appl. **258**, 110–130 (2001)

474. Henrion, R., Outrata, J.V.: Calmness of constraint systems with applications. Math. Program. B **104**, 437–464 (2005)

475. Henrion, R., Outrata, J.V.: On calculating the normal cone to a finite union of convex polyhedra. Optimization **57**(1), 57–78 (2008)

476. Henrion, R., Jourani, A., Outrata, J.V.: On the calmness of a class of multifunctions. SIAM J. Optim. **13**, 603–618 (2002)

477. Henrion, R., Outrata, J., Surowiec, T.: On the co-derivative of normal cone mappings to inequality systems. Nonlinear Anal. **71**(3–4), 1213–1226 (2009)

478. Hiriart-Urruty, J.-B.: On optimality conditions in non-differentiable programming. Math. Program. **14**(1), 73–86 (1978)

479. Hiriart-Urruty, J.-B.: Gradients généralisés de fonctions marginales. SIAM J. Contr. Optim. **16**, 301–316 (1978)

480. Hiriart-Urruty, J.-B.: New concepts in non differentiable programming. Bull. Soc. Math. France Mémoire **60**, 57–85 (1979)

481. Hiriart-Urruty, J.-B.: Tangent cones, generalized gradients and mathematical programming in Banach spaces. Math. Oper. Res. **4**, 79–97 (1979)

482. Hiriart-Urruty, J.-B.: Refinements of necessary optimality conditions in nondifferentiable programming. I. Appl. Math. Optim. **5**(1), 63–82 (1979)

483. Hiriart-Urruty, J.-B.: A note on the mean value theorem for convex functions. Boll. Un. Mat. Ital. B (5) **17**(2), 765–775 (1980)

484. Hiriart-Urruty, J.-B.: Mean value theorems in nonsmooth analysis. Numer. Funct. Anal. Optim. **2**(1), 1–30 (1980)

485. Hiriart-Urruty, J.-B.: Extension of Lipschitz functions. J. Math. Anal. Appl. **77**(2), 539–554 (1980)

486. Hiriart-Urruty, J.-B.: Refinements of necessary optimality conditions in nondifferentiable programming. II. Math. Program. Stud. **19**, 120–139 (1982)

487. Hiriart-Urruty, J.-B.: A short proof of the variational principle for approximate solutions of a minimization problem. Am. Math. Mon. **90**(3), 206–207 (1983)

488. Hiriart-Urruty, J.-B.: Miscellanies on nonsmooth analysis and optimization. In: Nondifferentiable Optimization: Motivations and Applications (Sopron, 1984), pp. 8–24. Lecture Notes in Econom. and Math. Systems, vol. 255. Springer, Berlin (1985)

489. Hiriart-Urruty, J.-B.: A general formula on the conjugate of the difference of functions. Can. Math. Bull. **29**(4), 482–485 (1986)

490. Hiriart-Urruty, J.-B.: From convex optimization to nonconvex optimization. Necessary and sufficient conditions for global optimality. In: Clarke, F.H., Demyanov, V.F., Giannessi, F. (eds.) Nonsmooth Optimization and Related Topics, pp. 219–239. Plenum, New York (1989)

491. Hiriart-Urruty, J.-B.: A note on the Legendre–Fenchel transform of convex composite functions. In: Nonsmooth Mechanics and Analysis, pp. 35–46. Adv. Mech. Math., vol. 12. Springer, New York (2006)

492. Hiriart-Urruty, J.-B.: Pot pourri of conjectures and open questions in nonlinear analysis and optimization. SIAM Rev. **49**(2), 255–273 (2007)

493. Hiriart-Urruty, J.-B.: Generalized differentiability, duality and optimization for problems dealing with differences of convex functions. In: Ponstein, J. (ed.) Convexity and Duality in Optimization. Lecture Notes in Econom. and Math. Systems, vol. 256, pp. 37–70 (1986)

494. Hiriart-Urruty, J.-B.: From convex optimization to nonconvex optimization. Necessary and sufficient conditions for global optimality. In: Clarke, F.H., Demyanov, V.F., Giannessi, F. (eds.) Nonsmooth Optimization and Related Topics, pp. 219–239. Ettore Majorana Internat. Sci. Ser. Phys. Sci., vol. 43. Plenum, New York (1989)

495. Hiriart-Urruty, J.-B., Imbert, C.: Les fonctions d'appui de la jacobienne généralisée de Clarke et de son enveloppe plénière. C.R. Acad. Sci. Paris I Math. **326**(11), 1275–1278 (1998)

496. Hiriart-Urruty, J.-B., Ledyaev, Y.S.: A note on the characterization of the global maxima of a (tangentially) convex function over a convex set. J. Convex Anal. **3**(1), 55–61 (1996)

497. Hiriart-Urruty, J.-B., Lemaréchal, C.: Convex Analysis and Minimization Algorithms. Part 1: Fundamentals. Grundlehren der Mathematischen Wissenschaften, vol. 305. Springer, Berlin (1993)

498. Hiriart-Urruty, J.-B., Lemaréchal, C.: Fundamentals of Convex Analysis. Grundlehren. Text editions. Springer, Berlin (2001)

499. Hiriart-Urruty, J.-B., Martínez-Legaz, J.-E.: New formulas for the Legendre–Fenchel transform. J. Math. Anal. Appl. **288**(2), 544–555 (2003)

500. Hiriart-Urruty, J.-B., Phelps, R.R.: Subdifferential calculus using ε-subdifferentials. J. Funct. Anal. **118**(1), 154–166 (1993)

501. Hiriart-Urruty, J.-B., Plazanet, Ph.: Moreau's theorem revisited. Anal. Non Linéaire. Ann. Inst. H. Poincaré **6**, 47, 325–338 (1989)

502. Hiriart-Urruty, J.-B., Thibault, L.: Existence et caractérisation de différentielles généralisées d'applications localement lipschitziennes d'un espace de Banach séparable dans un espace de Banach réflexif séparable. C.R. Acad. Sci. Paris A-B **290**(23), 1091–1094 (1980)

503. Hiriart-Urruty, J.-B., Torki, M.: Permanently going back and forth between the "quadratic world" and the "convexity world" in optimization. Appl. Math. Optim. **45**(2), 169–184 (2002)

504. Hiriart-Urruty, J.-B., Moussaoui, M., Seeger, A., Volle, M.: Subdifferential calculus without qualification conditions, using approximate subdifferentials: a survey. Nonlinear Anal. **24**(12), 1727–1754 (1995)

505. Hoffman, A.J.: On approximate solutions of systems of linear inequalities. J. Res. Nat. Bur. Stand. **49**, 263–265 (1952)

506. Holmes, R.B.: A Course on Optimization and Best Approximation. Lecture Notes in Mathematics, vol. 257. Springer, Berlin (1972)

507. Holmes, R.B.: Geometric Functional Analysis and Its Applications. Graduate Texts in Mathematics, vol. 24. Springer, New York (1975)

508. Hörmander, L.: Notions of Convexity. Reprint of the 1994 edition. Modern Birkhäuser Classics. Birkhäuser, Boston (2007). Applications. World Scientific, River Edge (1997)

509. Huang, L.R., Ng, K.F., Penot, J.-P.: On minimizing and stationary sequences in nonsmooth optimization. SIAM J. Optim. **10**(4), 999–1019 (2000)

510. Hyers, D.H., Isac, G., Rassias, T.M.: Topics in Nonlinear Analysis and Applications. World Scientific, River Edge (1997)

511. Ioffe, A.D.: Regular points of Lipschitz functions. Trans. Am. Math. Soc. **251**, 61–69 (1979)

512. Ioffe, A.D.: Sous-différentielles approchées de fonctions numériques. C.R. Acad. Sci. Paris **292**, 675–678 (1981)

513. Ioffe, A.D.: Calculus of Dini subdifferentials, Cahiers du Ceremade 8120, Université Paris IX Dauphine (1981)

514. Ioffe, A.D.: Nonsmooth analysis: Differential calculus of nondifferentiable mappings. Trans. Am. Math. Soc **266**(1), 1–56 (1981)

515. Ioffe, A.D.: On subdifferentiability spaces. Ann. New York Acad. Sci. **410**, 107–119 (1983)

516. Ioffe, A.D.: Subdifferentiability spaces and nonsmooth analysis. Bull. Am. Math. Soc. **10**, 87–90 (1984)

517. Ioffe, A.D.: Calculus of Dini subdifferentials of functions and contingent coderivatives of set-valued maps. Nonlinear Anal. **8**(5), 517–539 (1984)

518. Ioffe, A.D.: Necessary conditions in nonsmooth optimization. Math. Oper. Res. **9**(2), 159–189 (1984)

519. Ioffe, A.D.: On the theory of subdifferential. In: Hiriart-Urruty, J.B. (ed.) Fermat Days: Mathematics for Optimization. Math. Studies Series. North Holland, Amsterdam (1986)

520. Ioffe, A.D.: Approximate subdifferentials and applications. 1. The finite-dimensional theory. Trans. Am. Math. Soc **281**(1), 389–416 (1984) (submitted in 1982)

521. Ioffe, A.D.: Approximate subdifferentials and applications 2. Mathematika **33**, 111–128 (1986)

522. Ioffe, A.D.: On the local surjection property. Nonlinear Anal. Theor. Meth. Appl. **11**, 565–592 (1987)

523. Ioffe, A.D.: Global surjection and global inverse mapping theorems in Banach spaces. Ann. New York Acad. Sci. **491**, 181–189 (1987)

524. Ioffe, A.D.: Approximate subdifferentials and applications 3: the metric theory. Mathematika **36**(1), 1–38 (1989)

525. Ioffe, A.D.: Proximal analysis and approximate subdifferentials. J. Lond. Math. Soc. **41**(2), 175–192 (1990)

526. Ioffe, A.: A Lagrange multiplier rule with small convex-valued subdifferentials for non-smooth problems of mathematical programming involving equality and nonfunctional constraints. Math. Program. A **58**(1), 137–145 (1993)

527. Ioffe, A.: Separable reduction theorem for approximate subdifferentials. C.R. Acad. Sci. Paris I **323**(1), 107–112 (1996)

528. Ioffe, A.: Fuzzy principles and characterization of trustworthiness. Set-Valued Anal. **6**, 265–276 (1998)

529. Ioffe, A.D.: Variational methods in local and global non-smooth analysis. Notes by Igor Zelenko. In: Clarke, F.H., et al. (eds.) Nonlinear Analysis, Differential Equations and Control. Proceedings of the NATO Advanced Study Institute and séminaire de mathématiques supérieures, Montréal, Canada, 27 July–7 August 1998, NATO ASI Ser., Ser. C, Math. Phys. Sci., vol. 528, pp. 447–502. Kluwer, Dordrecht (1999)

530. Ioffe, A.D.: Codirectional compactness, metric regularity and subdifferential calculus. In: Théra, M. (ed.) Constructive, Experimental and Nonlinear Analysis, Limoges, France, 22–23 September 1999. CMS Conf. Proc., vol. 27, pp. 123–163. American Mathematical Society, Providence, RI (2000); publ. for the Canadian Mathematical Society

531. Ioffe, A.D.: Metric regularity and subdifferential calculus. Uspekhi Mat. Nauk **55**(3), 103–162 (2000). English transl.: Russ. Math. Surv. **55**(3), 501–558 (2000)

532. Ioffe, A.D.: On perturbation stability of metric regularity. Set-Valued Anal. **9**, 101–109 (2001)

533. Ioffe, A.: Towards metric theory of metric regularity. In: Lassonde, M. (ed.) Approximation, Optimization and Mathematical Economics. Proceedings of the 5th International Conference on Approximation and Optimization in the Caribbean, Guadeloupe, French West Indies, 29 March–2 April 1999, pp. 165–176. Physica-Verlag, Heidelberg (2001)

534. Ioffe, A.D.: Abstract convexity and non-smooth analysis. In: Kusuoka, S., et al. (eds.) Advances in Mathematical Economics, vol. 3, pp. 45–61. Springer, Tokyo (2001)

535. Ioffe, A.D.: On robustness of regularity properties of maps. Contr. Cybern. **32**, 543–555 (2003)

536. Ioffe, A.D.: On stability estimates for the regularity property of maps. In: Brézis, H., et al. (eds.) Topology Methods, Variational Methods, and Their Applications. Proceedings of the ICM 2002 Satellite Conference on Nonlinear Functional Analysis, Taiyuan, China, 14–18 August 2002, pp. 133–142. World Scientific, River Edge (2003)

537. Ioffe, A.D.: On regularity of convex multifunctions. Nonlinear Anal. **69**(3), 843–849 (2008)

538. Ioffe, A.D.: On regularity concepts in variational analysis. J. Fixed Point Theor. Appl. **8**(2), 339–363 (2010)
539. Ioffe, A.D.: Towards variational analysis in metric spaces: metric regularity and fixed points. Math. Program. B **123**, 241–252 (2010)
540. Ioffe, A.D.: Typical convexity (concavity) of Dini-Hadamard upper (lower) directional derivatives of functions on separable Banach spaces. J. Convex Anal. **17**(3–4), 1019–1032 (2010)
541. Ioffe, A.D.: On the theory of subdifferentials. Adv. Nonlinear Anal. **1**, 47–120 (2012)
542. Ioffe, A.D., Lewis, A.S.: Critical points of simple functions. Optimization **57**(1), 3–16 (2008)
543. Ioffe, A.D., Lucchetti, R.E.: Generic existence, uniqueness and stability in optimization problems. In: Di Pillo, G., Giannessi, F. (eds.) Nonlinear Optimization and Applications, vol. 2. Kluwer, Dordrecht (1998)
544. Ioffe, A.D., Lucchetti, R.E.: Generic well-posedness in minimization problems. Abstr. Appl. Anal. **4**, 343–360 (2005)
545. Ioffe, A., Lucchetti, R.E.: Typical convex program is very well posed. Math. Program. B **104**(2–3), 483–499 (2005)
546. Ioffe, A.D., Outrata, J.V.: On metric and calmness qualification conditions in subdifferential calculus. Set-Valued Anal. **16**, 199–227 (2008)
547. Ioffe, A.D., Penot, J.-P.: Subdifferentials of performance functions and calculus of coderivatives of set-valued mappings. Serdica Math. J. **22**, 257–282 (1996)
548. Ioffe, A.D., Sekiguchi, Y.: Regularity estimates for convex multifunctions. Math. Program. B **117**(1–2), 255–270 (2009)
549. Ioffe, A.D., Tikhomirov, V.M.: Theory of Extremal Problems. Nauka, Moscow (1974); English Translation, Studies in Mathematics and Its Applications, vol. 6. North Holland, Amsterdam (1979)
550. Ioffe, A.D., Tikhomirov, V.M.: Some remarks on variational principles. Math. Notes **61**(2), 248–253 (1997)
551. Ioffe, A.D., Zaslavski, A.J.: Variational principles and well-posedness in optimization and calculus of variations. SIAM J. Contr. Optim. **38**(2), 566–581 (2000)
552. Ioffe, A.D., Lucchetti, R.E., Revalski, J.P.: Almost every convex or quadratic programming problem is well-posed. Math. Oper. Res. **29**(2), 369–382 (2004)
553. Ioffe, A.D., Lucchetti, R.E., Revalski, J.P.: A variational principle for problems with functional constraints. SIAM J. Optim. **12**(2), 461–478 (2001/2002)
554. Jameson, G.J.O.: Convex series. Proc. Camb. Philos. Soc. **72**, 37–47 (1972)
555. Jameson, G.J.O.: Topology and Normed Spaces. Chapman and Hall, London (1974)
556. Janin, R.: Sur une classe de fonctions sous-linéarisables. C.R. Acad. Sci. Paris **277**, 265–267 (1973)
557. Janin, R.: Sensitivity for non convex optimization problems. In: Convex Analysis and Its Applications (Proc. Conf., Murat-le-Quaire, 1976), pp. 115–119. Lecture Notes in Econom. and Math. Systems, vol. 144. Springer, Berlin (1977)
558. Janin, R.: Sur des multiapplications qui sont des gradients généralisés. C.R. Acad. Sci. Paris **294**, 115–117 (1982)
559. Jeyakumar, V.: Asymptotic dual conditions characterizing optimality for convex programs. J. Optim. Theor. Appl. **93**, 153–165 (1997)
560. Jeyakumar, V., Luc, D.T.: Nonsmooth Vector Functions and Continuous Optimization. Springer, Berlin (2005)
561. Jeyakumar, V., Wolkowicz, H.: Generalizations of Slater's constraint qualification for infinite convex programs. Math. Program. **57** 85–102 (1992)
562. Jeyakumar, V., Lee, G.M., Dinh, N.: New sequential Lagrange multiplier conditions characterizing optimality without constraint qualifications for convex programs. SIAM J. Optim. **14**, 534–547 (2003)

563. Jeyakumar, V., Dinh, N., Lee, G.M.: A new closed cone constraint qualification for convex optimization. Applied Mathematics Research Report AMR04/8, School of Mathematics, University of New South Wales, Australia, 2004. http://www.maths.unsw.edu.au/applied/reports/amr08.html (2004)

564. Jeyakumar, V., Song, W., Dinh, N., Lee, G.M.: Liberating the subgradient optimality conditions from constraint qualifications. J. Global Optim. **36**, 127–137 (2006)

565. Jofré, A., Penot, J.-P.: A note on the directional derivative of a marginal function. Rev. Mat. Apl. **14**(2), 37–54 (1993)

566. Jofré, A., Penot, J.-P.: Comparing new notions of tangent cones. J. Lond. Math. Soc. **40**(2), 280–290 (1989)

567. Jofré, A., Rivera Cayupi, J.: A nonconvex separation property and some applications. Math. Program. **108**, 37–51 (2006)

568. Jofré, A., Thibault, L.: *D*-representation of subdifferentials of directionally Lipschitz functions. Proc. Am. Math. Soc. **110**(1), 117–123 (1990)

569. Jofré, A., Thibault, L.: *b*-subgradients of the optimal value function in nonlinear programming. Optimization **26**(3–4), 153–163 (1992)

570. Jofré, A., Thibault, L.: Proximal and Fréchet normal formulae for some small normal cones in Hilbert space. Nonlinear Anal. **19**(7), 599–612 (1992)

571. Jourani, A.: Regularity and strong sufficient optimality conditions in differentiable optimization problems. Num. Funct. Anal. Optim. **14**, 69–87 (1993)

572. Jourani, A.: Weak regularity of functions and sets in Asplund spaces. Nonlinear Anal. TMA **65**, 660–676 (2006)

573. Jourani, A., Thibault, L.: Approximate subdifferentials and metric regularity: the finite dimensional case. Math. Program. **47**, 203–218 (1990)

574. Jourani, A., Thibault, L.: A note on Fréchet and approximate subdifferentials of composite functions. Bull. Austral. Math. Soc. **49**(1), 111–115 (1994)

575. Jourani, A., Thibault, L.: Metric regularity for strongly compactly Lipschitzian mappings. Nonlinear Anal. Theor. Meth. Appl. **24**, 229–240 (1995)

576. Jourani, A., Thibault, L.: Metric regularity and subdifferential calculus in Banach spaces. Set-Valued Anal. **3**(1), 87–100 (1995)

577. Jourani, A., Thibault, L.: Verifiable conditions for openness and regularity of multivalued mappings in Banach spaces. Trans. Am. Math. Soc. **347**, 1255–1268 (1995)

578. Jourani, A., Thibault, L.: Extensions of subdifferential calculus rules in Banach spaces. Can. J. Math. **48**(4), 834–848 (1996)

579. Jourani, A., Thibault, L.: Qualification conditions for calculus rules of coderivatives of multivalued mappings. J. Math. Anal. Appl. **218**(1), 66–81 (1998)

580. Jourani, A., Thibault, L.: Chain rules for coderivatives of multivalued mappings in Banach spaces. Proc. Am. Math. Soc. **126**(5), 1479–1485 (1998)

581. Jourani, A., Thibault, L.: Coderivatives of multivalued mappings, locally compact cones and metric regularity. Nonlinear Anal. Theor. Meth. Appl. **35**(7A), 925–945 (1999)

582. Jules, F.: Sur la somme de sous-différentiels de fonctions continues inférieurement. Dissertationes Mathematicae, vol. 423. Polska Akad. Nauk, Warsaw (2003)

583. Jules, F., Lassonde, M.: Formulas for subdifferentials of sums of convex functions. J. Convex Anal. **9**(2), 519–533 (2002)

584. Kantorovich, L.V., Akilov, G.P.: Functional Analysis. Translated from the Russian, 2nd edn. Pergamon, Oxford–Elmsford (1982)

585. Katriel, G.: Mountain pass theorem and global homeomorphism theorems. Ann. Inst. H. Poincaré Anal. Non Linéaire **11**, 189–211 (1994)

586. Khanh, P.Q.: An induction theorem and general open mapping theorem. J. Math. Anal. Appl. **118**, 519–536 (1986)

587. Khanh, P.Q.: On general open mapping theorems. J. Math. Anal. Appl. **144**, 305–312 (1989)

588. Khanh, P.Q., Luc, D.T.: Stability of solutions in parametric variational relation problems. Set-Valued Anal. **16**(7–8), 1015–1035 (2008)

589. Khanh, P.Q., Quy, D.N.: A generalized distance and enhanced Ekeland's variational principle for vector functions. Nonlinear Anal. **73**(7), 2245–2259 (2010)
590. Klatte, D., Kummer, B.: Nonsmooth Equations in Optimization. Regularity, Calculus, Methods and Applications. Nonconvex Optimization and Its Applications, vol. 60. Kluwer, Dordrecht (2002)
591. Klein, E., Thompson, A.C.: Theory of Correspondences Including Applications to Mathematical Economics. Wiley, New York (1984)
592. Kočvara, M., Outrata, J.V.: A nonsmooth approach to optimization problems with equilibrium constraints. In: Complementarity and Variational Problems (Baltimore, MD, 1995), pp. 148–164. SIAM, Philadelphia (1997)
593. Křivan, V.: On the intersection of contingent cones. J. Optim. Theor. Appl. **70**(2), 397–404 (1991)
594. Kruger, A.Y.: Subdifferentials of Nonconvex Functions and Generalized Directional Derivatives. Mimeographied notes, VINITI Moscow 2661–77, p. 39 (1977) (in Russian)
595. Kruger, A.Y.: Generalized differentials of nonsmooth functions. Mimeographied notes, Belorussian State Univ. 1332–81, p. 64 (1981) (in Russian)
596. Kruger, A.Y.: Properties of generalized differentials. Sib. Math. J. **26**, 822–832 (1985)
597. Kruger, A.Ya.: A covering theorem for set-valued mappings. Optimization **19**(6), 763–780 (1988)
598. Kruger, A.Ya.: Strict ε, δ-subdifferentials and extremality conditions. Optimization **51**(3), 539–554 (2002)
599. Kruger, A.Ya.: On Fréchet subdifferentials. Optimization and related topics. J. Math. Sci. (NY) **116**(3), 3325–3358 (2003)
600. Kruger, A.Ya.: Stationarity and regularity of real-valued functions. Appl. Comput. Math. **5**(1), 79–93 (2006)
601. Kruger, A.Ya.: About regularity of collections of sets. Set-Valued Anal. **14**(2), 187–206 (2006)
602. Kruger, A.Ya.: About stationarity and regularity in variational analysis. Taiwanese J. Math. **13**(6A), 1737–1785 (2009)
603. Kruger, A.Ya., Mordukhovich, B.S.: Extremal points and the Euler equation in nonsmooth optimization problems. (Russian) Dokl. Akad. Nauk BSSR **24**(8), 684–687, 763 (1980)
604. Kummer, B.: Lipschitzian inverse functions, directional derivatives and applications in $C^{1,1}$ optimization. J. Optim. Theor. Appl. **70**, 561–582 (1991)
605. Kummer, B.: An implicit function theorem for $C^{0,1}$-equations and parametric $C^{1,1}$-optimization. J. Math. Anal. Appl. **158**, 35–46, (1991)
606. Kummer, B.: Metric regularity: characterizations, nonsmooth variations and successive approximation. Optimization **46**, 247–281 (1999)
607. Kuratowski, K.: Topologie I, II, Państwowe Wydawnictwo. Naukowe, Warsaw (1934); Academic Press, New York (1966)
608. Kurdila, A.J., Zabarankin, M.: Convex Functional Analysis, Systems and Control: Foundations and Applications. Birkhäuser, Basel (2005)
609. Lagrange, J.-L.: Leçons sur le Calcul des Fonctions. Paris (1804)
610. Gil de Lamadrid, J.: Topology of mappings and differentiation processes. Ill. J. Math. **3**, 408–420 (1959)
611. Lang, S.: Introduction to Differentiable Manifolds. Wiley, New York (1962)
612. Lang, S.: Analysis II. Addison-Wesley, Reading (1969)
613. Lasry, J.-M., Lions, P.-L.: A remark on regularisation in Hilbert spaces. Isr. J. Math. **55**(3), 257–266 (1986)
614. Lassonde, M.: Hahn-Banach theorems for convex functions. In: Ricceri, B., et al. (eds.) Minimax Theory and Applications. Proc. workshop, Erice, Italy, 30 September–6 October 1996. Nonconvex Optim. Appl. vol. 26, pp. 135–145. Kluwer, Dordrecht (1998)
615. Lassonde, M.: First-order rules for nonsmooth constrained optimization. Nonlinear Anal. Theor. Meth. Appl. **44A**(8), 1031–1056 (2001)

616. Lassonde, M.: Asplund spaces, Stegall variational principle and the RNP. Set-Valued Anal. **17**(2), 183–193 (2009)
617. Lassonde, M., Revalski, J.: Fragmentability of sequences of set-valued mappings with applications to variational principles. Proc. Am. Math. Soc. **133**, 2637–2646 (2005)
618. Lau, K.S.: Almost Chebyshev subsets in reflexive Banach spaces. Indiana Univ. Math. J. **27**, 791–795 (1978)
619. Laurent, P.-J.: Approximation et Optimisation. Hermann, Paris (1972)
620. Lebourg, G.: Perturbed optimization problems in Banach spaces. Analyse non convexe, Pau, 1977. Bull. Soc. Math. France Mémoire **60**, 95–111 (1979)
621. Lebourg, G.: Valeur moyenne pour gradients généralisés. C.R. Acad. Sci. Paris **281**, 795–798 (1975)
622. Lebourg, G.: Solutions en densité de problèmes d'optimisation paramétrés. C.R. Acad. Sci. Paris série A **289**, 79–82 (1979)
623. Lebourg, G.: Generic differentiability of Lipschitzian functions. Trans. Am. Math. Soc. **256**, 123–144 (1979)
624. Ledyaev, Yu.S., Zhu, Q.J.: Implicit multifunction theorems. Set Valued Anal. **7**, 209–238 (1999)
625. Lee, G.M., Tam, N.N., Yen, N.D.: Normal coderivative for multifunctions and implicit function theorems. J. Math. Anal. Appl. **338**, 11–22 (2008)
626. Lemaire, B.: Méthode d'optimisation et convergence variationelle. Travaux du Séminaire d'Analyse Convexe. Montpellier **17**, 9.1–9.18 (1986)
627. Lemaire, B.: The proximal algorithm. In: Penot, J.-P. (ed.) New Methods in Optimization and their Industrial Uses. Intern. Series Numer. Math. vol. 87, pp. 73–87. Birkhäuser, Basel (1989)
628. Lempio, F., Maurer, H.: Differential stability in infinite-dimensional nonlinear programming. Appl. Math. Optim. **6**(2), 139–152 (1980)
629. Lescarret, C.: Application "prox" dans un espace de Banach. C.R. Acad. Sci. Paris A **265**, 676–678 (1967)
630. Levy, A.B.: Implicit multifunction theorems for the sensitivity analysis of variational conditions. Math. Program. A **74**(3), 333–350 (1996)
631. Levy, A.B.: Nonsingularity conditions for multifunctions. Set-Valued Anal. **7**(1), 89–99 (1999)
632. Levy, A.B.: Sensitivity of solutions to variational inequalities on Banach spaces. SIAM J. Contr. Optim. **38**(1), 50–60 (1999)
633. Levy, A.B.: Calm minima in parameterized finite-dimensional optimization. SIAM J. Optim. **11**(1), 160–178 (2000)
634. Levy, A.B.: Lipschitzian multifunctions and a Lipschitzian inverse mapping theorem. Math. Oper. Res. **26**(1), 105–118 (2001)
635. Levy, A.B.: Solution stability from general principles. SIAM J. Contr. Optim. **40**, 209–238 (2001)
636. Levy, A.B.: Supercalm multifunctions for convergence analysis. Set-Valued Anal. **14**(3), 249–261 (2006)
637. Levy, A.B.: Constraint incorporation in optimization. Math. Program. A **110**(3), 615–639 (2007)
638. Levy, A.B., Mordukhovich, B.S.: Coderivatives in parametric optimization. Math. Program. **99**, 311–327 (2004)
639. Levy, A.B., Poliquin, R.A.: Characterizing the single-valuedness of multifunctions. Set-Valued Anal. **5**(4), 351–364 (1997)
640. Levy, A.B., Rockafellar, R.T.: Sensitivity analysis of solutions to generalized equations. Trans. Am. Math. Soc. **345**(2), 661–671 (1994)
641. Levy, A.B., Rockafellar, R.T.: Sensitivity of solutions in nonlinear programming problems with nonunique multipliers. In: Recent Advances in Nonsmooth Optimization, pp. 215–223. World Scientific, River Edge (1995)

642. Levy, A.B., Rockafellar, R.T.: Variational conditions and the proto-differentiation of partial subgradient mappings. Nonlinear Anal. **26**(12), 1951–1964 (1996)
643. Levy, A.B., Poliquin, R., Thibault, L.: Partial extensions of Attouch's theorem with applications to proto-derivatives of subgradient mappings. Trans. Am. Math. Soc. **347**(4), 1269–1294 (1995)
644. Levy, A.B., Poliquin, R.A., Rockafellar, R.T.: Stability of locally optimal solutions. SIAM J. Optim. **10**(2), 580–604 (2000)
645. Lewis, A.S.: Active sets, nonsmoothness, and sensitivity. SIAM J. Optim. **13**(3), 702–725 (2002)
646. Lewis, A.S.: Robust Regularization, Tech. report, School of ORIE, Cornell University, Ithaca, NY, 2002. Available online at http://people.orie.cornell.edu/aslewis/publications/2002.html (2002)
647. Lewis, A.S., Lucchetti, R.E.: Nonsmooth duality, sandwich, and squeeze theorems. SIAM J. Contr. Optim. **38**(2), 613–626 (2000)
648. Lewis, A.S., Pang, C.H.J.: Lipschitz behavior of the robust regularization. SIAM J. Contr. Optim. **48**(5), 3080–3104 (2009/2010)
649. Lewis, A.S., Pang, J.-S.: Error bounds for convex inequality systems. In: Generalized Convexity, Generalized Monotonicity: Recent Results (Luminy, 1996), pp. 75–110, Nonconvex Optim. Appl., vol. 27. Kluwer, Dordrecht (1998)
650. Lewis, A.S., Ralph, D.: A nonlinear duality result equivalent to the Clarke–Ledyaev mean value inequality. Nonlinear Anal. **26**(2), 343–350 (1996)
651. Lewis, A., Henrion, R., Seeger, A.: Distance to uncontrollability for convex processes. SIAM J. Contr. Optim. **45**(1), 26–50 (2006)
652. Li, C., Ni, R.: Derivatives of generalized distance functions and existence of generalized nearest points. J. Approx. Theor. **115**(1), 44–55 (2002)
653. Li, C., Ng, K.F.: On constraint qualification for an infinite system of convex inequalities in a Banach space. SIAM. J. Optim. **15**, 488–512 (2005)
654. Li, G., Ng, K.F.: Error bounds of generalized D-gap functions for nonsmooth and nonmonotone variational inequality problems. SIAM J. Optim. **20**(2), 667–690 (2009)
655. Li, Y.X., Shi, S.Z.: A generalization of Ekeland's ε-variational principle and of its Borwein-Preiss smooth variant. J. Math. Anal. Appl. **246**, 308–319 (2000)
656. Li, W., Singer, I.: Global error bounds for convex multifunctions. Math. Oper. Res. **23**, 443–462 (1998)
657. Li, S.J., Meng, K.W., Penot, J.-P.: Calculus rules of multimaps. Set-Valued Anal. **17**(1), 21–39 (2009)
658. Li, S.J., Penot, J.-P., Xue, X.W.: Codifferential calculus. Set-Valued Var. Anal. **19**(4), 505–536 (2011)
659. Li, C., Ng, K.F., Pong, T.K.: Constraint qualifications for convex inequality systems with applications in constrained optimization. SIAM J. Optim. **19**, 163–187 (2008)
660. Liapunov, A.A.: Sur les fonctions vecteurs complètement additives. Izvest. Akad. Nauk SSSR, Ser. Mat. **3**, 465–478 (1940)
661. Lin, P.K.: Strongly unique best approximation in uniformly convex Banach spaces. J. Approx. Theor. **56**, 101–107 (1989)
662. Lin, L.-J., Du, W.-S.: Ekeland's variational principle, minimax theorems and existence of nonconvex equilibria in complete metric spaces. J. Math. Anal. Appl. **323**(1), 360–370 (2006)
663. Lindenstrauss, J., Preiss, D. In: Fréchet differentiability of Lipschitz functions (a survey). In: Recent progress in functional analysis (Valencia, 2000), pp. 19–42, North Holland Math. Stud., vol. 189. North Holland, Amsterdam (2001)
664. Lindenstrauss, J., Preiss, D.: A new proof of Fréchet differentiability of Lipschitz functions. J. Eur. Math. Soc. **2**(3), 199–216 (2000)
665. Lindenstrauss, J., Matoušková, E., Preiss, D.: Lipschitz image of a measure-null set can have a null complement. Isr. J. Math. **118**, 207–219 (2000)

666. Lindenstrauss, J., Preiss, D., Tiser, J.: Fréchet differentiability of Lipschitz functions via a variational principle. J. Eur. Math. Soc. **12**(2), 385–412 (2010)
667. Lions, P.-L., Souganidis, P.E.: Differential games, optimal control and directional derivatives of viscosity solutions of Bellman's and Isaac's equations. SIAM J. Contr. Optim. **23**(4), 566–583 (1985)
668. Lindenstrauss, J.: On operators which attain their norms. Isr. J. Math. **3**, 139–148 (1963)
669. Loewen, Ph.D.: The proximal normal formula in Hilbert space. Nonlinear Anal. **11**(9), 979–995 (1987)
670. Loewen, Ph.D.: The proximal subgradient formula in Banach space. Can. Math. Bull. **31**(3), 353–361 (1988)
671. Loewen, Ph.D.: Limits of Fréchet normals in nonsmooth analysis. In: Ioffe, A., Marcus, M., Reich, S. (eds.) Optimization and Nonlinear Analysis, pp. 178–188. Longman, Essex (1992)
672. Loewen, Ph.D.: Optimal Control via Nonsmooth Analysis. CRM Proceedings and Lecture Notes, vol. 2. American Mathematical Society, Providence (1993)
673. Loewen, Ph.D.: A mean value theorem for Fréchet subgradients. Nonlinear Anal. Theor. Meth. Appl. **23**, 1365–1381 (1994)
674. Loewen, Ph.D., Wang, X.: A generalized variational principle. Can. J. Math. **53**(6), 1174–1193 (2001)
675. Loewen, Ph.D., Wang, X.: Typical properties of Lipschitz functions. R. Anal. Exchange **26**(2), 717–725 (2000/2001)
676. Loewen, Ph.D., Wang, X.: On the multiplicity of Dini subgradients in separable spaces. Nonlinear Anal. **58**(1–2), 1–10 (2004)
677. Loewen, Ph.D., Zheng, H.H.: Epi-derivatives of integral functionals with applications. Trans. Am. Math. Soc. **347**(2), 443–459 (1995)
678. Łojasiewicz, S.: Sur la géométrie semi et sous–analytique. Ann. Inst. Fourier **43**(5), 1575–1595 (1993)
679. López, M.A., Vercher, E.: Optimality conditions for nondifferentiable convex semi-infinite programming. Math. Program. **27**, 307–319 (1983)
680. Loridan, P., Morgan, J.: New results on approximate solutions in two-level optimization. Optimization **20**(6), 819–836 (1989)
681. Luc, D.T., Penot, J.-P.: Convergence of asymptotic directions. Trans. Am. Math. Soc. **353**(10), 4095–4121 (2001)
682. Lucchetti, R.: Hypertopologies and applications. In: Recent Developments in Well-Posed Variational Problems, pp. 193–209. Math. Appl., vol. 331. Kluwer, Dordrecht (1995)
683. Lucchetti, R.: Porosity of ill-posed problems in some optimization classes. In: Recent Advances in Optimization (Varese, 2002), pp. 119–130. Datanova, Milan (2003)
684. Lucchetti, R.: Convexity and well-posed problems. CMS Books in Mathematics/Ouvrages de Mathématiques de la S.M.C., vol. 22. Springer, New York (2006)
685. Lucchetti, R.: Some aspects of the connections between Hadamard and Tyhonov well-posedness of convex programs. Boll. Un. Mat. Ital. C (6) **1**(1), 337–345 (1982)
686. Lucchetti, R., Patrone, F.: On Nemytskii's operator and its application to the lower semicontinuity of integral functionals. Indiana Univ. Math. J. **29**(5), 703–713 (1980)
687. Lucchetti, R., Patrone, F.: Hadamard and Tyhonov well-posedness of a certain class of convex functions. J. Math. Anal. Appl. **88**(1), 204–215 (1982)
688. Lucchetti, R., Revalski, J.: Recent Developments in Well-Posed Variational Problems. Kluwer, Dordrecht (1995)
689. Lucchetti, R., Zolezzi, T.: On well-posedness and stability analysis in optimization. In: Mathematical Programming with Data Perturbations, pp. 223–251. Lecture Notes in Pure and Appl. Math., vol. 195. Dekker, New York (1998)
690. Lucet, Y.: What shape is your conjugate? A survey of computational convex analysis and its applications. SIAM Rev. **52**(3), 505–542 (2010)
691. Luo, Z.Q., Pang, J.S., Ralph, D.: Mathematical Programs with Equilibrium Constraints. Cambridge University Press, Cambridge (1996)
692. Luenberger, D.: Optimization by Vector Space Methods. Wiley, New York (1969)

693. Lyusternik, L.A.: On the conditional extrema of functionals. Mat. Sbornik **41**, 390–401 (1934)
694. Lyusternik, L.A., Sobolev, V.I.: Elements of Functional Analysis. Nauka, Moscow (1965)
695. Mäkelä, M.M., Neittaanmäki, P.: Nonsmooth optimization. Analysis and algorithms with applications to optimal control. World Scientific, River Edge (1992)
696. Mandelbrojt, S.: Sur les fonctions convexes. C.R. Acad. Sci. Paris **209**, 977–978 (1939)
697. Martínez-Legaz, J.-E., Penot, J.-P.: Regularization by erasement. Math. Scand. **98**, 97–124 (2006)
698. Matheron, G.: Random Sets and Integral Geometry. Wiley Series in Probability and Mathematical Statistics. Wiley, New York (1975)
699. Maurer, H., Zowe, J.: First and second order necessary and sufficient optimality conditions for infinite-dimensional programming problems. Math. Program. **16**(1), 98–110 (1979)
700. McShane, E.J.: The Lagrange multiplier rule. Am. Math. Mon. **80**, 922–925 (1973)
701. Mera, M.E., Morán, M., Preiss, D., Zajíček, L.: Porosity, σ-porosity and measures. Nonlinearity **16**(1), 247–255 (2003)
702. Michael, E.: Topologies on spaces of subsets. Trans. Am. Math. Soc. **71**, 152–182 (1951)
703. Michael, E.: Continuous selections I. Ann. Math. **63**, 361–381 (1956)
704. Michael, E.: Dense families of continuous selections. Fundamenta Math. **67**, 173–178 (1959)
705. Michal, A.D.: General differential geometries and related topics. Bull. Am. Math. Soc. **45**, 529–563 (1939)
706. Michal, A.D.: Differentials of functions with arguments and values in topological abelian groups. Proc. Natl. Acad. Sci. USA **26**, 356–359 (1940)
707. Michel, Ph., Penot, J.-P.: Calcul sous-différentiel pour des fonctions lipschitziennes et non lipschitziennes. C.R. Acad. Sci. Paris **298**, 684–687 (1984)
708. Michel, Ph., Penot, J.-P.: A generalized derivative for calm and stable functions. Diff. Integr. Equat. **5**, 189–196 (1992)
709. Milyutin, A.A., Osmolovskii, N.P.: Calculus of Variations and Optimal Control. Translations of Mathematical Monographs, vol. 180. American Mathematical Society, Providence (1998)
710. Minchenko, L.I.: Multivalued analysis and differential properties of multivalued mappings and marginal functions. Optimization and related topics, 3. J. Math. Sci. (N. Y.) **116**(3), 3266–3302 (2003)
711. Minoux, M.: Programmation Mathématique. Théorie et Algorithmes. Dunod, Paris (1983)
712. Meng, F., Zhao, G., Goh, M., De Souza, R.: Lagrangian-dual functions and Moreau–Yosida regularization. SIAM J. Optim. **19**(1), 39–61 (2008)
713. Mordukhovich, B.S.: Maximum principle in problems of time optimal control with nonsmooth constraints. J. Appl. Math. Mech. **40**, 960–969 (1976)
714. Mordukhovich, B.S.: Metric approximations and necessary optimality conditions for general classes of nonsmooth extremal problems. Sov. Math. Dokl. **22**, 526–530 (1980)
715. Mordukhovich, B.S.: Nonsmooth analysis with nonconvex generalized differentials and adjoint mappings. Dokl. Akad. Nauk BSSR **28**, 976–979 (1984)
716. Mordukhovich, B.S.: Approximation Methods in Problems of Optimization and Control. Nauk, Moscow (1988)
717. Mordukhovich, B.S.: Complete characterizations of openness, metric regularity, and Lipschitzian properties of multifunctions. Trans. Am. Math. Soc. **340**, 1–35 (1993)
718. Mordukhovich, B.S.: Variational Analysis and Generalized Differentiation, I: Basic Theory. Grundlehren der Math. Wissenschaften, vol. 330. Springer, Berlin (2006)
719. Mordukhovich, B.S.: Variational Analysis and Generalized Differentiation, II: Applications. Springer, Berlin (2006)
720. Mordukhovich, B.S., Outrata, J.V.: Coderivative analysis of quasi-variational inequalities with applications to stability and optimization. SIAM J. Optim. **18**(2), 389–412 (2007)
721. Mordukhovich, B.S., Shao, Y.: Differential characterizations of covering, metric regularity, and Lipschitzian properties of multifunctions between Banach spaces. Nonlinear Anal. Theor. Meth. Appl. **25**(12), 1401–1424 (1995)

722. Mordukhovich, B.S., Shao, Y.: Extremal characterizations of Asplund spaces. Proc. Am. Math. Soc. **124**, 197–205 (1996)
723. Mordukhovich, B.S., Shao Y.: Nonsmooth sequential analysis in Asplund spaces. Trans. Am. Math. Soc. **348**(4), 1235–1280 (1996)
724. Mordukhovich, B.S., Shao, Y.: Nonconvex coderivative calculus for infinite dimensional multifunctions. Set-Valued Anal. **4**, 205–236 (1996)
725. Mordukhovich, B.S., Shao, Y.: Stability of set-valued mappings in infinite dimension: point criteria and applications. SIAM J. Contr. Optim. **35**, 285–314 (1997)
726. Mordukhovich, B.S., Shao, Y.: Fuzzy calculus for coderivatives of multifunctions. Nonlinear Anal. Theor. Meth. Appl. **29**, 605–626 (1997)
727. Mordukhovich, B.S., Shao, Y.: Mixed coderivatives of set-valued mappings in variational analysis. J. Appl. Anal. **4**, 269–294 (1998)
728. Mordukhovich, B.S., Nam, N.M., Wang, B.: Metric regularity of mappings and generalized normals to set images. Set-Valued Var. Anal. **17**(4) 359–387 (2009)
729. Mordukhovich, B.S., Nam, N.M., Yen, N.D.: Subgradients of marginal functions in parametric mathematical programming. Math. Program. Ser. B **116**(1–2), 369–396 (2009)
730. Mordukhovich, B.S., Shao, Y., Zhu, Q.J.: Viscosity coderivatives and their limiting behavior in smooth Banach spaces. Positivity **4**, 1–39 (2000)
731. Mordukhovich, B.S., Wang, B.: Calculus of sequential normal compactness. J. Math. Anal. Appl. **283**, 63–84 (2003)
732. Mordukhovich, B.S., Wang, B.: Restrictive metric regularity and generalized differential calculus in Banach spaces. Int. J. Math. Math. Sci. **50**, 2650–2683 (2004)
733. Moreau, J.-J.: Fonctionnelles sous-différentiables. C.R. Acad. Sci. Paris **257**, 4117–4119 (1963)
734. Moreau, J.-J.: Proximité et dualité dans un espace hilbertien. Bull. Soc. Math. Fr. **93**, 273–299 (1965)
735. Moreau, J.-J.: Fonctionnelles Convexes. Collège de France, Paris (1966)
736. Mosco, U.: Convergence of convex sets and of solutions of variational inequalities. Adv. Math. **3**, 510–585 (1969)
737. Moulay, E., Bhat, S.P.: Topological properties of asymptotically stable sets. Nonlinear Anal. **73**(4), 1093–1097 (2010)
738. Nachi, K., Penot, J.-P.: Inversion of multifunctions and differential inclusions. Contr. Cybern. **34**(3), 871–901 (2005)
739. Nadler, S.B.: Hyperspaces of Sets. Dekker, New York (1978)
740. Nam, N.M.: Coderivatives of normal cone mappings and Lipschitzian stability of parametric variational inequalities. Nonlinear Anal. **73**(7), 2271–2282 (2010)
741. Nam, N.M., Wang, B.: Metric regularity, tangential distances and generalized differentiation in Banach spaces. Nonlinear Anal. **75**(3), 1496–1506 (2012)
742. Namioka, I., Phelps, R.R.: Banach spaces which are Asplund spaces. Duke Math. J. **42**, 735–750 (1975)
743. Nesterov, Y.: Introductory Lectures on Convex Optimization. A basic course. Applied Optimization, vol. 87. Kluwer, Boston (2004)
744. Neustadt, L.W.: An abstract variational theory with applications to a broad class of optimization problems. I. General theory. SIAM J. Contr. **4**, 505–527 (1966)
745. Neustadt, L.W.: A general theory of extremals. J. Comput. Syst. Sci. **3**, 57–92 (1969)
746. Neustadt, L.W.: Optimization. A Theory of Necessary Conditions. Princeton University Press, Princeton (1976)
747. Ng, K.F., Yang, W.H.: Regularities and relations to error bounds. Math. Prog. **99**, 521–538 (2004)
748. Ngai, H.V., Théra, M.: Metric inequality, subdifferential calculus and applications. Well-posedness in optimization and related topics (Gargnano, 1999). Set-Valued Anal. **9**(1–2), 187–216 (2001)
749. Ngai, H.V., Théra, M.: A fuzzy necessary optimality condition for non-Lipschitz optimization in Asplund spaces. SIAM J. Optim. **12**(3), 656–668 (2002)

750. Ngai, H.V., Théra, M.: Error bounds and implicit multifunction theorem in smooth Banach spaces and applications to optimization. Set-Valued Anal. **12**(1–2), 195–223 (2004)
751. Ngai, H.V., Théra, M.: Error bounds for convex differentiable inequality systems in Banach spaces. Math. Program. B **104**(2–3), 465–482 (2005)
752. Ngai, H.V., Théra, M.: Error bounds in metric spaces and application to the perturbation stability of metric regularity. SIAM J. Optim. **19**(1), 1–20 (2008)
753. Ngai, H.V., Théra, M.: Error bounds for systems of lower semicontinuous functions in Asplund spaces. Math. Program. B **116**(1–2), 397–427 (2009)
754. Nguyen, V.H., Strodiot, J.-J., Mifflin, R.: On conditions to have bounded multipliers in locally Lipschitz programming. Math. Program. **18**(1), 100–106 (1980)
755. Nijenhuis, A.: Strong derivatives and inverse mappings. Am. Math. Mon. **81**, 969–980 (1974)
756. Oettli, W., Théra, M.: Equivalents of Ekeland's principle. Bull. Aust. Math. Soc. **48**(3), 385–392 (1993)
757. Outrata, J.V.: On generalized gradients in optimization problems with set-valued constraints. Math. Oper. Res. **15**(4), 626–639 (1990)
758. Outrata, J.V.: Optimality conditions for a class of mathematical programs with equilibrium constraints. Math. Oper. Res. **24**(3), 627–644 (1999)
759. Outrata, J.V.: A generalized mathematical program with equilibrium constraints. SIAM J. Contr. Optim. **38**(5), 1623–1638 (2000)
760. Outrata, J.V., Jarušek, J.: On Fenchel dual schemes in convex optimal control problems. Kybernetika (Prague) **18**(1), 1–21 (1982)
761. Outrata, J.V., Jarušek, J.: Duality theory in mathematical programming and optimal control. Kybernetika (Prague) Suppl. **20/21**, 119 (1984/1985)
762. Outrata, J.V., Römisch, W.: On optimality conditions for some nonsmooth optimization problems over L_p spaces. J. Optim. Theor. Appl. **126**(2), 411–438 (2005)
763. Outrata, J.V., Sun, D.: On the coderivative of the projection operator onto the second-order cone. Set-Valued Anal. **16**(7–8), 999–1014 (2008). Erratum Set-Valued Var. Anal. **17**(3), 319 (2009)
764. Outrata, J.V., Kočvara M., Zowe, J.: Nonsmooth Approach to Optimization Problems with Equilibrium Constraints. Nonconvex Optimization and Its Applications, vol. 28. Kluwer, Dordrecht (1998)
765. Palais, R.S.: Critical point theory and the minimax principle. In: Global Analysis. Proc. Sympos. Pure Math., vol. XV, Berkeley, CA, 1968, pp. 185–212. American Mathematical Society, Providence (1970)
766. Palais, R.S.: When proper maps are closed. Proc. Am. Math. Soc. **24**, 835–836 (1970)
767. Palais, R.S., Smale, S.: A generalized Morse theory. Bull. Am. Math. Soc. **70**, 165–172 (1964)
768. Páles, Z.: General necessary and sufficient conditions for constrained optimum problems. Arch. Math. (Basel) **63**(3), 238–250 (1994)
769. Páles, Z.: Inverse and implicit function theorems for nonsmooth maps in Banach spaces. J. Math. Anal. Appl. **209**(1), 202–220 (1997)
770. Páles, Z., Zeidan, V.: Nonsmooth optimum problems with constraints. SIAM J. Contr. Optim. **32**(5), 1476–1502 (1994)
771. Páles, Z., Zeidan, V.: Infinite dimensional Clarke generalized Jacobian. J. Convex Anal. **14**(2), 433–454 (2007)
772. Páles, Z., Zeidan, V.: Generalized Jacobian for functions with infinite dimensional range and domain. Set-Valued Anal. **15**(4), 331–375 (2007)
773. Páles, Z., Zeidan, V.: Infinite dimensional generalized Jacobian: properties and calculus rules. J. Math. Anal. Appl. **344**(1), 55–75 (2008)
774. Páles, Z., Zeidan, V.: The core of the infinite dimensional generalized Jacobian. J. Convex Anal. **16**(2), 321–349 (2009)
775. Páles, Z., Zeidan, V.: Co-Jacobian for Lipschitzian maps. Set-Valued Var. Anal. **18**(1), 57–78 (2010)

776. Pallaschke, D., Rolewicz, S.: Foundations of Mathematical Optimization: Convex Analysis Without Linearity. Kluwer, Dordrecht (1997)
777. Pang, J.S.: Error bounds in mathematical programming. Math. Program. **79**, 299–332 (1997)
778. Peano, G.: Applicazioni Geometriche del Calcolo Infinitesimale. Fratelli Bocca, Torino (1887)
779. Penot, J.-P.: Calcul différentiel dans les espaces vectoriels topologiques. Studia Mathematica **47**(1), 1–23 (1972)
780. Penot, J.-P.: Sous-différentiels de fonctions numériques non convexes. C.R. Acad. Sci. Paris **278**, 1553–1555 (1974)
781. Penot, J.-P.: Continuité et différentiabilité des opérateurs de Nemytskii, Publications mathématiques de Pau, VIII 1–45 (1976)
782. Penot, J.-P.: Calcul sous-différentiel et optimisation. J. Funct. Anal. **27**, 248–276 (1978)
783. Penot, J.-P.: Inversion à droite d'applications non-linéaires. C.R. Acad. Sci. Paris **290**, 997–1000 (1980)
784. Penot, J.-P.: A characterization of tangential regularity. Nonlinear Anal. Theor. Meth. Appl. **5**(6), 625–643 (1981)
785. Penot, J.-P.: On regularity conditions in mathematical programming. Math. Program. Study **19**, 167–199 (1982)
786. Penot, J.-P.: Continuity properties of performance functions. In: Hiriart-Urruty, J.-B., Oettli, W., Stoer, J. (eds.) Optimization Theory and Algorithms. Lecture Notes in Pure and Applied Math., vol. 86, pp. 77–90. Marcel Dekker, New York (1983)
787. Penot, J.-P.: Compact nets, filters and relations. J. Math. Anal. Appl. **93**(2), 400–417 (1983)
788. Penot, J.-P.: Differentiability of relations and differential stability of perturbed optimization problems. SIAM J. Contr. Optim. **22**(4), 529–551 (1984). Erratum, ibid. (4), 260 (1988)
789. Penot, J.-P.: Open mapping theorems and linearization stability. Numer. Funct. Anal. Optim. **8**(1–2), 21–36 (1985)
790. Penot, J.-P.: Variations on the theme of nonsmooth analysis: another subdifferential. In: Nondifferentiable Optimization: Motivations and Applications (Sopron, 1984), pp. 41–54. Lecture Notes in Econom. and Math. Systems, vol. 255. Springer, Berlin (1985)
791. Penot, J.-P.: A characterization of Clarke's strict tangent cone via nonlinear semi-groups. Proc. Am. Math. Soc. **93**(1), 128–132 (1985)
792. Penot, J.-P.: The Drop theorem, the Petal theorem and Ekeland's variational principle. Nonlinear Anal. Theor. Meth. Appl. **10**(9), 459–468 (1986)
793. Penot, J.-P.: About linearization, conization, calmness, openness and regularity. In: Lakshmikhantham (ed.) Nonlinear Analysis, pp. 439–450. Marcel Dekker, New York (1987)
794. Penot, J.-P.: Metric regularity, openness and Lipschitzian behavior multifunctions. Nonlinear Anal. Theor. Meth. Appl. **13**(6), 629–643 (1989)
795. Penot, J.-P.: Topologies and convergences on the space of convex functions. Nonlinear Anal. Theor. Meth. Appl. **18**, 905–916 (1992)
796. Penot, J.-P.: Preservation of persistence and stability under intersections and operations. Part 1, Persistence, J. Optim. Theor. Appl. 79 (3), 525–550 (1993); Part 2, Stability, J. Optim. Theor. Appl. 79 (3), 551–561 (1993)
797. Penot, J.-P.: The cosmic Hausdorff topology, the bounded Hausdorff topology and continuity of polarity. Proc. Am. Math. Soc. **113**, 275–285 (1993)
798. Penot, J.-P.: Mean value theorems for small subdifferentials. J. Optim. Theor. Appl. **90**(3), 539–558 (1995)
799. Penot, J.-P.: Subdifferential calculus and subdifferential compactness. In: Sofonea, M., Corvellec, J.-N. (eds.) Proc. Second Catalan Days in Applied Mathematics, Font-Romeu-Odeillo, France, pp. 209–226. Presses Universitaires de Perpignan, Perpignan (1995)
800. Penot, J.-P.: Conditioning convex and nonconvex problems. J. Optim. Theor. Appl. **90**(3), 539–558 (1995)
801. Penot, J.-P.: Inverse function theorems for mappings and multimappings. South East Asia Bull. Math. **19**(2), 1–16 (1995)

802. Penot, J.-P.: Miscellaneous incidences of convergence theories in optimization and non-smooth analysis II: Applications to nonsmooth analysis. In: Du, D.Z., Qi, L., Womersley, R.S. (eds.) Recent Advances in Nonsmooth Optimization, pp. 289–321. World Scientific, Singapore (1995)

803. Penot, J.-P.: Generalized derivatives of a performance function and multipliers in mathematical programming. In: Guddat, J., Jongen, H.Th., Nozicka, F., Still, G., Twilt, F. (eds.) Parametric Optimization and Related Topics IV, Proceedings Intern. Conference Enschede, June 1995, pp. 281–298. Peter Lang, Frankfurt (1996)

804. Penot, J.-P.: Subdifferential calculus without qualification assumptions. J. Convex Anal. **3**(2), 1–13 (1996)

805. Penot, J.-P.: Favorable classes of mappings and multimappings in nonlinear analysis and optimization. J. Convex Anal. **3**(1), 97–116 (1996)

806. Penot, J.-P.: Metric estimates for the calculus of multimappings. Set-Valued Anal. **5**(4), 291–308 (1997)

807. Penot, J.-P.: Compactness property, openness criteria and coderivatives. Set-Valued Anal. **6**(4), 363–380 (1998) (preprint, Univ. of Pau, August 1995)

808. Penot, J.-P.: Well-behavior, well-posedness and nonsmooth analysis. Pliska Stud. Math. Bulgar. **12**, 141–190 (1998)

809. Penot, J.-P.: Central and peripheral results in the study of marginal and performance functions. In: Fiacco, A.V. (ed.) Mathematical Programming with Data Perturbations. Lectures Notes in Pure and Applied Math., vol. 195, pp. 305–337. Dekker, New York (1998)

810. Penot, J.-P.: Proximal mappings. J. Approx. Theor. **94**, 203–221 (1998)

811. Penot, J.-P.: On the minimization of difference functions. J. Global Optim. **12**, 373–382 (1998)

812. Penot, J.-P.: Delineating nice classes of nonsmooth functions. Pac. J. Optim. **4**(3), 605–619 (2008)

813. Penot, J.-P.: Cooperative behavior for sets and relations. Math. Meth. Oper. Res. **48**, 229–246 (1998)

814. Penot, J.-P.: Genericity of well-posedness, perturbations and smooth variational principles. Set-Valued Anal. **9**(1–2), 131–157 (2001)

815. Penot, J.-P.: Duality for anticonvex programs. J. Global Optim. **19**, 163–182 (2001)

816. Penot, J.-P.: The compatibility with order of some subdifferentials. Positivity **6**(4), 413–432 (2002)

817. Penot, J.-P.: A fixed point theorem for asymptotically contractive mappings. Proc. Am. Math. Soc. **131**(8), 2371–2377 (2003)

818. Penot, J.-P.: A metric approach to asymptotic analysis. Bull. Sci. Math. **127**, 815–833 (2003)

819. Penot, J.-P.: Multiplicateurs et analyse marginale. Matapli **70**, 67–78 (2003). http://smai.emath.fr/matapli/70/

820. Penot, J.-P.: Calmness and stability properties of marginal and performance functions. Numer. Funct. Anal. Optim. **25**(3–4), 287–308 (2004)

821. Penot, J.-P.: Differentiability properties of optimal value functions. Can. J. Math. **56**(4), 825–842 (2004)

822. Penot, J.-P.: Unilateral analysis and duality. In: Savard, G., et al. (eds.) GERAD, Essays and Surveys in Global Optimization, pp. 1–37. Springer, New York (2005)

823. Penot, J.-P.: Softness, sleekness and regularity properties in nonsmooth analysis. Nonlinear Anal. **68**(9), 2750–2768 (2008)

824. Penot, J.-P.: Gap continuity of multimaps. Set-Valued Anal. **16**(4), 429–442 (2008)

825. Penot, J.-P.: Error bounds, calmness and their applications in nonsmooth analysis. Contemp. Math. **514**, 225–247 (2010)

826. Penot, J.-P.: A short proof of the separable reduction theorem. Demonstratio Math. **43**(3), 653–663 (2010)

827. Penot, J.-P.: The directional subdifferential of the difference of two convex functions. J. Global Optim. **49**(3), 505–519 (2011)

828. Penot, J.-P., Ratsimahalo, R.: Subdifferentials of distance functions, approximations and enlargements. Acta Math. Sin. **23**(3), 507–520 (2006)
829. Penot, J.-P., Terpolilli, P.: Cônes tangents et singularités. C. R. Acad. Sci. Paris **296**, 721–725 (1983)
830. Penot, J.-P., Zălinescu, C.: Continuity of the Legendre-Fenchel transform for some variational convergences. Optimization **53**(5–6), 549–562 (2004)
831. Phelps, R.R.: Metric projections and the gradient projection method in Banach spaces. SIAM J. Contr. Optim. **23**(6), 973–977 (1985)
832. Phelps. R.R.: Convex Functions, Monotone Operators and Differentiability. Lecture Notes in Mathematics, vol. 1364. Springer, Berlin (1988)
833. Poliquin, R.A.: A characterization of proximal subgradient set-valued mappings. Can. Math. Bull. **36**(1), 116–122 (1993)
834. Poliquin, R.A., Rockafellar, R.T.: Amenable functions in optimization. In: Nonsmooth Optimization: Methods and Applications (Erice, 1991), pp. 338–353. Gordon and Breach, Montreux (1992)
835. Poliquin, R.A., Rockafellar, R.T.: A calculus of epi-derivatives applicable to optimization. Can. J. Math. **45**(4), 879–896 (1993)
836. Poliquin, R.A., Rockafellar, R.T.: Proto-derivative formulas for basic subgradient mappings in mathematical programming. Set convergence in nonlinear analysis and optimization. Set-Valued Anal. **2**(1–2), 275–290 (1994)
837. Poliquin, R.A., Rockafellar, R.T.: Prox-regular functions in variational analysis. Trans. Am. Math. Soc. **348**(5), 1805–1838 (1996)
838. Poliquin, R.A., Rockafellar, R.T.: Proto-derivatives of partial subgradient mappings. J. Convex Anal. **4**(2), 221–234 (1997)
839. Poliquin, R.A., Rockafellar, R.T.: A calculus of prox-regularity. J. Convex Anal. **17**(1), 203–210 (2010)
840. Poliquin, R.A., Rockafellar, R.T., Thibault, L.: Local differentiability of distance functions. Trans. Am. Math. Soc. **352**(11), 5231–5249 (2000)
841. Polovinkin, E.S., Smirnov, G.V.: Differentiation of multivalued mappings and properties of solutions of differential equations. Sov. Math. Dokl. **33**, 662–666 (1986)
842. Polyak, B.T.: Convexity of nonlinear image of a small ball with applications to optimization. Set-Valued Anal. **9**(1–2), 159–168 (2001)
843. Pourciau, B.H.: Analysis and optimization of Lipschitz continuous mappings. J. Optim. Theor. Appl. **22**(3), 311–351 (1977)
844. Pourciau, B.H.: Hadamard's theorem for locally Lipschitzian maps. J. Math. Anal. Appl. **85**(1), 279–285 (1982)
845. Pourciau, B.H.: Homeomorphisms and generalized derivatives. J. Math. Anal. Appl. **93**(2), 338–343 (1983)
846. Pourciau, B.H.: Global properties of proper Lipschitzian maps. SIAM J. Math. Anal. **14**(4), 796–799 (1983)
847. Pourciau, B.H.: Multiplier rules and the separation of convex sets. J. Optim. Theor. Appl. **40**(3), 321–331 (1983)
848. Pourciau, B.H.: Global invertibility of nonsmooth mappings. J. Math. Anal. Appl. **131**(1), 170–179 (1988)
849. Preiss, D.: Gâteaux differentiable functions are somewhere Fréchet differentiable. Rend. Circ. Mat. Palermo (2) **33**(1), 122–133 (1984)
850. Preiss, D.: Differentiability of Lipschitz functions on Banach spaces. J. Funct. Anal. **91**(2), 312–345 (1990)
851. Preiss, D., Zajíček, L.: Directional derivatives of Lipschitz functions. Isr. J. Math. **125**, 1–27 (2001)
852. Pritchard, G., Gürkan, G., Özge, A.Y.: A note on locally Lipschitzian functions. Math. Program. **71**, 369–370 (1995)
853. Pshenichnyi, B.N.: Necessary Conditions for an Extremum. Dekker, New York (1971); original Russian edition, Nauka, Moscow (1969)

854. Pshenichnii, B.N.: Leçons sur les jeux différentiels. Contrôle optimal et jeux différentiels, Cahier de l'INRIA **4**, 145–226 (1971)

855. Pták, V.: A quantitative refinement of the closed graph theorem. Czechoslovak Math. J. **24**, 503–506 (1974)

856. Pták, V.: A nonlinear subtraction theorem. Proc. Roy. Ir. Acad. Sci. A **82**, 47–53 (1982)

857. Qiu, J.-H.: Local completeness, drop theorem and Ekeland's variational principle. J. Math. Anal. Appl. **311**(1), 23–39 (2005)

858. Ramana, M.V., Tucel, L., Wolkowicz, H.: Strong duality for semi-definite programming. SIAM J. Optim. **7**, 644–662 (1997)

859. Reich, S., Zaslavski, A.J.: Convergence of generic infinite products of order-preserving mappings. Positivity **3**, 1–21 (1999)

860. Revalski, J.: Generic properties concerning well posed optimization problems. C. R. Acad. Bulgare Sci. **38**, 1431–1434 (1985)

861. Revalski, J.: Generic well posedness in some classes of optimization problems. Acta Univ. Math. Carol. Math. Phys. **28**, 117–125 (1987)

862. Revalski, J.: Well-posedness of optimization problems: a survey. In: Papini, P.L. (ed.) Functional Analysis and Approximation, pp. 238–255. Pitagora, Bologna (1988)

863. Revalski, J.: Hadamard and strong well-posedness for convex programs. SIAM J. Optim. **7**(2), 519–526 (1997)

864. Rifford, L.: Semiconcave control-Lyapunov functions and stabilizing feedbacks. SIAM J. Contr. Optim. **41**(3), 659–681 (2002)

865. Rifford, L.: A Morse–Sard theorem for the distance function on Riemannian manifolds. Manuscripta Math. **113**(2), 251–265 (2004)

866. Rifford, L.: Stratified semiconcave control-Lyapunov functions and the stabilization problem. Ann. Inst. H. Poincaré Anal. Non Linéaire **22**(3), 343–384 (2005)

867. Rifford, L.: Refinement of the Benoist theorem on the size of Dini subdifferentials. Comm. Pure Appl. Anal. **7**(1), 119–124 (2008)

868. Robinson, S.M.: Regularity and stability for convex multivalued functions. Math. Oper. Res. **1**, 130–143 (1976)

869. Robinson, S.M.: An implicit function theorem for a class of nonsmooth functions. Math. Oper. Res. **16**, 292–309 (1991)

870. Robinson, S.M.: Composition duality and maximal monotonicity. Math. Program. A **85**(1), 1–13 (1999)

871. Rockafellar, R.T.: Convex Analysis. Princeton University Press, Princeton (1970)

872. Rockafellar, R.T.: Conjugate Duality and Optimization. CBMS-NSF Regional Conf. Series in Applied Math. SIAM, Philadelphia (1974)

873. Rockafellar, R.T.: Clarke's tangent cones and the boundaries of closed sets in $\mathbb{R}^n$. Nonlinear Anal. **3**, 145–154 (1979)

874. Rockafellar, R.T.: Directionally Lipschitzian functions and subdifferential calculus. Proc. Lond. Math. Soc. **39**, 331–355 (1979)

875. Rockafellar, R.T.: Generalized directional derivatives and subgradients of nonconvex functions. Can. J. Math. **32**(2), 257–280 (1980)

876. Rockafellar, R.T.: The theory of subgradients and its applications to problems of optimization. Convex and Nonconvex Functions. Heldermann Verlag, Berlin (1986)

877. Rockafellar, R.T.: Lagrange multipliers and subderivatives of optimal value functions in nonlinear programming. Math. Program. Stud. **17**, 28–66 (1982)

878. Rockafellar, R.T.: Directional differentiability of the optimal value function in a nonlinear programming problem. Math. Program. Stud. **21**, 213–226 (1984)

879. Rockafellar, R.T.: Extensions of subgradient calculus with applications to optimization. Nonlinear Anal. Theor. Meth. Appl. **9**(7), 665–698 (1985)

880. Rockafellar, R.T.: Maximal monotone relations and the second derivatives of nonsmooth functions, Ann. Inst. H. Poincaré. Anal. Non Linéaire **2**, 167–184 (1985)

881. Rockafellar, R.T.: Nonsmooth analysis and parametric optimization. In: Methods of Nonconvex Analysis (Varenna, 1989), pp. 137–151. Lecture Notes in Mathematics, vol. 1446. Springer, Berlin (1990)

882. Rockafellar, R.T.: On a special class of convex functions. J. Optim. Theor. Appl. **70**(3), 619–621 (1991)
883. Rockafellar, R.T., Wets, R.J.-B.: Variational Analysis. Springer, New York (1998)
884. Rockafellar, R.T., Wolenski, P.R.: Convexity in Hamilton-Jacobi theory. I. Dynamics and duality. SIAM J. Contr. Optim. **39**(5), 1323–1350 (2000)
885. Rockafellar, R.T., Wolenski, P.R.: Convexity in Hamilton–Jacobi theory. II. Envelope representations. SIAM J. Contr. Optim. **39**(5), 1351–1372 (2000)
886. Rudin, W.: Functional Analysis, 2nd edn. International Series in Pure and Applied Mathematics. McGraw-Hill, New York (1991)
887. Saut, J.C., Témam, R.: Generic properties of nonlinear boundary value problems. Comm. Part. Differ. Equat. **4**(3), 293–319 (1979)
888. Saut, J.C.: Generic properties of nonlinear boundary value problems. In: Partial Differential Equations, vol. 10, pp. 331–351. Banach Center, Warsaw (1983)
889. Schaefer, H.H., Wolff, M.P.: Topological Linear Spaces, 2nd edn. Graduate Texts in Mathematics, vol. 3. Springer, New York (1999)
890. Schirotzek, W.: Nonsmooth Analysis. Universitext. Springer, Berlin (2007)
891. Severi, F.: Su alcune questioni di topologia infinitesimale. Ann. Polon. Soc. Math. **9**, 97–108 (1930)
892. Siegel, J.: A new proof of Caristi's fixed point theorem. Proc. Am. Math. Soc. **66**, 54–56 (1977)
893. Simons, S.: Minimax and Monotonicity. Lecture Notes in Mathematics, vol. 1693. Springer, Berlin (1998)
894. Simons, S., Zălinescu, C.: Fenchel duality, Fitzpatrick functions and maximal monotonicity. J. Nonlinear Convex Anal. **6**(1), 1–22 (2005)
895. Skripnik, I.V.: The application of Morse's methods to nonlinear elliptic equations (Russian). Dokl. Akad. Nauk SSSR **202**, 769–771 (1972)
896. Skripnik, I.V.: The differentiability of integral functionals (Ukrainian). Dopividi Akad. Nauk Ukrain RSR A 1086–1089 (1972)
897. Smale, S.: An infinite dimensional version of Sard's theorem. Am. J. Math. **87**, 861–866 (1965)
898. Smulian, V.L.: Sur la dérivabilité de la norme dans l'espace de Banach. Dokl. Akad. Nauk **27**, 643–648 (1940)
899. Song, W.: Calmness and error bounds for convex constraint systems. SIAM J. Optim. **17**, 353–371 (2006)
900. Sonntag, Y., Zalinescu, C.: Set convergence, an attempt of classification. Trans. Am. Math. Soc. **340**, 199–226 (1993)
901. Spingarn, J.E., Rockafellar, R.T.: The generic nature of optimality conditions in mathematical programming. Math. Oper. Res. **4**, 425–430 (1979)
902. Stegall, C.: Optimization of functions on certain subsets of Banach spaces. Math. Ann. **236**, 171–176 (1978)
903. Stegall, C.: The Radon–Nikodym property in conjugate Banach spaces II. Trans. Am. Math. Soc **264**, 507–519 (1981)
904. Strömberg, T.: On regularization in Banach spaces. Ark. Mat. **34**, 383–406 (1996)
905. Strömberg, T.: The operation of infimal convolution. Dissertationes Math. (Rozprawy Mat.) **352**, 58 (1996)
906. Takahashi, W.: Existence theorems generalizing fixed point theorems for multivalued mappings. In: Fixed Point Theory and Applications (Marseille 1989). Pitman Research Notes in Math., vol. 252, pp. 397–406. Longman, Harlow (1991)
907. Takahashi, W.: Nonlinear Functional Analysis. Yokohama publishers, Yokohama (2000)
908. Thibault, L.: Quelques propriétés des sous-différentiels de fonctions localement lipschitziennes. Travaux du Séminaire d'Analyse Convexe, vol. V, Exp. no. 16, p. 32. Univ. Sci. Tech. Languedoc, Montpellier (1975)
909. Thibault, L.: Problème de Bolza dans un espace de Banach séparable. C. R. Acad. Sci. Paris I Math. **282**, 1303–1306 (1976)

910. Thibault, L.: Subdifferentials of compactly Lipschitzian vector-valued functions. Travaux Sém. Anal. Convexe **8**(1, Exp. no. 5), 54 (1978)

911. Thibault, L.: Mathematical programming and optimal control problems defined by compactly Lipschitzian mappings, Sém. Anal. Convexe 8, Exp. no. 10 (1978)

912. Thibault, L.: Cônes tangents et épi-différentiels de fonctions vectorielles, Travaux Sém. Anal. Convexe **9**(2, Exp. no. 13), 38 (1979)

913. Thibault, L.: Subdifferentials of compactly Lipschitzian vector-valued functions. Ann. Mat. Pura Appl. **125**, 157–192 (1980)

914. Thibault, L.: Tangent cones and quasi-interiorly tangent cones to multifunctions. Trans. Am. Math. Soc. **277**(2), 601–621 (1983)

915. Thibault, L.: On subdifferentials of optimal value functions. SIAM J. Contr. Optim. **29**(5), 1019–1036 (1991)

916. Thibault, L.: A generalized sequential formula for subdifferentials of sums of convex functions defined on Banach spaces. In: Durier, R., et al. (eds.) Recent Developments in Optimization. Seventh French–German Conference on Optimization, Dijon, France, 27 June–2 July 1994. Lect. Notes Econ. Math. Syst., vol. 429, pp. 340–345. Springer, Berlin (1995)

917. Thibault, L.: Sequential convex subdifferential calculus and sequential Lagrange multipliers. SIAM J. Contr. Optim. **35**(4), 1434–1444 (1997)

918. Thibault, L.: Limiting convex subdifferential calculus with applications to integration and maximal monotonicity of subdifferential. In: Théra, M. (ed.) Constructive, Experimental and Nonlinear Analysis. CMS Conf. Proc., vol. 27, pp. 279–289. American Mathematical Society (AMS), Providence (2000); publ. for the Canadian Mathematical Society

919. Thibault, L., Zagrodny, D.: Enlarged inclusion of subdifferentials. Can. Math. Bull. **48**(2), 283–301 (2005)

920. Thibault, L., Zagrodny, D.: Subdifferential determination of essentially directionally smooth functions in Banach space. SIAM J. Optim. **20**(5), 2300–2326 (2010)

921. Tiba, D., Zălinescu, C.: On the necessity of some constraint qualification conditions in convex programming. J. Convex Anal. **11**(1), 95–110 (2004)

922. Tikhomirov, V.M.: Stories about Maxima and Minima. American Mathematical Society, Providence (1990). Translated from the Russian edition, Nauka (1986)

923. Toland, J.F.: Duality in nonconvex optimization. J. Math. Anal. Appl. **66**, 399–415 (1978)

924. Toland, J.F.: On subdifferential calculus and duality in non-convex optimization. Bull. Soc. Math. Fr. Suppl. Mém. (Proc. Colloq., Pau, 1977) **60**, 177–183 (1979)

925. Tran Duc Van, Tsuji M., Nguyen Duy Thai Son: The Characteristic Method and Its Generalizations for First-Order Nonlinear Partial Differential Equations. Chapman and Hall, Boca Raton, FL (2000)

926. Treiman, J.S.: Characterization of Clarke's tangent and normal cones in finite and infinite dimensions. Nonlinear Anal. **7**(7), 771–783 (1983)

927. Treiman, J.S.: Generalized gradients, Lipschitz behavior and directional derivatives. Can. J. Math. **37**(6), 1074–1084 (1985)

928. Treiman, J.S.: A new approach to Clarke's gradients in infinite dimensions. In: Nondifferentiable Optimization: Motivations and Applications (Sopron, 1984), pp. 87–93. Lecture Notes in Econom. and Math. Systems, vol. 255. Springer, Berlin (1985)

929. Treiman, J.S.: Generalized gradients and paths of descent. Optimization **17**, 181–186 (1986)

930. Treiman, J.S.: Clarke's gradients and epsilon-subgradients in Banach spaces. Trans. Am. Math. Soc. **294**(1), 65–78 (1986)

931. Treiman, J.S.: Shrinking generalized gradients. Nonlinear Anal. **12**(12), 1429–1450 (1988)

932. Treiman, J.S.: Finite-dimensional optimality conditions: *B*-gradients. J. Optim. Theor. Appl. **62**(1), 139–150 (1989)

933. Treiman, J.S.: An infinite class of convex tangent cones. J. Optim. Theor. Appl. **68**(3), 563–581 (1991)

934. Treiman, J.S.: The linear nonconvex generalized gradient and Lagrange multipliers. SIAM J. Optim. **5**(3), 670–680 (1995)

935. Treiman, J.S.: Lagrange multipliers for nonconvex generalized gradients with equality, inequality, and set constraints. SIAM J. Contr. Optim. **37**(5), 1313–1329 (1999)
936. Treiman, J.S.: The linear generalized gradient in infinite dimensions. Nonlinear Anal. Theor. Meth. A **48**(3), 427–443 (2002)
937. Troyanski, S.: On locally uniformly convex and differentiable norms in certain nonseparable Banach spaces. Stud. Math. **37**, 173–180 (1971)
938. Turinici, M.: Zhong's variational principle is equivalent with Ekeland's. Fixed Point Theor. **6**(1), 133–138 (2005)
939. Uderzo, A.: On a perturbation approach to open mapping theorems. Optim. Methods Soft. **25**(1), 143–167 (2000)
940. Uderzo, A.: Fréchet quasidifferential calculus with applications to metric regularity of continuous maps. Optimization **54**(4–5), 469–493 (2005)
941. Uhl, J.J.: The range of a vector-measure. Proc. Am. Math. Soc. **23**, 158–163 (1969)
942. Uhlenbeck, K.: Generic properties of eigenfunctions. Am. J. Math. **98**(4), 1059–1078 (1976)
943. Ursescu, C.: Multifunctions with closed convex graphs. Czech Math. J. **25**, 438–441 (1975)
944. Ursescu, C.: Linear openness of multifunctions in metric spaces. Int. J. Math. Sci. **2**, 203–214 (2005)
945. Valadier, M.: Contributions à l'Analyse Convexe. Thèse, Paris (1970)
946. Ver Eecke, P.: Fondements du Calcul Différentiel. Presses Universitaires de France, Paris (1983)
947. Ver Eecke, P.: Applications du Calcul Différentiel. Presses Universitaires de France, Paris (1985)
948. Vinter, R.B.: Optimal Control. Birkhäuser, Boston (2000)
949. Volle, M.: Some applications of the Attouch-Brézis conditions to closedness criterions, optimization and duality. Sem. Anal. Convexe Montpellier 22(16) (1992)
950. Wang, B.: The fuzzy intersection rule in variational analysis and applications. J. Math. Anal. Appl. **323**, 1365–1372 (2006)
951. Wang, X., Jeyakumar, V.: A sharp Lagrange multiplier rule for nonsmooth mathematical programming problems involving equality constraints. SIAM J. Optim. **10**(4), 1136–1148 (2000)
952. Ward, D.: Isotone tangent cones and nonsmooth optimization. Optimization **18**(6), 769–783 (1987)
953. Ward, D.: Chain rules for nonsmooth functions. J. Math. Anal. Appl. **158**, 519–538 (1991)
954. Ward, D.E., Borwein, J.M.: Nonconvex calculus in finite dimensions. SIAM J. Contr. Optim. **25**, 1312–1340 (1987)
955. Warga, J.: Optimal Control of Differential and Functional Equations. Academic Press, New York (1972)
956. Warga, J.: Fat homeomorphism and unbounded derivate containers. J. Math. Anal. Appl. **81**, 545–560 (1981)
957. Warga, J.: Optimization and controllability without differentiability assumptions. SIAM J. Contr. Optim. **21**(6), 837–855 (1983)
958. Warga, J.: Homeomorphisms and local C^1 approximations. Nonlinear Anal. **12**(6), 593–597 (1988)
959. Warga, J.: A necessary and sufficient condition for a constrained minimum. SIAM J. Optim. **2**(4), 665–667 (1992)
960. Ważewski, T.: Sur l'unicité et la limitation des intégrales des équations aux dérivées partielles du premier ordre. Rend. Acc. Lincei **17**, 372–376 (1933)
961. Weston, J.D.: A characterization of metric completeness. Proc. Am. Math. Soc. **64**, 186–188 (1977)
962. Wu, Z., Ye, J.J.: Sufficient conditions for error bounds. SIAM J. Optim. **12**(2), 421–435 (2001)
963. Wu, Z., Ye, J.J.: On error bounds for lower semicontinuous functions. Math. Program. A **92**(2), 301–314 (2002)

964. Wu, Z.: Equivalent reformulations of Ekeland's variational principle. Nonlinear Anal. Theor. Meth. Appl. **55**, 609–615 (2003)
965. Wu, Z., Ye, J.J.: First-order and second-order conditions for error bounds. SIAM J. Optim. **14**(3), 621–645 (2003)
966. Wu, Z., Ye, J.J.: Equivalence between various derivatives and subdifferentials of the distance function. J. Math. Anal. Appl. **282**, 629–647 (2003)
967. Ye, J.: Optimal strategies for bilevel dynamic problems. SIAM J. Contr. Optim. **35**(2), 512–531 (1997)
968. Ye, J.J.: Optimality conditions for optimization problems with complementarity constraints. SIAM J. Optim. **9**, 374–387 (1999)
969. Ye, J.J.: Constraint qualifications and necessary optimality conditions for optimization problems with variational inequality constraints. SIAM J. Optim. **10**, 943–962 (2000)
970. Ye, J.J.: Nondifferentiable multiplier rules for optimization and bilevel optimization problems. SIAM J. Optim. **15**, 252–274 (2004)
971. Ye, J.J.: Necessary and sufficient optimality conditions for mathematical programs with equilibrium constraints. J. Math. Anal. Appl. **307**, 350–369 (2005)
972. Ye, J.J., Ye, X.Y.: Necessary optimality conditions for optimization problems with variational inequality constraints. Math. Oper. Res. **22**, 977–997 (1997)
973. Ye, J.J., Zhu, D.L., Zhu, Q.J.: Exact penalization and necessary optimality conditions for generalized bilevel programming problems. SIAM J. Optim. **7**, 481–507 (1997)
974. Yen, N.D.: Hölder continuity of solutions to a parametric variational inequality. Appl. Math. Optim. **31**, 245–255 (1995)
975. Yost, D.: Asplund spaces for beginners. Acta Univ. Carol. Ser. Math. Phys. **34**, 159–177 (1993)
976. Yosida, K.: Functional Analysis. Springer, Berlin (1995)
977. Young, L.C.: Lectures on the Calculus of Variations and Optimal Control Theory. Saunders, Philadelphia (1969)
978. Zagrodny, D.: An example of bad convex function. J. Optim. Theor. Appl. **70**(3), 631–637 (1991)
979. Zagrodny, D.: Approximate mean value theorem for upper subderivatives. Nonlinear Anal. **12**, 1413–1428 (1988)
980. Zagrodny, D.: A note on the equivalence between the mean value theorem for the Dini derivative and the Clarke–Rockafellar derivative. Optimization **21**(2), 179–183 (1990)
981. Zajíček, L.: Porosity and σ-porosity. R. Anal. Exchange **13**, 314–350 (1987–1988)
982. Zălinescu, C.: On convex sets in general position. Lin. Algebra Appl. **64**, 191–198 (1985)
983. Zălinescu, C.: A comparison of constraint qualifications in infinite-dimensional convex programming revisited. J. Aust. Math. Soc. B **40**(3), 353–378 (1999)
984. Zălinescu, C.: Convex Analysis in General Vector Spaces. World Scientific, Singapore (2002)
985. Zangwill, W.I.: Nonlinear programming via penalty functions. Manag. Sci. **13**, 344–358 (1967)
986. Zaremba, S.C.: Sur les équations au paratingent. Bull. Sci. Math. **60**, 139–160 (1936)
987. Zeidler, E.: Nonlinear Functional Analysis and Its Applications. I. Fixed-Point Theory. Springer, New York (1986)
988. Zeidler, E.: Nonlinear Functional Analysis and Its Applications, III: Variational Methods and Optimization. Springer, New York (1985)
989. Zhang, R.: Weakly upper Lipschitzian multifunctions and applications to parametric optimization. Math. Program. **102**, 153–166 (2005)
990. Zhang, R., Treiman, J.: Upper-Lipschitz multifunctions and inverse subdifferentials. Nonlinear Anal. Theor. Meth. Appl. **24**(2), 273–286 (1995)
991. Zheng, X.Y., Ng, K.F.: The Fermat rule for multifunctions in Banach spaces. Math. Program. **104**, 69–90 (2005)
992. Zheng, X.Y., Ng, K.F.: Metric subregularity and constraint qualifications for convex generalized equations in Banach spaces. SIAM J. Optim. **18**(2), 437–460 (2007)

993. Zheng, X.Y., Ng, K.F.: Calmness for *L*-subsmooth multifunctions in Banach spaces. SIAM J. Optim. **19**(4), 1648–1673 (2008)
994. Zheng, X.Y., Ng, K.F.: Metric regularity of composite multifunctions in Banach spaces. Taiwanese J. Math. **13**(6A), 1723–1735 (2009)
995. Zhu, Q.J.: The equivalence of several basic theorems for subdifferentials. Set-Valued Anal. **6**, 171–185 (1998)
996. Zhu, Q.J.: Lower semicontinuous Lyapunov functions and stability. J. Nonlinear Convex Anal. **4**(3), 325–332 (2003)
997. Zhu, Q.J.: Nonconvex separation theorems for multifunctions, subdifferential calculus and applications. Set-Valued Anal. **12**, 275–290 (2004)
998. Zhu, J., Li, S.J.: Generalization of ordering principles and applications. J. Optim. Theor. Appl. **132**, 493–507 (2007)
999. Zolezzi, T.: Well-posedness criteria in optimization with application to calculus of variations. Nonlinear Anal. Theor. Meth. Appl. **25**, 437–453 (1995)
1000. Zolezzi, T.: Extended well-posedness of optimization problems. J. Optim. Theor. Appl. **91**, 257–268 (1996)
1001. Zolezzi, T.: Tikhonov regularization under epi-convergent perturbations. Ricerche Mat. **49**(suppl.), 155–168 (2000)
1002. Zolezzi, T.: Condition numbers theorems in optimization. SIAM J. Optim. **14**, 507–514 (2003)
1003. Zowe, J., Kurcyusz, S.: Regularity and stability for the mathematical programming problem in Banach spaces. Appl. Math. Optim. **5**, 49–62 (1979)

Index

J.-P. Penot, *Calculus Without Derivatives*, Graduate Texts in Mathematics 266,
DOI 10.1007/978-1-4614-4538-8, © Springer Science+Business Media New York 2013